DATE DUE

VOLUME FOUR HUNDRED AND NINETY-THREE

METHODS IN ENZYMOLOGY

Fragment-Based Drug Design

Tools, Practical Approaches, and Examples

METHODS IN ENZYMOLOGY

Editors-in-Chief

JOHN N. ABELSON AND MELVIN I. SIMON

Division of Biology
California Institute of Technology
Pasadena, California

Founding Editors

SIDNEY P. COLOWICK AND NATHAN O. KAPLAN

VOLUME FOUR HUNDRED AND NINETY-THREE

METHODS IN ENZYMOLOGY

Fragment-Based Drug Design

Tools, Practical Approaches, and Examples

EDITED BY

LAWRENCE C. KUO
Structural Biology
Johnson & Johnson Pharmaceutical
Research and Development, L.L.C.
Spring House, Pennsylvania, USA

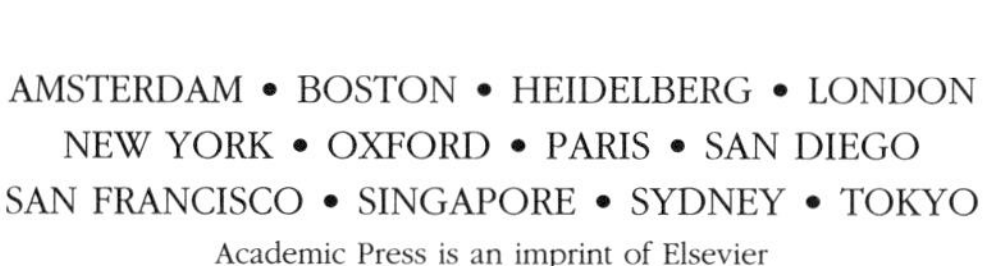

AMSTERDAM • BOSTON • HEIDELBERG • LONDON
NEW YORK • OXFORD • PARIS • SAN DIEGO
SAN FRANCISCO • SINGAPORE • SYDNEY • TOKYO
Academic Press is an imprint of Elsevier

Academic Press is an imprint of Elsevier
525 B Street, Suite 1900, San Diego, CA 92101-4495, USA
30 Corporate Drive, Suite 400, Burlington, MA 01803, USA
32 Jamestown Road, London NW1 7BY, UK

First edition 2011

ISBN: 978-0-12-381274-2
ISSN: 0076-6879

Printed and bound in United States of America
11 12 13 14 10 9 8 7 6 5 4 3 2 1

Contents

Contributors

Marta C. Abad
Structural Biology and Medicinal Chemistry, Johnson & Johnson Pharmaceutical Research and Development, L.L.C., Spring House, Pennsylvania, USA

Thomas B. Acton
Center for Advanced Biotechnology and Medicine, Department of Molecular Biology and Biochemistry, and Northeast Structural Genomics Consortium, Rutgers University, Piscataway, New Jersey, USA

Yasushi Amano
Advanced Genomics, Molecular Medicine Research Labs, Drug Discovery Research, Astellas Pharma Inc., Tsukuba, Ibaraki, Japan

Stephen Anderson
Center for Advanced Biotechnology and Medicine, Department of Molecular Biology and Biochemistry, and Northeast Structural Genomics Consortium, Rutgers University, Piscataway, New Jersey, USA

James Aramini
Center for Advanced Biotechnology and Medicine, Department of Molecular Biology and Biochemistry, and Northeast Structural Genomics Consortium, Rutgers University, Piscataway, New Jersey, USA

Jark Böttcher
Proteros biostructures GmbH, Am Klopferspitz 19, Martinsried, Germany

Darren W. Begley
Emerald BioStructures, Bainbridge Island, and Seattle Structural Genomics Center for Infectious Disease, Seattle, Washington, USA

William A. Buchwald
Center for Advanced Biotechnology and Medicine, Department of Molecular Biology and Biochemistry, and Northeast Structural Genomics Consortium, Rutgers University, Piscataway, New Jersey, USA

Colleen Ciccosanti
Center for Advanced Biotechnology and Medicine, Department of Molecular Biology and Biochemistry, and Northeast Structural Genomics Consortium, Rutgers University, Piscataway, New Jersey, USA

Miles Congreve
Heptares Therapeutics, Biopark, Welwyn Garden City, Hertfordshire, United Kingdom

Ken Conover
Center for Advanced Biotechnology and Medicine, Department of Molecular Biology and Biochemistry, and Northeast Structural Genomics Consortium, Rutgers University, Piscataway, New Jersey, USA

Douglas R. Davies
Emerald BioStructures, Bainbridge Island, and Seattle Structural Genomics Center for Infectious Disease, Seattle, Washington, USA

Renee L. DesJarlais
Structural Biology, Johnson & Johnson Pharmaceutical Research and Development, L.L.C., Spring House, Pennsylvania, USA

Hugh L. Eaton
Global Structural Chemistry, Merck Research Laboratories, Kenilworth, New Jersey, USA

Thomas E. Edwards
Emerald BioStructures, Bainbridge Island, and Seattle Structural Genomics Center for Infectious Disease, Seattle, Washington, USA

John Everett
Center for Advanced Biotechnology and Medicine, Department of Molecular Biology and Biochemistry, and Northeast Structural Genomics Consortium, Rutgers University, Piscataway, New Jersey, USA

Francis Figaroa
Leiden Institute of Chemistry, ZoBio and Leiden University, Einsteinweg 55, Leiden, The Netherlands

Anthony M. Giannetti
Genentech Inc., 1 DNA Way, South San Francisco, California, USA

Alan C. Gibbs
Structural Biology and Medicinal Chemistry, Johnson & Johnson Pharmaceutical Research and Development, L.L.C., Spring House, Pennsylvania, USA

Keith Hamilton
Center for Advanced Biotechnology and Medicine, Department of Molecular Biology and Biochemistry, and Northeast Structural Genomics Consortium, Rutgers University, Piscataway, New Jersey, USA

Robert C. Hartley
Emerald BioStructures, Bainbridge Island, and Seattle Structural Genomics Center for Infectious Disease, Seattle, Washington, USA

Yuanpeng Janet Huang
Center for Advanced Biotechnology and Medicine, Department of Molecular Biology and Biochemistry, and Northeast Structural Genomics Consortium, Rutgers University, Piscataway, New Jersey, USA

Roderick E. Hubbard
Vernalis (R&D) Ltd., Granta Park, Cambridge, and YSBL & HYMS, University of York, Heslington, York, United Kingdom

Haleema Janjua
Center for Advanced Biotechnology and Medicine, Department of Molecular Biology and Biochemistry, and Northeast Structural Genomics Consortium, Rutgers University, Piscataway, New Jersey, USA

Anja Jestel
Proteros biostructures GmbH, Am Klopferspitz 19, Martinsried, Germany

Reiner Kiefersauer
Proteros biostructures GmbH, Am Klopferspitz 19, Martinsried, Germany

Anthony E. Klon
Department of Design, Ansaris, Four Valley Square, Blue Bell, Pennsylvania, USA

Zenon D. Konteatis
Department of Design, Ansaris, Four Valley Square, Blue Bell, Pennsylvania, USA

Gregory Kornhaber
Center for Advanced Biotechnology and Medicine, Department of Molecular Biology and Biochemistry, and Northeast Structural Genomics Consortium, Rutgers University, Piscataway, New Jersey, USA

James K. Kranz
Biopharmaceutical Technologies, GlaxoSmithKline Biopharmaceutical Research and Development, Upper Merion, Pennsylvania, USA

Stephan Krapp
Proteros biostructures GmbH, Am Klopferspitz 19, Martinsried, Germany

Lawrence C. Kuo
Structural Biology, Johnson & Johnson Pharmaceutical Research and Development, L.L.C., Spring House, Pennsylvania, USA

James Lanter
Medicinal Chemistry, Johnson & Johnson Pharmaceutical Research and Development, L.L.C., Spring House, Pennsylvania, USA

Jessica Lau
Center for Advanced Biotechnology and Medicine, Department of Molecular Biology and Biochemistry, and Northeast Structural Genomics Consortium, Rutgers University, Piscataway, New Jersey, USA

Dong Yup Lee
Center for Advanced Biotechnology and Medicine, Department of Molecular Biology and Biochemistry, and Northeast Structural Genomics Consortium, Rutgers University, Piscataway, New Jersey, USA

Christopher A. Lepre
Structural Biology, Vertex Pharmaceuticals, Incorporated, Cambridge, Massachusetts, USA

Gaohua Liu
Center for Advanced Biotechnology and Medicine, Department of Molecular Biology and Biochemistry, and Northeast Structural Genomics Consortium, Rutgers University, Piscataway, New Jersey, USA

Betsy L. Lytle
Department of Biochemistry, Medical College of Wisconsin, Milwaukee, Wisconsin, USA

Lichung Ma
Center for Advanced Biotechnology and Medicine, Department of Molecular Biology and Biochemistry, and Northeast Structural Genomics Consortium, Rutgers University, Piscataway, New Jersey, USA

Melissa Maglaqui
Center for Advanced Biotechnology and Medicine, Department of Molecular Biology and Biochemistry, and Northeast Structural Genomics Consortium, Rutgers University, Piscataway, New Jersey, USA

Lei Mao
Center for Advanced Biotechnology and Medicine, Department of Molecular Biology and Biochemistry, and Northeast Structural Genomics Consortium, Rutgers University, Piscataway, New Jersey, USA

Fiona H. Marshall
Heptares Therapeutics, Biopark, Welwyn Garden City, Hertfordshire, United Kingdom

Till Maurer
Department of Structural Biology, Genentech Inc., South San Francisco, California, USA

Siavash Meshkat
Discovery Technologies Department, Ansaris, Four Valley Square, Blue Bell, Pennsylvania, USA

Gaetano T. Montelione
Center for Advanced Biotechnology and Medicine, Department of Molecular Biology and Biochemistry, and Northeast Structural Genomics Consortium, Rutgers University, and Department of Biochemistry, Robert Wood Johnson

Medical School, University of Medicine and Dentistry of New Jersey, Piscataway, New Jersey, USA

James B. Murray
Vernalis (R&D) Ltd., Granta Park, Cambridge, United Kingdom

Peter J. Myler
Seattle Biomedical Research Institute, and Seattle Structural Genomics Center for Infectious Disease, Seattle, Washington, USA

David G. Myszka
Department of Biochemistry, University of Utah, Salt Lake City, Utah, USA

Susanna Nagel
Proteros biostructures GmbH, Am Klopferspitz 19, Martinsried, Germany

Lars Neumann
Proteros Biostructures GmbH, Am Klopferspitz 19, Martinsried, Germany

Tatsuya Niimi
Chemistry for Leads, Chemistry Research Labs, Drug Discovery Research, Astellas Pharma Inc., Tsukuba, Ibaraki, Japan

Kazuki Ohno
Chemistry for Leads, Chemistry Research Labs, Drug Discovery Research, Astellas Pharma Inc., Tsukuba, Ibaraki, Japan

Masaya Orita
Chemistry for Leads, Chemistry Research Labs, Drug Discovery Research, Astellas Pharma Inc., Tsukuba, Ibaraki, Japan

Michael H. Parker
Structural Biology and Medicinal Chemistry, Johnson & Johnson Pharmaceutical Research and Development, L.L.C., Spring House, Pennsylvania, USA

Dayaban Patel
Center for Advanced Biotechnology and Medicine, Department of Molecular Biology and Biochemistry, and Northeast Structural Genomics Consortium, Rutgers University, Piscataway, New Jersey, USA

Francis C. Peterson
Department of Biochemistry, Medical College of Wisconsin, Milwaukee, Wisconsin, USA

Rebecca L. Rich
Department of Biochemistry, University of Utah, Salt Lake City, Utah, USA

Paolo Rossi
Center for Advanced Biotechnology and Medicine, Department of Molecular Biology and Biochemistry, and Northeast Structural Genomics Consortium, Rutgers University, Piscataway, New Jersey, USA

Seema Sahdev
Center for Advanced Biotechnology and Medicine, Department of Molecular Biology and Biochemistry, and Northeast Structural Genomics Consortium, Rutgers University, Piscataway, New Jersey, USA

Celine Schalk-Hihi
Structural Biology, Johnson & Johnson Pharmaceuticals Research and Development, LLC, Spring House, Pennsylvania, USA

Ritu Shastry
Department of Biochemistry, Robert Wood Johnson Medical School, University of Medicine and Dentistry of New Jersey, Piscataway, New Jersey, USA

Gregg Siegal
Leiden Institute of Chemistry, ZoBio and Leiden University, Einsteinweg 55, Leiden, The Netherlands

John C. Spurlino
Structural Biology, Johnson & Johnson Pharmaceutical Research and Development, LLC, Pennsylvania, USA

Bart L. Staker
Emerald BioStructures, Bainbridge Island, and Seattle Structural Genomics Center for Infectious Disease, Seattle, Washington, USA

Stefan Steinbacher
Proteros biostructures GmbH, Am Klopferspitz 19, Martinsried, Germany

Holger Steuber
Proteros biostructures GmbH, Am Klopferspitz 19, Martinsried, Germany

Lance J. Stewart
Emerald BioStructures, Bainbridge Island, and Seattle Structural Genomics Center for Infectious Disease, Seattle, Washington, USA

Zhihua Sui
Medicinal Chemistry, Johnson & Johnson Pharmaceutical Research and Development, L.L.C., Spring House, Pennsylvania, USA

G. V. T. Swapna
Center for Advanced Biotechnology and Medicine, Department of Molecular Biology and Biochemistry, and Northeast Structural Genomics Consortium, Rutgers University, Piscataway, New Jersey, USA

Yeufeng Tang
Center for Advanced Biotechnology and Medicine, Department of Molecular Biology and Biochemistry, and Northeast Structural Genomics Consortium, Rutgers University, Piscataway, New Jersey, USA

Saichiu Tong
Center for Advanced Biotechnology and Medicine, Department of Molecular Biology and Biochemistry, and Northeast Structural Genomics Consortium, Rutgers University, Piscataway, New Jersey, USA

Brett A. Tounge
Structural Biology and Medicinal Chemistry, Johnson & Johnson Pharmaceutical Research and Development, L.L.C., Spring House, Pennsylvania, USA

Dirk Ullmann
Proteros Biostructures GmbH, Am Klopferspitz 19, Martinsried, Germany

Wesley C. Van Voorhis
Department of Medicine, University of Washington, and Seattle Structural Genomics Center for Infectious Disease, Seattle, Washington, USA

Brian F. Volkman
Department of Biochemistry, Medical College of Wisconsin, Milwaukee, Wisconsin, USA

Konstanze von König
Proteros biostructures GmbH, Am Klopferspitz 19, Martinsried, Germany

Dongyan Wang
Center for Advanced Biotechnology and Medicine, Department of Molecular Biology and Biochemistry, and Northeast Structural Genomics Consortium, Rutgers University, Piscataway, New Jersey, USA

Huang Wang
Center for Advanced Biotechnology and Medicine, Department of Molecular Biology and Biochemistry, and Northeast Structural Genomics Consortium, Rutgers University, Piscataway, New Jersey, USA

Masaichi Warizaya
Advanced Genomics, Molecular Medicine Research Labs, Drug Discovery Research, Astellas Pharma Inc., Tsukuba, Ibaraki, Japan

Daniel F. Wyss
Global Structural Chemistry, Merck Research Laboratories, Kenilworth, New Jersey, USA

Rong Xiao
Center for Advanced Biotechnology and Medicine, Department of Molecular Biology and Biochemistry, and Northeast Structural Genomics Consortium, Rutgers University, Piscataway, New Jersey, USA

Xuqing Zhang
Structural Biology and Medicinal Chemistry, Johnson & Johnson Pharmaceutical Research and Development, L.L.C., Spring House, Pennsylvania, USA

Li Zhao
Center for Advanced Biotechnology and Medicine, Department of Molecular Biology and Biochemistry, and Northeast Structural Genomics Consortium, Rutgers University, Piscataway, New Jersey, USA

Joshua J. Ziarek
Department of Biochemistry, Medical College of Wisconsin, Milwaukee, Wisconsin, USA

Jinming Zou
Department of Design, Ansaris, Four Valley Square, Blue Bell, Pennsylvania, USA

Preface

There has been a plethora of technological innovations since the late 1980s that have influenced the way research is pursued in the biotechnology and pharmaceutical arena. Investigators are now applying ever more elaborate methods to elucidate the molecular basis underlying the biology of human diseases and to tackle discovery of new medicine. A few of these new technologies have fundamentally altered the means by which we look for hits as well as the routes we use to evolve hits to leads. Notably, protein crystallography was applied in the early 1990s to provide a structure-based approach to optimize drug leads and at about the same time high-throughput automated screening was introduced to evaluate at an unprecedented speed the effect of compounds on protein targets. Both were adopted swiftly as standard operating procedures by the pharmaceutical industry. In contrast, the use of very low molecular weight compounds, known as fragments and introduced by Abbott Laboratories in the mid-1990s, has only gained widespread acceptance in recent years.

A typical screening exercise searches a compound library with the aim of finding inhibitors against a protein target. Traditional high-throughput screening does not always offer hits of sufficient quality to be progressed into a lead compound. Fragment-based approaches utilize low molecular weight compounds that are associated with favorable physiochemical and pharmacokinetic properties. Hits derived from low molecular weight compounds offer a viable and an orthogonal entry point to finding lead compounds that in principle and by design shun unappealing pharmacophores. The use of fragments is now accepted as a "legitimate" starting point in the discovery of new medical entities as therapeutics. There are numerous, excellent reviews on Fragment-Based Drug Design (FBDD). For those new to the field, there is a need for comprehensive walk-through protocols with which one can embark readily on this creative approach to complement traditional screening methodologies. This *Methods in Enzymology* volume offers tools, practical approaches, and hit-to-lead examples on how to conduct FBDD screens. The chapters in this volume are written by experts in the field to cover methods that have proven to be successful with a focus on how to mount a successful FBDD campaign. The chapters include computational techniques, nuclear magnetic resonance, surface plasma resonance, thermal shift and enzyme kinetic assays, protein crystallography, and medicinal chemistry. Emphasis is placed on practical aspects and effective progression of lead generation to include sample preparations of fragments, proteins, protein crystals, and G-protein coupled receptors. Explicit examples are given on how to generate leads from low-affinity fragment hits.

I want to thank all the authors; it is solely their contributions that render this volume possible. I want to thank Paul Prasad Chandramohan and Zoe Kruze of Elsevier Publishing Company for their guidance and patience throughout all stages of putting this volume together. Thanks are also due to Drs. John Abelson and Melvin Simon for their support in selecting the timely topic. I am grateful to my wife and my children for their patience during my time spent on this volume.

Lawrence C. Kuo

Methods in Enzymology

Volume I. Preparation and Assay of Enzymes
Edited by Sidney P. Colowick and Nathan O. Kaplan

Volume II. Preparation and Assay of Enzymes
Edited by Sidney P. Colowick and Nathan O. Kaplan

Volume III. Preparation and Assay of Substrates
Edited by Sidney P. Colowick and Nathan O. Kaplan

Volume IV. Special Techniques for the Enzymologist
Edited by Sidney P. Colowick and Nathan O. Kaplan

Volume V. Preparation and Assay of Enzymes
Edited by Sidney P. Colowick and Nathan O. Kaplan

Volume VI. Preparation and Assay of Enzymes *(Continued)*
Preparation and Assay of Substrates
Special Techniques
Edited by Sidney P. Colowick and Nathan O. Kaplan

Volume VII. Cumulative Subject Index
Edited by Sidney P. Colowick and Nathan O. Kaplan

Volume VIII. Complex Carbohydrates
Edited by Elizabeth F. Neufeld and Victor Ginsburg

Volume IX. Carbohydrate Metabolism
Edited by Willis A. Wood

Volume X. Oxidation and Phosphorylation
Edited by Ronald W. Estabrook and Maynard E. Pullman

Volume XI. Enzyme Structure
Edited by C. H. W. Hirs

Volume XII. Nucleic Acids (Parts A and B)
Edited by Lawrence Grossman and Kivie Moldave

Volume XIII. Citric Acid Cycle
Edited by J. M. Lowenstein

Volume XIV. Lipids
Edited by J. M. Lowenstein

Volume XV. Steroids and Terpenoids
Edited by Raymond B. Clayton

Volume XVI. Fast Reactions
Edited by Kenneth Kustin

Volume XVII. Metabolism of Amino Acids and Amines (Parts A and B)
Edited by Herbert Tabor and Celia White Tabor

Volume XVIII. Vitamins and Coenzymes (Parts A, B, and C)
Edited by Donald B. McCormick and Lemuel D. Wright

Volume XIX. Proteolytic Enzymes
Edited by Gertrude E. Perlmann and Laszlo Lorand

Volume XX. Nucleic Acids and Protein Synthesis (Part C)
Edited by Kivie Moldave and Lawrence Grossman

Volume XXI. Nucleic Acids (Part D)
Edited by Lawrence Grossman and Kivie Moldave

Volume XXII. Enzyme Purification and Related Techniques
Edited by William B. Jakoby

Volume XXIII. Photosynthesis (Part A)
Edited by Anthony San Pietro

Volume XXIV. Photosynthesis and Nitrogen Fixation (Part B)
Edited by Anthony San Pietro

Volume XXV. Enzyme Structure (Part B)
Edited by C. H. W. Hirs and Serge N. Timasheff

Volume XXVI. Enzyme Structure (Part C)
Edited by C. H. W. Hirs and Serge N. Timasheff

Volume XXVII. Enzyme Structure (Part D)
Edited by C. H. W. Hirs and Serge N. Timasheff

Volume XXVIII. Complex Carbohydrates (Part B)
Edited by Victor Ginsburg

Volume XXIX. Nucleic Acids and Protein Synthesis (Part E)
Edited by Lawrence Grossman and Kivie Moldave

Volume XXX. Nucleic Acids and Protein Synthesis (Part F)
Edited by Kivie Moldave and Lawrence Grossman

Volume XXXI. Biomembranes (Part A)
Edited by Sidney Fleischer and Lester Packer

Volume XXXII. Biomembranes (Part B)
Edited by Sidney Fleischer and Lester Packer

Volume XXXIII. Cumulative Subject Index Volumes I-XXX
Edited by Martha G. Dennis and Edward A. Dennis

Volume XXXIV. Affinity Techniques (Enzyme Purification: Part B)
Edited by William B. Jakoby and Meir Wilchek

VOLUME LIII. Biomembranes (Part D: Biological Oxidations)
Edited by SIDNEY FLEISCHER AND LESTER PACKER

VOLUME LIV. Biomembranes (Part E: Biological Oxidations)
Edited by SIDNEY FLEISCHER AND LESTER PACKER

VOLUME LV. Biomembranes (Part F: Bioenergetics)
Edited by SIDNEY FLEISCHER AND LESTER PACKER

VOLUME LVI. Biomembranes (Part G: Bioenergetics)
Edited by SIDNEY FLEISCHER AND LESTER PACKER

VOLUME LVII. Bioluminescence and Chemiluminescence
Edited by MARLENE A. DELUCA

VOLUME LVIII. Cell Culture
Edited by WILLIAM B. JAKOBY AND IRA PASTAN

VOLUME LIX. Nucleic Acids and Protein Synthesis (Part G)
Edited by KIVIE MOLDAVE AND LAWRENCE GROSSMAN

VOLUME LX. Nucleic Acids and Protein Synthesis (Part H)
Edited by KIVIE MOLDAVE AND LAWRENCE GROSSMAN

VOLUME 61. Enzyme Structure (Part H)
Edited by C. H. W. HIRS AND SERGE N. TIMASHEFF

VOLUME 62. Vitamins and Coenzymes (Part D)
Edited by DONALD B. MCCORMICK AND LEMUEL D. WRIGHT

VOLUME 63. Enzyme Kinetics and Mechanism (Part A: Initial Rate and Inhibitor Methods)
Edited by DANIEL L. PURICH

VOLUME 64. Enzyme Kinetics and Mechanism
(Part B: Isotopic Probes and Complex Enzyme Systems)
Edited by DANIEL L. PURICH

VOLUME 65. Nucleic Acids (Part I)
Edited by LAWRENCE GROSSMAN AND KIVIE MOLDAVE

VOLUME 66. Vitamins and Coenzymes (Part E)
Edited by DONALD B. MCCORMICK AND LEMUEL D. WRIGHT

VOLUME 67. Vitamins and Coenzymes (Part F)
Edited by DONALD B. MCCORMICK AND LEMUEL D. WRIGHT

VOLUME 68. Recombinant DNA
Edited by RAY WU

VOLUME 69. Photosynthesis and Nitrogen Fixation (Part C)
Edited by ANTHONY SAN PIETRO

VOLUME 70. Immunochemical Techniques (Part A)
Edited by HELEN VAN VUNAKIS AND JOHN J. LANGONE

Volume 89. Carbohydrate Metabolism (Part D)
Edited by Willis A. Wood

Volume 90. Carbohydrate Metabolism (Part E)
Edited by Willis A. Wood

Volume 91. Enzyme Structure (Part I)
Edited by C. H. W. Hirs and Serge N. Timasheff

Volume 92. Immunochemical Techniques (Part E: Monoclonal Antibodies and General Immunoassay Methods)
Edited by John J. Langone and Helen Van Vunakis

Volume 93. Immunochemical Techniques (Part F: Conventional Antibodies, Fc Receptors, and Cytotoxicity)
Edited by John J. Langone and Helen Van Vunakis

Volume 94. Polyamines
Edited by Herbert Tabor and Celia White Tabor

Volume 95. Cumulative Subject Index Volumes 61–74, 76–80
Edited by Edward A. Dennis and Martha G. Dennis

Volume 96. Biomembranes [Part J: Membrane Biogenesis: Assembly and Targeting (General Methods; Eukaryotes)]
Edited by Sidney Fleischer and Becca Fleischer

Volume 97. Biomembranes [Part K: Membrane Biogenesis: Assembly and Targeting (Prokaryotes, Mitochondria, and Chloroplasts)]
Edited by Sidney Fleischer and Becca Fleischer

Volume 98. Biomembranes (Part L: Membrane Biogenesis: Processing and Recycling)
Edited by Sidney Fleischer and Becca Fleischer

Volume 99. Hormone Action (Part F: Protein Kinases)
Edited by Jackie D. Corbin and Joel G. Hardman

Volume 100. Recombinant DNA (Part B)
Edited by Ray Wu, Lawrence Grossman, and Kivie Moldave

Volume 101. Recombinant DNA (Part C)
Edited by Ray Wu, Lawrence Grossman, and Kivie Moldave

Volume 102. Hormone Action (Part G: Calmodulin and Calcium-Binding Proteins)
Edited by Anthony R. Means and Bert W. O'Malley

Volume 103. Hormone Action (Part H: Neuroendocrine Peptides)
Edited by P. Michael Conn

Volume 104. Enzyme Purification and Related Techniques (Part C)
Edited by William B. Jakoby

VOLUME 206. Cytochrome P450
Edited by MICHAEL R. WATERMAN AND ERIC F. JOHNSON

VOLUME 207. Ion Channels
Edited by BERNARDO RUDY AND LINDA E. IVERSON

VOLUME 208. Protein–DNA Interactions
Edited by ROBERT T. SAUER

VOLUME 209. Phospholipid Biosynthesis
Edited by EDWARD A. DENNIS AND DENNIS E. VANCE

VOLUME 210. Numerical Computer Methods
Edited by LUDWIG BRAND AND MICHAEL L. JOHNSON

VOLUME 211. DNA Structures (Part A: Synthesis and Physical Analysis of DNA)
Edited by DAVID M. J. LILLEY AND JAMES E. DAHLBERG

VOLUME 212. DNA Structures (Part B: Chemical and Electrophoretic Analysis of DNA)
Edited by DAVID M. J. LILLEY AND JAMES E. DAHLBERG

VOLUME 213. Carotenoids (Part A: Chemistry, Separation, Quantitation, and Antioxidation)
Edited by LESTER PACKER

VOLUME 214. Carotenoids (Part B: Metabolism, Genetics, and Biosynthesis)
Edited by LESTER PACKER

VOLUME 215. Platelets: Receptors, Adhesion, Secretion (Part B)
Edited by JACEK J. HAWIGER

VOLUME 216. Recombinant DNA (Part G)
Edited by RAY WU

VOLUME 217. Recombinant DNA (Part H)
Edited by RAY WU

VOLUME 218. Recombinant DNA (Part I)
Edited by RAY WU

VOLUME 219. Reconstitution of Intracellular Transport
Edited by JAMES E. ROTHMAN

VOLUME 220. Membrane Fusion Techniques (Part A)
Edited by NEJAT DÜZGÜNEŞ

VOLUME 221. Membrane Fusion Techniques (Part B)
Edited by NEJAT DÜZGÜNEŞ

VOLUME 222. Proteolytic Enzymes in Coagulation, Fibrinolysis, and Complement Activation (Part A: Mammalian Blood Coagulation Factors and Inhibitors)
Edited by LASZLO LORAND AND KENNETH G. MANN

Volume 223. Proteolytic Enzymes in Coagulation, Fibrinolysis, and Complement Activation (Part B: Complement Activation, Fibrinolysis, and Nonmammalian Blood Coagulation Factors)
Edited by Laszlo Lorand and Kenneth G. Mann

Volume 224. Molecular Evolution: Producing the Biochemical Data
Edited by Elizabeth Anne Zimmer, Thomas J. White, Rebecca L. Cann, and Allan C. Wilson

Volume 225. Guide to Techniques in Mouse Development
Edited by Paul M. Wassarman and Melvin L. DePamphilis

Volume 226. Metallobiochemistry (Part C: Spectroscopic and Physical Methods for Probing Metal Ion Environments in Metalloenzymes and Metalloproteins)
Edited by James F. Riordan and Bert L. Vallee

Volume 227. Metallobiochemistry (Part D: Physical and Spectroscopic Methods for Probing Metal Ion Environments in Metalloproteins)
Edited by James F. Riordan and Bert L. Vallee

Volume 228. Aqueous Two-Phase Systems
Edited by Harry Walter and Göte Johansson

Volume 229. Cumulative Subject Index Volumes 195–198, 200–227

Volume 230. Guide to Techniques in Glycobiology
Edited by William J. Lennarz and Gerald W. Hart

Volume 231. Hemoglobins (Part B: Biochemical and Analytical Methods)
Edited by Johannes Everse, Kim D. Vandegriff, and Robert M. Winslow

Volume 232. Hemoglobins (Part C: Biophysical Methods)
Edited by Johannes Everse, Kim D. Vandegriff, and Robert M. Winslow

Volume 233. Oxygen Radicals in Biological Systems (Part C)
Edited by Lester Packer

Volume 234. Oxygen Radicals in Biological Systems (Part D)
Edited by Lester Packer

Volume 235. Bacterial Pathogenesis (Part A: Identification and Regulation of Virulence Factors)
Edited by Virginia L. Clark and Patrik M. Bavoil

Volume 236. Bacterial Pathogenesis (Part B: Integration of Pathogenic Bacteria with Host Cells)
Edited by Virginia L. Clark and Patrik M. Bavoil

Volume 237. Heterotrimeric G Proteins
Edited by Ravi Iyengar

Volume 238. Heterotrimeric G-Protein Effectors
Edited by Ravi Iyengar

Volume 256. Small GTPases and Their Regulators (Part B: Rho Family)
Edited by W. E. Balch, Channing J. Der, and Alan Hall

Volume 257. Small GTPases and Their Regulators (Part C: Proteins Involved in Transport)
Edited by W. E. Balch, Channing J. Der, and Alan Hall

Volume 258. Redox-Active Amino Acids in Biology
Edited by Judith P. Klinman

Volume 259. Energetics of Biological Macromolecules
Edited by Michael L. Johnson and Gary K. Ackers

Volume 260. Mitochondrial Biogenesis and Genetics (Part A)
Edited by Giuseppe M. Attardi and Anne Chomyn

Volume 261. Nuclear Magnetic Resonance and Nucleic Acids
Edited by Thomas L. James

Volume 262. DNA Replication
Edited by Judith L. Campbell

Volume 263. Plasma Lipoproteins (Part C: Quantitation)
Edited by William A. Bradley, Sandra H. Gianturco, and Jere P. Segrest

Volume 264. Mitochondrial Biogenesis and Genetics (Part B)
Edited by Giuseppe M. Attardi and Anne Chomyn

Volume 265. Cumulative Subject Index Volumes 228, 230–262

Volume 266. Computer Methods for Macromolecular Sequence Analysis
Edited by Russell F. Doolittle

Volume 267. Combinatorial Chemistry
Edited by John N. Abelson

Volume 268. Nitric Oxide (Part A: Sources and Detection of NO; NO Synthase)
Edited by Lester Packer

Volume 269. Nitric Oxide (Part B: Physiological and Pathological Processes)
Edited by Lester Packer

Volume 270. High Resolution Separation and Analysis of Biological Macromolecules (Part A: Fundamentals)
Edited by Barry L. Karger and William S. Hancock

Volume 271. High Resolution Separation and Analysis of Biological Macromolecules (Part B: Applications)
Edited by Barry L. Karger and William S. Hancock

Volume 272. Cytochrome P450 (Part B)
Edited by Eric F. Johnson and Michael R. Waterman

Volume 273. RNA Polymerase and Associated Factors (Part A)
Edited by Sankar Adhya

Volume 293. Ion Channels (Part B)
Edited by P. Michael Conn

Volume 294. Ion Channels (Part C)
Edited by P. Michael Conn

Volume 295. Energetics of Biological Macromolecules (Part B)
Edited by Gary K. Ackers and Michael L. Johnson

Volume 296. Neurotransmitter Transporters
Edited by Susan G. Amara

Volume 297. Photosynthesis: Molecular Biology of Energy Capture
Edited by Lee McIntosh

Volume 298. Molecular Motors and the Cytoskeleton (Part B)
Edited by Richard B. Vallee

Volume 299. Oxidants and Antioxidants (Part A)
Edited by Lester Packer

Volume 300. Oxidants and Antioxidants (Part B)
Edited by Lester Packer

Volume 301. Nitric Oxide: Biological and Antioxidant Activities (Part C)
Edited by Lester Packer

Volume 302. Green Fluorescent Protein
Edited by P. Michael Conn

Volume 303. cDNA Preparation and Display
Edited by Sherman M. Weissman

Volume 304. Chromatin
Edited by Paul M. Wassarman and Alan P. Wolffe

Volume 305. Bioluminescence and Chemiluminescence (Part C)
Edited by Thomas O. Baldwin and Miriam M. Ziegler

Volume 306. Expression of Recombinant Genes in Eukaryotic Systems
Edited by Joseph C. Glorioso and Martin C. Schmidt

Volume 307. Confocal Microscopy
Edited by P. Michael Conn

Volume 308. Enzyme Kinetics and Mechanism (Part E: Energetics of Enzyme Catalysis)
Edited by Daniel L. Purich and Vern L. Schramm

Volume 309. Amyloid, Prions, and Other Protein Aggregates
Edited by Ronald Wetzel

Volume 310. Biofilms
Edited by Ron J. Doyle

Volume 328. Applications of Chimeric Genes and Hybrid Proteins (Part C: Protein–Protein Interactions and Genomics)
Edited by Jeremy Thorner, Scott D. Emr, and John N. Abelson

Volume 329. Regulators and Effectors of Small GTPases (Part E: GTPases Involved in Vesicular Traffic)
Edited by W. E. Balch, Channing J. Der, and Alan Hall

Volume 330. Hyperthermophilic Enzymes (Part A)
Edited by Michael W. W. Adams and Robert M. Kelly

Volume 331. Hyperthermophilic Enzymes (Part B)
Edited by Michael W. W. Adams and Robert M. Kelly

Volume 332. Regulators and Effectors of Small GTPases (Part F: Ras Family I)
Edited by W. E. Balch, Channing J. Der, and Alan Hall

Volume 333. Regulators and Effectors of Small GTPases (Part G: Ras Family II)
Edited by W. E. Balch, Channing J. Der, and Alan Hall

Volume 334. Hyperthermophilic Enzymes (Part C)
Edited by Michael W. W. Adams and Robert M. Kelly

Volume 335. Flavonoids and Other Polyphenols
Edited by Lester Packer

Volume 336. Microbial Growth in Biofilms (Part A: Developmental and Molecular Biological Aspects)
Edited by Ron J. Doyle

Volume 337. Microbial Growth in Biofilms (Part B: Special Environments and Physicochemical Aspects)
Edited by Ron J. Doyle

Volume 338. Nuclear Magnetic Resonance of Biological Macromolecules (Part A)
Edited by Thomas L. James, Volker Dötsch, and Uli Schmitz

Volume 339. Nuclear Magnetic Resonance of Biological Macromolecules (Part B)
Edited by Thomas L. James, Volker Dötsch, and Uli Schmitz

Volume 340. Drug–Nucleic Acid Interactions
Edited by Jonathan B. Chaires and Michael J. Waring

Volume 341. Ribonucleases (Part A)
Edited by Allen W. Nicholson

Volume 342. Ribonucleases (Part B)
Edited by Allen W. Nicholson

Volume 343. G Protein Pathways (Part A: Receptors)
Edited by Ravi Iyengar and John D. Hildebrandt

Volume 344. G Protein Pathways (Part B: G Proteins and Their Regulators)
Edited by Ravi Iyengar and John D. Hildebrandt

VOLUME 381. Oxygen Sensing
Edited by CHANDAN K. SEN AND GREGG L. SEMENZA

VOLUME 382. Quinones and Quinone Enzymes (Part B)
Edited by HELMUT SIES AND LESTER PACKER

VOLUME 383. Numerical Computer Methods (Part D)
Edited by LUDWIG BRAND AND MICHAEL L. JOHNSON

VOLUME 384. Numerical Computer Methods (Part E)
Edited by LUDWIG BRAND AND MICHAEL L. JOHNSON

VOLUME 385. Imaging in Biological Research (Part A)
Edited by P. MICHAEL CONN

VOLUME 386. Imaging in Biological Research (Part B)
Edited by P. MICHAEL CONN

VOLUME 387. Liposomes (Part D)
Edited by NEJAT DÜZGÜNEŞ

VOLUME 388. Protein Engineering
Edited by DAN E. ROBERTSON AND JOSEPH P. NOEL

VOLUME 389. Regulators of G-Protein Signaling (Part A)
Edited by DAVID P. SIDEROVSKI

VOLUME 390. Regulators of G-Protein Signaling (Part B)
Edited by DAVID P. SIDEROVSKI

VOLUME 391. Liposomes (Part E)
Edited by NEJAT DÜZGÜNEŞ

VOLUME 392. RNA Interference
Edited by ENGELKE ROSSI

VOLUME 393. Circadian Rhythms
Edited by MICHAEL W. YOUNG

VOLUME 394. Nuclear Magnetic Resonance of Biological Macromolecules (Part C)
Edited by THOMAS L. JAMES

VOLUME 395. Producing the Biochemical Data (Part B)
Edited by ELIZABETH A. ZIMMER AND ERIC H. ROALSON

VOLUME 396. Nitric Oxide (Part E)
Edited by LESTER PACKER AND ENRIQUE CADENAS

VOLUME 397. Environmental Microbiology
Edited by JARED R. LEADBETTER

VOLUME 398. Ubiquitin and Protein Degradation (Part A)
Edited by RAYMOND J. DESHAIES

Volume 399. Ubiquitin and Protein Degradation (Part B)
Edited by Raymond J. Deshaies

Volume 400. Phase II Conjugation Enzymes and Transport Systems
Edited by Helmut Sies and Lester Packer

Volume 401. Glutathione Transferases and Gamma Glutamyl Transpeptidases
Edited by Helmut Sies and Lester Packer

Volume 402. Biological Mass Spectrometry
Edited by A. L. Burlingame

Volume 403. GTPases Regulating Membrane Targeting and Fusion
Edited by William E. Balch, Channing J. Der, and Alan Hall

Volume 404. GTPases Regulating Membrane Dynamics
Edited by William E. Balch, Channing J. Der, and Alan Hall

Volume 405. Mass Spectrometry: Modified Proteins and Glycoconjugates
Edited by A. L. Burlingame

Volume 406. Regulators and Effectors of Small GTPases: Rho Family
Edited by William E. Balch, Channing J. Der, and Alan Hall

Volume 407. Regulators and Effectors of Small GTPases: Ras Family
Edited by William E. Balch, Channing J. Der, and Alan Hall

Volume 408. DNA Repair (Part A)
Edited by Judith L. Campbell and Paul Modrich

Volume 409. DNA Repair (Part B)
Edited by Judith L. Campbell and Paul Modrich

Volume 410. DNA Microarrays (Part A: Array Platforms and Web-Bench Protocols)
Edited by Alan Kimmel and Brian Oliver

Volume 411. DNA Microarrays (Part B: Databases and Statistics)
Edited by Alan Kimmel and Brian Oliver

Volume 412. Amyloid, Prions, and Other Protein Aggregates (Part B)
Edited by Indu Kheterpal and Ronald Wetzel

Volume 413. Amyloid, Prions, and Other Protein Aggregates (Part C)
Edited by Indu Kheterpal and Ronald Wetzel

Volume 414. Measuring Biological Responses with Automated Microscopy
Edited by James Inglese

Volume 415. Glycobiology
Edited by Minoru Fukuda

Volume 416. Glycomics
Edited by Minoru Fukuda

Volume 417. Functional Glycomics
Edited by Minoru Fukuda

Volume 418. Embryonic Stem Cells
Edited by Irina Klimanskaya and Robert Lanza

Volume 419. Adult Stem Cells
Edited by Irina Klimanskaya and Robert Lanza

Volume 420. Stem Cell Tools and Other Experimental Protocols
Edited by Irina Klimanskaya and Robert Lanza

Volume 421. Advanced Bacterial Genetics: Use of Transposons and Phage for Genomic Engineering
Edited by Kelly T. Hughes

Volume 422. Two-Component Signaling Systems, Part A
Edited by Melvin I. Simon, Brian R. Crane, and Alexandrine Crane

Volume 423. Two-Component Signaling Systems, Part B
Edited by Melvin I. Simon, Brian R. Crane, and Alexandrine Crane

Volume 424. RNA Editing
Edited by Jonatha M. Gott

Volume 425. RNA Modification
Edited by Jonatha M. Gott

Volume 426. Integrins
Edited by David Cheresh

Volume 427. MicroRNA Methods
Edited by John J. Rossi

Volume 428. Osmosensing and Osmosignaling
Edited by Helmut Sies and Dieter Haussinger

Volume 429. Translation Initiation: Extract Systems and Molecular Genetics
Edited by Jon Lorsch

Volume 430. Translation Initiation: Reconstituted Systems and Biophysical Methods
Edited by Jon Lorsch

Volume 431. Translation Initiation: Cell Biology, High-Throughput and Chemical-Based Approaches
Edited by Jon Lorsch

Volume 432. Lipidomics and Bioactive Lipids: Mass-Spectrometry–Based Lipid Analysis
Edited by H. Alex Brown

Volume 433. Lipidomics and Bioactive Lipids: Specialized Analytical Methods and Lipids in Disease
Edited by H. Alex Brown

Volume 434. Lipidomics and Bioactive Lipids: Lipids and Cell Signaling
Edited by H. Alex Brown

Volume 435. Oxygen Biology and Hypoxia
Edited by Helmut Sies and Bernhard Brüne

Volume 436. Globins and Other Nitric Oxide-Reactive Protiens (Part A)
Edited by Robert K. Poole

Volume 437. Globins and Other Nitric Oxide-Reactive Protiens (Part B)
Edited by Robert K. Poole

Volume 438. Small GTPases in Disease (Part A)
Edited by William E. Balch, Channing J. Der, and Alan Hall

Volume 439. Small GTPases in Disease (Part B)
Edited by William E. Balch, Channing J. Der, and Alan Hall

Volume 440. Nitric Oxide, Part F Oxidative and Nitrosative Stress in Redox Regulation of Cell Signaling
Edited by Enrique Cadenas and Lester Packer

Volume 441. Nitric Oxide, Part G Oxidative and Nitrosative Stress in Redox Regulation of Cell Signaling
Edited by Enrique Cadenas and Lester Packer

Volume 442. Programmed Cell Death, General Principles for Studying Cell Death (Part A)
Edited by Roya Khosravi-Far, Zahra Zakeri, Richard A. Lockshin, and Mauro Piacentini

Volume 443. Angiogenesis: *In Vitro* Systems
Edited by David A. Cheresh

Volume 444. Angiogenesis: *In Vivo* Systems (Part A)
Edited by David A. Cheresh

Volume 445. Angiogenesis: *In Vivo* Systems (Part B)
Edited by David A. Cheresh

Volume 446. Programmed Cell Death, The Biology and Therapeutic Implications of Cell Death (Part B)
Edited by Roya Khosravi-Far, Zahra Zakeri, Richard A. Lockshin, and Mauro Piacentini

Volume 447. RNA Turnover in Bacteria, Archaea and Organelles
Edited by Lynne E. Maquat and Cecilia M. Arraiano

VOLUME 448. RNA Turnover in Eukaryotes: Nucleases, Pathways and Analysis of mRNA Decay
Edited by LYNNE E. MAQUAT AND MEGERDITCH KILEDJIAN

VOLUME 449. RNA Turnover in Eukaryotes: Analysis of Specialized and Quality Control RNA Decay Pathways
Edited by LYNNE E. MAQUAT AND MEGERDITCH KILEDJIAN

VOLUME 450. Fluorescence Spectroscopy
Edited by LUDWIG BRAND AND MICHAEL L. JOHNSON

VOLUME 451. Autophagy: Lower Eukaryotes and Non-Mammalian Systems (Part A)
Edited by DANIEL J. KLIONSKY

VOLUME 452. Autophagy in Mammalian Systems (Part B)
Edited by DANIEL J. KLIONSKY

VOLUME 453. Autophagy in Disease and Clinical Applications (Part C)
Edited by DANIEL J. KLIONSKY

VOLUME 454. Computer Methods (Part A)
Edited by MICHAEL L. JOHNSON AND LUDWIG BRAND

VOLUME 455. Biothermodynamics (Part A)
Edited by MICHAEL L. JOHNSON, JO M. HOLT, AND GARY K. ACKERS (RETIRED)

VOLUME 456. Mitochondrial Function, Part A: Mitochondrial Electron Transport Complexes and Reactive Oxygen Species
Edited by WILLIAM S. ALLISON AND IMMO E. SCHEFFLER

VOLUME 457. Mitochondrial Function, Part B: Mitochondrial Protein Kinases, Protein Phosphatases and Mitochondrial Diseases
Edited by WILLIAM S. ALLISON AND ANNE N. MURPHY

VOLUME 458. Complex Enzymes in Microbial Natural Product Biosynthesis, Part A: Overview Articles and Peptides
Edited by DAVID A. HOPWOOD

VOLUME 459. Complex Enzymes in Microbial Natural Product Biosynthesis, Part B: Polyketides, Aminocoumarins and Carbohydrates
Edited by DAVID A. HOPWOOD

VOLUME 460. Chemokines, Part A
Edited by TRACY M. HANDEL AND DAMON J. HAMEL

VOLUME 461. Chemokines, Part B
Edited by TRACY M. HANDEL AND DAMON J. HAMEL

VOLUME 462. Non-Natural Amino Acids
Edited by TOM W. MUIR AND JOHN N. ABELSON

VOLUME 463. Guide to Protein Purification, 2nd Edition
Edited by RICHARD R. BURGESS AND MURRAY P. DEUTSCHER

VOLUME 464. Liposomes, Part F
Edited by NEJAT DÜZGÜNEŞ

VOLUME 465. Liposomes, Part G
Edited by NEJAT DÜZGÜNEŞ

VOLUME 466. Biothermodynamics, Part B
Edited by MICHAEL L. JOHNSON, GARY K. ACKERS, AND JO M. HOLT

VOLUME 467. Computer Methods Part B
Edited by MICHAEL L. JOHNSON AND LUDWIG BRAND

VOLUME 468. Biophysical, Chemical, and Functional Probes of RNA Structure, Interactions and Folding: Part A
Edited by DANIEL HERSCHLAG

VOLUME 469. Biophysical, Chemical, and Functional Probes of RNA Structure, Interactions and Folding: Part B
Edited by DANIEL HERSCHLAG

VOLUME 470. Guide to Yeast Genetics: Functional Genomics, Proteomics, and Other Systems Analysis, 2nd Edition
Edited by GERALD FINK, JONATHAN WEISSMAN, AND CHRISTINE GUTHRIE

VOLUME 471. Two-Component Signaling Systems, Part C
Edited by MELVIN I. SIMON, BRIAN R. CRANE, AND ALEXANDRINE CRANE

VOLUME 472. Single Molecule Tools, Part A: Fluorescence Based Approaches
Edited by NILS G. WALTER

VOLUME 473. Thiol Redox Transitions in Cell Signaling, Part A Chemistry and Biochemistry of Low Molecular Weight and Protein Thiols
Edited by ENRIQUE CADENAS AND LESTER PACKER

VOLUME 474. Thiol Redox Transitions in Cell Signaling, Part B Cellular Localization and Signaling
Edited by ENRIQUE CADENAS AND LESTER PACKER

VOLUME 475. Single Molecule Tools, Part B: Super-Resolution, Particle Tracking, Multiparameter, and Force Based Methods
Edited by NILS G. WALTER

VOLUME 476. Guide to Techniques in Mouse Development, Part A Mice, Embryos, and Cells, 2nd Edition
Edited by PAUL M. WASSARMAN AND PHILIPPE M. SORIANO

VOLUME 477. Guide to Techniques in Mouse Development, Part B Mouse Molecular Genetics, 2nd Edition
Edited by PAUL M. WASSARMAN AND PHILIPPE M. SORIANO

VOLUME 478. Glycomics
Edited by MINORU FUKUDA

VOLUME 479. Functional Glycomics
Edited by MINORU FUKUDA

VOLUME 480. Glycobiology
Edited by MINORU FUKUDA

VOLUME 481. Cryo-EM, Part A: Sample Preparation and Data Collection
Edited by GRANT J. JENSEN

VOLUME 482. Cryo-EM, Part B: 3-D Reconstruction
Edited by GRANT J. JENSEN

VOLUME 483. Cryo-EM, Part C: Analyses, Interpretation, and Case Studies
Edited by GRANT J. JENSEN

VOLUME 484. Constitutive Activity in Receptors and Other Proteins, Part A
Edited by P. MICHAEL CONN

VOLUME 485. Constitutive Activity in Receptors and Other Proteins, Part B
Edited by P. MICHAEL CONN

VOLUME 486. Research on Nitrification and Related Processes, Part A
Edited by MARTIN G. KLOTZ

VOLUME 487. Computer Methods, Part C
Edited by MICHAEL L. JOHNSON AND LUDWIG BRAND

VOLUME 488. Biothermodynamics, Part C
Edited by MICHAEL L. JOHNSON, JO M. HOLT, AND GARY K. ACKERS

VOLUME 489. The Unfolded Protein Response and Cellular Stress, Part A
Edited by P. MICHAEL CONN

VOLUME 490. The Unfolded Protein Response and Cellular Stress, Part B
Edited by P. MICHAEL CONN

VOLUME 491. The Unfolded Protein Response and Cellular Stress, Part C
Edited by P. MICHAEL CONN

VOLUME 492. Biothermodynamics, Part D
Edited by MICHAEL L. JOHNSON, JO M. HOLT, AND GARY K. ACKERS

VOLUME 493. Fragment-Based Drug Design
Tools, Practical Approaches, and Examples
Edited by LAWRENCE C. KUO

SECTION ONE

TOOLS

CHAPTER ONE

Designing a Diverse High-Quality Library for Crystallography-Based FBDD Screening

Brett A. Tounge *and* Michael H. Parker

Contents

Abstract

A well-chosen set of fragments is able to cover a large chemical space using a small number of compounds. The actual size and makeup of the fragment set is dependent on the screening method since each technique has its own practical limits in terms of the number of compounds that can be screened and requirements for compound solubility. In this chapter, an overview of the general requirements for a fragment library is presented for different screening platforms. In the case of the FBDD work at Johnson & Johnson Pharmaceutical Research and Development, L.L.C., our main screening technology is X-ray crystallography. Since every soaked protein crystal needs to be diffracted and a protein structure determined to delineate if a fragment binds, the size of our

Structural Biology and Medicinal Chemistry, Johnson & Johnson Pharmaceutical Research and Development, L.L.C., Spring House, Pennsylvania, USA

Methods in Enzymology, Volume 493
ISSN 0076-6879, DOI: 10.1016/B978-0-12-381274-2.00001-7

initial screening library cannot be a rate-limiting factor. For this reason, we have chosen 900 as the appropriate primary fragment library size. To choose the best set, we have developed our own mix of simple property ("Rule of 3") and "bad" substructure filtering. While this gets one a long way in terms of limiting the fragment pool, there are still tens of thousands of compounds to choose from after this initial step. Many of the choices left at this stage are not drug-like, so we have developed an *FBDD Score* to help select a 900-compound set. The details of this score and the filtering are presented.

1. Introduction

The typical collection, or deck, used for high throughput screening (HTS) is comprised mostly of compounds that have $\geq$15 nonhydrogen atoms. As an example, the size distribution of our compound collection at Johnson & Johnson Pharmaceutical Research and Development, L.L.C. shows an average nonhydrogen atom count of ~30 (Fig. 1.1). Compounds smaller than this, fragments, tend to have lower absolute affinities, and thus are difficult to detect in routine HTS campaigns. However, over the past decade, there has been a considerable amount of effort put into making use of these smaller compounds (Chen and Hubbard, 2009; Chessari and Woodhead, 2009; de Kloe Gerdien *et al.*, 2009; Fischer and Hubbard, 2009; Murray and Rees, 2009; Orita *et al.*, 2009b; Schulz and Hubbard, 2009; Wang *et al.*, 2009). These efforts have been driven by the recognition that fragments offer several unique properties relative to typical "drug-size" molecules. Since smaller compounds are less complex, they have a higher

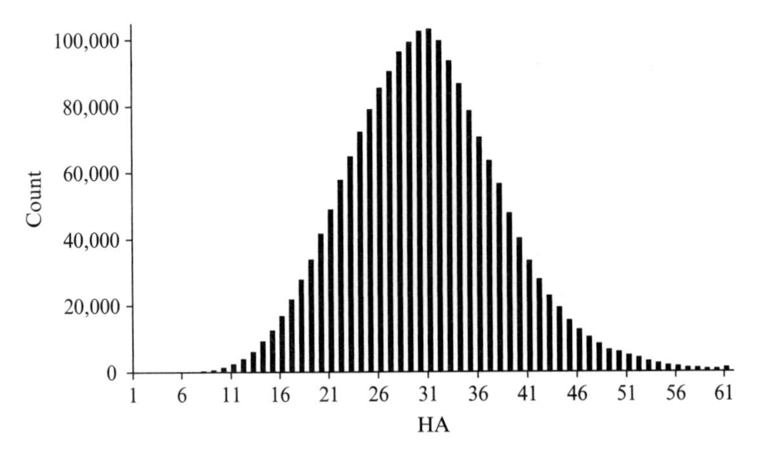

Figure 1.1 The size distribution of the Johnson & Johnson Pharmaceutical Research and Development, L.L.C. compound collection is centered on ~30 nonhydrogen atoms (HA). The typical HTS screen uses compounds with $\geq$15 nonhydrogen atoms.

probability of binding, tend to bind with higher ligand efficiencies, cover greater chemical space with fewer compounds, and, depending on how the fragments are chosen, have no built-in bias for a particular target.

As one moves from large complex drug-size molecules to smaller fragments, there are fewer constraints that need to be met for a compound to bind to a target protein. For example, for a protein to accommodate three hydrogen bonding partners while maintaining good van der Waals contacts requires a very specific binding pocket and ligand geometry. However, many proteins can accommodate a single hydrogen bond acceptor attached to a phenyl ring. As a result, the probability of binding goes up as the size of the ligand goes down (Hann *et al.*, 2001).

Support for this theory can be seen in the high ligand efficiencies found for smaller molecules. Ligand efficiency is generically defined as $\Delta\Delta G_{binding}/HA$, where HA is the nonhydrogen count. Numerous papers have been published showing that smaller molecules bind with higher ligand efficiencies (Fig. 1.2; Abad-Zapatero, 2007; Bembenek *et al.*, 2009; Hopkins Andrew *et al.*, 2004; Nissink, 2009). This is directly related to the argument made above for binding probability. Smaller, less complex molecules have fewer constraints that need to be met when binding and thus they typically have a better "fit" to the protein.

The efficiency of chemical space coverage afforded by using fragments is illustrated in Fig. 1.3. There are ~33 unique six-membered rings in the Comprehensive Medicinal Chemistry database (Bemis and Murcko, 1996,

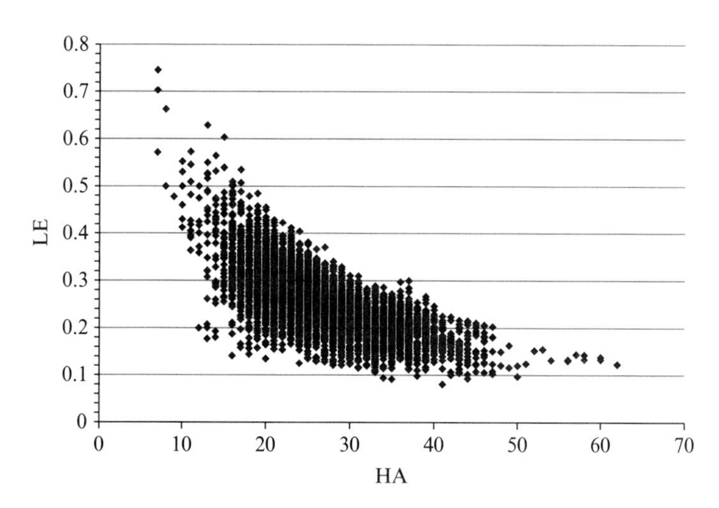

Figure 1.2 Ligand efficiency (LE) shows a precipitous decline between 10 and 25 nonhydrogen atoms (HA). We have extracted the affinity data used in this plot from the BindingDB database developed at the University of Maryland Biotechnology Institute (Liu *et al.*, 2007).

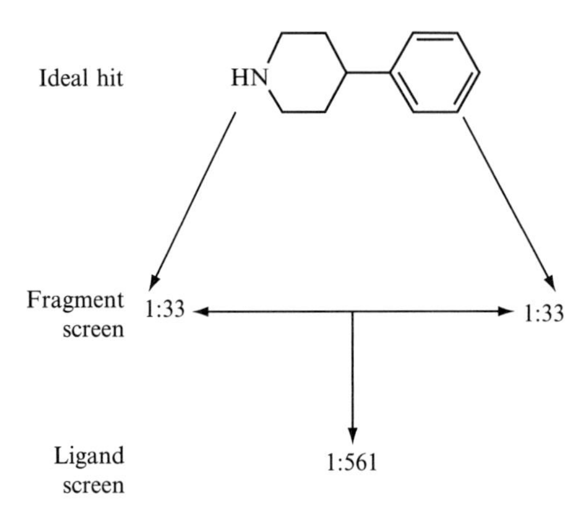

Figure 1.3 By breaking compounds into smaller substituents, one is able to cover chemical space more efficiently. In this case, by screening the piperidine and phenyl rings as fragments, the ideal two-ring system could be mapped by screening just 33 compounds. Enumerating all 33 possible six-membered rings into a two-ring system would result in 561 compounds that would need to be screened.

1999). For just a simple two-ring system, that would result in 561 unique combinations (assuming a single attachment point for each ring). As a result, to find the "ideal" hit, all 561 compounds would need to be screened. If one takes the fragments individually, only 33 compounds would need to be screened to find the best two rings.

Finally, one of the biggest advantages for fragment screening is that you are not starting with compound libraries that are already biased for specific target classes. For example, many corporate compound collections are heavily populated with ligands designed to hit kinases. The biased ligands will likely not work well for very divergent protein classes. In contrast, by breaking ligands into their component fragments, the resulting library will have broader applicability in screening.

2. Library Requirements for Different Screening Methods

For all these reasons, fragment-based drug discovery (FBDD) has greatly expanded over the past decade. In particular, there has been considerable effort put into developing screening technologies suited to detecting weak binders. The methodologies used fall into three basic categories: traditional biophysical screening, NMR, and X-ray crystallography (Barker *et al.*, 2007; Blaney *et al.*, 2006; Dalvit, 2009; Danielson, 2009;

Hartshorn *et al.*, 2005; Jhoti, 2007; Jhoti *et al.*, 2007; Orita *et al.*, 2009b; Perspicace *et al.*, 2009). Each of these methods represents tradeoffs between throughput and structural information content. As a result, the library design criteria are screening method-dependent.

2.1. Traditional biophysical screening methods

Traditional biophysical methods (e.g., surface plasma resonance, Thermofluor®) represent the high end in terms of throughput. For most of these methods, running $\geq 100{,}000$ compounds for a screen is routine. However, the very high throughput methods represent the other end of the spectrum in terms of binding site information content. Traditional biophysical methods simply tell you whether or not a binding event occurs, but not where on the protein it occurred. As a result, these methods are often used as a prescreen to select ligands for NMR or X-ray crystallography. In addition, compound solubility is a crucial aspect of the library design for these methods.

2.2. NMR screening

Ligand-detected NMR-based screening falls in the middle in terms of both structural information and throughput. Typical fragment screening libraries for NMR are 20,000–50,000 in size. Ligand-detected NMR methods offer limited information in terms of binding. At best, the pharmacophore for the ligand can be defined using saturation transfer difference-based methods which tell you what part of the ligand is in contact with the protein (Mayer and Meyer, 1999). Protein chemical shift perturbation-based NMR methods offer more structural information, but are lower throughput ($\sim 10{,}000$ compounds; Hajduk *et al.*, 1999). In addition, since the backbone resonance assignments must be done, chemical shift perturbation techniques are limited to proteins ≤ 40 kDa in molecular weight. Finally, as in the methods mentioned above, compound solubility is crucial for NMR-based screening.

2.3. X-ray screening

While X-ray screening represents the low end in terms of typical throughput (~ 1000) and it can only be applied in cases where a robust X-ray structure can be produced, it offers the highest information content. Once a hit is found, the exact binding location and orientation is known, which allows for more direct follow-up chemistry. In addition, we have found that in practice, compound solubility is not a limiting factor. As long as we can obtain a sufficient concentration in the buffer solution for soaking, any precipitate simply provides a source for additional compound as fragments soak into the crystal lattice.

3. Library Design for X-Ray Screening

For various reasons that are outlined in other chapters of this book, X-ray crystallography was chosen as the main screening platform for our FBDD program. As mentioned above, the low throughput of this screening method forces one to balance screening coverage with time to screen. In practice, we settle on a 900-compound screening set. This allows for sufficient chemical space coverage and allows the screen to be accomplished in a reasonable time frame. Various methods have been published for making such a selection. We will review some of the more common techniques as well as present our unique metric for final compound selection.

3.1. Property filters

The overall process for compound selection is summarized in Fig. 1.4. Our initial pool of candidate fragments is drawn from both our in-house library and commercial sources. This rather large list can be quickly reduced by applying simple property filters. The most commonly used filter is the "Rule of 3" (Congreve *et al.*, 2003). For the initial property filtering step, we use a modified rule of three set,

1. $5 \leq$ nonhydrogen atoms ≤ 15
2. Hydrogen bond acceptors ≤ 3
3. Hydrogen bond donors ≤ 3
4. $1 \leq$ number of rings ≤ 3
5. Number of unspecified stereo centers $= 0$

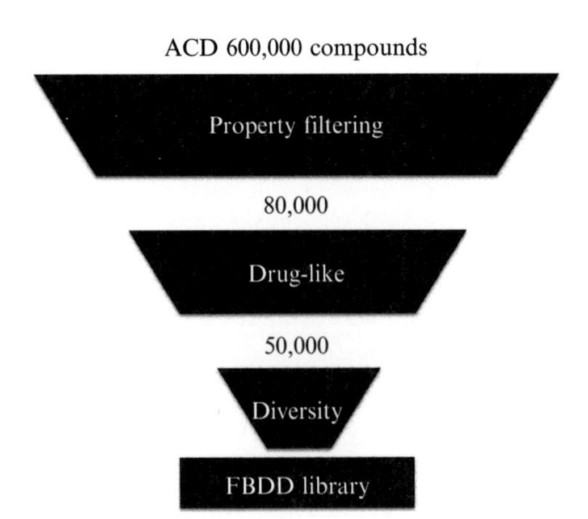

Figure 1.4 Overview of the general filtering process used to build FBDD libraries. The compound counts after each stage is based on a filtering of the ACD.

coupled with a substructure search to remove unwanted functional groups (Table 1.1).

As an example, the above filtering is applied to the Available Chemicals Directory (ACD; Symyx Solutions, Inc., 2007a). After running the property filtering on the initial set (632,790 compounds), 101,952 compounds remain. The substructure filtering then brings the list down to 77,078.

3.2. Fragment ranking—*FBDD Score*

At this point, several methods are available to rank and select fragments (Baurin *et al.*, 2004; Blomberg *et al.*, 2009; Brewer *et al.*, 2008; Colclough *et al.*, 2008; Jacoby *et al.*, 2003; Mercier *et al.*, 2006; Orita *et al.*, 2009a; Schuffenhauer *et al.*, 2005). While diversity metrics could be used, they tend to pick out non-drug-like compounds since the rough property filtering outlined above does not remove these compounds. Instead, it is important at this stage to introduce a method to select the more desirable compounds. For this, we use a metric we call *FBDD Score* that captures size, chemical complexity, and drug-likeness.

For drug-likeness, we have adapted a previously published method used to rank reagents for combinatorial chemistry libraries (Tounge and Reynolds, 2004). For each fragment molecule, we calculate a subsimilarity to each compound in a drug-like database (the Comprehensive Medicinal Chemistry database) and keep the highest subsimilarity score (Symyx Solutions, Inc., 2007b). The score (CMCSubSim) is computed using the following general formula:

$$\text{CMCSubSim} = \frac{\text{Total number of keys in the target that match keys in the probe}}{\text{Keys}_\text{P}} \tag{1.1}$$

where Keys_P is the total number of unique keys in the probe molecule (i.e., the fragment). In our implementation, the Extended Connectivity with a path length of 4, ECFP_4, descriptor keys are used (Rogers and Hahn, 2010). The CMCSubSim score ranges from 1 to 0. A fragment molecule whose entire structure is found in one or more compounds from the reference database is considered drug-like and will have a CMCSubSim = 1.

The next element of the score is designed to capture the complexity of the molecule. The goal with this aspect of the score is to bias our set of fragments to those with a simple pharmacophore (e.g., limit the number of functional groups (amines, acids, halogens, etc.) on a given fragment). This keeps the probability of binding high since fewer constraints need to be met. To capture this, we use the simple metric of calculating the percentage of heteroatoms in the fragment. This component of the score is simply

Table 1.1 A substructure filtering of the initial fragment pool is run to eliminate all compounds that have the functional groups listed

1-2 Diketones	Aldehydes	Itrazole-like	Diazo	Azides
Acetylene	Alkyltriene	Long chain aliphatics (C7+)	Nitroxide	Nitro
Nitrile	Alkyl halide	Carbamoyl halides	Peroxide	*S*-Oxide
Acetals	α-Halo carbonyls	Carbodiimides	Hypoiodate-like	Phosphorhalide
Acid chlorides	Anhydrides	Disulfide oxides	Isocyanates	Diphosphide
Acrylates	Anthracenes	1,4-Pentadiene-3-diphenylmethylene	Isothiocyanates	Sulfonyl halides
Acyl halides	Aziridines	Fullerenes	Methylenehydrazines	Thiocarbamoyl halides
Acylhydrazinones	Oxiridines	Halogen-heteroatom	Diamine (N_chain_N)	1,1,2-Trimercaptoethylene
Adamantanes	Miconazole-like	Hemes	*N*-hydroxy	Thioyl halides

$$\text{PercHetAtom} = \frac{\text{HA} - \text{Number of carbon atoms}}{\text{HA}} \tag{1.2}$$

where HA is the nonhydrogen atom count.

The final element of the score is added to bias our selections to smaller molecules. This is accomplished by adding a penalty score which is defined as

$$\text{SizePenalty} = \frac{\text{HA} - 6}{9} \tag{1.3}$$

This score applies no penalty when the atom count is 6 and −1 for fragments with 15 nonhydrogen atoms. This aspect of the score can be adjusted depending on the size bias one wants to apply.

All these components are combined into the final *FBDD Score* as follows:

$$\textit{FBDD Score} = \text{CMCSubSim} - \frac{\text{SizePenalty}}{2} - \frac{\text{PercHetAtom}}{2} \tag{1.4}$$

In this combination of the terms, both the size and complexity terms have been divided by 2 in order to give more weight to the drug-like term. An example of the score can be found in Fig. 1.5. Once applied to the entire property-filtered fragment set, the score allows one to pick out drug-like, small compounds with simple pharmacophores. In practice, all compounds with an *FBDD Score* $\geq \sim 0.0$ are kept (Fig. 1.6). For example, scoring and filtering the remaining 77,078 compounds from the above ACD filtering using a cutoff of ≥ 0 leaves 53,515 compounds from which to do the final library selection.

3.3. Diversity

Once the *FBDD Score* filtering is done, diversity can be used to select the final number of fragments needed for the screen. At this point, many different algorithms are available. The details of the final selection of our fragment library for X-ray screening set are outlined in Section 4 of this chapter.

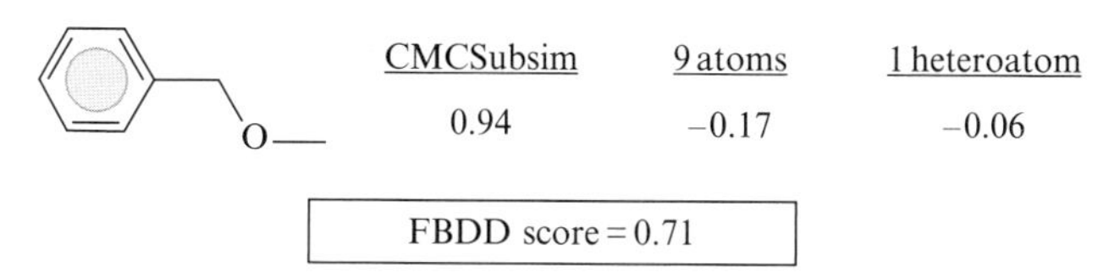

Figure 1.5 The *FBDD Score* is composed of three terms and helps to rank fragments. In the example shown, this fragment ranks high for drug-likeness (CMCSubSim) and has low penalties for size (−0.17) and complexity (−0.06).

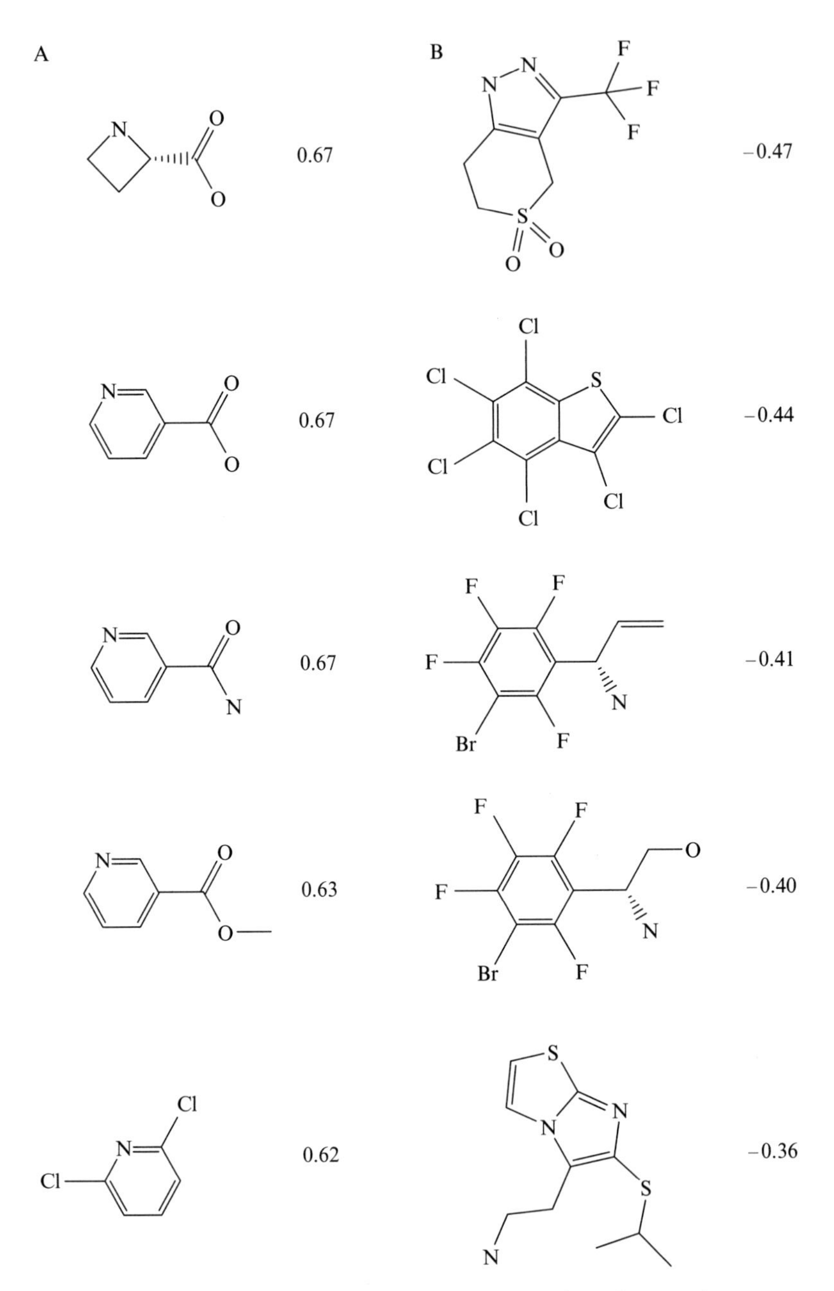

Figure 1.6 The *FBDD Score* provides a robust scoring algorithm to select out more desirable fragments. The examples shown are a selection of high (A) and low (B) scoring fragments from the ACD.

4. Implementation

4.1. X-ray primary screening library

Given our choice of X-ray crystallography as our primary screening platform, a non-target-biased initial screening set of 900 compounds needs to be selected (Primary Screening Library). The initial pool of candidate compounds is a mix of commercial sources (ACD) and our corporate collection. In total, this entailed ~2.5 million compounds. The first step is to apply the property and substructure filtering outlined above. This initial filtering leaves a pool of ~110,000 compounds.

After the property filtering, the *FBDD Score* is calculated. The score ranges from –0.47 to 1.0 for this set, with ~74,000 compounds having a score greater than zero. In practice, most of the compounds chosen for the primary screen deck have an *FBDD Score* of greater than zero. However, some compounds that scored lower than 0.0 are kept. These compounds offer a unique functionality in terms of 3D shape. For example, some bridged and spiro systems that had an *FBDD Score* of approximately –0.3 are included.

The final step is to use a diversity metric to select the final set of 900 compounds. To ensure we cover the desired functional groups, this selection is done in two steps. First, substructure filtering is used to divide the set into six groups (Hartshorn *et al.*, 2005):

1. Carboxylic acids
2. Amines
3. Amidines
4. Alcohols
5. Amides
6. Other

Second, from each of these groups, a diverse set of fragments are chosen. This is done using the "Diverse Molecule" selection tool in Pipeline Pilot (ECFP_4 fingerprint; Accelrys, 2009). For each subgroup, ~150 compounds are chosen to establish the final screening deck of 900 fragments. A represented set of this final selection can be found in Fig. 1.7.

4.2. Quantity and purity

Once the final selection is complete, the compounds are all ordered as neat samples from either vendors or from our internal collection. Before entering into the FBDD library, quality control of purity is run on each sample to ensure a purity of ≥95%. All compounds are stocked at ≥50 mg providing a supply to cover multiple years of screening.

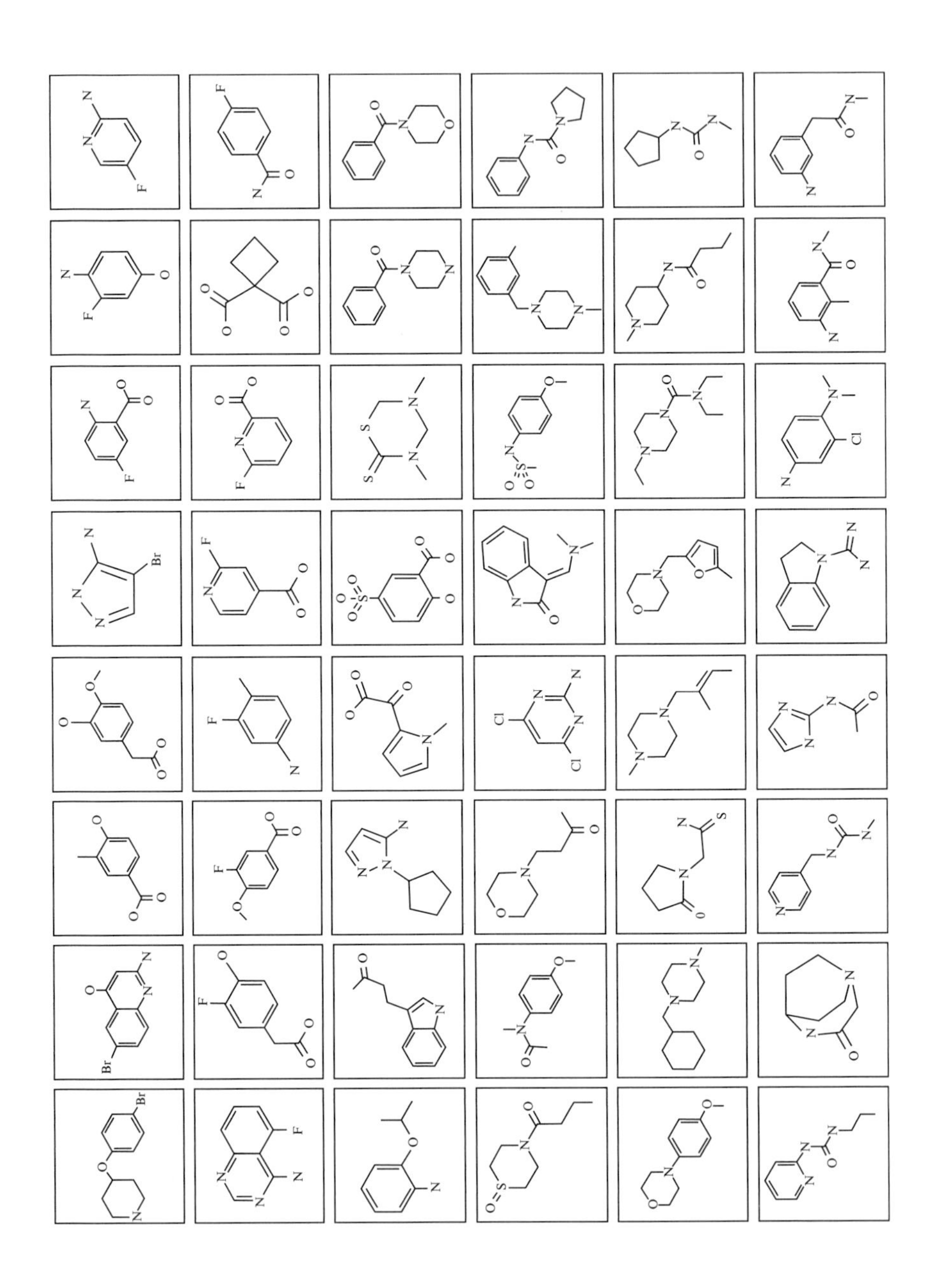

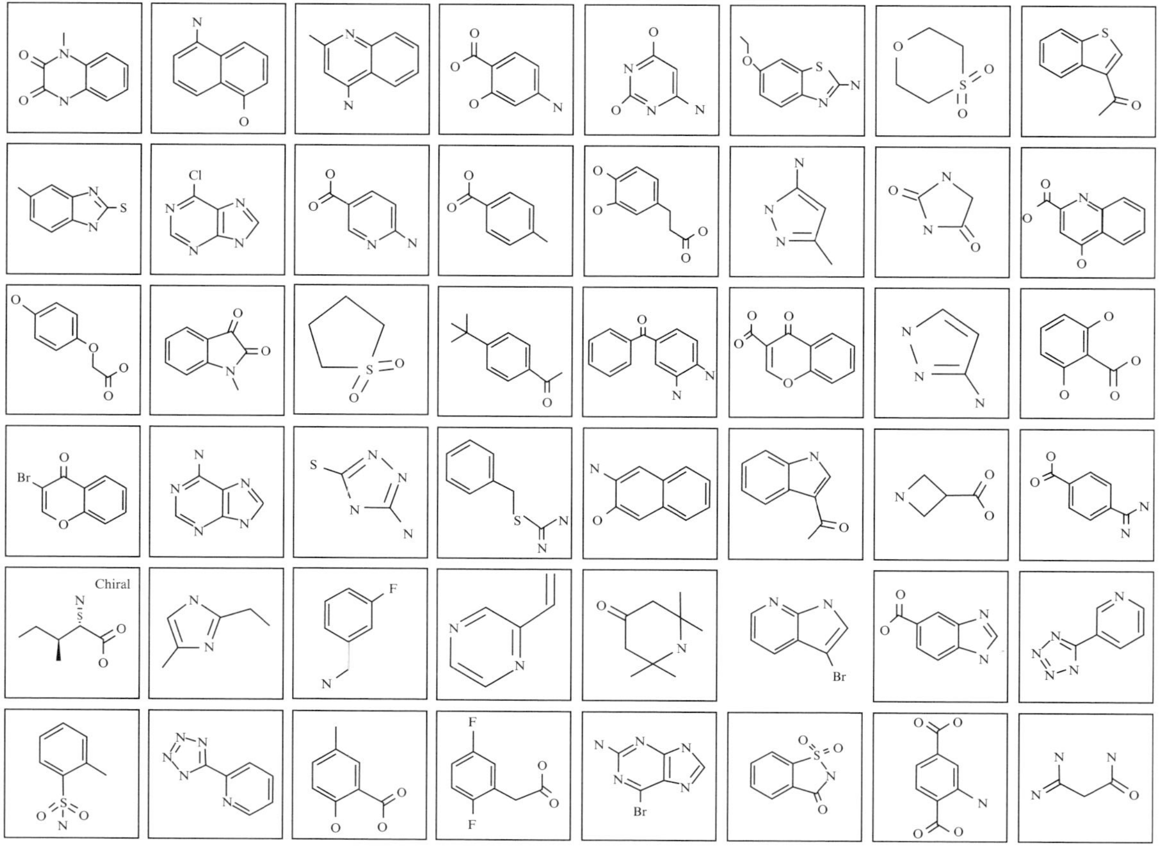

Figure 1.7 (Continued).

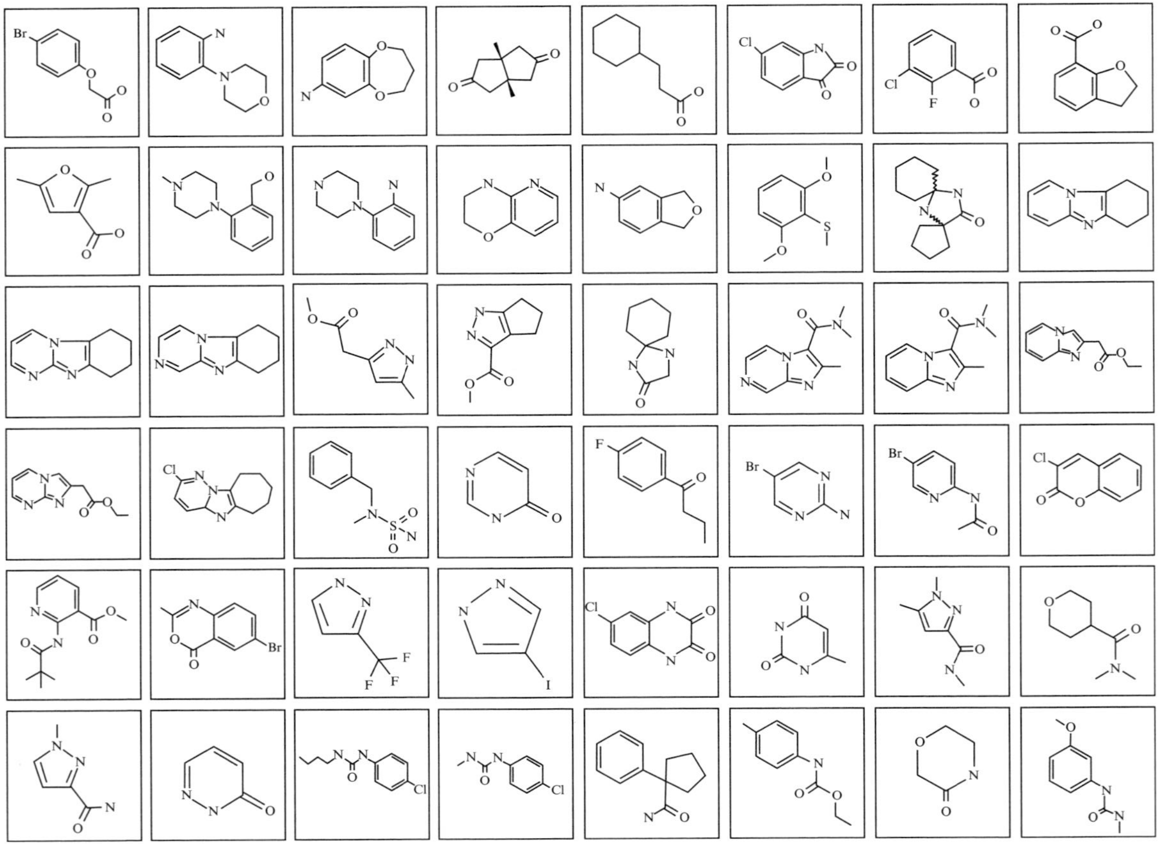

Figure 1.7 A representative sample of the final FBDD Primary Library.

4.3. Clustering for plating

In order to increase throughput for the crystallography studies, all the primary library compounds are grouped into clusters of five for screening. In order to avoid the need for deconvolution after finding a hit well, the compounds are grouped into shape-similar sets. This differs from the more routinely used method of grouping by dissimilarity. A detailed discussion of the ideas behind the shape-similar grouping choice can be found in Chapter 13 (Spurlino, 2011). In brief, one of its main advantages is that it allows us to quickly determine the pharmacophore of the binding fragments.

The clustering of the compound is done using the Pipeline Pilot platform (Fig. 1.8). The clustering algorithm first splits the fragments into compounds with neutral, negative, and positively charged groups. Within the groups, the compounds are clustered into groups of approximately five using the "Cluster Molecules" tool to determine the cluster center molecules. Since the "Cluster Molecules" component only allows you to define an average cluster size, this step is used only to find the most dissimilar compounds that can then be used as the seed molecules for the next component, "Equipartition Molecules." This component finds the four closest molecules to the seed and groups them to form the final clusters of five. For all these steps, the following descriptors are used: Pipeline Pilot functional class fingerprints (path length 4), Pipeline Pilot extended connectivity class fingerprints (path length 4), calculated logP (AlogP; Ghose *et al.*, 1998), number of hydrogen bond acceptors, number of hydrogen bond donors, MDL public keys, number of atoms, and number of rings.

5. Conclusions

As outlined above, the requirements for an FBDD library in terms of size, and in some cases, physical properties such as solubility, differ depending on the screening methods being used. For X-ray crystallography-based screening, throughput is low, so the library size is small. This puts tighter constraints on how a library is chosen. In particular, it is important to select molecules that have no more than 15 nonhydrogen atoms, have a simple pharmacophore, and are drug-like. While very useful, "Rule of 3" type property filtering does not capture enough chemical information to fully define this set. To get to the final selection, the *FBDD Score* was developed. It provides a robust scoring metric to bias fragment selections to simple, in terms of pharmacophore, and drug-like compounds.

One criterion that must be met independent of the screening method is a strict purity quality control cutoff. All compounds must be $\geq$95% pure before becoming part of the screening deck. Poor purity could lead

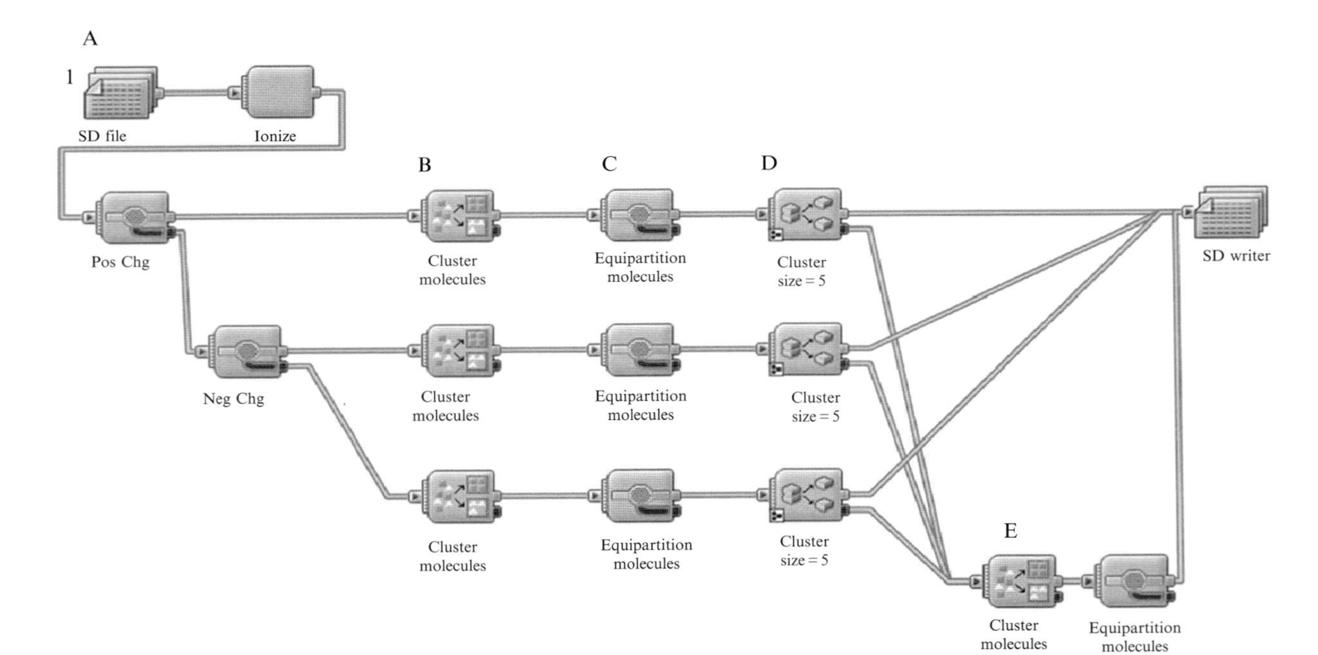

Figure 1.8 Pipeline Pilot protocol for creating the shape-similar wells of five compounds. The compounds are sorted by charge state (step A) and then piped into a clustering component (step B) where they are grouped into clusters of approximately five to find the cluster centers. To enforce the fixed cluster size of five, the compounds are next piped to the equipartition component (step C). This component finds the four closest molecules to the seed (from step B) and groups them to form the final clusters of five. Finally, a check is done to make sure all compounds are grouped into wells of five (steps D and E).

to false-positive hits that are close to impossible to track down. This can waste valuable time and money. In the end, one needs a high-quality fragment pool at the start of the FBDD process.

REFERENCES

Abad-Zapatero, C. (2007). Ligand efficiency indices for effective drug discovery. *Expert Opin. Drug Discov.* **2,** 469–488.

Accelrys (2009). *Pipeline Pilot, 7.5.* Accelrys, San Diego, CA.

Barker, J., Hesterkamp, T., Schade, M., and Whittaker, M. (2007). Fragment screening: Biochemical assays versus NMR. *Innov. Pharm. Technol.* **23,** 19–22.

Baurin, N., Aboul-Ela, F., Barril, X., Davis, B., Drysdale, M., Dymock, B., Finch, H., Fromont, C., Richardson, C., Simmonite, H., and Hubbard, R. E. (2004). Design and characterization of libraries of molecular fragments for use in NMR screening against protein targets. *J. Chem. Inf. Comput. Sci.* **44,** 2157–2166.

Bembenek, S. D., Tounge, B. A., and Reynolds, C. H. (2009). Ligand efficiency and fragment-based drug discovery. *Drug Discov. Today* **14,** 278–283.

Bemis, G. W., and Murcko, M. A. (1996). The properties of known drugs. 1. Molecular frameworks. *J. Med. Chem.* **39,** 2887–2893.

Bemis, G. W., and Murcko, M. A. (1999). Properties of known drugs. 2. Side chains. *J. Med. Chem.* **42,** 5095–5099.

Blaney, J., Nienaber, V., and Burley, S. K. (2006). Fragment-based lead discovery and optimization using X-ray crystallography, computational chemistry, and high-throughput organic synthesis. *Meth. Princ. Med. Chem.* **34,** 215–248.

Blomberg, N., Cosgrove, D. A., Kenny, P. W., and Kolmodin, K. (2009). Design of compound libraries for fragment screening. *J. Comput. Aided Mol. Des.* **23,** 513–525.

Brewer, M., Ichihara, O., Kirchhoff, C., Schade, M., Whittaker, M. (2008). Assembling a fragment library. *In* "Fragment-Based Drug Discovery: A Practical Approach," (E. Zartler, M. Shapiro, eds.), pp39–62. John Wiley & Sons, Ltd, United Kingdom.

Chen, I. J., and Hubbard, R. E. (2009). Lessons for fragment library design: Analysis of output from multiple screening campaigns. *J. Comput. Aided Mol. Des.* **23,** 603–620.

Chessari, G., and Woodhead, A. J. (2009). From fragment to clinical candidate—A historical perspective. *Drug Discov. Today* **14,** 668–675.

Colclough, N., Hunter, A., Kenny, P. W., Kittlety, R. S., Lobedan, L., Tam, K. Y., and Timms, M. A. (2008). High throughput solubility determination with application to selection of compounds for fragment screening. *Bioorg. Med. Chem.* **16,** 6611–6616.

Congreve, M., Carr, R., Murray, C., and Jhoti, H. (2003). A 'rule of three' for fragment-based lead discovery? *Drug Discov. Today* **8,** 876–877.

Dalvit, C. (2009). NMR methods in fragment screening: Theory and a comparison with other biophysical techniques. *Drug Discov. Today* **14,** 1051–1057.

Danielson, U. H. (2009). Fragment library screening and lead characterization using SPR biosensors. *Curr. Top. Med. Chem. (Sharjah, United Arab Emirates)* **9,** 1725–1735.

de Kloe Gerdien, E., Bailey, D., Leurs, R., and de Esch Iwan, J. P. (2009). Transforming fragments into candidates: Small becomes big in medicinal chemistry. *Drug Discov. Today* **14,** 630–646.

Fischer, M., and Hubbard, R. E. (2009). Fragment-based ligand discovery. *Mol. Interv.* **9,** 22–30.

Ghose, A. K., Viswanadhan, V. N., and Wendoloski, J. J. (1998). Prediction of hydrophobic (lipophilic) properties of small organic molecules using fragment methods: An analysis of AlogP and CLogP methods. *J. Phys. Chem. A* **102,** 3762–3772.

Hajduk, P. J., Gerfin, T., Boehlen, J.-M., Haeberli, M., Marek, D., and Fesik, S. W. (1999). High-throughput nuclear magnetic resonance-based screening. *J. Med. Chem.* **42,** 2315–2317.

Hann, M. M., Leach, A. R., and Harper, G. (2001). Molecular complexity and its impact on the probability of finding leads for drug discovery. *J. Chem. Inf. Comput. Sci.* **41,** 856–864.

Hartshorn, M. J., Murray, C. W., Cleasby, A., Frederickson, M., Tickle, I. J., and Jhoti, H. (2005). Fragment-based lead discovery using X-ray crystallography. *J. Med. Chem.* **48,** 403–413.

Hopkins Andrew, L., Groom Colin, R., and Alex, A. (2004). Ligand efficiency: A useful metric for lead selection. *Drug Discov. Today* **9,** 430–431.

Jacoby, E., Davies, J., and Blommers, M. J. J. (2003). Design of small molecule libraries for NMR screening and other applications in drug discovery. *Curr. Top. Med. Chem. (Hilversum, Netherlands)* **3,** 11–23.

Jhoti, H. (2007). Fragment-based drug discovery using rational design. *Ernst Schering Found. Symp. Proc.* **3,** 169–185.

Jhoti, H., Cleasby, A., Verdonk, M., and Williams, G. (2007). Fragment-based screening using X-ray crystallography and NMR spectroscopy. *Curr. Opin. Chem. Biol.* **11,** 485–493.

Liu, T., Lin, Y., Wen, X., and Jorissen, R. N. (2007). BindingDB: A web-accessible database of experimentally determined protein-ligand binding affinities. *Nucleic Acids Res.* **35,** D198–D201.

Mayer, M., and Meyer, B. (1999). Characterization of ligand binding by saturation transfer difference NMR spectroscopy. *Angew. Chem. Int. Ed.* **38,** 1784–1788.

Mercier, K. A., Germer, K., and Powers, R. (2006). Design and characterization of a functional library for NMR screening against novel protein targets. *Comb. Chem. High Throughput Screening* **9,** 515–534.

Murray, C. W., and Rees, D. C. (2009). The rise of fragment-based drug discovery. *Nat. Chem.* **1,** 187–192.

Nissink, J. W. M. (2009). Simple size-independent measure of ligand efficiency. *J. Chem. Inf. Model.* **49,** 1617–1622.

Orita, M., Ohno, K., and Niimi, T. (2009a). Two golden ratio' indices in fragment-based drug discovery. *Drug Discov. Today* **14,** 321–328.

Orita, M., Warizaya, M., Amano, Y., Ohno, K., and Niimi, T. (2009b). Advances in fragment-based drug discovery platforms. *Expert Opin. Drug Discov.* **4,** 1125–1144.

Perspicace, S., Banner, D., Benz, J., Muller, F., Schlatter, D., and Huber, W. (2009). Fragment-based screening using surface plasmon resonance technology. *J. Biomol. Screen.* **14,** 337–349.

Rogers, D., and Hahn, M. (2010). Extended-connectivity fingerprints. *J. Chem. Inf. Model.* **50,** 742–754.

Schuffenhauer, A., Ruedisser, S., Marzinzik, A. L., Jahnke, W., Blommers, M., Selzer, P., and Jacoby, E. (2005). Library design for fragment based screening. *Curr. Top. Med. Chem. (Sharjah, United Arab Emirates)* **5,** 751–762.

Schulz, M. N., and Hubbard, R. E. (2009). Recent progress in fragment-based lead discovery. *Curr. Opin. Pharmacol.* **9,** 615–621.

Spulino, J. (2011). Fragment Screening Purely with Protein Crystallography. *In* "Fragment Based Drug Design Tools, Practical Approaches, and Examples," (L. Kuo, ed.), Vol. 493. Elsevier, San Diego.

Symyx Solutions, Inc. (2007a). Available Chemicals Directory. Symyx Solutions, Inc, San Diego, CA.

Symyx Solutions, Inc. (2007b). Comprehensive Medicinal Chemistry. Symyx Solutions, Inc, San Diego, CA.

Tounge, B. A., and Reynolds, C. H. (2004). Defining privileged reagents using subsimilarity comparison. *J. Chem. Inf. Comput. Sci.* **44,** 1810–1815.

Wang, X., Yang, Q., and You, Q. (2009). Fragment-based drug discovery. *Zhongguo Yaoke Daxue Xuebao* **40,** 289–296.

CHAPTER TWO

Preparation of Protein Samples for NMR Structure, Function, and Small-Molecule Screening Studies

Thomas B. Acton,* Rong Xiao,* Stephen Anderson,* James Aramini,* William A. Buchwald,* Colleen Ciccosanti,* Ken Conover,* John Everett,* Keith Hamilton,* Yuanpeng Janet Huang,* Haleema Janjua,* Gregory Kornhaber,* Jessica Lau,* Dong Yup Lee,* Gaohua Liu,* Melissa Maglaqui,* Lichung Ma,* Lei Mao,* Dayaban Patel,* Paolo Rossi,* Seema Sahdev,* Ritu Shastry,† G. V. T. Swapna,* Yeufeng Tang,* Saichiu Tong,* Dongyan Wang,* Huang Wang,* Li Zhao,* *and* Gaetano T. Montelione*,†,1

Contents

* Center for Advanced Biotechnology and Medicine, Department of Molecular Biology and Biochemistry, and Northeast Structural Genomics Consortium, Rutgers University, Piscataway, New Jersey, USA
† Department of Biochemistry, Robert Wood Johnson Medical School, University of Medicine and Dentistry of New Jersey, Piscataway, New Jersey, USA
1 Corresponding author.

Methods in Enzymology, Volume 493
ISSN 0076-6879, DOI: 10.1016/B978-0-12-381274-2.00002-9

Abstract

In this chapter, we concentrate on the production of high-quality protein samples for nuclear magnetic resonance (NMR) studies. In particular, we provide an in-depth description of recent advances in the production of NMR samples and their synergistic use with recent advancements in NMR hardware. We describe the protein production platform of the Northeast Structural Genomics Consortium and outline our high-throughput strategies for producing high-quality protein samples for NMR studies. Our strategy is based on the cloning, expression, and purification of 6×-His-tagged proteins using T7-based *Escherichia coli* systems and isotope enrichment in minimal media. We describe 96-well ligation-independent cloning and analytical expression systems, parallel preparative scale fermentation, and high-throughput purification protocols. The 6×-His affinity tag allows for a similar two-step purification procedure implemented in a parallel high-throughput fashion that routinely results in purity levels sufficient for NMR studies (>97% homogeneity). Using this platform, the protein open reading frames of over 17,500 different targeted proteins (or domains) have been cloned as over 28,000 constructs. Nearly 5000 of these proteins have been purified to homogeneity in tens of milligram quantities (see Summary Statistics, http://nesg.org/statistics.html), resulting in more than 950 new protein structures, including more than 400 NMR structures, deposited in the Protein Data Bank. The Northeast Structural Genomics Consortium pipeline has been effective in producing protein samples of both prokaryotic and eukaryotic origin. Although this chapter describes our entire pipeline for producing isotope-enriched protein samples, it focuses on the major updates introduced during the last 5 years (Phase 2 of the National Institute of General Medical Sciences Protein Structure Initiative). Our advanced automated and/or parallel cloning, expression, purification, and biophysical screening technologies are suitable for implementation in a large individual laboratory or by a small group of collaborating investigators for structural biology, functional proteomics, ligand screening, and structural genomics research.

1. Introduction

The production of high-quality protein samples is critical to success in structural biology and drug discovery. During the second phase of the Protein Structure Initiative, the Northeast Structural Genomics Consortium (NESG; http://www.nesg.org) was one of the four Large Scale Centers funded by the National Institutes of Health and National Institute of General Medical Sciences. The goal of these centers was to determine the three-dimensional atomic-level structures of hundreds of novel proteins and protein domains and leverage this novel structural information to allow three-dimensional modeling of thousands of additional proteins (or protein domains). Another major goal of these centers was to develop and refine new technologies for high-throughput protein production, X-ray crystallography, NMR spectroscopy, structural bioinformatics, and related supporting infrastructure. The Protein Structure Initiative has worked to advance the field of biology through the dissemination of three-dimensional structural information on important protein domain families, production of protein expression systems and protocols, and by providing improved technology for protein sample preparation and structural analysis.

The protein sample production pipeline of the NESG has been previously described in detail (Acton *et al.*, 2005; Xiao *et al.*, 2010). Unlike nucleic acid-based genomics studies, where the macromolecules share common biophysical traits and standardized preparation techniques, proteins have a wide range of biophysical properties, making high-throughput production and sample purification significantly more challenging. Adding further complexity is the need to produce selenomethionine-labeled proteins for X-ray crystallography studies and isotope-enriched samples for nuclear magnetic resonance (NMR) studies. One of the unique features of the NESG pipeline is the ability to produce protein samples suitable for both structural determination strategies. Indeed, the NESG has produced a similar number of three-dimensional structures by each method. However, in this chapter, we concentrate on the production of high-quality protein samples for NMR studies. In particular, we provide an in-depth description of recent advances in the production of NMR samples and their synergistic use with recent advancements in NMR hardware. These include NMR microprobe technology allowing determination of protein structures from 100 μg quantities of protein (Aramini *et al.*, 2007), microprobe screening of small quantities of proteins in order to assess amenability to NMR structural determination, buffer optimization for protein stability and/or spectral quality (Rossi *et al.*, 2010), wheat germ cell-free protein translation for the production of protein samples (Vinarov *et al.*, 2006; Zhao *et al.*, 2010), and condensed-phase expression technologies allowing reduced costs in preparing isotope-enriched protein samples (Schneider *et al.*, 2009, 2010).

The NESG high-throughput cloning, protein expression, and protein purification pipeline is primarily based on *E. coli* T7 expression systems (Studier and Moffatt, 1986), utilizing 6×-His tags to allow a similar purification strategy to be utilized for proteins with diverse biophysical characteristics. Overall, this approach has proven to be a highly productive, efficient, and inexpensive method to produce the quantities of protein required for structural studies (Graslund *et al.*, 2008a). We also describe our strategies for target selection, construct optimization, ligation-independent cloning, analytical scale expression and solubility screening, midi-scale expression, purification and biophysical characterization, and large-scale protein sample production. The protein targets of the NESG project are both full-length proteins and domain constructs; we describe experimental and prediction methods for identifying disordered regions of proteins and how this information is used to design protein constructs suitable for NMR studies.

The current output of the NESG protein production facility on a weekly basis includes cloning and expression screening of over 100 protein targets, fermenting on a preparative (1–2 L) scale 50–75 expression constructs, and purifying in tens of milligram quantities roughly 30–40 targets for biophysical characterization, including NMR and/or crystallization screening. This platform can be readily implemented by traditional structural biology laboratories, biotechnology industry, and various proteomics and functional genomics projects since it is scalable, portable, and largely comprised of commercially available equipment.

2. Bioinformatics Infrastructure and Target Curation

During the second phase of the Protein Structure Initiative program, a centralized bioinformatics committee selected and distributed protein targets to the Large Scale Centers (Dessailly *et al.*, 2009). These generally constituted broadly conserved protein domain families for which no structural representative was yet available (BIG families), very large protein domain families with limited structural coverage (MEGA families), and domain families selected from metagenomic projects (META families) such as the human gut microbiome project (Gill *et al.*, 2006). The overall goal of targeting large protein domain families is to provide the greatest novel leverage of structure space per target (Liu *et al.*, 2007; Nair *et al.*, 2009). Fortuitously, this allows for a highly effective pan-genomic targeting strategy, taking advantage of the sequence differences and their concomitant biophysical characteristics within a domain family to identify and pursue the members most amenable for structure determination (Acton *et al.*, 2005; Liu *et al.*, 2004; Punta *et al.*, 2009). Each Large Scale Center also defined a biomedical theme; for example, the NESG is pursuing structural analysis of large numbers of proteins from the Human Cancer Pathway Protein Interaction Network (Huang *et al.*, 2008). These proteins are involved in cancer-associated signaling pathways and biological processes, together with their associated protein–protein interaction partners (http://nesg.org:9090/HCPIN/). In addition, the biomedical community nominates targets to the Protein Structure Initiative (http://www.nesg.org/target_nomination.html), which distributes these "Community Nominated Targets" to the various Large Scale Centers. Irrespective of the varied source of protein targets, the focus of the NESG remains the domain families represented in eukaryotic proteomes, including families that have exclusively eukaryotic members (e.g., the Ubiquitin Domain family) and families that have both eukaryotic and prokaryotic members (e.g., the START Domain family).

One of the main tasks of the Protein Structure Initiative and structural genomics as a whole is to increase the efficiency of protein three-dimensional structure production, ranging from target selection to automated Protein Data Bank depositions. With respect to protein production, experimental and bioinformatics studies have been undertaken to identify the parameters and procedures that correlate with success, such as high levels of protein solubility or "clone to Protein Data Bank deposition rates" (Dyson *et al.*, 2004; Goh *et al.*, 2004; Graslund *et al.*, 2008b; Slabinski *et al.*, 2007). The NESG has developed numerous bioinformatics tools to take advantage of pan-genomic targeting and identify the members of a protein domain family that are most amenable to protein production and structure determination. It is now well established that variation in protein sequence within a protein domain family can greatly affect its biophysical properties and therefore its success in protein production. The NESG has collected a vast data set on the behavior of proteins from diverse sources and families (both eukaryotic and prokaryotic) with the important fact that they were all prepared in a similar fashion. This has allowed NESG researchers to identify primary sequence traits that correlate with (i) high levels of protein expression (*E*) and solubility (*S*) in our bacterial expression systems (P_{ES}; unpublished results), (ii) greater probability of crystal structure determination based on protein sequence (P_{XS}; Price *et al.*, 2009) (http://nmr.cabm.rutgers.edu:8080/PXS/calculatePXS.jsp), and (iii) greater probability of amenability to NMR structure determination (P_{NMR}; unpublished results). Using our extensive list of over 175 Reagent Genomes (fully sequenced archeal, bacterial, and eukaryotic genomes and the corresponding genetic material for cloning) and these tools, we identify members of a protein family that are most likely to proceed to structure determination. This allows us to select several (4–6) proteins from each family for protein production in an effort to take advantage of the pan-genomic targeting strategy without pursuing large numbers of targets from a given protein family.

Beyond increasing efficiency by enriching our protein production pipeline with amenable targets, another major enhancement to our pipeline in

Protein Structure Initiative-2 is the NESG Construct Optimization Software. Highly homogeneous protein samples with minimal numbers of disordered residues are generally more amenable for successful protein crystallization and structure determination by X-ray crystallography. Although this chapter is focused on sample preparation for NMR studies, which can often be used successfully to study even fully disordered proteins, disordered segments of proteins can promote aggregation and deleteriously affect NMR spectral quality. In addition, a large percentage of targets (particularly human and other eukaryotic targets) are within multidomain proteins, which often misfold in prokaryotic systems (Netzer and Hartl, 1997). Of even more importance, many multidomain proteins exceed the size limitations for high-throughput NMR structural determination techniques. Domain parsing can be used to circumvent these significant issues. However, it is extremely challenging to predict the protein subsequence that will produce a soluble well-behaved protein, particularly with domains for which the three-dimensional structure is not yet known. This arises from problems with accurately predicting the domain boundaries and locations of disordered residues, and how to use such information to design an open reading frame that produces expression and solubility in the T7 system. Currently, our approach is to take advantage of our high-throughput platform and produce several alternative constructs, varying the termini of a targeted domain, followed by experiments to identify the protein subsequence with the best behavior. The success of this strategy has been reported by the NESG and others (Chikayama *et al.*, 2010; Graslund *et al.*, 2008b; Xiao *et al.*, 2010).

The NESG construct optimization software uses reports from the DisMeta Server (http://www-nmr.cabm.rutgers.edu/bioinformatics/disorder/), a metaserver that generates a consensus analysis of eight sequence-based disorder predictors to identify regions that are likely to be disordered. It also identifies predicted secretion signal peptides using SignalP (Bendtsen *et al.*, 2004), transmembrane segments by TMHMM (Krogh *et al.*, 2001), possible metal binding sites (Bertini *et al.*, 2010), secondary structure by PROFsec (Rost *et al.*, 2004) and PSIPred (McGuffin *et al.*, 2000), and interdomain disordered linkers (Fig. 2.1A). The data from these prediction servers, along with multiple sequence alignments of homologous proteins and hidden Markov models characteristic of the targeted protein domain families (Dessailly *et al.*, 2009), are used to predict possible structural domain boundaries. Based on this information, the software generates nested sets of alternative constructs for full-length proteins, multidomain constructs, and single-domain constructs. Thus for a single targeted region, we generally design multiple open reading frames varying the N and/or C-terminal sequences (Fig. 2.1B). These alternative constructs often possess significantly better expression, solubility, and biophysical behavior than their full-length parent sequences, increasing the likelihood of success in crystallization and the efficiency of structure

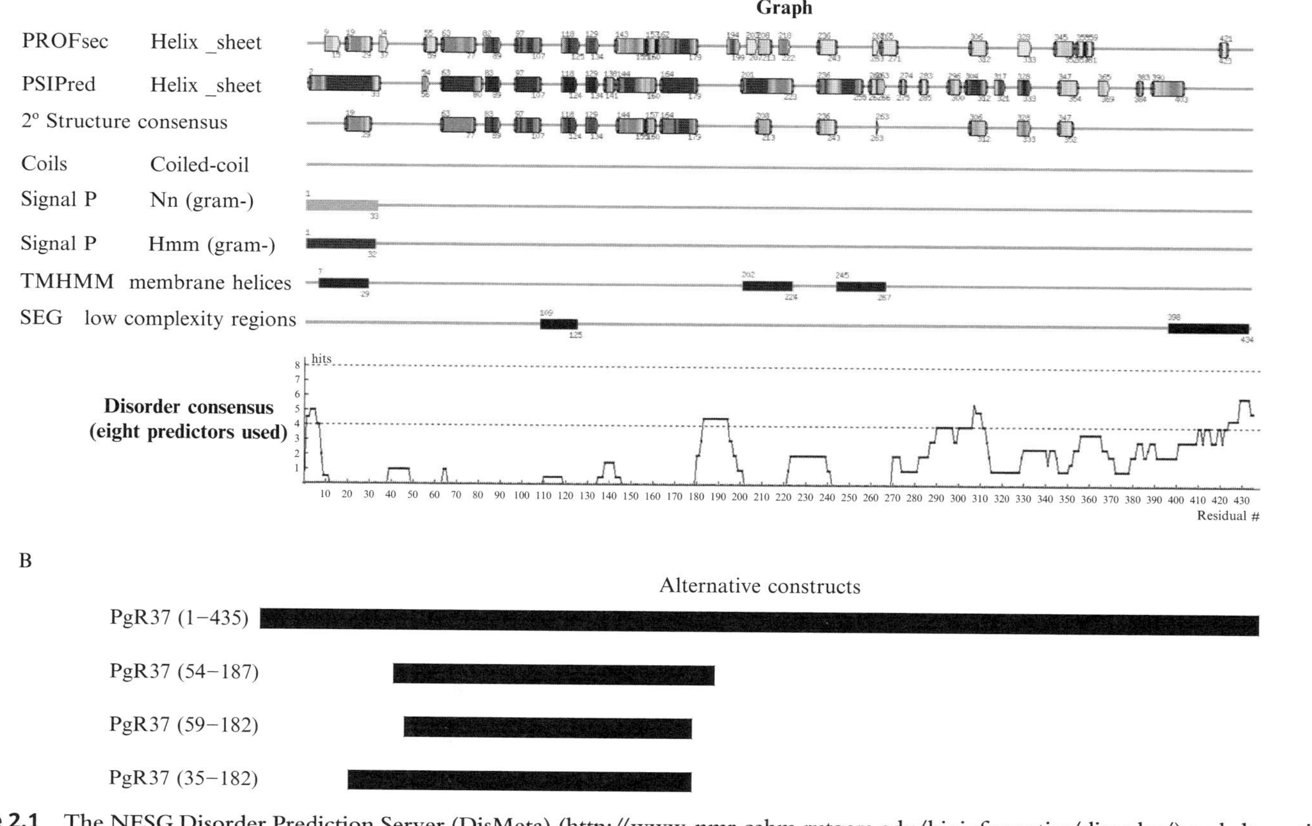

Figure 2.1 The NESG Disorder Prediction Server (DisMeta) (http://www-nmr.cabm.rutgers.edu/bioinformatics/disorder/) and alternative constructs produced by the Construct Optimization Software for the *Porphyromonas gingivalis* protein Q7MX54 (NESG ID: PgR37). (A) Output of the DisMeta server including prediction of secondary structure (PROFsec, PSIPred), parallel coiled coil regions (COIL; Lupas *et al.*, 1991), signal peptides (SignalP), transmembrane helices (TMHHM), low complexity regions (SEG; Wootton and Federhen, 1996), and the

production. In addition, it generates domain-sized regions that are amenable to high-throughput NMR studies, allowing access to proteins that would otherwise be too large to study by NMR.

An example of our domain parsing/construct optimization approach is shown in Fig. 2.1A. Consensus analysis of several disorder prediction algorithms (see Disorder Consensus panel) suggests that the C-terminal half of the 434-residue protein from *Porphyromonas gingivalis* Q7MX54 (NESG ID: PgR37) contains disordered regions. Cloning and expression analysis of the full-length protein in our bacterial expression system results in no detectable expression, supporting the disorder prediction based on the fact that proteins with significant disorder are often degraded in the *E. coli* cell. The construct optimization software generated alternative constructs based on the previously outlined criteria and the database of protein families that includes their annotations and multiple sequence alignments (DUF477) domain boundary (residues 59–182). These are depicted in Fig. 2.1B. The two expression constructs comprised of residues 54–187 and 59–182 also did not express at detectable levels. However, a slightly longer construct (residues 35–182) was highly expressed and soluble and ultimately allowed the structure of this targeted domain to be solved by NMR (Protein Data Bank ID:2KW7). Interestingly, the three-dimensional structure reveals the presence of two short helical regions between residues 38–48 and a β-strand for residues 54–56. These helices and the β-strand are tightly packed against each other and to other regions of the protein. The loss of these interactions in the shorter constructs likely destabilizes the protein, leading to degradation in the expression host.

Alternative constructs generated by the Construct Optimization Software are reviewed by a bioinformatics expert and, if approved, are entered into the NESG Protein Laboratory Information Management System (PLIMS). This JAVA-based Oracle database provides a detailed protein production data model following lab activities on a step-by-step basis. PLIMS is a web-based application consisting of four main modules: (i) Target registration and management, (ii) Molecular biology and analytical-scale protein expression, (iii) Large-scale fermentation, and (iv) Protein purification. PLIMS captures all pertinent information during the protein sample production process, interfacing where possible with robotics. It utilizes bar codes, personal digital assistants, and wireless technology. Key data from PLIMS is then uploaded to the Internet-accessible NESG SPINE Structure Production Database (Bertone *et al.*, 2001; Goh *et al.*, 2003) to be shared across the

---

Disorder Consensus plot showing the number of disorder prediction algorithms (0–8) predicting disorder versus the protein residue number in linear order N- to C-terminus. (B) Schematic representation of construct optimization for NESG protein target PgR37 from PFAM domain family DUF477, including full-length, residues 54–187, residues 59–182, and residues 35–182. Only the 35–182 construct produced a soluble-expressed protein and ultimately, an NMR structure (Protein Data Bank ID:2KW7).

consortium and with public databases, including the Protein Structure Initiative TargetDB (http://www.sbkb.org/; Chen *et al.*, 2004).

The final step before wet laboratory work can commence is the design of polymerase chain reaction (PCR) primers. The DNA sequences for the proposed alternative constructs are generated by the PLIMS database, usually in 96-well format. These sequences are entered into the freely available web-based software Primer Prim'er software (http://www.nesg.org/primer_primer) for automated primer design (Everett *et al.*, 2004). Vector-specific PCR primer sets are designed to amplify and insert targeted regions into a vector of choice. The NESG has designed the "NESG Multiplex Vector Kit," a series of vectors with a common multiple cloning site designed to minimize the number of nonnative residues while adding a short 6×-His tag (Acton *et al.*, 2005). These expression vectors are available from the Protein Structure Initiative Materials Repository (http://psimr.asu.edu/). Although affinity tags are generally required for high-throughput purification protocols (Crowe *et al.*, 1994; Sheibani, 1999), many commercial vector systems contain numerous nonnative residues (which are likely disordered) in their extended tags. Such large, disordered purification tags can produce large sharp peaks in the NMR spectrum that interfere with NMR studies. Primer Prim'er supports classical restriction endonuclease cloning, viral recombination cloning strategies such as Gateway Cloning (Invitrogen; Hartley *et al.*, 2000), as well as the InFusion (Clonetech)-based cloning most commonly used in the NESG sample production pipeline. It can also be easily modified to support other cloning strategies. In this task, Primer Prim'er generates ORF-specific primers with regions of vector overlap for use with the ligation-independent cloning strategy, as discussed in detail below. The software then organizes the primers into 96-well format, separating forward and reverse primers onto two different plates but in corresponding wells. Additionally, the user can array the primers in a variety of fashions, grouping primer sets on an expected size of amplification (increasing, decreasing, staggered, etc.), template source (e.g., all *E. coli* targets grouped together), etc. (Everett *et al.*, 2004). The arrayed primer sets are entered into PLIMS, which generates order forms for an oligonucleotide vendor.

3. Ligation-Independent High-Throughput Cloning and Analytical Scale-Expression Screening

3.1. Ligation-independent cloning and automated vector construction

The first step in the high-throughput cloning is the procurement of template DNA for the PCR amplification of the coding sequence regions selected by the construct design process. Unlike oligonucleotide primers that are chemically synthesized and easily available from a variety of vendors at inexpensive rates, obtaining a high-quality PCR template has a number of issues. For prokaryotic targets, the NESG has organized a collection of genomic templates for over 150 reagent genomes. In many cases, Whole Genome Amplification by Multiple Displacement Amplification has been used to generate template genomic DNA for our prokaryotic reagent genomes (Acton *et al.*, 2005; Xiao *et al.*, 2010). However, in recent years, we have had growing emphasis on production of human and other eukaryotic proteins. As *E. coli* does not have the splicing machinery to allow the use of eukaryotic genomic DNA as PCR template for the expression of genes containing introns, cDNA must be used.

3.1.1. Whole genome amplification by multiple displacement amplification

Although the complete genomic sequenes of over 1200 prokaryotic organism have been elucidated, with even more in progress, genomic DNA preparations are commercially available for only a small fraction ($\sim$10%). Recent methods that predict success in expression, solubility, crystallization, and NMR spectral quality, based on primary sequence (Price *et al.*, 2009; Slabinski *et al.*, 2007), often identify a prokaryotic member of a given protein domain family as amenable to structural determination. However, genomic DNA preparations from these organisms are often not available, and extracting genomic DNA from these organisms can be problematic. To circumvent these issues and increase project efficiency by expanding our reagent genome list, we have implemented Whole Genome Amplification by Multiple Displacement Amplification utilizing phi29 DNA polymerase (Kvist *et al.*, 2007; Lasken, 2007, 2009), to produce microgram quantities of genomic DNA suitable for use as cloning template (Dean *et al.*, 2002). Specifically, a small aliquot of freeze-dried cells from the ATCC (American Type Culture Collection), most sequenced prokaryotic strains are available from this resource, are used as a template for Multiple Displacement Amplification. We have generated high molecular weight genomic DNA preparations for over 30 new prokaryotic genomes from organisms ranging from gut metagenomic bacteria (Gill *et al.*, 2006; Qin *et al.*, 2010) to extremophiles using this high-fidelity technique, greatly expanding the range of proteomes that we can target (Acton *et al.*, 2011).

3.1.2. RT-PCR

Numerous cDNA sources are commercially available ranging from cDNA clone pools, full-length verified clones (length verified by sequencing), and full-length, fully sequenced cDNA clones such as the ORFeome collaboration clones (Open Biosystems; Lamesch *et al.*, 2007; Rual *et al.*, 2004). Only the latter type of cDNA clone is flawless as a template source with respect to sequence, while the others may contain PCR-based errors and polymorphisms. All individual clone libraries have logistical issues; they must be purchased (at considerable cost), archived, and rearrayed before use as PCR template. To circumvent these problems, we have developed protocols for reverse transcriptase-mediated generation of cDNA from PolyA RNA extracted from various eukaryotic organisms (RT-PCR). cDNA pools are generated from commercially available polyadenylated mRNA preparations from various tissues, cell types, and developmental stages (PolyA$^+$ RNA, Clonetech), including a considerable number of tumor cells and human cell lines. The PolyA$^+$ RNA is incubated with oligo-dT or random primers and MMLV reverse transcriptase to generate cDNA pools, which are then combined and used as a common template that is added to each PCR reaction with target-specific primers much like using bacterial genomic DNA. Although this approach may generate clones with polymorphisms (a problem also when using cDNA clone collections that are not full-length and fully sequenced), we find it effective in terms of cost, amenability to high-throughput manipulations, and PCR efficiency. Indeed, our PLIMS database indicates that 88% of GC rich (> 59% GC content) and 96% of lower GC content RT-PCR amplification products are of the correct size. This strategy has proved successful for cloning genes suitable for protein sample production from *Homo sapiens*, *Bos taurus*, *Mus musculus*, *Rattus norvegicus*, and *Aribidopsis thaliana* and from other eukaryotic organisms.

3.1.3. Vector construction by ligation-independent cloning

The NESG originally developed a restriction endonuclease/T4 DNA ligase-based 96-well cloning strategy, using our Multiplex Cloning Vector Set and a Qiagen BioRobot 8000 automated liquid handling device (Acton *et al.*, 2005). The vector system minimizes the number of nonnative residues in the open reading frame while adding a 6×-His tag and common polylinker sequences for purification and cross compatibility, respectively. Although the vector set comprises nine different vectors in three different reading frames, the majority of NESG cloned targets reside in one of the three NESG-modified T7 expression vector derivatives: pET15_NESG, pET21_NESG, or pET15TEV_NESG, coding for proteins with N- (MGHHHHHHSH-), C- (-LEHHHHHH), or N- (MGHHHHHHEN-LYFQSH-) 6×-His affinity purification tags, respectively. The latter expression vector contains a TEV protease cleavage site for the removal of the 6×-His tag (Nallamsetty *et al.*, 2004).

During Protein Structure Initiative-2, InFusion-based ligation-independent cloning was introduced to increase cloning efficiency and throughput. Like other ligation-independent cloning systems, InFusion is more efficient, less time-consuming, and requires less technical skill than ligase-dependent cloning (Aslanidis and de Jong, 1990; Haun and Moss, 1992). No vector modification is necessary, allowing complete compatibility with the NESG Multiplex Expression Vectors, keeping the number of nonnative residues in the expressed construct to a minimum. The InFusion enzyme only requires the addition of a 15 base pair tail to each of the gene-specific PCR primers for a given target ORF. As depicted in the lower center portion of Fig. 2.2, these base pairs are identical to the 5′ and 3′ regions of the vector multicloning site, respectively (Zhu *et al.*, 2007); the 15 base pair region of overlap (identical double-stranded DNA sequence) at the 5′ end is shown in gray and the 3′ end is shown in black on both the vector and PCR amplification product (insert). The vector is cleaved at the region of overlap to produce linear vector DNA. The InFusion enzyme recognizes the identical ends of the two linear DNA strands (vector and insert) and through its strand displacement and exonuclease activity, promotes the pairing and resecting of the DNA (Marsischky and LaBaer, 2004). Transformation of the resulting DNA complex into *E. coli* results in ligated circular DNA clones.

Vector preparation for InFusion cloning is nearly identical to restriction endonuclease cloning (Acton *et al.*, 2005). To minimize nonnative residues, the vector is cleaved at the outermost restriction sites in the multicloning site, *Nde*I, followed by *Xho*I restriction endonuclease digestion. The linearized vector DNA is purified by agarose gel electrophoresis, followed by gel extraction and normalized to 8 ng/μL.

3.1.4. High-throughput construction of expression plasmids

Having addressed important cloning-related issues, alternative PCR template strategies, primer design, and the InFusion cloning approach provide the framework for a detailed description of the high-throughput cloning procedure. Figure 2.2 outlines each step of the high-throughput vector construction pipeline and illustrates the contributions of the PLIMS system and the BioRobot 8000 automation in our 96-well high-throughput cloning. A detailed protocol of the entire process can be downloaded (http://www-nmr.cabm.rutgers.edu/labdocuments/proteinprod/index.htm), including a description of custom Qiasoft 4.1 programs developed in-house to perform the automated steps. Briefly, PLIMS-generated 96-well primer plates are provided by the vendor (Eurofin MWG Operon) at 50 μ*M* concentration. Forward and reverse primers for each specific ORF (identical wells on two separate 96-well blocks) are placed on the BioRobot. An ABI thermocycler-compatible 96-well PCR plate is chilled (4 °C) on a temperature-controlled BioRobot 8000 slot, and the eight-channel pipette head transfers 46 μL of an appropriate PCR reaction mix (dNTPs, high-fidelity thermostable DNA polymerase and buffer, template DNA, and nuclease-free H_2O) to each well in a 96-well PCR plate. Multiple PCR reaction mixes are prepared for each source of template DNA (cDNA, prokaryotic genomic DNA, etc.). The BioRobot then transfers 100 pmol of the appropriate forward and reverse primers from the primer blocks into the corresponding well for each target in the PCR plate. The PCR plate is covered with a MicroAmp™ 96-Well Full Plate Cover (Applied

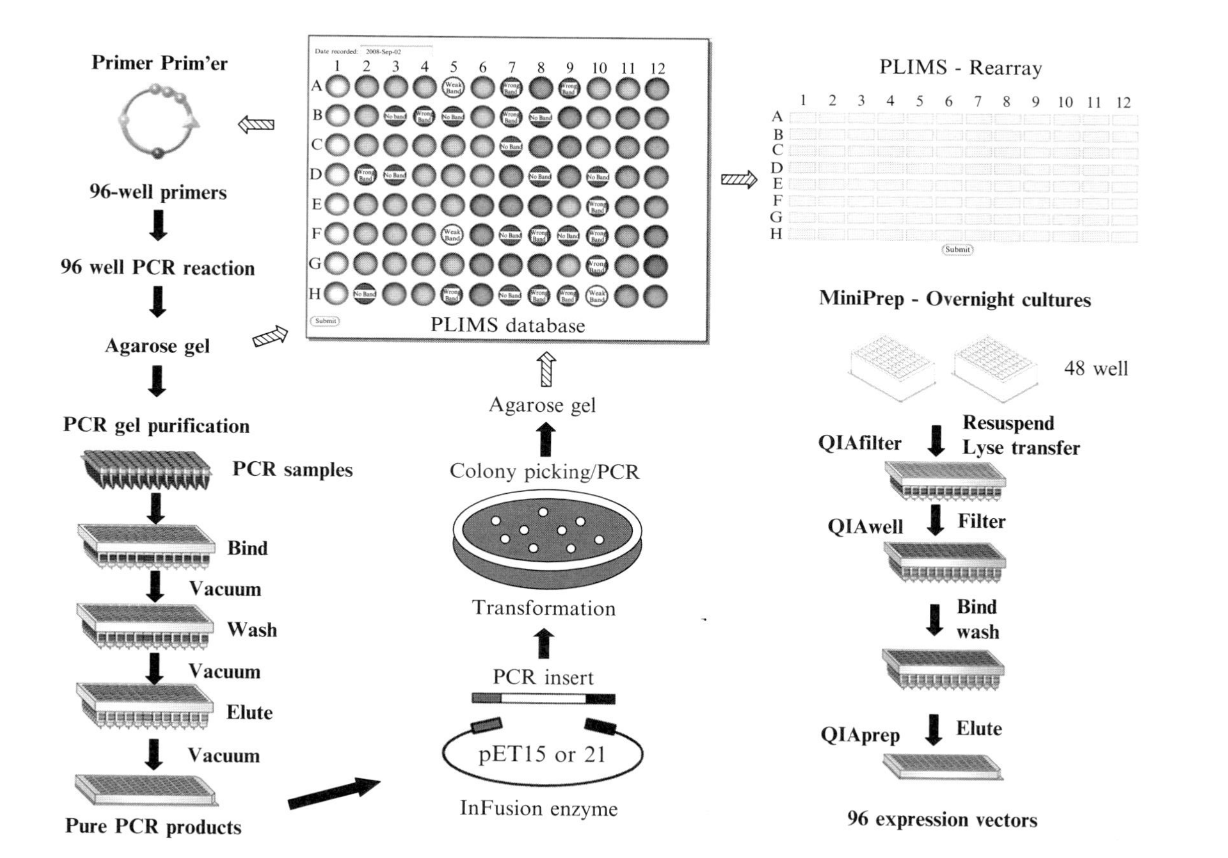

Figure 2.2 Schematic of the PLIMS-directed high-throughput cloning process using the Qiagen BioRobot 8000. The PLIMS Database, utilizing a 96-well graphical user interface (GUI), generates DNA sequences corresponding to the protein regions selected by the NESG Construct Optimization Software. Primer Prim'er generates primer sets in 96-well format, each with the necessary 15 base pair overhangs for Infusion ligation-independent cloning. 96-well PCR reactions are performed, and amplification products are visualized and separated by agarose gel

Biosystems) and transferred to a Veriti® (Applied Biosystems) PCR thermocycler for amplification. Primer Prim'er generates primers with similar melting temperatures allowing thermocycling parameters to be standardized across the plate. Each cycle contains a 10 s, 94 °C melting step, a 30 s annealing step (50–55 °C), and a 3 min 68 °C elongation step (~1 min per kilobase pair). An annealing temperature step increase after 10 rounds of amplification (57–60 °C) is included for the final 25 cycles to adjust for the increased stability derived from the added recombination sites base pairs (Xiao *et al.*, 2010). Advantage® HF2 Polymerase (Clonetech) is normally used for the initial PCR of most targets. This enzyme preparation has proved extremely robust, with very high fidelity. However, human target genes, as well as some prokaryotic strains added to our Reagent Genome List during Protein Structure Initiative-2, may have high GC content. As the GC content increases, PCR amplification becomes increasingly more challenging. Alternative thermostable DNA polymerases (Advantage® GC 2 Polymerase; Clonetech) have proved to be very useful for such GC-rich PCR amplifications. Various concentrations of DMSO are also evaluated to produce PCR products of the correct size and of sufficient quantity for high-throughput cloning. Although higher error rates will occur with these GC-rich PCR conditions, efforts must be made (often opposing) to adjust buffer and annealing temperature conditions to maximize fidelity while increasing the likelihood of obtaining amplification product.

Following thermocycling, 10× DNA Loading Dye is added to each well of the PCR plate using the BioRobot 8000. A Matrix EXP 8-Channel Pipette (Thermo Scientific) with an adjustable tip spacing is used to load the PCR products on a 2% agarose gel for separation and visualization. The results are documented by Alpha Imager (Cell Biosciences) documentation, entered into the PLIMS database, and successful PCR amplifications are identified (correct size and acceptable yield). DNA fragments for the successful PCR reactions are excised from the gel with a SafeXtractor (5 Prime) and transferred to the appropriate well of a 96-well S-Block (Qiagen). The BioRobot 8000 performs an automated 96-well gel extraction with a protocol developed using Buffer QG from the Qiagen Gel Extraction Kit and a QIAquick 96-well column PCR Cleanup plate. An aliquot (1 μL) of the

---

electrophoresis. The results are archived in PLIMS and correct bands excised and purified using a custom 96-well-gel extraction protocol on the BioRobot. Purified amplification products contain 15-base pair regions of homology with each end of the linearized vector (shown as gray (5′) and black (3′) extensions on the PCR insert and vector in the lower center of the figure). Addition of the Infusion enzyme for pairing and resecting the region of overlap is followed by *E. coli* transformation. Colony PCR/agarose gel electrophoresis identifies the correct clones and the data is entered into PLIMS, which rearrays the colony PCR template plates for inoculation of 48-well blocks and an automated 96-well miniprep protocol on the BioRobot.

resulting purified PCR products (~40 μL/well) is transferred to the appropriate well of a fresh PCR plate. A reaction mix for 96-well ligation-independent cloning is then prepared by adding 96 μL of the above-treated vector (8 ng/μL) with 288 μL of sterile deionized water. The reaction mix is used to rehydrate 24 wells from the Infusion Dry Down PCR Cloning Kit (96-well). A considerable cost savings is achieved by diluting the InFusion reaction fourfold over vendor suggestion (4 reactions per well or 384 reactions per plate). The InFusion reaction mix is aliquoted to each well (4 μL) and the plate covered and incubated for 15 min at 37 °C, followed by 15 min at 50 °C; the reaction mix is then flash-frozen (−80 °C). The plate is thawed on ice and the 96 InFusion reactions (resected and paired vector and insert) are transformed into *E. coli* cells using a 24-well format robotic transformation procedure. A single microliter of the ligation-independent cloning product is robotically transferred to the corresponding well of a fresh 96-well PCR plate containing 10 μL of XL-10 Gold® Ultracompetent Cells (Agilent) chilled at 0 °C on the robot deck. Following incubation for 30 min at 0 °C, a manual heat shock step (1 min at 42 °C) is performed, SOC (100 μL) is added to each well, and the plate is incubated at 37 °C for 1 h. The robotic eight-channel pipette head transfers the entire content of each well to a corresponding well in one of the four 24-well blocks. The platform shaker distributes the transformation reactions via 5–10 (3-mm-diameter) glass beads over 2 mL of Luria Broth (LB) medium/agar with ampicillin (100 μg/mL) contained in each well. Following overnight incubation at 37 °C, two colonies per ORF are harvested for colony PCR using primers flanking the multiple cloning site. Colonies arising from an empty vector are rare since the InFusion enzyme does not have ligase activity, and self-ligation by host enzymes appears inefficient with the minute overhangs produced by restriction digest. The colony PCR reactions are loaded on a 2% agarose gel and the results are documented in the PLIMS database. Correct transformants are transferred to a PLIMS-directed well of a 48-well block (Qiagen) containing 2 mL of LB/ampicillin and grown overnight on a platform shaker (37 °C, 210 rpm). Cells are harvested by centrifugation (3000×*g*-force for 10 min) and the media is discarded. Plasmid DNA is isolated using a completely automated QIAprep 96 Turbo Miniprep Kit and BioRobot 8000 procedure. Both the overnight culture and miniprep DNA are archived in an NESG Reagent Repository. Using the InFusion ligation-independent cloning method, we have cloned over 20,000 constructs of some 9000 unique protein targets (multiple alternative constructs per target) into pET expression vectors.

3.2. Analytical scale expression screening

We have developed a microtiter plate-based high-throughput analytical scale protein expression system to screen the numerous expression constructs produced by high-throughput cloning. The overall goal of this assay is to identify protein constructs or homologs that will produce highly expressed, soluble samples for structural analysis when produced by preparative fermentation (1–3 L) in minimal media suitable for isotope enrichment or selenomethione labeling. Although nearly 100 new constructs are produced in each cloning set, only 30–50% of the clones (depending on the source of the protein) will express soluble protein at the levels needed for NMR studies. Such high attrition rates exclude performing large-scale fermentation on every construct due to high costs and inefficiency. We have therefore developed the plate-based strategy to evaluate expression (*E*) and solubility (*S*) in a high-throughput fashion while maintaining the highly aerated growth conditions found in later fermentation efforts.

An outline of this procedure is presented in Fig. 2.3. It starts with transformation into a codon-enhanced *E. coli* expression strain. For each construct, at least two isolates are pursued. Often, this identifies a clone with more favorable characteristics. Typically, codon-enhanced strains BL21 (DE3)-Gold (Agilent), harboring the pMgK plasmid that encodes for rare tRNA codons of Arg (AGA, AGG) and Isoleucine (ATA), or BL21-CodonPlus(DE3)-RIPL (Agilent) that additionally supplies rare tRNA codons for Pro (CCC) and Leu (CTA), are used. Briefly, a robotic transformation protocol is performed, 1 μL of each expression plasmid is transferred to the corresponding well of a PCR plate (on ice) containing 10 μL of competent cells. Following a 30 -min incubation period, the plate is transferred to a PCR thermocycler for a 1 -min heat shock at 42 °C. The cells are transferred to the corresponding well of one of the four Falcon Multiwell 24-well tissue culture plates (Becton-Dickinson) containing 2 mL of LB agar (with appropriate antibiotics and 0.5% glucose) and 5–10 (3 mm) borosilicate glass beads per well. The latter is used to distribute the cells much like the transformation step during ligation-independent cloning. Following overnight growth, individual colonies are inoculated into the corresponding well of a 96-well S-block (Qiagen) containing 0.5 mL of selective LB medium per well. The plate is covered with AirPore Tape (Qiagen) and incubated for 6 h at 37 °C on an Innova platform shaker (New Brunswick Scientific) rotating at roughly 210 rpm. A simple microplate freezer rack can be modified and attached to the shaker platform to provide a low-cost method to simultaneously secure up to 24-deep-well microplate blocks for shaking incubation (see Fig. 2.3). Following the allotted time, the BioRobot transfers 10 μL of each well into the corresponding well of a fresh 96-well S-block containing 0.5 mL (per well) of MJ9 minimal media (Jansson *et al.*, 1996). The block is covered with AirPore Tape and incubated overnight (37 °C at 210 rpm). We have found that growth under minimal media conditions differs significantly from growth in rich media (e.g., LB), often affecting expression and solubility behavior. Therefore, analytical-scale expression utilizes the same minimal media as preparative-scale fermentation in an effort to maximize

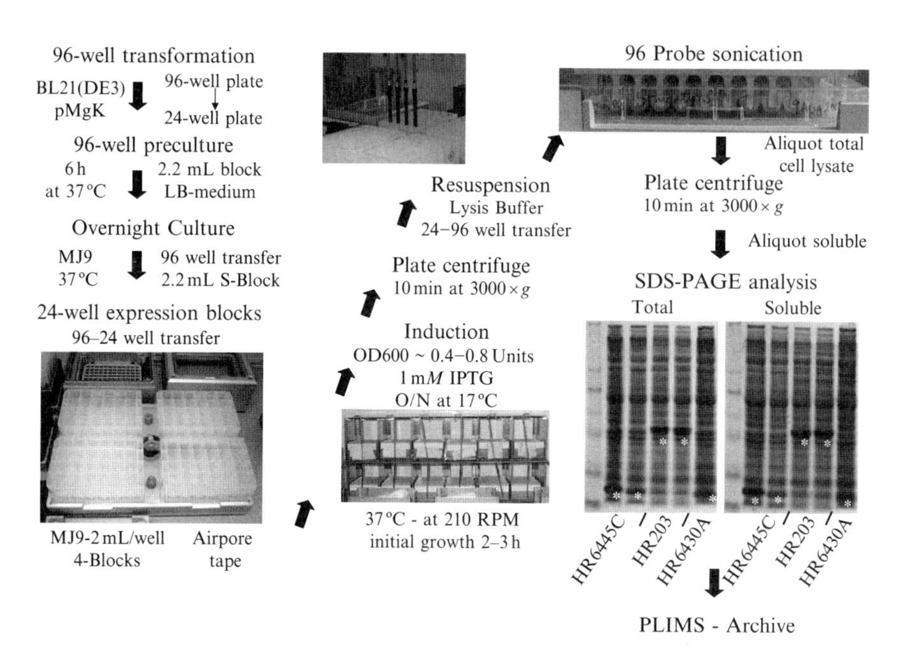

Figure 2.3 High-throughput analytical scale protein expression screening using robotic methods. A 96-well transformation protocol is performed on the BioRobot. Colonies are transferred to an LB containing 2.2-mL 96-well S-Block (Qiagen), followed by overnight subculturing in MJ9 media (all manipulations performed on the BioRobot deck, shown in lower left). Plates are agitated in a microtiter plate freezing rack attached to a platform shaker (shown in the lower middle). A 1:20 dilution into four 24-well blocks (Qiagen) is performed and cells are grown to mid-log phase and induced overnight at 17 °C. Following overnight incubation, cells are harvested by centrifugation (3000×*g*-force, 10 min) and resuspended in 100 μL of lysis buffer and transferred to a 96-well Round Bottom plate (Greiner). Following sonication (Qsonix 96 probe with cell lysate shown in upper right), a 30-μL aliquot of the total cellular lysate (Tot) is transferred to a new plate. The remainder is centrifuged for 10 min at 3000×*g*-force, and a 30-μL aliquot of the supernatant (Sol) is transferred to a new plate. SDS-PAGE analysis of equal amounts of total cell extract (left) or soluble cell extract (right): lane 1, SDS-PAGE standard (Precision Plus, Bio-Rad), lanes 2 and 3 NESG target HR6654C (residues 207–289, 11 kDa), lanes 4 and 5 NESG target HR203 (residues 18–247, 26 kDa), and lane 6 NESG target HR6430A (residues 287–370, 10.4 kDa). The constructs of all the three human proteins produce soluble over-expressed proteins of correct size (highlighted with an asterisk).

reproducibility between the two expression scales. The MJ9 minimal media allows for either isotope enrichment or selenomethionine labeling in preparative-scale fermentation.

Following overnight incubation, the BioRobot performs a 1:20 dilution of the saturated growth into one of the four 24-square-well blocks (10 mL maximum volume/well) containing 2 mL of MJ9 media, preserving well assignment. Each block is covered with Airpore Tape and incubated at

37 °C (210 rpm) until a mid-log phase (2–3 h growth, 0.5–1.0 OD_{600} units) is reached. The 24-well blocks are cooled and isopropyl thiogalactopyranoside (IPTG, 1 m*M* final concentration) is added to each well to induce expression. The block is resealed with fresh AirPore Tape and incubated overnight at 17 °C with shaking at 210 rpm. The low-temperature incubation often aids in producing soluble proteins (Shirano and Shibata, 1990), while the vigorous shaking with gas permeable tape allows for greater aeration rates like those that we obtain in our Midi-scale fermentor (described below) or large-scale fermentation in baffled flasks. Following overnight induction, cells are harvested by centrifugation; the pellets are resuspended in a lysis buffer (50 m*M* NaH_2PO_4, 300 m*M* NaCl, 10 m*M* 2-mercaptoethanol) and robotically transferred to a 96-well round bottom plate (Greiner). A chilled 96-probe sonicator (Qsonica, LLC.) is used for cell disruption, with a 12-min cycle time (30 s bursts at 18 W followed by 30 s cooling periods). Total and soluble portions of the cell lysate are then analyzed by SDS-PAGE (see Fig. 2.3 for details). Expression (E) and solubility (S) are scored, each on a scale of 0 (none) to 5 (max); that is, the $E \times S$ (or ES value) ranges from 0 to 25. All data are documented in the PLIMS system, and 96-well glycerol stocks of the constructs in the expression strain are archived in the NESG Reagent Repository.

Based on the expression and solubility levels obtained from subsequent large-scale (1–3 L) fermentations, we have defined $ES > 11$ as a "usability score." Expression constructs meeting this criterion generally provide enough material (5–50 mg/L) from preparative-scale fermentation and protein purification to allow high-throughput structural determination techniques. In spite of the challenges in producing similar growth conditions in plate format and 1–2 mL volume versus 1000 mL volumes, after considerable efforts, we have achieved 90% agreement between the different expression scales. This allows accurate prediction of large-scale expression yields.

4. Midi-Scale Protein Expression and Purification

A retrospective analysis of the NESG expression constructs produced during the second phase of the Protein Structure Initiative indicates that roughly one-third of all expression constructs possess expression and solubility levels (i.e., $ES > 11$) consistent with the requirements for large-scale fermentation. However, a large fraction of these samples prove intractable at later steps in the protein production pipeline. For example, roughly one-half of the proteins produced in the last 2 years are not monodisperse in solution and contain some level of aggregation. Analysis of NESG data indicates that crystallization success rates are dramatically increased more than 10-fold for monodisperse protein samples in comparison with those polydisperse or aggregated (Price *et al.*, 2009). Protein aggregation is also deleterious for NMR studies, often resulting in protein precipitation or negatively affecting NMR spectra. Approximately one-half of protein samples prepared for HSQC screening provide "good" or "promising" HSQC spectra and are amenable to structural determination by NMR.

Since about one-half of purified proteins have been intractable for structural studies due to their biophysical properties, we have developed a *high-throughput Midi-scale Protein Production Pipeline* which provides a biophysical characterization of small quantities of protein constructs before investing in more labor-intensive large-scale expression and purification.

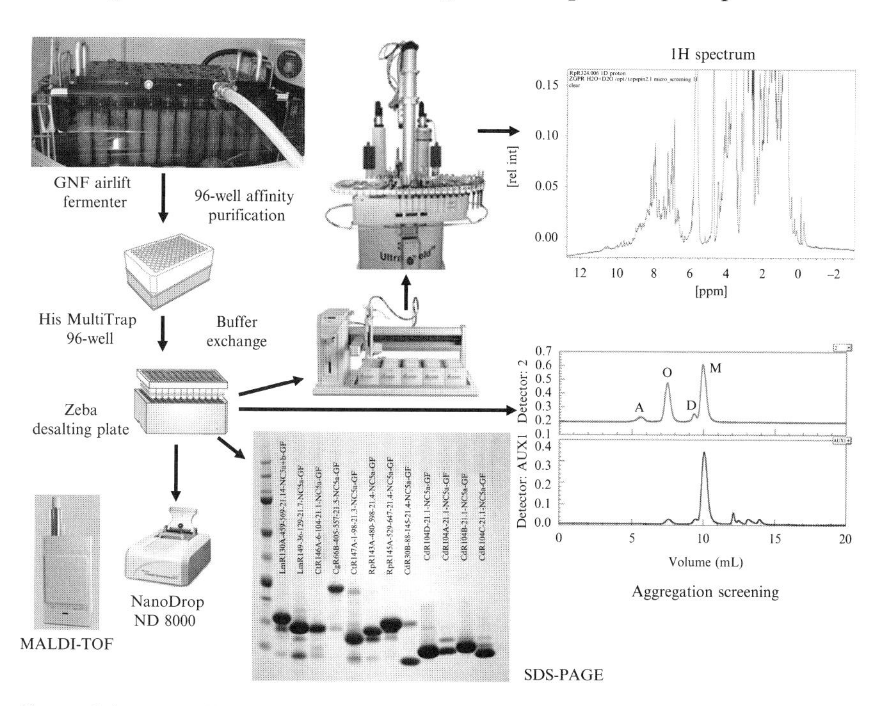

Figure 2.4 96-well midi-scale protein expression, purification, and characterization. This system utilizes (i) an Airlift Fermentation System (Genomics Institute of the Novartis Research Foundation - GNF) with O_2 aeration at 60-mL scale; (ii) a His MultiTrap HP 96-well plate (GE Healthcare) for Ni^{2+}-affinity protein purification; and (iii) Zeba^{TM} 96-well desalting spin plate (Thermo Scientific) for buffer exchange. Analytical protein chemistry and biophysical screening steps include target validation by MALDI-TOF mass spectrometry, concentration determination by a NanoDrop ND-8000 Spectrophotometer, homogeneity analysis by SDS-PAGE, aggregation screening by analytical gel filtration with static light scattering, and NMR screening (sample tubes loaded with Gilson 215-based automation) using a 1.7-mm micro cryo NMR probe and automated sample changer. Insets show representative results of ^{1}H NMR Screening (top right) and aggregation screening with static light scattering (lower right, M, monomer; D, dimer; O, oligomer; A, soluble aggregate, as explained in the text).

This system, outlined in Fig. 2.4, utilizes (i) a GNF Airlift Fermentation System (GNF Systems) with O_2 aeration allowing for 96 simultaneous high cell density fermentations; (ii) a His MultiTrap HP plate (GE Healthcare) for 96 parallel immobilized metal affinity chromatography (IMAC) purifications; and (iii) Zeba^{TM} 96-well desalting spin plate (Thermo Scientific) for buffer exchange. Typical yields of 0.2–1.0 mg of protein per 60 mL fermentation are achieved, which is sufficient for a series of analytical protein chemistry steps, including: (i) aggregation screening by analytical gel filtration with static light scattering, (ii) homogeneity analysis by SDS-PAGE analysis, (iii) target validation by MALDI-TOF mass spectrometry, (iv) concentration determination by a NanoDrop ND-8000 spectrophotometer, and (v) 1D ^{1}H NMR screening using a 1.7-mm micro cryo NMR probe (35 μL sample volume; Rossi *et al.*, 2010). Identification of aggregated/polydisperse proteins and proteins exhibiting poor-quality 1D ^{1}H NMR spectra avoids scale-up of intractable protein targets, and allows us to concentrate finite resources for isotope labeling and protein purification on targets amenable to NMR studies or crystallization screening, greatly increasing project efficiency

4.1. Midi-scale fermentation with the GNF Airlift Fermentation System

To produce sufficient quantities of protein for biophysical characterization in a high-throughput fashion, we have adapted an Airlift Fermentation System, developed by the Genomics Institute of the Novartis Research Foundation (GNF), to our Midi-scale pipeline. This system uses 96 (100 mL) test tubes, each with a fermentation capacity of ~60 mL of media. A common manifold with 96 canulae is used to deliver oxygen to each tube for both cell growth, and to provide agitation for cell suspension and nutrient mixing. Temperature regulation is controlled using a water bath with a refrigerated/heated water circulator (VWR Scientific). Using rich TB media (Peti *et al.*, 2005), this system routinely reaches cell densities in the range of 15–20 OD_{600} units, corresponding to a quarter of the final cell mass obtained from 1 L of our large-scale protein expression in minimal media (3–5 OD_{600} units). Briefly, this procedure starts with PLIMS generating a target pool of constructs with high *ES* values (>11). After selecting 96 targets, robotic transfer from the appropriate glycerol stock to a PLIMS-directed well in a 96-well S-block is performed. Each well contains 500 μL of TB media with ampicillin and kanamycin and is then covered with AirPore tape. Following incubation (37 °C, 210 rpm) for 6 h, 100 μL from each well is transferred to a 100-mL test tube in the corresponding position of the GNF fermentor. Each tube contains 3 mL of TB, Antifoam Y-30 Emulsion (Sigma), and appropriate antibiotics. The rack is covered with aluminum foil and placed on an Innova shaking platform. Following overnight growth at 37 °C, 57 mL of fresh TB/antifoam/antibiotics are added to each of the 96 tubes. The air intake manifold is inserted into the tube rack and the entire system is relocated into a water bath preheated to 37 °C. Using the manifold and its canulae, 100% oxygen is distributed to each well at a flow rate of ~3.5 cfm. We have found that the dual-functioning canulae, providing both oxygenation and lift (large bubble size), necessitate a high percentage of oxygen addition for the greatest yield. *This is potentially dangerous, and strong system ventilation must be used for safety.* When OD_{600} reaches 5–6 units, each tube receives IPTG (1 m*M* final concentration) and Antifoam Y-30 Emulsion through ports in the Airlift Fermentation System manifold. Concurrently, the water bath temperature is decreased to 17 °C using the refrigerated water circulator. Following 16 h of incubation at this temperature and aeration with 100% oxygen, an aliquot is taken from each well to assay final cell density and for SDS-PAGE analysis of expression and solubility levels. The resulting data is documented and processed in the PLIMS database. The remaining contents of each tube are transferred to an appropriately labeled (PLIMS-generated well position and bar-code) 50-mL conical tube and centrifuged for 20 min at 6000×*g*-force. The media is poured off and the pellets are flash-frozen and stored at −80 °C.

4.2. Ni^{2+}-affinity protein purification using 96-well IMAC plates

The Midi-scale fermentation system produces sufficient protein levels for carrying out numerous biophysical characterization techniques. To purify these proteins to levels suitable for these techniques, we have also developed a Ni^{2+}-based IMAC step in plate format. This takes advantage of the 6×-His tag incorporated into each construct. Briefly, the cell pellets are thawed and resuspended in a lysis buffer containing 1× Cell Lytic B (Sigma), 500 μg/mL lysozyme (Sigma), 100 units/mL RNAse (Sigma), 100 units/mL DNAse (Sigma), 40 m*M* imidazole, and Complete Protease Inhibitor Cocktail (Roche). The resuspension mix is incubated for 30 min at 37 °C. Each 50-mL tube is then centrifuged at 3000 rpm (1600×*g*-force) for 30 min to clear the cell debris. Two milliliters of each resulting supernatant is transferred to an empty 2.2-mL deep-well plate (Qiagen S-block). A Liquidator96 (Rainin) is used to transfer 400 μL from each well to the corresponding well of a His MultiTrap™ HP 96-well plate (GE Healthcare) equilibrated with lysis buffer. The plate is centrifuged for 4 min at 100×*g*-force and the flow-through is discarded. This process is repeated, loading the entire 2 mL of cell lysate into each respective well. The IMAC plate is subjected to three wash steps consisting of 500 μL of lysis buffer (per well) containing 40 m*M* imidazole (pH 7.5), followed by centrifugation at 500×*g*-force for 2 min. Proteins are eluted from the Nickel Sepharose with a lysis buffer containing 300 m*M* imidazole (pH 7.5) per well. Seventy-five microliters of this solution is added to each well and followed by incubation at room temperature for 10 min. The step is completed by centrifugation at 100×*g*-force for 4 min. The entire volume of Ni^{2+}-affinity-purified proteins are then immediately transferred to an equilibrated Zeba™ 96-well desalting spin plate (Thermo Scientific). Following centrifugation for 2 min at 1000×*g*-force, the proteins are ready for biophysical characterization.

4.3. Biophysical characterization of the midi-scale-generated proteins

4.3.1. Target validation by MALDI-TOF mass spectrometry

At this point in the NESG Protein Production Pipeline, DNA sequence verification of expression constructs has generally not been performed. Therefore, we have implemented a quality-control step using MADLI-TOF mass spectrometry (MS) to identify the protein molecular weight of the protein produced by the expression vector in each well. Samples are prepared by mixing 1 μL of the protein sample from each well with 10 μL of sinapinic acid matrix solution (10 mg/mL sinapinic acid in 50% acetonitrile/ 50% 0.1% TFA) in the corresponding well of a PCR plate. One microliter of this solution is transferred to the appropriate well position on an Opti-TOF Sample Plate (AB SCIEX) and spectra are collected for each protein spot (corresponding to a well position) on a MALDI-TOF/TOF (4800 Plus MALDI TOF/TOF™ Analyzer, AB SCIEX) in single TOF mode. The spectrum of each well is compared to the expected size of the purified protein; species differing from their expected mass by greater than 500 Da likely represent invalid, processed, or proteolyzed targets. Biophysical analysis continues on those samples with aberrant molecular weight; however, these expression constructs and purified proteins are subjected to DNA sequence and liquid chromatography–mass spectrometry analysis for verification of the cloned DNA sequence and protein product, respectively. These mass spectrometry data are all archived in the SPINE database (Goh *et al.*, 2003).

4.3.2. NanoDrop and SDS-PAGE for assaying protein concentration and homogeneity

To measure protein concentration in standard non-high-throughput applications, protein samples are transferred to cuvettes or capillary tubes for spectrophotometric measurement of absorbance at 280 nm. In many high-throughput projects, absorbance measurements are performed on solutions in microtiter plates. However, we have found that neither of these methods is well suited for the Midi-scale pipeline since the samples are of limited volume and large in number. To circumvent these issues, we have incorporated a NanoDrop™ 8000 spectrophotometer into the NESG Protein Production Pipeline. This instrument accurately measures absorbance values of sample volumes <2 μL without dilution. Briefly, a Multichannel Pipette (Rainin) is used to simultaneously transfer single columns (eight samples, e.g., A1–H1) of the purified proteins from the PCR plate to a linear array of eight pedestals. The pedestals are arrayed with standard 96-well plate spacing, each with optics for absorbance measurement. Using this system, a full plate of 96 samples can be measured in less than 6 min. The protein concentration in each well is calculated automatically using its respective extinction coefficient (calculated by the PLIMS system). The concentration of each protein is recorded in the SPINE database and used in the aggregation screening analysis (described below), in the preparation of NMR samples, and to determine process yield.

An important quality-control step in protein purification is to assess the homogeneity of the preparation. For the larger proteins in the NESG X-ray crystallography protein production pipeline, we utilize a LabChip® 90 system (Caliper) for this task. Although this system is ideal for high-throughput procedures, it is technologically limited to proteins greater than 12 kDa and cannot be used for some of the smaller proteins produced in the NMR protein production pipeline. For these smaller proteins, homogeneity is assessed by SDS-PAGE analysis. NuPage® Novex® Bis–Tris Mini-Gels (Invitrogen) are utilized for ease of sample preparation, excellent resolution, long shelf life, and high reproducibility. Samples are prepared by aliquoting 2 μL of the purified protein into a fresh PCR plate containing 5.5 μL of deionized water, 0.5 μL of DTT (1 m*M*), and 2.5 μL of NuPage LDS Sample Buffer per well. The samples are mixed and heated at 70 °C for 10 min. Gels are loaded using an eight-channel expandable pipette (Matrix EXP Pipette, Thermo Scientific) that reduces from normal 96-well spacing to that of the SDS-PAGE gel. Gels are fixed and stained using GelCode Blue (Thermo Scientific), documented (Alphaimager™, Cell Biosciences™), and archived in SPINE.

4.3.3. Aggregation screening to detect protein mass distribution

Numerous studies have shown that proteins that are monodisperse in solution are more likely to produce diffraction quality crystals than polydisperse or aggregated samples (Ferre-D'Amare and Burley, 1994, 1997; Klock *et al.*, 2008; Price *et al.*, 2009). Roughly one-half of NESG structures are determined by X-ray crystallography. Therefore, to increase project efficiency, an aggregation screening step was implemented to characterize purified proteins by analytical gel filtration, followed by multiangle static light scattering. This allows for accurate measurement of the distribution of oligomers and/or aggregates in a protein sample. Although this system was originally implemented for X-ray crystallography samples, probing the oligomerization state of NMR samples is also valuable for assessing the methods to be used in NMR studies; for example, monomer and homodimers require different data collection and analysis strategies, and higher order multimers may be beyond the size constraints for high-throughput NMR structural determination. These aggregation screening data also predict NMR sample stability (proteins with significant aggregation may precipitate before a full data set is collected). Aggregation screening is carried out using an Agilent 1200 series HPLC system with an 8 × 300 mm Shodex KW-802.5 HPLC size-exclusion column (Showa Denko K.K.), connected in line with a miniDAWN TREOS detector (Wyatt technologies) and a Optilab rEX Refractometer (Wyatt Technology). Protein samples are loaded from the automated 96-well sample changer and injected into the HPLC system equilibrated in a candidate NMR buffer (see Section 4.3.4). Following separation by size exclusion, the light-scattering properties of each successive protein species in solution are simultaneously measured at three different angles (45°, 90°, and 135°) and their refractive index is detected. The analysis of the light scattering and refractive index data provides the shape-independent weight-average molecular mass of each species and their relative distributions. As shown in Fig. 2.4, the light-scattering trace for the NESG human protein target HR3580C indicates peaks corresponding to monomer (M), dimer (D), higher oligomers (O), and aggregates (A) of the protein (panel-marked aggregation screening). The bottom trace shows the refractive index indicating that the majority of mass is contained as a monomer. However, analysis of the data indicates that ~75% of the mass is monomeric (22.9 kDa) in nature. Although generally monomeric and therefore within the size-limits of high-throughput NMR assignment and structural analysis, the significant amount of aggregated protein observed suggests that further buffer optimization, construct optimization, or other "salvage" efforts are required before promotion to large-scale fermentation and purification.

4.3.4. High-throughput micro cryo probe screening by 1D ^{1}H-NMR

The NESG Microscale Protein NMR Sample-Screening Pipeline has been described in detail (Rossi *et al.*, 2010). It takes advantage of NMR microprobe technology and its inherent ability to function with relatively low quantities of protein. Typically, 10–200 μg of protein in a volume of 35 μL is sufficient for screening with a Bruker 600 MHz TXI 1.7-mm micro cryo probe. This probe is well suited for the yields generated by our Midi-scale Pipeline. Briefly, the 96 purified proteins are buffer-exchanged into an appropriate NMR buffer using a Zeba™ 96-well desalting spin plate (Thermo Scientific), as described in Section 7.1. The initial NMR buffer is selected based on the isoelectric point (p*I*) of the protein; typically, 20 m*M* MES, 100 m*M* NaCl, 5 m*M* $CaCl_2$, 10 m*M* DTT, and 0.02% NaN_3 at pH 6.5, or alternatively, 20 m*M* ammonium acetate, 100 m*M* NaCl, 5 m*M* $CaCl_2$, 10 m*M* DTT, and 0.02% NaN_3 at pH 4.5. Aliquots are then transferred to 1.7-mm SampleJet Tubes (Bruker) using a Gilson 215 Liquid Handler. Only 1D ^{1}H-NMR spectra are collected, since the rich TB broth in the Midi-scale Pipeline does not allow for isotope enrichment. However, this screen can detect dispersion of amide protons and upfield-shifted methyl protons, indicative of aromatic and methyl stacking (folded protein core). Protein constructs exhibiting tractable traits from the Midi-scale biophysical screening (correct molecular weight, monodisperse monomers or low molecular weight dimers, disperse amide, and methyl proton resonance frequencies) are more than likely amenable for structure determination by NMR. These are subsequently scaled up for fermentation and purification with isotope enrichment.

5. Preparative-Scale Fermentation

The recent advent of NMR microprobe technology has allowed for structural determination using relatively minute amounts (<100 μg) of protein sample (Aramini *et al.*, 2007). However, micro cryo NMR probes have lower sensitivity than 5-mm cyro probes, which utilize significantly greater amounts of protein (>5 mg). We have designed our process for preparative-scale or large-scale protein expression to produce these yields while optimizing cost and throughput, and incorporating the flexibility to use essentially the same pipeline to produce samples for both NMR studies and X-ray crystallography. This strategy is based on parallel fermentations in 2.5-L baffled Ultra Yield™ Fernbach flasks (Thomson Instrument Company) with agitation on low-cost platform shakers (Innova 2300, New Brunswick Scientific) housed in controlled temperature rooms (Acton *et al.*, 2005). The Ultra Yield™ flasks, with advanced 6-baffle design, are capable of supporting high cell density (Brodsky and Cronin, 2006), while their compact design allows for up to 15 parallel fermentations per platform shaker. Together with buffer- and vitamin-enriched MJ9 minimal media (Jansson *et al.*, 1996), this system of flasks and shakers achieves cell density and protein expression levels consistent with the needs of a high-throughput structural genomics pipeline. MJ9 media allows for enrichment with ^{15}N, ^{13}C, and/or ^{2}H isotopes for NMR structural studies, or with selenomethionine for single and multiple anomalous diffraction X-ray crystallography (Hendrickson, 1991). For most NMR studies undertaken by the NESG, an NC5 (100% ^{15}N, 5% ^{13}C) sample is first produced for 2D ^{1}H–^{15}N HSQC screening (Rossi *et al.*, 2010). NC5 samples are fermented in MJ9 media supplemented with uniformly (U)–$^{15}NH_4$ salts as the sole nitrogen source, while the glucose (sole carbon source) is 95% natural abundance and 5% (U)–^{13}C-enriched. The NC5 samples are also used for stereo-specific assignment of isopropyl methyl groups (Neri *et al.*, 1989). For targets providing good-quality HSQC spectra, (U)–^{13}C, (U)–^{15}N-enriched proteins are then produced using MJ9 media supplemented with (U)–^{13}C glucose and (U)–$^{15}NH_4$ salts as the sole source of carbon and nitrogen, respectively.

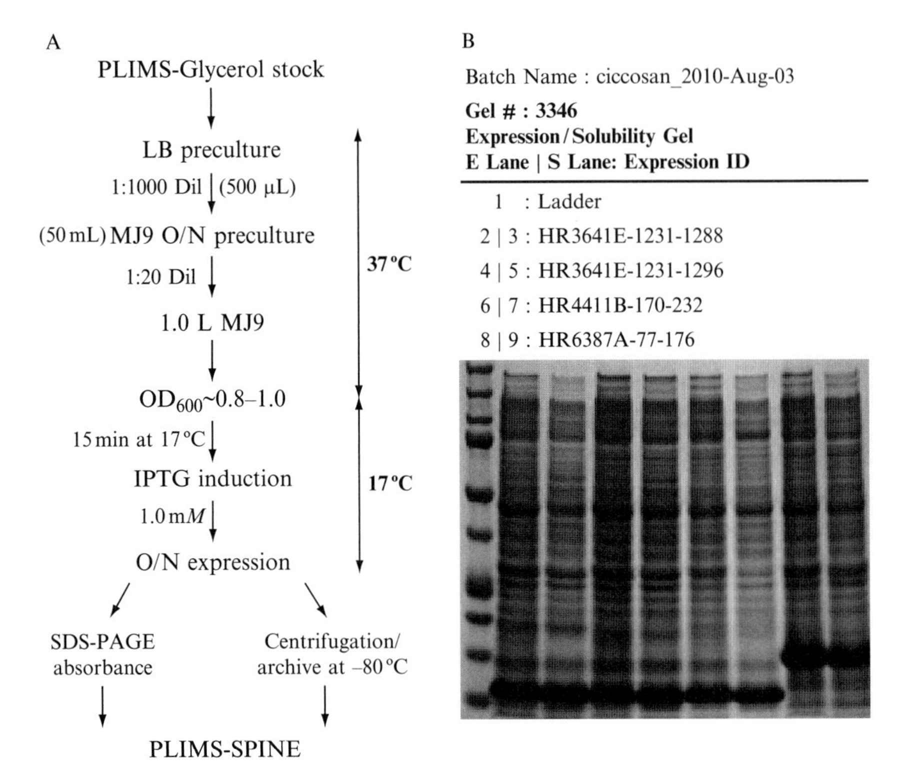

Figure 2.5 Preparative-scale fermentation. (A) Flow chart of the preparative-scale fermentation process. The PLIMS database identifies targets suitable for fermentation and their expression glycerol stock location. The precultures and initial log-phase growth are carried out at 37 °C and overnight induction is performed at 17 °C. (B) SDS-PAGE analysis of a fermentation with four human protein targets showing Total Expressed (E—lanes 2, 4, 6, and 8) and Soluble (S—lanes 3, 5, 7, 9) extracts. The intensity of the targeted protein band in the total extract is used to estimate the total expression level (*E*), and the intensity of the band in the S extract is used to estimate the portion of the expressed protein that is soluble (*S*).

The fermentation process (outlined in Fig. 2.5) begins with the PLIMS database generating a pool of protein expression constructs that pass analytical scale ($ES > 11$) and Midi-scale biophysical characterization, and meet a variety of project-defined criteria (e.g., structural uniqueness). The appropriate glycerol stock plate and well position for each selected construct is reported (each plate has a unique bar code identification), and the plates are retrieved from −80 °C storage. An aliquot from each selected well is robotically transferred to 500 μL of LB media with ampicillin and kanamycin or chloramphenicol [BL21(DE3)-Gold pMgK or BL21-CodonPlus (DE3)-RIPL, respectively] in an S-block and incubated for 6 h at 37 °C. An aliquot (50 μL) of this rich media preculture is then used to inoculate

50 mL of MJ9 minimal media in a 250-mL flask and incubated overnight at 37 °C on an Innova shaker (250 rpm). The entire volume of overnight culture is then used to inoculate a 2.5-L Ultra Yield™ flask containing 1.0 L of MJ9 (preheated to 37 °C) supplemented with the appropriate isotopes and selective agents. The cultures are then incubated at 37 °C (shaking at 250 rpm) until the OD_{600} reaches 0.6–0.8 units. The 2.5-L Ultra Yield™ flasks containing the cultures are then transferred to an Innova 2300 Shaker in a 17 °C constant temperature room. Following equilibration at this temperature (~10 min), protein expression is induced with IPTG (1 m*M* final concentration). Incubation with vigorous shaking (250 rpm) at this temperature continues overnight with a total induction time of 16 h. Before harvest, 250 μL of Antifoam Y-30 Emulsion is added to each flask to disperse any accumulated bubbles that may interfere with harvest. Before any further manipulations, a 500-μL aliquot of cells from each culture is transferred to a microcentrifuge tube for (i) determining final cell density (OD_{600}), which typically ranges from 3 to 5 OD units; (ii) SDS-PAGE analysis of expression (*E*) and solubility (*S*); and (iii) quality control by DNA sequence analysis. The remaining culture is transferred into a HarvestLine System Liner (Beckman Coulter) contained in a specialized 1000-mL polycarbonate centrifuge bottle and cap (J-Lite PC-1000, Beckman Coulter). The cells are harvested by centrifugation at 9000×*g*-force for 30 min using an Avanti J-26 XP centrifuge (Beckman Coulter). Following centrifugation, the supernatant is discarded and the HarvestLine System Liner bags are removed from the centrifuge bottle and archived at −80 °C in a PLIMS-directed location. SDS-PAGE analysis of expression and solubility is performed in a manner similar to analytical scale expression (Fig. 2.5). The fermentation data is uploaded in the PLIMS database for archival, including automated gel labeling. A subset of this information is transferred to the web-based SPINE database (Bertone *et al.*, 2001; Goh *et al.*, 2003) for web-based access across the NESG consortium and/or with the public databases

6. Preparative-Scale Purification

One of the greatest challenges in producing highly homogenous protein preparations (suitable for NMR structure determination and/or crystallization) in a high-throughput manner is the diverse biophysical characteristic of the targets themselves. The addition of a short 6×-His tag (Porath, 1992) utilized by the NESG "Multiplex Vector System" addresses this issue by allowing the same affinity purification to be utilized across our diverse target set. The 6×-His tag and the associated Ni^{2+}-based IMAC along with automated four-module ÄKTAxpress systems (GE Healthcare) serve as the foundation of the NESG purification pipeline. Although the IMAC purification achieves a high degree of purity for most bacterially overexpressed proteins, a size-exclusion chromatography step is performed to produce the purity levels necessary for crystallization and NMR studies. Each module in the ÄKTAxpress system is fitted with four separate HisTrap HP columns (GE Healthcare) and one size-exclusion column per module allowing four separate completely automated affinity/gel filtration two-step purifications, or 16 total purifications per each four-module system, in less than 12 h.

Briefly, the SPINE database generates a list of fermented targets available for purification and their corresponding −80 °C storage locations. The database also reports associated protein target characteristics critical for the purification effort, including protein molecular weight with tag for theexact isotope enrichment strategy, protein isoelectric point, and extinction coefficient among others. The centrifuge bags of targets selected for purification are thawed on ice, 30 mL of binding buffer (50 m*M* Tris–HCl, 500 m*M* NaCl, 40 m*M* imidazole, 1 m*M* TCEP, and 0.02% NaN_3, pH 7.5) containing a protease inhibitor cocktail (Complete Protease Inhibitor Cocktail, Roche) is pipetted into each bag, and the pellet is resuspended by simple hand manipulation of the cell paste. The bag contents are then transferred to a prechilled stainless steel 125-mL beaker (Vollrath®) and sonicated in an ice water bath using a Dual Horn 3/4″ probe (Qsonica, LLC) for 10 min with 30-s bursts, followed by 30-s cooling periods (30 s on/30 s off). The cell debris is cleared by centrifugation at 27,000×*g*-force for 40 min, followed by filtration (0.2 μm). The supernatant is then loaded onto one of the 16 positions on the ÄKTAxpress system and a two-step automated purification protocol is performed using the preinstalled default settings (AF-GF). Specifically, a HisTrap HP IMAC column (5 mL) and Superdex 75 26/60 gel filtration column are run in a linear series. The 6×-His-tagged proteins are eluted from the HisTrap column using five column volumes of elution buffer (50 m*M* Tris–HCl, 500 m*M* NaCl, 500 m*M* imidazole, and 0.02% NaN_3 at pH 7.5) at a flow rate of 4 mL/minute. The ÄKTAxpress system monitors the elution profile (A_{280}) and collects major peaks into internal storage loops, which are then automatically injected onto the Superdex 75 gel filtration column equilibrated in one of the two NMR buffers: (i) 20 m*M* MES, 100 m*M* NaCl, 5 m*M* $CaCl_2$, 10 m*M* DTT, and 0.02% NaN_3 at pH 6.5, or alternatively (ii) 20 m*M* ammonium acetate, 100 m*M* NaCl, 5 m*M* $CaCl_2$, 10 m*M* DTT, and 0.02% NaN_3 at pH 4.5, based on the protein's theoretical p*I*. The elution is monitored by A_{280} detection and major peaks emanating from the gel filtration column are collected in 96-well blocks. The resulting fractions are analyzed by SDS-PAGE and pooled followed by concentration using Amicon ultrafiltration concentrators (Millipore). Molecular weight validation by MALDI-TOF mass spectrometry, homogeneity analysis by SDS-PAGE, aggregation screening by analytical gel filtration with static light scattering, and concentration determination is performed on each sample. The NMR sample preparation is completed with the addition of Complete Protease Inhibitor Cocktail (Roche), 10% $^{2}H_2O$, and 50 μM DSS (4,4-dimethyl-4-silapentane-1-sulfonic acid) as an internal NMR reference (Markley *et al.*, 1998).

For NMR microprobe data collection, aliquots (35 μL) are transferred to 1.7-mm SampleJet Tubes (Bruker), using a Gilson 215 Liquid Handler. For samples destined for data collection using 5-mm probes, 300-μL aliquots are transferred into 4- or 5-mm Shigemi tubes. The ÄKTAxpress purification trace, MALDI-TOF, aggregation screening, sample concentration, and other data are archived into the SPINE database. SPINE also serves to direct and track shipping of samples to NESG NMR spectroscopists around the country for data collection and structural determination. The database coordinates this effort with bar code-based registration of shipment tubes and automatically tracks shipments through the FedEx database.

7. Salvage Strategies

Depending on the source of the protein target, up to 70% of expression constructs will fail to express soluble proteins at a level consistent with preparation of structurable samples. Many of these are high value targets. Therefore, alternative strategies and expression hosts have been explored in an attempt to rescue these targets. In addition, many proteins proceed through the pipeline providing samples that are nearly structurable, but are not quite suitable for high-throughput structural determination techniques. The NESG Protein Production Pipeline has explored and developed "salvage" techniques to enhance the behavior of these proteins in attempts to provide samples more amenable to NMR studies. Some of the most successful salvage strategies are described below.

7.1. NMR buffer optimization

During Protein Structure Initiative-2, roughly one-half of all proteins purified for NMR screening produce 2D ^{1}H–^{15}N HSQC spectra that were scored as "Good" or "Promising" (Rossi *et al.*, 2010; Snyder *et al.*, 2005). The former are sufficient for high-throughput NMR structural determination. However, the "Promising" spectra are marginal in quality and often cannot be used for resonance assignment or structural analysis. In the case of proteins with "Good" spectra, a significant proportion of these proteins will suffer sample stability issues (e.g., slow precipitation) in the time frame of complete NMR data acquisition (4–10 days). Although targets are screened in one of the two different buffers, based on avoiding close proximity to the estimated protein p*I*, neither of these buffer conditions may be optimal. We have found that alternative buffer conditions can often improve NMR spectral quality, reduce aggregation and sample precipitation, and improve sample stability (Rossi *et al.*, 2010). To screen for conditions that promote stability or better-quality spectra, we have implemented a high-throughput buffer optimization procedure. Purified proteins with good spectra and problematic precipitation or "Promising" spectra are thawed on ice. For each row of a Zeba^{TM} 96-well desalting spin plate (Thermo Scientific), the corresponding microcolumns are first washed with one of the 12 alternative buffers (e.g., listed in Table 2.1), by centrifuging the plates for 2 min at 1000×*g*-force to remove the storage buffer. Then, each microcolumn is equilibrated with the corresponding buffer by eluting

Table 2.1 Buffer optimization for NMR studies

Buffer ID	pH	Buffer formula
MJ001	6.5	20 m*M* MES[a], 100 m*M* NaCl, 5 m*M* $CaCl_2$, 10 m*M* DTT[b], 0.02% NaN_3, Complete[c], 10% D_2O
MJ002	5.5	20 m*M* NH_4OAc, 100 m*M* NaCl, 5 m*M* $CaCl_2$, 10 m*M* DTT[b], 0.02% NaN_3, Complete[c], 10% D_2O
MJ003	4.5	20 m*M* NH_4OAc, 100 m*M* NaCl, 5 m*M* $CaCl_2$, 10 m*M* DTT[b], 0.02% NaN_3, Complete[c], 10% D_2O
MJ004	5	50 m*M* NH_4OAc, 10 m*M* DTT[b], 50 m*M* arginine, 0.02% NaN_3, Complete[c], 10% D_2O
MJ005	5	50 m*M* NH_4OAc, 10 m*M* DTT[b], 5% CH_3CN, 0.02% NaN_3, Complete[c], 10% D_2O
MJ006	6	50 m*M* MES[a], 10 m*M* DTT[b], 50 m*M* arginine, 0.02% NaN_3, Complete[c], 10% D_2O
MJ007	6	50 m*M* MES[a], 10 m*M* DTT[b], 5% CH_3CN, 0.02% NaN_3, Complete[c], 10% D_2O
MJ008	6.5	25 m*M* Na_2PO_4, 450 m*M* NaCl, 10 m*M* DTT[b], 20 m*M* $ZnSO_4$, Complete[c], 0.02% NaN_3, 10% D_2O
MJ009	6.5	20 m*M* MES[a], 100 m*M* NaCl, 5% CH_3CN, 10 m*M* DTT[b], 0.02% NaN_3, Complete[c], 10% D_2O
MJ010	6.5	20 m*M* MES[a], 100 m*M* NaCl, 50 m*M* Arginine, 10 m*M* DTT[b], 0.02% NaN_3, Complete[c], 10% D_2O
MJ011	6.5	20 m*M* MES[a], 100 m*M* NaCl, 1% Zwitter[d], 10 m*M* DTT[b], 0.02% NaN_3, Complete[c], 10% D_2O
MJ012	6.5	20 m*M* MES[a], 100 m*M* NaCl, 50 m*M* $ZnSO_4$, 10 m*M* DTT[b], 0.02% NaN_3 Complete[c], 10% D_2O

[a] MES: 2-(*N*-morpholino)ethanesulfonic acid (Sigma).
[b] DTT: DL-dithiothreitol (Sigma).
[c] Complete: Complete Protease Inhibitor Cocktail (ROCHE).
[d] Zwitter: ZWITTERGENT® 3-12 (CALBIOCHEM).

with 250 μL of buffer transferred from a 12 partition Microplate Reservoir (Seahorse Bioscience) using a 12-channel pipette, followed by centrifugation for 2 min at 1000×*g*-force. This step is repeated three times until each well has received 1000 μL of the appropriate buffer. A candidate-purified protein is then transferred (~70 μL) to each well to sample all the 12 conditions. Buffer-exchanged samples are collected following centrifugation for 2 min at 1000×*g*-force. The eluate containing the buffer-exchanged protein is then loaded into 1.7-mm NMR microprobe tubes using a Gilson 215 Liquid Handler. Each tube is stored at room temperature and scored by visual inspection for precipitation after 10 days.

1D ^{1}H-NMR spectra with solvent presaturation are acquired for each of the buffer conditions, including the slowly precipitating samples. 2D ^{1}H–^{15}N-HSQC spectra are acquired only on samples that qualify as "well folded" based on the dispersed amide protons and the upfield-shifted methyl protons in the 1D spectrum. The 2D ^{1}H–^{15}N-HSQC spectra under different buffer conditions are then overlayed and assessed. The scoring is now based on the solubility as well as "foldedness." It has been observed that certain buffer conditions that provide excellent solubility provide spectra characteristic of partially or totally unfolded proteins. It is important to note that the presence or absence of some precipitation is not correlated with spectral quality, as some precipitation may occur without any significant loss of NMR signal. The buffer conditions providing the best-quality 2D HSQC NMR spectra and possessing sufficient sample stability are identified, and future protein samples are prepared in this buffer. A more detailed description of sample preparation for buffer optimization has been previously described (Rossi *et al.*, 2010). This same technique is also used for screening for ligands that bind to the protein and for those that can improve the quality of NMR spectra.

7.2. Construct optimization using amide hydrogen deuterium exchange with mass spectrometry (HDX-MS) detection

One of the keys to the NESG Construct Optimization Software (and the associated increase in project efficiency) lies in the identification of predicted disordered regions by the DisMeta server and elimination of these residues from the protein construct. In spite of these efforts, expert analysis of the 2D ^{1}H–^{15}N-HSQC spectra generated by the NESG high-throughput NMR Microprobe Screening Pipeline can often identify regions of disorder in some proteins with "Good" or "Promising" spectra. Although it is possible, in some cases, to identify these regions following resonance assignment, this approach is time-consuming. For this reason, we have implemented HDX-MS to experimentally identify these disordered regions in a high-throughput manner (Englander, 2006; Sharma *et al.*, 2009; Woods and Hamuro, 2001). HDX-MS studies, outlined in Fig. 2.6A, are based on

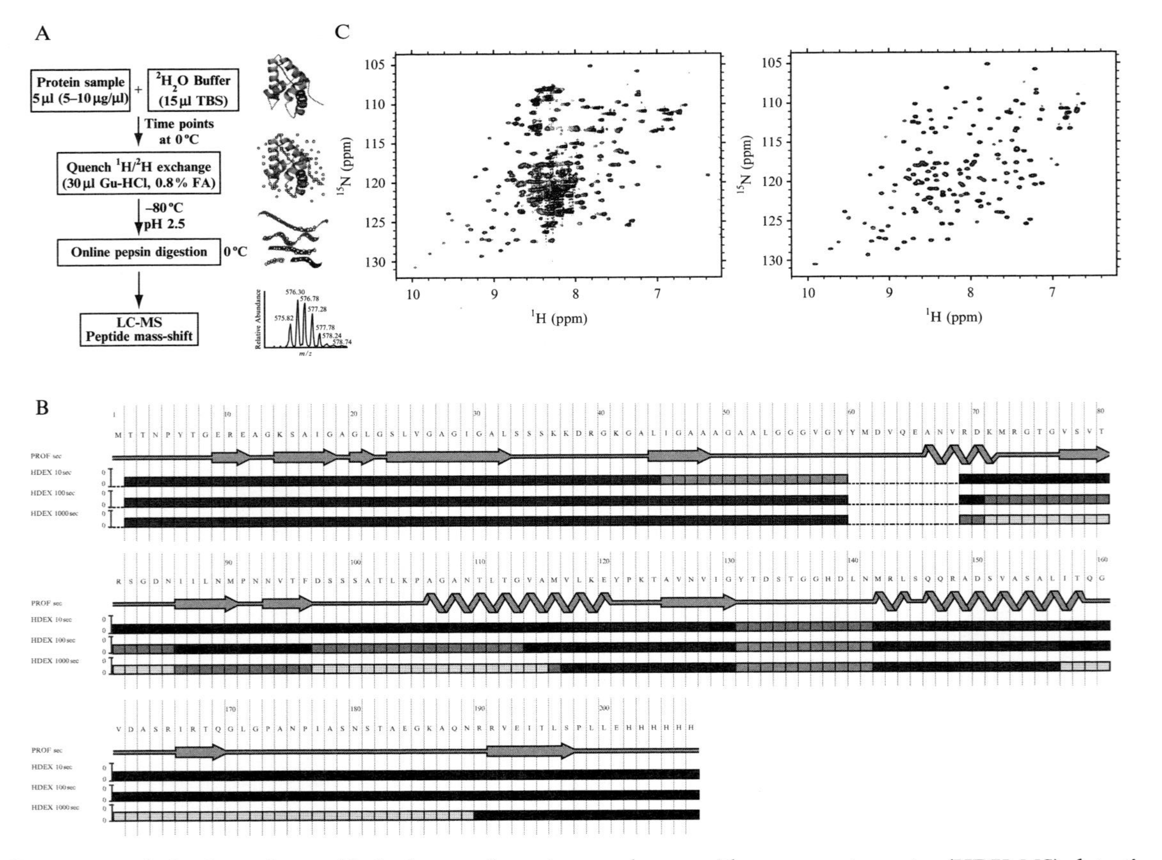

Figure 2.6 Construct optimization using amide hydrogen–deuterium exchange with mass spectrometry (HDX-MS) detection. (A) Schematic of the HDX-MS process. Proteins are mixed with $^{2}H_{2}O$ and the exchange is quenched at three time points (e.g., 10, 100, and 1000 s) by lowering the pH and temperature. The samples are then subjected to a pepsin column for digestion, followed by reverse-phase

the concept that backbone amide protons in disordered regions are solvent-accessible and therefore exchange with solvent deuterium (^2H_2O) at a faster rate than backbone amide protons in less solvent-accessible ordered regions. The degree of exchange over various time intervals is assessed by quenching the exchange kinetics by lowering the pH and temperature, fragmenting the protein by pepsin proteolysis, and measuring the mass of the resulting fragments by mass spectrometry. Briefly, a 5-μL aliquot of purified protein (25–50 μg) is mixed with 15 μL of ^2H_2O and incubated at 0 °C. At several time intervals, 30 μL of Quench Solution (1.0 M guanidine hydrochloride, 0.5% formic acid, ~pH 2.5) is added to stop the exchange reaction and partially unfold the protein (increasing efficiency of pepsin cleavage). These samples are then immediately flash-frozen at −80 °C. Samples are then thawed on ice and subjected to limited proteolysis (pepsin column liquid chromatography). The resulting peptides are separated by reverse phase liquid chromatography using a C18 column and acetonitrile gradient. The eluate is analyzed using an electrospray-linear ion-trap mass spectrometer (ESI-MS) and the resulting peptides and their masses are identified. Peptides with greater mass (higher deuterium exchange) compared to the fully protonated control are identified. The results are depicted graphically as a heat map (Fig. 2.6B); residues in peptides with the greatest amount of mass increase (disordered regions) are represented with red boxes. Residues in peptides showing little or no mass increase (ordered regions) are represented by blue boxes. This information allows design of alternative constructs (removing disordered regions) with improved crystallization properties (Pantazatos *et al.*, 2004; Spraggon *et al.*, 2004) and providing spectra more amenable for NMR assignment and structure determination (Sharma *et al.*, 2009), as shown in Fig. 2.6C.

7.3. Wheat germ cell-free protein expression

The NESG has made considerable gains in the area of eukaryotic expression and protein production during Protein Structure Initiative-2. However, analysis of the expression and solubility data derived from the NESG pipeline during this time period reveals that the eukaryotic targets are still recalcitrant in comparison to prokaryotic proteins. Overall, 42% of bacterial targets produce constructs with *ES* values greater than 11. By contrast, only 34% of the human

---

chromatography for peptide separation, and the degree of amide exchange prior to quenching is assessed by ESI-MS. (B) Results of the HDX-MS analysis for *E. coli* yiaD (NESG target ID: ER553). Primary sequence is shown at the top, followed by PROFsec secondary structure prediction and heat map of the HDX-MS results at 10, 100, and 1000 s reaction intervals. Note that the numbering does not include the N-terminal lipoprotein signal sequence that is cleaved from the mature protein. (C) ^1H–^{15}N-HSQC spectra of full-length (left) and construct-optimized (right) *E. coli* yiaD (59–199).

targets produce constructs that meet our usability score criteria. Clearly, the production of human and other eukaryotic proteins in *E. coli* remains challenging, and alternative expression systems are needed to allow the pursuit of high value human targets proven intractable in bacterial expression systems. Eukaryotic cell-free systems often permit successful production of proteins that undergo proteolysis or accumulate in inclusion bodies during bacterial expression (Vinarov *et al.*, 2006). Although the yields produced by wheat germ cell-free systems can be challenging for crystallization studies, recent advances in NMR microprobe technologies allow for structural determination using relatively small protein quantities (< 100 μg; Aramini *et al.*, 2007) that can be readily produced by wheat germ cell-free expression systems even when they cannot be made in *E. coli* cell expression systems.

In order to exploit this technology together with the "NESG Multiplex Vector Set," we have modified the Promega TnT wheat germ cell-free vector to allow ligase-independent cloning of the same PCR products used in our normal T7-based cloning pipeline. Specifically, we have modified the wheat germ cell-free expression vector pTSHQn from Promega to be compatible with our pET15_NESG vector and contain the identical N-terminal 6×-His tag and multicloning site sequences. This allows for the direct comparison of the same protein sequence expressed in both *E. coli* and wheat germ cell-free system, while also allowing a single PCR amplification product to be cloned into both vectors/expression systems. We have used this modified wheat germ cell-free system to study the efficacy of this approach with 66 nonsecreted human targets that were problematic (both in expression and/or solubility levels) in the prokaryotic expression system (Zhao *et al.*, 2010). Following wheat germ cell-free expression, we have found that ~70% of the insolubly expressed (in pET15_NESG) proteins were solubly expressed in wheat germ cell-free system. Overall, 52% of the human target proteins (34 of the 66) were solubly expressed in the wheat germ cell-free system. Although the yields of purified protein produced in this wheat germ cell-free system are limited (100–500 μg levels), NMR microprobe technology and selective labeling strategies such as SAIL (*S*tereo *A*rray *I*sotope *L*abeling; Aramini *et al.*, 2007; Kainosho *et al.*, 2006) enable wheat germ cell-free protein expression as a valuable tool for preparing samples for NMR studies.

7.4. Total gene synthesis and codon optimization

One possible reason for the low success rates with human and eukaryotic proteins in producing expression constructs meeting our usability criteria may relate to codon usage or other translation-related factors. Numerous studies have been published linking the differences in frequency of codon usage in *E. coli* (and tRNA pool) and the effects on expression of recombinant proteins of human or other origin with differing codon usage patterns (Ikemura, 1985; Kanaya *et al.*, 1999). Other studies have shown that codon usage is not the critical factor, rather the stability of mRNA secondary structure near the ribosomal binding site is the overriding factor (Kudla *et al.*, 2009), or a combination of both (Supek and Muc, 2010). Total gene synthesis allows for codons and secondary structure to be optimized to maximize translation (Tian *et al.*, 2004). To explore this technology and its efficacy for high-throughput protein production, some human genes have been optimized and synthesized for expression in *E. coli* (Codon Devices). We have compared the expression and solubility (*ES*) score of the optimized constructs versus natural sequences, and natural sequences with various codon-enhanced strains (Acton and Montelione, unpublished results). These studies indicate that codon optimization can be very effective in improving both expression and solubility of human proteins in *E. coli* expression hosts, providing a means to produce proteins that otherwise fail in bacterial expression systems.

8. *E. coli* Single Protein Production System for Isotopic Enrichment

Perdeuteration is an invaluable approach to the study of proteins by NMR. This is especially true for larger proteins (Gardner and Kay, 1998; Kay and Gardner, 1997; Rosen *et al.*, 1996; Venters *et al.*, 1995) and membrane proteins (Fernandez *et al.*, 2004; Hiller *et al.*, 2008). ^{2}H-, ^{13}C-, ^{15}N-enriched protein samples are also required for certain strategies utilizing rapid, fully automated analysis of small protein structures (Tang *et al.*, 2010; Zheng *et al.*, 2003). Although perdeuteration is a valuable tool, the preparation of deuterated samples can be costly; isotope costs for preparing ^{2}H-,^{13}C-,^{15}N-enriched samples range from \$1500 to \$3000 L^{-1}. *E. coli* expression, the preferred host for most protein production for NMR studies, is often inhibited by $^{2}H_2O$, and time-consuming, laborious acclimation steps must be employed for effective growth in deuterium (Venters *et al.*, 1995). Together, these issues have limited the use of predeuteration in NMR studies. However, the use of the condensed single-protein production method in *E. coli* (Schneider *et al.*, 2010; Suzuki *et al.*, 2005, 2006, 2007) alleviates many of these problems, including the cost barrier.

Briefly, the condensed single protein production system utilizes MazF endoribonuclease overexpression (of the MazE/MazF addiction module) to induce bacteriostasis. The toxin cleaves mRNA at ACA triplets, inhibiting the translation of *E. coli* proteins and causing the cells to enter into a quiescent state. Heterologous genes can be expressed at high levels under these conditions by the removal of ACA triplets in their coding sequence by total gene synthesis or site-directed mutagenesis. Since the cells are in a state of bacteriostasis, they (i) are not negatively affected by media containing deuterium and

(ii) can be condensed 40-fold or more while producing heterologous proteins for several days. The condensation provides cost savings by decreasing the volume of deuterated media used, and concomitantly the amounts of isotope-labeled carbon and nitrogen sources and/or other precursors.

Recently, two modified condensed single-protein production systems have been developed. Both are engineered to decrease or delay the expression of the heterologous gene before the addition of ^2H_2O containing media and hence produce recombinant protein with higher levels of perdeuteration. This is achieved in the condensed single protein production (tet) system (Schneider *et al.*, 2009) using a dual induction system, IPTG for MazF and anhydrous tetracycline, inducing the *tet* O_1 operator for the heterologous protein in a modified Cold-Shock Vector (Takara Biosciences). Alternatively, IPTG-inducible MazF variants were produced that lack tryptophan or histidine residues (Vaiphei *et al.*, 2010). Expression of these variants and heterologous genes (cloned into a Cold Shock Vector) in Trp or His auxotrophic strains does not allow the production of the recombinant protein until Trp or His is added to the media. Briefly, protein-coding regions for a target are produced lacking ACA triplets and cloned into an appropriate Cold-Shock Vector. *E. coli* strains harboring a plasmid with an IPTG-inducible MazF gene are transformed with the expression construct. Transformants are grown in 1 L of minimal media until OD_{600} reaches ~0.5 units and cold-shocked at 15 °C. IPTG is added to induce MazF expression for 2–3 h. Either before or after IPTG induction, the culture is centrifuged to pellet the cells, the supernatant is discarded, and the cells are condensed (resuspended in 10–100-fold decreased volume) in deuterated media. Finally, heterologous protein expression is induced (anhydrotetracycline) or permitted (addition of labeled Trp or His) and incubation at 15 °C is performed for at least 16 h.

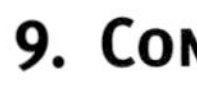

9. Conclusions

The main goal of this chapter is to inform the NMR community of the tools and strategies used by the NESG for high-throughput protein production of protein samples for NMR studies. The current weekly capacity of the NESG protein sample production pipeline is roughly 100 new expression constructs and analytical scale expression screens, 96 midi-scale fermentations/purifications, 72-preparative scale fermentations, and purification of 30–40 proteins in the tens of milligrams scale. In the last 5 years, this pipeline has produced over 21,000 expression constructs. Some 7000 of these have *ES* values meeting our usability criteria, with nearly 4000 successful protein purifications, resulting in over 750 Protein Data Bank depositions (in addition to more than 200 PDB depositions in the first-phase Protein Structure Initiative-1 project). The platform has provided samples allowing determination of NMR resonance assignments and three-dimensional structures for more than 400 proteins or protein domains. It is designed to work with commercially available yet relatively inexpensive equipment such as Qiagen BioRobot 8000 automation systems and plasticware, inexpensive platform shakers, and ÄKTAxpress FPLC systems among others. This approach allows for (i) scalability, as higher throughput can be attained through the acquisition of additional equipment or personnel, and (ii) duplication in industrial, core facilities or small consortiums of structural biology labs. The tools outlined here, including DisMeta and Primer Prim'er, and our detailed protocols for high-throughput cloning, expression, and purification, are freely available (http://www-nmr.cabm.rutgers.edu/labdocuments/proteinprod/index.htm) and will hopefully prove to be a valuable resource for the biological research community.

ACKNOWLEDGMENTS

We thank Profs. C. Arrowsmith, G. DeTitta, W. Hendrickson, J. Hunt, M. Gerstein, M. Inouye, M. Kennedy, J. Marcotrigiano, B. Rost, T. Szyperski, and L. Tong, along with all the current and former members of the Rutgers Protein Production Team and all members of the NESG Consortium, for their valuable advice in the development of the NESG Protein Sample Production Platform. This work was supported by a grant from the National Institute of General Medical Sciences Protein Structure Initiative U54-GM074958 and U54-GM094597 (to G. T. M.).

REFERENCES

Acton, T. B., *et al.* (2005). Robotic cloning and protein production platform of the northeast structural genomics consortium. *Methods Enzymol.* **394,** 210–243.

Acton, T. B., *et al.* (2011). An Alternative Method for Generating Genomic DNA PCR Template and Expansion of 'Reagent Genomes'. (in preparation).

Aramini, J. M., *et al.* (2007). Microgram-scale protein structure determination by NMR. *Nat. Methods* **4,** 491–493.

Aslanidis, C., and de Jong, P. J. (1990). Ligation-independent cloning of PCR products (LIC-PCR). *Nucleic Acids Res.* **18,** 6069–6074.

Bendtsen, J. D., *et al.* (2004). Improved prediction of signal peptides: SignalP 3.0. *J. Mol. Biol.* **340,** 783–795.

Bertini, I., *et al.* (2010). The annotation of full zinc proteomes. *J. Biol. Inorg. Chem.* **15,** 1071–1078.

Bertone, P., *et al.* (2001). SPINE: An integrated tracking database and data mining approach for identifying feasible targets in high-throughput structural proteomics. *Nucleic Acids Res.* **29,** 2884–2898.

Brodsky, O., and Cronin, C. N. (2006). Economical parallel protein expression screening and scale-up in *Escherichia coli*. *J. Struct. Funct. Genomics* **7,** 101–108.

Chen, L., *et al.* (2004). TargetDB: A target registration database for structural genomics projects. *Bioinformatics* **20,** 2860–2862.

Chikayama, E., *et al.* (2010). Mathematical model for empirically optimizing large scale production of soluble protein domains. *BMC Bioinform.* **11,** 113.

Crowe, J., *et al.* (1994). 6xHis-Ni-NTA chromatography as a superior technique in recombinant protein expression/purification. *Methods Mol. Biol.* **31,** 371–387.
Dean, F. B., *et al.* (2002). Comprehensive human genome amplification using multiple displacement amplification. *Proc. Natl. Acad. Sci. USA* **99,** 5261–5266.
Dessailly, B. H., *et al.* (2009). PSI-2: Structural genomics to cover protein domain family space. *Structure* **17,** 869–881.
Dyson, M. R., *et al.* (2004). Production of soluble mammalian proteins in *Escherichia coli*: Identification of protein features that correlate with successful expression. *BMC Biotechnol.* **4,** 32.
Englander, S. W. (2006). Hydrogen exchange and mass spectrometry: A historical perspective. *J. Am. Soc. Mass Spectrom.* **17,** 1481–1489.
Everett, J. K., *et al.* (2004). Primer Prim'r: A web based server for automated primer design. *J. Funct. Struct. Genomics* **5,** 13–21.
Fernandez, C., *et al.* (2004). NMR structure of the integral membrane protein OmpX. *J. Mol. Biol.* **336,** 1211–1221.
Ferre-D'Amare, A. R., and Burley, S. K. (1994). Use of dynamic light scattering to assess crystallizability of macromolecules and macromolecular assemblies. *Structure* **2,** 357–359.
Ferre-D'Amare, A. R., and Burley, S. K. (1997). Dynamic light scattering in evaluating crystallizability of macromolecule. *Methods Enzymol.* **276,** 157–166(Academic Press, New York).
Gardner, K. H., and Kay, L. E. (1998). The use of ^{2}H, ^{13}C, ^{15}N multidimensional NMR to study the structure and dynamics of proteins. *Annu. Rev. Biophys. Biomol. Struct.* **27,** 357–406.
Gill, S. R., *et al.* (2006). Metagenomic analysis of the human distal gut microbiome. *Science* **312,** 1355–1359.
Goh, C. S., *et al.* (2003). SPINE 2: A system for collaborative structural proteomics within a federated database framework. *Nucleic Acids Res.* **31,** 2833–2838.
Goh, C. S., *et al.* (2004). Mining the structural genomics pipeline: Identification of protein properties that affect high-throughput experimental analysis. *J. Mol. Biol.* **336,** 115–130.
Graslund, S., *et al.* (2008a). Protein production and purification. *Nat. Methods.* **5,** 135–146.
Graslund, S., *et al.* (2008b). The use of systematic N- and C-terminal deletions to promote production and structural studies of recombinant proteins. *Protein Expr. Purif.* **58,** 210–221.
Hartley, J. L., *et al.* (2000). DNA cloning using in vitro site-specific recombination. *Genome Res.* **10,** 1788–1795.
Haun, R. S., and Moss, J. (1992). Ligation-independent cloning of glutathione S-transferase fusion genes for expression in *Escherichia coli*. *Gene* **112,** 37–43.
Hendrickson, W. A. (1991). Determination of macromolecular structures from anomalous diffraction of synchrotron radiation. *Science* **254,** 51–58.
Hiller, M., *et al.* (2008). [2, 3-(13)C]-labeling of aromatic residues–getting a head start in the magic-angle-spinning NMR assignment of membrane proteins. *J. Am. Chem. Soc.* **130,** 408–409.
Huang, Y. J., *et al.* (2008). Targeting the human cancer pathway protein interaction network by structural genomics. *Mol. Cell. Proteomics* **7,** 2048–2060.
Ikemura, T. (1985). Codon usage and tRNA content in unicellular and multicellular organisms. *Mol. Biol. Evol.* **2,** 13–34.
Jansson, M., *et al.* (1996). High-level production of uniformly ^{15}N- and ^{13}C-enriched fusion proteins in *Escherichia coli*. *J. Biomol. NMR* **7,** 131–141.
Kainosho, M., *et al.* (2006). Optimal isotope labelling for NMR protein structure determinations. *Nature* **440,** 52–57.
Kanaya, S., *et al.* (1999). Studies of codon usage and tRNA genes of 18 unicellular organisms and quantification of Bacillus subtilis tRNAs: Gene expression level and species-specific diversity of codon usage based on multivariate analysis. *Gene* **238,** 143–155.
Kay, L. E., and Gardner, K. H. (1997). Solution NMR spectroscopy beyond 25 kDa. *Curr. Opin. Struct. Biol.* **7,** 722–731.
Klock, H. E., *et al.* (2008). Combining the polymerase incomplete primer extension method for cloning and mutagenesis with microscreening to accelerate structural genomics efforts. *Proteins* **71,** 982–994.

Krogh, A., *et al.* (2001). Predicting transmembrane protein topology with a hidden Markov model: Application to complete genomes. *J. Mol. Biol.* **305,** 567–580.

Kudla, G., *et al.* (2009). Coding-sequence determinants of gene expression in *Escherichia coli. Science* **324,** 255–258.

Kvist, T., *et al.* (2007). Specific single-cell isolation and genomic amplification of uncultured microorganisms. *Appl. Microbiol. Biotechnol.* **74,** 926–935.

Lamesch, P., *et al.* (2007). hORFeome v3.1: A resource of human open reading frames representing over 10,000 human genes. *Genomics* **89,** 307–315.

Lasken, R. S. (2007). Single-cell genomic sequencing using Multiple Displacement Amplification. *Curr. Opin. Microbiol.* **10,** 510–516.

Lasken, R. S. (2009). Genomic DNA amplification by the multiple displacement amplification (MDA) method. *Biochem. Soc. Trans.* **37,** 450–453.

Liu, J., *et al.* (2004). Automatic target selection for structural genomics on eukaryotes. *Proteins* **56,** 188–200.

Liu, J., *et al.* (2007). Novel leverage of structural genomics. *Nat. Biotechnol.* **25,** 849–851.

Lupas, A., *et al.* (1991). Predicting coiled coils from protein sequences. *Science* **252,** 1162–1164.

Markley, J. L., *et al.* (1998). Recommendations for the presentation of NMR structures of proteins and nucleic acids–IUPAC-IUBMB-IUPAB Inter-Union Task Group on the standardization of data bases of protein and nucleic acid structures determined by NMR spectroscopy. *Eur. J. Biochem.* **256,** 1–15.

Marsischky, G., and LaBaer, J. (2004). Many paths to many clones: A comparative look at high-throughput cloning methods. *Genome Res.* **14,** 2020–2028.

McGuffin, L. J., Bryson, K., and Jones, D. T. (2000). The PSIPRED protein structure prediction server. *Bioinformatics* **16,** 404–405.

Nair, R., *et al.* (2009). Structural genomics is the largest contributor of novel structural leverage. *J. Struct. Funct. Genomics* **10,** 181–191.

Nallamsetty, S., *et al.* (2004). Efficient site-specific processing of fusion proteins by tobacco vein mottling virus protease in vivo and in vitro. *Protein Expr. Purif.* **38,** 108–115.

Neri, D., *et al.* (1989). Stereospecific nuclear magnetic resonance assignments of the methyl groups of valine and leucine in the DNA-binding domain of the 434 repressor by biosynthetically directed fractional 13C labeling. *Biochemistry* **28,** 7510–7516.

Netzer, W. J., and Hartl, F. U. (1997). Recombination of protein domains facilitated by co-translational folding in eukaryotes. *Nature* **388,** 343–349.

Pantazatos, D., *et al.* (2004). Rapid refinement of crystallographic protein construct definition employing enhanced hydrogen/deuterium exchange MS. *Proc. Natl. Acad. Sci. USA* **101,** 751–756.

Peti, W., *et al.* (2005). Towards miniaturization of a structural genomics pipeline using micro-expression and microcoil NMR. *J. Struct. Funct. Genomics* **6,** 259–267.

Porath, J. (1992). Immobilized metal ion affinity chromatography. *Protein Expr. Purif.* **3,** 263–281.

Price, W. N., 2nd, *et al.* (2009). Understanding the physical properties that control protein crystallization by analysis of large-scale experimental data. *Nat. Biotechnol.* **27,** 51–57.

Punta, M., *et al.* (2009). Structural genomics target selection for the New York consortium on membrane protein structure. *J. Struct. Funct. Genomics* **10,** 255–268.

Qin, J., *et al.* (2010). A human gut microbial gene catalogue established by metagenomic sequencing. *Nature* **464,** 59–65.

Rosen, M. K., *et al.* (1996). Selective methyl group protonation of perdeuterated proteins. *J. Mol. Biol.* **263,** 627–636.

Rossi, P., *et al.* (2010). A microscale protein NMR sample screening pipeline. *J. Biomol. NMR* **46,** 11–22.

Rost, B., *et al.* (2004). The PredictProtein server. *Nucleic Acids Res.* **32,** W321–W326.

Rual, J. F., *et al.* (2004). Human ORFeome version 1.1: A platform for reverse proteomics. *Genome Res.* **14,** 2128–2135.

Schneider, W. M., *et al.* (2009). Independently inducible system of gene expression for condensed single protein production (cSPP) suitable for high efficiency isotope enrichment. *J. Struct. Funct. Genomics* **10,** 219–225.

Schneider, W. M., *et al.* (2010). Efficient condensed-phase production of perdeuterated soluble and membrane proteins. *J. Struct. Funct. Genomics* **11,** 143–154.

Sharma, S., *et al.* (2009). Construct optimization for protein NMR structure analysis using amide hydrogen/deuterium exchange mass spectrometry. *Proteins* **76,** 882–894.

Sheibani, N. (1999). Prokaryotic gene fusion expression systems and their use in structural and functional studies of proteins. *Prep. Biochem. Biotechnol.* **29,** 77–90.

Shirano, Y., and Shibata, D. (1990). Low temperature cultivation of *Escherichia coli* carrying a rice lipoxygenase L-2 cDNA produces a soluble and active enzyme at a high level. *FEBS Lett.* **271,** 128–130.

Slabinski, L., *et al.* (2007). XtalPred: A web server for prediction of protein crystallizability. *Bioinformatics* **23,** 3403–3405.

Snyder, D. A., *et al.* (2005). Comparisons of NMR spectral quality and success in crystallization demonstrate that NMR and X-ray crystallography are complementary methods for small protein structure determination. *J. Am. Chem. Soc.* **127,** 16505–16511.

Spraggon, G., *et al.* (2004). On the use of DXMS to produce more crystallizable proteins: Structures of the T. maritima proteins TM0160 and TM1171. *Protein Sci.* **13,** 3187–3199.

Studier, F. W., and Moffatt, B. A. (1986). Use of bacteriophage T7 RNA polymerase to direct selective high-level expression of cloned genes. *J. Mol. Biol.* **189,** 113–130.

Supek, F., and Muc, T. (2010). On relevance of codon usage to expression of synthetic and natural genes in *Escherichia coli. Genetics* **185,** 1129–1134.

Suzuki, M., *et al.* (2005). Single protein production in living cells facilitated by an mRNA interferase. *Mol. Cell* **18,** 253–261.

Suzuki, M., *et al.* (2006). Bacterial bioreactors for high yield production of recombinant protein. *J. Biol. Chem.* **281,** 37559–37565.

Suzuki, M., *et al.* (2007). Single protein production (SPP) system in *Escherichia coli. Nat. Protoc.* **2,** 1802–1810.

Tang, Y., *et al.* (2010). Fully automated high-quality NMR structure determination of small (2)H-enriched proteins. *J. Struct. Funct. Genomics* **11,** 223–232.

Tian, J., *et al.* (2004). Accurate multiplex gene synthesis from programmable DNA microchips. *Nature* **432,** 1050–1054.

Vaiphei, S. T., *et al.* (2010). Use of amino acids as inducers for high-level protein expression in the single-protein production system. *Appl. Environ. Microbiol.* **76,** 6063–6068.

Venters, R. A., *et al.* (1995). High-level $^{2}H/^{13}C/^{15}N$ labeling of proteins for NMR studies. *J. Biomol. NMR* **5,** 339–344.

Vinarov, D. A., *et al.* (2006). Wheat germ cell-free platform for eukaryotic protein production. *FEBS J.* **273,** 4160–4169.

Woods, V. L., Jr., and Hamuro, Y. (2001). High resolution, high-throughput amide deuterium exchange-mass spectrometry (DXMS) determination of protein binding site structure and dynamics: Utility in pharmaceutical design. *J. Cell. Biochem. Suppl.* **37,** 89–98.

Wootton, J. C., and Federhen, S. (1996). Analysis of compositionally biased regions in sequence databases. *Methods Enzymol.* **266,** 554–571.

Xiao, R., *et al.* (2010). The high-throughput protein sample production platform of the Northeast Structural Genomics Consortium. *J. Struct. Biol.* **172,** 21–33.

Zhao, L., *et al.* (2010). Engineering of a wheat germ expression system to provide compatibility with a high throughput pET-based cloning platform. *J. Struct. Funct. Genomics* **11,** 201–209.

Zheng, D., *et al.* (2003). Automated protein fold determination using a minimal NMR constraint strategy. *Protein Sci.* **12,** 1232–1246.

Zhu, B., *et al.* (2007). In-fusion assembly: Seamless engineering of multidomain fusion proteins, modular vectors, and mutations. *Biotechniques* **43,** 354–359.

CHAPTER THREE

Key Factors for Successful Generation of Protein–Fragment Structures: Requirement on Protein, Crystals, and Technology

Jark Böttcher, Anja Jestel, Reiner Kiefersauer, Stephan Krapp, Susanna Nagel, Stefan Steinbacher, *and* Holger Steuber

Contents

Abstract

In the past two decades, fragment-based approaches have evolved as a predominant strategy in lead discovery. The availability of structural information on the interaction geometries of binding fragments is key to successful structure-guided fragment-to-lead evolution. In this chapter, we illustrate methodological advances for protein–fragment crystal structure generation in order to offer

Proteros biostructures GmbH, Am Klopferspitz 19, Martinsried, Germany

Methods in Enzymology, Volume 493
ISSN 0076-6879, DOI: 10.1016/B978-0-12-381274-2.00003-0

general lessons on the importance of fragment properties and the most appropriate crystallographic setup to evaluate them. We analyze elaborate protocols, methods, and clues applied to challenging complex formation projects. The results should assist medicinal chemists to select the most promising targets and strategies for fragment-based crystallography as well as provide a tutorial to structural biologists who attempt to determine protein–fragment structures.

Abbreviation

FBDD Fragment-based drug discovery

1. Introduction

Within the last decade, fragment-based drug discovery (FBDD) has evolved as a new paradigm for the design of small-molecule lead compounds and an alternative to high-throughput screening well documented in the literature (Congreve *et al.*, 2008; Coyne *et al.*, 2010; Erlanson, 2006; Erlanson and Hansen, 2004; Hajduk and Greer, 2007; Murray and Blundell, 2010; Rees *et al.*, 2004). To generate molecules with desired drug-like properties, fragments require significant design efforts in the context of growing, linking, or merging strategies (Carr *et al.*, 2005). In the course of the fragment-to-lead optimization, the availability of structural information on fragment–target interactions is of utmost importance in order to avoid "blind" puzzling in chemical space.

Due to their small size, generally, a lower number of directed interactions with the target protein are formed by fragments, often resulting in lower affinities, compared to more elaborate ligands (K_d or IC_{50} < 10 μM). Typical fragments may be characterized by the rule of three: molecular weight <300 Da, hydrogen bond donors $\leq$3, hydrogen bond acceptors $\leq$3, and clogP $\leq$3 and have affinities in the range of 1 μM to several millimolars (Congreve *et al.*, 2003). In addition to weak binding, nonspecific interactions or binding in multiple conformations may represent a major problem in their crystallographic characterization. In the case of high-affinity ligands, the predominant technique is cocrystallization of the ligand and the target protein (Hassell *et al.*, 2007). For low-affinity fragments, cocrystallization is often hampered by the high concentration of fragments necessary to afford full occupancy at the binding site to yield an interpretable difference electron density. In addition, the number of fragments needed to undergo structural analysis is disproportionately higher than in traditional structure-based drug discovery projects (Hann *et al.*, 2001). The large number of fragments with a wide range of physicochemical properties would require specific adjustments of cocrystallization conditions for each single fragment. Alternative approaches of employing crystals of the apo protein or of a protein prebound with a weak inhibitor are better suited to generate cocrystals of protein–fragment complexes. Organic solvents necessary to increase the solubility of the fragments to give rise to higher ligand concentrations are better tolerated by preformed crystals.

To illustrate general principles that we have observed in various projects, we have generated a set of protein–fragment structures from the Protein Data Bank (PDB) by applying a ligand molecular weight filter via Relibase (Hendlich *et al.*, 2003) to select those crystal structures containing only ligands with a molecular weight of 70–350 Da. As a second selection criterion, hits passing the molecular weight filter must result from a FBDD campaign that has generated at least three protein–fragment structures. Some exceptions from this second criterion have been made to include crystal structures containing fragment-type ligands with drug-like properties for comparison, for example, estrogen receptor beta, cyclooxygenase 1, and phenylethanolamine *N*-methyl transferase. A total 244 protein–fragment structures from targets listed in Table 3.1 are analyzed; these structures are also discussed in the context of technologies applied and the lessons learned at our company.

2. General Target Properties and Protein Sources in Protein–Fragment Crystallography

FBDD initially focused on proteases and protein kinases (Table 3.1) in line with their general amenability to X-ray crystallography and their relevance to drug design. Meanwhile, it has been shown that this approach is generic and can be extended to a broad variety of other proteins interacting with either small molecules or biological macromolecules such as proteins or nucleic acids (Coyne *et al.*, 2010). Almost all studies published so far have generated leads for globular and soluble targets. The size of the target protein appears not to be restricted for X-ray crystallography (e.g., yeast proteasome with a molecular weight of 700 kDa, unpublished in-house results). Taking the recent progress in the structural biology of G-protein-coupled receptors and ion channels (Doyle *et al.*, 1998; Rasmussen *et al.*, 2007) into account, it will be interesting to see whether the approach can also be extended to these target classes.

Table 3.1 Overview of targets and associated references evaluated in this contribution

Kinases	Other targets
p38 kinase (11) (Gill *et al.*, 2005; Hartshorn *et al.*, 2005; Wang *et al.*, 1997)	Cytochrome *c* peroxidase (15) (Brenk *et al.*, 2006; Musah *et al.*, 2002)
ERK2 (3) (Aronov *et al.*, 2007)	Cyp 51 *Mycobacterium tuberculosis* (3) (Podust *et al.*, 2007)
CDK2 (4) (Anderson *et al.*, 2008; Hartshorn *et al.*, 2005; Howard *et al.*, 2009; Lawrie *et al.*, 1997; Rosenblatt *et al.*, 1993)	COX1 (4) (Selinski *et al.*, 2001)
Aurora A (5) (Howard *et al.*, 2009)	NO-Synthase (4) (Fedorov *et al.*, 2003; Ghosh *et al.*, 2001)
JAK2 (4) (Antonysamy *et al.*, 2009; Howard *et al.*, 2009)	Leukotriene A4 Hydrolase (20) (Davies *et al.*, 2009)
PKA-PKB chimera (7) (Davies *et al.*, 2007; Engh *et al.*, 1996; Saxty *et al.*, 2007)	Estrogen receptor β (7) (Malamas, *et al.*, 2004; Manas *et al.*, 2004)
Proteases	β-lactamase (14) (Chen & Shoichet, 2009; Teotico *et al.*, 2009)
BACE (11) (Godemann *et al.*, 2009; Murray *et al.*, 2007; Patel *et al.*, 2004; Wang *et al.*, 2010)	Nucleoside 2-Deoxyribosyltransferase *Trypanosoma brucei* (4) (Bosch *et al.*, 2006)
Thrombin (10) (Howard *et al.*, 2006; Katz *et al.*, 2001)	IL-2 (2) (Arkin *et al.*, 2003; Thanos *et al.*, 2003)
Trypsin (11) (Katz *et al.*, 2001; Toyota *et al.*, 2001)	Protein Tyrosine phosphatase 1b (PTP 1B) (9) (Andersen *et al.*, 2000; Hartshorn *et al.*, 2005; Iversen *et al.*, 2000; Szczepankiewicz *et al.*, 2003)
MMP-12 (3) (Holmes *et al.*, 2009)	RNAse A (1) (Hartshorn *et al.*, 2005; Leonidas *et al.*, 1997)
MMP-13 (3) (Engel *et al.*, 2006)	PDE4 (7) (Card *et al.*, 2004)
Urokinase-type plasminogen activator (9) (Frederickson *et al.*, 2008; Katz *et al.*, 2001)	Pantothenate synthase *M. tuberculosis* (9) (Hung *et al.*, 2009)
HIV Protease (6) (Perryman *et al.*, 2010)	HSP90 (11) (Brough *et al.*, 2009; Huth *et al.*, 2007)
DPP IV (4) (Engel *et al.*, 2006)	Biotin carboxylase (Mochalkin *et al.*, 2009)
MetAP (5) (Schiffmann *et al.*, 2005; Wang *et al.*, 2007)	Human uracil DNA glycosylase (4) (Chung *et al.*, 2009)
Thermolysin (4) (Englert *et al.*, 2010; English *et al.*, 2001)	Prostaglandin D2 synthase (6) (Hohwy *et al.*, 2008)
	HCV NS5B (6) (Antonysamy *et al.*, 2008)
	Ricin A (4) (Carra *et al.*, 2007)
	Carbonic anhydrase (3)(Scott *et al.*, 2009)
	PNMT (McMillan *et al.*, 2004)

Numbers in brackets indicate the number of protein–fragment structures analyzed for each target.

The nature of the targeted site or molecular interaction has significant impact on the suitability of a target for FBDD. Only well-defined cavities typical for enzymes are regarded as highly druggable. Protein–protein interactions are generally seen as very challenging because of the often flat and featureless contact surfaces that lack defined pockets for binding. Not surprisingly, published results on FBDD application to protein–protein interactions are not well represented in the literature—only one example, the interfering of interactions between IL2 and IL2 receptors, has been reported. In that case, tethering via covalent linkage of the fragment to the target protein has been successfully applied to obtain protein–fragment structures to guide the design for the evolution of a known inhibitor (Braisted *et al.*, 2003; Hyde *et al.*, 2003; Thanos *et al.*, 2003).

From a practical standpoint, a FBDD project requires reliable access to sufficient amounts of the target protein. An evaluation of the sources of protein for the preparation of the protein–fragment structures included in our data set is given in Fig. 3.1. The protein sources are classified according to the most common expression systems, *Escherichia coli*, baculovirus-infected insect cells, and other sources, and compared to the plethora of all PDB-deposited structures. As expected, the clear majority of protein–fragment structures have been generated using protein expressed in and purified from *E. coli* that is not different from all PDB-deposited structures (see Fig. 3.1; Joachimiak, 2009). Baculovirus/insect

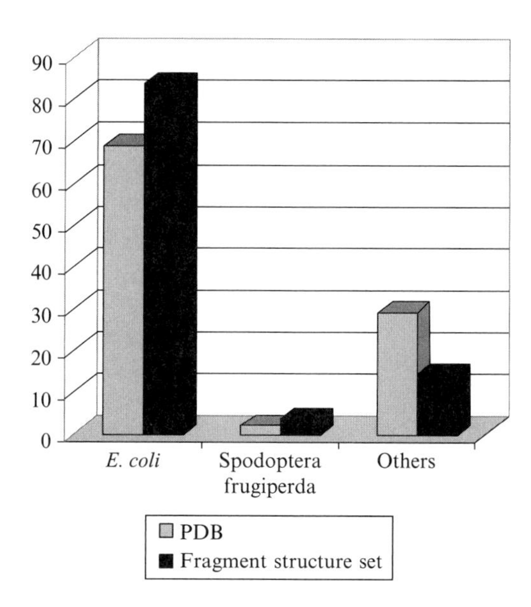

Figure 3.1 Protein sources used for fragment structures compared to all PDB-deposited structures.

cell expression systems play a similar role for the generation of protein–fragment structures as for all PDB structures. In practice, the amount of a protein required for a FBDD crystallography project may vary significantly from several milligrams to 100 mg or more, depending on the efficiency of crystal generation.

3. Fragment Properties and the Crystallization Setup

3.1. Robust crystallization

Crystallography can either be used as a semi-high-throughput method to directly identify fragments by screening medium-sized fragment libraries or applied to fragments preselected by various methods, including surface plasmon resonance, NMR-based techniques, fluorescence correlation spectroscopy, or virtual fragment screening. For protein crystallography, we can make use of advances in automation, X-ray brilliance, crystal and data handling, as well as computational resources. The application of cocktail screening proposed in Wim Hols Lab book in 1990 (Verlinde *et al.*, 2009) offers an additional strategy to decrease the number of data sets that have to be collected in a fragment-based project. The conversion even of preselected fragment hits into crystal structures is still characterized by a high attrition rate due to the need for a large number of crystals that must be generated in a highly reproducible manner. In practice, crystallization conditions are only rarely established specifically for fragment-based crystallography, as in the case of β-secretase (BACE). Here, a well-reproducible setup for achievement of strongly diffracting and soakable crystals containing high crystallographic symmetry has been identified (Patel *et al.*, 2004). Treatment of the crystals with stabilizing cross-linking agents such as glutaraldehyde should also be considered to make crystal forms more amenable to soaking procedures (Andersen *et al.*, 2009; Mattos *et al.*, 2006). In summary, a FBDD crystallography projects may easily require several dozens of crystals and therefore, an appropriate crystal system has to be established in the initial project phase.

3.2. Fragment affinity

The affinity range of fragments that has been successfully translated into protein–fragment structures is 100 n*M*–10 m*M* (Fig. 3.2). Few fragments possessing lower affinities have been crystallized in complex with a protein target, mainly due to an extremely high concentration needed in the molar range. There are, however, interesting exceptions where a fragment shows no affinity in an enzymatic assay but nevertheless reveals binding in an

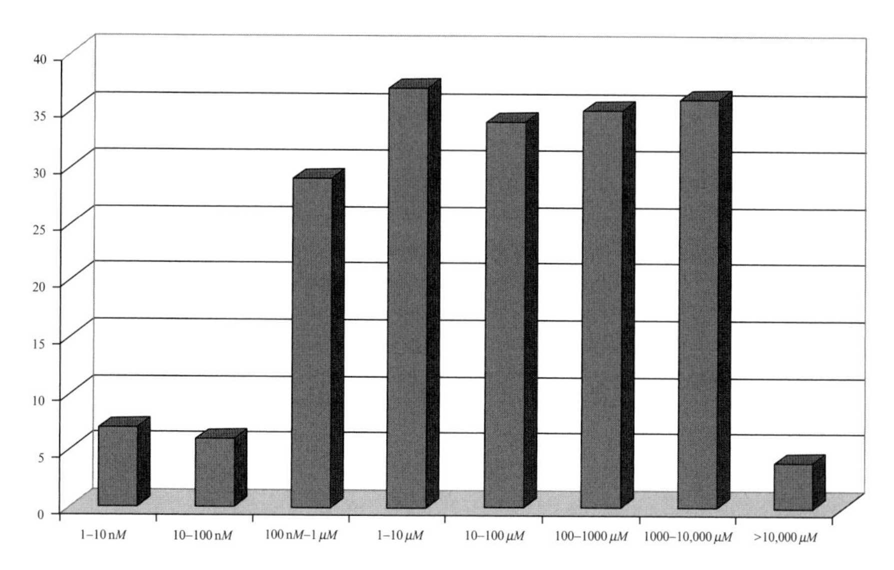

Figure 3.2 Distribution of fragment affinities encountered for the set of PFS analyzed in this chapter.

X-ray costructure as observed for thrombin (Howard *et al.*, 2006) and RNAse A (Hartshorn *et al.*, 2005). This discrepancy might be due to either an insufficient concentration range covered in the enzymatic assay or distinct binding sites for the fragment and substrate/probe molecule. These observations support the use of crystallographic cocktail screening compared to other methods of prescreening.

In some rare cases, small, fragment-like compounds may have tight binding properties as observed for natural compounds or their derivatives such as estradiol-type derivatives as ligands of the estradiol receptor (Malamas *et al.*, 2004), and nicotine or epibatidine as ligands of the nicotinic acetylcholine receptor (Hansen *et al.*, 2005). The origin of this extraordinary high ligand efficiency obviously involves natural evolution of these targets toward binding of their fragment-type ligands over millions of years. The use of a fragment library assembled from building blocks of natural compounds ("Fragments of Life"®) attempts to benefit from this concept (Davies *et al.*, 2009).

Mapping of active site hot spots with organic solvent additives such as phenol, acetone, and acetonitrile have also been reported (English *et al.*, 2001; Mattos *et al.*, 2006). Successful complex formation requires unusually high concentrations, as in the case of thermolysin with up to 70 and 80% of acetone or acetonitrile. Consistent with the binding mode of the parent inhibitor HONH-BAGN (Holmes and Matthews, 1981), the phenol fragment is observed to bind into the S1′ pocket at a concentration of 50 m*M*.

3.3. Crystallization buffer and precipitants

A major concern to obtain fragment–protein structures is the composition of the crystallization buffer needed to obtain the cocrystals. It is rarely possible to perform fragment screening under crystallization conditions, due to the increased viscosity and other effects encountered in the crystallization medium (Chung, 2007). The crystallization buffer usually not only deviates from the solution screening conditions in salt, ionic strength, and pH but also in the presence of precipitating agents required to achieve crystallization. Hence, the influence of crystallization buffer components on fragment binding should be considered when selecting a specific crystal system. Crystallization conditions commonly contain one or more precipitants to include, for example, inorganic salts, organic solvents, long-chain polymers (>1000 Da), low molecular-weight polymers (<1000 Da), and nonvolatile organic compounds (see Fig. 3.3; McPherson, 1990). The widespread use of high molecular-weight polyethylene glycol (PEG), low molecular-weight polymers, and nonvolatile organic compounds for crystallization is consistent with the notion that these precipitants can act as solubilizing agents for hydrophobic chemical compounds (Hassell *et al.*, 2007), giving rise to higher concentrations in the crystallization buffer of hydrophobic, otherwise poorly soluble fragments. Thus, if possible, these conditions should be generally preferred compared to salt conditions. Interestingly, electron density for PEG molecules is only rarely observed in

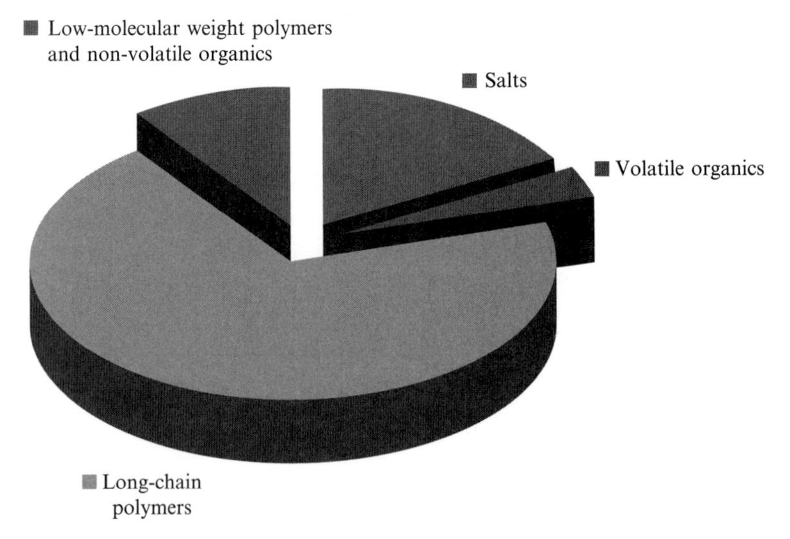

Figure 3.3 Usage of precipitants in successful protein–fragment structure determination. (See Color Insert.)

X-ray maps, suggesting no, or only limited, firm interactions of PEG with the protein to interfere with fragment binding.

3.4. Influence of pH

The influence of pH on a protein–ligand structure has been well documented. For example, a dramatic change in binding geometry was detected in crystallographic analysis for factor Xa inhibitor bound to trypsin when the complex was crystallized at different pH (Stubbs *et al.*, 2002). In order to minimize the influence of pH, a similar pH should be applied in both screening and structural analyses. This is particularly relevant for fragments that exhibit only a few interactions with a target protein. A change in the protonation state of either a protein or ligand functional group can cause a significant affinity change. The change of protonation on the ligand can also alter the solubility of the ligand. In the case of the serine proteases trypsin, thrombin, and urokinase-type plasminogen, a detailed analysis was performed using fragments with high affinity (Katz *et al.*, 2001). Changes in pH under otherwise unchanged solvent conditions caused changes in inhibitor solubility by one order of magnitude, with a simultaneous loss of affinity by about the same magnitude. For weak-binding fragments, this loss in affinity would hamper a successful structure determination. For trypsin, the structures of a weak-binding inhibitor determined at the pH values of 7 and 8 showed two different binding modes (Stubbs *et al.*, 2002). In the worst case, determination of a protein–fragment structure at the "wrong" pH could lead to a misguided optimization strategy.

The aspartic protease BACE is probably one of the most studied targets in FBDD. The enzyme exhibits its optimum activity at pH 4.5 (Vassar *et al.*, 1999). Given that the established crystallization condition has a pH of around 6 and that most fragments are found to interact with the carbonyl of the two catalytic aspartate residues via a protonated nitrogen, several strategies have been conducted to lower the crystallization pH to near 4.5 (Patel *et al.*, 2004). The scrutiny of endothiapepsin (which crystallizes at lower pH) as a surrogate target for BACE to overcome the affinity discrepancy between the pH values of screening and crystallographic conditions underscores the importance of pH when screening hits are evaluated in cocrystal structures (Geschwindner *et al.*, 2007).

A similar example has been observed for a scenario established to probe molecular docking to a designed, charged binding site embedded into cytochrome *c* peroxidase (Brenk *et al.*, 2006). In order to validate the docking results, crystal structures were determined. The set of putative binders comprised phenol which did not yield a significant occupancy after soaking with 50 m*M* fragment at neutral pH. Surprisingly, a similar soak at pH 4.5 produced an interpretable electron density for phenol present at 65% occupancy. The origin of this apparent discrepancy can easily be understood by inspection of the complex structure in which the phenol hydroxyl group donates a hydrogen bond to a neighboring peptide carbonyl group.

3.5. Flexibility in binding sites

The generation of protein–fragment structures strongly benefits from the availability of soakable crystal forms which allow the use of apo crystals for the complexation process. Such a workflow is the method of choice not only because it reduces efforts, compared to the alternative of cocrystallization, but also because it often represents the only available option when either small molecule or solvent additives interfere with crystallization of the protein target.

For soakable crystal forms, two main prerequisites should be fulfilled. The site of interest has to be accessible via a solvent channel in the crystal, and sufficient protein conformational flexibility should be inherent at the binding site to respond to induced-fit processes. However, these prerequisites are not necessarily fulfilled due to either target-specific or crystal-form specific challenges. The binding site of the protein may be completely buried in the crystal or altered upon crystallization. Sometimes, stabilizing ligands facilitate crystallization, as observed for nuclear receptors such as estrogen receptor beta (Fig. 3.4A; Malamas *et al.*, 2004) and phenylethanolamine *N*-methyl transferase (Fig. 3.4B; McMillan *et al.*, 2004). For such targets, cocrystallization would be the method of choice rather than soaking or ligand-exchange soaking attempts. Crystallographic screening, as usually performed by cocktail soaking, would likely be impossible for these targets.

Another unfavorable situation detrimental to soaking is observed if the active site is fixed by crystal packing in a conformation noncompetent to ligand binding. This case is exemplified in Fig. 3.5 on Bcl-XL (Muchmore *et al.*, 1996). To resolve this issue for FBDD, either NMR screening and determination of the binding properties has been performed (Petros *et al.*, 2006), or a more favorable crystal form needs to be established (Lee *et al.*, 2007; unpublished in-house results).

There are various examples demonstrating that even the smallest fragments of urea-like size are able to stabilize a distinct binding pocket geometry compared to the unbound state (Carra *et al.*, 2007). Even though there is an ongoing debate whether induced-fit is indeed "induced" or only obtained by conformational selection of the most favorable binding conformer by the small-molecule compound, a crystal form suitable for fragment soaking should account for appropriate mobility required to accommodate the ligand. A study on Ricin A demonstrates, in particular, that even fragments of a molecular weight as low as 60 g mol^{-1} are able to cause significant active site shifts such as the change of tyrosine $\chi 1$ and $\chi 2$ angles, strongly influencing the active site architecture. Other examples

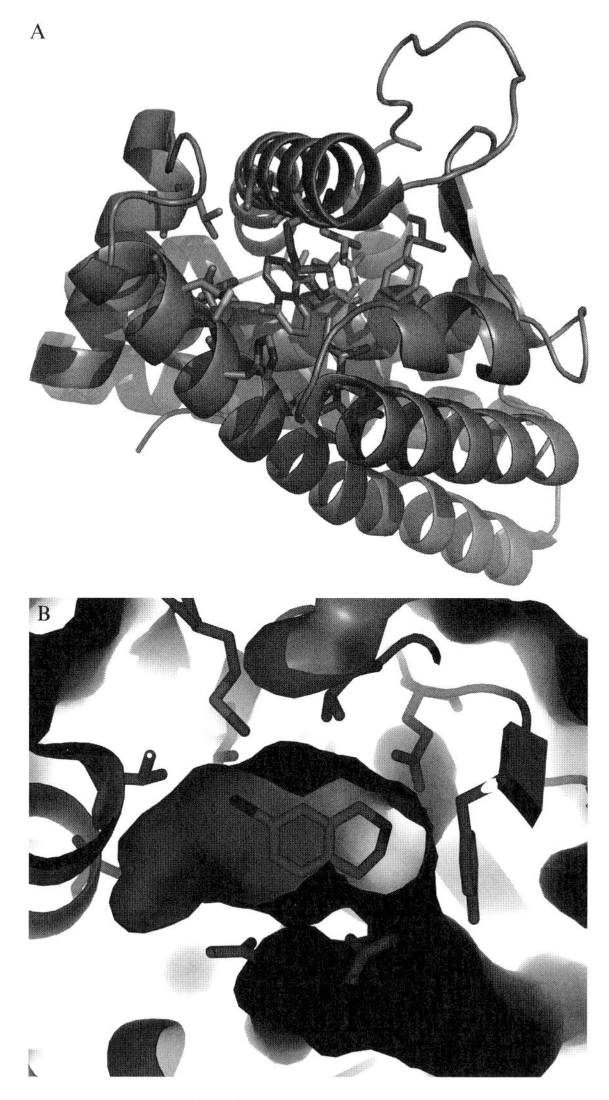

Figure 3.4 Representations of buried binding pockets not suitable for complex formation by soaking. (A) Estrogen receptor beta; (B) phenylethanolamin *N*-methyl transferase. The protein binding pocket is shown in blue, the buried ligand as orange sticks. (See Color Insert.)

of induced-fit processes caused by fragments are observed for BACE (Patel *et al.*, 2004) and cytochrome *c* peroxidase (Brenk *et al.*, 2006). Although BACE and HIV protease both represent aspartic proteases of therapeutic interest and share certain similarity with respect to their active site

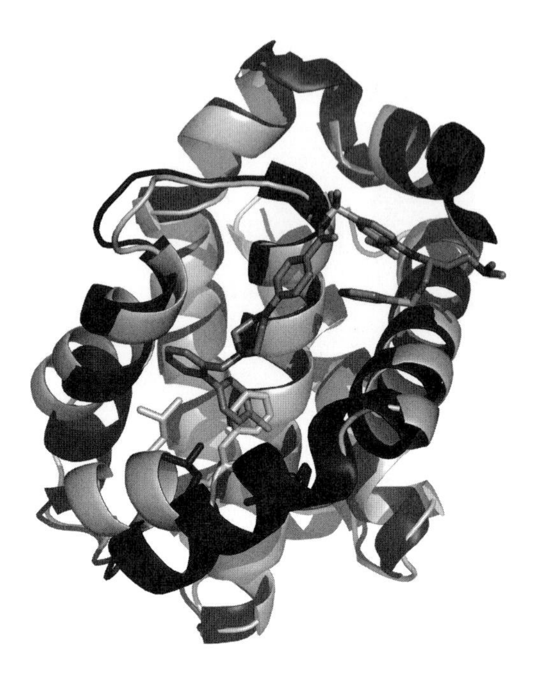

Figure 3.5 Superposition of Bcl-XL crystal structures in an unbound state (PDB number 1MAZ, gray) and a ligand-bound state (PDB number 2YXJ, blue, ligand shown as sticks in light blue). The conformation observed for the unbound protein in space group $P4_12_12$ strongly interferes with ligand binding, while the crystal form observed in space group $P2_12_12_1$ exhibits a pronounced binding capability. (See Color Insert.)

architecture, a larger number of BACE–fragment structures is available but no HIV protease–fragment structure (except those containing fragments bound to exosites; Perryman *et al.*, 2010). A likely explanation is the difference in the dynamic properties of the flap regions of both proteases which depend also on the crystal form. While the flap region of BACE in crystal forms successfully used possesses sufficient mobility to bind soaked ligands even in a crystalline state (Godemann *et al.*, 2009; Patel *et al.*, 2004; Fig. 3.6), this flexibility is not observed for HIV protease crystals obtained in an unbound state. Here, the open-flap conformation, as observed in $P4_12_12$ (Heaslet *et al.*, 2007), is trapped by crystal contacts which prevent accommodation of small-molecule entities. Obviously, the open-flap conformation observed for this crystal form so far prohibits its use in FBDD. In other crystal forms (e.g., Pillai *et al.*, 2001), the flap regions are only to a lesser extent involved in crystal contacts, and visual inspection of crystal packing allows one to understand their mobility. This example shows that the availability of "apo" crystals is necessary but insufficient to afford a favorable soaking system, and further requirements need to be fulfilled.

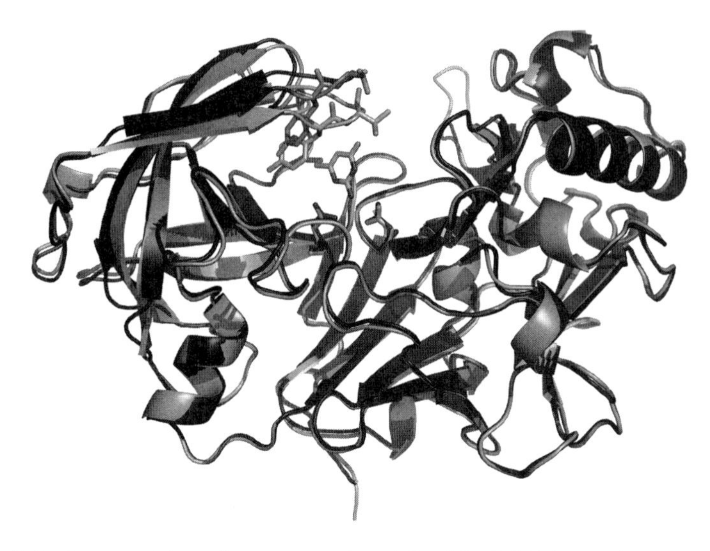

Figure 3.6 Superposition of monomers A and C of the BACE crystal structure (PDB number 3HVG; Godemann *et al.*, 2009). While monomer C (dark blue) represents the unbound state with the flap region in open conformation, monomer A (green) exhibits a closed conformation interacting with the bound fragment (orange). (See Color Insert.)

3.6. Optimization of fragment occupancy

High occupancy of a fragment at the site of interest is a necessary, but not in all cases a sufficient, prerequisite for an unambiguous determination of the binding mode. As reported for cyclophilin 3 inhibited by a Pro–Ala dipeptide (Wu *et al.*, 2001), it has been shown that (1) the occupancy of a ligand at the protein active site in the crystal correlates well with the ligand concentration the protein crystal is exposed to and (2) the affinity to the protein active site in a crystal is comparable to that observed in biochemical assays. These observations are of fundamental importance for a rational protein–fragment complexation process. It is generally assumed that the ligand concentration should be chosen at least 10 times higher than the expected binding affinity to obtain a reasonable ($\sim$90%) occupancy in the protein crystal (Chung, 2007; Danley, 2006). However, as evidenced by the day-by-day laboratory experience, most of the ligands are not sufficiently soluble under the established crystallization condition (despite the solubilizing character of certain precipitants as addressed above) and require the addition of organic additives in order to reach higher concentrations. Even though such additives can be detrimental to hydrophobicity-driven binding ("antihydrophobic solvent effects"; see Bartlett *et al.*, 2002), this method is the most popular one in fragment crystallography as well as in protein complexation with lead-like ligands. The effect of organic solvents on crystal stability can be judged by exposure of crystals to the soaking composition in the absence of a fragment. Various studies make use of dimethylsulfoxide in the concentration range of 2.5–10% (v/v) and a fragment concentration of up to 10 m*M*. The true fragment concentration in a soaking experiment may be overestimated due to unobserved or unreported precipitation when diluting from dimethylsulfoxide stock solutions. In the case of extremely poor ligand affinities but high compound solubilities, the issue can simply be resolved by using a sufficient excess of fragments, as reported for the Ricin A–fragment complexes. Here, acetamide and urea derivatives of three-digit millimolar affinity could be converted to protein–fragment structures by soaking in up to 2 *M* fragment concentrations for 48 h (Carra *et al.*, 2007). A striking result has been described for RNAse A where structure determination at 200 m*M* soaking concentration of a nucleotide-like fragment that had no biochemically detectable activity could be performed (Hartshorn *et al.*, 2005). However, in certain studies, more elaborate complexation protocols have been established to fulfill target-specific requirements as described below.

If there is convincing evidence that the properties of a crystallization condition do not provide the optimum condition for protein–ligand complex formation, it might help to choose a more favorable buffer system. In order to stabilize crystals during the soaking process, a popular method is to increase the PEG concentration in the soaking solution (compared to the original crystallization condition) and to reduce the amount of salt to avoid detrimental effects toward fragment solubility. This approach was performed for BACE crystals, first grown in a solution of 20% PEG 5000 MME, 200 m*M* sodium citrate, and 200 m*M* ammonium iodide and subsequently transferred to a solution of 30% PEG 5000 MME, 100 m*M* sodium citrate, and 220 m*M* ammonium iodide (Patel *et al.*, 2004). A similar situation was encountered for pantothenate synthase. In this case, a sulfate ion originating from the crystallization condition (containing 100–150 m*M* lithium sulfate) was removed from the crystal active site to provide access for fragment binding. This was achieved by "washing" the crystals in a similar buffer containing lithium chloride instead of lithium sulfate (Hung *et al.*, 2009). Remarkably, as a further peculiarity in this study, a soak of a two-compound cocktail was successfully performed—two fragments were soaked simultaneously in order to occupy neighboring pockets, and this structural knowledge could be exploited in terms of fragment linking.

3.7. Copurification, cocrystallization, and exchange of tool compounds

Various proteins require either the presence of ligands during purification and crystallization, for example, inhibitors of proteases to prevent autocatalytic digestion prior to crystallization, or crystals form only in the presence of a particular stabilizing inhibitor. Clearly, this requirement is detrimental to ligand exchange with weak-binding fragments and makes further steps necessary to remove the originally bound ligand. Such a case, combined with a complete change of the main precipitant needed for crystallization of the protein–ligand complex, has been reported for blood coagulation factor VIIa. In this case, the protein was supplemented with benzamidine to prevent autoproteolysis and crystals were grown using 2 *M* ammonium sulfate, with 15% glycerol as the main precipitant. The crystals were transferred to a washing and soaking buffer containing 25% PEG 6000 and 15% ethylene glycol, and the benzamidine ligand was successfully removed. These manipulations likely contributed to a loss in resolution from 1.69 Å for the benzamidine bound to 2.44 Å for the ligand-free factor VIIa structure (Sichler *et al.*, 2002).

Cocrystallization of a protein with a relatively weakly binding tool compound offers another route to the determination of protein–fragment structures. For protein kinases, the popular substrate analog AMP-PNP-$MnCl_2$ offers the advantage in that its affinity drops significantly after removing the divalent metal ion with ethylenediaminetetraacetic acid, as described for protein kinase B (Davies *et al.*, 2007).

Crystallographic cocktail screening, for example, working with mixtures of fragments, requires more elaborated soaking protocols. In most cases, six to eight preferably shape-diverse compounds are mixed as dimethylsulfoxide stocks and used for soaking at 10–50 m*M* final concentration in the soaking drop. When successful, in terms of positive difference electron density at the site of interest, the soak is usually repeated with the expected single compound for validation. A more sophisticated protocol has been developed for highly soluble fragments (Davies *et al.*, 2009)—mixtures are prepared as methanol stocks, spotted onto the drop chambers of the crystallization plates, and after evaporation of the methanol, the remaining dry powder is resolubilized using reservoir solution, leading to a soaking solution lacking any organic solvent that might interfere with crystal integrity.

An approach to reduce the amount of organic solvent required to dissolve hydrophobic components consists of the use of solubilizers such as cyclodextrin derivatives; for example, addition of beta-cyclodextrine to the PEG-containing soaking solution assists in the complexation of a fragment-type ligand to the active site of aldose reductase (Steuber *et al.*, 2006). This approach basically resembles the use of hydroxypropyl cyclodextrine-coated screening plates to avoid the presence of dimethylsulfoxide in biochemical assays (Benson *et al.*, 2005).

Although cocrystallization is usually not suitable for high throughput, it has been applied successfully in the past. This approach can be done either by addition of the fragment ligand to the crystallization drop or to the protein solution at any stage of the protein preparation. The first, more common strategy has been performed for HSP90 (Huth *et al.*, 2007), estrogen receptor beta (Malamas *et al.*, 2004), and phenylethanolamine *N*-methyl transferase (McMillan *et al.*, 2004). The second strategy is exemplified by addition of the fragment during concentration of estrogen receptor protein and by dialysis of cyclooxygenase against 1 m*M* inhibitor solution (Selinski *et al.*, 2001). Cocrystallization is necessary when the binding site of the protein targets is either deeply buried or nonaccessible via solvent channels in a soaking-based setup.

Similarly, protein–fragment structure generation using the tethering approach requires more elaborate protocols, for example, a covalent linkage of the fragment to the protein. A successful example has been described for IL-2 (Hyde *et al.*, 2003). A recombinantly introduced cysteine residue served as an anchor for covalent linkage. However, a complex procedure was required. As the cysteine of interest was found to undergo formation of a mixed disulfide with cystamine, initially, a reduction step with 10 m*M* β-mercaptoethanol was performed for 4 h at room temperature. After removal of β-mercaptoethanol and subsequent concentration of the protein, the protein solution was incubated with a fragment susceptible for tethering and a drug-like tool compound overnight at 4 °C. Following an additional desalting step and confirmation of complex formation by LC–MS, the protein was crystallized for structural analysis of tethered fragment and bound tool compound.

4. Quality of Crystallographic Data

Over 90% of the protein–fragment structures summarized in Table 3.1 have a resolution of 2.5 Å or better, and more than 50% have a resolution better than 2.0 Å. Only six crystal structures have a resolution of 2.75 Å or worse, and five of these data sets have been collected on an in-house rotating anode X-ray source and would likely have yielded better resolution at synchrotron sources (Fig. 3.7). Interestingly, out of a total of 244 PDB entries, ~100 have been refined using data collected at in-house rotating anode X-ray sources. Thus, about 60% of the data sets have been collected at synchrotron sources combining high data quality with short turnaround times. The high resolution is mandatory for a correct interpretation of the resulting electron densities (Davis *et al.*, 2003, 2008). It should be noted in this context that soaking procedures that are convenient for the evaluation of a large number of fragments often interfere with crystal quality. Therefore, special attention should be paid to the establishment of stabilizing soaking buffers.

Caused by poor affinity and/or solubility properties, the occupancy of the fragment might be significantly lower than 100%. This characteristic is

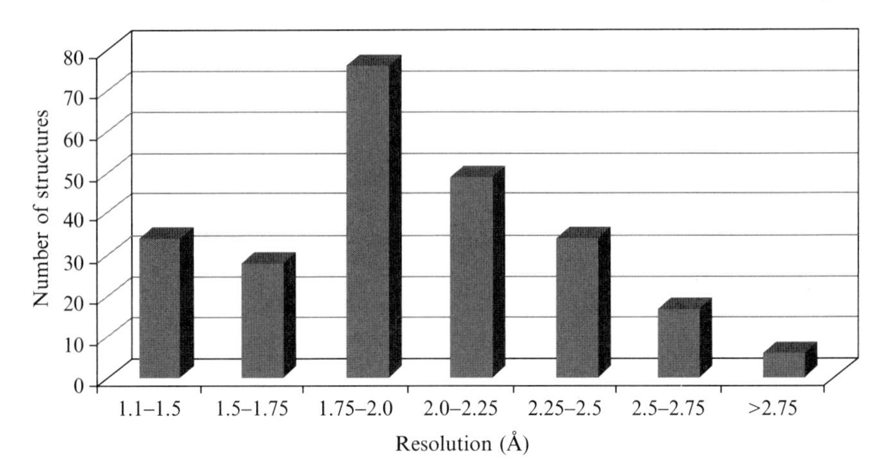

Figure 3.7 Bar diagram depicting the number of protein–fragment structures versus crystallographic resolution.

usually associated with various undesired effects such as the difficultly in interpreting the difference electron densities, occurrence of negative electron density, and elevated *B*-values after inclusion of the small-molecule ligand to the model. Accordingly, as for common protein–ligand structures, temperature factors are significantly higher for the small-molecule entity compared to the protein environment and might originate from a reduced occupancy and require careful treatment and inspection. It is assumed that a minimum of 25–30% occupancy is required to observe an interpretable electron density map (Wu *et al.*, 2001). The PDB entry 3FUD provides an example for such a constellation representative for various similar cases in the PDB. In this protein–fragment structure of leucotriene A4 hydrolase (Davies *et al.*, 2009), the average *B*-value of the ligand atoms amounts to 57.6 Å^2, while for the average *B*-value of the protein main chain and side chain atoms, values of 19.0 and 20.1 Å^2, respectively, are observed. Interestingly, in this crystal structure, the fragment molecule also possesses a significantly higher than average *B*-value compared to all water molecules included to the model (average *B*-value, 22.3 Å^2; maximum B-value, 43.8 Å^2). These observations suggest only a partial occupancy; however, explicit refinement of this value requires sufficient data quality and usage of appropriate programs such as SHELX (Sheldrick, 2008). Nevertheless, a less clear-cut electron density or higher *B*-values might also originate from multiple binding modes or increased mobility of certain parts of the model, giving rise, at least to some extent, to an entropy-driven thermodynamic fingerprint of the binding properties (Gerlach *et al.*, 2007).

5. Application of the Free Mounting System and the Picodropper Technology to Improve Ligand Occupancy

The limited solubility of fragments combined with low affinity often results in a low occupancy at the binding site. Electron densities are often distorted by partially occupied water molecules with overlapping binding sites. To improve the occupancy of fragments at their binding site and to overcome crystal damage due to soaking procedures, we have developed the Picodropper^{TM} technology[1] based on the Free Mounting System^{TM} (FMS). By means of the Picodropper technology, picoliter-sized drops of a ligand solution can be applied to a protein crystal. The FMS is used to mount protein crystals in a humidified airstream without a surrounding liquid phase (Fig. 3.8). FMS experiments have shown that many crystal systems gain physical stability in the absence of any surrounding mother liquor, the presence of which is unavoidable in conventional soaking procedures. Hence, crystals often tolerate higher concentrations of organic solvents (such as dimethylsulfoxide) when mounted on the FMS, which is beneficial to achieving higher concentrations of poorly soluble fragments.

Using this setup, protein crystals can also be exposed to subtle humidity changes. The rearrangements of crystal contacts occurring during controlled hydration or dehydration are often associated with a significant improvement of the diffraction quality (Kiefersauer *et al.*, 2000). The humid airstream can also be supplemented with volatile components like ethanol or even dimethylsulfoxide to control their concentration, or ammonia to change the pH. The compound solubility may even be optimized by changing the temperature of the humid airstream of the FMS by ± 10 °C. A clear advantage is that the diffraction quality of crystals can be continuously checked throughout the Picodropper procedure.

To quantify the amount of a fragment solution applied to a crystal, a video system continually records projection areas of the crystal. Each application of a single, picoliter-sized drop to the protein crystal results in a peak of the projection area (Fig. 3.9). Drops can either be released with a fixed frequency, or a certain change of projection area can be used as a trigger threshold. In this way, dimethylsulfoxide can be allowed to evaporate in the humid airstream, and it does not accumulate to destructive levels, thus preserving the diffraction quality of the crystal.

For illustration purposes, a crystal of BACE was used to demonstrate the time-resolved diffusion of a colored ligand within the crystal by applying the Picodropper^{TM} technology. Picoliter-sized drops (2 pl drop size) of a 5 m*M*

[1] Patents WO 2005/017236 A1 and EP 1 505 179 A1.

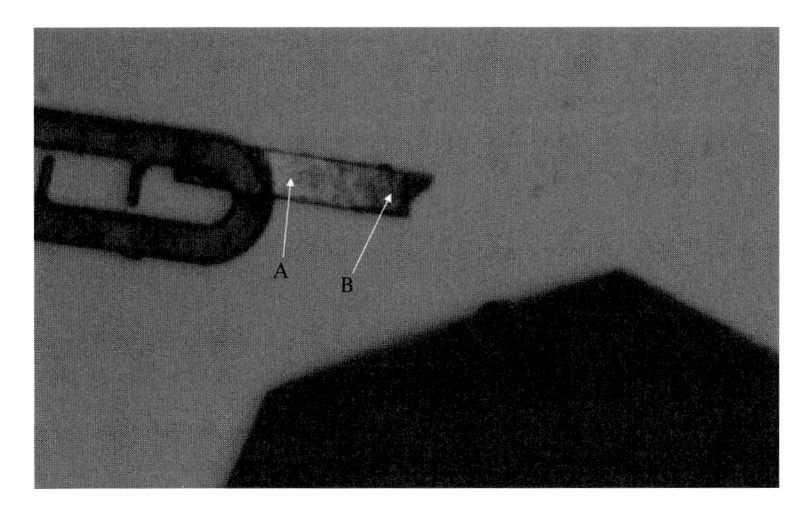

Figure 3.8 BACE crystal on a MicroLoops E^{TM} support (MiTeGen, Jena Bioscience). The extended dark object in the lower right corner is the outlet of the Picodropper in the site view. Drops strike the crystal at point B. To track diffusion of the dye into the crystal, the grayscale value was measured at point A over the course of the experiment.

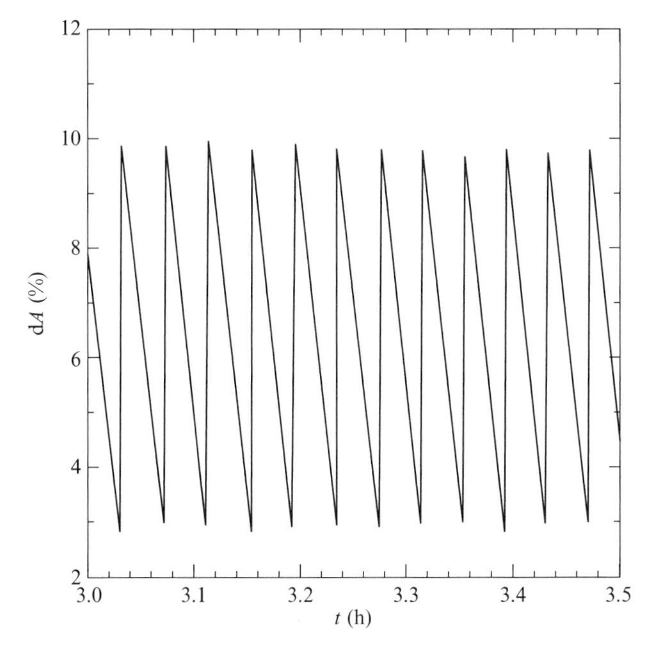

Figure 3.9 Representative case of drop detection by the video system. Each peak represents a single drop to impact on the crystal. Decrease in projection area is due to evaporation of the dimethylsulfoxide from the crystal surface.

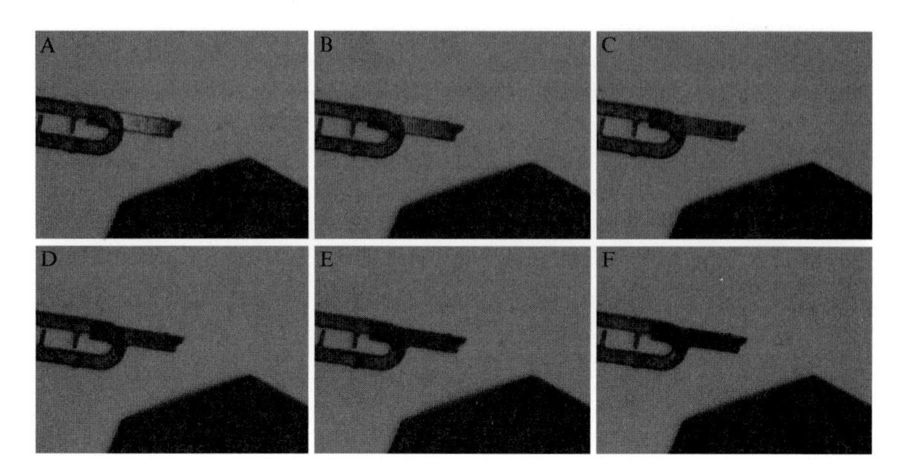

Figure 3.10 Six points in time during dropping procedure: (A) 29 min, (B) 5.8 h, (C) 11.7 h, (D) 22.8 h, (E) 1.4 days, (F) 2.3 days.

solution of methylene blue in 100% dimethylsulfoxide were applied to the crystal (1000 pl volume) in a frequency of 0.1 Hz during a period of 2 days. Images were taken automatically every minute to track diffusion and accumulation of the dye in the crystal. To visualize diffusion of methylene blue from the location of impact to a distant location in the crystal, we collected grayscale values at point of measurement A (distance between point A and point B is ~40 μm) at different points in time (Fig. 3.8). Figure 3.10 shows the diffusion and accumulation of the dye in the crystal during Picodropper procedure. Grayscale values vary from white at the maximum to black at the minimum intensity (Figs. 3.10 and 3.11).

During the first 1.3 days of the Picodropper procedure, the gray scale at point A decreased rapidly (continuous line) as methylene blue diffused into the crystal and coloring of the entire crystal increased continuously. After about 2 days, the grayscale value at point A reached a plateau as it was already small and did not shift proportionally to increasing concentrations of methylene blue anymore. During the Picodropper procedure, no precipitation of methylene blue was observed on the surface of the crystal. Knowledge of the size of the unit cell, the number of binding sites per unit cell, volume of the crystal, concentration of the stock solution, drop size, and the number of applied drops allowed the calculation of the excess of compound molecules applied with respect to the number of binding sites (Table 3.2). In the present case, the excess of compound with respect to the number of active sites is calculated to be 14-fold at the end of the Picodropper procedure.

This method of applying a substance dissolved in an organic solvent by means of small picoliter-sized drops allows to accumulate an excess of a

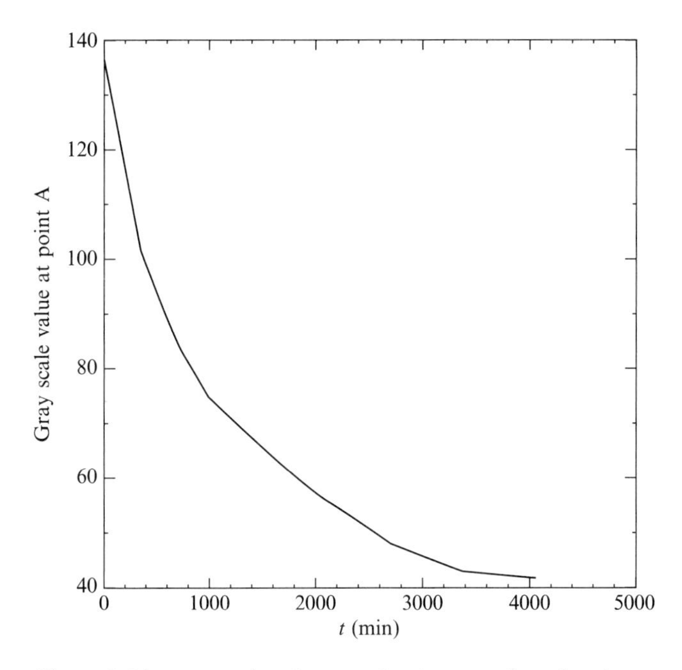

Figure 3.11 Grayscale value at point A versus dropping time.

Table 3.2 Controlled increase of ligand excess within a crystal treated by the Picodropper experiment

Image number (minutes of dropping)	29	350	699	1369	2039	3379
Number of drops	174	2100	4194	8214	12,234	20,274
Excess of methylene blue with respect to active sites present in the crystal	0.02	1.4	2.9	5.7	8.5	14
Grayscale value at position A	135	102	85	68	57	43

compound in the crystal in a stepwise manner. Via drop size and application frequency, the level of the solvent in the crystal can be balanced with respect to the needs of the crystal system (tolerance of organic solvents) and the applied substance (solubility at different levels of organic solvents). Furthermore, the organic solvent content in the crystal can be controlled by supplementing the main gas stream with volatile organic solvents. In this way, a combination of FMS and Picodropper technology provides an additional strategy for complexation beyond the classical approaches.

6. Conclusions

Within the last two decades, the approach of FBDD has evolved to a concept nowadays widely and successfully used for the generation of novel lead compounds. Here, a collective of 244 protein–fragment structures deposited in the PDB has been evaluated in the light of desirable target properties, fragment features, and methodological options for overcoming challenges in fragment-based crystallography. In addition to the type of interaction site to be addressed, crystallographic features such as crystal form, resolution obtained, crystallization conditions, and requirement of tool compounds have to be evaluated upon selection of putative targets for FBDD. The pH of the complexation solution and its components turn out to have an important influence on the outcome of the study. Most fragments successfully crystallized have a reported affinity range between 1 and 10,000 μmol. However, in this context, many methodological aspects are typically not explicitly considered or experimentally checked in fragment-based lead discovery projects. The influence of the components of the complexation solution on the affinity of the fragments is disregarded in most of the studies. Various contributions make use of organic solvent additives such as dimethylsulfoxide to the soaking process for solubilizing fragments with otherwise insufficient solubility properties. However, the influence of those additives on the dielectric constant in this medium and the thermodynamic forces driving the binding event of the fragment to the crystal active site remain mostly ignored. These aspects would deserve more experimental efforts to obtain a more profound understanding. Even though the complex formation process in a crystal sometimes resembles a "black box," there are various parameters to be modified in order to successfully determine protein–fragment X-ray structures.

Acknowledgments

The authors thank Ms. Birgit Liebner, Dr. Timothy Woodcock, Dr. Martin Augustin, and Dr. Dirk Ullmann for helpful discussions and critical review of the chapter.

References

Andersen, H. S., Iversen, L. F., Jeppesen, C. B., Branner, S., Norris, K., Rasmussen, H. B., Møller, K. B., and Møller, N. P. (2000). 2-(oxalylamino)-benzoic acid is a gneral, competitive inhibitor of protein-tyrosine phophatases. *J. Biol. Chem.* **275,** 7101–7108.

Andersen, O. A., Schönfeld, D. L., Toogood-Johnson, I., Felicetti, B., Albrecht, C., Fryatt, T., Whittaker, M., Hallett, D., and Barker, J. (2009). Cross-linking of protein crystals as an aid in the generation of binary protein-ligand crystal complexes, exemplified by the human PDE10a-papaverine structure. *Acta Crystallogr. D* **65,** 872–874.

Anderson, M., Andrews, D. M., Barker, A. J., Brassington, C. A., Breed, J., Byth, K. F., Culshaw, J. D., Finlay, M. R., Fisher, E., McMiken, H. H., Green, C. P., Heaton, D. W., *et al.* (2008). Imidazoles: SAR and development of a potent class of cyclin-dependent kinase inhibitors. *Bioorg. Med. Chem. Lett.* **18,** 5487–5492.

Antonysamy, S. S., Aubol, B., Blaney, J., Browner, M. F., Giannetti, A. M., Harris, S. F., Hébert, N., Hendle, J., Hopkins, S., Jefferson, E., Kissinger, C., Leveque, V., *et al.* (2008). Fragment-based discovery of hepatitis C virus NS5b RNA polymerase inhibitors. *Bioorg. Med. Chem. Lett.* **18,** 2990–2995.

Antonysamy, S. S., Hirst, G., Sprengler, P., Stappenbeck, F., Steensma, R., Wilson, M., and Wong, M. (2009). Fragment-based discovery of JAK-2 inhibitors. *Bioorg. Med. Chem. Lett.* **19,** 279–282.

Arkin, M. R., Randal, M., DeLano, W. L., Hyde, J., Luong, T. N., Oslob, J. D., Raphael, D. R., Taylor, L., Wang, J., McDowell, R. S., Wells, J. A., and Braisted, A. C. (2003). Binding of small molecules to an adaptive protein-protein interface. *PNAS* **100,** 1603–1608.

Aronov, A. M., Baker, C., Bernis, G. W., Cao, J., Chen, G., Ford, P. J., Germann, U. A., Green, J., Hale, M. R., Jacobs, M., Janetka, J. W., Maltais, F., *et al.* (2007). Flipped out: Structure-guided design of selective pyrazolylpyrrole ERK inhibitors. *J. Med. Chem.* **50,** 1280–1287.

Bartlett, P. A., Yusuff, N., Rico, A. C., and Lindvall, M. K. (2002). Antihydrophobic solvent effects: An experimental probe for the hydrophobic contribution to enzyme-inhibitor binding. *J. Am. Chem. Soc.* **124,** 3853–3857.

Benson, N., Boyd, H. F., Everett, J. R., Fries, J., Gribbon, P., Haque, N., Henco, K., Jessen, T., Martin, W. H., Mathewson, T. J., Sharp, R. E., Spencer, R. W., *et al.* (2005). NanoStore: A concept for logistical improvements of compound handling in high-throughput screening. *J. Biomol. Screen.* **10,** 573–580.

Bosch, J., Robien, M. A., Mehlin, C., Boni, E., Riechers, A., Buckner, F. S., Van Voorhis, W. C., Myler, P. J., Worthey, E. A., DeTitta, G., Luft, J. R., Lauricella, A., *et al.* (2006). Using fragment cocktail crystallography to assist inhibitor design of Trypanosoma brucei nucleoside 2-deoxyribosyltransferase. *J. Med. Chem.* **49,** 5939–5946.

Braisted, A. C., Oslob, J. D., Delano, W. L., Hyde, J., McDowell, R. S., Waal, N., Yu, C., Arkin, M. R., and Raimundo, B. C. (2003). Discovery of a potent small molecule IL-2 inhibitor through fragment assembly. *J. Am. Chem. Soc.* **125,** 3714–3715.

Brenk, R., Vetter, S. W., Boyce, S. E., Goodin, D. B., and Shoichet, B. K. (2006). Probing molecular docking in a charged model binding site. *J. Mol. Biol.* **357,** 1449–1470.

Brough, P. A., Barril, X., Borgognoni, J., Chene, P., Davies, N. G., Davis, B., Drysdale, M. J., Dymock, B., Eccles, S. A., Garcia-Echeverria, C., Fromont, C., Hayes, A., *et al.* (2009). Combining hit identification strategies: Fragment-based and in silico approaches to orally active 2-aminothieno[2, 3-d]pyrimidine inhibitors of the Hsp90 molecular chaperone. *J. Med. Chem.* **52,** 4794–4809.

Card, G. L., England, B. P., Suzuki, Y., Fong, D., Powell, B., Lee, B., Luu, C., Tabrizizad, M., Gillette, S., Ibrahim, P. N., Artis, D. R., Bollag, G., *et al.* (2004). Structural basis for the activity of drugs that inhibit phosphodiesterases. *Structure* **12,** 2233–2247.

Carr, R. A. E., Congreve, M., Murray, C. W., and Rees, D. C. (2005). Fragment-based lead discovery: Leads by design. *Drug Discov. Today* **10,** 987–992.

Carra, J. H., McHugh, C. A., Mulligan, S., Machiesky, L. M., Soares, A. S., and Millard, C. B. (2007). Fragment-based identification of determinants of conformational and spectroscopic change at the ricin active site. *BMC Struct. Biol.* **7,** 72.

Chen, Y., and Shoichet, B. K. (2009). Molecular docking and ligand specificity in fragment-based inhibitor discovery. *Nat. Chem. Biol.* **5,** 358–364.

Chung, C.-W. (2007). The use of biophysical methods increases success in obtaining liganded crystal structures. *Acta Crystallogr. D* **63,** 62–71.

Chung, S., Parker, J. B., Bianchet, M., Amzel, L. M., and Stivers, J. T. (2009). Impact of linker strain and flexibility in the design of a fragment-based inhibitor. *Nat. Chem. Biol.* **5,** 407–413.

Congreve, M., Carr, R., Murray, C., and Jhoti, H. (2003). A 'Rule of Three' for fragment-based lead discovery? *Drug Discov. Today* **8,** 876–877.

Congreve, M., Chessari, G., Tisi, D., and Woodhead, A. J. (2008). Recent developments in fragment-based drug discovery. *J. Med. Chem.* **51,** 3661–3680.

Coyne, A. G., Scott, D. E., and Abell, C. (2010). Drugging challenging targets using fragment-based approaches. *Curr. Opin. Chem. Biol.* **14,** 299–307.

Danley, D. E. (2006). Crystallization to obtain protein-ligand complexes for structure-aided drug design. *Acta Crystallogr. D* **62,** 569–575.

Davies, T. G., Verdonk, M. L., Graham, B., Saalau-Bethell, S., Hamlett, C. C. F., McHardy, T., Collins, I., Garrett, M. D., Workman, P., Woodhead, S. J., Jhoti, H., and Barford, D. (2007). A structural comparison of inhibitor binding to PKB, PKA and PKA-PKB chimera. *J. Mol. Biol.* **367,** 882–894.

Davies, D. R., Mamat, B., Magnusson, O. T., Christensen, J., Haraldsson, M. H., Mishra, R., Pease, B., Hansen, E., Singh, J., Zembower, D., Kim, H., Kiselyov, A. S., *et al.* (2009). Discovery of Leukotriene A4 hydrolase inhibitors using metabolomics biased fragment crystallography. *J. Med. Chem.* **52,** 4694–4715.

Davis, A. M., Teague, S. J., and Kleywegt, G. J. (2003). Application and limitations of X-ray crystallographic data in structure-based ligand and drug design. *Angew. Chem. Int. Ed.* **42,** 2718–2736.

Davis, A. M., St-Gallay, S. A., and Kleywegt, G. J. (2008). Limitations and lessons in the use of X-ray structural information in drug design. *Drug Discov. Today* **13,** 831–841.

Doyle, D. A., Morais, C. J., Pfuetzner, R. A., Kuo, A., Gulbis, J. M., Cohen, S. L., Chait, B. T., and MacKinnon, R. (1998). The structure of the potassium channel: Molecular basis of K+ conduction and selectivity. *Science* **280,** 69–77.

Engel, M., Hoffmann, T., Manhart, S., Heiser, U., Chambre, S., Huber, R., Demuth, H. U., and Bode, W. (2006). Rigidity and flexibility of dipeptidyl peptidase IV: Crystal structures of and docking experiments with DPIV. *J. Mol. Biol.* **355,** 768–783.

Engh, R., Girod, A., Kinzel, V., Huber, R., and Bossemeyer, D. (1996). Crystal structures of catalytic subunit of cAMP-dependent protein kinase in complex with isoquinolinesulfonyl protein kinase inhibitors H7, H8, and H89. Structural implications for selectivity. *J. Biol. Chem.* **271,** 26157–26164.

Englert, L., Biela, A., Zayed, M., Heine, A., Hangauer, D., and Klebe, G. (2010). Displacement of disordered water molecules from hydrophobic pocket creates enthalpic signature: binding of phosphonamidate to the S(1)'-pocket of thermolysin. *Biochim. Biophys. Acta* **1800,** 1192–1202.

English, A. C., Groom, C. R., and Hubbard, R. E. (2001). Experimental and computational mapping of the binding surface of a crystalline protein. *Protein Eng.* **14,** 47–59.

Erlanson, D. A. (2006). Fragment-based lead discovery: A chemical update. *Curr. Opin. Biotechnol.* **17,** 643–652.

Erlanson, D. A., and Hansen, S. K. (2004). Making drugs on proteins: Site-directed ligand discovery for fragment-based lead assembly. *Curr. Opin. Chem. Biol.* **8,** 399–406.

Fedorov, R., Ghosh, D. K., and Schlichting, I. (2003). Crystal structures of cyanide complexes of P450cam and the oxgenase domain of inducile nitric oxide synthase-structural models of the short-lived oxygen complexes. *Arch. Biochem. Biophys.* **409,** 25–31.

Frederickson, M., Callaghan, O., Chessari, G., Congreve, M., Cowan, S. R., Matthews, J. E., McMenamin, R., Smith, D. M., Vinković, M., and Wallis, N. G. (2008). Fragment-based discovery of mexiletine derivatives as orally bioavailable inhibitors of urokinase-type plasminogen activator. *J. Med. Chem.* **51,** 183–186.

Gerlach, C., Smolinski, M., Steuber, H., Sotriffer, C. A., Heine, A., Hangauer, D. G., and Klebe, G. (2007). Thermodynamic inhibition profile of a cyclopentyl and a cyclohexyl derivative towards thrombin: The same but for different reasons. *Angew. Chem. Int. Ed.* **46,** 8511–8514.

Geschwindner, S., Olsson, L.-L., Albert, J. S., Deinum, J., Edwards, P. D., de Beer, T., and Folmer, R. H. A. (2007). Discovery of a Novel Warhead against beta-secretase through fragment-based lead generation. *J. Med. Chem.* **50,** 5903–5911.

Ghosh, D. K., Rashid, M. B., Crane, B., Taskar, V., Mast, M., Misukonis, M. A., Weinberg, J. B., and Eissa, N. T. (2001). Characterization of key residues in the subdomain encoded by exons 8 and 9 of human inducible nitric oxide synthase: A critical role for Asp-280 in substrate binding and subunit interactions. *PNAS* **98,** 10392–10397.

Gill, A. L., Frederickson, M., Cleasby, A., Woodhead, S. J., Carr, M. G., Woodhead, A. J., Walker, M. T., Congreve, M. S., Devine, L. A., Tisi, D., O'Reilly, M., Seavers, L. C., *et al.* (2005). Identification of novel p38alpha MAP kinase inhibitors using fragment-based lead generation. *J. Med. Chem.* **48,** 414–426.

Godemann, R., Madden, J., Krämer, J., Smith, M., Fritz, U., Hesterkamp, T., Barker, J., Höppner, S., Hallett, D., Cesura, A., Ebneth, A., and Kemps, J. (2009). Fragment-based discovery of BACE inhibitors using functional assays. *Biochemistry* **48,** 10743–10751.

Hajduk, P. J., and Greer, J. (2007). A decade of fragment-based drug design: Strategic advances and lessons learned. *Nat. Rev. Drug Discov.* **6,** 211–219.

Hann, M. M., Leach, A. R., and Harper, G. (2001). Molecular complexity and its impact on the probability of finding leads for drug discovery. *J. Chem. Inf. Comput. Sci.* **41,** 856–864.

Hansen, S. B., Sulzenbacher, G., Huxford, T., Marchot, P., Taylor, P., and Bourne, Y. (2005). Structure of *Aplysia* AChBP complexes with nicotinic agonists and antagonists reveal distictive binding interfaces and confirmations. *EMBO J.* **24,** 3635–3646.

Hartshorn, M. J., Murray, C. W., Cleasby, A., Frederickson, M., Tickle, I. J., and Jhoti, H. (2005). Fragment-based Lead discovery using X-ray crystallography. *J. Med. Chem.* **48,** 403–413.

Hassell, A. M., An, G., Bledsoe, R. K., Bynum, J. M., Carter, H. L., III, Deng, S.-J. J., Gampe, R. T., Grisard, T. E., Madauss, K. P., Nolte, R. T., Rocque, W. J., Wang, L., *et al.* (2007). Crystallization of protein-ligand complexes. *Acta Crystallogr. D* **63,** 72–79.

Heaslet, H., Rosenfeld, R., Giffin, M., Lin, Y. C., Tam, K., Torbett, B. E., Elder, J. H., McRee, D. E., and Stout, C. D. (2007). Conformational flexibility in the flap domains of ligand-free HIV protease. *Acta Crystallogr. D* **63,** 866–875.

Hendlich, M., Bergner, A., Günther, J., and Klebe, G. (2003). Relibase: Design and development of a database for comprehensive analysis of protein-ligand interactions. *J. Mol. Biol.* **326,** 607–620.

Hohwy, M., Spadola, L., Lundquist, B., Hawtin, P., Dahmén, J., Groth-Clausen, I., Nilsson, E., Persdotter, S., von Wachenfeldt, K., Folmer, R. H., and Edman, K. (2008). Novel prostaglandin D synthase inhibitors generated by fragment-based drug design. *J. Med. Chem.* **51,** 2178–2186.

Holmes, M. A., and Matthews, B. W. (1981). Binding of hydroxamic acid inhibitors to crystalline thermolysin suggests a pentacoordinate zinc intermediate in catalysis. *Biochemistry* **20,** 6912–6920.

Holmes, I. P., Gaines, S., Watson, S. P., Lorthioir, O., Walker, A., Baddeley, S. J., Herbert, S., Egan, D., Convery, M. A., Singh, O. M., Gross, J. W., Strelow, J. M., *et al.* (2009). The identification of beta-hydroxy carboxylic acids as selective MMP-12 inhibitors. *Bioorg. Med. Chem. Lett.* **19,** 5760–5763.

Howard, N., Abell, C., Blakemore, W., Chessari, G., Congreve, M., Howard, S., Jhoti, H., Murray, C. W., Seavers, L. C. A., and van Montfort, R. L. M. (2006). Application of fragment screening and fragment linking to the discovery of novel thrombin inhibitors. *J. Med. Chem.* **49,** 1346–1355.

Howard, S., Berdini, V., Boulstridge, J. A., Carr, M. G., Cross, D. M., Curry, J., Devine, L. A., Early, T. R., Fazal, L., Gill, A. L., Heathcote, M., Maman, S., *et al.* (2009). Fragment-based discovery of the pyrazol-4-yl urea (AT9283), a multitargeted kinase inhibitor with potent aurora kinase activity. *J. Med. Chem.* **52,** 379–388.

Hung, A. W., Silvestre, H. L., Wen, S., Ciulli, A., Blundell, T. L., and Abell, C. (2009). Application of fragment growing and fragment linking to the discovery of inhibitors of mycobacterium tuberculosis panththenate synthase. *Angew. Chem. Int. Ed.* **48,** 8452–8456.

Huth, J. R., Park, C., Petros, A. M., Kunzer, A. R., Wendt, M. D., Wang, X., Lynch, C. L., Mack, J. C., Swift, K. M., Judge, R. A., Chen, J., Richardson, P. L., *et al.* (2007). Discovery and design of novel HSP90 inhibitors using multiple fragment-based design strategies. *Chem. Biol. Drug Des.* **70,** 1–12.

Hyde, J., Braisted, A. C., Randal, M., and Arkin, M. R. (2003). Discovery and characterization of cooperative ligand binding in the adaptive region of interleukin-2. *Biochemistry* **42,** 6475–6483.

Iversen, L. F., Andersen, H. S., Branner, S., Mortensen, S. B., Peters, G. H., Norris, K., Olsen, O. H., Jeppesen, C. B., Lundt, B. F., Ripka, W., Møller, K. B., and Møller, N. P. (2000). Structure-based design of a low molecular weight, nonphosphorus, nonpeptide, and highly selective inhibitor of protein-tyrosine phosphatase 1B. *J. Biol. Chem.* **275,** 10300–10307.

Joachimiak, A. (2009). High-throughput crystallography for structural genomics. *Curr. Opin. Struct. Biol.* **19,** 573–584.

Katz, B. A., Elrod, K., Luong, C., Rice, M. J., Mackman, R. L., Sprengeler, P. A., Spencer, J., Hataye, J., Janc, J., Link, J., Litvak, J., Rai, R., *et al.* (2001). A novel serine protease inhibition motif involving a multi-centered short hydrogen bonding network at the active site. *J. Mol. Biol.* **307,** 1451–1486.

Kiefersauer, R., Than, M. E., Dobbek, H., Gremer, L., Melero, M., Strobl, S., Dias, J. M., Soulimane, T., and Huber, R. (2000). A novel free-mounting system for protein crystal transformation and improvement of diffraction power by accurately controlled humidity change. *J. Appl. Cryst.* **33,** 1223–1230.

Lawrie, A. M., Noble, M. E., Tunnah, P., Brown, N. R., Johnson, L. N., and Endicott, J. A. (1997). Protein kinase inhibition by staurosporine revealed in details of the molecular interaction with CDK2. *Nat. Struct. Biol.* **4,** 796–801.

Lee, E. F., Czabotar, P. E., Smith, B. J., Deshayes, K., Zobel, K., Colman, P. M., and Fairlie, W. D. (2007). Crystal structure of ABT-737 complexed with Bcl-XL: Implications for selectivity of antagonists of the Bcl-2 family. *Cell Death Differ.* **14,** 1711–1713.

Leonidas, D. D., Shapiro, R., Irons, L. I., Russo, N., and Acharya, K. R. (1997). Crystal structures of ribonuclease A complexes with 5′-diphosphoadenosine 3′-phosphate and 5′-diphosphoadenosine 2′phosphate at 1.7 A resolution. *Biochemistry* **36,** 5578–5588.

Malamas, M. S., Manas, E. S., McDevitt, R. E., Gunawan, I., Xu, Z. B., Collini, M. D., Miller, C. P., Dinh, T., Henderson, R. A., Keith, J. C., Jr., and Harris, H. A. (2004). Design and synthesis of aryl diphenolic azoles as potent and selective estrogen receptor-b ligands. *J. Med. Chem.* **47,** 5021–5040.

Manas, E. S., Unwalla, R. J., Xu, Z. B., Malamas, M. S., Miller, C. P., Harris, H. A., Hsiao, C., Akopian, T., Hum, W. T., Malakian, K., Wolfrom, S., Bapat, A., *et al.* (2004). Structure-based design of estrogen receptor-beta selective ligands. *J. Am. Chem. Soc.* **126,** 15106–15119.

Mattos, C., Bellamacina, C. R., Peisach, E., Pereira, A., Vitkup, D., and Petsko, G. A. Ringe, D. (2006). Multiple solvent crystal structures: Probing binding sites, plasticity and hydration. *J. Mol. Biol.* **357,** 1471–1482.

McMillan, F. M., Archbold, J., McLeish, M. J., Caine, J. M., Criscione, K. R., Grunewald, G. L., and Martin, J. L. (2004). Molecular recognition of sub-micromolar inhibitors by the epinephrine-synthesizing enzyme phenylethanolamine N-methyltransferase. *J. Med. Chem.* **47,** 37–44.

McPherson, A. (1990). Current approaches to macromolecular crystallization. *Eur. J. Biochem.* **189,** 1–23.

Mochalkin, I., Miller, J. R., Narasimhan, L., Thanabal, V., Erdman, P., Cox, P. B., Prasad, J. V., Lightle, S., Huband, M. D., and Stover, C. K. (2009). Discovery of antibacterial biotin carboxylase inhibitors by virtual screening and fragment-based approaches. *ACS Chem. Biol.* **4,** 473–483.

Muchmore, S. W., Sattler, M., Liang, H., Meadows, R. P., Harlan, J. E., Yoon, H. S., Nettesheim, D., Chang, B. S., Thompson, C. B., Wong, S. L., Ng, S. L., and Fesik, S. W. (1996). X-ray and NMR structure of human Bcl-XL, an inhibitor of programmed cell death. *Nature* **381,** 335–341.

Murray, C. W., and Blundell, T. L. (2010). Structural biology in fragment-based drug design. *Curr. Opin. Struct. Biol.* **20,** 1–11.

Murray, C. W., Callaghan, O., Chessari, G., Cleasby, A., Congreve, M., Frederickson, M., Hartshorn, M. J., McMenamin, R., Patel, S., and Wallis, N. (2007). Application of fragment screening by X-ray crystallography to beta-secretase. *J. Med. Chem.* **50,** 1116–1123.

Musah, R. A., Jensen, G. M., Bunte, S. W., Rosenfeld, R. J., and Goodin, D. B. (2002). Artificial protein cavities as specific ligand-binding templates: Characterization of an engineered heterocyclic cation-binding site that preserves the evolved specificity of the parent protein. *J. Mol. Biol.* **315,** 845–857.

Patel, S., Vuillard, L., Cleasby, A., Murray, C. W., and Yon, J. (2004). Apo and inhibitor complex structures of BACE ([beta]-secretase). *J. Mol. Biol.* **343,** 407–416.

Perryman, A. L., Zhang, Q., Soutter, H. H., Rosenfeld, R., McRee, D. E., Olson, A. J., Elder, J. E., and Stout, C. D. (2010). Fragment-based Screen against HIV protease. *Chem. Biol. Drug Des.* **75,** 257–268.

Petros, A. M., Dinges, J., Augeri, D. J., Baumeister, S. A., Betebenner, D. A., Bures, M. G., Elmore, S. W., Hajduk, P. J., Joseph, M. K., Landis, S. K., Nettesheim, D. G., Rosenberg, S. H., *et al.* (2006). Discovery of a potent inhibitor of the antiapoptotic protein Bcl-XL from NMR and Parallel synthesis. *J. Med. Chem.* **49,** 656–663.

Pillai, B., Kannan, K. K., and Hosur, M. V. (2001). 1.9 Å X-ray study shows closed flap conformation in crystals of tethered HIV-1 PR. *Proteins* **43,** 57–64.

Podust, L. M., von Kries, J. P., Eddine, A. N., Kim, Y., Yermalitskaya, L. V., Kuehne, R., Ouellet, H., Warrier, T., Alteköster, M., Lee, J. S., Rademann, J., Oschkinat, H., *et al.* (2007). Small-molecule scaffolds for CYP51 inhibitors identified by high-throughput screening and defined by X-ray crystallography. *Antimicrob. Agents Chemother.* **5,** 3915–3923.

Rasmussen, S. G., Choi, H. J., Rosenbaum, D. M., Kobilka, T. S., Thian, F. S., Edwards, P. C., Burghammer, M., Ratnala, V. R., Sanishvili, R., Fischetti, R. F., Schertler, G. F., Weis, W. I., *et al.* (2007). Crystal structure of the human beta2 adrenergic G-protein-coupled receptor. *Nature* **450,** 383–387.

Rees, D. C., Congreve, M., Murray, C. W., and Carr, R. (2004). Fragment-based lead discovery. *Nat. Rev. Drug Discov.* **3,** 660–672.

Rosenblatt, J., De Bondet, H., Jancarik, J., Morgan, D. O., and Kim, S. H. (1993). Purification and crystallization of human cyclin-dependent kinase 2. *J. Mol. Biol.* **230,** 1317–1319.

Saxty, G., Woodhead, S. J., Berdini, V., Davies, T. G., Verdonk, M. L., Wyatt, P. G., Boyle, R. G., Barford, D., Downham, R., Garrett, M. D., and Carr, R. A. (2007). Identification of inhibitors of protein kinase B using fragment-based lead discovery. *J. Med. Chem.* **50,** 2293–2296.

Schiffmann, R., Heine, A., Klebe, G., and Klein, C. D. (2005). Metal ions as cofactors for the binding of inhibitors to methionine aminopeptidase: A critical view of the relevance of in vitro metalloenzyme assays. *Angew. Chem. Int. Ed. Engl.* **44,** 3620–3623.

Scott, A. D., Phillips, C., Alex, A., Flocco, M., Bent, A., Randall, A., O'Brien, R., Damian, L., and Jones, L. H. (2009). Thermodynamic optimization in drug discovery: A case study using carbonic anhydrase inhibitors. *ChemMedChem.* **4,** 1985–1989.

Selinski, B. S., Gupta, K., Sharkey, C. T., and Loll, P. J. (2001). Structural analysis of NSAID binding by prostaglandine H2-synthase: Time-dependent and time-independent inhibitors elicit identical enzyme conformations. *Biochemistry* **40,** 5172–5180.

Sheldrick, G. M. (2008). A short history of SHELX. *Acta Crystallogr. A* **64,** 112–122.

Sichler, K., Banner, D. W., DÁrcy, A., Hopfner, K. P., Huber, R., Bode, W., Kresse, G. B., Kopetzki, E., and Brandstetter, H. (2002). Crystal structures of uninhibited factor VIIa link its cofactor and substrate-assisted activation to specific interactions. *J. Mol. Biol.* **322,** 591–603.

Steuber, H., Zentgraf, M., Podjarny, A., Heine, A., and Klebe, G. (2006). High-resolution crystal structure of aldose reductase complexed with the novel sulfonyl-pyridazinone inhibitor exhibiting an alternative active site anchoring group. *J. Mol. Biol.* **356,** 45–56.

Stubbs, M. T., Reyda, S., Dullweber, F., Möller, M., Klebe, G., Dorsch, D., Mederski, W. W., and Wurziger, H. (2002). pH-Dependent binding modes observed in trypsin crystals: Lessons for structure-based drug design. *Chembiochem* **3,** 246–249.

Szczepankiewicz, B. G., Liu, G., Hajduk, P. J., Abad-Zapatero, C., Pei, Z., Xin, Z., Lubben, T. H., Trevillyan, J. M., Stashko, M. A., Ballaron, S. J., Liang, H., Huang, F., *et al.* (2003). Discovery of a potent, selective protein tyrosine phosphatase 1B inhibitor using a linked-fragment strategy. *J. Am. Chem. Soc.* **125,** 4087–4096.

Teotico, D. G., Babaoglu, K., Rocklin, G. J., Ferreira, R. S., Giannetti, A. M., and Shoichet, B. K. (2009). Docking for fragment inhibitors of AmpC beta-lactamase. *PNAS* **106,** 7455–7460.

Thanos, C. D., Randal, M., and Wells, J. A. (2003). Potent small-molecule binding to a dynamic hot spot on IL-2. *J. Am. Chem. Soc.* **125,** 15280–15281.

Toyota, E., Ng, K. K., Sekizaki, H., Itoh, K., Tanizawa, K., and James, M. N. (2001). X-ray crystallographic analyses of complexes between bovine beta-trypsin and Schiff base copper(II) or iron(III) chelates. *J. Mol. Biol.* **305,** 471–479.

Vassar, R., Bennett, B. D., Babu-Khan, S., Kahn, S., Mendiaz, E. A., Denis, P., Teplow, D. B., Ross, S., Amarante, P., Loeloff, R., Luo, Y., Fisher, S., *et al.* (1999). {Beta}-Secretase cleavage of Alzheimer's amyloid precursor protein by the transmembrane aspartic protease BACE. *Science* **286,** 735–741.

Verlinde, C. L. M. J., Fan, E., Shibata, S., Zhang, Z., Sun, Z., Deng, W., Ross, J., Kim, J., Xiao, L., Arakaki, T. L., Bosch, J., Caruthers, J. M., *et al.* (2009). Fragment-based cocktail crystallography by the medical structural genomics of pathogenic protozoa consortium. *Curr. Top. Med. Chem.* **9,** 1678–1687.

Wang, Z., Harkins, P. C., Ulevitch, R. J., Han, J., Cobb, M. H., and Goldsmith, E. J. (1997). The structure of mitogen-activated protein kinase p38 at 2.1-A resolution. *PNAS* **94,** 2327–2332.

Wang, T. G., Mantei, R. A., Kawai, M., Tedrow, J.S., Barnes, D. M., Wang, J., Zhang, Q., Lou, P., Garcia, L. A., Bouska, J., Yates, M., Park, C., *et al.* (2007). Lead optimization of methionine aminopeptidase-2 (MetAP2) inhibitors containing sulfonamides of 5,6-disubstituted anthranilic acids. *Bioorg. Med. Chem. Lett.* **17,** 2817–2822.

Wang, Y. S., Strickland, C., Voigt, J. H., Kenedy, M. E., Beyer, B. M., Senior, M. M., Smith, E. M., Nechuta, T. L., Madison, V. S., Czarniecki, M., McKittrick, B. A., Stamford, A. W., *et al.* (2010). Application of fragment-based NMR screening, X-ray crystallography, structure-based design, and focused chemical library design to identify novel microM leads for the development of nM BACE-1 (beta-site APP cleaving enzyme 1) inhibitors. *J. Med. Chem.* **53,** 942–950.

Wu, S.-y., Dornan, J., Kontopidis, G., Taylor, P., and Walkinshaw, M. D. (2001). The first direct determination of a ligand binding constant in protein crystals. *Angew. Chem. Int. Ed.* **40,** 582–586.

CHAPTER FOUR

Predicting the Success of Fragment Screening by X-Ray Crystallography

Douglas R. Davies, Darren W. Begley, Robert C. Hartley, Bart L. Staker, *and* Lance J. Stewart

Contents

Abstract

Fragment screening using X-ray crystallography is a method that can provide direct three-dimensional readouts of the structures of protein–small molecule complexes for lead development and fragment-based drug discovery. With current technology, an amenable crystal form can be screened crystallographically against a library of 1000–2000 fragments in 1–2 weeks. We have performed over a dozen crystallographic screening campaigns using our own compound collection called Fragments of Life™ (FOL). While the majority of our fragment screening campaigns have generated multiple hits, some unexpectedly turned out to be nonproductive, either yielding no bound ligands, or only those thought to be inadequate for lead development. In this chapter, we have attempted to identify one or more parameters which could be used to

Emerald BioStructures, Bainbridge Island, Washington, USA

Methods in Enzymology, Volume 493
ISSN 0076-6879, DOI: 10.1016/B978-0-12-381274-2.00004-2

predict whether a crystallized protein target would be a good candidate for fragment hit discovery. Here, we describe the parameters of crystals from 18 fragment screening campaigns, including six unsuccessful targets. From this analysis, we have concluded that there are no parameters that are absolutely predictive of fragment screening success. However, we do describe a parameter we have termed pocket factor which provides a statistically significant variance between nonproductive targets and productive targets shown to bind fragments. The pocket factor is calculated using a novel method of consensus scoring from three distinct pocket-finding algorithms, and the results may be used to prioritize targets for fragment screening campaigns based on an initial crystal structure.

1. Introduction

The past decade has seen an emergence of numerous biophysical techniques employed for screening and evaluation of fragment binding to biological molecules (Carr and Jhoti, 2002; Congreve *et al.*, 2008; Erlanson *et al.*, 2004; Hajduk and Greer, 2007; Shuker *et al.*, 1996). X-ray crystallography is a critical component of the fragment-based drug design toolkit, with an unmatched ability to rapidly provide a direct three-dimensional readout of the interactions between small-molecule fragment and macromolecular target (Nienaber *et al.*, 2000). Continuing developments in automation, X-ray optics, and detector sensitivity have improved the throughput of X-ray data collection to the point where crystallography can be routinely used as a primary screening method for fragments (Albert *et al.*, 2007; Davies *et al.*, 2009). We have conducted many fragment screens at Emerald BioStructures, using pools of fragments, various crystal soaking techniques, and X-ray diffraction, to search for novel small-molecule binding entities. Success in this endeavor means obtaining one or more fragment-bound complexes through random screening of an all-purpose, chemically diverse compound library against a single target. In developing our own fragment screening methods, we have found that success is highly variable, depending on the target and its crystal form. Given this experience, we wondered whether or not the ultimate success of a crystallographic fragment screen could be predicted based on the properties of the crystal itself. Knowledge of such factors in advance of fragment screening could greatly facilitate target selection and help prioritize those targets with the best chances for fragment screening campaigns.

To address this question, we examined a variety of parameters that can be readily obtained from an initial (*apo*) crystal structure, including crystallization conditions, solvent content, and size of solvent channels. We also used several publicly available computational tools to see whether the "druggability" (Hopkins and Groom, 2002) of a given target can predict the likelihood of success in obtaining fragment-bound structures through X-ray crystallographic screening. Described in this chapter are both "productive" and "nonproductive" targets, in terms of generating complex structures through crystallographic fragment screening. The dataset consists of targets from 18 different screening campaigns, including 10 from our infectious disease structural genomics and drug discovery pipelines, four internal proprietary drug discovery targets, and four external targets previously described in the scientific literature. We describe our analysis methods and summarize our results, concluding that none of the parameters which we examined in detail provide an absolute predictor of crystallographic fragment screening success. However, some trends were observed which distinguish productive from nonproductive targets, and may furnish opportunities to prioritize targets among many potential candidates for fragment screening.

2. Fragment Screening and Druggability

Over the past two decades, fragment-based methods for drug discovery have gone from virtually nonexistent to widespread throughout industry and academia. The majority of fragment-based screening protocols apply one or more biophysical screening techniques to separate fragment-sized molecules (Congreve *et al.*, 2003) which bind a target from those that do not. One strategy for following up initial fragment hits is through structural characterization of small-molecule binding modes relative to a macromolecular target. Recent advances in home source X-ray generators, robotics, and software applications have enabled the use of X-ray crystallography as both a primary screen technique and a swift, robust method for determining fragment-bound structures (Bosch *et al.*, 2006; Carr and Jhoti, 2002; Hartshorn *et al.*, 2005; Nienaber *et al.*, 2000; Pflugrath, 1999; Rees *et al.*, 2004). For well-diffracting, readily obtainable crystals, over 200 data sets can be collected in less than a week, representing over 1500 fragments tested against a single target. This was the case for one infectious disease target (IspF from *Burkholderia psuedomallei*, detailed below) where a Rigaku FR-E+SuperBright rotating anode X-ray generator featuring two Saturn 944+ CCD X-ray detectors with VariMax optics and ACTOR robotic sample mounters were used to complete a primary screen of a ~1450 compound library in 5 days (Begley *et al.*, manuscript in preparation). Thus, screening directly by X-ray crystallography has the advantage of producing a hit simultaneously with a ligand-bound structure, which can serve as a starting point for structure-based drug design.

Much has already been written on techniques for designing and assembling libraries suitable for fragment screening by various biophysical methods

(Albert *et al.*, 2007; Alex and Flocco, 2007; Barelier *et al.*, 2010; Baurin *et al.*, 2004; Blomberg *et al.*, 2009; Chen and Shoichet, 2009; Duarte *et al.*, 2007; Hajduk *et al.*, 2000; Hubbard *et al.*, 2007; Jacoby *et al.*, 2003; Makara, 2007; Rees *et al.*, 2004; Schuffenhauer *et al.*, 2005; Siegel and Vieth, 2007; Zartler and Shapiro, 2008). Most fragment libraries are either intended to be general use, covering sufficient chemical space for screening diverse targets, or focused, with small molecules selected for binding to a specific target or target family. Different biophysical methods also require a degree of library tailoring, such as resonance overlap in NMR spectroscopy versus shape complementarity in X-ray diffraction. Studies at Emerald BioStructures use a library of our own design called the Fragments of Life™ (FOL) (Davies *et al.*, 2009). The FOL collection is a general, all-purpose fragment screening compound library with components that generally fall into three categories: natural metabolites, including molecules known to be present in livings cells; derivatives of metabolites, such as isosteres or heteroatom derivatives of natural metabolites; and synthetic biaryl molecules, whose energy-minimized structures tend to mimic peptidic α-, β-, and γ-turns (Biros *et al.*, 2007; Robinson, 2008; Saraogi and Hamilton, 2008). In the light of the extensive previous literature on fragment libraries themselves, we have set aside questions of privileged scaffolds and other ligand-specific issues for this book chapter. Here, we will instead focus on the biologic macromolecular targets employed in fragment studies.

Regardless of the biophysical methodology used, a fragment screening campaign is a significant investment in resources, requiring specialized equipment, expert technicians, and highly pure materials, prior to any actual experimentation. Therefore, choosing a suitable target is of paramount importance and has generated much discussion on "druggability" as a key factor in target validation (Chen and Hubbard, 2009; Hajduk *et al.*, 2005a,b; Hopkins and Groom, 2002). Of course, one can select a target based on the previous success of others with the same or a structurally homologous protein. HSP90 and MAPK14 are examples of specific targets proven to be amenable to fragment screening, providing sufficient diversity among fragment hits to develop novel lead candidates (Brough *et al.*, 2009; Gill *et al.*, 2005). However, many assumptions about the likelihood of a given target yielding fragment hits based on its inherent biology do not necessarily hold. For instance, metabolic enzymes which catalyze reactions with hydrophobic small molecules would be more likely to yield fragment hits than targets which participate in protein–protein interactions, based purely on chemical similarity between fragments and the native substrate. However, there are many examples in the literature of small molecules developed from fragment screening which disrupt protein–protein interactions, such as the chaperone HSP90, signaling factors in the B-cell lymphoma 2 family, and others (Chen *et al.*, 2007; Oltersdorf *et al.*, 2005; Shuker *et al.*, 1996; Stebbins *et al.*, 2007; Wells and McClendon, 2007). Likewise, not all enzymes that act upon small molecules lend themselves toward fragment screening, such as the nonproductive targets from our structural genomics pipeline involved in fatty acid synthesis and carbohydrate metabolism. Thus, although druggability is of primary concern when conducting drug discovery investigations for a new or unproven target, the basic biology of a target does not dictate its suitability for fragment screening in advance of experimentation.

3. Fragment Screening Campaigns in This Review

In an attempt to better understand which targets are most suitable for fragment studies by X-ray crystallography, we surveyed 18 screening campaigns to determine whether predictive parameters could be gleaned from an initial or *apo* crystal structure. Of these, 12 were deemed to be "productive" screening campaigns yielding structures of bound fragments and providing opportunities for further development. The remaining six were designated "nonproductive," meaning targets either did not yield any complex structures, or furnished fragments bound to a crystal packing interface with no immediate utility for inhibitor design. These case studies were drawn from 12 targets screened at Emerald BioStructures, both proprietary and nonproprietary work, and four literature cases as detailed below.

3.1. Targets from the Seattle Structural Genomics Center for Infectious Disease

As a part of the Seattle Structural Genomics Center for Infectious Disease (SSGCID) (www.ssgcid.org), we have been on track to solve over 500 X-ray crystal structures from infectious disease targets between 2008 and 2012 (Myler *et al.*, 2009; Van Voorhis *et al.*, 2009). This National Institute of Allergy and Infectious Disease-funded initiative applies structural genomics to assist in structure-based drug discovery for Category A–C agents and other emergent and reemerging infectious diseases. As part of that effort, we generate multiple ligand-bound structures of selected proteins. To our knowledge, the SSGCID is the first structural genomics project to incorporate fragment screening into its pipeline.

The SSGCID targets that proved to be productive from fragment screening included four metabolic enzymes (2C-methyl-D-erythritol 2,4-cyclodiphosphate synthase from *Burkholderia pseudomallei* (IspF); glutaryl-CoA dehydrogenase from *B. pseudomallei* (GCD); the glycolytic enzyme phosphoglyceromutase from *B. pseudomallei* (GCM); and β-ketoacyl synthase from *Brucella melitensis* (KAS I)) and the periplasmic domain of the risS sensor protein from *B. pseudomallei* (risS) (Table 4.1). In these cases, fragment screening was carried out by soaking preformed *apo* crystals in crystallization solution with 180 fragment cocktails, each containing up to 8 fragments at approximately 6.25 m*M* in the drop. Hit rates for fragment screens with these targets ranged between 1% and 6%. The nonproductive SSGCID targets included three metabolic enzymes (inorganic pyrophosphatase from *B. pseudomallei* (PPA); 3-oxoacyl-(acyl-carrier-protein) synthase III from *B. pseudomallei* (KASIII), and triosephosphate isomerase from *Mycobacterium tuberculosis* (TPI)) and the large C-terminal domain of polymerase basic protein 2 from the H1N1 (swine flu) influenza virus (PB2) (Table 4.1). Three of these nonproductive targets were identified after a small prescreening effort with approximately 13% of the FOL library. When zero hits are detected after such a prescreen, it is presumed that the overall hit rate for a target will be sufficiently low as to warrant discontinuation of additional screening investigations.

One nonproductive target from the SSGCID pipeline (PPA) was originally the subject of two noncrystallographic screens: an STD-NMR screen carried out with a library of 520 fragments, over 90% of which were present in the FOL library; and an SPR screen carried out using 384 FOL fragments. From the putative hits by STD-NMR and SPR, a list of 20 consensus fragments were selected and either soaked into PPA crystals or cocrystallized with PPA protein. Of these, only two were observed to be bound in the X-ray crystal structure, and both bound in a hydrophobic pocket at a twofold crystallographic biological dimer interface 19 Å away from the substrate binding site. While such fragment hits could possibly inspire the development of compounds that modulate the oligomerization state of inorganic pyrophosphatase, the target was deemed to be nonproductive due to lack of active site binding fragments identified.

3.2. Drug discovery targets at Emerald BioStructures

In addition to targets from our SSGCID pipeline, we employed both proprietary and nonproprietary drug discovery targets from our in-house fragment studies. Previously, we reported the application of FOL library screening to human leukotriene A4 hydrolase (LTA4H). In these studies, numerous fragments were found to interact with the long hydrophobic substrate binding cleft featuring a Zn^{2+} metal active site and recapitulated binding modes of previously reported LTA4H inhibitors (Davies *et al.*, 2009). In this regard, LTA4H is considered to be a highly productive fragment target. The results of proprietary fragment screens are confidential, but the experience gained from these efforts is informative for the current discussion. Accordingly, we have carried out a full analysis on four proprietary fragment screening targets, withholding only certain parameters from publication that might be used to identify the target. Two of the examples,

Table 4.1 Target identifiers, crystal conditions, and other crystallographic data on productive and nonproductive proteins surveyed for fragment screening suitability

					Monomer		Oligomeric state		Crystal conditions					Crystal parameters			
	Target	PDB	E.C. (if appropriate)	Native organism	Chain length	MW (Da)	Biologic	ASU	pH	Precipitant	Salt	Buffer	Other components	RMSD (Å)	Target volume ($Å^3$)	Solvent content (%)	Solvent channel diameter (Å)
Productive targets	DHNA	1rrw	4.1.2.25	*S. aureus*	121	13,770	4	8	5.6	30% MPD	200 m*M* AMOAc	Citrate	N.A.	2.2	47,121	41	18
	GCD	3eom	1.3.99.7	*B. pseudomallei*	395	43,117	4	4	7.5	20% PEG 3000	200 m*M* NaCl	HEPES	N.A.	1.96	73,278	39	31
	HSP90	2wi1	N.A.	*H. sapiens*	236	26,810	1	1	6.5	25% PEG 2000	200 m*M* $MgCl_2$ or 0.8 *M* Na formate	Cacodylate	N.A.	2.3	19,980	58	26
	IspF	3f0d	4.6.1.12	*B. pseudomallei*	162	17,175	3	6	8	20% PEG 4000	200 m*M* NaCl	Tris	5 m*M* $ZnCl_2$	2.05	42,379	45	34
	KASI	3lrf	2.3.1.41	*B. melitensis*	407	43,311	2	1	8.5	20% PEG 3350	200 m*M* Na malonate	BTP	N.A.	1.6	75,944	44	25
	LTA4H	1h19	3.3.2.6	*H. sapiens*	611	69,285	1	1	6.5	13% PEG 8000	100 m*M* Na acetate	Imidazole	5 m*M* YbCl3	1.58	60,417	49	19
	MAPK14	1wbo	2.7.11.24	*H. sapiens*	360	41,344	1	1	7	18% PEG 8000	200 m*M* MgOAc	HEPES	N.A	2.16	34,641	56	28
	PDE4B/D	1y2b	3.1.4.17	*H. sapiens*	349	81,888	2	1	7	19–24% PEG 3350	N.A.	Bis–Tris	10% Isopropanol, 28–34% EG	1.4	65,420	50	20
	PGM	3ezn	5.4.2.1	*B. pseudomallei*	249	27,893	2	2	7.5	30% PEG 600	N.A.	MES	5% PEG 1000, 10% Glycerol	2.1	45,673	40	27
	PRP1	PRP1	N.A.	N.A.	N.A.	N.A.	1	1	8.5	0.8 *M* Na citrate	200 m*M* NaCl	Tris	2.5% PEG 4000	1.7	38,670	55	30
	PRP4	PRP4	N.A.	N.A.	N.A.	N.A.	1	1	7.5	PEG 550 MME	200 m*M* $CaCl_2$	HEPES	N.A.	1.85	28,083	62	33
	risS	3lr0	2.7.13.3	*B. pseudomallei*	445	49,572	2	1	4.2	25% PEG 1000	200 m*M* Li_2SO_4	Phos/Citrate	N.A.	1.9	19,572	46	21
Nonproductive targets	KASIII	3gwa	2.3.1.180	*B. pseudomallei*	344	37,060	2	2	7	15% PEG 3350	N.A.	Succinate	N.A.	1.6	63,498	46	28
	PB2	3khw	N.A.	Influenza A virus	759	85,835	1	2	7	25% PEG 1500	N.A.	Propionic acid, Cacodylate, BTP	N.A.	2.1	18,938	43	15
	PPA	3eiz	3.6.1.1	*B. pseudomallei*	175	19,161	6	1	5.4	20% PEG 3500	100 m*M* NaSCN	Imidazole	N.A.	1.75	104,132	39	22
	PRP2	PRP2	N.A.	N.A.	N.A.	N.A.	2	2	8.5	PEG 4000	$MgCl_2$	Tris	N.A.	1.5	25,898	40	30
	PRP3	PRP3	N.A.	N.A.	N.A.	N.A.	3	6	7.5	PEG 8000	1 *M* Na formate	N.A.	N.A.	1.8	40,674	50	31
	TPI	3gvg	5.3.1.1	*M. tuberculosis*	261	27,403	2	2	N.D.	20% PEG 3350	200 m*M* ammonium citrate	N.A.	N.A.	1.55	47,319	47	24

designated PRP1 and PRP4, were chosen from successful fragment-screening campaigns that yielded multiple protein–fragment complexes. Two other examples, PRP2 and PRP3, came from a more speculative project where neither target yielded any verified fragment cocrystal structures.

3.3. Literature targets

To reduce the bias inherent from screening targets against the same library of fragment molecules, we included four external examples of fragment screening campaigns from the scientific literature. Two targets were screened by soaking preformed crystals into pools of fragments as a primary detection method. These were dihydroneopterin aldolase (DHNA) from *Staphylococcus aureus* done at Abbott Laboratories (Nienaber *et al.*, 2000; Sanders *et al.*, 2004) and human p38 alpha (MAPK14) conducted at Astex Therapeutics (Gill *et al.*, 2005; Hartshorn *et al.*, 2005). Two additional targets were the focus of fragment-based drug design approaches which involved different screening methods, followed by cocrystallization trials with hits to derive structural data. The catalytic domains of human phosphodiesterases 4B and D (PDE4B/D) were screened by scintillation assay at Plexxikon (Card *et al.*, 2005), while human heat-shock protein 90 (HSP90) was screened at Vernalis using a panel of ligand-observe NMR experiments as well as *in silico* screening methods (Brough *et al.*, 2009; Wright *et al.*, 2004) to obtain novel fragment hits. In all cases, teams used distinct focused and all-purpose libraries in their screening efforts, making it difficult to calculate meaningful, relative hit rates across this target series. Nevertheless, all the four targets are well-known, structurally characterized proteins which have been successfully used in fragment-based methods to obtain initial hits for drug design and development, and are therefore classified as productive for our comparative analysis. Further literature analysis did not afford additional suitable nonproductive targets, presumably because negative results are rarely published in great detail.

4. Crystallization Conditions of Fragment Targets

Protein crystallization is considered an art as much as a science by some, since it is virtually impossible to accurately predict crystallization conditions from the primary sequence of a protein. A typical crystallization project begins with random sparse matrix screening, utilizing tens to hundreds of conditions, with only one or a few combinations of precipitant, salts, buffers, and additives yielding useful crystals. Thus, the crystallographer is often at the mercy of these conditions, with limited latitude to change the composition of the solutions used for crystallographic fragment screening. In analyzing specific crystallization conditions for targets of fragment studies, the crystallant itself can present obvious challenges. For instance, an alternate crystal form of phosphoglyceromutase (GCM) was discovered, which grew out of tacsimate, a mixture of several carboxylic acids. Under these conditions, malonate, a close chemical analog of the native substrate, was observed to be bound in the enzyme active site. Malonate is unlikely to be displaced by any weakly binding fragment soaked or cocrystallized under these conditions, thus making these crystal conditions unsuitable for fragment studies. Another more subtle example of interference from crystallization conditions was found for the proprietary target PRP4. This target was very productive but yielded no carboxylic acid-containing fragments upon an initial screen. This result was unexpected, since acid compounds had been discovered to have activity against the target in an independent assay (data not shown). The crystallization conditions included 200 m*M* calcium chloride, and it was hypothesized that the calcium ion was decreasing the solubility of the acid-containing fragments, thus preventing their binding. Accordingly, subsequent replacement of calcium chloride with a different chloride salt for soaking experiments allowed the discovery of carboxylic acid fragments bound to the target.

It is clear that certain crystallization conditions required for the growth of high diffraction quality crystals preclude their use in fragment screening by X-ray crystallography. Conditions that affect fragment solubility or that utilize high concentrations of small molecules with the potential to bind the site of interest, are not amenable to fragment screening campaigns. Further examination of general crystallization conditions for productive and nonproductive targets are summarized in Table 4.1. No single condition identifies any of these crystals as potentially productive or nonproductive prior to experimental testing. Furthermore, we were unable to discern any trends for pH, precipitant, or salts, other than the more extreme cases outlined above. Interestingly, the pH range for productive crystals in our analysis was larger than that for nonproductive crystals; the average crystallization pH for both target classes is near neutrality.

5. Solvent Content and Solvent Channels

When conducting fragment studies by X-ray crystallography, the crystallization conditions and the crystal lattice itself may have a significant impact on the ability of fragments to bind to the protein target. The two most common methods for exposing small molecules to crystals are soaking and cocrystallization. Soaking requires a preexisting crystal, either an *apo* crystal or a crystal with a weakly bound ligand that can be displaced by screening molecules. As fragments are typically low-affinity binders, ligand displacement is rarely attempted in the context of a fragment screening campaign. Our crystallography-based fragment screening is done by soaking preformed protein crystals into solution drops containing mixtures of small molecules (Davies *et al.*, 2009). In extreme cases, the crystal packing arrangement may trap the protein in a conformational state unable to bind ligands, or occlude fragment binding sites by symmetry-related protein molecules. Many of these issues can be discerned by a careful examination of the symmetry-related molecules surrounding the asymmetric unit of an initial crystal structure. However, we did not know if solvent content, solvent channels, or other morphological characteristics of the crystals themselves lent any indication toward their suitability for fragment screening studies.

We analyzed our productive and nonproductive proteins to see if there were any trends in the solvent content of *apo* protein crystals predictive of the outcome of fragment screening campaigns. The majority of protein crystals amenable to medium- or high-resolution diffraction data contain solvent-filled channels comprising 30–65% of their total volume (Matthews, 1968). Our limited dataset covered most of this range, with 39–62% solvent content for both our productive and nonproductive fragment screening protein targets. The oligomeric states observed in the crystallographic asymmetric units of these targets ranged from monomers to octamers, consisting of polypeptide chains from 121 to 759 residues in length. Four out of the 12 productive targets and none of the six nonproductive targets had over 50% solvent content, while both groups were represented at the lowest crystal solvent percentage (Table 4.1). The four productive proteins with over 50% solvent content were monomers, both in terms of the asymmetric unit and the biological molecule. However, we did not observe a clear trend with the data available. We concluded that employing a $>50\%$ solvent content cutoff may eliminate crystal forms for many productive targets, including the majority of our productive candidates.

Small-molecule fragments must diffuse through solvent channels in order to successfully "soak" into the binding pockets of preformed crystals without destroying the crystal in the process. This requires that the narrowest part of the solvent channels be adequately wide for fragments to travel through them. The typical maximum width of a fragment in our library is 10 Å at its widest point across nonrotatable bonds (data not shown). Given that most fragments are asymmetrical in shape and have on average at least one rotatable bond, it is clear that most fragments could traverse a solvent channel that is more constricted than 10 Å. With this in mind, we estimated the maximum diameter of solvent channels between crystallographic monomers in each asymmetric unit of our set of proteins to see if there was any connection between solvent channel diameter and successful fragment screening.

To calculate the size of the largest solvent channel in a crystal form, we used the program AREAIMOL from CCP4 (CCP4, 1994). High point density AREAIMOL was recompiled from source as needed, increasing the parameters MAXNET, MAXBIN, MXNNBR, and MXPNT (CCP4, 1994). Symmetry and translation operations were applied to a PDB coordinate file which had been stripped of all heteroatoms (except for di- and trivalent metal atoms) to generate a model of the crystal lattice. Starting with a radius of 1.4 Å, the AREAIMOL probe radius was tested, then increased by 2.0 Å steps until the contact area between the probe and the protein monomer at the center of the simulated lattice dropped to zero. The largest radius with a nonzero contact area was iteratively determined and taken to be the radius of the largest solvent channel in the crystal. To increase accuracy, a point density of 100 pt/Å^2 was used to accommodate the largest probes.

The majority of targets possessed maximum diameters of 20 Å or more, a value in keeping with the apparent pore diameters of chemically cross-linked protein crystals calculated from macromolecular porosimetry (Vilenchik *et al.*, 1998). None of our survey target proteins would prevent the majority of 10 Å-wide fragments in the FOL library from passing through them at their widest point. However, the range of maximal diameters across productive targets reflected that of nonproductive targets and in no meaningful way connected solvent channel diameter to success rates with fragment screening by X-ray crystallography. Solvent channel diameters also did not correlate with the oligomeric state of the protein, either in terms of the biologic molecule or the asymmetric unit (Table 4.1). Thus, except in the most extreme cases, neither the overall solvent channel content nor the diameter of solvent channels between crystallographic protein monomers for an *apo* crystal structure give an advanced indication of fragment screening suitability.

6. Pocket Predictions for Fragment Screening Targets

6.1. Predicting ligand-binding sites

Computational tools exist which utilize geometry, energetic calculations, or some combination thereof to derive from crystal structures predictions about ligand-binding sites (Charifson *et al.*, 1999; Hendlich *et al.*, 1997; Laurie and Jackson, 2005; Le Guilloux *et al.*, 2009; Schmidtke *et al.*, 2010; Shoichet *et al.*, 1993). Previous analysis of these tools have shown the ability of such methods to rank order true ligand-binding sites among the highest scored sites (Chen and Hubbard, 2009; Hajduk *et al.*, 2005a,b; Hopkins and Groom, 2002). Although useful in determining statistical trends for ligand binding and druggability, it was unclear from the literature how effective any one method would be in determining the suitability of a single target for fragment screening studies by X-ray crystallography. We therefore applied some of these tools to our survey proteins to see if anything about predicted ligand-binding pockets might indicate the likelihood of success in crystallographic fragment screening. From our experience, we knew which pockets on which protein targets could bind fragments after the fact. But, we desired a method which could be used for new, untested structures, for which fragment screening data would not yet be available. In addition, a set of tools was needed to obtain physical and chemical characteristics of known ligand-binding pockets and to assess which (if any) parameters might be significant determinants for binding fragments.

To conduct this investigation, we performed *in silico* analyses of our 18 productive and nonproductive targets using three different, publicly available pocket prediction methods: Pocket-Finder (P-Pocket), Q-SiteFinder (Q-Pocket), and F-Pocket. Each computational tool uses a unique set of algorithms to predict and rank order potential ligand-binding pockets on a 3D protein structure. The P-Pocket algorithm uses a spatial/directional approach based on LIGSITE (Hendlich *et al.*, 1997), which effectively pushes a neutral sphere through a protein structure while taking note of its surroundings. After performing this along several different axes, the software generates protein cavities likely to bind ligands based on certain adjacent space requirements. The Q-Pocket (Laurie and Jackson, 2005) method positions and clusters methyl probes to the protein surface, followed by calculation and ranking of likely pockets based on predicted binding energies for such hydrophobic clusters. The F-Pocket (Le Guilloux *et al.*, 2009; Schmidtke *et al.*, 2010) approach detects protein cavities by scanning, categorizing, clustering, and ranking sets of alpha spheres which can be drawn within a 3D protein structure.[1] In applying each of the three methods to our target set, we included transition metal ions (e.g., zinc) and prosthetic groups (e.g., heme) as part of the protein, while removing salts, waters, and all other small molecules. We assigned ligand status to small molecules of six nonhydrogen atoms or larger and the spots where they bound as experimentally validated ligand-binding pockets.

Consensus pockets (C-Pockets) were identified by overlaying F-, P-, and Q-Pocket vertices for a given target, followed by visual determination of those with approximately 50% or more overlap between all the three prediction methods (Fig. 4.1). Since P-Pocket vertex representations tended to cover the most volume and mask results generated by the other two methods, P-Pockets were visualized using dotted spheres for overlays and consensus scoring. For P- and Q-Pocket, data were obtained using their respective web

[1] Alpha spheres are defined as any sphere drawn in space which has four atoms on its surface, but does not contain any atoms, and are quickly generated using Voronoi vertices (see Le Guilloux *et al.*, 2009).

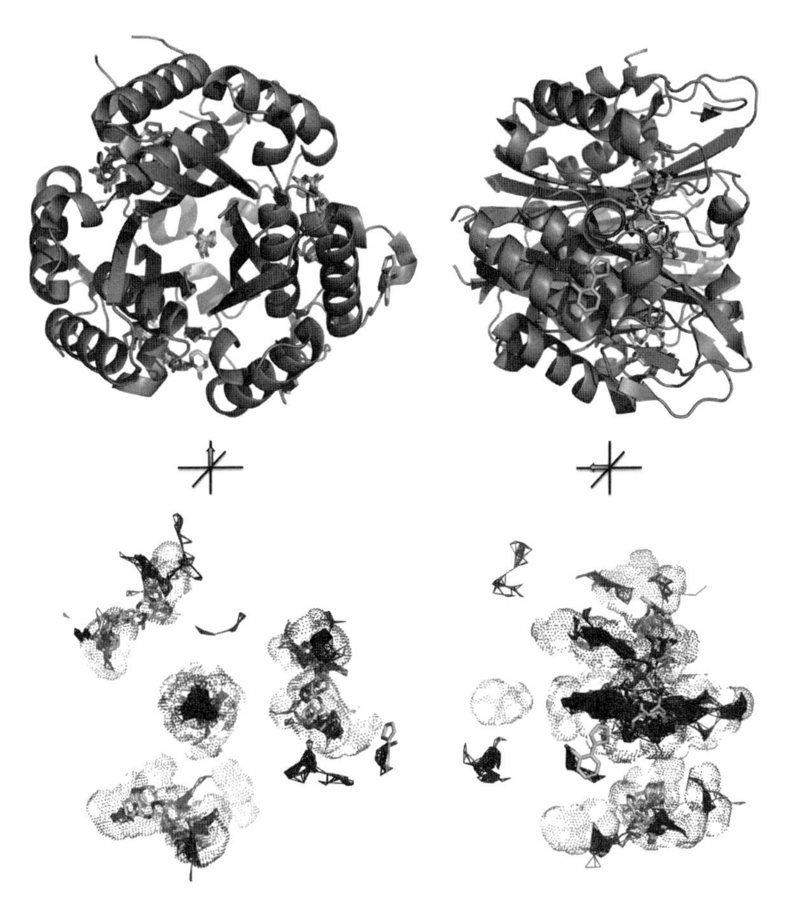

Figure 4.1 (Above) Two ribbon views of the IspF trimer with bound ligands from several crystal structures. (Below) Spatial arrangement of predicted pockets determined by P-Pocket (dots), Q-Pocket (dark gray sticks), and F-Pocket (light gray sticks) relative to known ligand-binding sites (Hendlich *et al.*, 1997; Laurie and Jackson, 2005; Le Guilloux *et al.*, 2009; Schmidtke *et al.*, 2010). Figure generated with PyMol (DeLano, 2008) from PDB IDs 3KE1, 3JVH, and 3MBM.

servers, while F-Pocket was compiled and run locally to generate nonnormalized statistics for each pocket. Once a consensus C-Pocket had been identified, we used F-Pocket data to interpret the chemical and physical parameters associated with each pocket. Mean B-factors were calculated independently from F-Pocket, using pocket atoms assigned by F-Pocket and B-values listed in the original RCSB PDB files. Mean B-factors were calculated for each entire protein, as well as for individual atoms and residues which wholly or partially comprise a given predicted pocket. Estimates of overall target volume were calculated by Q-Pocket.

6.2. Pocket prediction and consensus C-Pocket methodology

Both ligand-bound and *apo* crystal structures were available as representative models for novel fragment screening candidates. In comparing ligand-bound to *apo* crystals for four different proteins (two productive, two nonproductive), the three computational methods gave nearly identical pocket statistics, revealing no bias of protein complexes over *apo* structures (data not shown). Moreover, all three algorithms failed to find one or more experimentally validated ligand-binding pockets for several productive and nonproductive targets, indicating that the presence of a ligand does not influence these computational predictors. We therefore used both *apo* and complex structures with small molecules removed together in our qualitative assessment of predicted protein pockets.

Each computational method generates a set of vertices corresponding to the centers of spheres which fill each protein cavity, in addition to the set of protein atoms which line the pocket. Both sets of coordinates can be visualized using PyMol (DeLano, 2008) or another PDB file-compatible viewer. When results from all three methods are overlaid on a given target, each algorithm reveals certain tendencies which lead to different kinds of pocket sets (Fig. 4.1). Typically, P-Pocket generated very large and spherically shaped pockets, often connecting and combining distinct cavities into a single site, while ignoring smaller, solvent-exposed binding surfaces. Conversely, Q-Pocket often split a P- or an F-Pocket into two distinct sites and tended to return a greater number of small, shallow pockets, including solvent-accessible hydrophobic binding patches on protein surfaces. F-Pocket generated both extended pocket networks and small pockets for each target, some of which appeared too small or inaccessible for typical fragments to bind.

The percentage of F, P, and Q pockets found to contain actual ligands ranged from 16 to 26% across productive targets and 6–8% in nonproductive targets (Fig. 4.2). When judged against known sites on *apo* structures or complex structures with ligands removed, no single method proved superior in predicting ligand-binding sites. True druggability sites are often associated with increased concavity and complexity (Hajduk *et al.*, 2005b); although all three methods returned better scores for larger, more complex sites, none was individually consistent in doing so, particularly for proven ligand-binding sites. However, in 29 out of 47 cases, P-, Q-, and F-Pocket all discovered the same sites where fragments were known to bind, regardless of score or rank. These consensus pockets, where F-, P-, and Q-Pocket results overlap, proved more accurate at predicting actual ligand-binding sites for both productive (38%) and nonproductive (15%) targets than any single method alone (Fig. 4.2). Since the majority of experimentally validated ligand-binding sites showed consensus among the three techniques, we used consensus (C-Pockets) as our benchmark determinant.

6.3. C-Pocket, F-Pocket, and pocket factor analysis

Using consensus scoring, we obtained 68 productive and 20 nonproductive consensus C-Pockets for our set of targets, of which 26 and 3 are experimentally proven ligand-binding sites, respectively (Fig. 4.2). Many C-Pockets include unlikely spots for small-molecule binding and pockets that have not been experimentally demonstrated to bind ligands from diverse fragment screening or other studies (Table 4.2). Both P- and

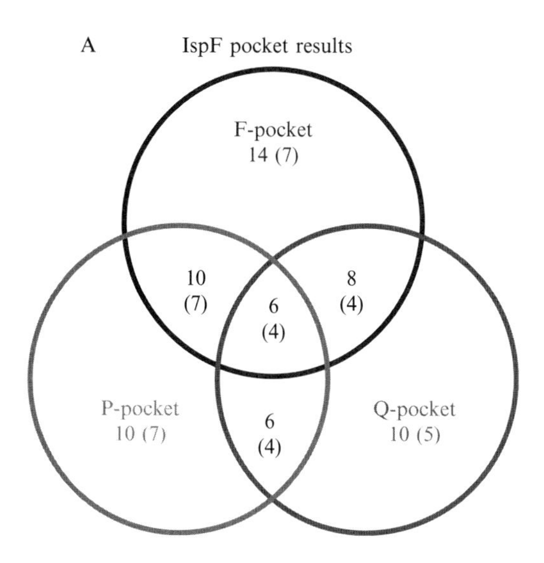

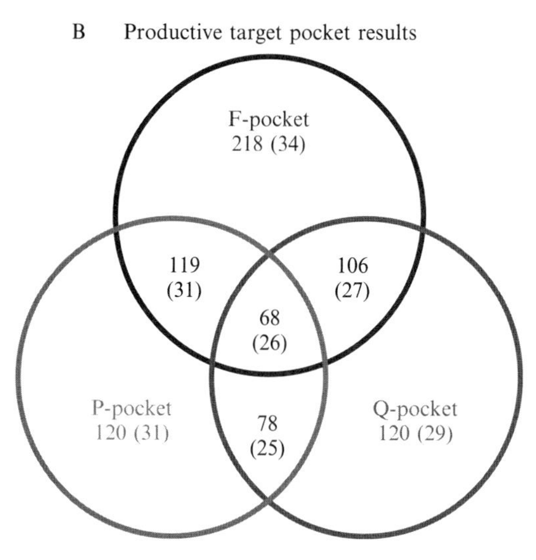

Figure 4.2 (Continued).

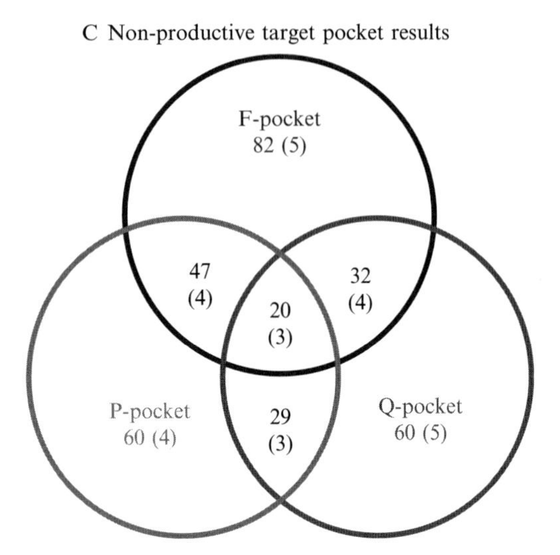

Figure 4.2 F-, P-, Q-Pocket total for (A) IspF from *Burkholderia pseudomallei*, (B) all productive targets, and (C) all nonproductive targets in the survey. Predicted pockets with $\geq$50% spatial overlap between F-, P-, and Q-Pocket results are designated as consensus C-Pockets. Brackets indicate numbers of predicted pockets experimentally proven to bind ligands.

Q-Pocket algorithms were written to generate 10 pockets per target, regardless of the size or oligomeric state of the protein crystal, while F-Pocket is written to return any number of pockets, depending on its standard scoring function. Thus, the total number of possible C-Pockets is dependent on the number of F-Pockets generated and therefore on the overall size of the biological unit analyzed. To account for size differences, we calculated a "pocket factor" for each biological molecule, taking the total number of C-Pockets per volume of protein (in 100 nm^3). Interestingly, the average pocket factor for productive proteins was twice that of nonproductive proteins (Table 4.2). With one exception in each category, productive proteins gave a pocket factor greater than 10, while all nonproductive targets returned values less than eight. Thus, there appears to be a statistically significant trend ($p = 0.024$) toward greater overall pocket character for productive versus nonproductive proteins across our limited survey.

Correlating "pocket factor" with the likelihood of success in fragment screening is intuitive: the more fragment-friendly pockets per cubic nanometer of protein, the greater the number of potential fragment binding sites and the higher the chances for obtaining screening hits. But this crude determination of pocket character does *not* correlate with biologically

Table 4.2 Total number of known ligand-binding pockets, F-Pockets, C-Pockets, C-Pockets with ligands, and calculated pocket factors for each productive and nonproductive protein target

PDB	Known sites	Total F-Pockets	Total C-Pockets	Pocket factor[a]	C-Pockets with ligands
Productive					
3lr0	1 (2)	10	5	26	1
1wbo	1	15	7	20	1
PRP4	N.A.	11	4	14	N.A.
3f0d	8	14	6	14	4
1h19	5	19	8	13	4
PRP1	N.A.	13	5	13	N.A.
1rrw	1 (4)	28	6	13	4
3ezn	1 (2)	13	5	11	2
3eom	3 (12)	44	8	11	7
1y2b	2 (4)	19	7	11	2
2wi1	1	4	2	10	1
3lrf	1	22	5	6.6	0
	Sum	212	68		26
	Min	4	2	7	
	Max	44	8	26	
	Mean	18	6	14	
	SD	10	2	5	
Nonproductive					
3khw	1	10	3	16	0
3eiz	3 (18)	18	8	7.7	1
PRP3	N.A.	12	3	7.4	N.A.
3gvg	1 (2)	12	3	6.3	2
PRP2	N.A.	4	1	3.9	N.A.
3gwa	0	22	2	3.1	0
	Sum	78	20		3
	Min	4	1	3	
	Max	22	8	16	
	Mean	13	3	7	
	SD	6	2	5	

Pocket factor confidence level (prod. vs. nonprod.): $p = 0.024$ (>98%).
[a] Pocket factor = (no. consensus C-Pockets/target volume [100 nm^3]).

relevant active sites, nor the number of pockets actually observed to bind ligands. The *apo* structure of IspF from *B. pseudomallei* generated six C-Pockets, of which only four were among the eight experimentally proven binding sites. Although one of the three computational methods found each of the remaining four ligand-binding pockets, these were excluded using our C-Pocket consensus criteria (Fig. 4.2). Conversely, we obtained three C-Pockets for a polymerase-binding protein from influenza which generated a "pocket factor" of 16, despite the lack of experimental crystal soaking hits. Just as C-Pocket scoring does not accurately predict every validated binding pocket, employing a strict pocket-factor cutoff based on our limited dataset has the potential to include poor targets and exclude good ones (Table 4.2). Bearing that in mind, targets with higher pocket factors appear to do well, which might prove a useful benchmark in rank-ordering or prioritizing a large number of targets queued for fragment screening.

Geometrically, there is nothing distinct about F-Pockets for a productive versus a nonproductive target (Fig. 4.3). However, F-Pocket returns a large number of properties associated with each predicted ligand-binding pocket, including volume, hydrophobicity, polarity, solvent-accessible surface area, as well as "druggability" and other overall scores (Le Guilloux *et al.*, 2009). We therefore mined these parameters between the 68 productive and 20 nonproductive C-Pockets to search for any trends associated with fragment screening suitability. In doing so, we did not discover any meaningful or statistically significant trends related to productive fragment binding. The only mean values not returning the null hypothesis between productive and nonproductive targets were polarity and mean local hydrophobic density of alpha spheres (Table 4.3). In both cases, mean values for productive C-Pockets were greater than those for nonproductive targets, with most of the proven ligand-binding pockets at the high end of the scale. Thus, a

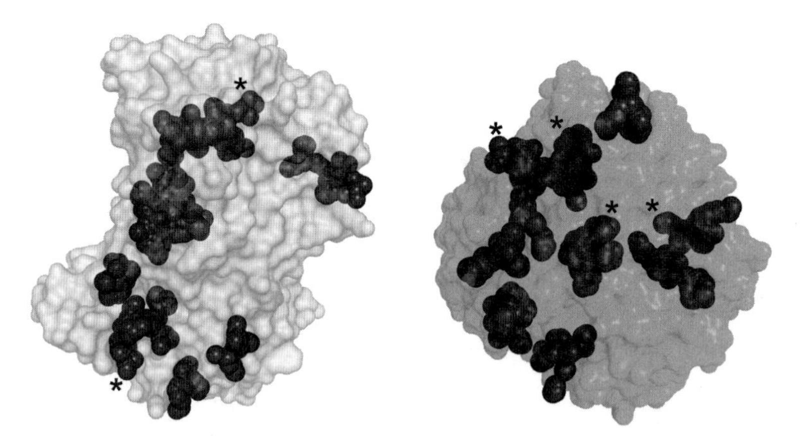

Figure 4.3 Results for triosephosphate isomerase, a nonproductive target (left) which yielded 10 predicted binding pockets, and 2C-methyl-D-erythritol 2,4-cyclodiphosphate synthase, a productive target (right) which yielded 14 predicted pockets by F-Pocket (Le Guilloux *et al.*, 2009; Schmidtke *et al.*, 2010). Predicted sites which are known to bind ligands are denoted with asterisks. Figure generated using PyMol (DeLano, 2008) from PDB IDs 3GVG and 3F0D.

Table 4.3 F-Pocket parameter statistics for productive and nonproductive protein targets

	Productive			Nonproductive			Productive vs. nonproductive	
	Min	Max	Avg	Min	Max	Avg	p	CI
F-Pocket score	0	47	21	5	36	18	0.21	79
Druggability score	0.01	0.82	0.20	0.03	0.80	0.26	0.41	59
Hydrophobicity score	−18	66	24	−2	55	30	0.11	89
Volume score	3.00	5.22	3.94	3.44	5.10	4.06	0.21	79
Polarity score	3.00	31.00	10.19	1.00	15.00	7.90	0.07	93
Charge score	−3.00	4.00	0.13	−2.00	2.00	−0.05	0.64	36
Flexibility score	0.05	0.65	0.30	0.00	0.59	0.22	0.06	94
B-factor (protein)	10.64	60.57	27.03	16.92	37.21	21.78	0.02	98
B-factor (pocket atoms)	9.92	66.66	28.59	6.90	36.10	18.50	0.0003	>99
B-factor (pocket residues)	9.25	65.66	26.95	7.02	38.33	17.02	0.0003	>99
DIFF (Bprot−Batom)	−11.16	22.72	1.56	−10.12	6.77	−3.28	0.002	>99
DIFF (Bprot−Bres)	−10.60	15.27	−0.08	−10.00	2.72	−4.77	0.0004	>99
Total no. α-spheres	36	534	119	37	538	104	0.57	43
Total SASA	61	1969	414	74	1429	337	0.35	65
Total polar SASA	14	930	203	25	701	173	0.47	53
Total apolar SASA	34	1039	210	50	728	164	0.26	74
Total volume	182	3550	829	139	2578	651	0.23	77
Mean α-sphere density	2.34	15.01	5.83	2.39	17.58	5.37	0.57	43
Mean local hydrophobic density	2	80	29	4	41	21	0.02	98
Mean α-sphere radius	3.38	4.42	3.77	3.39	3.99	3.71	0.23	77
Mean α-sphere SA	0.38	0.60	0.49	0.41	0.53	0.48	0.22	78
Percent apolar α-spheres	6%	85%	40%	11%	70%	37%	0.50	50
Percent polar atoms	22%	59%	41%	27%	50%	41%	0.88	12
Mass-centered α-sphere max distance	5.89	40.32	14.59	5.31	43.76	13.19	0.51	49

higher score correlated to more hydrophobic *and* hydrophilic (i.e., polar) character for strong fragment-binding pockets by these indicators, which is at best an indication of increased molecular complexity at these sites. However, we could not extract a meaningful cutoff value for either parameter without eliminating many productive targets and including several nonproductive targets. We also could not devise any sensible conclusions from these two factors which would identify untested targets for fragment studies.

B-factors are temperature factors calculated by crystallographic refinement to estimate the thermal motion of individual atoms in a structure. B-factors for atoms defined by F-Pocket associated with each C-Pocket were calculated and compared to the average B-factor across the entire target, without waters or other ligands. Contrary to our expectations, nonproductive protein C-Pockets gave lower B-factor scores, relative to the entire protein, than did C-Pockets for productive targets (Table 4.3). Some of the productive targets used in the analysis were ligand-bound complexes for which atoms lining the pocket were expected to have less internal motion than the (mostly) unbound protein surfaces found among targets in nonproductive group. However, on the whole, nonproductive targets generated more stable, less disordered *apo* pockets than did productive proteins, even when taking the entire residue into account for a given pocket atom (Table 4.3). If we limit our analysis to only experimentally validated, ligand-binding C-Pockets, there are only three pockets for the nonproductive class, making comparisons of proven pockets between the two groups rather problematic, if not altogether contradictory. Like the other F-Pocket parameters, the range of B-factor differences between individual pocket atoms and the whole protein is evenly dispersed across both groups, making it impossible to use B-factor differences to discern which targets would be most amenable to fragment screening. Thus, with our limited dataset, we cannot find any reliable cutoff value for a given F-Pocket parameter or B-factor difference which clearly separates productive from nonproductive targets.

7. Summary

We conducted a detailed analysis of 18 fragment screening targets and found no single "red-flag" parameter or characteristic which would preclude a given protein for such studies prior to preliminary experimental investigations. The data presented herein include unsuccessful targets which often go unreported in the literature, but which can provide invaluable learning experiences for making decisions about resource allocation for large-scale multitarget drug discovery projects. Overall, our analysis does show trends which correlate the size, hydrophobicity, polarity, and the overall number of predicted pockets with increasing numbers of fragment hits identified by

X-ray crystallography. We have also shown how calculation of a "pocket factor" based on consensus pocket predictions from publicly available prediction algorithms can provide a statistically significant variance between productive and nonproductive targets. Although our pocket factor is not absolutely predictive, the trend associated with higher hit rates might prove a useful factor in prioritizing targets for fragment screening.

In this chapter, we have attempted to identify one or more biophysical parameters which could be used to predict if a protein target would be a good candidate for fragment screening by X-ray crystallography. To do this, we categorized targets as either nonproductive, yielding no relevant fragment-bound structures, or productive, yielding structures with bound fragments that provide opportunities for further development. In practical terms, this is a somewhat arbitrary distinction. If a highly prized drug target only yields a single hit after fragment screening, it might very well be worth the effort if that hit is amenable to chemical elaboration. Conversely, targets known to bind many fragments at promiscuous and biologically irrelevant sites may possess higher screening hit rates, but generate leads of little value. The likelihood of obtaining hits by fragment screening is only one of the many important factors which must be weighed in deciding which targets to carry forward. In our experience, the best test of a given target in its ability to bind fragments is through collection of preliminary experimental data. One prudent strategy for allocating resources against multiple targets is to conduct an initial screen using a subset of molecules from a fragment library. However, in the case where a crystallized target is of high importance for drug discovery efforts, it seems equally sensible to conduct a full library screen, to ensure that no hits are left uncovered despite what prediction methods are used to initially assess the target.

ACKNOWLEDGMENTS

We thank Drs. Alex S. Kiselyov and John M. McCall for their helpful comments in reading this chapter. We further acknowledge and greatly appreciate the efforts of the entire team at SSGCID, without whom our accomplishments would not have been possible. Support for this research was funded by the National Institute of Allergy and Infectious Disease under Federal Contract No. HHSN272200700057C.

REFERENCES

Albert, J. S., Blomberg, N., Breeze, A. L., Brown, A. J., Burrows, J. N., Edwards, P. D., Folmer, R. H., Geschwindner, S., Griffen, E. J., Kenny, P. W., Nowak, T., Olsson, L. L., *et al.* (2007). An integrated approach to fragment-based lead generation: Philosophy, strategy and case studies from AstraZeneca's drug discovery programmes. *Curr. Top. Med. Chem.* **7,** 1600–1629.

Alex, A. A., and Flocco, M. M. (2007). Fragment-based drug discovery: What has it achieved so far? *Curr. Top. Med. Chem.* **7,** 1544–1567.

Barelier, S., Pons, J., Gehring, K., Lancelin, J. M., and Krimm, I. (2010). Ligand specificity in fragment-based drug design. *J. Med. Chem.* **53,** 5256–5266.

Baurin, N., Aboul-Ela, F., Barril, X., Davis, B., Drysdale, M., Dymock, B., Finch, H., Fromont, C., Richardson, C., Simmonite, H., and Hubbard, R. E. (2004). Design and characterization of libraries of molecular fragments for use in NMR screening against protein targets. *J. Chem. Inf. Comput. Sci.* **44,** 2157–2166.

Begley, D. W., Hartley, R. C., Davies, D. R., Edwards, T. E., Leonard, J. T., Abendroth, J., Burris, C. A., Bhandari, J. Myler, P. J., and Stewart, L. J. Leveraging Structure Determination with Fragment Screening for Infectious Disease Drug Targets. *J. Struct. Funct. Genomics.* (manuscript in preparation).

Biros, S. M., Moisan, L., Mann, E., Carella, A., Zhai, D., Reed, J. C., and Rebek, J., Jr. (2007). Heterocyclic alpha-helix mimetics for targeting protein-protein interactions. *Bioorg. Med. Chem. Lett.* **17,** 4641–4645.

Blomberg, N., Cosgrove, D. A., Kenny, P. W., and Kolmodin, K. (2009). Design of compound libraries for fragment screening. *J. Comput. Aided Mol. Des.* **23,** 513–525.

Bosch, J., Robien, M. A., Mehlin, C., Boni, E., Riechers, A., Buckner, F. S., Van Voorhis, W. C., Myler, P. J., Worthey, E. A., DeTitta, G., Luft, J. R., Lauricella, A., *et al.* (2006). Using fragment cocktail crystallography to assist inhibitor design of Trypanosoma brucei nucleoside 2-deoxyribosyltransferase. *J. Med. Chem.* **49,** 5939–5946.

Brough, P. A., Barril, X., Borgognoni, J., Chene, P., Davies, N. G., Davis, B., Drysdale, M. J., Dymock, B., Eccles, S. A., Garcia-Echeverria, C., Fromont, C., Hayes, A., *et al.* (2009). Combining hit identification strategies: Fragment-based and in silico approaches to orally active 2-aminothieno[2, 3-d]pyrimidine inhibitors of the Hsp90 molecular chaperone. *J. Med. Chem.* **52,** 4794–4809.

Card, G. L., Blasdel, L., England, B. P., Zhang, C., Suzuki, Y., Gillette, S., Fong, D., Ibrahim, P. N., Artis, D. R., Bollag, G., Milburn, M. V., Kim, S. H., *et al.* (2005). A family of phosphodiesterase inhibitors discovered by cocrystallography and scaffold-based drug design. *Nat. Biotechnol.* **23,** 201–207.

Carr, R., and Jhoti, H. (2002). Structure-based screening of low-affinity compounds. *Drug Discov. Today* **7,** 522–527.

CCP4 (1994). The CCP4 suite: Programs for protein crystallography. *Acta Crystallogr. D Biol. Crystallogr.* **50,** 760–763.

Charifson, P. S., Corkery, J. J., Murcko, M. A., and Walters, P. W. (1999). Consensus scoring: A method for obtaining improved hit rates from docking databases of three-dimensional structures into proteins. *J. Med. Chem.* **42,** 5100–5109.

Chen, I. J., and Hubbard, R. E. (2009). Lessons for fragment library design: Analysis of output from multiple screening campaigns. *J. Comput. Aided Mol. Des.* **23,** 603–620.

Chen, Y., and Shoichet, B. K. (2009). Molecular docking and ligand specificity in fragment-based inhibitor discovery. *Nat. Chem. Biol.* **5,** 358–364.

Chen, J., Zhang, Z., Stebbins, J. L., Zhang, X., Hoffman, R., Moore, A., and Pellecchia, M. (2007). A fragment-based approach for the discovery of isoform-specific p38alpha inhibitors. *ACS Chem. Biol.* **2,** 329–336.

Congreve, M., Carr, R., Murray, C., and Jhoti, H. (2003). A 'rule of three' for fragment-based lead discovery? *Drug Discov. Today* **8,** 876–877.

Congreve, M., Chessari, G., Tisi, D., and Woodhead, A. J. (2008). Recent developments in fragment-based drug discovery. *J. Med. Chem.* **51,** 3661–3680.

Davies, D. R., Mamat, B., Magnusson, O. T., Christensen, J., Haraldsson, M. H., Mishra, R., Pease, B., Hansen, E., Singh, J., Zembower, D., Kim, H., Kiselyov, A. S., *et al.* (2009). Discovery of leukotriene A4 hydrolase inhibitors using metabolomics biased fragment crystallography. *J. Med. Chem.* **52,** 4694–4715.

DeLano, W. L. (2008). The PyMOL Molecular Graphics System. DeLano Scientific LLC, Palo Alto, CA, USA.

Duarte, C. D., Barreiro, E. J., and Fraga, C. A. (2007). Privileged structures: A useful concept for the rational design of new lead drug candidates. *Mini Rev. Med. Chem.* **7,** 1108–1119.

Erlanson, D. A., McDowell, R. S., and O'Brien, T. (2004). Fragment-based drug discovery. *J. Med. Chem.* **47,** 3463–3482.

Gill, A. L., Frederickson, M., Cleasby, A., Woodhead, S. J., Carr, M. G., Woodhead, A. J., Walker, M. T., Congreve, M. S., Devine, L. A., Tisi, D., O'Reilly, M., Seavers, L. C., *et al.* (2005). Identification of novel p38alpha MAP kinase inhibitors using fragment-based lead generation. *J. Med. Chem.* **48,** 414–426.

Hajduk, P. J., and Greer, J. (2007). A decade of fragment-based drug design: Strategic advances and lessons learned. *Nat. Rev. Drug Discov.* **6,** 211–219.

Hajduk, P. J., Bures, M., Praestgaard, J., and Fesik, S. W. (2000). Privileged molecules for protein binding identified from NMR-based screening. *J. Med. Chem.* **43,** 3443–3447.

Hajduk, P. J., Huth, J. R., and Fesik, S. W. (2005a). Druggability indices for protein targets derived from NMR-based screening data. *J. Med. Chem.* **48,** 2518–2525.

Hajduk, P. J., Huth, J. R., and Tse, C. (2005b). Predicting protein druggability. *Drug Discov. Today* **10,** 1675–1682.

Hartshorn, M. J., Murray, C. W., Cleasby, A., Frederickson, M., Tickle, I. J., and Jhoti, H. (2005). Fragment-based lead discovery using X-ray crystallography. *J. Med. Chem.* **48,** 403–413.

Hendlich, M., Rippmann, F., and Barnickel, G. (1997). LIGSITE: Automatic and efficient detection of potential small molecule-binding sites in proteins. *J. Mol. Graph. Model.* **15** (359–363), 389.

Hopkins, A. L., and Groom, C. R. (2002). The druggable genome. *Nat. Rev. Drug Discov.* **1,** 727–730.

Hubbard, R. E., Chen, I., and Davis, B. (2007). Informatics and modeling challenges in fragment-based drug discovery. *Curr. Opin. Drug Discov. Devel.* **10,** 289–297.

Jacoby, E., Davies, J., and Blommers, M. J. (2003). Design of small molecule libraries for NMR screening and other applications in drug discovery. *Curr. Top. Med. Chem.* **3,** 11–23.

Laurie, A. T., and Jackson, R. M. (2005). Q-SiteFinder: An energy-based method for the prediction of protein-ligand binding sites. *Bioinformatics* **21,** 1908–1916.

Le Guilloux, V., Schmidtke, P., and Tuffery, P. (2009). Fpocket: An open source platform for ligand pocket detection. *BMC Bioinformatics* **10,** 168.

Makara, G. M. (2007). On sampling of fragment space. *J. Med. Chem.* **50,** 3214–3221.

Matthews, B. W. (1968). Solvent content of protein crystals. *J. Mol. Biol.* **33,** 491–497.

Myler, P. J., Stacy, R., Stewart, L., Staker, B. L., Van Voorhis, W. C., Varani, G., and Buchko, G. W. (2009). The Seattle Structural Genomics Center for Infectious Disease (SSGCID). *Infect. Disord. Drug Targets* **9,** 493–506.

Nienaber, V. L., Richardson, P. L., Klighofer, V., Bouska, J. J., Giranda, V. L., and Greer, J. (2000). Discovering novel ligands for macromolecules using X-ray crystallographic screening. *Nat. Biotechnol.* **18,** 1105–1108.

Oltersdorf, T., Elmore, S. W., Shoemaker, A. R., Armstrong, R. C., Augeri, D. J., Belli, B. A., Bruncko, M., Deckwerth, T. L., Dinges, J., Hajduk, P. J., Joseph, M. K., Kitada, S., *et al.* (2005). An inhibitor of Bcl-2 family proteins induces regression of solid tumours. *Nature* **435,** 677–681.

Pflugrath, J. W. (1999). The finer things in X-ray diffraction data collection. *Acta Crystallogr. D Biol. Crystallogr.* **55,** 1718–1725.

Rees, D. C., Congreve, M., Murray, C. W., and Carr, R. (2004). Fragment-based lead discovery. *Nat. Rev. Drug Discov.* **3,** 660–672.

Robinson, J. A. (2008). Beta-hairpin peptidomimetics: Design, structures and biological activities. *Acc. Chem. Res.* **41,** 1278–1288.

Sanders, W. J., Nienaber, V. L., Lerner, C. G., McCall, J. O., Merrick, S. M., Swanson, S. J., Harlan, J. E., Stoll, V. S., Stamper, G. F., Betz, S. F., Condroski, K. R., Meadows, R. P., *et al.* (2004). Discovery of potent inhibitors of dihydroneopterin aldolase using CrystaLEAD high-throughput X-ray crystallographic screening and structure-directed lead optimization. *J. Med. Chem.* **47,** 1709–1718.

Saraogi, I., and Hamilton, A. D. (2008). alpha-Helix mimetics as inhibitors of protein-protein interactions. *Biochem. Soc. Trans.* **36,** 1414–1417.

Schmidtke, P., Le Guilloux, V., Maupetit, J., and Tuffery, P. (2010). fpocket: Online tools for protein ensemble pocket detection and tracking. *Nucleic Acids Res.* **38**(Suppl), W582–W589.

Schuffenhauer, A., Ruedisser, S., Marzinzik, A. L., Jahnke, W., Blommers, M., Selzer, P., and Jacoby, E. (2005). Library design for fragment based screening. *Curr. Top. Med. Chem.* **5,** 751–762.

Shoichet, B. K., Stroud, R. M., Santi, D. V., Kuntz, I. D., and Perry, K. M. (1993). Structure-based discovery of inhibitors of thymidylate synthase. *Science* **259,** 1445–1450.

Shuker, S. B., Hajduk, P. J., Meadows, R. P., and Fesik, S. W. (1996). Discovering high-affinity ligands for proteins: SAR by NMR. *Science* **274,** 1531–1534.

Siegel, M. G., and Vieth, M. (2007). Drugs in other drugs: A new look at drugs as fragments. *Drug Discov. Today* **12,** 71–79.

Stebbins, J. L., Zhang, Z., Chen, J., Wu, B., Emdadi, A., Williams, M. E., Cashman, J., and Pellecchia, M. (2007). Nuclear magnetic resonance fragment-based identification of novel FKBP12 inhibitors. *J. Med. Chem.* **50,** 6607–6617.

Van Voorhis, W. C., Hol, W. G., Myler, P. J., and Stewart, L. J. (2009). The role of medical structural genomics in discovering new drugs for infectious diseases. *PLoS Comput. Biol.* **5,** e1000530.

Vilenchik, L. Z., Griffith, J. P., St. Clair, N., Navia, M. A., and Margolin, A. L. (1998). Protein crystals as novel microporous materials. *J. Am. Chem. Soc.* **120,** 4290–4294.

Wells, J. A., and McClendon, C. L. (2007). Reaching for high-hanging fruit in drug discovery at protein-protein interfaces. *Nature* **450,** 1001–1009.

Wright, L., Barril, X., Dymock, B., Sheridan, L., Surgenor, A., Beswick, M., Drysdale, M., Collier, A., Massey, A., Davies, N., Fink, A., Fromont, C., *et al.* (2004). Structure-activity relationships in purine-based inhibitor binding to HSP90 isoforms. *Chem. Biol.* **11,** 775–785.

Zartler, E., and Shapiro, M. (2008). Fragment-based drug discovery: A practical approach. John Wiley & Sons, Hoboken, N.J.

CHAPTER FIVE

Fragment Screening of Stabilized G-Protein-Coupled Receptors Using Biophysical Methods

Miles Congreve,* Rebecca L. Rich,† David G. Myszka,†
Francis Figaroa,‡ Gregg Siegal,‡ *and* Fiona H. Marshall*

Contents

Abstract

Biophysical studies with G-protein-coupled receptors (GPCRs) are typically very challenging due to the poor stability of these receptors when solubilized from the cell membrane into detergent solutions. However, the stability of a GPCR can be greatly improved by introducing a number of point mutations into the protein sequence to give a stabilized receptor or StaR®. Here, we present the utility of StaRs for biophysical studies and the screening of fragment libraries. Two case studies are used to illustrate the methods: first, the screening of a library of fragments by surface plasmon resonance against the adenosine A_{2A} receptor StaR, demonstrating how very small and weakly active xanthine

* Heptares Therapeutics, Biopark, Welwyn Garden City, Hertfordshire, United Kingdom
† Department of Biochemistry, University of Utah, Salt Lake City, Utah, USA
‡ Leiden Institute of Chemistry, ZoBio and Leiden University, Einsteinweg 55, Leiden, The Netherlands

Methods in Enzymology, Volume 493
ISSN 0076-6879, DOI: 10.1016/B978-0-12-381274-2.00005-4

fragments can be detected binding to the protein on chips; second, the screening and detection of fragment hits of a larger fragment library in an NMR format called TINS (target-immobilized NMR screening) against the β_1 adrenergic StaR.

1. Introduction

G-protein-coupled receptors (GPCRs) are one of the most important classes of protein due to their critical role in cell signaling in response to neurotransmitters, hormones, and other signaling molecules. Historically, GPCRs have been successfully targeted to develop a wealth of small-molecule and biological drugs across many therapeutic areas and are still an intense area of interest today. To date, most GPCR drug discovery programs have relied on cell-based assay systems combined with high-throughput screening of large compound libraries. While this approach has been successful for many GPCRs, more recent targets of interest, which include peptide receptors, are proving less tractable to traditional lead-generation methods. Given the success that fragment screening and structure-based drug discovery methods have had for soluble targets such as kinases and proteases, it would be of great benefit to apply these to GPCRs (Chessari and Woodhead, 2009; Congreve *et al.*, 2008). However, biophysical assays and structural techniques are extremely challenging for membrane proteins and these have so far precluded biophysical fragment-based screening methods to discover new chemistry starting points. These limitations are largely due to the difficulties in isolating pure stable protein in functionally folded forms and in sufficient quantities for use as reagents. GPCRs are typically highly unstable when extracted from the cell membrane, making them particularly difficult to isolate and purify for screening with biophysical methods. Popular fragment screening methods such as nuclear magnetic resonance (NMR) spectroscopy or surface plasmon resonance (SPR) have generally proved particularly difficult with GPCRs, with only a few of the chemokine receptors successfully studied by SPR to date (Navratilova *et al.*, 2005, 2006).

A solution to the stability and purification problems is the production of StaR®s (stabilized receptors): GPCRs engineered to contain a small number of point mutations that greatly improve their thermostability (Magnani *et al.*, 2008; Robertson *et al.*, 2010; Serrano-Vega *et al.*, 2008; Shibata *et al.*, 2009). As part of the thermostabilization process, the StaR is trapped in either the agonist or inverse agonist/antagonist conformation in a complex with a suitable ligand. The two antagonist StaRs employed in the studies described herein are the β_1 adrenergic receptor M36 (which contains mutations R68S, M90V, Y227A, A282L, F327A, F338M) (Serrano-Vega *et al.*, 2008; Warne *et al.*, 2008) and the adenosine A_{2A} receptor Rant22 (which contains mutations A54L, T88A, K122A, and V239A) (Magnani *et al.*, 2008; Robertson *et al.*, 2010). Using these StaRs trapped *in situ*, the advantages of using NMR and SPR to screen small molecules against GPCRs are exemplified in this chapter. StaRs are highly suitable for both methods due to their much improved stability in detergent solutions and for SPR, they are unparalleled because either the purified or crude receptor preparations can be utilized. Described below are conditions for solubilization, purification, and obtaining binding information for receptor-binding partners including low-molecular weight fragments of low affinity. Approaches for preparing active receptor biosensor surfaces and active receptor in the target-immobilized NMR screening (TINS) format are discussed. Two case studies of fragment screens are reported that serve to illustrate the potential of StaRs for fragment-based drug discovery for GPCRs.

2. Case Study 1: Biacore Fragment Screen of Adenosine A_{2A} StaR

Using the adenosine A_{2A} G-protein-coupled stabilized receptor system, we demonstrate the ability to resolve receptor–ligand interactions using optimized assay conditions. In addition, the advantages of using SPR-based biosensors to monitor the activity of receptor preparations, determine binding constants, and screen panels of analytes are outlined. Methods to extract active receptor from cell membranes (both simple solubilization and solubilization/affinity purification protocols); approaches for tethering StaRs on biosensor surfaces; and high-resolution kinetic and equilibrium analyses, as well as higher throughput screening assays that serve to identify hits from a fragment library, are reported. The ability to immobilize high concentrations of active functional protein on SPR chips has allowed a screen of a small focused fragment library based on xanthine structures. This library contained a range of very low-molecular weight (136–194 Da) molecules which included known binders to the A_{2A} receptor such as caffeine. This study provided proof of concept for using StaRs together with SPR for GPCR fragment screening.

2.1. Materials and methods

2.1.1. Reagents and instrumentation

Studies were performed at 10 °C using Biacore 2000 and S51 optical biosensors equipped with NTA sensor chips and equilibrated with running buffer (20 m*M* Tris–HCl, 350 m*M* NaCl, 0.1% dodecyl maltoside (DDM), 5% DMSO, pH 7.8). Antagonist compounds were purchased from Sigma and Tocris, detergents from Anatrace, and general laboratory reagents from Sigma and Fisher Scientific.

2.1.2. Receptor preparation

For biophysical studies, receptors, including a C-terminal His-10 tag, were expressed in Trichoplusia ni (Tni) cells using the FastBac expression system (Invitrogen), as previously described (Robertson *et al.*, 2010). All protein purification steps were carried out at 4 °C. The solubilized material was applied to Ni-NTA superflow cartridge (Qiagen) and then eluted with a linear gradient (5–400 m*M*) of imidazole in buffer supplemented with 0.15% *n*-decyl-β-maltoside (DM). Receptor protein was detected with an online detector to monitor absorbance A_{280}, and column fractions were collected and analyzed by SDS-PAGE. Fractions containing the ca. 35 kDa protein were pooled and concentrated using a YM50 Amicon ultrafiltration membrane. The protein sample was then applied to a 10/30 S200 size exclusion column and fractions pooled and concentrated as before to a final concentration of 10 mg/ml and stored at −80 °C. The protein concentration was determined using a detergent compatible Bradford assay (Bio-Rad).

2.1.3. Capture and activity of purified receptor

An NTA chip was preconditioned with three 1-min pulses of 350 m*M* EDTA in running buffer and charged for 3 min with 500 μ*M* Ni^{2+} in running buffer. Purified receptor was diluted approximately 100× in running buffer and injected across an NTA surface to achieve capture levels of >7000 resonance units (RU). After capture, the surface was washed for 2 h with running buffer to allow all antagonists added during the purification step to fully dissociate from the receptor. Afterward, activity of the captured receptor was evaluated using 500 n*M* xanthine amine congener (XAC).

2.1.4. Capture and activity of solubilized receptor

A frozen aliquot of crude membrane preparation was diluted with an equal volume of solubilization buffer (40 m*M* Tris, 1.5% decyl maltoside (DM), pH 7.4), sonicated for six 1-s pulses, rotated gently at 4 °C for 50 min, and centrifuged for 5 min at 14,000 rpm. Crude cell supernatant (diluted 1/3 in running buffer) was injected across a preconditioned, charged NTA surface to capture receptor to initial densities of ~10,000 RU. The surface was washed for 2–3 h with running buffer to remove nonspecifically bound supernatant debris from the chip surface and to allow full dissociation of any receptor-bound antagonist. Activity of the captured receptor was evaluated using 300 n*M* XAC.

2.1.5. Kinetic characterization of $A_{2A}R$ antagonists

For the kinetic analyses, five antagonists of 286–428 Da were each tested in triplicate in threefold dilution series for binding to $A_{2A}R$. The receptor surface was washed with buffer for 3–60 min after each compound injection to avoid using a surface regeneration step. For the kinetic screening studies, these five compounds were each tested at one concentration (ranging from 300 to 2500 n*M*).

2.1.6. Fragment screening against $A_{2A}R$

A panel of xanthine derivatives and unrelated compounds ranging in size from 136 to 194 Da were tested at 200 μ*M* for binding to solubilized $A_{2A}R$ captured to a density of ~8000 RU on an NTA sensor surface. 8-Cyclopentyl-1,3-dipropylxanthine (DPCPX) was included as a positive control (at 1 μ*M*) and was tested at every eighth injection.

2.1.7. Equilibrium analysis of screening hits against $A_{2A}R$

Each hit from the fragment screen was tested in replicate in a twofold dilution series for binding to a ~8000-RU solubilized $A_{2A}R$ surface.

2.1.8. Data processing and analysis

All responses were double-referenced (Myszka, 1999). For kinetic analyses, data were globally fit to a 1:1 interaction model to obtain binding parameters. For equilibrium analyses, the responses at equilibrium were plotted against analyte concentration and fit to a simple binding isotherm. All data processing and analysis were performed using Scrubber 2 (BioLogic Software Pty Ltd).

2.2. Results

2.2.1. NTA capture of purified $A_{2A}R$ StaR

Figure 5.1 depicts the NTA capture and XAC-binding test of an affinity-purified, C-terminally His_{10}-tagged $A_{2A}R$ StaR. The receptor was readily captured to a relatively high density, and the flat response at $t > 1200$ s in Fig. 5.1 indicated that the capture of $A_{2A}R$-His10 was very stable. Furthermore, the response obtained for XAC (428 Da) suggested that this captured receptor preparation was active enough to detect small-molecule binding. Together, these data demonstrated that NTA capture of this purified $A_{2A}R$-His_{10} StaR produced receptor surfaces that may be used for detailed characterization of $A_{2A}R$ antagonist compounds.

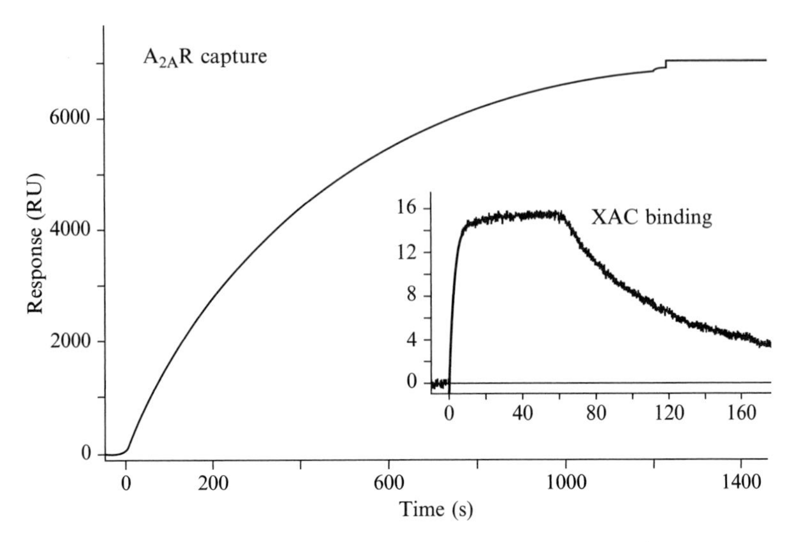

Figure 5.1 Capture and activity of purified A_{2A}R StaR. Main panel: NTA capture of A_{2A}R-His_{10}. Inset: binding response for 500 nM XAC.

2.2.2. Kinetic analyses of antagonists binding to purified A_{2A}R StaR

Figure 5.2 highlights the kinetic characterization of five A_{2A}R antagonists. For each of these compounds (which ranged in size from 296 to 438 Da), binding was readily detected, the responses were concentration-dependent, and the triplicate analyses of each compound overlay well. In addition, each data could be globally fit to a 1:1 interaction model to obtain binding parameters. This series of antagonists displayed a range of association and dissociation rate constants that produced a ~10,000-fold span in affinities, which agree well with the affinities reported from cell-based assays (Robertson *et al.*, 2010).

2.2.3. Capturing of solubilized StaRs

Having established that the biosensor was able to monitor antagonist binding to surface-tethered StaRs, the efficiency of the assay was optimized. Toward increasing the sampling throughput, the effects (on both receptor capture and activity) of omitting the affinity-purification step was examined. Figure 5.3 shows the NTA capture of the A_{2A}R StaR from cell membrane fractions. High-density capture was achieved simply by injecting the crude solubilization supernatant across the sensor surface. But, after capture of A_{2A}R preparations, the signal showed some decay. Since the A_{2A} receptor itself is stably captured (as shown in Fig. 5.1), the postcapture drift observed here was most likely due to the washing away of material in the crude supernatant that bound nonspecifically to the NTA surface.

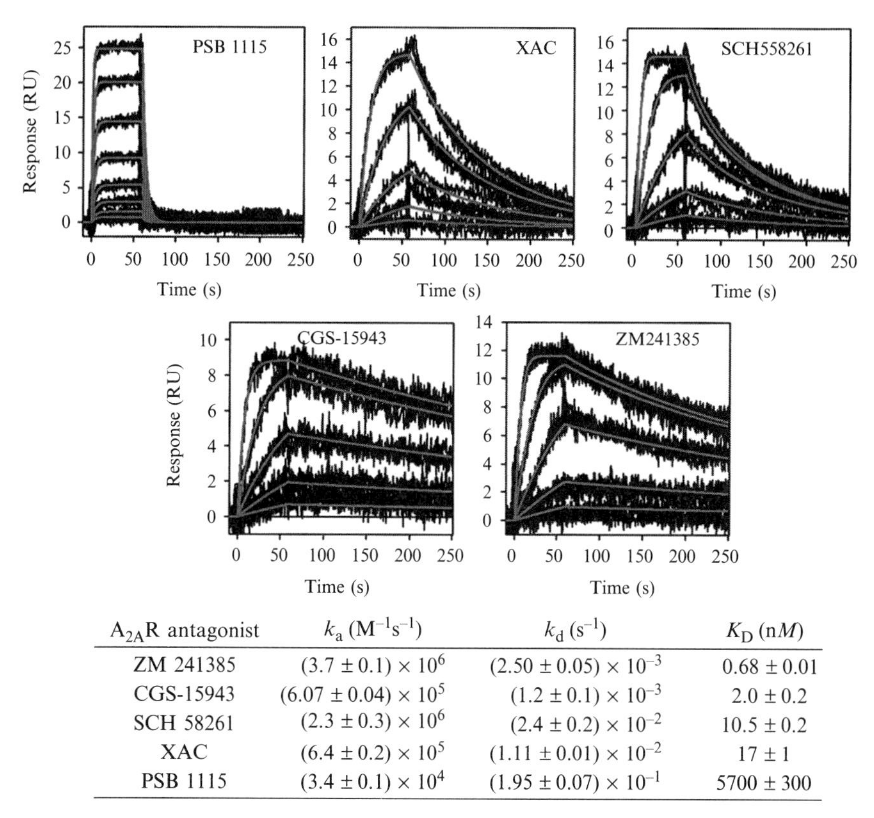

$A_{2A}R$ antagonist	k_a ($M^{-1}s^{-1}$)	k_d (s^{-1})	K_D (n*M*)
ZM 241385	$(3.7 \pm 0.1) \times 10^6$	$(2.50 \pm 0.05) \times 10^{-3}$	0.68 ± 0.01
CGS-15943	$(6.07 \pm 0.04) \times 10^5$	$(1.2 \pm 0.1) \times 10^{-3}$	2.0 ± 0.2
SCH 58261	$(2.3 \pm 0.3) \times 10^6$	$(2.4 \pm 0.2) \times 10^{-2}$	10.5 ± 0.2
XAC	$(6.4 \pm 0.2) \times 10^5$	$(1.11 \pm 0.01) \times 10^{-2}$	17 ± 1
PSB 1115	$(3.4 \pm 0.1) \times 10^4$	$(1.95 \pm 0.07) \times 10^{-1}$	5700 ± 300

Figure 5.2 Full kinetic analyses of five antagonists (286–428 Da) binding to purified $A_{2A}R$. Triplicate responses of each analyte concentration are overlaid with the fit of a 1:1 interaction model, with determined binding parameters listed in the table.

After washing the crude $A_{2A}R$ surface with running buffer for several hours, the signal was found to stabilize. Furthermore, XAC binding (Fig. 5.3, inset) demonstrated that solubilization alone produced an active $A_{2A}R$ preparation that could be used to examine antagonist binding.

2.2.4. Kinetic screening of antagonists against solubilized $A_{2A}R$ StaR

Having established the feasibility of preparation of viable $A_{2A}R$ surfaces from crude cell supernatants, improving the efficiency of this assay even more by increasing the analyte throughput was examined. Using a kinetic screening mode, analytes were each tested at one concentration and these single sensorgrams were fitted to obtain estimates of the binding parameters. Figure 5.4 shows the responses that were obtained from a kinetic screen of the five $A_{2A}R$ antagonists characterized in Fig. 5.2.

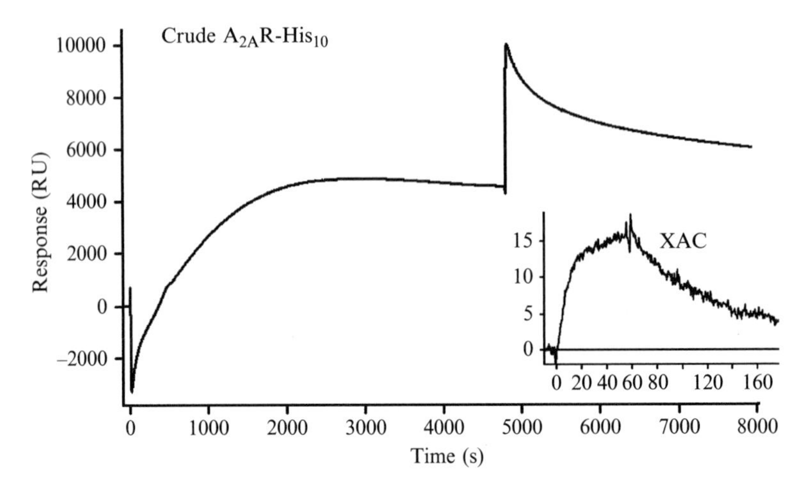

Figure 5.3 NTA capture of $A_{2A}R$-His_{10} from crude cell supernatant (main panel) and binding response for 300 n*M* XAC (inset).

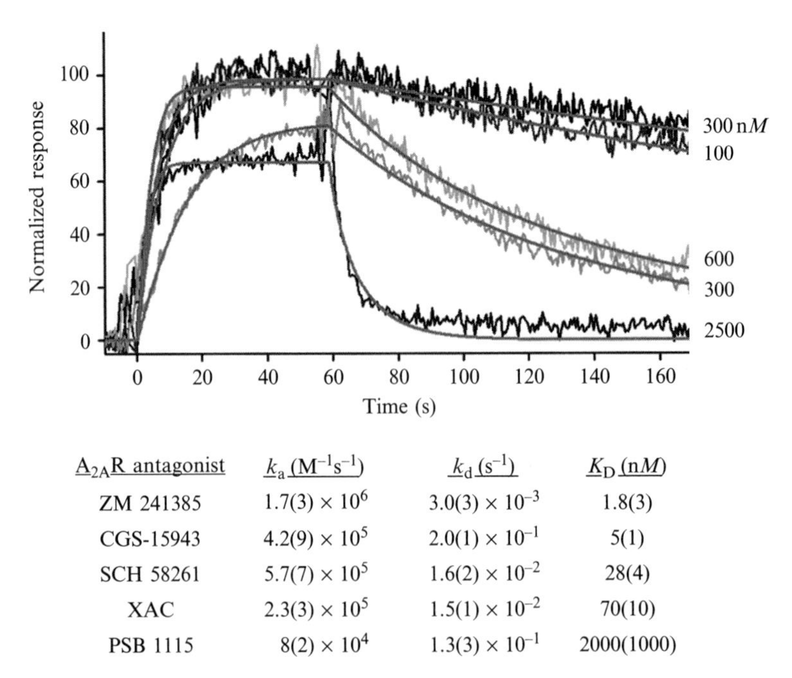

$A_{2A}R$ antagonist	k_a ($M^{-1}s^{-1}$)	k_d (s^{-1})	K_D (n*M*)
ZM 241385	$1.7(3) \times 10^6$	$3.0(3) \times 10^{-3}$	1.8(3)
CGS-15943	$4.2(9) \times 10^5$	$2.0(1) \times 10^{-1}$	5(1)
SCH 58261	$5.7(7) \times 10^5$	$1.6(2) \times 10^{-2}$	28(4)
XAC	$2.3(3) \times 10^5$	$1.5(1) \times 10^{-2}$	70(10)
PSB 1115	$8(2) \times 10^4$	$1.3(3) \times 10^{-1}$	2000(1000)

Figure 5.4 Responses for five small-molecule antagonists (286–428 Da) tested at 100 n*M*–2.5 μ*M*) for binding to $A_{2A}R$. Binding parameters obtained from the fit of a 1:1 interaction model (shown as red lines) are listed in the table. (For interpretation of the references to color in this figure legend, the reader is referred to the Web version of this chapter.)

The binding parameters determined from this kinetic screen are reasonable estimates of those determined from the kinetic analysis (Fig. 5.2). This kinetic screening approach allowed the rapid identification of binders as well as ranking of compounds based on their kinetics and affinity, and/or the prescreening of compounds before a detailed kinetic analysis. Since the GPCR preparations were found to slowly lose activity over several days, the higher throughput of this approach means testing of larger compound libraries against a single GPCR surface is possible.

2.2.5. Fragment screening against $A_{2A}R$ StaR

Figure 5.5A shows the responses for a compound library (136–194 Da) screened against $A_{2A}R$. Replicate analyses of a positive control (DPCPX, black lines) were included throughout the screen; the overlay of these replicates demonstrated that the receptor did not lose activity during this screen. Not unexpectedly, only a small subset (~10%) of the samples showed receptor binding significantly above background. These binding data are summarized in the trend plot (Fig. 5.5B), in which the responses are plotted against analyte number. Again, the reproducibility of the binding of the positive control is obvious, as are the signals from the few compounds that appear to be hits for $A_{2A}R$. From this trend plot, eight samples were identified as potential hits.

To confirm the hits identified from this screen, each of these eight compounds were rerun in a concentration series (Fig. 5.6A). From an equilibrium analysis (Fig. 5.6B), these hits were determined to have $A_{2A}R$ affinities ranging from ~10 μM to 5 mM. The detailed analyses of these eight samples confirmed that (1) the screen hits were indeed $A_{2A}R$ binders and (2) the SPR screen could reliably detect hits that are fragment-sized compounds and/or particularly weak binders.

2.3. Discussion of results

Here, we demonstrated the viability of using SPR approaches for characterizing GPCR StaRs. With SPR, we can establish the effectiveness of StaR capture on the sensor surface and confirm the activity of the receptor (both affinity-purified and extracted from crude cell supernatants). In addition, SPR can be used in a variety of formats, from kinetic and equilibrium analyses, which provide rate and affinity information, to screening assays for discriminating between small-molecule antagonists and for identifying hits in a fragment library that are worth pursing in downstream drug discovery efforts.

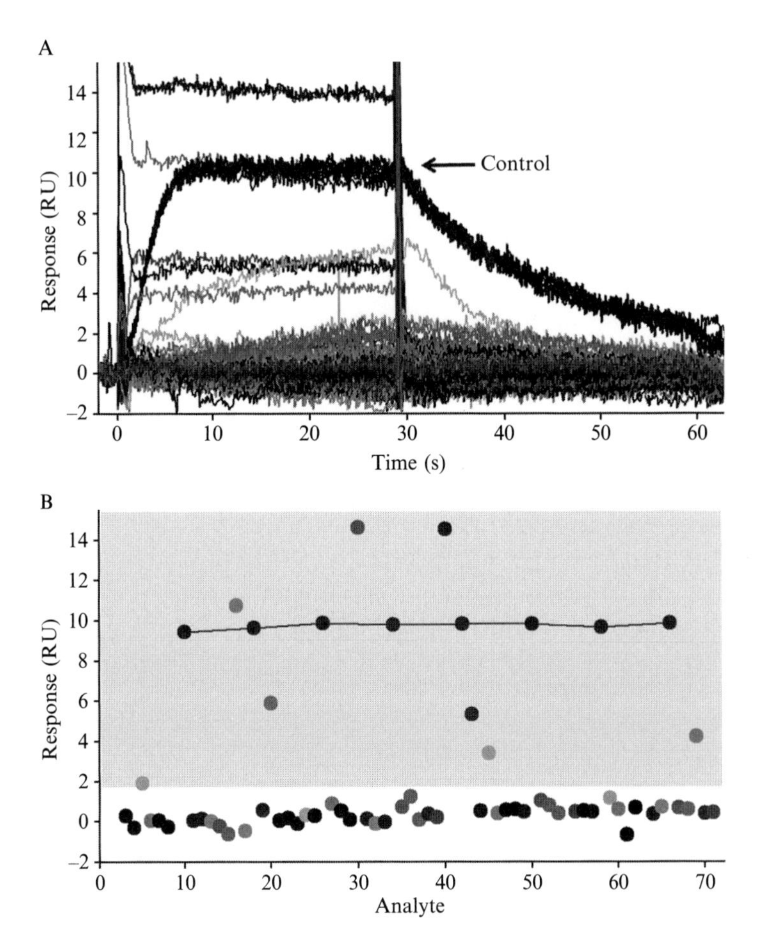

Figure 5.5 Fragment screen against $A_{2A}R$. (A) Responses for samples (tested at 200 μ*M*) and replicates of the control compound (DPCPX at 1 μ*M*; shown in black) binding to $A_{2A}R$. (B) Trend plot of the screening responses. Reproducibility of the DPCPX replicates is indicated by the line and hits are highlighted by the shaded box.

3. Case Study 2: TINS NMR Fragment Screen of β_1-Adrenergic StaR

NMR methods for ligand screening have had little or no success when applied to membrane proteins. The hurdle arises primarily from two issues: limited availability of the protein and significant levels of nonspecific partitioning of compounds into the media used to solubilize the protein. Here, we demonstrate the potential of the combination of stabilized receptors and

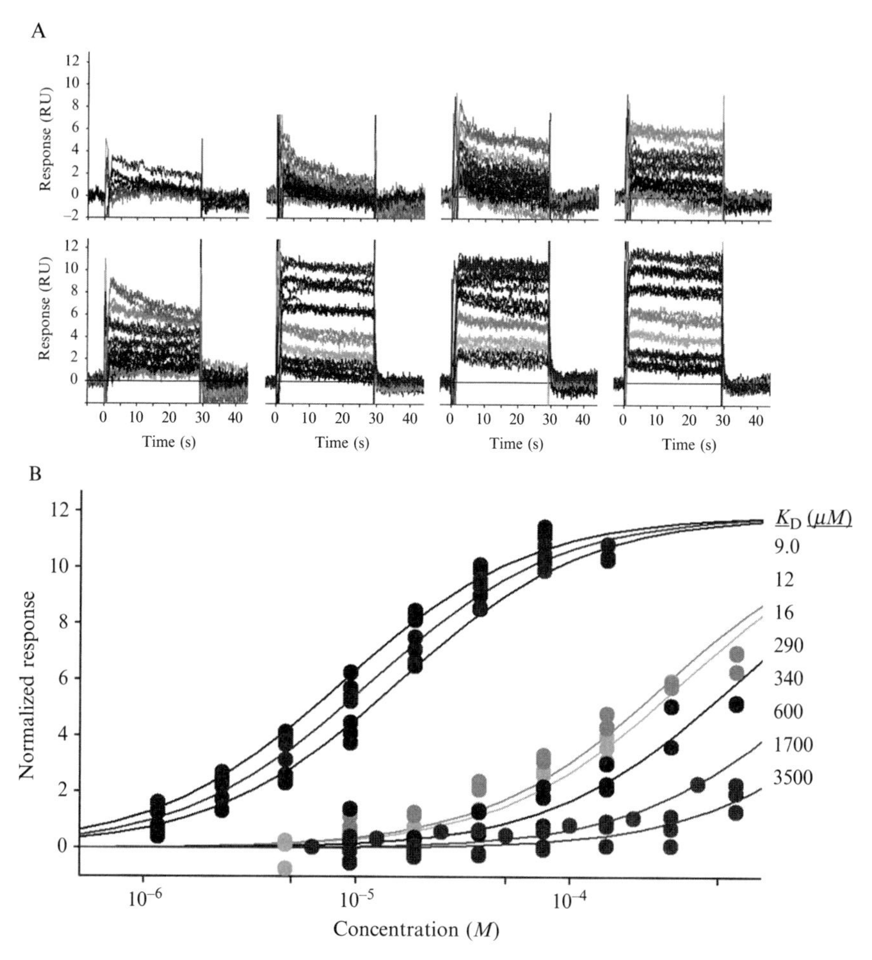

Figure 5.6 Equilibrium analyses of hits against $A_{2A}R$. (A) Responses for twofold dilution series of each hit. (B) Responses at equilibrium (t = 5–25 s) are plotted against concentration and fit to a simple binding isotherm to obtain the affinities listed on the right.

TINS for efficient screening of Rule of Three compliant fragment libraries (Congreve *et al.*, 2003). Methods to functionally immobilize StaRs as well as the stability of the immobilized preparations are discussed. Proof-of-concept data are subsequently provided that demonstrate the successful use of NMR to detect binding of a fragment to a GPCR. Based on these results, a limited screen of approximately 575 fragments was performed. The results of the screen and biochemical validation assays are presented. As a first step toward understanding the binding site of fragment ligands and gaining insight into their molecular mechanism of action, we present the results of competition binding studies using known ligands. Together, the data underscore the usefulness of StaRs and TINS for drug discovery efforts focused on GPCRs.

3.1. Materials and methods

3.1.1. Protein purification

The β_1-adrenergic StaR (β_1AR) and OmpA were expressed and purified, as previously reported (Arora *et al.*, 2000; Warne *et al.*, 2003). Both proteins have a 6x-His tag at the N- or C-terminus for affinity purification. Successful refolding of OmpA from inclusion bodies was monitored by SDS-PAGE analysis (Arora *et al.*, 2000). Both β_1AR and OmpA were buffer-exchanged to PBS (0.1 *M* sodium phosphate, 0.15 *M* sodium chloride, pH 7.2) containing 15 m*M* dodecylphosphocholine (DPC) for OmpA or 0.1% decyl maltoside (DM) for β_1AR.

3.1.2. Protein immobilization

DM-solubilized β_1AR was immobilized on Actigel ALD resin (Sterogene, Carlsbad, CA, USA) via Schiff's base chemistry using the manufacturer's protocol. After overnight incubation at 4 °C, residual unreacted aldehydes on the resin were blocked by addition of 50 m*M* d_{11}-Tris buffer. The same procedure was repeated for DPC-solubilized OmpA. Quantification of immobilized protein was monitored by absorption of the supernatant at 280 nm before and after immobilization, and by SDS-PAGE with a known standard curve and band volume analysis. This data indicated that a final concentration of 50 μ*M* of immobilized β_1AR and 75 μ*M* OmpA was achieved (pmol protein/ml settled bed volume), equating to a 90% and 80% yield. Subsequently, the buffer of both protein samples was exchanged to PBS containing 10 m*M* *n*-dodecyl-β-D-maltopyranoside (DDM) for ligand screening experiments.

3.1.3. Target-immobilized NMR screening

Immobilized, DDM-solubilized β_1AR and OmpA were each packed into a separate cell of a dual-cell sample holder. Mixes of the 579 fragments were made by 200-fold dilution of a 100 m*M* stock of each compound in d_6-DMSO such that the final DMSO concentration was never greater than 5%. Upon injection of each mix into the dual-cell sample holder, flow was stopped and spatially selective Hadamard spectroscopy was used to acquire a 1D ^{1}H spectrum of each sample separately. A CPMG T2 filter of 80 ms was used to remove residual broad resonances from the sepharose resin. To maintain the proper fold of each protein, the screen was performed at 15 °C, and 200 μ*M* DDM was included in the buffer (PBS in D_2O) used to wash the fragment mixes from the sample holder. Periodic injection of dopamine was used as a positive control.

3.1.4. β_1AR activity assays

The activity of TINS hits was characterized using radioligand competition binding studies. Any hits that bind to the orthosteric binding site would be expected to have activity in such an assay if they are sufficiently potent. Such an assay may also pick up effects of allosteric modulators. Whole cell binding assays were conducted on Tni insect cells expressing the stabilized β_1AR-m23 (50,000 cells/well), in a final volume of 0.2 ml of PBS buffer and 4% DMSO with compounds at either a single concentration of 500 μM or 8 × 0.5 log unit dilutions and a final [^{3}H]-dihydroalprenolol concentration of 2 nM. Dobutamine was included as a control. After incubation for 60 min at room temperature, assays were terminated by rapid filtration through 96-well GF/B UniFilter plates presoaked with 0.5% PEI, followed by washing with 5 × 0.25 ml ddH_2O. Plates were dried, 50 μl SafeScint was added per well, and bound radioactivity was measured using a Packard Microbeta counter. Data were analyzed using GraphPad Prism v5 and normalized as "% specific binding," and IC_{50} values were calculated. All fragments from the TINS screen that were designated as positive for binding were assayed in the radioligand displacement assay at 500 μM.

3.2. Functional immobilization and stability

As a first step toward TINS ligand screening, we assessed the ability to immobilize functional micelle solubilized β_1AR and the stability of such an immobilized preparation (Fig. 5.7A). We followed a simple approach that was successfully applied to the *Escherichia coli* membrane proteins, DsbB and OmpA (Fruh *et al.*, 2010), based on direct covalent attachment using Schiff base chemistry. The DDM-solubilized β_1AR was efficiently immobilized (80%), and subsequently, unreacted aldehydes on the resin were blocked using deuterated Tris buffer. The immobilized protein was assayed for functionality by binding of [^{3}H]-dihydroalprenolol, a well-characterized, high-affinity ligand that binds with 1:1 stoichiometry. Initial measurement of [^{3}H]-dihydroalprenolol binding after immobilization indicated that nearly 100% of the immobilized receptor was functional. Even after 4 days, nearly 50% of the immobilized receptor remained functional for high-affinity ligand binding if stored at temperatures up to 10 °C.

In any assay in which the target is reused to assess ligand binding, it is convenient to have a known ligand that can be readily removed without denaturing the protein as a tool compound. Accordingly, we investigated the affinity of both dopamine and dobutamine for β_1AR, two compounds on the biosynthetic route to epinephrine. Both dopamine and dobutamine inhibited dihydroalprenolol binding with IC_{50}s of 60 and 0.60 μM, respectively (data not shown). In order to determine the feasibility of detecting

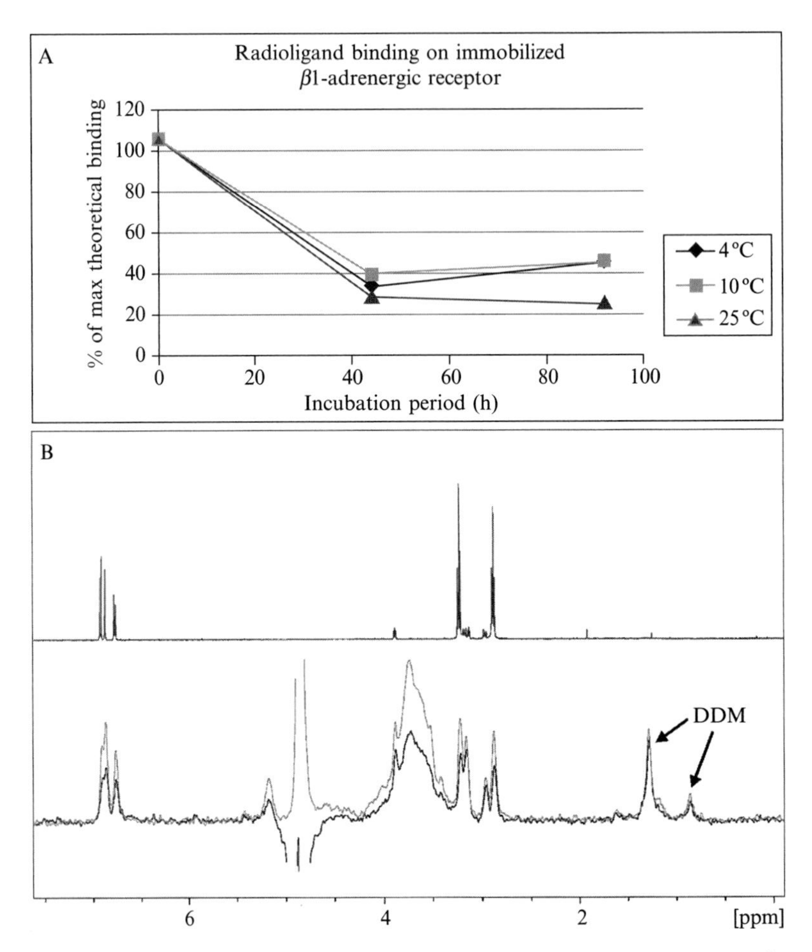

Figure 5.7 (A) Functional immobilization and stability of micelle-solubilized β_1AR. β_1AR was covalently bound to Sepharose resin (see text for details). Functionality was assessed by binding of ^{3}H-dihydroalprenolol upon storage of the protein at the indicated temperature and for the indicated time periods. (B) Use of TINS to detect weakly binding ligands. A solution of 500 μM dopamine was injected into the cell containing immobilized β_1AR (blue) and OmpA (red) and a ^{1}H spectrum of each is presented. The green spectrum is that of dopamine in solution. The NMR signals from DDM are indicated. (See Color Insert.)

weak ligand binding to immobilized β_1AR, we assayed for dopamine binding using the TINS assay. TINS (Vanwetswinkel *et al.*, 2005) uses a reference to cancel out nonspecific binding, thereby reducing the hit rate of false positives. The immobilized proteins, target and reference, are packed into separate cells of a dual-cell sample holder (Marquardsen *et al.*, 2006) and placed inside the magnet. Previously, we had determined that OmpA was an appropriate reference for micelle-solubilized membrane proteins and therefore, we have used it for all studies involving StaRs. To assay for binding, compounds are injected, typically in mixtures, into both cells simultaneously and flow is stopped at a predetermined point. A spatially selective, one-dimensional ^{1}H Hadamard experiment (Murali *et al.*, 2006) is used to acquire the NMR spectrum of the compounds in solution. Since the NMR relaxation of a spin is approximately 1000 times more efficient in the solid state than in solution, binding of a ligand to the immobilized protein results in the complete disappearance of the magnetization from that molecule in the NMR spectrum. Accordingly, binding is detected as a simple reduction in the amplitude of the signals of a compound in the presence of the target with respect to those in the presence of the reference (Fig. 5.7B). We assayed for specific binding of dopamine to immobilized β_1AR using immobilized OmpA as the reference. As seen in Fig. 5.7B, the amplitude of all of the NMR signals from dopamine is significantly reduced in the presence of β_1AR, while the signals derived from soluble DDM are the same in both samples. This data suggests that TINS is capable of detecting weak but specific binding typical of fragments.

3.3. Fragment screening profile and biochemical validation

Based on the successful detection of dopamine binding to immobilized β_1AR, we decided to perform a screen of a limited portion of the ZoBio fragment collection. The goals of the project were to test the stability of the immobilized β_1AR to repeated injections of mixtures of fragments and to assess the robustness of detection of specific ligands. Accordingly, 579 fragments were assayed for binding in mixtures of, on average, approximately 4.5 compounds each. In order to assess the physical integrity of the immobilized samples, control experiments are routinely performed at the beginning and periodically throughout a screen. Including controls, slightly more than 200 cycles of compound application and washing were performed. The most convenient measure of binding in TINS is the ratio of the amplitude of the NMR signals for each compound in the presence of the target to that in the presence of the reference, which we refer to as the T/R ratio. Figure 5.8A presents the T/R ratio of dopamine at various points during the screen. One can clearly see that the apparent specific binding of dopamine to β_1AR decreases over the 4 days required to perform the screen, which was done at 15 °C. We have used binding of dopamine as a proxy for the functionality of the immobilized β_1AR StaR, where interestingly, the decay in TINS is essentially identical to the decay observed when the samples are simply stored. This suggests that the decay is a stochastic function of the intrinsic stability of the protein and not a result of the cycles of compound application and washing.

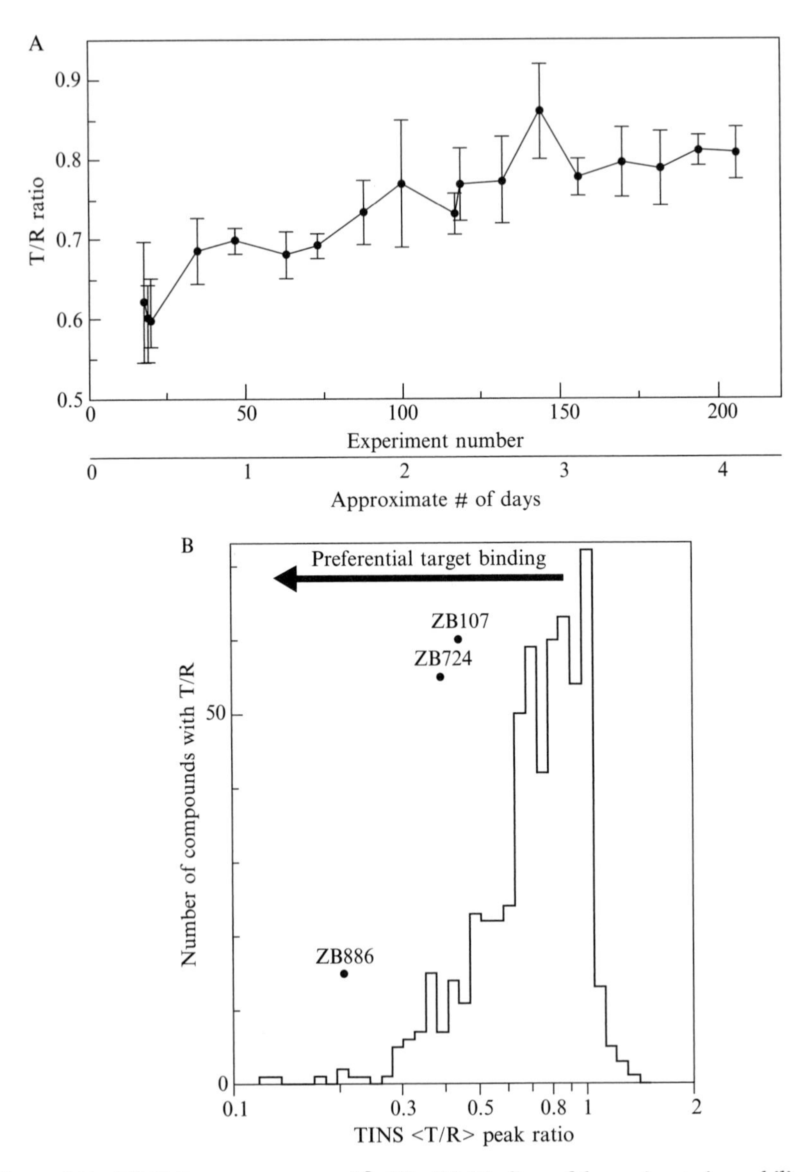

Figure 5.8 TINS fragment screen of β_1AR. (A) Binding of dopamine to immobilized β_1AR before, during, and after the screen (see text for definition of T/R ratio). (B) Profile of the complete screen. The number of fragments with the indicated T/R ratio is shown. The T/R ratio of three fragment hits that have been biochemically characterized is indicated.

In order to differentiate hits from nonhits, we analyze a profile of the entire screen (Fig. 5.8B). In this plot, the T/R ratio for each fragment has been bucketed and the number of fragments in each bucket is presented as a histogram. As expected, the bulk of the fragments has T/R ratios close to 1, indicating that they do not preferentially bind to either the target or reference. Similarly, there are very few fragments with T/R greater than 1 confirming our earlier work, indicating that OmpA lacks significant small-molecule binding sites. In contrast, there are large numbers of fragments with T/R ratios significantly less than 1. Moreover, there is a range of values displayed reflecting the various affinities and specificity of the fragments in the collection for β_1AR. In order to make a selection, a discontinuity in the histogram is chosen and that value is used as a cutoff. Such a discontinuity can be found at a T/R ratio of about 0.35, which would result in 29 fragments defined as hits. It was noted in this work, as well as previous work on a different membrane protein (Fruh *et al.*, 2010), that the T/R ratios were considerably smaller than we had observed for soluble proteins. Therefore, a very conservative hit selection was made using a T/R ratio of about 0.5, which resulted in 79 fragments being defined as hits. We assessed whether the decay in the amount of functional β_1AR affected our ability to detect ligand binding by investigating a potential relationship between the hits and the cycle in which they were analyzed. While a decline in the number of hits found per cycle was not found, it is entirely possible that the selection became more stringent as the screen progressed (i.e., the same fragment might have resulted in a lower T/R if assessed earlier in the screen).

As an initial step toward validating specific binding of the fragment hits to β_1AR, the ability of each of the fragment hits from the screen to displace [^{3}H]-dihydroalprenolol was assessed using a whole cell assay with wild-type receptor. Dihydroalprenolol has a K_d of approximately 5 nM and the fragments were titrated from 500 nM to 5 mM. Eleven of the 79 hits elicited a significant response in the radioligand displacement assay (at least 30% displacement), strongly suggesting a specific and biologically relevant interaction between the fragment and β_1AR. Subsequently, the 11 fragments were reassayed using membrane preparations devoid of any intrinsic nucleotide. The curves for three selected hits, with IC_{50} values ranging from 5 mM to 5 μM, are presented in Fig. 5.9. The Hill coefficient, in conjunction with the TINS data, indicates that these fragments reversibly bind β_1AR with an approximate 1:1 stoichiometry.

3.4. TINS competition binding

The data presented above have shown that the combination of TINS with StaR protein can be used to screen a fragment library and to detect compounds that are weakly interacting with the receptor in a biologically meaningful manner. Inspired by the results from the biochemical radioligand displacement assay, the feasibility of performing competition binding experiments using TINS was assessed as a means to both validate and

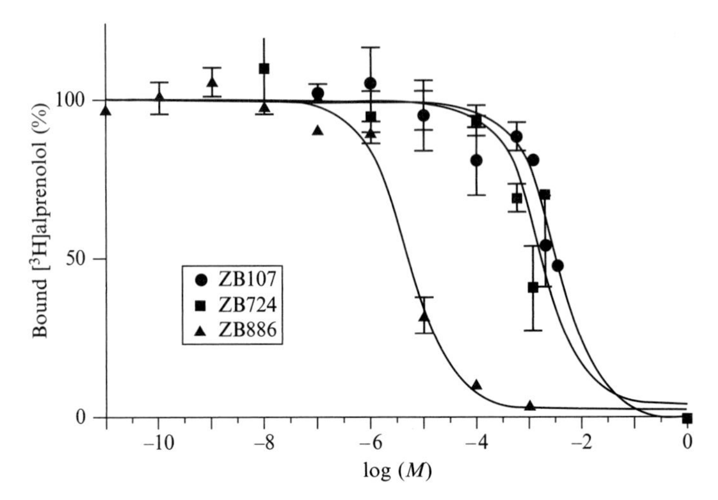

Figure 5.9 Biochemical characterization of fragments hits. Fragments were assayed for their ability to displace the well-characterized, high-affinity (5 n*M*) ligand [^{3}H]-dihydroalprenolol. The IC_{50} curves for three selected fragments from the screen (see Fig. 5.8) are shown. The calculated IC_{50}s range from approximately 5 μ*M* to 5 m*M*.

determine the binding site of hits from a TINS screen. The antagonist propranolol was selected as a known ligand with high-affinity (3 n*M*) and well-characterized binding site. Figure 5.10 shows the binding of dopamine to immobilized β_1AR in the presence and absence of propranolol using TINS. As expected, inclusion of 40 μ*M* propranolol has nearly completely abrogated binding of dopamine to β_1AR. Thirty of the 79 hits from the TINS screen were assayed for binding ± propranolol. Interestingly, only about one third of the hits showed a reduction in binding in the presence of propranolol and none were completely abrogated. A possible interpretation of this data is that some of the fragment hits target a separate or only partially overlapping binding site on β_1AR. However, at this early stage, this is only an intriguing hypothesis. In a second, ongoing project on a StaR, we have clear evidence for allosteric ligands from a TINS screen (to be published elsewhere).

3.5. Discussion of results with TINS

Here, we have shown the potential of TINS and StaRs to enable fragment-based drug discovery on GPCRs. The flexibility of the TINS immobilization approach allows careful characterization of the functionality of immobilized receptor before, during, and after screening. The sensitivity of TINS allows detection of ligands with a wide range of affinities, thereby minimizing false negatives. Given the stability or StaRs, it is possible to screen a large portion of

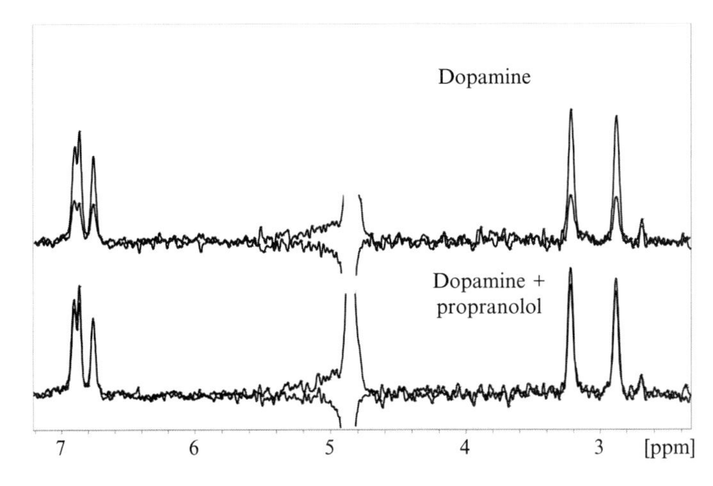

Figure 5.10 Competition binding studies for fragment validation and characterization. Two pairs of overlaid spectra are shown. The lower pair shows binding of dopamine to immobilized β_1AR. The upper pair shows the lack of binding of dopamine to immobilized β_1AR in the presence of the high-affinity orthosteric ligand propranolol.

a fragment collection on a single protein sample, although one must be careful to monitor the functionality of the immobilized target. Since, in general, StaRs can be more readily produced than their wild-type analog, having to use more than one sample to screen the collection is not necessarily a limiting step. Furthermore, it may be possible to functionally immobilize biotinylated StaRs from crude cell lysates, similar to the procedure for SPR. Such a procedure has already been used to immobilize and screen unpurified soluble proteins. Lastly, the potential to discover allosteric modulators of GPCRs is quite intriguing.

4. Discussion

Fragment screening at high concentrations in binding or cellular assays has been attempted for GPCR targets but is very limited in the sensitivity and reliability of the results and also in the ability to distinguish between false positive and useful hits (Albert *et al.*, 2007). The results described here establish that it is possible to screen libraries of fragments against GPCRs and to identify low-affinity hits using StaRs with biophysical screening platforms. The use of such methods allows the identification of very weak binders which are unlikely to be identified using conventional cell-based assays. It is necessary to include suitable controls in all the methods described to eliminate false positives which may be merely nonspecific protein binders. This is readily done by including an alternative membrane protein such as another receptor run in parallel during the screen. The availability of a range of assays also gives the option to screen fragment hits in an orthogonal assay format to triage hits and only focus on progression of useful chemotypes. StaR reagents make fragment-based drug discovery now a viable possibility and open up the potential for identification of new GPCR modulators to receptors that have proved intractable to high-throughput screening across the pharmaceutical industry and also to identify ligands that bind to new binding sites that would be silent in radiochemical binding assays.

StaRs are stabilized into a single chosen receptor conformation, typically the antagonist or agonist conformation (Magnani *et al.*, 2008; Serrano-Vega *et al.*, 2008). StaRs selectively bind ligands that recognize that conformation; that is, antagonists typically bind with higher affinity to antagonist StaRs and agonists bind with higher affinity to agonist StaRs compared to binding to the wild-type receptor. The reason for this is that the stabilization process is carried out in the presence of a ligand which selects a specific conformation, and one of the factors that contribute to the additional thermal stability is the trapping of this bound receptor conformation and reduction in receptor flexibility. This opens up the intriguing possibility of identifying low-affinity fragments that bind selectively to the higher energy agonist state of the receptor that can then be optimized to give functional agonists. This would be almost impossible with wild-type receptors, as low-affinity fragments would be highly unlikely to be detectable in a functional cellular assay, and in a binding assay, any fragments that might be detectable binders are far more likely to bind to the ground state antagonist conformation.

Evolution of fragment hits into lead compounds relies on an understanding of the binding site and the mode of binding of the fragment in that site. Optimization is then possible using structure-based design approaches, where new functionality is added to better fill the site and to pick up further polar interactions. Until recently, there has been very limited structural information available for GPCRs; indeed, the only structure known from this family was that of the visual pigment rhodopsin (Palczewski *et al.*, 2000). Knowledge of how ligands interacted with receptors was limited to models based on homology with rhodopsin or from site-directed mutagenesis experiments. These limitations have precluded optimization of fragment hits using a rational approach. However, in the past 3 years, a range of different technological developments have resulted in the structures of three new GPCRs, all of which are important drug targets; the β1 and β2 adrenergic receptors and the adenosine A_{2A} receptor (Cherezov *et al.*, 2007; Jaakola *et al.*, 2008; Rasmussen *et al.*, 2007; Warne *et al.*, 2008).

StaR proteins are highly suitable reagents for structural biology studies and in our hands have been useful in solving multiple protein–ligand complexes for several GPCR systems. Critically, it has been possible to show that ligands of low affinity (>10 μ*M*) can be crystallized and their structures solved bound to the GPCR in the required conformation (unpublished data). These developments show the potential of StaRs to be used to solve receptor–fragment complexes in the future for use in fragment-based drug discovery for rational, structure-based optimization.

In summary, the studies described herein demonstrating the potential of StaRs for use in fragment-based drug discovery and other developments that will be published elsewhere on their utility for protein–ligand structure solution by X-ray crystallography, give great promise that fragment-based drug discovery is now possible using all of the tool box available for soluble proteins, such as kinases and proteases. Although at an early stage today, the power of fragment-based drug discovery and structure-based drug design applied to GPCRs may deliver important breakthroughs that lead to new therapeutic agents over the coming years.

ACKNOWLEDGMENTS

This work was funded in part by NIH grant GM 071697 (awarded to D. G. M.). The authors thank Malcolm Weir and Andrei Zhukov for proof-reading this chapter. Thanks also to Andrei Zhukov for technical discussions relating to this work.

REFERENCES

Albert, J. S., Blomberg, N., Breeze, A. L., Brown, A. J., Burrows, J. N., Edwards, P. D., Folmer, R. H., Geschwindner, S., Griffen, E. J., Kenny, P. W., Nowak, T., Olsson, L. L., *et al.* (2007). An integrated approach to fragment-based lead generation: Philosophy, strategy and case studies from AstraZeneca's drug discovery programmes. *Curr. Top. Med. Chem.* **7,** 1600–1629.

Arora, A., Rinehart, D., Szabo, G., and Tamm, L. K. (2000). Refolded outer membrane protein A of *Escherichia coli* forms ion channels with two conductance states in planar lipid bilayers. *J. Biol. Chem.* **275,** 1594–1600.

Cherezov, V., Rosenbaum, D. M., Hanson, M. A., Rasmussen, S. G., Thian, F. S., Kobilka, T. S., Choi, H. J., Kuhn, P., Weis, W. I., Kobilka, B. K., and Stevens, R. C. (2007). High-resolution crystal structure of an engineered human β_2-adrenergic G protein-coupled receptor. *Science* **318,** 1258–1265.

Chessari, G., and Woodhead, A. J. (2009). From fragment to clinical candidate—A historical perspective. *Drug Discov. Today* **14,** 668–675.

Congreve, M., Carr, R., Murray, C., and Jhoti, H. (2003). A 'rule of three' for fragment-based lead discovery? *Drug Discov. Today* **8,** 876–877.

Congreve, M., Chessari, G., Tisi, D., and Woodhead, A. J. (2008). Recent developments in fragment-based drug discovery. *J. Med. Chem.* **51,** 3661–3680.

Fruh, V., Zhou, Y., Chen, D., Loch, C., Ab, E., Grinkova, Y. N., Verheij, H., Sligar, S. G., Bushweller, J. H., and Siegal, G. (2010). Application of fragment-based drug discovery to membrane proteins: Identification of ligands of the integral membrane enzyme DsbB. *Chem. Biol.* **17,** 881–891.

Jaakola, V. P., Griffith, M. T., Hanson, M. A., Cherezov, V., Chien, E. Y., Lane, J. R., Ijzerman, A. P., and Stevens, R. C. (2008). The 2.6 angstrom crystal structure of a human A_{2A} adenosine receptor bound to an antagonist. *Science* **322,** 1211–1217.
Magnani, F., Shibata, Y., Serrano-Vega, M. J., and Tate, C. G. (2008). Co-evolving stability and conformational homogeneity of the human adenosine A_{2a} receptor. *Proc. Natl. Acad. Sci. USA* **105,** 10744–10749.
Marquardsen, T., Hofmann, M., Hollander, J. G., Loch, C. M., Kiihne, S. R., Engelke, F., and Siegal, G. (2006). Development of a dual cell, flow-injection sample holder, and NMR probe for comparative ligand-binding studies. *J. Magn. Reson.* **182,** 55–65.
Murali, N., Miller, W. M., John, B. K., Avizonis, D. A., and Smallcombe, S. H. (2006). Spectral unraveling by space-selective Hadamard spectroscopy. *J. Magn. Reson.* **179,** 182–189.
Myszka, D. G. (1999). Improving biosensor analysis. *J. Mol. Recognit.* **12,** 279–284.
Navratilova, I., Sodroski, J., and Myszka, D. G. (2005). Solubilization, stabilization, and purification of chemokine receptors using biosensor technology. *Anal. Biochem.* **339,** 271–281.
Navratilova, I., Dioszegi, M., and Myszka, D. G. (2006). Analyzing ligand and small molecule binding activity of solubilized GPCRs using biosensor technology. *Anal. Biochem.* **355,** 132–139.
Palczewski, K., Kumasaka, T., Hori, T., Behnke, C. A., Motoshima, H., Fox, B. A., Le Trong, I., Teller, D. C., Okada, T., Stenkamp, R. E., Yamamoto, M., and Miyano, M. (2000). Crystal structure of rhodopsin: A G protein-coupled receptor. *Science* **289,** 739–745.
Rasmussen, S. G., Choi, H. J., Rosenbaum, D. M., Kobilka, T. S., Thian, F. S., Edwards, P. C., Burghammer, M., Ratnala, V. R., Sanishvili, R., Fischetti, R. F., Schertler, G. F., Weis, W. I., *et al.* (2007). Crystal structure of the human β_2 adrenergic G-protein-coupled receptor. *Nature* **450,** 383–387.
Robertson, N., Jazayeri, A., Errey, J., Baig, A., Hurrell, E., Zhukov, A., Langmead, C. J., Weir, M., and Marshall, F. H. (2011). The properties of thermostabilised G protein-coupled receptors (StaRs) and their use in Drug Discovery. *Neuropharmacology* **60,** 36–44.
Serrano-Vega, M. J., Magnani, F., Shibata, Y., and Tate, C. G. (2008). Conformational thermostabilization of the β_1-adrenergic receptor in a detergent-resistant form. *Proc. Natl. Acad. Sci. USA* **105,** 877–882.
Shibata, Y., White, J. F., Serrano-Vega, M. J., Magnani, F., Aloia, A. L., Grisshammer, R., and Tate, C. G. (2009). Thermostabilization of the neurotensin receptor NTS1. *J. Mol. Biol.* **390,** 262–277.
Vanwetswinkel, S., Heetebrij, R. J., van Duynhoven, J., Hollander, J. G., Filippov, D. V., Hajduk, P. J., and Siegal, G. (2005). TINS, target immobilized NMR screening: An efficient and sensitive method for ligand discovery. *Chem. Biol.* **12,** 207–216.
Warne, T., Chirnside, J., and Schertler, G. F. (2003). Expression and purification of truncated, non-glycosylated turkey beta-adrenergic receptors for crystallization. *Biochim. Biophys. Acta* **1610,** 133–140.
Warne, T., Serrano-Vega, M. J., Baker, J. G., Moukhametzianov, R., Edwards, P. C., Henderson, R., Leslie, A. G., Tate, C. G., and Schertler, G. F. (2008). Structure of a β_1-adrenergic G-protein-coupled receptor. *Nature* **454,** 486–491.

CHAPTER SIX

Using Computational Techniques in Fragment-Based Drug Discovery

Renee L. DesJarlais

Contents

Abstract

Fragment-based drug discovery has emerged over the past 15 years as an effective lead discovery paradigm that is complementary to traditional high-throughput screening. The starting point for fragment-based drug discovery is the identification of low-molecular weight, typically low-affinity compounds that bind to a target of interest. These fragments can then be elaborated by growing or linking to create compounds with high affinity and selectivity. A wide variety of techniques from the computational chemistry tool chest can be applied in a fragment-based project. The computational tools are equally useful in combination with experimental-binding determination or in a completely *in silico* design procedure. This chapter will outline these techniques, their utility, and their validation in the design of novel lead compounds.

Structural Biology, Johnson & Johnson Pharmaceutical Research and Development, L.L.C., Spring House, Pennsylvania, USA

Methods in Enzymology, Volume 493
ISSN 0076-6879, DOI: 10.1016/B978-0-12-381274-2.00006-6

1. Introduction

A wide variety of computational methods can be applied throughout a fragment-based campaign. A general schematic of the FBDD process is shown in Fig. 6.1. Initially, one must identify a set of fragments to be screened, a target to be screened, and a method for detecting fragment binding to the target. With these inputs, one then embarks on a cycle of testing fragments for binding, selecting fragments for elaboration, growing or linking fragments, synthesizing the designed compounds, and back to evaluating the next-generation molecules for binding to the target. Typically, two or three cycles of design are required to identify a compound that would qualify as a lead to be optimized through standard medicinal chemistry. Computational chemistry tools can be used in the design of the initial fragment library, the identification of fragments that bind to a target, the selection of fragments for expansion, and in the design of subsequent compounds. Many of the computational methods that have been applied to FBDD are proprietary code or academic code and are not generally accessible. This chapter will focus on the use of methods that are widely available.

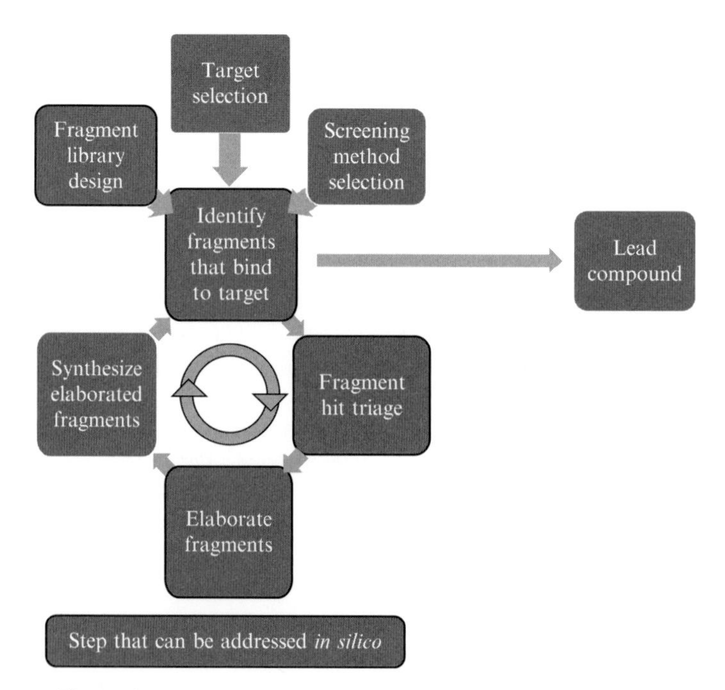

Figure 6.1 Generic fragment-based drug discovery flowchart.

2. Fragment Library Design

The choice of compounds for any screening campaign has a clear effect on the outcome. One seeks a diverse library of fragments that when combined or elaborated will produce molecules compatible with optimization into drugs. The first step in identifying the molecules that make up a fragment screen is to choose a source of molecules. Typically, the source of compounds would be a commercial collection, for example, a vendor catalog, a compilation of commercial sources such as the ZINC database (Irwin and Shoichet, 2005), or a proprietary corporate collection. In order to computationally evaluate the source molecules, their structures and data need to be captured in a machine-readable data format. Both structure data files[1] and simplified molecular input line entry specification (SMILES™) files[2] are formats containing two-dimensional connectivity information and are accepted by most commercial software (see Table 6.1 for software companies). For the purposes of this discussion, the set of molecules from which the fragments will be chosen is referred to as the set of available molecules.

The set of available molecules now needs to be filtered for those that meet the fragment definition. The molecules in a fragment screen are small typically less than 250–300 Da in molecular weight. Depending upon the screening strategy, fragments may need to be soluble at high concentrations in order to detect weak binding. In analogy with Lipinski's "Rule of Five"

Table 6.1 Major software companies and their official Web sites

Software company/academic labs	Web site
Accelrys	www.accelrys.com
BioSolveIT	www.biosolveit.com
Cambridge Crystallographic Data Centre	www.ccdc.cam.ac.uk
Chemical Computing Group, Inc.	www.chemcomp.com
Daylight Chemical Information Systems, Inc.	www.daylight.com
OpenEye Scientific Software	www.eyesopen.com
Schrödinger, LLC	www.schrodinger.com
Tripos, L.C.	www.tripos.com

[1] For a description of structure data file format, commonly known as SD or SDF format, see the Symyx Solutions, Inc. web site: www.symyx.com/downloads/public/ctfile/ctfile.pdf.

[2] For a description of SMILES™ file format see the Daylight Chemical Information Systems, Inc. web site: www.daylight.com/dayhtml/doc/theory/theory.smiles.html.

(Lipinski *et al.*, 2001) for predicting compounds likely to be permeable, Congreve *et al.* (2003) have proposed a "Rule of Three" for fragments to be screened. The "Rule of Three" was derived from an analysis of hits from FBDD screens using a variety of techniques. The rule states that fragments should be less than 300 Da in molecular weight, possess less than or equal to three hydrogen bond acceptors, less than or equal to three hydrogen bond donors, and cLogP less than or equal to three. Their analysis indicates that additional filters restricting rotatable bonds to less than or equal to three and polar surface area to less than or equal to 60 $Å^2$ would provide compounds more likely to be fragment hits (Congreve *et al.*, 2003). The features embodied in the Rule of Three have been widely accepted, and most researchers use some version of the rule as part of the selection criteria for their fragment library. All of these properties can be readily calculated from a structure data file or SMILES™ file with commercial computational chemistry software packages.

Several studies have shown a linear relationship between molecular size and affinity, with every nonhydrogen atom contributing ~0.3 to the pK_d of the compound (Hajduk, 2006; Kuntz *et al.*, 1999). Hajduk (2006) points out the clear implications of this correlation. If the goal for a project is single-digit nanomolar affinity ($pK_d > 8$) in a final compound that is less than 500 Da in molecular weight (~35 nonhydrogen atoms), then the K_d of a 300 Da (~21 nonhydrogen atoms) fragment starting point must be at least 30 μ*M* ($pK_d = 4.5$). This fact, plus the higher probability of finding hits with smaller molecules (Hann *et al.*, 2001), suggests that skewing the fragments in the set to molecular weights of 250 Da, or lower, is advisable. The set of molecules remaining after the nonfragment molecules are eliminated will be referred to as the set of available fragments.

It is advisable to filter the set of available fragments to eliminate those that are reactive, contain known toxicophores, or functionality that would make them an undesirable starting point from a medicinal chemistry point of view. The last of these characteristics will be, to some extent, subjective and should be decided upon with input of the local chemistry team. Examples of substructures from each of these categories are shown in Table 6.2.

Having eliminated known reactive and undesirable functionality, the remaining task is to limit the set to a feasible number for the screening methods. For a virtual screen, one will likely be able to examine all remaining fragments. For resource intensive screens, such as X-ray crystallography, the number will need to be limited to less than 1000. The main consideration in reducing the number is to maintain maximum diversity while retaining drug-likeness. Chemical diversity is assessed using traditional methods to compare molecular similarity. The common similarity metrics use some type of two-dimensional fingerprints. Fingerprints are bit

Table 6.2 Examples of substructural filters for undesirable functional groups

Reactive substructures 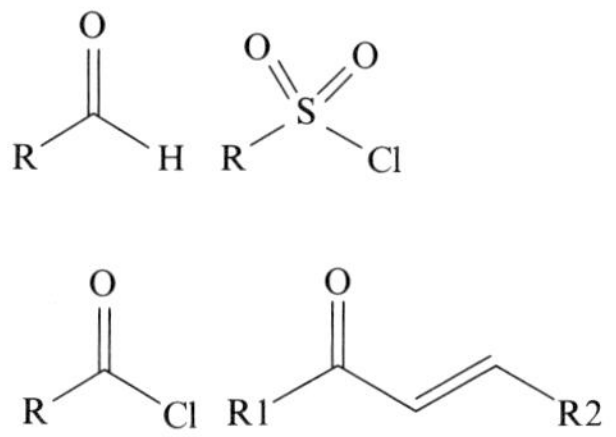	Toxicophores 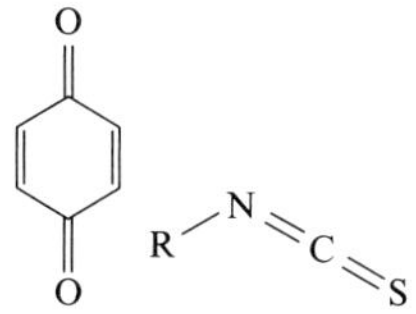	Undesirable substructures 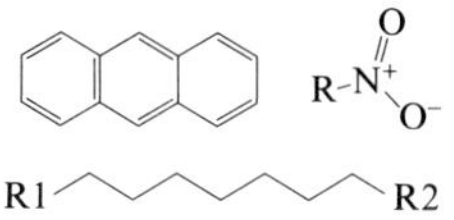

Table 6.3 Molecular fingerprint example

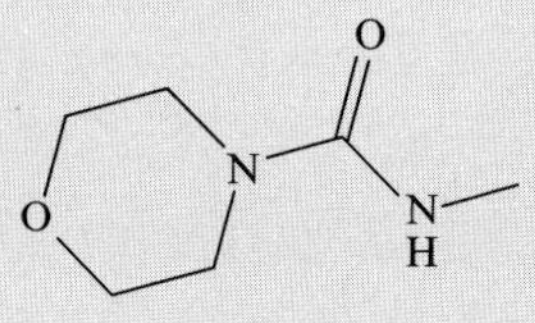

Zero-bond paths	C (6), N (2), O (2)
One-bond paths	C–C (2), C–N (5), C–O (2), C=O (1)
Two-bond paths	C–C–N (2), C–C–O (2), C–N–C (4), C–O–C (1), N–C–N (1), N–C=O (2)
Three-bond paths	CNCN (3), CNC=O (3), CNCC (4), NCCO (2), CCOC (2)

The number in parentheses indicates the number of times the bond path occurs in the molecule.

strings that encode the presence or absence of certain substructures, elements, bonded fragments, or chemical environment. The bit string is a set of 1's and 0's allowing facile comparison. Commonly used fingerprints include *Molecular Design Limited*, or MDL keys (Durant *et al.*, 2002), *Daylight* fingerprints (Shemetulskis *et al.*, 1996; www.daylight.com), and *Extended Connectivity Fingerprints* (Rogers and Hahn, 2010; www.accelrys.com). As an example of the information encoded in a molecular fingerprint, the bonded paths of atropine from zero to three bonds are shown in Table 6.3. Any of these fingerprints capture the atom and bond-type information as well as the local chemical environment around each atom and offer a good choice for the user designing an FBDD library. The choice of method can be one of convenience. A common metric to compare fingerprints is the *Tanimoto* coefficient (Tanimoto, 1957), which is defined as the number of bits the fingerprints have in common divided by the sum of the number of bits unique to molecule 1, the number of bits unique to molecule 2, and the number of bits in common. This metric varies between 1 for identical molecules and 0 for molecules that do not have any elements in common. All the major software companies have implemented fingerprint methods and algorithms for selecting a diverse subset.

Various investigators include an assessment of drug-likeness either as another filter before making their diverse selection or in conjunction with a diversity assessment. Drug-likeness is not a characteristic that can be calculated directly from a commercial software package. In order to select fragments that are drug-like, some groups have taken the approach of dissecting known drugs into their component fragments (Hartshorn *et al.*, 2005; Kolb and Caflisch, 2006). Figure 6.2 illustrates how a particular drug molecule might be broken down into fragments. Once the set of know drugs has been dissected and the unique fragments identified, the fragments can be retrieved from the available fragments list. An alternative is to assess the substructural similarity of fragments to a set of known drugs (Chapter 1, Tounge and Parker, 2011). The substructural similarity is a special case of the fingerprint similarity metric discussed above. In this case,

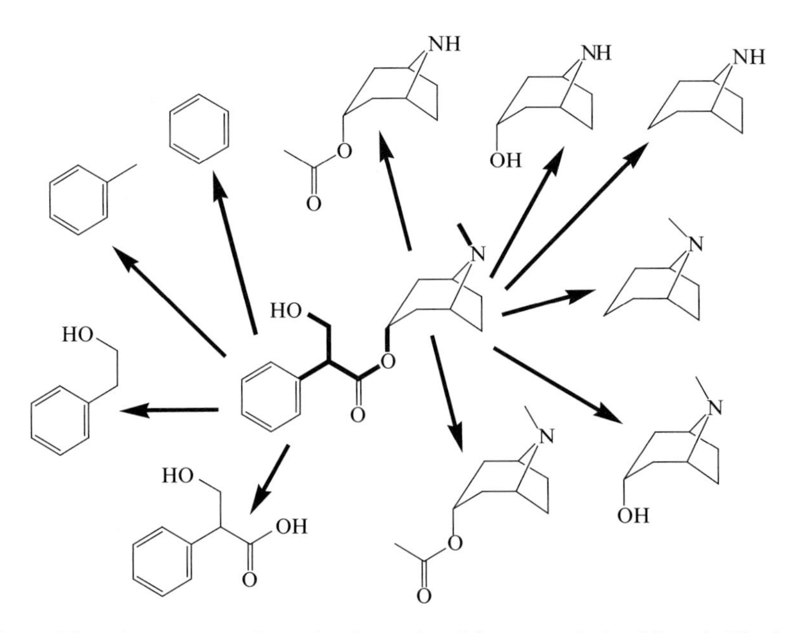

Figure 6.2 The structure of atropine (center) and fragments derived from it. The bold bonds in the atropine structure indicate the bonds broken to generate the fragments.

the subsimilarity metric is the number of bits in common between the fragment and the drug molecule divided by the number of bits total in the fragment (Tounge and Reynolds, 2004). The value of subsimilarity will vary from 1 where the fragment is a substructure of the drug and 0 where the fragment has no elements in common with the drug. Table 6.4 illustrates the subsimilarity and classic similarity values for a set of fragments when compared to atropine using AtomType/Path fingerprints of length 4 as calculated in *PipelinePilot* V7.5 (Hassan *et al.*, 2007; www.accelrys.com).

3. *In Silico* Fragment Screening

Experimental fragment screening requires significant investment in equipment and resources. In addition to access to an X-ray beam line, high-field nuclear magnetic resonance, or surface plasmon resonance equipment, expertise in the interpretation of these specialized data is important for obtaining the best results. Although *in silico* screening also requires expertise to appropriately set up computational experiments and interpret results, the investment in computer hardware and software is much less than required for X-ray, nuclear magnetic resonance, or surface plasmon resonance detection methods. Another practical advantage for *in silico* methods is that they can be applied to any system for which a high-quality three-dimensional model is available. Computational screening provides an opportunity for structure-based optimization that is lacking with an affinity technique such as surface plasmon resonance used alone and without the limitations of X-ray crystallography (protein crystals) or nuclear magnetic resonance (molecular weight and solubility). Taken together, these attributes make screening with a computational method an attractive choice for many groups and many targets. An *in silico* screen may be used as a prescreen for an experimental technique or may drive several cycles of optimization, resulting in molecules expected to provide binding hits with similar affinity as hits from a high-throughput screen.

3.1. Preparation of the protein structure

To begin an *in silico* screen, one must have a crystal structure or a high-quality homology model of the target protein. A protein structure from X-ray crystallography typically does not include hydrogens and may have some ambiguities with respect to the identity of atoms in the amide side chains of asparagine and glutamine. Side-chain protonation, placement of hydroxyl hydrogens, and the tautomeric state of histidine must all be carefully considered. The commercial packages all have a series of dialog boxes to help the user through the process of adding hydrogen atoms, insuring correct protonation states, and assigning tautomers and amide

Table 6.4 Subsimilarity and similarity values for a set of fragments compared to atropine

Target structure

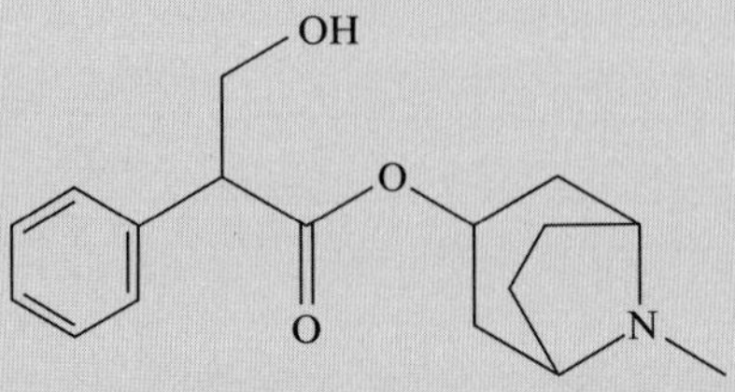

Fragment structure	Subsimilarity SA/(SA+SB)	Similarity SA/(SA+SB+SC)	SA	SB	SC
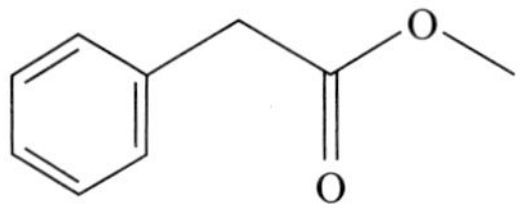	1.00	0.49	42	0	44
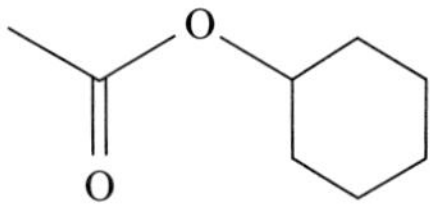	1.00	0.41	35	0	51
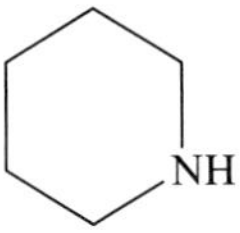	1.00	0.23	20	0	66
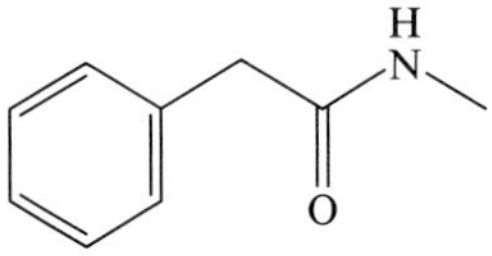	0.66	0.29	29	15	57
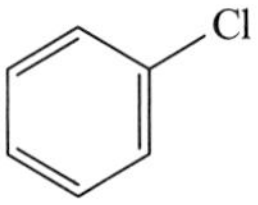	0.43	0.06	6	8	80
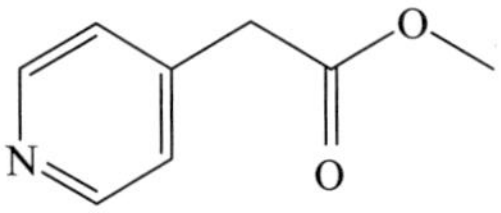	0.75	0.43	43	14	43
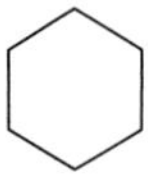	1.00	0.06	5	0	81

SA, number of bits in common between target and fragment; SB, number of bits unique to fragment; SC, number of bits unique to target.

side-chain rotomers that optimally satisfy internal hydrogen bonds. Incorrect placement of hydrogen atoms or heteroatoms can significantly alter the electrostatic environment in a binding site, leading to incorrect placement and/or ranking of fragments. An example of the kind of correction that may be required in preparing a protein for further computation is shown in Fig. 6.3. Figure 6.3A shows a close-up of the hydrogen bonding environment of asparagine 44 and glutamine 57 in hen egg white lysozyme (Datta *et al.*, 2001; PDBID 1jis). As deposited in the PDB (Berman *et al.*, 2000; www.pdb.org), the side chain NH_2 of glutamine 57 makes hydrogen bonds with the backbone carbonyl oxygens of alanine 42 and glycine 54. This arrangement leaves the side-chain oxygen of glutamine 57 only 2.7 Å from the side-chain oxygen of asparagine 44, resulting in an unfavorable interaction for these two hydrogen bond acceptors. Rotation of the χ^2 dihedral angle of asparagine 44 removes this unfavorable interaction and replaces it with a hydrogen bond between the side-chain NH_2 of asparagine 44 and the side-chain oxygen of glutamine 57 (Fig. 6.3B).

At present, efficient *in silico* fragment screening requires use of a rigid target structure. The possibility exists that the X-ray structure or initial homology model is not the optimum conformation for compound binding. Proteins are flexible and a single conformation may not be appropriate for binding all ligands. Unfortunately, it is not possible simply to view a single protein structure and know whether it is the best for ligand binding. A combination of protein molecular dynamics and limited fragment docking has proved

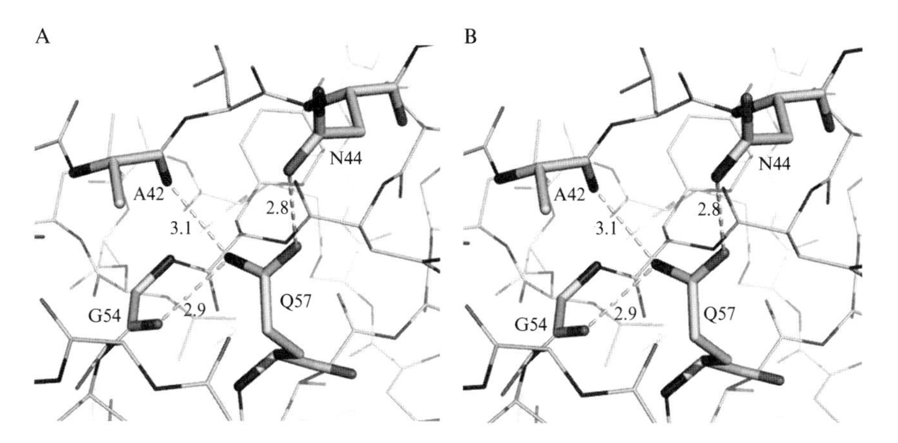

Figure 6.3 (A) The structure of hen egg white lysozyme (Datta *et al.*, 2001; PDBID 1jis) original atom positions as deposited in the PDB (Berman *et al.*, 2000; www.pdb.org) for asparagine 44 showing unfavorable interactions with glutamine 57. (B) The optimized atom positions for asparagine 44 with the amide side chain flipped making a favorable hydrogen bond with glutamine 57. Carbon atoms are colored light gray, oxygen atoms are colored medium gray, and nitrogen atoms are colored black.

useful in exploring target conformations and selecting a conformation for subsequent *in silico* screening. Ekonomiuk *et al.* (2009) docked several small probe molecules (benzene, methylguanidine, and 2-phenylimidazoline) into conformations of West Nile virus NS3 protease generated in a molecular dynamics simulation. The conformation for which the probe molecules were predicted to bind with lowest energy as evaluated with SEED (Majeux *et al.*, 1999, 2001) was selected for use in a standard virtual screen. Their results using the molecular dynamics-generated protein structure indicate that this procedure allowed the group to identify inhibitors that would have been missed if the X-ray structure had been used. If conformational flexibility is known or suspected, it is well worth the effort to perform molecular dynamics and examine conformations seen in the simulation. All the major software vendors offer molecular dynamics capability.

3.2. Screening for bound fragments

Once one or a small set of protein conformations have been identified for use as target structures, the user must choose a method for placing and ranking fragments. Table 6.5 lists some of the programs for fragment screening that are generally available from commercial and academic sources. The choice of computational technique for an *in silico* fragment screen is often governed by what software is available and personal experience. The success of a given *in silico* screen is likely to depend more on the careful setup of the experiment than the particular method. Strategies for selecting parameters and evaluating whether the computational results are reliable are discussed in Section 3.3.

3.3. Choosing parameters wisely

Whenever possible, it is important to calibrate your computational method with known ligands and fragments thereof. Such calibration calculations are not an exercise that will bias the end results. Instead, an initial calibration study will allow the user to choose parameters appropriate for the target of interest as well as to provide guidance with respect to how or whether scores relate to affinity for this system. It is important to include fragments as part of the set for calibration, as parameters best for identifying fragment binders may not be the same as for docking lead- or drug-sized molecules. This process is analogous to altering experimental conditions to pick up weak binders. For Glide, two separate studies have shown the simpler *Standard-Precision*, or SP, scoring scheme to perform better in fragment placement than the more elaborate *Extra-Precision*, or XP, scoring scheme (Kawatka *et al.*, 2009; Sandor *et al.*, 2010).

Parameters for a computational screen will typically fall into two categories: those that influence how many poses are explored (sampling) and

Table 6.5 Major companies/academic groups and their software that can be used in fragment screening

Software company or academic lab	Products	Type of calculation	Literature reference
Accelrys	LUDI	*De novo* design	Böhm (1992)
	MCSS	Ensemble molecular mechanics	Miranker and Karplus (1991)
	LigandFit	Docking	Venkatachalam *et al.* (2003)
BioSolveIT	FlexX	Docking	Rarey *et al.* (1996)
	FlexNovo	*De novo* design	Degen and Rarey (2006)
Cambridge Crystallographic Data Centre	GOLD	Docking	Jones *et al.* (1997), Verdonk *et al.* (2003)
Chemical Computing Group, Inc.	Multifragment search	Ensemble molecular mechanics	MOE (2009)
	Ligand–Receptor Docking	Docking	
OpenEye Scientific Software	FRED	Docking	McGann *et al.* (2003)
Schrödinger, LLC	GLIDE	Docking	Friesner *et al.* (2004), Halgren *et al.* (2004)
Tripos, L.P.	Surflex-Dock	Docking	Jain (2003)
UCSF	DOCK	Docking	Kuntz *et al.* (1982, 1994)

those that influence the ranking of those poses. Adjusting the sampling is typically a tradeoff between insuring that the orientation and conformation space of the fragment around the target is covered and the time that it takes to perform the calculation. Because methods have different algorithms, it is not possible to present specific settings. A good approach would be to test different sampling parameters with several fragments with different balances of polar versus nonpolar character. As sampling increases, you expect to see a decrease in energy of the best poses eventually leveling off with more sampling. Knowing how much sampling is necessary to find the lowest energy poses, one can then decide on the amount of sampling that will give the best orientations in an acceptable time.

Setting parameters for ranking or scoring ligands can be trickier. Most scoring functions have been parameterized for replicating the crystallographically determined binding mode, for correlating with experimentally determined binding energies, or both. It is advisable to start with default parameter sets or parameter sets that are known to have performed well on your target class.[3] If there are known ligands for the target, calibration studies can be conducted on these molecules and related fragments to insure that known poses receive high ranking and/or whether correlation with known binding affinities can be achieved. Several studies have found better performance in identifying a binding mode than in predicting binding affinity (Cummings *et al.*, 2005; Warren *et al.*, 2006) when drug-sized molecules are used in an *in silico* screen.

3.4. What to watch out for

Results from a computational study should always be given a "gut check" to assess reasonableness. One should be skeptical of a method that returns molecules with poor geometries, fragments that clash with the target, or fragments that have clear mismatches of their chemistry with the target, for example, the juxtaposition of a hydrophilic group with a hydrophobic subsite. Any such discrepancies suggest that parameters are incorrectly set. While such checks might be informed by interactions of known ligands, one should be wary of introducing constraints for specific interactions either in the ranking or as a filter after the initial computational screen. This type of constraint biases the search toward known ligands and lessens the power of FBDD to find new starting points.

4. Hit Triage

Depending on the target, a fragment screen may produce many hits and these hits may identify several sites on the protein surface. When multiple fragments' binding sites are found, one must choose how to prioritize the sites' follow-up as well as prioritizing the fragments in them. The known binding or active site typically takes priority, since assays to detect the biological effect of binding to that site will be established. For other sites, two questions need to be addressed: Is the site of a size and complexity to bind a drug-sized molecule and will binding at this site produce a relevant biological effect? The first of these questions is related to the issue of "druggability." The study by Cheng *et al.* (2007) presents a druggability index that attempts to predict maximal affinity achievable at a

[3] For example, in the program GOLD (Jones et al., 1997; Verdonk et al., 2003), one can choose parameter sets optimized to provide the best performance for particular target classes such as kinases or proteases.

given binding site. The amount of hydrophobic surface area and a measure of curvature of the site are used to distinguish easily druggable targets from those that are more difficult. While one can argue with the specific formalism, this definition of druggability captures the common wisdom that a good small-molecule binding site will be a cleft capable of binding a drug-sized molecule and have a balance of hydrophobic and hydrophilic surface to drive desolvation of a ligand and specific binding. One should look for these properties in assessing a potential binding site. The question of biological relevance is only addressable with experiment. Such experiments need to be carefully planned, as the full functionality of a target may not be detectable with an *in vitro* activity assay.

Once the site or sites are chosen, other ligand-specific considerations come into play in choosing the fragments for further study. The concept that some molecules make more efficient use of their constituent atoms in interacting with a target has a long history (Andrews *et al.*, 1984; Kuntz *et al.*, 1999). This idea has been formalized as ligand efficiency, which can be defined as the binding affinity divided by the number of nonhydrogen atoms in the ligand (Hopkins *et al.*, 2004). It has become an important parameter driving decisions in fragment-based efforts and in medicinal chemistry more generally (Carr *et al.*, 2005; Leeson and Springthorpe, 2007), spawning several refinements and extensions of the concept (Abad-Zapatero and Metz, 2005; Leeson and Springthorpe, 2007; Reynolds *et al.*, 2008). An overview of various ligand efficiency implementations can be found in Chapter 15 (Orita *et al.*, 2011). All things being equal, the principle embodied in ligand efficiency says that smaller is better. As discussed above for fragments to be able to be optimized to nanomolar binders, it is essential to select the maximal binding in the minimal molecular weight. Examining ligand efficiency makes this sort of prioritization simple. In general, ligand efficiency values of $\sim$0.3 pK_d (or pIC_{50}) per nonhydrogen atom would be a reasonable starting point. One should expect this value to be approximately constant as the fragment is evolved toward a lead (Hajduk, 2006).

Assuming that the initial fragment collection has excluded reactive and non-drug-like elements from the beginning, further property filtering will not be necessary at this stage. Some groups choose to focus on fragments that make interactions with the target previously shown to be important in ligand binding (Hartshorn *et al.*, 2005; Law *et al.*, 2009; Teotico *et al.*, 2009). Bias based on known interactions can increase the likelihood that newly designed molecules will share structural similarity and potentially the liabilities of previously identified compounds. If the identified fragments are sufficiently distinct from the known ligands, such bias is an acceptable compromise between novelty of the designed compounds and being able to quickly discover ligands with good affinity.

5. Hit Follow-Up

There are two basic ways to follow-up on fragment hits: select existing compounds for testing that include the fragment as a substructure or design compounds that build onto or join fragment hits. The former is really a way to use a fragment screen to create a more efficient high-throughput screen. One is limited to finding molecules that already exist in a company inventory or are purchasable. Testing of existing compounds may be a practical approach for groups where both synthetic capabilities and screening capabilities are limiting. To fully take advantage of the knowledge gained from a fragment screen, it is desirable to design and synthesize compounds specifically targeting your site of interest. The design of compounds increases greatly the chances that they will be patentable.

5.1. Selection of existing compounds

Several groups have been successful in using standard substructure searching and similarity searching methods to follow-up on fragment hits (see Section 2 for a description of similarity searching methods). The Vertex SHAPES method explicitly includes following up their hits from nuclear magnetic resonance with known compounds identified based on substructure and similarity searching around the fragment hits (Lepre, 2007). Law *et al.* (2009) discuss their application of FBDD to lead discovery on several targets. For many of these targets, substructure or similarity searching of known compounds is part of the process leading to the successful identification of active compounds based on initial fragment hits.

5.2. Design of compounds for follow-up

Design and synthesis of compounds based on a fragment hit is more resource-intensive than choosing compounds from an existing set, but it holds the potential advantage of getting the project directly into a patentable chemical space. The buildup of a fragment hit into a lead compound ready for optimization typically takes the form of extension of the fragment or linking of fragments. These approaches may be used in any combination. Computational tools can assist with each of these processes, although often design is carried out by interactive examination of the target binding site and the poses of one or more fragments.

Table 6.6 lists commercially available compound design tools categorized based on their approach, either growing or linking. The growing tools take fragments one at a time and attempt append atoms or groups to make additional favorable interactions with the target. Linking tools look at the set

Table 6.6 Major companies and their software that can be used in fragment expansion

Software company/ academic lab	Product	Type of calculation	Literature reference
Accelrys	LUDI	Growing	Böhm (1992)
	HOOK	Linking	Eisen *et al.* (1994)
BioSolveIT	FlexNovo	Growing	Degen and Rarey (2006)
Chemical Computing Group, Inc.	LigX	Growing	MOE (2009)
Tripos, L.P.	EA-Inventor	Growing	Liu *et al.* (2007)

of fragments for pairs or sets that are positioned so that they can easily be joined. The majority of the tools are of the growing type reflecting the predominant use of this approach. Although the theoretical improvement in free energy from an appropriate linking of two fragments is large, the chance of finding two fragments at the appropriate geometry is low, and small changes needed to accommodate the link might disrupt the binding of one or both fragments, negating some of the theoretical benefits of linking them.

An elaborated fragment needs to be synthesizable and the synthesis should be facile. There will be little support from medicinal chemistry colleagues for a long synthesis when starting from weak binders. Although some academic and pharmaceutical industry groups have developed software to determine whether a molecule can be easily made (Gillett *et al.*, 1995; Vinkers *et al.*, 2003), the assessment is generally left to synthetic chemists working in close collaboration with computational chemists. At this point, there is no commercial software that can be relied upon to evaluate how easily a given molecule could be synthesized.

As the design process moves toward more drug-like molecules, it is advisable to reassess the molecular properties. The team should avoid designing in any functional groups associated with toxicity. Rule of five parameters (Lipinski *et al.*, 2001) as well as the number of rotatable bonds and polar surface area should be kept in a reasonable range for lead-like molecules. Table 6.7 lists the values of these parameters associated with permeable drugs (Lipinski *et al.*, 2001; Veber *et al.*, 2002). It is important to keep an eye on ligand efficiency once affinity or activity is being assessed. These considerations are the same as in any hit-to-lead or lead optimization program.

Table 6.7 Computed parameters and values associated with permeable drugs

Parameter	Desired value	Reference
Number of hydrogen bond donors (NH, OH)[a]	<5 drugs	Lipinski *et al.* (2001)
Number of hydrogen bond acceptors (N, O)[a]	<10	Lipinski *et al.* (2001)
Molecular weight[a]	<500 Da	Lipinski *et al.* (2001)
logP[a]	<5	Lipinski *et al.* (2001)
Number of rotatable bonds	<10	Veber *et al.* (2002)
Polar surface area	<140 Å^2	Veber *et al.* (2002)

[a] Part of Lipinski's Rule of Five.

6. Iteration

Once a first round of molecules is selected or designed and synthesized, the FBDD cycle goes back to evaluation of the interaction between each new molecule and the target. The interaction may be assessed experimentally or predicted computationally. Typically, two to three cycles of design are necessary to discover a molecule that could be classified as a lead compound and prompt a medicinal chemistry lead optimization effort.

7. Conclusion

Computational searching methods can provide an easy entry into the world of fragment-based design. Calculations can typically be run on existing hardware and software is readily available. Although the literature offers few examples of purely computational FBDD campaigns, there are reports of hits found computationally that can be developed into lead molecules of sufficient interest to spur a lead optimization campaign (e.g., Böhm *et al.*, 2000 and Teotico *et al.*, 2009). In addition, computational methods can play a key role in selecting a set of fragments and in the structure-based elaboration of a fragment hit identified by experimental screening whatever the screening method may be. As our ability to accurately assess binding energies improves, computational FBDD will meet with increasing success.

REFERENCES

Abad-Zapatero, C., and Metz, J. T. (2005). Ligand efficiency indices as guideposts for drug discovery. *Drug Discov. Today* **10,** 464–469.

Andrews, P., Craik, D. J., and Martin, J. L. (1984). Functional group contributions to drug–receptor interactions. *J. Med. Chem.* **27,** 1648–1657.

Berman, H. M., Westbrook, J., Feng, Z., Gilliland, G., Bhat, T. N., Weissig, H., Shindyalov, I. N., and Bourne, P. E. (2000). The Protein Data Bank. *Nucleic Acids Res.* **28,** 235–242.

Böhm, H.-J. (1992). The computer program LUDI: A new simple method for the *de novo* design of enzyme inhibitors. *J. Comput. Aided Mol. Des.* **6,** 61–78.

Böhm, H.-J., Boehringer, M., Bur, D., Gmuender, H., Huber, W., Klaus, W., Kostrewa, D., Kuehne, H., Luebbers, T., Meunier-Keller, N., and Mueller, F. (2000). Novel inhibitors of DNA gyrase: 3D structure based biased needle screening, hit validation by biophysical methods, and 3D guided optimization. A promising alternative to random screening. *J. Med. Chem.* **43,** 2664–2674.

Carr, R. A., Congreve, M., Murray, C. W., and Rees, D. C. (2005). Fragment-based lead discovery: Leads by design. *Drug Discov. Today* **10,** 987–992.

Cheng, A. C., Coleman, R. G., Smyth, K. T., Cao, Q., Soulard, P., Caffrey, D. R., Salzberg, A. C., and Huang, E. S. (2007). Structure-based maximal affinity model predicts small-molecule druggability. *Nat. Biotech.* **25,** 71–75.

Congreve, M., Carr, R., Murray, C., and Jhoti, H. (2003). A 'Rule of Three' for fragment-based lead discovery? *Drug Discov. Today* **8,** 876–877.

Cummings, M. D., DesJarlais, R. L., Gibbs, A. C., Mohan, V., and Jaeger, E. P. (2005). Comparison of automated docking programs as virtual screening tools. *J. Med. Chem.* **48,** 962–976.

Datta, S., Biswal, B. K., and Vijayan, M. (2001). The effect of stabilizing additives on the structure and hydration of proteins: A study involving tetragonal lysozyme. *Acta Crystallogr. D* **57,** 1614–1620.

Degen, J., and Rarey, M. (2006). Flexnovo: Structure-based searching in large fragment spaces. *ChemMedChem* **1,** 854–868.

Durant, J. L., Leland, B. A., Henry, D. R., and Nourse, J. G. (2002). Reoptimization of MDL keys for use in drug discovery. *J. Chem. Inf. Comput. Sci.* **42,** 1273–1280.

Eisen, M. B., Wiley, D. C., Karplus, M., and Hubbard, R. E. (1994). HOOK: A program for finding novel molecular architectures that satisfy the chemical and steric requirements of a macromolecule binding site. *Proteins* **19,** 199–221.

Ekonomiuk, D., Su, X.-C., Ozawa, K., Bodenreider, C., Lim, S. P., Otting, G., Huang, D., and Caflisch, A. (2009). Flaviviral protease inhibitors identified by fragment-based library docking into a structure generated by molecular dynamics. *J. Med. Chem.* **52,** 4860–4868.

Friesner, R. A., Banks, J. L., Murphy, R. B., Halgren, T. A., Klicic, J. J., Mainz, D. T., Repasky, M. P., Knoll, E. H., Shelley, M., Perry, J. K., Shaw, D. E., Francis, P., *et al.* (2004). Glide: A new approach for rapid, accurate docking and scoring. 1. Method and assessment of docking accuracy. *J. Med. Chem.* **47,** 1739–1749.

Gillett, V. J., Myatt, G., Zsoldos, Z., and Johnson, A. P. (1995). SPROUT, HIPPO and CAESA: Tools for de novo structure generation and estimation of synthetic accessibility. *Perspect. Drug Discov. Des.* **3,** 34–50.

Hajduk, P. J. (2006). Fragment-Based drug design: How big is too big? *J. Med. Chem.* **49,** 6972–6976.

Halgren, T. A., Murphy, R. B., Friesner, R. A., Beard, H. S., Frye, L. L., Pollard, W. T., and Banks, J. L. (2004). Glide: A new approach for rapid, accurate docking and scoring. 2. Enrichment factors in database screening. *J. Med. Chem.* **47,** 1750–1759.

Hann, M. M., Leach, A. R., and Harper, G. (2001). Molecular Complexity and its impact on the probability of finding leads for drug discovery. *J. Chem. Inf. Comput. Sci.* **41,** 856–864.

Hartshorn, M. J., Murray, C. W., Cleasby, A., Frederickson, M., Tickle, I. J., and Jhoti, H. (2005). Fragment-based lead discovery using X-ray crystallography. *J. Med. Chem.* **48,** 403–413.

Hassan, M., Brown, R. D., Varma-O'Brien, S., and David Rogers, D. (2007). Cheminformatics analysis and learning in a data pipelining environment. *Mol. Divers.* **10,** 283–299.

Hopkins, A. L., Groom, C. R., and Alex, A. (2004). Ligand efficiency: A useful metric for lead selection. *Drug Discov. Today* **9,** 430–431.

Irwin, J. J., and Shoichet, B. K. (2005). ZINC—A free database of commercially available compounds for virtual screening. *J. Chem. Inf. Model.* **45,** 177–182.

Jain, A. N. (2003). Surflex: Fully automatic flexible molecular docking using a molecular similarity-based search engine. *J. Med. Chem.* **46,** 499–511.

Jones, G., Willet, P., Glen, R. C., Leach, A. R., and Taylor, R. (1997). Development and validation of a genetic algorithm for flexible docking. *J. Mol. Biol.* **267,** 727–748.

Kawatka, S., Wang, H., Czerminski, R., and Joseph-McCarthy, D. (2009). Virtual fragment screening: An exploration of various docking and scoring protocols for fragments using glide. *J. Comput. Aided Mol. Des.* **23,** 527–539.

Kolb, P., and Caflisch, A. (2006). Automatic and efficient decomposition of two-dimensional structures of small molecules for fragment-based high-throughput docking. *J. Med. Chem.* **49,** 7384–7392.

Kuntz, I. D., Blaney, J. M., Oatley, S. J., Langridge, R., and Ferrin, T. E. (1982). A geometric approach to macromolecule–ligand interactions. *J. Mol. Biol.* **161,** 269–288.

Kuntz, I. D., Meng, E. C., and Shoichet, B. K. (1994). Structure-based molecular design. *Acc. Chem. Res.* **27,** 117–123.

Kuntz, I. D., Chen, K., Sharp, K. A., and Kollman, P. A. (1999). The maximal affinity of ligands. *Proc. Natl. Acad. Sci. USA* **96,** 9997–10002.

Law, R., Barker, O., Barker, J. J., Hesterkamp, T., Godemann, R., Andersen, O., Fryatt, T., Courtney, S., Hallett, D., and Whittaker, M. (2009). The multiple roles of computational chemistry in fragment-based drug design. *J. Comput. Aided Mol. Des.* **23,** 459–473.

Leeson, P. D., and Springthorpe, B. (2007). The influence of drug-like concepts on decision-making in medicinal chemistry. *Nat. Rev. Drug Discov.* **6,** 881–890.

Lepre, C. (2007). Fragment-based drug discovery using the SHAPES method. *Expert Opin. Drug Discov.* **2,** 1555–1566.

Lipinski, C. A., Lombardo, F., Dominy, B. W., and Feeney, P. J. (2001). Experimental and computational approaches to estimate solubility and permeability in drug discovery and development settings. *Adv. Drug Deliv. Rev.* **46,** 3–26.

Liu, Q., Masek, B., Smith, K., and Smith, J. (2007). Tagged fragment method for evolutionary structure-based de novo lead generation and optimization. *J. Med. Chem.* **50,** 5392–5402.

Majeux, N., Scarsi, M., Apostolakis, J., Ehrhardt, C., and Caflisch, A. (1999). Exhaustive docking of molecular fragments on protein binding sites with electrostatic solvation. *Proteins* **37,** 88–105.

Majeux, N., Scarsi, M., and Caflisch, A. (2001). Efficient electrostatic solvation model for protein-fragment docking. *Proteins* **42,** 256–268.

McGann, M., Almond, H., Nicholls, A., Grant, J. A., and Brown, F. (2003). Gaussian docking functions. *Biopolymers* **68,** 76–90.

Miranker, A., and Karplus, M. (1991). Functionality maps of binding sites: A multiple copy simultaneous search method. *Proteins* **11,** 29–34.

MOE—The Molecular Operating Environment (2009). Software available from Chemical Computing Group Inc., 1010 Sherbrooke Street West, Suite 910, Montreal, Canada H3A 2R7. http://www.chemcomp.com.

Orita, M., Ohno, K., Warizaya, M., Amano, Y., and Niimi, T. (2011). Lead generation and examples: Opinion regarding how to follow up hits. *Methods Enzymol.* **493,** 383–422.

Rarey, M., Kramer, B., Lengauer, T., and Klebe, G. A. (1996). Fast flexible docking method using an incremental construction algorithm. *J. Mol. Biol.* **261,** 470–489.

Reynolds, C. R., Tounge, B. A., and Bembenek, S. D. (2008). Ligand binding efficiency: trends, physical basis, and implications. *J. Med. Chem.* **51,** 2432–2438.

Rogers, D., and Hahn, M. (2010). Extended-connectivity fingerprints. *J. Chem. Inf. Model.* **50,** 742–754.

Sandor, M., Kiss, R., and Keseru, G. M. (2010). Virtual fragment docking by glide: A validation study on 190 protein–fragment complexes. *J. Chem. Inf. Model.* **50,** 1165–1172.

Shemetulskis, N. E., Weininger, D., Blankley, C. J., Yang, J. J., and Humblet, C. (1996). Stigmata: An algorithm to determine structural commonalities in diverse datasets. *J. Chem. Inf. Comput. Sci.* **36,** 862–871.

Tanimoto, T. T. (1957). IBM Internal Report, 17th Nov. IBM Corp., Armonk, New York.

Teotico, D. G., Babaoglu, K., Rocklin, G. J., Ferreira, R. S., Giannetti, A. M., and Shoichet, B. K. (2009). Docking for fragment inhibitors of AmpCβ-lactamase. *Proc. Natl. Acad. Sci. USA* **106,** 7455–7460.

Tounge, B. A., and Parker, M. H. (2011). Designing a diverse high-quality library for crystallography-based FBDD screening. *Methods Enzymol.* **493,** 1–20.

Tounge, B. A., and Reynolds, C. H. (2004). Defining privileged reagents using subsimilarity comparison. *J. Chem. Inf. Comput. Sci.* **44,** 1810–1815.

Veber, D. F., Johnson, S. R., Cheng, H.-Y., Smith, B. R., Ward, K. W., and Kopple, K. D. (2002). Molecular properties that influence the oral bioavailability of drug candidates. *J. Med. Chem.* **45,** 2615–2623.

Venkatachalam, C. M., Jiang, X., Oldfield, T., and Waldman, M. (2003). Ligand-Fit: A novel method for the shape-directed rapid docking of ligands to protein active sites. *J. Mol. Graph. Model.* **21,** 289–307.

Verdonk, M. L., Cole, J. C., Hartshorn, M. J., Murray, C. W., and Taylor, R. D. (2003). Improved protein–ligand docking using GOLD. *Proteins* **52,** 609–623.

Vinkers, H. M., de Jonge, M. R., Daeyaert, F. F. D., Heeres, J., Koymans, L. M. H., van Lenthe, J. H., Lewi, P. J., Timmerman, H., Van Aken, K., and Janssen, P. A. J. (2003). SYNOPSIS: Synthesize and optimize system in silico. *J. Med. Chem.* **46,** 2765–2773.

Warren, G. L., Andrews, C. W., Capelli, A.-M., Clarke, B., LaLonde, J., Lambert, M. H., Lindvall, M., Nevins, N., Semus, S. F., Senger, S., Tedesco, G., Wall, I. D., *et al.* (2006). A critical assessment of docking programs and scoring functions. *J. Med. Chem.* **49,** 5912–5931.

SECTION TWO

PRACTICAL APPROACHES

CHAPTER SEVEN

How to Avoid Rediscovering the Known

Lawrence C. Kuo

Contents

Abstract

For anyone who has participated in a screening exercise in a pharmaceutical or biotech setting with the aim to discover hits against protein targets, it is evident that a single screening exercise does not always offer hits of sufficient quality to be progressed into a lead compound. Often, more than one screen is needed. The premise in conducting a new screening exercise is to find "better" hits that are chemically and pharmacologically more attractive. As we move into challenging, new target classes, the need of new methods to find tractable hits is ever more urgent and the availability of differentiated backup compounds is crucial to sustain a clinical program. The obvious alternate routes to conduct the new screen include an improved compound library, a larger compound library, and different or more sensitive detection methods. As many of us have experienced firsthand, repeating a screen without drastically changing the chemical nature of the compound library or information content of the readout likely will offer only variations of the original hits. This chapter describes the strategies to adopt in fragment-based lead discovery to avoid rediscovering the known.

Structural Biology, Johnson & Johnson Pharmaceutical Research & Development, L.L.C., Spring House, Pennsylvania, USA

Methods in Enzymology, Volume 493
ISSN 0076-6879, DOI: 10.1016/B978-0-12-381274-2.00007-8

1. Why FBDD?

With the continued development and refinement of biophysical techniques to conduct fragment-based drug discovery (FBDD), using low-molecular weight scaffolds that are "lead like"[1] with favorable physiochemical and ADME/PK[2] properties offers a viable and "orthogonal" entry point to finding hits that in principle and by design shun unappealing pharmacophores. The logical questions are what molecular fragments to use in a screen, which technique to adopt as a first-in, and how to follow up the primary hits. A most pertinent question is what kind of readout information is orthogonal to the high-throughput screen (HTS) already conducted for a given target. Besides starting with compounds in the 100–200 MW range, how does an FBDD screen distinguish itself as one that is worthy of the time and cost?

Many analytical biophysical approaches have been adapted for FBDD, some to medium- or high-throughput mode. The list includes biochemical assays, nuclear magnetic resonance (NMR), protein crystallography, surface plasma resonance, mass spectrometry,[3] isothermal titration calorimetry, and capillary electrophoresis.[4] As no two biophysical methods will likely offer an identical set of hits,[5] one may wish to consider using several screening methods to catch as many hits as possible before progressing to the next stage of "sprouting" a fragment hit. The reason to run a second screen in an FBDD endeavor is different from that in an HTS. Because of the enormous size of an HTS compound library, and hence the raw number of hits, a second and sometimes a third follow-up screen by means of a different detection method is used to filter out false positives. In the case of an FBDD screen, one should apply a different second screen to avoid false negatives, particularly if the fragment library is small (in the range of a few hundred to a few thousand compounds).

It is understandable that when FBDD is used to augment an HTS, the speed of completing a fragment screen could be important to keep the

[1] Following Lipinski's "Rule of Five" (Lipinski *et al.*, 2001) that provides a guide for the development of orally bioavailable compounds, Congreve *et al.* (2003) have suggested a "Rule of Three" for the design of fragment libraries based on an analysis of a diverse set of fragment hits against a range of targets. The "Rule of Three" includes molecular weight <300, the number of hydrogen bond donor is ≤ 3, the number of hydrogen acceptor is ≤ 3, and logP or calculated logP is ≤ 3.

[2] ADME is absorption, distribution, metabolism, and excretion; PK is pharmacokinetics.

[3] Mass spectrometry has been applied to chemistry to progress fragments (Poulsen and Kruppa, 2008), but to the extent of its application to discover hits in a primary screening exercise, its throughput is too low to be the first-in choice.

[4] For an overview, see Orita et al. (2009).

[5] Preliminary experiments in our laboratory using enzyme activity to detect fragment binding to the enzyme ketohexokinase suggest that the primary hits from the same fragment library are significantly nonoverlapping with those found in a pure protein crystallography screen (Sun *et al.*, 2010). This observation suggests that two different approaches offer two different sets of hits. We plan to continue the analysis with the application of other detection approaches such as mass spectrometry, saturation transfer difference NMR, and surface plasma resonance to offer a comprehensive view.

chemistry team from moving onto another target. Nevertheless, the speed of a screen should be weighed against the informational content of the outcome. A "fast" screen usually offers only an apparent affinity as the readout. A screen that offers higher sensitivity or specific binding information in addition to simply an apparent affinity may take longer to conduct, but the content of the resulting data could be more than compensating. Thus, for every target, particularly one that has been screened before, one should consider the extent to which binding information is needed in order to support progression of a hit to a pharmacologically attractive and tight-binding lead, rather than the speed of coming up with a new list of hit compounds. As molecular fragments are low-affinity binders, generally with K_D, or IC_{50}, in the high micromolar to low millimolar range, how one wishes to progress the hits also determines what type of fragment screen one should undertake.

2. Types of Screen

There are two types of fragment screens to consider—one detects binding and the other reveals binding interactions. Most biophysical techniques being applied to fragment screening offer various degrees of affinity information (e.g., enzyme assays, surface plasma resonance, calorimetry, ligand-detected NMR[6]). Protein-detected NMR[7] and X-ray crystallography are currently the only methods that offer binding interaction information. Understandably, the first group is faster than the latter to screen a primary molecular fragment library. In all FBDD efforts, three-dimensional information about binding interactions is ultimately needed to progress a hit toward a lead. One could consider the affinity-guided FBDD screens as ones to be partnered up with a structure-based discovery design approach. Alternatively, one could screen a primary library of molecular fragments at the start with either protein-detected NMR or protein crystallography to go after binding interactions at atomic resolution. Using binding interactions derived from a bound structure as the screening approach is less popular for obvious reasons. First of all, the time needed to prepare a target to sustain a fragment screen is usually in the 6–12 months range. For protein-detected NMR, one needs to label the protein with ^{2}H, ^{13}C, or ^{15}N to provide a large supply of stable protein and have access to a high-field magnet. For protein crystallography, the protein needs to be crystallized, the supply of high-resolution diffracting single crystals must be robust, and the protein crystals must be stable when soaked in molecular fragment solutions.

[6] For example, nuclear Overhauser effect and saturation transfer difference NMR.
[7] For example, heteronuclei single quantum correlation NMR.

For both, data collecting and processing cannot be prohibitively rate limiting. However, the payback is enormous in terms of data content; one obtains upfront data on binding location as well as binding interactions at atomic resolution to guide the design of the next round of molecular fragments.

If affinity readout is the only choice, for example, the target protein is too large for protein-detected NMR or single crystals are unavailable, one can apply the tools of computer-assisted modeling to provide structural information to understand how hits bind.[8] Structural information is absolutely necessary to progress fragment hits.

If there is no suitable homologous structure to provide a molecular model, FBDD could be of low value because the probability to progress a hit toward a lead will then be entirely dependent on affinity. Using affinity solely to guide drug discovery has been successful in numerous HTS campaigns in which tight-binding hits are uncovered and ADME/PK friendly leads are generated. However, for molecular fragments in the 100–200 MW range, it is unlikely that there will be compelling and differentiating affinity data to rank the hits; the margin of error could be 1–2 orders of magnitude. In other words, the affinity readout could be entirely misleading. Basically, the hit progression exercise will fall back entirely on prior knowledge from previous HTS campaigns and one's prejudice of which fragment is more likely to bear fruit.

If structural information is available to discern binding interactions, one can screen with affinity readouts as the first-in, add hits via a subsequent screen with a different method, determine the bound ligand-target structures (via protein crystal structure or NMR protein structure) or the binding interactions via computational chemistry, and design a set of new molecular fragments to pin down target-specific interactions per structural information. The steps to progress a fragment hit are not really different from that for an HTS hit. So what should be different? What should one pay attention to? How should one avoid recapitulating the path of an HTS?

3. Size of a Fragment

As progression of an initial hit in potency toward a lead usually entails an increase in molecular weight and lipophilicity, the size of a molecular fragment should be biased toward its being small and polar. Hajduk (2006) showed in a defragmentation analysis of 18 drug leads (average molecular weight equals 463) that when various portions of each compound were

[8] For discussion on application of computational chemistry to FBDD, see Law *et al.* (2009). See also Chapter 6 by DesJarlais (2011) in this volume and reviews cited within.

sequentially removed to yield the simplest core (average molecular weight equals 224), generating a total of 73 fragments in the process, there was a reasonable linear relationship between the logarithmic value of affinity (K_D) and molecular weight. Using these data, a predictive plot of the expected molecular weight of a final optimized inhibitor with a potency of <10 nM is obtained for fragments of various size and potency as starting point. Capping the maximal molecular weight at 500 for the eventual lead compound, the starting fragments should not be of molecular weight >250 unless their potency is <30 μM.

The study by Hajduk (2006) is in agreement with an analysis by Kuntz *et al.* (1999) on the relationship between size and potency for a set of 160 highly optimized leads. Both studies show roughly a -0.3 kcal mol^{-1} contribution to the $\Delta\Delta G_{binding}$ per nonhydrogen atom added onto a core fragment of around 10 nonhydrogen atoms; the final optimized inhibitor approaches a limiting $\Delta\Delta G_{binding}$ of around -12 kcal mol^{-1} with a total number of 20–30 nonhydrogen atoms.[9] The study by Kuntz *et al.* (1999) shows that beyond 30 nonhydrogen atoms, the increase in energetic gain per nonhydrogen atom drops precipitously to zero.

These analyses strongly imply that the method applied to FBDD must be suitable to detect binding (K_D or IC_{50}) in the high micromolar to low millimolar range and that the molecular fragments should be <250 MW. If the method applied is aimed at detecting binding in the low-micromolar range and the compound library contains a good percentage of fragments >250 MW, the screen would be insufficiently "orthogonal" to augment a traditional HTS screen and its purpose to add value in finding non-HTS-like hits and leads would be largely lost.

4. Choice of Fragments

The primary set of molecular fragments to screen should be one that offers simple and less complex compounds yet samples a wide chemical diversity. Application of chemical diversity and molecular complexity in selecting compounds for a screen has been eloquently elaborated in a number of review articles.[10] Favorable properties to consider in selecting fragments include physiochemical properties, ADME/PK characteristics,

[9] For the 160 highly optimized leads analyzed by Kuntz *et al.* (1999), the initial five nonhydrogen atoms offer approximately -1.5 kcal mol^{-1} contribution to $\Delta\Delta G_{binding}$ and thereafter about -0.3 kcal mol^{-1} per nonhydrogen atom up to ~30 nonhydrogen atoms. Thus, for an optimized lead compound with 30 nonhydrogen atoms, the total $\Delta\Delta G_{binding}$ is close to 12 kcal mol^{-1}, an energetic that translates into an apparent $K_D \cong 1$ nM.

[10] See Patterson *et al.* (1996), Hann *et al.* (2001, 2006), and Hesterkamp and Whittaker (2008).

drug-like scaffolds,[11] lead-like scaffolds, functional groups that are synthesis friendly, and solubility. Among the preferred properties, solubility is an important one to note. As weak binding is the norm for the initial hits, the fragments chosen should be highly soluble to afford use at a high concentration to compensate for low affinity.

The size of the library depends to some degree on the screening method of choice as well as how exhaustive a search one prefers. It ranges from single- to double-digit thousand ranges. In using small scaffolds with high ligand efficiency, the freedom to "grow" and "merge" is greatly expanded; a relatively small number of molecules can provide the coverage of chemical space equivalent to the large libraries routinely employed for HTS.[12] Didactically, a small library may offer fewer hits and a less defined chemical boundary, but the emphasis is placed on novelty if not a more adventurous starting point. If structural guidance is readily available, out of a small (say 500–1000 fragments) yet highly diverse compound library, the customary 2–5% hit rate found in fragment screens offers an excellent springboard to prompt hit progression.

One should emphasize chemical diversity in any screen, *particularly* if one knows intimately the binding properties of a target, to steer clear of introducing bias. The primary library for a fragment screen should be one applicable to different types of targets. Because certain chemical moieties are known to provide favorable affinity and ADME/PK properties in the context of an inhibitor for several families of target, it has been argued by some that this prior knowledge should be built into the design of molecular fragment libraries. For example, indane, isoquinoline, and amide bond mimetic tend to bind proteases while quinazoline, amino pyrimidine, and amino-pyrazine tend to bind kinases. With this view, one would screen kinases with a kinase-directed fragment library and proteases with a protease-directed fragment library. The presumed impetus for such an approach is that the binding location of fragments would remain unaltered to provide an anchor when merged or evolved into a larger, higher affinity molecule. This logic is counterproductive in a FBDD campaign—if the components of a larger molecule do not, or are not designed to, reside at the same location as the individual components themselves, there is no need for one to prepare a biased library. However, if components of a larger molecule do, or are designed to, reside roughly at the same location of the individual fragments themselves, then one would be expected to recapitulate in building lead compounds similar to previously known ligands. Target-focused functional groups should be avoided in any fragment library.

[11] The pros and cons regarding a rigid use of "drug-like" properties as a filter in the design of a fragment library have been discussed (Lepre, 2001; Oprea *et al.*, 2001; Teague *et al.*, 1999).

[12] For an in-depth discussion, see Fink and Reymond (2007).

5. Fragment Progression

A couple of studies have addressed the question of whether small fragment molecules bind to protein targets at the same location and with similar conformation as seen for the eventual lead compound they constitute. Findings of these analyses impact the decision of how one should follow up hits. By deconstructing a series of 9 Bcl-x_L inhibitors, Barelier *et al.* (2010) analyzed 22 fragments to identify their binding interactions with the protein and found no conservation of binding interactions between the fragments and the lead molecule. Their fragments interact with their preferred binding site which can be different from the site they occupy when they are part of the lead molecule. The same authors have also shown that the affinity of these fragments to the protein is not additive. Similar findings have been revealed in a study by Babaoglu and Shoichet (2006). Fragmentation by these authors of a known β-lactamase inhibitor reveals that none of the individual fragments binds in a manner seen in the context of the larger inhibitor. In an FBDD-directed study to generate lead molecules from fragment hits for different classes of targets, researchers at Johnson & Johnson Pharmaceutical Research & Development, LLC have observed that while many fragments approximately retain their binding interactions as well as location when progressed toward a lead, some do not.[13] Collectively, these studies suggest that in "sprouting" a hit toward a lead, it is better not to be fixated too firmly on an observed binding interaction. The binding juxtaposition of fragments could change during hit progression. There is no guarantee of conservation in binding conformation of the fragments that spawn the eventual lead molecule.

Often, the resolution of X-ray crystal structure of a protein complex is in the mid 2 Å range. For a fragment with a molecular weight of 100–250, this resolution is sufficient to define its location unambiguously but not necessarily its binding interaction or even orientation. For the first iteration of fragment progression, it is unnecessary to try to define the binding interactions at a higher resolution or with analogs that include added functional groups. Instead, it would be much more advantageous to apply an observed interaction somewhat loosely to test more chemotype derivatives for each subpocket of a binding site. It is not only more fruitful but also more gratifying to find surprises. Some useful general rules for hit progression are

[13] Using atomic resolution binding interactions as a guide, one would expect less movement of a fragment when it is being "sprouted" into a larger compound if each iteration of elaboration is by design to be "constrained" by the three-dimensional structure of the bound compound.

1. Emphasize chemical diversity.
2. Apply "fuzziness" when viewing bound interactions so as to sample many possibilities.
3. Add functional groups iteratively, but no more than 6–8 heavy atoms at a time.
4. Do not use target-specific "preferred" moieties.
5. Guide prejudicially with new structural information.
6. Deemphasize affinity data and do not rank order hits.
7. Consider lipophilicity to introduce interactions.[14]
8. Evaluate potency and ADME/PK properties against off-target effects only in the latter part of hit progression.

6. Conclusion

Compounds fail in the early development stage of a drug discovery campaign for various unforeseeable reasons. The availability of differentiated backup leads is always highly desirable. Nonetheless, far too often minimal changes in chemical content are made from the original lead for backup compounds. Thus, early in a drug discovery program, it is highly desirable to conduct additional screens intended to provide orthogonal outcomes. With the coming of age of sufficiently sensitive and fast throughput biophysical techniques, it is not a matter of whether FBDD can offer leads with IC_{50} in the single-digit nanomolar range but rather how differently it can do so when compared to traditional HTS. It is important to not screen fragments with the same technique one applies to HTS and/or previous lead optimization to avoid running into the same set of target- or method-specific issues, known or unknown. The speed to find and transform a hit into a lead should be a lesser worry than the type of information being extracted at the different stages of an FBDD effort for the evolution of a truly different chemical entity. The key is to follow a distinct scheme in fragment evolution, elaboration, and optimization from what has been applied in a HTS. The imperative endeavor is to chart in a FBDD exercise a new course of hit-to-lead progression uncompromised by "prior" knowledge.

[14] The need to screen fragments at high concentrations, usually in the 1–20 m*M* range (Abad *et al.*, 2011; Barker *et al.*, 2006; Baurin *et al.*, 2004; Hubbard *et al.*, 2007), predisposes the selection of fragments with high aqueous solubility, either by choice per logP ("Rule of Three") or by necessity.

REFERENCES

Abad, M. C., Zhang, X., and Gibbs, A. C. (2011). Structure density relationship based FBDD. *Methods Enzymol.* **493,** 489–510.

Babaoglu, K., and Shoichet, B. K. (2006). Deconstructing fragment-based inhibitor discovery. *Nat. Chem. Biol.* **2,** 720–723.

Barelier, S., Pons, J., Marcillat, O., Lancelin, J.-M., and Krimm, I. (2010). Fragment-based deconstruction of Bcl-x_L inhibitors. *J. Med. Chem.* **56,** 2577–2588.

Barker, J., Courtney, S., Hesterkamp, T., Ullmann, D., and Whittaker, M. (2006). Fragment screening by biochemical assay. *Expert Opin. Drug Discov.* **1,** 225–236.

Baurin, N., Aboul-Ela, F., Barril, X., Davis, B., Drysdale, M., Dymock, B., Finch, H., Fromont, C., Richardson, C., Simmonite, H., and Hubbard, R. E. (2004). Design and characterization of libraries of molecular fragments for use in NMR screening against protein targets. *J. Chem. Inf. Comput. Sci.* **44,** 2157–2166.

Congreve, M., Carr, R., Murray, C., and Jhoti, H. (2003). A "Rule of Three" for fragment-based lead discovery? *Drug Discov. Today* **8,** 876–877.

DesJarlais, R. L. (2011). Using computational techniques in fragment-based drug discovery. *Methods Enzymol.* **493,** 137–156.

Fink, T., and Reymond, J.-L. (2007). Virtual exploration of the chemical universe up to 11 atoms of C, N, O, F: Assembly of 26.4 million structures (110.9 million stereoisomers) and analysis for new ring systems, stereochemistry, physicochemical properties, compound classes, and drug discovery. *J. Chem. Inf. Model.* **47,** 342–353.

Hajduk, P. J. (2006). Fragment-based drug design: How big is too big? *J. Med. Chem.* **49,** 6972–6976.

Hann, M. M., Leach, A. R., and Harper, G. (2001). Molecular complexity and its impact on the probability of finding leads for drug discovery. *J. Chem. Inf. Comput. Sci.* **41,** 856–864.

Hann, M. M., Leach, A. R., Burrows, J. N., and Griffen, E. (2006). Lead discovery and the concepts of complexity and lead-likeness in the evolution of drug candidates. *Compr. Med. Chem. II* **4,** 435–458.

Hesterkamp, T., and Whittaker, M. (2008). Fragment-based activity space: Smaller is better. *Curr. Opin. Chem. Biol.* **12,** 260–268.

Hubbard, R. E., Davis, B., Chen, I., and Drysdale, M. J. (2007). The SeeDs approach: Integrating fragments into drug discovery. *Curr. Top. Med. Chem.* **7,** 1568–1581.

Kuntz, I. D., Chen, K., Sharp, K. A., and Kollman, P. A. (1999). The maximal affinity of ligands. *Proc. Natl. Acad. Sci. USA* **96,** 9997–10002.

Law, R., Barker, O., Barker, J. J., Hesterkamp, T., Godemann, R., Andersen, O., Fryatt, T., Courtney, S., Hallett, D., and Whittaker, M. (2009). The multiple roles of computational chemistry in fragment-based drug design. *J. Comput. Aided Mol. Des.* **23,** 459–473.

Lepre, C. A. (2001). Library design for NMR-based screening. *Drug Discov. Today* **6,** 133–140.

Lipinski, C. A., Lombardo, F., Dominy, B. W., and Feeney, P. J. (2001). Experimental and computational approaches to estimate solubility and permeability in drug discovery and development settings. *Adv. Drug Deliv. Rev.* **46,** 3–26.

Oprea, T. I., Davis, A. M., Teague, S. J., and Leeson, P. D. (2001). Is there a difference between leads and drugs? A historical perspective. *J. Chem. Inf. Comput. Sci.* **41,** 1308–1315.

Orita, M., Warizaya, M., Amano, Y., Ohno, K., and Niimi, T. (2009). Advances in fragment-based drug discovery platforms. *Expert Opin. Drug Discov.* **4,** 1125–1144.

Patterson, D. E., Cramer, R. D., Ferguson, A. M., Clark, R. D., and Weinberger, L. E. (1996). Neighborhood behavior: A useful concept for validation of "molecular diversity" descriptors. *J. Med. Chem.* **39,** 3049–3059.

Poulsen, S.-A., and Kruppa, G. H. (2008). In situ fragment-based medicinal chemistry: Screening by mass spectrometry. *In* "Fragment-Based Drug Discovery: A Practical Approach," (E. R. Zartler and M. J. Shapiro, eds.), pp. 159–198. John Wiley & Sons, Ltd., UK.

Sun, W., Struble, G., and Kuo, L. C. (2010). Unpublished observations.

Teague, S. J., Davis, A. M., Leeson, P. D., and Oprea, T. (1999). The design of leadlike combinatorial libraries. *Angew. Chem. Int. Ed.* **38,** 3743–3747.

CHAPTER EIGHT

From Experimental Design to Validated Hits: A Comprehensive Walk-Through of Fragment Lead Identification Using Surface Plasmon Resonance

Anthony M. Giannetti

Contents

Genentech Inc., 1 DNA Way, South San Francisco, California, USA

Methods in Enzymology, Volume 493
ISSN 0076-6879, DOI: 10.1016/B978-0-12-381274-2.00008-X

Abstract

The detection and characterization of fragment binding requires the use of technologies with extreme sensitivity to observe the binding interactions of low-affinity and low-molecular weight compounds to proteins. A number of methods have emerged capable of providing fragment hits to project teams including, but certainly not limited to, NMR, X-ray crystallography, and surface plasmon resonance (SPR). SPR-based biosensors are sufficiently sensitive and high throughput to provide complete fragment screens on libraries of several thousand compounds in just a few weeks per target. Biosensors provide quantitative binding information for ranking fragments by affinity and ligand efficiency and can support ongoing quantitative structure–activity efforts during fragment hit-to-lead development. The combination of speed and binding quantitation makes SPR a valuable technology in pharmaceutical fragment-based drug discovery and development. Successful implementation of SPR biosensors in fragment efforts requires specialized methods for instrument preparation, assay development, primary compound handling, primary screening, confirmation testing, and data analysis. In this chapter, each of these topics is discussed in detail with general best practices for maintaining the highest throughput while maximizing data quality.

1. Introduction: Biophysical Principles of Surface Plasmon Resonance

The use of optical biosensors for pharmaceutical applications has risen since the introduction of commercially available instruments, primarily by Biacore Inc. in 1990. In the biopharmaceutical arena, their use has primarily been restricted to protein–protein interactions, but in the past 10 years, they have gained significant ground in small-molecule applications. While many different biosensors are on the market, the most widely used employ the technique of surface plasmon resonance (SPR). There are many commercial implementations of SPR. For a recent extensive list of vendors, see Rich and Myszka (2008). In a typical SPR experiment, a thin gold film, deposited on glass, is excited on one side by plane polarized light to induce an evanescent plasmon wave that radiates ~150 nm away from surface of the film. The presence of any matter, and more specifically the dipole moment of that matter that give rise to its refractive index, entering the region of the plasmon wave will alter the wave's resonance frequency and change the critical angle for total internal reflection of the incident light. The change in this angle, which is directly proportional to the amount of matter entering the field, is read in real time as a change in response, measured in resonance units, which are often referred to as response units, (1 RU = 1 pg protein/mm^2; Stenberg *et al.*, 1991) to resolve the kinetics of the movement of the matter in and out of the detection volume. Many methods have been developed to attach molecules, such as proteins, nucleic acids, and even small molecules, to the gold surface (Lofas and McWhirter, 2006). Usually a polydimethylsiloxane-based microfluidic system is used to stamp out a set of flow cells on the gold surface allowing different targets to be immobilized in separate regions on a single chip. Built-in microfluidics allow for precise control over each flow cell alone and in combination, as well as injection of binding partners and running buffers over the surfaces. When binding partners are flowed over the derivatized surface the interactions between the pair of molecules results in a rise in response. The fluidics then switches to inject running buffer over the surface and dissociation is observed. Global analysis of the shape of the response curves for multiple concentrations of the injected molecule yields the on-rate (k_a expressed as $M^{-1}s^{-1}$), off-rate (k_d expressed as s^{-1}), and by their ratio the dissociation constant (K_D expressed as M) (Fig. 8.1). Some interactions, especially weak ones, equilibrate too quickly to be resolved by the instrument's sampling frequency (typically 10 Hz) and kinetic data are not obtainable. In these cases, the equilibrium binding level for the different concentrations of flowed binding partner is plotted versus the concentration to reveal the K_D (Fig. 8.1D).

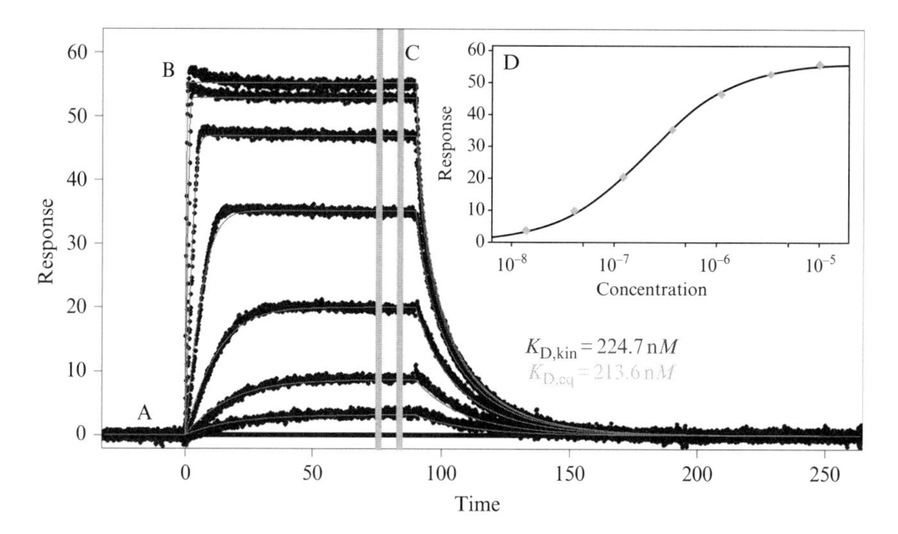

Figure 8.1 Experimental SPR sensorgram from a Biacore T100 with data, black points, overlaid with a 1:1 Langmuir binding model kinetic fit (red). The top concentration is 10 μM and injections are related by threefold dilutions. (A) Preinjection baseline and (B) association phase. Compound is injected at $t = 0$ s and continues for 90 s. (C) Injection ends and dissociation is observed back to baseline levels. (D) Plot of the equilibrium response calculated as the average of the response values between the green bars versus the log of the sample concentration in molar. The $K_{D,kin}$ is taken as the ratio of the off-rate divided by the on-rate and $K_{D,eq}$ is derived from the fit in (D). The lowest concentration injection represents 6.2% surface occupancy and is at exactly three times the standard deviation of the noise. (See Color Insert.)

Many reviews have been dedicated to best practices for obtaining high-quality data using biosensors, especially for the more traditional uses of the technology such as protein–protein and protein–nucleic acid interactions (Myszka, 1999; Myszka *et al.*, 2003). Many of the methods described, such as double referencing, chip preconditioning, and routine instrument maintenance, are critical to obtain quality data on small-molecule interactions. However, small molecules, unlike proteins, are associated with a host of additional technical challenges including, but not limited to, working with high refractive index solvents such as DMSO, compound solubility limitations in both DMSO and aqueous solvents, compound aggregation and promiscuous binding, nonspecific binding, high-density coupling of target to surface, high numbers of samples, low-molecular weight compounds, and extremely low-affinity interactions. These last three are particularly relevant to fragment-based lead discovery efforts where libraries consist of compounds with very low-molecular weights (generally <300 Da), and affinities ranging from 10 μM to 20 mM, meaning there will be little compound binding even at high concentrations. In the next sections, discussion of solutions to each of the challenges faced by the SPR users in the fragment discovery environment will be described based on experiences performing $\sim$30 fragment screening campaigns on $\sim$20 different protein targets.

2. Preparing the Instrument

Every type of biosensor will require its own unique procedures for maintenance and cleaning, and proper preparation of the instrument for fragment experiments should be discussed with the instrument manufacturer. The descriptions below will generally apply to the Biacore family of biosensor products but can be extended within reason to other types of instruments. The goal of these procedures is to maximize sensitivity and minimize systematic (e.g., drift) and nonsystematic (e.g., air bubbles) artifacts in the data.

To function properly, an instrument needs to be serviced regularly, ideally twice a year, to maintain integrity of the integrated microfluidic, pneumatic, and robotic subsystems that are generally not available for user-level service. However, user vigilance needs to be maintained to watch for gradual or sudden decay of these subsystems, such as frequent running of the air compressor (air bladder crack), sticky or inoperative valves on the integrated microfluidic cartridge that controls the individual experimental flow cells, and misaligned auto sampler (e.g., not piercing the center of wells on a 384-well plate) that can lead to clogged or bent injection needles.

For user-level service, Biacore suggests running the Desorb procedure at least weekly, which would involve washing the fluidics alternately with sodium dodecylsulfate and high-pH glycine buffer. This procedure was originally developed to strip away protein deposits that can accumulate throughout the instrument when regularly used in a protein–protein interaction environment where generally only one to a few different proteins are tested in the course of an experiment. In contrast, small-molecule screening experiments only expose the instrument to protein solutions for a few minutes and instead runs hundreds to thousands of different compounds through the fluidics, all with different aqueous solubility and stability properties, through the fluidics. Heterogeneous compound deposits are generally resistant to cleaning agents such as detergent and high-/low-pH buffers and require more extreme measures for their removal. Generally, running the Sanitize procedure, which thoroughly washes the fluidics with a 5% solution of sodium hypochlorite, is sufficient to restore an instrument to high-level operation. As a modification to Biacore's Sanitize protocol, it is recommended to first conduct the Sanitize protocol, followed by priming with water two to three times, followed by priming with a concentrated neutral buffer to rapidly restore to a neutral pH. The recommended procedure of priming and slowly rinsing with unbuffered water and letting stand overnight is time consuming and often insufficient to return the fluidic paths to a neutral pH. Priming in buffer will return the instrument to a functional state in a few minutes and allow for a more rapid setup of the next experiment.

Although keeping the fluidics clean is critical for obtaining high-quality data, over cleaning can be a problem. The harsher conditions of the Sanitize procedure can prematurely age the microfluidics cartridge. In addition, the constant use of DMSO in small-molecule work, often at 5%, helps continuously clean the fluidics. Thus, strict adherence to cleaning schedules may be unnecessary and waste time. Typically, the quality of data slowly decays over time as the need to clean increases and can be monitored daily by evaluating the quality of the blank buffer injections included for double referencing (see Section 9). An instrument in need of cleaning will show a large degree of shape in the blank injections, often in the form of large positive or negative values and sometimes significant variance and drift among the replicates. A highly functioning instrument will produce blank buffer injections exhibiting some, but minimal, systematic shape and high agreement among replicates. It is advised to clean the instrument before the variance between replicate buffer injections becomes significant. This is a more useful metric for when to clean an instrument as it prevents overcleaning if the instrument is not accumulating compound deposits and will indicate the need for cleaning if a particular bad compound or set of compounds is injected and fouls the fluidics system significantly. Fortunately, when the variance starts to increase, there will be little to no impact on the final reduced data as long as the instrument is cleaned before the variance becomes too significant. In practice, cleaning is not necessary more than every 8–12 weeks for small-molecule work where DMSO is included in all of the experimental running buffers.

3. Surface Preparation

The small size and weak affinity of fragments means the SPR signal for fragment-based experiments will be very close to the detection limit of modern instrumentation. Thus, minimizing noise and maximizing instrument stability are critical to successful fragment screens, and both can be achieved by proper preconditioning and normalization of each new sensor chip. Most biosensor chips are not flat gold surfaces but are covalently coated with a layer of dextran gel (Lofas, 1995). Chips arrive from the manufacturer in a dry nitrogen atmosphere and the dextran is dehydrated. Priming the system in buffer begins the hydration process, which may not complete for some time. As the gel swells mass moves away from the surface of the chip translating to a downward drift in the baseline and can continue for hours. The swelling can be rapidly completed by exposing the new chip to duplicate 12-s injections each of 50 m*M* NaOH, 10 m*M* HCl, 0.1% sodium dodecylsulfate, and 0.085% phosphoric acid. When a neutral running buffer is used, each agent will be removed by the continuous buffer flow that occurs while the instrument prepares each subsequent injection. At the end of these cycles, the baseline will be 50–100 RU lower than at the start. The next step involves normalization and is a premade program available in all biosensor instruments. This procedure corrects for the slight reflectance differences of the gold surface between different chips and flow cells. Usually 70% glycerol is pumped through the flow cells to saturate the SPR signal allowing direct observation of the reflectance properties of the surfaces. Internal software corrects for the differences observed. Some users prefer to perform this step after protein coupling because the chip will be closer to the experimental state, but many proteins can be damaged by the exposure to unbuffered 70% glycerol for 5–10 min. While the normalization protocol will attempt to rinse the glycerol from the fluidics, a prime into running buffer is recommended to be sure it is fully removed. Once these preparatory procedures are complete, protein can be coupled to the surface.

4. Target Immobilization

SPR technology is usually described as label free, meaning it requires no addition of radioactive tracers or fluorescent tags. However, the fundamental "label" required in all SPR-based experiments is the attachment of the target to the hydrogel. Originally, it was thought that biosensors lacked sufficient sensitivity to detect the mass of a small-molecule binding a large protein and thus some approaches have been developed that involve the coupling of a small molecule to the surface and monitoring the binding of proteins injected over the small-molecule surface (e.g., see Geschwindner *et al.*, 2007). Indeed, some theoretical calculations, which frequently neglect that the refractive index increment for small molecules is frequently higher than that for proteins giving compounds more signal per Dalton than protein (Davis and Wilson, 2000), have been presented (Dalvit, 2009; Hamalainen *et al.*, 2008) estimating that SPR's utility, especially on proteins heavier than 60 kDa, is limited for fragments, even at high concentration. However, with the development of improved instrumentation, the methods described here, and the use of highly pure and active protein (Bukhtiyarova *et al.*, 2004), fragment binding to proteins as large as 170 kDa has been reliably detected and an example is shown in Fig. 8.2. The data have sufficient window (R_{max} ~ 40 RU) to suggest that small-molecule binding to even larger proteins and/or smaller compounds is possible. Thus, this chapter will only be concerned with a protein-immobilized fragment screening format.

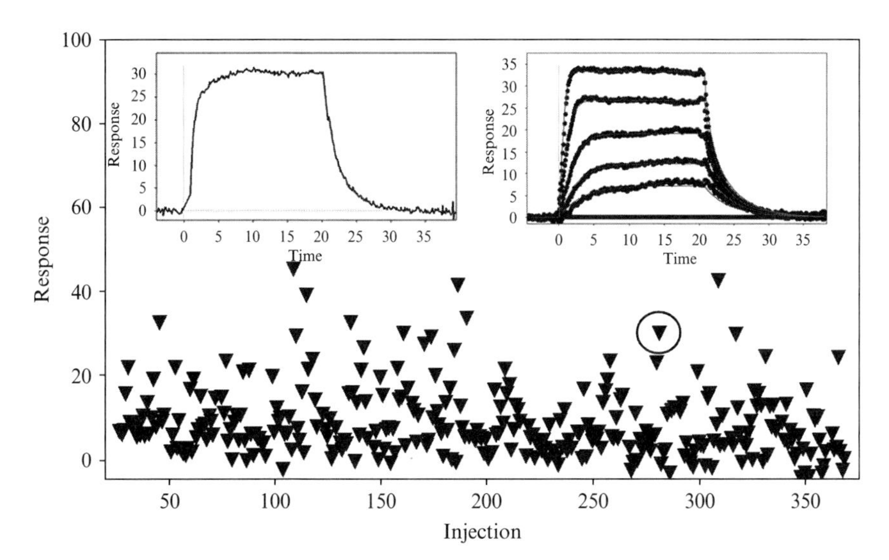

Figure 8.2 Fragments tested on a 170-KDa protein. Equilibrium binding levels in RU for each fragment (triangles) are plotted versus the injection cycle number. The average response level for this screen is somewhat less than when tested on smaller proteins; however, sufficient window is available to identify binders. The sensorgram for the circled compound is a fragment with molecular weight of ~320 Da, and the sensorgram generating this point is shown in the upper left. Note the rapid approach to equilibrium and return to baseline binding after the injection. This fragment was confirmed in a dose–response experiment during secondary confirmation and has a $K_{D,kin}$ of 2.05 μM, and R_{max} is ~40 RU indicating a large enough window to detect much smaller compounds (< 100 Da) and/or study larger proteins.

While the term "immobilization" has been used to describe the process of attaching proteins to the SPR chip, it is somewhat inaccurate and has led to misunderstandings and heated debate about the relevance of obtaining data on proteins that are "not in solution" or completely immobile. This has been addressed by numerous comparison studies (e.g., see Day *et al.*, 2002). While attaching proteins to a flat gold surface is an option, it is almost never done. Instead, the gold chips are prederivatized by the manufacturer with a dextran hydrogel. The dextran can be chemically activated by injecting a 1:1 mixture of 100 m*M* *N*-hydroxysuccinimide (NHS) and 400 m*M* 1-ethyl-3-(3-dimethylaminopropyl)-carbodiimide for 7 min at 30 μl/min to make it reactive with primary amino groups on proteins (Johnsson *et al.*, 1991). Protein, often at 50 μg/mL, is injected at 10 μL/min, and the reaction monitored in real time and stopped when a sufficient immobilization level is reached. When using this "amine coupling method," protein is more precisely described as tethered to a matrix that, at the molecular level, moves around in solution. Thus, protein motions are restricted to the detection volume of the plasmon wave but without making them perfectly still. Thus, they retain the bulk of their rotational and translational degrees of freedom. Many experiments have shown that binding parameters of properly amine-coupled proteins match those measured on completely free protein (Day *et al.*, 2002; Navratilova *et al.*, 2007; Papalia *et al.*, 2006, 2008; Rich *et al.*, 2009).

Amine coupling requires preparing the protein in a low-salt buffer at least one pH unit below the protein isoelectric point. Care should be taken not to expose the protein to denaturing conditions and to minimize exposure time to the coupling buffer. Taking measures to enhance stability and/or prevent reactions with an exposed active-site lysine, such as coupling in the presence of an inhibitor (Casper *et al.*, 2004) may also be used. Damage to the protein via amine coupling will always remain a formal possibility so careful testing and characterization of the surfaces is a critical part of assay development (covered in Section 6). However, amine coupling is considered extremely convenient because it is often rapid, requires minimal protein, and avoids the need for any additional tagging or labeling. Note that primary amine containing buffers such as Tris cannot be used as they will block the surface. Once protein has been coupled to a sufficient density unused amine-reactive groups on the dextran must be neutralized. Traditionally, this is accomplished with three 1-min injections of 1 *M* ethanolamine, pH 8. Some proteins are not stable in this buffer, so a gentler method is to mix the ethanolamine 1:1 with the running buffer and inject twice for 30 s, thus maintaining the pH and salt combinations favorable to each protein system.

Many other methods for immobilization exist for proteins not amenable to amine coupling. The next most straightforward is limiting biotinylation. In this approach, a substoichiometric amount of biotinylation agent is mixed with the protein (usually, 0.5:1 agent:protein, though lower may be necessary and higher can be possible) and incubated on ice for 2–3 h. The mix is then thoroughly desalted, preferably using size exclusion chromatography as gravity desalting columns do not always completely eliminate the free biotin. The recovered protein can be injected over chips prederivatized with and avidin of choice, such as streptavidin or neutravidin, streptavidin or neutravidin to yield a very stable captured protein surface. The most commonly used biotinylation agent is sulfosuccinimidyl-6′-(biotinamido)-6-haxanamido hexanoate (sulfo-NHS-LC-LC-biotin; Pierce catalog #21338) that reacts with primary amine groups in proteins over a range of pH values. This agent is even compatible with Tris buffers. The spacer arm in this agent allows the target protein to move ~30 Å away from the avidin capture molecule giving the protein even more room to translate and rotate as if it were free in solution.

If more specific biotinylation is desired, then an Avi-tag can be introduced either at the N- or at the C-terminus of the protein. The Avi-tag (GLNDIFEAQKIEWHE; Beckett *et al.*, 1999) is specifically biotinylated on the underlined lysine by biotin ligase. Purified protein can be biotinylated using purified biotin ligase, biotin, and ATP; however, this introduces two additional steps in the purification (biotinylation and desalting) and not all proteins are compatible with the optimal ligase buffer. Biotinylation and purification efficiency can be enhanced by coexpression of ligase and target protein resulting in upwards of 100% biotinylated protein extracted from the cells (Duffy *et al.*, 1998). The resulting purified protein can be easily captured, even when at very low concentration by extending capture times, to an avidin-based biosensor surface. When a tagging approach is used it is generally advisable to prepare two constructs with the tag at the N- or C-terminus in the event that the tag interferes with expression, stability, or activity of the protein. Examiniation of known crystal structures of the target protein can help guide the choice of terminus and linker length. For 10 assays developed in house using Avi-tagged protein, only one showed a preference for a particular terminus. In the case of oligomeric proteins, Avi-tagging is a preferred strategy because tags from different monomers can be captured by neighboring avidin molecules in the coupled dextran hydrogel to keep the monomers localized near each other and prevent monomer dissociation from the surface. In fact, multimeric Avi-tagged proteins can make more stable surfaces than monomeric proteins due to avidity effects arising from the interactions of multiple tags on the multimer with multiple avidin molecules.

Neutravidin and streptavidin contain a small-molecule binding site that can bind fragments and interfere with the results. Therefore, once the targets have been captured, it is advisable to block all surfaces with biotin. While the free biotin will slowly dissociate over time, it is so slow that few open binding sites will become available during the screen for any fragment binding at the biotin sites to be detected. Blocking the biotin binding sites is best achieved using soluble biotin analogs such as amino-PEG-biotin

(Pierce catalog #21346) or biocytin (Pierce catalog #28022). Usually a solution of the blocking agent is mixed with running buffer and two 30-s injections are made over all surfaces at 30 μl/min. Biotin capture after the first injection will be noted as an upward baseline shift on all flow cells. If no additional shift is observed after the second injection, then the surfaces are considered blocked.

A variety of other methodologies for tethering proteins to biosensors exist but are less useful in the small molecule, and especially fragment screening, arena. Capture with coupled antitarget or antitag antibodies, while highly effective in protein–protein work, generally does not allow for the highly dense surfaces required for fragment work. It can also complicate fragment identification because fragments will also bind to the antibodies. While careful deconvolution of antibody versus target binding may be possible, the additional complications can be avoided by using one of the previously discussed methods. GST-tags are also to be avoided as GST binds many fragments as well. Thiol-coupling through endogenous, engineered, or chemically added thiol groups on the protein surface can generate dense surfaces suitable for fragment work, but in general, thiol-coupling is a tedious and time-consuming coupling procedure especially if cysteine mutations are introduced. Some users have had success adding several repeats of poly-histidine sequences at the N- or C-terminus for use in capturing to Ni^{2+}-NTA chips. This has the advantage that protein can be easily removed from the surface with EDTA stripping, the chip recharged with an injection of nickel ions, and then fresh protein captured. Indeed, successful SPR fragment screens have been executed rebuilding the chip after testing each fragment. The additional steps required in each cycle, however, will consume significant time and have an impact on throughput. Additional care needs to be taken with the false positives and negatives that may arise with fragments containing carboxylic acids, imidazole rings, or other functionalities that can bind to the nickel instead of the captured protein. Several repeats of the poly-histidine sequence are necessary as the affinity of the Ni-NTA/poly-histidine interaction is low micromolar (Lofas and McWhirter, 2006).

Most modern instruments come bundled with control software containing menu-driven wizards that can be useful in helping the new user become familiar with standard chip-building procedures. However, these wizards often consume considerably more time and reagent than necessary to build a surface because they work sequentially over individual flow cells rather than in parallel across all surfaces. Additionally, the wizard may not allow the user to fix a problem (e.g., the injection time was insufficient to achieve the desired coupling density and another injection is necessary) and instead often automatically block the surface with ethanolamine or biotin preventing additional protein capture. To prevent premature surface blocking, blocking agents should never be placed in the machine until sufficient target density has been achieved on every flow cell. With experience, the user can manually perform all chip preconditioning, normalization, activation, coupling, capture, and blocking procedures in less than an hour, and the user can often perform additional tasks such as buffer preparation during the long priming, normalization, or capture steps. Because of the ease in making chips, it is advisable to never reuse chips. Once undocked, a chip begins to collect dust which can then contaminate the microfluidics if redocked with the instrument. It is simpler to rebuild the chip rather than clean or service the instrument. Additionally, many proteins are insufficiently stable to maintain a working surface for more than 1–2 days.

For small molecule, especially fragment work, it is necessary to work at high protein coupling densities. However, it may not be necessary to work at the highest densities possible and usually anywhere from 5000 to 10,000 RU of protein is sufficient. Some experimentation with control compounds will be necessary for every target to establish if the window is sufficiently large for a screen. Usually maximal signal at compound saturation of 30–100 RU is sufficient with modern instruments to reliably detect and characterize fragment binding. Densities up to 20,000 RU on standard chips can be achieved, but higher density is challenging due to molecular crowding effects. In general, it is not advisable to go as high as possible because the protein concentration on chip gets very high (e.g., 5000 RU of a 50-kDa protein is ~660 μ*M*) and aggregation could become a problem. In addition, if a capture method is used, the baseline drift rate will become more significant at higher coupling densities resulting in a longer time until the baseline stabilizes enough to screen.

At high coupling densities, SPR experiments can suffer from an artifact known as mass transport. This occurs for samples where the association rate is highly relative to the injected molecule's diffusion rate, and rebinding to free binding sites affects a molecule's ability to leave the system once dissociated. Thus, the true k_a and k_d become obscured by diffusion limitations. This effect is sometimes observed in small-molecule experiments, but does not affect fragment binding because fragments are generally too weak to be subject to the rebinding effects and kinetics are rarely observed at all. It is likely that coupling density-related artifacts, which can play a role when fragments are developed into more potent leads, will be minimized as sensitivity increases in future generations of instruments.

Success of biophysical binding experiments is tightly tied to protein purity and homogeneity. Unlike many biochemical assays, biophysical binding experiments can detect binding to inactive or partially denatured protein and impurities. Even partially denatured proteins, with exposed hydrophobic cores, can nonspecifically bind compounds, especially fragments. Thus, impurities or lack of homogeneity can contribute significantly to the signal and results in the identification of false positives (e.g., see Fig. 8.3 and Section 8). It is therefore critical to ensure protein preparations

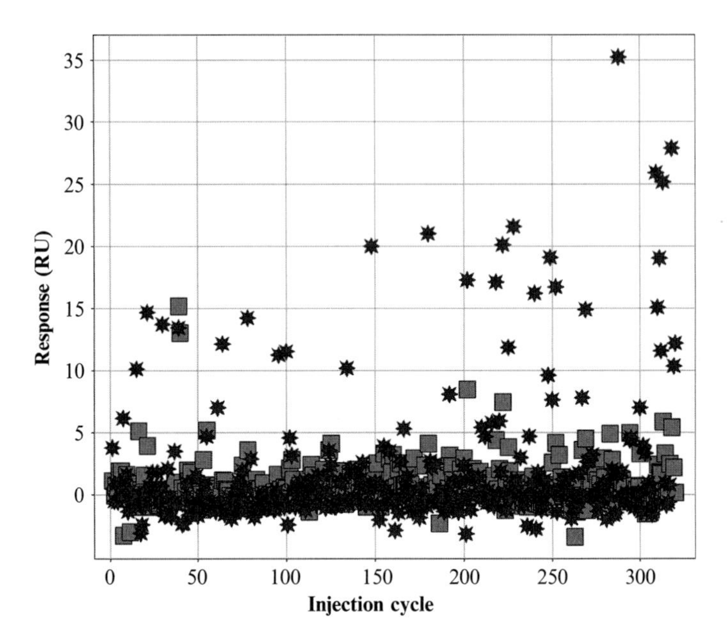

Figure 8.3 Equilibrium binding level plot for 320 fragments tested against high- and low-quality protease preparations. Protease with low-specific activity (<10%) was captured to a neutravidin surface and screened (red stars) yielding an apparent hit rate of ~16.4%. A new batch of protein was made and purified to higher homogeneity and specific activity of ~90%. The same fragments were retested (gray squares) and give a more reasonable hit rate of ~6.4%, indicating the better protein preparation is binding fewer compounds nonspecifically.

destined for SPR testing are free of contaminants that can capture to the surface (e.g., biotinylated host proteins in a cell lysate containing Avi-tagged protein or aggregates). In addition, testing protein-specific activity, or performing active-site titrations, is important to establish that only homogenous, well-behaved, protein is being used. The extra stringency of protein homogeneity is balanced by the advantage of the binding assay that inactive forms of proteins, such as nonphosphorylated kinases, zymogen forms of proteases, or active-site mutants, can be tested despite the lack of biochemical activity. In these cases, measurement of the binding level of a control compound, protein, or peptide at saturation compared with the binding level expected from the target coupling density and molecular weight (Danielson, 2009) can be used as a rough indication of protein integrity. While high-quality protein is a requirement, SPR biosensors only require small quantities of it to achieve dense surfaces. In practice, only a few micrograms are needed for each round of coupling or capture. As little as 0.5 mg of quality protein can be sufficient to support assay development, screening, and additional SAR support.

5. Buffer and Compound Preparation

The SPR effect is sensitive to all molecules passing through the plasmon wave. Thus, care needs to be taken in buffer preparation as small differences in salt concentration, additives, and especially high refractive index components, such as DMSO, will all have an effect on the signal. Careless buffer preparation means that specific binding may be hard to distinguish from bulk solvent shifts.

Typical instruments and proteins are stable in buffers with 5% DMSO. Thus, when preparing 1 L of running buffer, add all buffer components except the DMSO, dilute to 975 mL, and mix. Remove 15 mL and label as buffer-DMSO. Then add 50 mL of dimethyl sulfoxide (DMSO) to make 1 L of buffer + DMSO and filter through a 0.2-μm filter. DMSO-resistant filters, such as nylon or Teflon, and housings made of polypropylene, such as the Zapcaps-CR (Whatman, #10443421), must be used to prevent DMSO leaching filter and plastic components into the buffer. For screening plates, additional buffer-DMSO will be needed, so instead dilute with water to 1 L, then remove 50 mL of buffer-DMSO, and add 50 mL of DMSO. It is advisable to degas buffer+DMSO solutions whether or not the instrument has an in-line degasser. Maintain the vacuum after filtering, swirl the buffer, and strike the bottle flat on a table to induce cavitation and out-gassing. The buffer can be placed in a sonicating water bath under vacuum as well. While an in-line degasser helps prevent bubbles from developing in the microfluidics over the course of a day due to in-gassing of the running buffer, this same buffer will be used to make and dilute samples that will be injected and could introduce bubbles during the association phase. Use degassed buffer for making samples and in-gassing will be minimized by sealing the samples.

Small-molecule samples are almost invariably handled as DMSO stock solutions. DMSO has a very high refractive index such that a 0.1% variation in DMSO concentration between the sample and running buffer translates to 100 RUs of SPR signal (2002). Because DMSO solutions rapidly absorb water from the atmosphere, it is impossible to maintain a constant DMSO concentration from sample to sample, especially if plates of compounds are handled and opened more than once. This variation can be minimized by purchasing smaller bottles of DMSO (<500 mL), minimizing time of air exposure, and making all buffers and compound dilutions for a given experiment from the same bottle of DMSO. However, fragment libraries pass through many steps in compound management groups and maintaining constant DMSO concentrations for an entire library is impossible. These differences in DMSO concentration will affect the observed signals in the SPR experiment because of the excluded volume effect (Navratilova and

Myszka, 2006). Briefly, protein coupled in one cell takes up volume within the plasmon detection region that is still open in the reference flow cell where no protein is coupled. Thus, during an injection, more DMSO molecules will pass through the plasmon wave in the reference cell than the experimental cell and, after reference subtraction, produce an incorrect representation of binding differences between the experimental and reference cell that is proportionate to the DMSO difference between the sample and running buffers. This amount will be different for every injection. While this difference is minimized when concentrated DMSO stocks are diluted with fresh DMSO, and concentration series are diluted with running buffer, the effect is very significant when screening through plates of hundreds of compounds, each of which may have been handled differently in compound management. A solvent correction procedure can be used to correct for this effect. Take two 1.5 mL samples of running buffer and adjust one up ~1.5% in DMSO concentration by adding 100% DMSO, and the other sample down in DMSO concentration by ~1% using the buffer-DMSO sample. Make at least three but up to six linear serial dilutions by mixing these two buffers together in different proportions and add the five to eight total solutions to a plate or tubes in the instrument. These will be injected as additional samples during the experiment to construct a standard curve that is best fit with a second-order polynomial. The curve is plotted as the response on the reference flow cell versus the response differences between the reference and experimental cells. Individual samples in the experiment are plotted based on the bulk signal differences between the experimental and reference flow cells observed at the start of the injection and applied to the standard curve to derive how much additional signal should be added or subtracted from the association phase data to eliminate the DMSO contributions (Fig. 8.4). More details of this method and the exclusion effect can be found in (Frostell-Karlsson *et al.*, 2000). Fortunately, all modern SPR data analysis packages contain a routine to handle this correction automatically. Some users prefer to repeat the DMSO standard injections at even intervals throughout the experiment while others prefer to run the DMSO standards in duplicate or triplicate and scattered throughout the run. Testing of various methods has shown that these are both redundant and testing the samples once at the beginning of an experiment is sufficient to apply over the entire experiment. Replicates usually overlay exactly or closely enough that there is no effect on the final applied correction. If it is found at the end of the experiment that the curve was insufficiently wide, or some other error was made, a new curve can be made immediately and applied to the dataset retroactively. It is important that any injection lying outside of the samples used to create the standard curve be eliminated from the dataset going forward. Extrapolation of the curve is inexact and will result in calculation of an incorrect binding level.

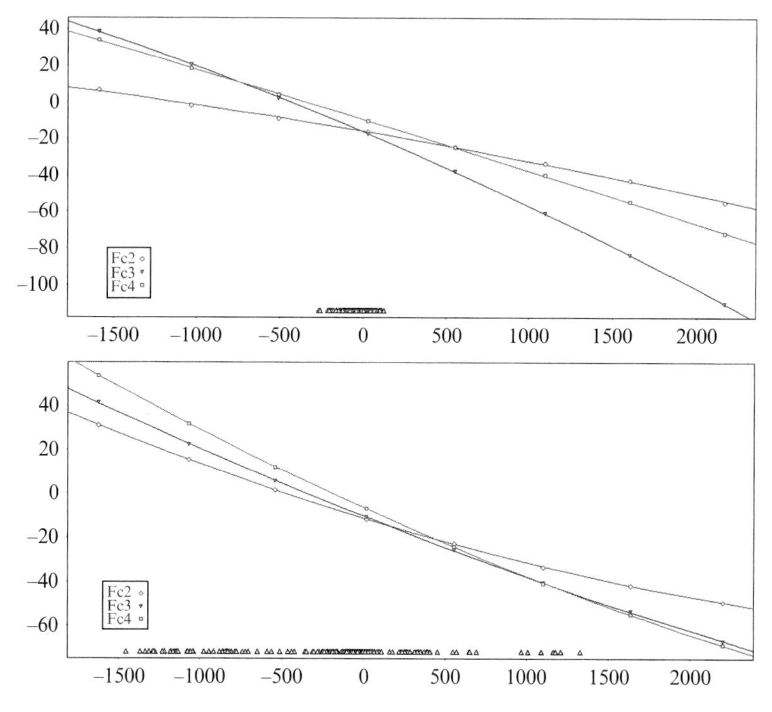

Figure 8.4 Example solvent correction curves from fragment screening. Each panel represents a screen of 320 fragments on three proteins. The X-axis is the bulk response value observed on the target flow cell and the Y-axis is the difference in response between the target and reference flow cells. Eight solvent correction solutions were used to construct the standard curve. Variation in sample DMSO levels is shown by plotting a triangle on the X-axis representing the bulk shift in response on the target surface due to solvent mismatch. Due to differences in capture levels and target weights, the shape of the correction curves varies between days and flow cells and is rarely linear. In (A), the dispersion of DMSO concentrations in the fragment plate is low but is very high in (B). All points are within the correction curve and can be retained for analysis. (See Color Insert.)

6. Assay Development

Once sufficiently pure and active protein has been obtained then attempts to couple or capture the protein to a sensor chip can begin. If possible, more than one coupling method should be attempted, and if tags are being used they should be tried both at the N- or C-terminus. Testing of the assay quality is best approached using small-molecule control compounds in the affinity range from approximately 100 nM to 10 μM. These compounds are likely to exhibit kinetics, which are useful for a better determination of K_D

during the early stages of assay development. Generally, compounds in this affinity range dissociate to baseline in 1–2 min speeding up cycle times and assay development. In addition, they will not need a regeneration strategy, which involves searching for a buffer that can strip the small molecule from the surface without damaging the protein. Surface regeneration for tight binding small molecules can be challenging and time consuming, and different conditions may be needed for different small molecules, even on the same target (Papalia *et al.*, 2008). A panel of 20–50 compounds from diverse scaffolds with known IC_{50} or K_I values is ideal to establish a correlation between binding and inhibition. A large panel is also useful, as some compounds, especially from the literature, will often be insoluble or will suffer from promiscuous binding (McGovern *et al.*, 2002; Section 12). For targets with a few or no known binding ligands, limited fragment screening (Section 9) can often be successful in identifying compounds that can serve as controls.

It is advised that the panel of candidate control compounds be tested under a variety of buffer conditions early in assay development. The SPR fragment screening assay requires proteins to be sufficiently stable for at least 24 h and ideally over many days. Certain combinations of buffer components may interact with the SPR instrumentation in ways that affect the data quality, and these may even be different for different instrument types, even within a vendor's product line. For example, note the difference in data quality between the sensorgrams shown in Fig. 8.5A and B. The only variable between these experiments was a difference in the detergent. While the kinetics, K_D, and amplitude of the data are equivalent, the injections in Fig. 8.5A appear much noisier and may reflect that, under this particular combination of buffer components, the Brij-35 is not as well tolerated as the Tween-20 and is causing problems with tubing or microfluidics, but not the protein. In contrast, Fig. 8.5C shows clean data with kinetics and K_D similar to Fig. 8.5A and B, but with a 2.5-fold larger amplitude indicating that the protein prefers dithiothreitol (DTT), and possibly Triton. In Fig. 8.5D, the combination of DTT and Tween-20 enhances the window by an additional twofold indicating a protein preference for these additives. By systematic variation of single buffer components, the user can eventually determine a condition that gives a high signal, low random noise, good reproducibility, and high agreement between IC_{50} and K_D. Note that for some systems, agreement with IC_{50} may not be expected if the form of the protein is not the same as the enzyme assay. In these cases, a similar rank ordering or a small 3–5 shift may be tolerated. For most systems, however, the agreement between pIC_{50} and pK_D is excellent with a slope close to 1 and R^2 values near and sometimes above 0.9 (Fig. 8.6).

For ~30 different SPR assays developed in house, 5% DMSO was tolerated in all cases except one, where it had the effect of lowering the signal intensity but did not affect K_D. While care should be taken to look for effects of DMSO, proteins are generally significantly less sensitive to it than

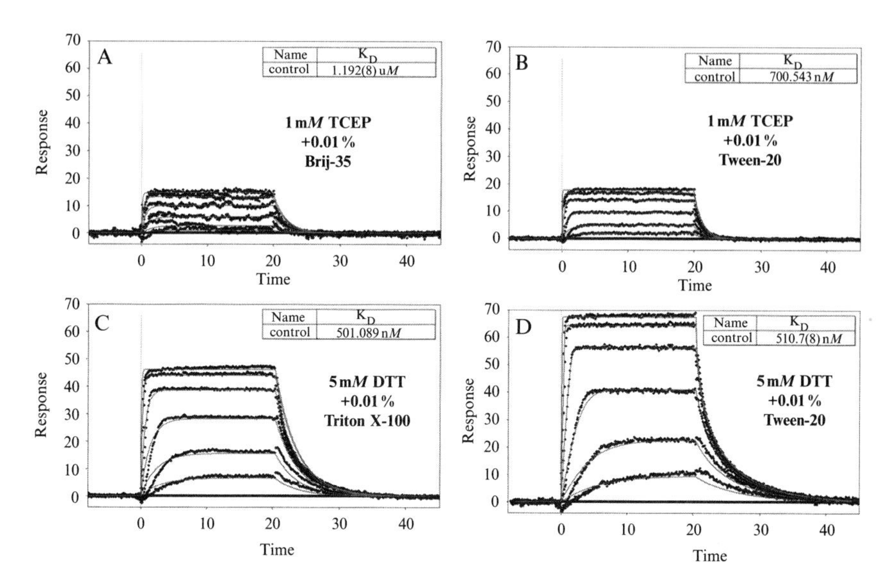

Figure 8.5 Buffer testing cycle during kinase assay development. In (A), the combination of Brij-35 and 1-mM TCEP results in low signal and waviness in the association phase data that is resolved in (B) when the detergent is changed to Tween-20 indicating a buffer more suitable to the instrument fluidics. In (C), the combination of Triton X-100 and DTT maintains artifact-free sensorgrams while increasing the amplitude, indicating the protein prefers either Triton or DTT. In (D), the combination of DTT and Tween yields the highest signals and clean sensorgrams free of fluidic artifacts. The K_D determined in each experiment is listed in the inset and agrees well with the K_I of 1.23 μM.

pH, detergent, and type and concentration of reducing agent. Some assays have shown better signal and stability when DTT is increased from the standard 5 to 50 mM. Regarding salts, it is not recommended to run binding assays in no salt, as this significantly raises the chances for nonspecific binding. Many assays use 100–150 mM NaCl, and higher or lower NaCl rarely has any significant effect on control compound affinity.

A final consideration for assay development is temperature. Most SPR instruments can vary the analysis temperature between ~4 and 40 °C. Generally, initial assay development and screening are performed at 20 °C, as the instruments and protein are often more stable and better behaved than at 25 °C. Many proteins can be significantly stabilized by lowering the temperature. However, great care should be taken in interpreting K_D values determined at extremes of temperature as binding affinity for some compounds can be very temperature dependent resulting in large K_D changes. In addition, the off-rate can be significantly slowed resulting in a control dissociating too slowly to be useful as a screening control (Navratilova and Hopkins, 2010).

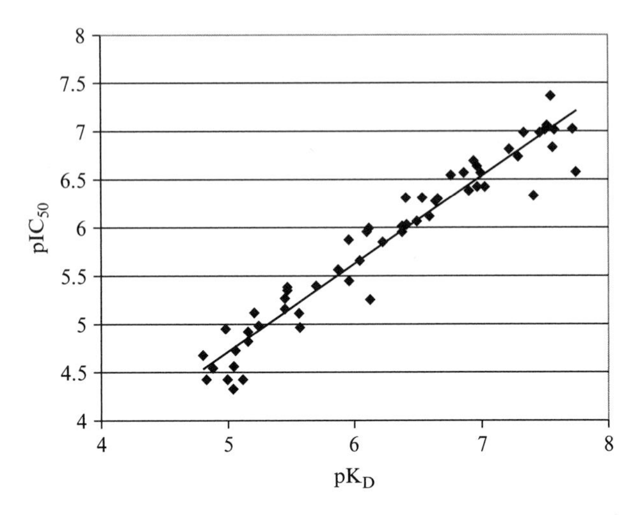

Figure 8.6 Plot of the pK_D versus pIC_{50} for 61 compounds tested against a protease. The slope of the fitted line is 0.91 and the $R^2 = 0.94$ indicating strong agreement between binding and inhibition. In general, K_D values will be slightly more potent than IC_{50} values for competitive compounds because of the lack of substrate competition in the binding experiment.

7. Aligning the SPR Assay with Crystallography

Once a coupling strategy, controls, and buffer system have been developed, it is worth considering a final round of buffer testing to align the assay more closely with the crystallization conditions for the target. Successful fragment development is heavily dependent on obtaining high-resolution fragment/protein crystal structures. Differences in conditions between the SPR screening and crystallization assays can significantly decrease the chance of subsequently obtaining structures of the fragment hits. Thus, it is recommended that some attempt be made to translate the screening conditions to more closely represent the crystallization conditions, accounting both for the contributions from the protein buffer and precipitant in the crystallization experiment. SPR instruments will not tolerate 25% PEG-8000 or 4 *M* ammonium sulfate, as those solutions have a high refractive index and/or viscocity out of range for modern SPR instruments. However, 1% PEG-3350 and 2–3 *M* salt can be easily tolerated and often provide a significant reduction in nonspecific or additive sensitive binding (Perspicace *et al.*, 2009). In some cases, compounds that bound the target in the SPR assay, but could not be cocrystallized, were retested in the presence of 1% PEG-3350. These compounds either no longer bound, or in one case became a promiscuous binder. Thus matching additives as much as reasonably possible is critical for improved crystallographic success with fragments. In most cases, a change in buffer pH of 1–2 units will not adversely affect control compound binding and can be tried to improve alignment with crystallization. Similarly, if SPR is used to characterize high-throughput screen (HTS) hits, or to compare K_D and IC_{50}, alignment of the buffers to match the HTS or enzyme assay conditions should be attempted.

8. Pilot Screening

The final test before the assay is ready for fragment screening is to perform a pilot fragment screen. In this experiment, a single fragment plate is tested with the screening protocol, and the reproducibility and decay rate of the control binding level and apparent hit rate are assessed. If the control binding level decays too quickly, or if the hit rate is significantly higher than expected for the target class, additional refinements to the protein construct, surface preparation method, control choice, or buffers may be needed. Figure 8.3 shows the same fragment plate tested on a protein lot with ~10% specific activity and apparent hit rate of ~15% (expected <5% for this target class). Based on these results, the protein purification procedure was refined to yield a protein preparation with >90% specific activity. The same fragment plate was retested to yield a hit rate of ~5%. These data suggest that the less active protein bound compounds nonspecifically resulting in significant false positives. A plate of DMSO or identical control compound dilutions can also be used to assess the reproducibility and long-term stability of the surface.

9. Setting Up the Fragment Screen

Executing a fragment screen, especially on larger libraries (e.g., 3000–6000 compounds), requires achieving significant screening throughput without compromising data quality. Because low molecular weight, low-affinity fragment binding signals are near the detection limits of modern instrumentation, compromises in data quality cannot be tolerated. Shortcuts such as low-quality injections, low data collection rates, improper referencing procedures, or inadequate washing can yield results that are difficult or impossible to interpret and can increase false positive and negative rates. Fortunately, almost all of the techniques described here to enhance data quality have a minimal impact on speed, but give a significant improvement in data quality. In addition, most SPR instruments allow the testing of more than one target at a time, allowing for the possibility of multiplexing screening activities. Of course, screening multiple targets requires that all proteins being screened are compatible with a single buffer system. This is often achievable, especially with kinases, but even proteins from different families can sometimes be paired with only minor modification to the conditions.

Figure 8.7A shows an example of a fragment screening platemap using a Biacore T100 sample racks color coded by the type of solution located in each position. The roles of each solution type in the screen are discussed below. This setup can usually be adapted to other SPR platforms with minor modifications, including ones with parallel injection modes.

Startup cycles: SPR instruments, like other sensitive analytical instruments, need to perform some cycles of sample injections before the machine equilibrates and starts producing stable data. The early cycles can be contaminated by extra baseline drift and injection artifacts. Performing eight injections of running buffer (Fig. 8.7A, yellow) will allow the baseline to stabilize and prevent these artifacts from affecting data. These cycles will be discarded during data reduction. Some SPR instruments are programmed with a special startup-cycle function that draws running buffer directly from the buffer tank instead of a plate thereby preventing the use of positions in the 384-well plate or rack for the discarded injections. However, it is important that the machine carry out all the activities it will perform when testing fragment samples, and thus the user should consider having these cycles sampled from the 384-well plate.

Solvent correction: Normally, five solvent correction samples are sufficient to correct a wide range of buffer/DMSO mismatches. Given the convenience of splitting a single 16-well column in a 384-well plate into startup cycles and solvent correction solutions, Fig. 8.7A (magenta) shows the solvent correction extended to eight points, providing finer sampling of the correction curve. Because of this an operator can consider widening the DMSO concentration range of the correction solutions more than normal. This is especially useful for fragment screening where plates may have been replicated using robotics and possibly exposed to air for long periods of time, increasing DMSO/water content dispersion. Figure 8.4A shows the solvent correction curves for two plates from a screening library. Both plates were formulated at the same time, but one has a small range of DMSO variation (Fig. 8.4A) while the other shows a large amount of variation (Fig. 8.4B) indicating differences in how they were handled by compound management.

Buffer blanks: 16 wells (cyan) are dedicated to replicates of the running buffer to be used for double referencing during data collection. No matter how well prepared an instrument is, the injections of buffer blanks will never be perfectly flat, may exhibit drift, and will reveal contributions from systematic injection and valve-switching artifacts that complicate the shape of sample injections. Double referencing removes systematic artifacts by recording the injection of buffer identical to the running buffer and subtracting it from sample data. Typically, it is performed after the subtraction

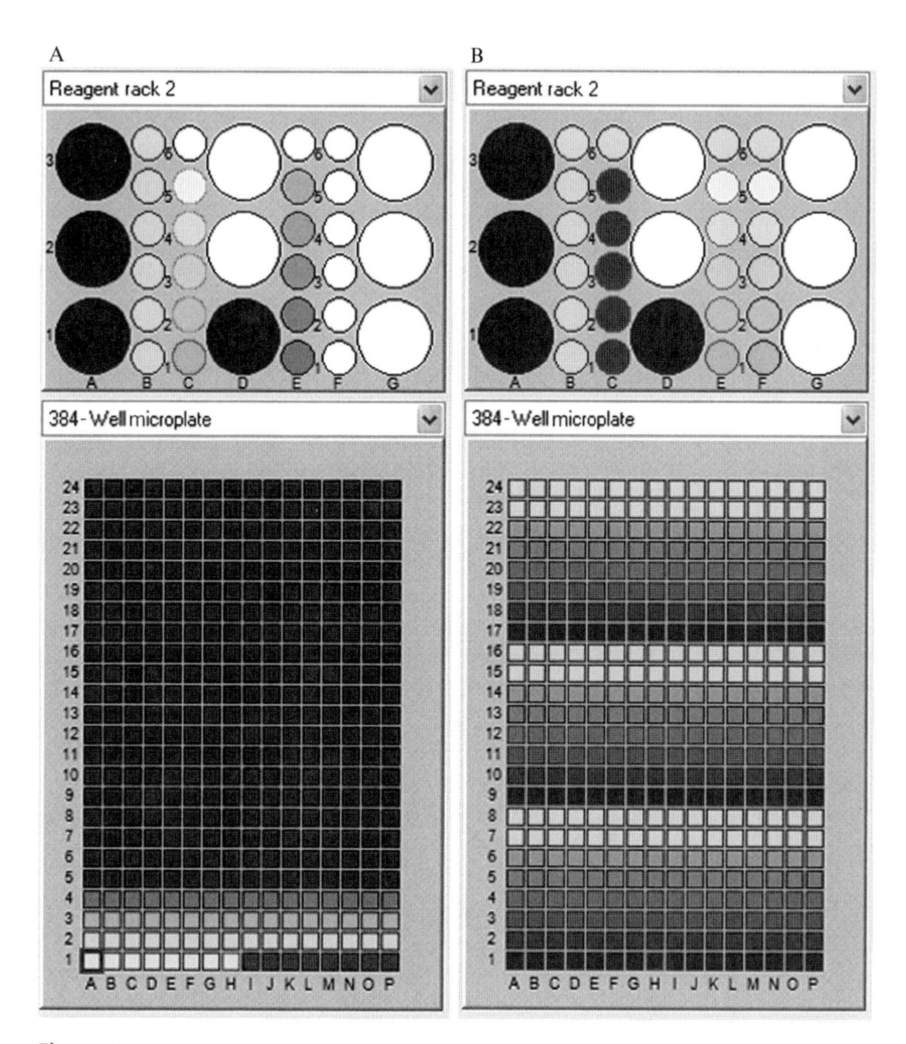

Figure 8.7 Example platemaps for fragment screening and characterization. Maps are specific to the Biacore T100 SPR platform but can be adapted to other systems. (A) Screening platemap with positions colored by sample type. Yellow: Startup cycles, magenta: solvent correction, cyan: running buffer used for double referencing, green: first control compounds, orange: second control compound, red: fragment, and blue: wash solution. Often the wash solution is simply the running buffer but can be different for various projects. In the reagent rack, dilution series of the controls are indicated by a color gradient in the green and orange samples. The blue samples are the blanks for double referencing the controls dilutions. (B) Colored as in (A) but with the layout for a dose–response plate. Forty-eight fragments (red) are shown in dilution with two buffer blanks for double referencing. There is only sufficient space for two dilution series of one control compound to be tested at the beginning and end of the run. If an additional control is required, then two fragment positions can be used for the second control reducing throughput to 46 fragments/machine/day. (See Color Insert.)

of contributions from the reference slow cell. Software packages that support double referencing on screening data, such as Scrubber2 (BioLogic Software, Campbell, Australia; http://www.biologic.com/au), offer the option of subtracting the average of all blank injection, or subtracting the blank injection closest in time to a given sample injection. Generally, it is best to subtract the closest blank as systematic artifacts can change over time.

Some users advocate skipping the blank injections and the double referencing procedure because of the extra time costs it brings to the screen and argue any systematic artifact should affect all injections equally. In practice, the additional injections represent $< 4\%$ of the total screening time, and frequently the systematic artifacts drift over the course of a single day's screen, which can last up to 23 h. It is reasonable to expect that subtle behaviors in a sensitive instrument can drift over such a long time period and the artifacts at the end of the screen can be different than those at the beginning. Indeed, Fig. 8.8 shows the equilibrium response values for two fragment plates screened on the same machine using freshly neutravidin-captured target protein. The data are presented both single referenced (subtraction of only the reference channel) in Fig. 8.8B and D and double referenced (subtraction of the reference flow cell followed by subtraction of the closest buffer blank Fig. 8.8A and C). All 16 blank injections used for the double referencing are shown in the insets. In Fig. 8.8A, the inset buffer blanks exhibit a clear negative response of ~ 12 RU during the injection phase. The magnitude of the drop decreases over the course of the experiment. The plot of binding response versus injection cycle for the entire plate exhibits a constant downward drift in the baseline when double referencing is not used (Fig. 8.8B). However, when double referencing is implemented as shown in Fig. 8.8A, the drift is essentially eliminated, the baseline is more flat, true hits no longer have a negative response value, and the results are easier to interpret. Figure 8.8D shows the data for a different fragment plate where the buffer blanks and nonreferenced data more closely aligned and exhibit less drift. Nevertheless, application of double referencing still provides some improvement of the baseline. Statistical analysis of the baseline shows that the double referencing improves the standard deviation and reduces skew by one-third. It is interesting to note that the only difference between these two datasets is the contents of the fragment plate used, indicating the double referencing controls not just for systematic drift in the instrument itself, but for artifacts contributed by the collection of compounds tested, possibly by nonspecific reference binding, sticking to microfluidics, etc. Thus, while double referencing does add an additional 16 injection cycles to a single experiment, the significant enhancement of data quality it provides can make the difference between a usable and unusable dataset.

Controls: A round of SPR-based fragment screening currently takes $\sim$18–23 h (depending on instrument) and reuses the same protein and reference surfaces the entire time. Inclusion of a control injection at regular

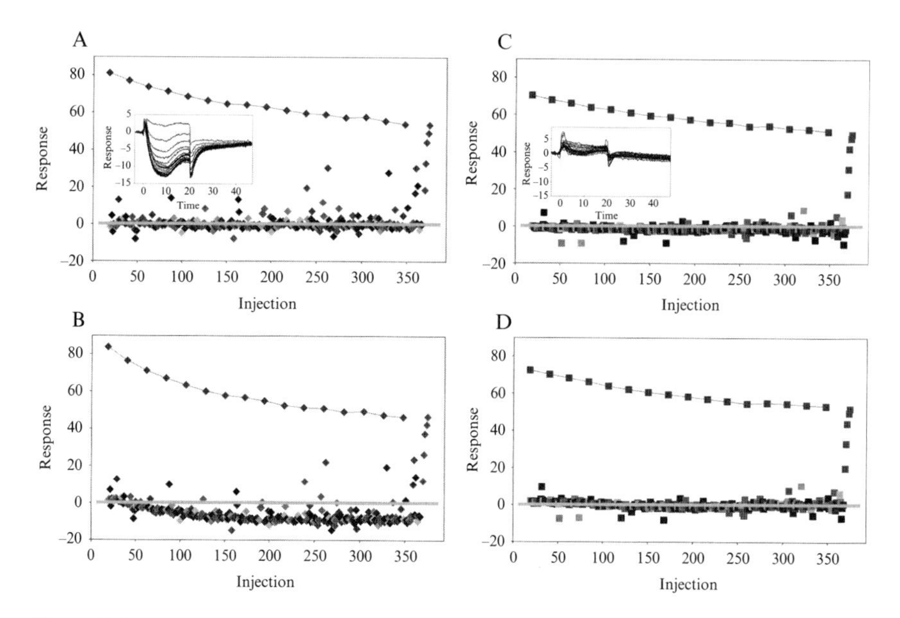

Figure 8.8 Effect of double referencing. Equilibrium response values for two different fragment plates on the same target and machine (but fresh surface) are plotted with (A) and (C) and without (B) and (D) double referencing. The 16 buffer blank injections from each screen are shown in the insets. The zero RU line is highlighted in green. Connected red dots are control replicates, magenta samples at the end are the control dilution series at the end of the run, and the rest of the points are individual fragment samples. The lack of double referencing in (B) significantly distorts the baseline and allows some fragment hits to have negative RU values, as compared with the double referenced data in (A). The improvements in the data in (C) are hard to visualize but statistical analysis reveals significant improvement in standard deviation and skew (see text). (See Color Insert.)

intervals throughout the assay is required to observe the assay window, the rate of active-surface decay, and changes in the control kinetic parameters and K_D over the course of the experiment. Typically, 16 replicates evenly distributed over a single 384-well plate (Fig. 8.7A, green and orange; every 20 fragment samples and run after the blank injection) are more than sufficient. Almost all protein surfaces show a loss of the active fraction with time, most likely due to unfolding. Fortunately, this loss can generally be corrected for, even with fairly significant decay (see Section 13). In addition, all capture systems, including biotin/streptavidin, will show a relatively constant decay due to loss of protein from the surface that can also be normalized and corrected with control replicates. Software permitting, all control replicates can be independently kinetically fit to determine the on/off rate and K_D to look for sudden changes in the binding constant or R_{max} that may indicate surface damage. While this is almost never observed, it can be worth checking to ensure high-quality data.

SPR fragment screens require very high target coupling densities (typically 5000–10,000 RU) making it difficult or impossible to use native ligands such as protein-binding partners, peptides, or nucleic acids as screening controls. When flowed over high-density surfaces, these types of binding partners tend to exhibit significant artifacts and/or stick to the chip permanently because of avidity and rebinding effects. These classes of molecules can be very useful on low-density surfaces during the early stages of assay development, especially when optimizing buffers. Thus, the ideal screening control is a compound that is soluble at concentrations ~5- to 10-fold higher than its K_D, so the binding response level at surface saturation can be established, displays detectable off-rate kinetics, and dissociates to baseline in fewer than 90 s so as to not block up binding sites or lengthen the screen. Compounds with these properties generally have K_D values between 100 nM and 10 μM and can be identified during initial assay development. In cases of novel targets or novel target classes where no known ligands are available at all, the high hit rates expected in most fragment screens (3–15%) can be leveraged to search for a compound from the fragment library that can serve as a workable control. In this approach, a small fragment screen of up to 1000 compounds can be run without controls. The data are carefully and manually inspected to identify compounds that appear to bind cleanly. As many putative binders as possible are tested in a dose–response experiment to identify the best binders. This approach has yielded potent (<100 μM) and selective compounds for all targets where it has been attempted. For example, two proteases were screened in parallel against ~400 fragments (Fig. 8.9). Visually, confirmed putative binders were tested, and the dose–response analysis of the most potent and selective hits for each protein is shown. While this process rarely leads to controls more potent than ~30 μM, those are still sufficient for controlling and executing a screen on the full library.

Most SPR instruments are capable of screening more than one target and frequently three compatible proteins can be screened in parallel. Note that the platemap in Fig. 8.7A only contains space for two sets of control replicates. This is because multiplexed screens can be accelerated if controls with the right binding profiles across the various targets can be identified. For example, if a single compound can be found that binds two or even all three targets, then the screening speed can be accelerated by using one control injection to satisfy two or three surfaces. In other cases, it is possible to identify two compounds that are perfectly selective for each of two targets and mix them together so that both are run in a single injection. This is especially easy if the targets are different classes (e.g., kinase and polymerase). By choosing nonselective controls, or mixtures of highly selective controls, screens of two and three targets can be carried out using only one or two sets of control replicates accelerating screening speed.

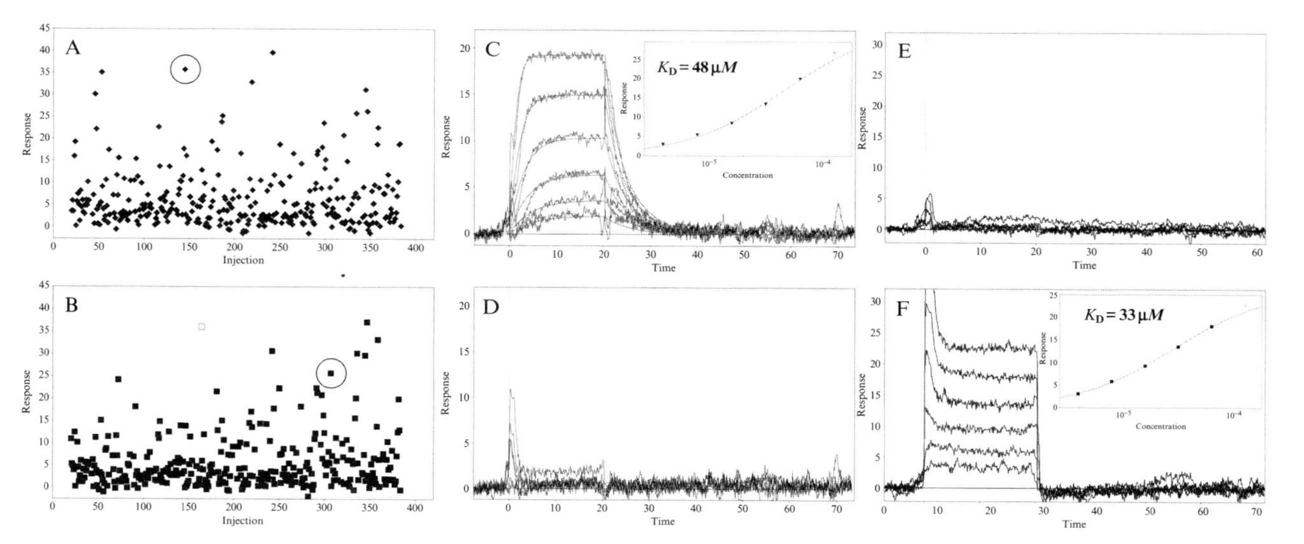

Figure 8.9 Using fragment screens to identify control compounds. A single fragment plate was screened against two proteases in parallel and the binding response plotted versus injection cycle (A and B). All compounds binding above the noise level were tested in a dose–response confirmation experiment against both proteins at 125 μ*M* in a twofold dilution series. The dose–response for the circled compound in (A) is shown for each protein in (C) and (D). The dose–response for the circled compound in (B) is shown in (E) and (F). Equilibrium fits are shown in the insets in (C) and (F). Both of these selective compounds are soluble up to saturation and were used to control fragment screens against each of these targets. The kinetics observed in (C) fit well to a 1:1 binding model and yield $K_{D,\text{kinetic}} = 51.3\ \mu M$, very close to the equilibrium value shown. Compounds this weak that yield observable kinetics are rare but do occur. The molecular mechanism that gives rise to the slower off-rates cannot be discerned from these data alone and are not discussed further.

Control dose–response: It is advisable to include a dilution series of the control compound(s) at the end of screen (shown as green and orange color gradients in Fig. 8.7A with cyan buffer blanks for double referencing). The dose–response curve based on these samples can be used to establish that the K_D for the control has not drifted by the end of the screen and to decide if there is sufficient signal window, based on the R_{max}, to reuse the surface for another day. In addition, collection of additional replicate measurements of the control K_D will be useful during the analysis of the complete screen (Section 13) to establish the K_D with more confidence and analyze the day-to-day performance of the screen.

Fragments: The red squares in Fig. 8.7A represent the fragment samples. In this format, 320 fragments are contained in the plate. This is the preferred layout wherein each fragment would be screened once at a single concentration. Some users screen each sample at two or more concentrations to boost confidence that the observed binding is real and to reduce the number of false positives passed to the more rigorous confirmation phase (Navratilova and Hopkins, 2010). However, for duplicate testing to improve total screening efficiency on a 6000 member library would require that more than 750 compounds (12.5% of the total library) be identified as false positives to make up for the extra 6000 injections at the second concentration. Fragment screen hit rates rarely exceed 10% indicating that inclusion of even a large number of false positives from the primary screen in the confirmation phase is more efficient than duplicate primary screening, especially for larger libraries.

Pooling or "cocktailing" fragments and screening the mixtures improve throughput for NMR and crystallographic screening (Jhoti *et al.*, 2007) but offer no improvement in SPR screening efficiency. Unlike NMR and crystallographic screens, which can readily determine the identity of binders from a mixture of fragments, SPR methodology only detects that binding occurred but cannot provide the identity of the binder from a mixture. As a result, every compound in a well containing one or more binders would have to be retested. Since the probability that a given well contains a hit will be multiplied by the number of fragments/well, the number of hitting wells in a singleton or pooled fragment screen will be statistically identical. The result is more compounds will always have to be moved into secondary confirmation in the multiplex versus singleton format, negating savings made during the primary screen. Library pooling in this format will only realize savings when hit rates are less than ~2.5%, as may be expected in an HTS campaign. However, fragment hit rates are often above 5% and can be as high as 15%, meaning pooling in SPR fragment screening will result in 30–300% more effort than the singleton approach.

The choice of compound concentration for the primary screen varies among users with some preferring millimolar concentrations (Hamalainen *et al.*, 2008) and others preferring concentrations around 100 μ*M*

(Navratilova and Myszka, 2006). Higher screening concentrations result in a higher occupancy of protein binding sites and larger binding signals making all hits, especially weakly interacting ones, easier to detect. However, as concentration rises the probability increases that compounds in aqueous buffers will bind nonspecifically to the target or reference, bind to secondary or tertiary binding sites, form promiscuously binding colloidal aggregates (Feng and Shoichet, 2006), or simply precipitate. Compound precipitation can be largely avoided by prescreening the compounds in a solubility assay during library quality control. Nonspecific binding can be reduced through the inclusion of detergents, ionic strength, and additives such as polyethylene glycol, but none are completely sufficient or as effective as lowering compound concentration. Enzymatic assays often include carrier protein such as bovine serum albumin or bovine gamma globulin to reduce nonspecific inhibition, and while these proteins can be tolerated by SPR instrumentation when no reducing agents are used, it is suggested their use be avoided as a large fraction of fragments are likely to bind to them increasing the false negative rate.

Promiscuous binding refers to the property of some compounds to form colloidal aggregates, like micelles, with diameters ranging from 30 to 400 nm (McGovern *et al.*, 2002). The compound aggregates are capable of binding protein with high stoichiometry and affinity resulting in binding signals in the SPR (or inhibition in enzyme assays) due to nonspecific, nondrug-like interactions. These molecules cannot be developed further into leads or drugs and must be identified and removed from hit lists before presentation to chemistry. More details on the use of SPR for detection and characterization of promiscuous binders are given in Section 12 and included references. A pernicious feature of promiscuous binders is that this behavior can be buffer- and target-dependent and observations have been made that a given compound may aggregate and bind promiscuously to one target but not to another (Giannetti *et al.*, 2008). Thus, despite efforts to purge them from libraries, promiscuously binding compounds will be a feature of all screens because every screen will involve different proteins and/or buffers. Some methods to remove promiscuous binders from SPR screening libraries have been proposed where an SPR fragment screen is performed, sensorgrams consistent with promiscuous binding are noted, the compound removed from the library, and the entire screen performed again (Hamalainen *et al.*, 2008). This approach doubles the time for a screen and places a burden on compound management groups to re-plate the library, which may not be compatible with project timelines in the pharmaceutical setting. While a library prescreen may remove some promiscuous compounds likely to offend in every screen, the changing nature of the targets and buffers means new bad actors will be observed in the next screen requiring a prescreen and library reformat for every screening campaign. In addition, observations from many screens have revealed that most promiscuous binders do not hit in every screen, or even most screens, but are observed infrequently and the set of observed promiscuous binders is fairly unique to each screen. Table 8.1 shows the number of times a given compound behaved promiscuously in 10 different screens. No promiscuous compound was identified in all 10 screens and only 18 were identified in 5 or more screens. In contrast, ~200 compounds gave sensorgrams consistent with promiscuous binding in four screens suggesting that removal of the most frequently poorly behaved compounds still leaves behind a significantly large number that can still aggregate or bind nonspecifically to targets not yet tested.

Ultimately, relative to true fragment hits, promiscuous binders represent a small fraction of fragment libraries and large efforts to identify and eliminate them from the library are disproportionate to the size of the problem. Table 8.2 compares the percentage of identified promiscuous binders with the percentage of validated fragment hits for 13 screens. In general, promiscuous compounds represent fewer than 2% of the total library, with half of the screens identifying fewer than 1% (or ~3 promiscuous compounds per plate), while fragment hits are usually in excess of 5%. Only one screen had a disproportionate number of promiscuous binders and was performed at an extremely low pH and very low ionic strength, far from the conditions where initial solubility testing was performed during library quality control.

In general, the probability of a compound forming a colloidal aggregate in aqueous buffers has been observed across multiple libraries to rise dramatically at and above 500 μ*M* (Hamalainen *et al.*, 2008; Navratilova and Hopkins, 2010), and screens performed below 500 μ*M* give cleaner results.

Table 8.1 Number of compounds identified as promiscuous binders in one or more screens of a ~2300 member fragment library

Frequency of compound identification as a promiscuous binder	Number of compounds
1	144
2	36
3	15
4	4
5	6
6	4
7	6
8	2
9	0
10	0

No compounds are badly behaved in all screens and most are identified only once.

Table 8.2 Fraction of promiscuous binders versus fraction of validated hits in 13 fragment screens of the same library

Screen ID	Promiscuous binders (%)	Hits (%)
1	6.35	2.88
2	2.33	6.65
3	2.24	10.76
4	2.16	2.63
5	1.52	14.91
6	1.52	5.00
7	1.52	13.70
8	1.02	5.05
9	0.68	4.62
10	0.64	3.26
11	0.38	12.54
12	0.30	9.19
13	0.13	6.18

In all but one case, there were significantly more hits than promiscuously binding compounds.

Lower concentration screening is feasible because the first phase of screening only requires identifying potential binders, not observing their binding at saturation. Modern SPR instruments generally have <0.5 RU short-term noise making the threefold signal-to-noise level equivalent to only ~5% compound occupancy on a typically dense protein surface (e.g., lowest concentration injection in Fig. 8.1). If a screen was conducted at 100 μ*M*, and 5% binding taken as a cutoff, then the weakest hits that could be typically detected are ~2 m*M*. Fragments much weaker than this will generally have a ligand efficiency too low to be of interest and represent significant challenges to obtaining occupied protein cocrystal structures necessary to initiate structure-based drug design. The affinity ranges of validated hits from 13 screens are listed in Table 8.3 and were initially identified from screens using 100 μ*M* fragment in the primary screen and 200 μ*M* in the dose–response confirmation phase.

10. Executing the Screen

Single use fragment screening plates are replicated from master plates with 320 compounds/384-well plate and contain 5 μL of 2 m*M* compound. A buffer-DMSO solution (95 μL) is added and mixed. Controls, buffer blanks, startup solutions, and control dilution series are racked as shown in Fig. 8.7A. This racking pattern is ideal for single injection instruments but may need modification for multiplexed instruments. The highest sample

Table 8.3 Affinity ranges of hits from 20 screens across four target classes

Target class	Most potent K_D (μ*M*)	Least potent K_D (μ*M*)
Cytokine	30	5000
Kinase	0.32	2800
Kinase	0.91	1090
Kinase	0.94	3520
Kinase	1.1	1180
Kinase	1.8	1700
Kinase	2.3	1990
Kinase	4.7	2600
Kinase	10.5	2010
Kinase	15.6	2900
Kinase	64	3400
Kinase	72	2400
Kinase	123	1230
Kinase	70	3900
Polymerase	65	4000
Protease	2.9	2300
Protease	44	2010
Protease	207	1900
Protease	210	3900
Protease	240	2900

throughput that maintains high data quality is necessary to combat surface decay and allow for time to prepare the instrument the next day to continue to screen. A screen as shown, using two controls, requires 408 injection cycles. Thus, every 1-s reduction in time per injection cycle reduces the experimental run by ~7 min. Typically an association phase time of 20 s is used, followed by a dissociation measurement for 10 s. This is sufficient time for well-behaved fragments because equilibration and complete dissociation to baseline are nearly instantaneous. The extra time is used to note whether the sensorgram exhibits nonideal behavior. If a chosen control needs extra time to dissociate to baseline, then the dissociation phase for just that control, and the buffer blanks, can be lengthened while maintaining short times for the fragment samples. If the blank is not extended, then double referencing cannot be properly performed across the entire dissociation phase. Note that most instruments continue to provide buffer flow over the surfaces during needle wash steps, so additional dissociation phase data can continue to be collected, albeit with more systematic noise, beyond the end of the programmed dissociation phase to complete control dissociation and save time.

Injection-based SPR instruments all exhibit signal artifacts that occur at the beginning and end of injections. In addition, the longer the compound solution stays in the sample loop the more likely the shape and quality of the data will degrade due to nonspecific binding to the tubing or diffusion of the sample plug. Thus, it is recommended to use the highest flow rate allowed by the instrument and sample availability, often 100 μL/min, to obtain cleaner data. Even with high-flow rates, data points will be lost around the injection stop and start. For this reason, and to aid in sensorgram review (Section 12), it is recommended to use the highest data collection rate the instrument supports. Typically this is around 10 Hz. Some users advocate low-data collection rates to minimize the size of data files, especially for large screens, but this comes at the expense of obtaining the highest quality data required to differentiate weak binding from noise. Data collected at lower rates can be much harder to interpret, making it difficult to identify promiscuous binders, and resulting in additional false positives and negatives.

While it does add significant time to the injection cycle, the use of extra wash procedures between injection cycles (e.g., extraclean routine in Biacore products) is critical. Washing reduces the probability and magnitude of compound carryover from injection to injection. A wash in the running buffer is typically sufficient; though making a wash solution by adding an equal volume of 100% DMSO to a sample of running buffer can provide more thorough cleaning. The stability of the wash solution container and cap to high DMSO percentage solutions should be tested as some rubber caps and tubing have been observed to slowly dissolve in such solutions and deposit substances leached from the plastic onto the biosensor surface. Most instruments offer the ability to perform a carryover injection where the injection needle aspirates a small amount of running buffer and injects it over the surface revealing if additional compound is still stuck to the fluidics. While useful, this adds significant time to each sample and is unnecessary, as most samples do not carry over. Instead, signs of carryover will be obvious during data reduction and any negative effects on the subsequent sample can be managed during data evaluation and hit selection (Section 12).

11. Primary Screen Data Reduction

Proper reduction of raw biosensor data has been extensively discussed elsewhere (Myszka, 1999) and will not be shown in detail here. Briefly, raw data need to be zeroed a few seconds before the start of the injection and the useful data cropped. Alignment in time such that all injections start at the same time is critical and software utilizing algorithms to identify the injection start times work very well. Some flow systems such as those in the Biacore A100, Biacore 4000, and S51 have all the surfaces arranged such that the sample injection hits all surfaces simultaneously. However, injection start times from other cycles generally do not align in time exactly, especially for samples further away from the injection needle where more time is needed to retrieve the sample from the plate. With 100-ms sampling resolution, proper time alignment is critical to minimize the injection artifacts. After subtracting the reference surface from each injection, solvent correction is performed and is implemented in a straightforward way by all modern data reduction programs such as those provided with modern SPR instrument and third party software packages such as the Scrubber2 (Bio-Logic Software). Most screening libraries will contain a few compounds that consistently fall outside the solvent correction calibration range, even with the additional widening of the standard curve. The fragments outside the range must be rejected from further consideration at this step.

Inspection of the blanks used for double referencing is critical to identify any blank that might be an outlier. When using the screening method described, a corrupted blank will affect the 20 fragments around it. Experimentation with deleting and adding back blanks while observing the effect on the results on samples close to that blank, as well as across the whole plate, may be necessary to optimize the data. While this is usually not the case, and frequently all the blanks agree well with each other, this can become critical when dealing with a large number of poorly behaved compound samples or unstable proteins. Sometimes an unusual blank is correcting for a systematic behavior and needs to be retained.

12. Data Quality Control and Extraction of the Equilibrium Binding Level

Once data have been properly reduced, the equilibrium binding level for every injection is determined by averaging the observed response near the end of the injection. The time window that is chosen should apply to all injections in the experiment. Typically a 2-s window (20 data points at 10 Hz) from $t = 16$–18 s for a 20-s association phase is sufficient to obtain an R_{eq} (equilibrium response) value that is not affected by the injection stop, or compounds with a drifting signal, such as promiscuous binders. Care should be taken to look across all the injections in the chosen time window for artifacts from bad injections, such as a random jump in signal for one or more samples, which could result in an erroneous R_{eq} value. Choosing a different region of the association phase for response calculation that is free of such artifacts is generally sufficient to circumvent the problem.

It is critical to check the data quality at the end of each plate before proceeding with the next day of screening in case of a sudden change in quality, systematic problem, unexpected surface decay, unstable control, etc. This is best tested by looking at a plot of the R_{eq} value versus the injection order (Fig. 8.8A). The shape of the baseline, drift in the positive control, apparent hit rate, and binding level from the control dose–response at the end of the experiment can be rapidly assessed and compared with previous runs. Next the control dose–response should be fit. Ideally the control has observable on- and off-rates, and these rates and K_D are comparable with the historic data from prior screening plates and assessments made during assay development. In addition, the rates and K_D can be individually determined for the 16 control replicates to look for any systematic change that could indicate adverse assay performance.

Additional quality control of the primary screening data can be performed by examining response values taken at other points in the injection cycle. For example, the binding level assessed 5–10 s after the end of the injection can reveal slowly dissociating compounds, such as promiscuous binders. The value of the baseline at the injection start before zeroing all injections can reveal sudden jumps in the baseline consistent with a large amount of undissociated compound, usually a promiscuous binder. Undissociated compound aggregates generally do not adversely affect the quality of the protein surface (Giannetti *et al.*, 2008) and zeroing the data eliminates contributions from baseline shifts. Only significant changes in sensorgram quality, signal intensity, and K_D or kinetic parameters of the control replicates are cause for concern.

The report point plots described above can be time consuming to create and interpret. In addition, they are significant simplifications of the high content information contained within the original sensorgram. A faster and more thorough way to validate the data is to visually inspect all the sensorgrams from the run looking for the hallmarks of promiscuous binding and other deviations from the expected characteristics of well-behaved fragments. This will also reveal additional information about machine status such as slow changes in injection quality or unexpected systematic data features, such as carryover. In practice, with the right software tools, a review of all sensorgrams for a complete plate takes fewer than 5 min, owing in part to promiscuous compounds being a small proportion of the total tested compounds. Samples yielding undesirable sensorgrams are flagged and not characterized further. User time for visual sensorgram validation is well spent as every compound flagged saves ~20 min of machine time in the dose–response evaluation, as well as user time eliminating or attempting to fit dose–response data from ill-behaved compounds. The visual inspection rate can be further enhanced when multiple surfaces are being used simultaneously by coloring each surface differently and then overlaying the sensorgrams in a single plot. Compounds that behave poorly on all surfaces versus just one or two surfaces are more easily spotted this way. Though rare, it is possible for "sticky" compounds to affect the quality of one or two compounds tested after it, due to carryover. In these rare cases, the user can inspect the subsequent data to see if there is evidence of well-behaved fragment binding convoluted with the behavior of the sticky compound and decide if it should get a "second chance" by manually selecting it for inclusion in secondary confirmation testing.

Almost all screens result in sensorgrams with large negative values where the binding curve looks "flipped" upside-down and the user is left to interpret "negative binding." This shape is generally the result of significant compound binding to the reference surface, but less well to the target surfaces (i.e., the target is blocking nonspecific binding sites on the surface). In these cases, any real binding has been obscured by the nonspecific interactions and the compound must be discarded from further characterization. It is unlikely that the compound has specific binding to the target, but if it does, it would have to be revealed by another technology because reference subtraction is fundamental to the SPR experiment.

13. Scaling and Normalization of Primary Screening Data

After the primary screen and data validation activities are completed, additional data correction and scaling must be applied before actives, which are unconfirmed potential binders, set can be selected for dose-response confirmation experiments. The binding signal in the SPR is a convolution of multiple parameters including target coupling density, target-binding activity, target molecular weight, sample versus running buffer mismatch, and compound/target affinity. To accurately select hits, data must be corrected for contributions from the first four factors so the fifth can be clearly determined. For example, if one plate was tested on a surface with 10,000 RU of target coupled, and the next plate tested on a new surface where it was only possible to couple 8000 RU, then the signals from the second plate will be reduced 20% relative to the first. The second plate may contain fragments with more potent binding than fragments on the first plate, but the 20% lower RU values may obscure this. In addition to interplate experimental variations, intraplate variations due to surface decay can also impact binding signals. For example, in a screen where 50% surface activity is lost, a compound binding with 10 RU early in the run is actually showing weaker binding than a compound measured at 10 RU at the end of the run. Similarly, how can the performance of a given compound in the screen on one target be compared with a different target protein run in parallel or in a different screen if only the RU level is compared? Differences in surface densities and molecular weights of the two targets will complicate comparisons based on response level. Small variations in buffer preparation day-to-day will affect the baseline such that all injections may have a systematic shift above or below the baseline by several RU.

Therefore, the screening data need to be corrected plate-by-plate for drift, coupling density, target molecular weight, target activity, and baseline shifts. Data from all the controls can be used to normalize for the first four factors, and a simple adjustment applied for the fifth.

The first correction that needs to be applied is for the drift in the controls. Control drift is the sum of losses of protein from the surface and decay in the fraction of active surface. The shape and magnitude of the drift can vary across plates. To perform the drift correction, extract the 16 control replicates from each plate and plot their R_{eq} value versus their cycle number (Fig. 8.10A). A third-order polynomial is fit to the 16 points, and the user can opt to eliminate obvious outliers from the fit. Because multiple processes contribute to surface decay, linear fits, second-order polynomials, and exponential decay functions are usually insufficient to model the decay. In addition, long-period fluctuations arising from systematic drifts in the instrument can often be observed and corrected for by the third-order fit. The third-order polynomial also allows for the variance amongst the control injections to be maintained, whereas experiments with fourth-order polynomials resulted in fitting of the noise and should not be used. Once the equation $Y = \mathrm{A}x^3+\mathrm{B}x^2+\mathrm{C}x+\mathrm{D}$ is determined for the control replicates, then every data point in the plate should be scaled by

$$R_{\mathrm{drift}} = \frac{R_{\mathrm{eq}}D}{Ax^3 + Bx^2 + Cx + D},$$

where R_{eq} is the equilibrium binding level in RU, x is the cycle number, and A–D are the coefficients from the nonlinear regression fitting of the controls. The resulting data are corrected for the drift and normalized to the Y-intercept, D, from the regression (Fig. 8.10B). While the data could be scaled to the R_{eq} value of the first control replicate, occasionally that data point can be corrupted by a bad injection and has to be discarded. Using the Y-intercept of the fit provides a systematic and nonarbitrary scaling point. This drift correction procedure has the advantage of scaling the fragment data continuously over the run, unlike a percent-of-control approach that would require binning fragments and scaling them to their nearest control.

Buffer mismatches between sample buffer and running buffer can contribute to the baseline noise scattering about a value other than zero, usually plus or minus 2 RU, though larger has been observed. For proper hit selection across multiple plates, the baseline needs to be adjusted around a fixed value of zero. To do this, note that most samples (85–97%) are nonbinding and the bulk of the tested fragments must be scattering around a common zero-binding level. Calculating the median value from R_{drift} of the points representing the obvious baseline (i.e., excluding the controls, obvious hits, and samples with very large positive or negative values due to nonspecific binding) will reveal how far shifted the entire dataset is due to

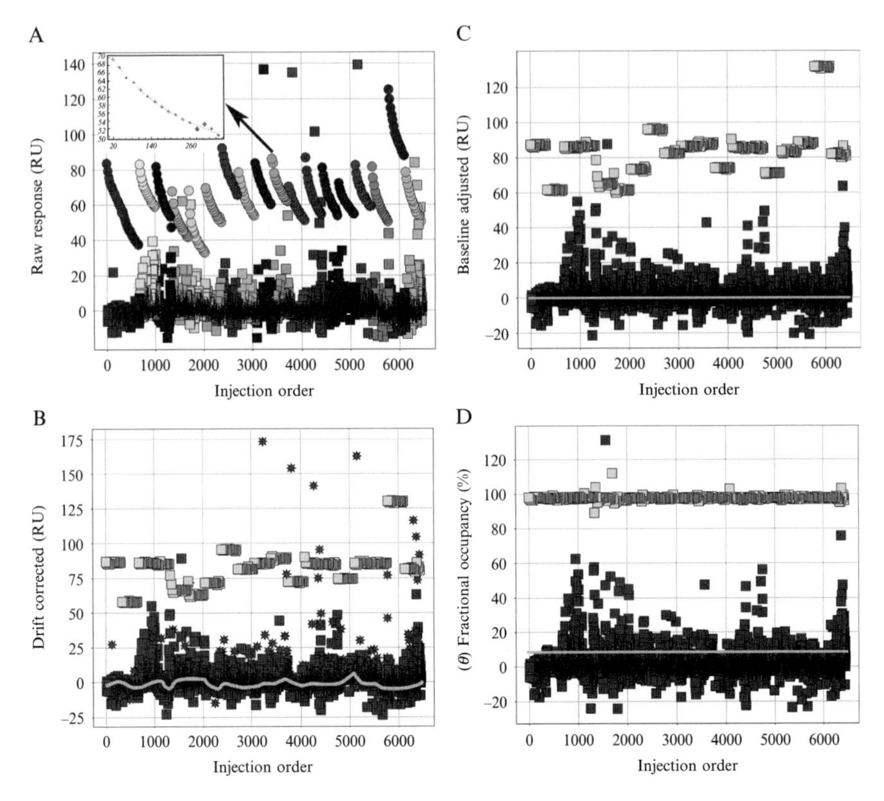

Figure 8.10 Scaling and normalization of primary fragment screening data. The plots represent a screen of ~6000 fragments (19 plates) on a single target. (A) Raw equilibrium response values (R_{eq}) from the instrument for all fragment and control replicate injections are shown versus the injection order. Points are colored by individual screening plate. Dose–response injections at the end of each plate are omitted for clarity. The inset shows a plot of the R_{eq} versus injection cycle number for the 16 replicate controls for the plate colored in gray. The green line represents the fit of a third-order polynomial to model the control decay. Two points highlighted in diamond boxes are statistical outliers and are excluded from the fit. (B) Screening data after correction for drift in the controls. Fragment samples shown in magenta, controls in cyan, and promiscuous binders identified from visual inspection of the sensorgrams are shown as stars and are excluded going forward. The green line trends the average of the baseline showing variance from day to day due to small buffer mismatches. (C) Data adjusted on a per-plate basis so the baseline scatters evenly around zero RU (green line). (D) Screen adjusted from RU to fractional occupancy, expressed as percentage, per the calculations described in Section 13. The average and standard deviation of the controls and baseline are 97.6 ± 1.33% ($N = 304$) and 0.17 ± 2.80% ($N = 5439$) yielding a Z′ factor (Zhang *et al.*, 1999) of 0.95. The three-sigma level used for initial hit selection is shown with a green line drawn at 8.4% and selects 344 compounds (5.7%). This initial hit rate is not officially quoted as only confirmed compounds from the dose–response analysis will qualify as true hits. (See Color Insert.)

buffer mismatch. That median value is then subtracted across the entire dataset to yield R_{med} and the baseline is now observed to scatter evenly around zero (Fig. 8.10C).

After these two adjustments, the data are still reported in RU. The RU value is affected by target coupling density, activity, molecular weight, and binding level. If the first three parameters are known the dataset could be scaled by those parameters, but rarely is activity precisely known or constant, and careful recording of the surface density would be required for every experiment. Instead, the response level of the control replicates, which have a known concentration and affinity, can be used to convert from RU to fractional surface occupancy (θ), where 100% would be binding to all available specific binding sites. This eliminates the need to consider density and ignores any inactive fraction of surface protein. Scaling to fractional occupancy, θ is accomplished by

$$\theta = \frac{R_{\text{med}}[100C/(C + K_{\text{D}})]}{\bar{R}_{\text{med,control}}},$$

where C is the control concentration, K_{D} is the control's equilibrium dissociation constant determined by averaging all replicate determinations of the control from the dose–response test at the end of each plate, R_{med} is a given compound's drift and baseline corrected value in RU, and $\bar{R}_{\text{med,control}}$ is the average value of the 16 control replicates (excluding obvious outliers) after the median baseline adjustment. Once applied, the controls will scatter by their noise level around the value representing the expected surface occupancy for their testing concentration and K_{D}. For example, if the control replicates are run at five times the K_{D} then their surface occupancy is 83.3%. Now individual fragments can be compared across different plates, and even different targets, on the same absolute scale (Fig. 8.10D). Some compounds may receive a θ greater than 100. The user will have to inspect the sensorgram and decide if this represents a previously unflagged promiscuous binder, is a multisite binder, or has a high refractive index increment relative to the control compound (Davis and Wilson, 2000). In practice, compounds with θ values below 150 often confirm, but only rarely when θ is greater than 150%.

14. Primary Screen Active Selection

Using the θ values, calculate the standard deviation of the baseline scatter and initially select the compounds with θ values greater than three times the standard deviation for confirmation. This can be performed on the entire screen, or on a plate-by-plate basis, and represents binders that are statistically likely to be real. In contrast, active selections made using arbitrary RU-based cutoffs on noncorrected or normalized data will result in more false positives and negatives due to the influence of the many experimental factors the corrections described in Section 13 eliminate. If the capacity for follow-up activities exceeds the initial hit list, then additional compounds can be added by lowering the bar. Many more real hits can be identified by selecting compounds in the noise, though they are generally weaker and the confirmation rate will be lower. If the number of hits exceeds follow-up capacity then testing cluster representatives or eliminating known or uninteresting scaffolds can reduce the list to a more manageable number. This is more effective than raising the cutoff threshold as chemical diversity may be lost if only the tightest binders are retained.

15. Collecting Dose–Response Hit Confirmation Data

Actives identified in the primary screen need to be confirmed in a secondary SPR experiment. False positives in the primary screening data can arise for many reasons including noise, bad injections, wrong compound identity, experimental errors, and certain types of promiscuous binding that can only be revealed by multiple concentration data. Thus, before compounds can be confirmed as hits, SPR testing of the compounds at multiple concentrations is required to confirm binding, as well as determine a K_D for affinity ranking and ligand efficiency calculation. Fortunately, the same assay format for the primary screen can be used for the confirmation stage by modifying only the plate map. Figure 8.7B shows a typical plate map for this experiment containing startup cycles, solvent correction solutions, duplicate dilution series of up to two controls to be run before and after the fragment samples, and 48 fragments to be tested at 6 concentrations, with two buffer blanks per fragment for double referencing.

After startup and solvent correction cycles are completed, one set of controls is run. The second set is run after all 48 fragments have been tested. These "early" and "late" controls will be useful in establishing the stability and reliability of the surface over the course of testing the fragments by revealing the degree of drift in the R_{max}, as well as any decay in the K_D of the control. Some decay in the R_{max} is expected due to slow loss of protein when using capture systems, and/or from protein decay due to unfolding, oxidation, proteolysis, etc. Up to 50% and occasionally 75% loss in R_{max} can be managed during data fitting but it is often around 10–30%, while some systems show almost no loss at all. Unlike the R_{max}, decay in the K_D generally cannot be tolerated or corrected. While some K_D drift has been tracked to control compounds that were chemically unstable in aqueous buffer, stuck to plastic, or were slowly aggregating and precipitating, these are rare and are typically discovered and replaced with better compounds during assay development. With suitable control compounds, K_D drift may indicate protein damage due to a surface ageing too long, mistake in buffer preparation, or surface exposure to a test compound that denatured the target (though these are exceptionally rare). Even a twofold weakening of the K_D can be considered a significant loss. Figure 8.11 shows the K_D of early and late controls for an SPR assay driving SAR for a project team over more than a year. Note that the standard deviations for the ~120 independent K_D determinations of the before and late control are ~15% of the average K_D value, well within the normal variance observed in SPR experiments (Myszka *et al.*, 2003).

In the platemap in Fig. 8.7B, 48 fragments are arrayed such that each fragment is tested once at six concentrations. Two wells of running buffer for double referencing are interspersed throughout the collection of multiple concentration data. Arguments have been made (Myszka, 1999) for sampling from the plate randomly with respect to both sample and concentration to increase confidence in the data, and some instruments even support an automated random sampling mode. However, the benefits of this approach are outweighed in fragment characterization by several factors. The random

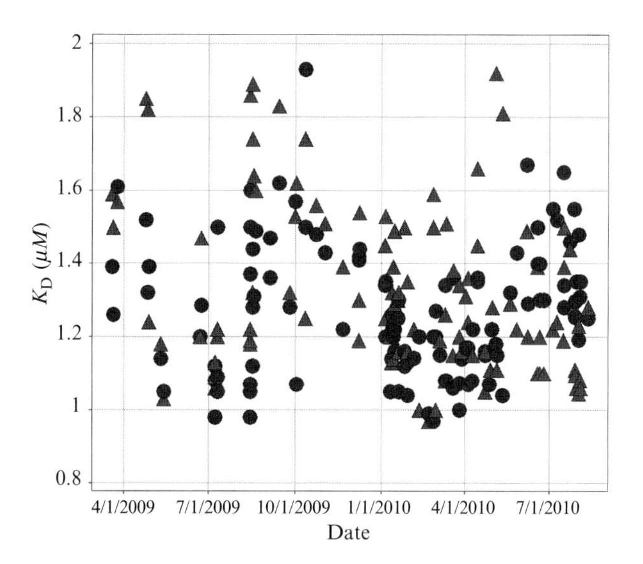

Figure 8.11 Reproducibility of "early" and "late" controls over ~18 months. The K_D values for 112 pairs of "early" and "late" controls used in a routine SPR-based SAR assay are plotted on a linear scale on the *Y*-axis versus the date of the collection. Note the *Y*-axis ranges from 0.8 to 2 μ*M*. The early controls are shown as blue circles and the late controls, at the end of the plate, are depicted as red triangles. The average and standard deviation for the late and early controls are 1.27 ± 0.18 μ*M* and 1.32 ± 0.22 μ*M* (*N* = 112 for each set). (See Color Insert.)

sampling approach was developed more around protein–protein interactions, which are less susceptible to drift over time than small-molecule work where small changes in surface stability can have a large impact on the ability to determine the K_D. Thus, sampling the high concentrations of one compound early, and the low concentration later, can result in a discontinuity in the data. Some badly behaved compounds are sticky and carry over into the next injection. While this usually invalidates only the compound in question, if it carries over into other samples it can corrupt their data. Run randomly, a sticky compound could interfere with six other compounds, whereas in a sequential mode this effect is minimized to only the next compound, and often only the lowest one or two concentrations. Having the samples collected randomly can make troubleshooting of spurious sensorgrams difficult and complicate other aspects of data postprocessing.

Running the six concentrations of each compound from lowest to highest concentration will minimize the impact of compounds that exhibit poor behavior, especially at higher concentrations, such as concentration-dependent aggregation, sticking to plastic, nonspecific reference binding, etc., and allow for the maximum number of unaffected data points per compound to be collected. Running one of the two blanks at the beginning of the series, and second in the middle, distributes all the blanks such that all but the second and fifth concentration are adjacent in run order to a blank for more effective double referencing. This requires the data reduction software to be able to use the nearest blank as opposed to the blanks assigned to the same compound identifier, as the first blank for the second compound, will also serve as a double reference for the last injection of the first compound, etc.

16. Dose–Response Data Reduction and Quality Control

After data collection, the sensorgrams for each compound must be visually inspected for quality. While many promiscuous binders were removed by sensorgram inspection during the primary screen, some types of promiscuous binding, such as concentration-dependent aggregation (Giannetti *et al.*, 2008), are not visible until tested in a dose–response format. Other types of poor behavior such as nonspecific reference binding (upside-down curves), nonbinding, and high stoichiometry of binding may also be observed and these should be removed before fitting. In general, a well-behaved fragment rises to its equilibrium binding level in 1 or 2 s, the equilibrium phase of the sensorgram starts completely flat until the end of the injection, and then dissociates completely to baseline within 1–5 s. A clear dose–response is visible when at least two concentrations are resolved above the noise. All injection in the series should have the same qualitative sensorgram shape. Changes in the shape at higher concentrations are indicative of solubility/stability problems with the fragment that may ultimately be compounded at the very high concentrations used in crystallographic experiments. While stringent curation of the dose–response curves at this stage can result in some false negatives, it is preferable in order to identify fragments with the highest chances of crystallographic success. A challenge of using very high compound concentrations (1–5 m*M*) as the maximal concentration for measuring K_D is the poor solubility of many compounds at these concentration. Contributions to the binding signal from nonspecific binding to the target and/or reference surface can complicate data analysis and invalidate many of the high-concentration points.

17. Global Analysis for K_D Determination

However, in the SPR experiment, the direct observation of the K_D is not necessary to determine the K_D. This is due in part to the low noise and high reproducibility of most SPR instruments, but largely because all compounds tested in the dose–response experiment as outlined will share the same target surface. This eliminates one term in the binding equation

$$R_{obs} = R_{max}\left[\frac{C}{(C + K_D)}\right],$$

where R_{obs} is the observed response, R_{max} is the maximal response, or top of the curve, C is the injected concentration, and K_D is the equilibrium binding constant. Because R_{max} is shared across all compounds in the experiment, including the controls, one value for this parameter can be used for the entire dataset. One method for fitting the data is to determine R_{max} from the early and late controls using standard fitting procedures and apply the average R_{max} value to all compounds yielding a data-to-parameter ratio for each fragment of up to 6:1. Another method, using software capable of global analysis, is to allow one free R_{max} term for the entire dataset, and one K_D per compound. In this way, every compound contributes some information to the R_{max} parameter and the total data-to-parameter ratio is 5.9:1. The shape of the binding curve is uniquely defined by just the R_{max} and K_D, so even curves with no observed inflection point can be well fit with fixed R_{max} to estimate the K_D in a reliable way. The fits can be further enhanced by controlling for variation in the compound molecular weights by first scaling the response values by dividing by the compound's mass and multiplying by 100 (Fig. 8.13 and legend). Table 8.4 lists the K_D values determined for 21 compounds in a single experiment in

Table 8.4 K_D of fragment binding determined with the locked R_{max} method

Compound	$K_D(6)$	$K_D(5)$	$K_D(4)$	$K_D(3)$	$K_D(2)$	$K_D(1)$
Early control	1.15	1.15	1.15	1.15	1.15	1.152
Late control	1.18	1.18	1.18	1.18	1.18	1.84
Fragment	82.30	82.30	79.30	75.90	74.80	74
Fragment	105.30	99.30	95.00	90.10	88.30	88
Fragment	108.30	105.50	104.80	99.20	91.90	90
Fragment	138.40	142.10	143.30	144.00	159.00	161
Fragment	144.00	152.60	146.40	142.10	148.00	138
Fragment	177.30	177.30	176.60	172.00	171.00	174
Fragment	186.60	200.80	192.00	184.00	**Fail**	**Fail**
Fragment	206.80	205.90	202.00	197.00	187.00	174
Fragment	217.00	216.60	212.00	202.00	198.00	202
Fragment	258.00	258.00	255.00	257.00	244.00	260
Fragment	280.00	280.00	261.00	234.00	226.00	230
Fragment	290.00	275.00	275.00	280.00	334.00	360
Fragment	323.00	326.00	311.00	291.00	**Fail**	**Fail**
Fragment	367.00	365.00	373.00	360.00	349.00	320
Fragment	373.00	320.00	323.00	322.00	370.00	480
Fragment	464.00	464.00	456.00	454.00	**840.00**	**Fail**
Fragment	466.00	466.00	469.00	481.00	580.00	**Fail**
Fragment	527.00	510.00	507.00	520.00	**Fail**	**Fail**
Fragment	534.00	534.00	542.00	500.00	570.00	470
Fragment	595.00	545.00	537.00	530.00	**Fail**	**Fail**
Fragment	1460.00	1380.00	1580.00	2000.00	**Fail**	**Fail**

The K_D in μM is listed followed in parentheses by the number of data points included in the fit. Data points were excluded from highest to lowest concentration. The "early" and "late" controls were collected as five-point dose–response curves. Samples are sorted from lowest to highest potency and span the typical range for a fragment hit confirmation experiment. The greatest shift (1.8-fold) relative to $K_D(6)$ or failed fits are highlighted in bold.

this way. The K_D values were initially determined using all six concentrations and an R_{max} value determined from the controls. To test the robustness of the fitting method, the R_{max} was maintained, but all data points for a single concentration were deleted, remaining points fit, and new K_D values assigned. Note that no compounds exhibit a significant change until the three highest concentrations were deleted. Furthermore, 70% of compounds still yield a K_D consistent with the full six-point determination using only the lowest two concentration data points, with only one additional compound lost when the data were further reduced to only the lowest concentration data point. Further evidence of the robustness of this approach is shown in Fig. 8.12. Six-point binding curves determined in separate experiments were collected with a top compound concentration of either 250 or 2500 μM. This allows for the comparison of K_D values

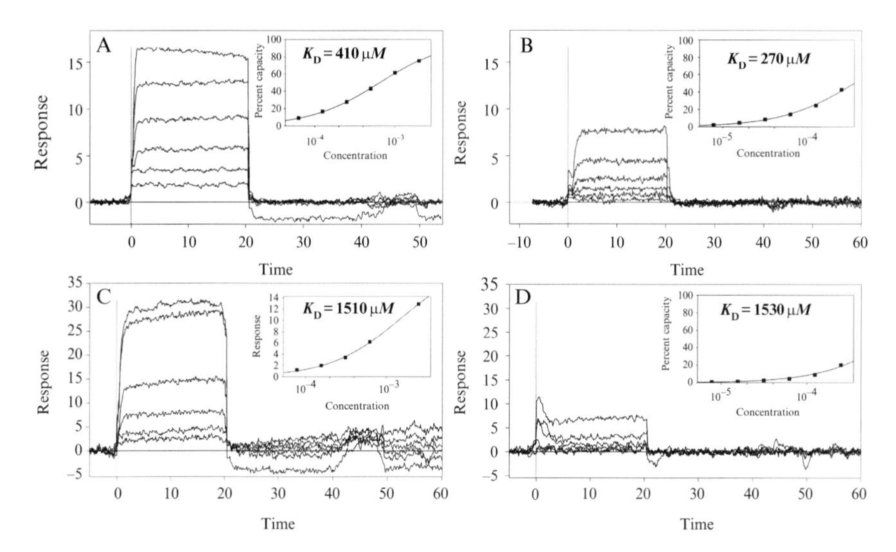

Figure 8.12 Compound K_D determination using high- (free R_{max}) and low-concentration (locked R_{max}) approaches. Two compounds (A and B or C and D) were tested at either 2500 μ*M* (A and C) or 250 μ*M* (B and D) top concentrations in a twofold dilution series. Data in (A) and (B) were fit with a free K_D and R_{max} to determine the fit. One bad injection had to be excluded in (C). (B) and (D) were fit with a free K_D but the R_{max} was locked to the average of the early and late control R_{max}. The variance in K_D values for each compound determined by both methods is within the normal range for biosensor experiments (Myszka *et al.*, 2003).

determined by extrapolation or by direct observation of the inflection. The values determined using both methods agree extremely well (<25% variation). In the original experiment, 43 compounds were tested at both concentrations. K_D values for all 43 compounds could be determined when starting at 250 μ*M*. When tested at 2500 μ*M*, 17 compounds precipitated either in the aqueous solvents or in the high-concentration DMSO stocks that were made separately for this experiment. Thus, while K_D values determined using a locked R_{max} method are properly considered extrapolations, the method allows for a much greater number of K_D values to be estimated and also avoids collecting erroneous K_D values corrupted by compounds that are insoluble as concentrated DMSO stocks and lead to an underestimate of the K_D by being at a lower concentration than expected. The rank ordering determined in this way is highly precise and useful for prioritizing fragments for follow-up studies in structural biology and can reveal any emergent SAR present in the fragment set. Example fragment dose–response data from a screen using this approach are shown in Fig. 8.13A–E.

The binding signal in the SPR experiment is due, in part, to the refractive index increment of the compound, which can vary from compound to

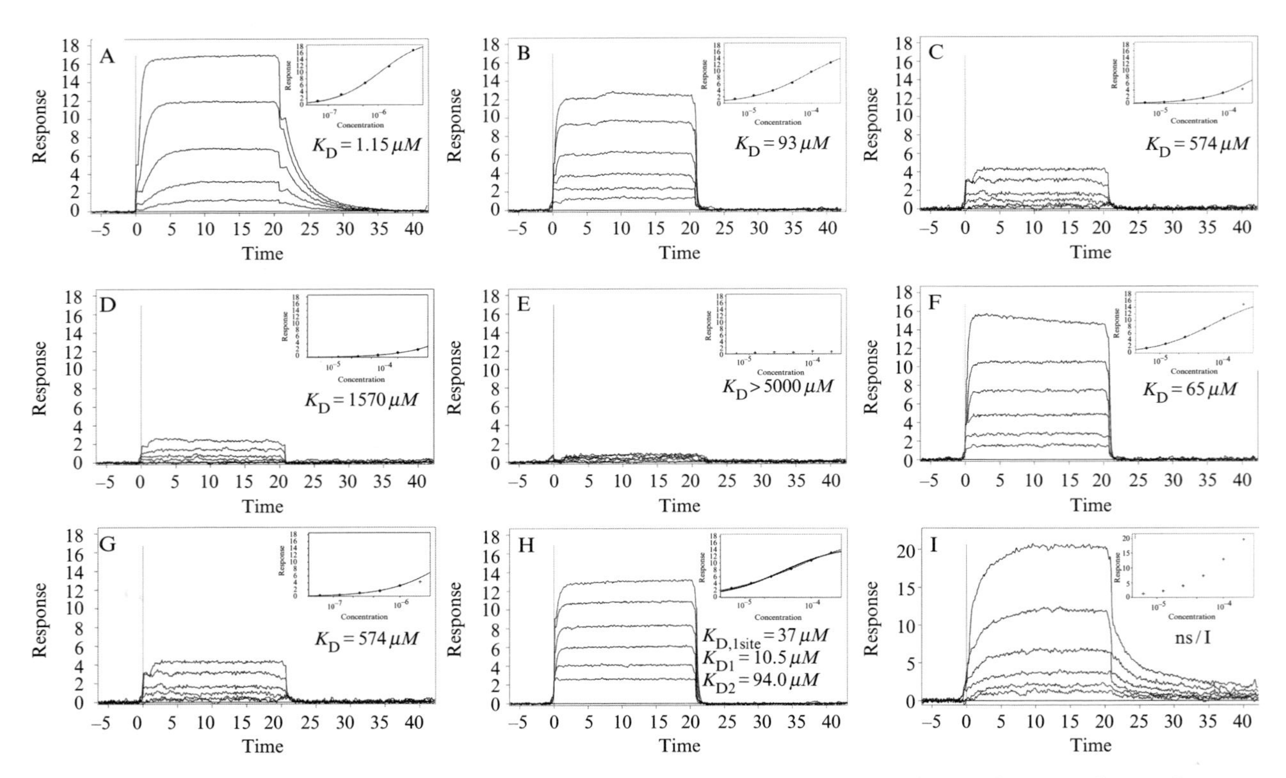

Figure 8.13 Representative sensorgrams of commonly observed behaviors in dose–response hit confirmation from a fragment screen. Except for the control compound in (A), all compounds were tested at 200 μ*M* with a twofold dilution factor. All sensorgrams were normalized by dividing by the molecular weight of the compound and multiplying by 100 RU. For example, the control R_{max} before

compound, especially when halogenated. A survey of RI variations in small molecules has been performed and found that the value does not vary by more than twofold (Davis and Wilson, 2000). Thus, after initial fitting, all dose–response curves should be manually inspected to see if the R_{max} for a given compound needs to be raised or lowered, but not by more than twofold. In practice, these R_{max} adjustments have a less than twofold effect on the final K_D, but need to be performed to ensure the highest quality rank ordering. While checking the curves it may occasionally be noted that a fragment has a K_D value more potent than the highest tested concentration.

---

normalization is 71 RU and after is 20 RU. (A) Early control showing a clean dose–response, kinetics, and an equilibrium K_D fits to a one-site model with free K_D and R_{max} parameters. (B) This fragment's K_D is more potent than the top concentration tested (200 μM) and has been fit with free K_D and R_{max} terms as in (A), and the R_{max} is not significantly different from the control R_{max}. (C) This fragment binds more weakly than the top tested concentration (200 μM) and has been fit with a free K_D and an R_{max} locked to the value determined in (A). (D) A fragment even weaker than in (C) fit with a locked R_{max}. Multiple injections are resolved above the baseline noise, are quite flat at equilibrium, and return quickly to the baseline, and the 1:1 fit matches the data quite well, thus meriting a fit. (E) Fewer than two concentrations resolve above the baseline and cannot be fit. (F) The top concentration injection must be rejected to obtain a quality fit to the dose–response curve. Note that the shape of the highest concentration injection is slightly different than the others, indicating an issue with solubility, second-site binding, or nonspecific binding. The remaining five concentrations can still be used to obtain a clear K_D. (G) The highest concentration injection has a response value lower than expected from a fit to the dose–response curve and is almost identical to the 100 μM data point. The top concentration point is rejected and may have a lower response than expected because of solubility problems above 100 μM. Sometimes this point can be and even lower response level than the next lower concentration. A K_D is still reported but a note about potential solubility problems at concentrations higher than the highest concentration point retained in the fit is made. (H) This compound is a two-site binder. The 1:1 fit to the data (thick line) underestimates response at the low concentration, overestimates it at the higher concentrations, and underestimates the highest concentration. A fit to a two-site model (thin line) fits the data more precisely. When two-site binding is well fit, the geometric mean of K_{D1} and K_{D2} (31 μM) will closely match the K_D determined from the 1:1 model (37 μM). (I) Dose–response curve failing strict quality control. This fragment shows a slow rise to equilibrium and a slow decay back to baseline. The kinetics cannot be fit by a 1:1 model. The equilibrium binding level has exceeded the R_{max} and shows no approach to saturation. In fact, the space between the highest and second highest concentration injections is greater than between the second and third highest concentrations indicating the system is far from saturation. Thus, this is labeled as a nonstoichiometric irreversible binder because it has and will continue to exceed 1:1 stoichiometry and, while dissociation does return to baseline, the time required to do so is too large for a fragment. While this is a more mild-case of poor behavior and significantly worse sensorgrams have been observed (Giannetti *et al.*, 2008), there are much better behaved compounds more likely to yield occupied crystal structures that should get higher priority than a more marginal one like this.

In these cases, the R_{max} can be locally fit to see if it can be restrained by the data. This allows a more precise fit for that compound, and usually the new R_{max} is within 20–30% of the globally applied one. However, if the new R_{max} is significantly lower (more than twofold) than the R_{max} determined from the control, the curve may be an artifact where nonspecific binding has coincidentally taken the shape of a sigmoidal binding curve but is not reflective of true binding. For example, it is unlikely that a fragment will exhibit an R_{max} of 5 RU when the control has an R_{max} of 50 RU. These samples should be excluded from further consideration or very carefully considered in light of additional data.

Fragments often have the ability to recognize multiple binding sites on proteins. Occasionally, a binding curve is recorded that does not fit well to a 1:1 binding model. If the sensorgrams are well behaved, then one can consider using a two-site binding model. These models add an additional term to the binding equation yielding an $R_{max,2}$ and a K_{D2}. Testing the quality of the fit and its parameters to two-site binders is important as fitting two-site data with a one-site model will significantly shift the K_D away from the true K_{D1} making the compound appear significantly weaker than it really is (Fig. 8.13H). As two-site binders are frequently the most potent fragments in a screen, forcing a one-site fit could result in inappropriate deprioritization of the best binders.

Even though quality control has been performed at the level of the sensorgram data, the quality of the fits to the data, as well as choices about exclusion of points from the fit are also necessary. Frequently, the top concentration data point may fit poorly to the binding isotherm due to either inadequate solvent correction or small refractive index differences between that sample and the running buffer. Those differences are diluted away in the lower concentrations if dilutions are made with the running buffer. Additionally, the top concentration could contain a contribution due to small amounts of nonspecific binding or the beginnings of second-site binding and may need to be excluded from the fit (Fig. 8.13D). Occasionally, the top one or two concentrations will fall below the fit indicating that the compound concentration is lower than expected due to either aggregation or precipitation (Fig. 8.13G). This phenomenon is also observed in HTS IC_{50} data. In these cases, a K_D may still be reported, but a note should be made about compound solubility being limited greater than the highest concentration data point retained in the fit.

The dilution factor used in the experiments shown in Fig. 8.13 is twofold. While threefold allows a greater sampling of concentration ranges and is preferred when working with more potent systems, the more useful information for fragment K_D determination comes from the higher concentration data and twofold, or even 1.5-fold dilutions are recommended.

In practice, the number of fragment hits that need K_D determination is higher than modern instrument throughput allows and so dose–response

Table 8.5 K_D reproducibility from independent experiments

ragment ID	K_D FL inactive replicate 1	K_D FL active replicate 1	K_D kinase domain replicate 1	K_D FL inactive replicate 2	K_D FL active replicate 2	K_D kinase domain replicate 2	FL inactive shift	FL active shift	Kinase domain shift
1	6.2	6.3	6.3	13	12.1	11.8	**2.10**	1.92	1.87
2	12.1	10.2	11.8	13	15	14	1.07	1.47	1.19
3	32	25	40	34	31	33	1.06	1.24	0.83
4	38	36	39	40	45	38	1.05	1.25	0.97
5	43	39	47	91	92	62	2.12	**2.36**	1.32
6	52	50	55	100	98	100	1.92	1.96	1.82
7	84	82	89	90	84	102	1.07	1.02	1.15
8	90	83	112	157	125	181	1.74	1.51	1.62
9	97	75	155	121	92	157	1.25	1.23	1.01
10	101	78	116	121	82	136	1.20	1.05	1.17
11	104	82	128	110	95	113	1.06	1.16	0.88
12	113	96	140	134	107	155	1.19	1.11	1.11
13	122	112	119	199	225	289	1.63	**2.01**	**2.43**
14	134	152	123	135	143	127	1.01	0.94	1.03
15	166	158	195	182	213	252	1.10	1.35	1.29
16	184	176	203	212	275	275	1.15	1.56	1.35
17	200	178	265	227	235	363	1.14	1.32	1.37
18	209	202	226	274	306	398	1.31	1.51	1.76
19	216	214	264	246	287	349	1.14	1.34	1.32
20	225	203	285	230	179	240	1.02	0.88	0.84
21	271	252	308	332	371	458	1.23	1.47	1.49
22	470	460	520	507	577	681	1.08	1.25	1.31

Twenty two compounds were tested in duplicate on three forms of a kinase (full length, full length activated, and kinase domain only). Duplicates were run on different plates, different days, different machines, and by different users. Reported K_Ds were fit with the locked R_{max} method. K_Ds are reported in μ*M*. Four compounds with more than a twofold shift between replicates are highlighted in bold.

curves are usually collected as singletons rather than in duplicate or triplicate, which could extend the follow-up time for a fragment set from 1–2 weeks to nearly 1–2 months. This approach is possible because most SPR instruments have extremely good reproducibility. Table 8.5 shows the results of 22 compounds tested in duplicate on three forms of a kinase, full-length, full-length activated, and kinase-domain only. Duplicates were seeded into full dose–response plates containing different fragment sets and tested over multiple days, instruments, and users. No K_D determination made using the locked R_{max} procedure for the 66 pairs of interactions differs by more than twofold indicating a high level of reproducibility using the extrapolated K_D method.

18. Conclusion

SPR methods have the advantage of being able to screen large fragment libraries using small amounts of protein to deliver fragment hits as weak as 5 m*M* in short time frames. One dedicated single injection-mode instrument, such as a Biacore T100, can screen 5000 compounds and provide dose–response confirmation data on up to 15% of that library in approximately 1 month. In addition, most SPR instruments are multiplexed allowing multiple targets to be screened simultaneously thereby increasing turnaround speed to project teams. Given the screening speed, sensitivity, and ability to drive SAR down to biochemical potency, SPR methods represent a robust way to initiate and support a fragment-based lead identification approach. Advances in instrument sensitivity, multiplexing targets and injections, and various ancillary method improvements are proceeding and will help this technology continue to provide value to drug discovery programs.

ACKNOWLEDGMENTS

I wish to thank many of my current and former colleagues who have supported the development of these methods including Ken Brameld, Michelle Browner, Michael Dillon, Seth Harris, Chris Heise, Carolyn Jackson, Andreas Kuglstatter, Bruce Koch, Kevin Lindquist, and Eric Sjogren. Brandon Bravo and Keith Pitts especially have been instrumental in the continuous refinement and automation of these methods, and I thank them for their continuous dedication to the collection of high-quality SPR data.

David Myszka has been a good friend and collaborator, and I thank him for the many discussions and collaborations, especially on fragment work, over the past six years. I extend special thanks to Tom Morton, the developer of the Scrubber software package, for continuing to rapidly refine and enhance that software to support industrial applications of SPR technology, especially around fragment screening.

I am indebted to Cristina Lewis, Michael Bradshaw, and Rebecca Rich for their help reviewing, editing, and commenting on the chapter.

REFERENCES

(2002). Protocol for Measuring Small Molecule Interactions Using Biacore: A Practical Guide to Experimental Design and Data Evaluation. Biacore Symposium 2002, Chicago, Illinois. pp. 1–17.

Beckett, D., Kovaleva, E., *et al.* (1999). A minimal peptide substrate in biotin holoenzyme synthetase-catalyzed biotinylation. *Protein Sci.* **8**(4), 921–929.

Bukhtiyarova, M., Northrop, K., *et al.* (2004). Improved expression, purification, and crystallization of p38alpha MAP kinase. *Protein Expr. Purif.* **37**(1), 154–161.

Casper, D., Bukhtiyarova, M., *et al.* (2004). A Biacore biosensor method for detailed kinetic binding analysis of small molecule inhibitors of p38alpha mitogen-activated protein kinase. *Anal. Biochem.* **325**(1), 126–136.

Dalvit, C. (2009). NMR methods in fragment screening: Theory and a comparison with other biophysical techniques. *Drug Discov. Today* **14**(21–22), 1051–1057.

Danielson, U. H. (2009). Fragment library screening and lead characterization using SPR biosensors. *Curr. Top. Med. Chem.* **9**(18), 1725–1735.

Davis, T. M., and Wilson, W. D. (2000). Determination of the refractive index increments of small molecules for correction of surface plasmon resonance data. *Anal. Biochem.* **284**(2), 348–353.

Day, Y. S., Baird, C. L., *et al.* (2002). Direct comparison of binding equilibrium, thermodynamic, and rate constants determined by surface- and solution-based biophysical methods. *Protein Sci.* **11**(5), 1017–1025.

Duffy, S., Tsao, K. L., *et al.* (1998). Site-specific, enzymatic biotinylation of recombinant proteins in *Spodoptera frugiperda* cells using biotin acceptor peptides. *Anal. Biochem.* **262**(2), 122–128.

Feng, B. Y., and Shoichet, B. K. (2006). Synergy and antagonism of promiscuous inhibition in multiple-compound mixtures. *J. Med. Chem.* **49**(7), 2151–2154.

Frostell-Karlsson, A., Remaeus, A., *et al.* (2000). Biosensor analysis of the interaction between immobilized human serum albumin and drug compounds for prediction of human serum albumin binding levels. *J. Med. Chem.* **43**(10), 1986–1992.

Geschwindner, S., Olsson, L. L., *et al.* (2007). Discovery of a novel warhead against beta-secretase through fragment-based lead generation. *J. Med. Chem.* **50**(24), 5903–5911.

Giannetti, A. M., Koch, B. D., *et al.* (2008). Surface plasmon resonance based assay for the detection and characterization of promiscuous inhibitors. *J. Med. Chem.* **51**(3), 574–580.

Hamalainen, M. D., Zhukov, A., *et al.* (2008). Label-free primary screening and affinity ranking of fragment libraries using parallel analysis of protein panels. *J. Biomol. Screen.* **13**(3), 202–209.

Jhoti, H., Cleasby, A., *et al.* (2007). Fragment-based screening using X-ray crystallography and NMR spectroscopy. *Curr. Opin. Chem. Biol.* **11**(5), 485–493.

Johnsson, B., Lofas, S., *et al.* (1991). Immobilization of proteins to a carboxymethyldextran-modified gold surface for biospecific interaction analysis in surface plasmon resonance sensors. *Anal. Biochem.* **198**(2), 268–277.

Lofas, S. (1995). Dextran modified self-assembled monolayer surfaces for use in biointeraction analysis with surface-plasmon resonance. *Pure Appl. Chem.* **67**(5), 829–834.

Lofas, S., and McWhirter, A. (2006). The Art of Immobilization for SPR Sensors. Surface Plasmon Resonance Based Sensors. New York, Springer, 117–151.

McGovern, S. L., Caselli, E., *et al.* (2002). A common mechanism underlying promiscuous inhibitors from virtual and high-throughput screening. *J. Med. Chem.* **45**(8), 1712–1722.

Myszka, D. G. (1999). Improving biosensor analysis. *J. Mol. Recognit.* **12**(5), 279–284.

Myszka, D. G., Abdiche, Y. N., *et al.* (2003). The ABRF-MIRG'02 study: Assembly state, thermodynamic, and kinetic analysis of an enzyme/inhibitor interaction. *J. Biomol. Tech.* **14**(4), 247–269.

Navratilova, I., and Hopkins, A. L. (2010). Fragment screening by surface plasmon resonance. *Acs Med. Chem. Lett.* **1**(1), 44–48.

Navratilova, I., and Myszka, D. (2006). Investigating biomolecular interactions and binding properties using SPR biosensors. *In* "Surface Plasmon Resonance Based Sensors," (J. Homola, ed.), pp. 155–176. Springer, New York.

Navratilova, I., Papalia, G. A., *et al.* (2007). Thermodynamic benchmark study using Biacore technology. *Anal. Biochem.* **364**(1), 67–77.

Papalia, G. A., Leavitt, S., *et al.* (2006). Comparative analysis of 10 small molecules binding to carbonic anhydrase II by different investigators using Biacore technology. *Anal. Biochem.* **359**(1), 94–105.

Papalia, G. A., Giannetti, A. M., *et al.* (2008). Thermodynamic characterization of pyrazole and azaindole derivatives binding to p38 mitogen-activated protein kinase using Biacore T100 technology and van't Hoff analysis. *Anal. Biochem.* **383**(2), 255–264.

Perspicace, S., Banner, D., *et al.* (2009). Fragment-based screening using surface plasmon resonance technology. *J. Biomol. Screen.* **14**(4), 337–349.

Rich, R. L., and Myszka, D. G. (2008). Grading the commercial optical biosensor literature-Class of 2008: 'The Mighty Binders'. *J. Mol. Recognit.* **23**(1), 1–64.

Rich, R. L., Papalia, G. A., *et al.* (2009). A global benchmark study using affinity-based biosensors. *Anal. Biochem.* **386**(2), 194–216.

Stenberg, E., Persson, B., *et al.* (1991). Quantitative-determination of surface concentration of protein with surface-plasmon resonance using radiolabeled proteins. *J. Colloid Interface Sci.* **143**(2), 513–526.

Zhang, J. H., Chung, T. D., *et al.* (1999). A simple statistical parameter for use in evaluation and validation of high throughput screening assays. *J. Biomol. Screen.* **4**(2), 67–73.

CHAPTER NINE

Practical Aspects of NMR-Based Fragment Screening

Christopher A. Lepre

Contents

Abstract

NMR spectroscopy is a popular and highly versatile screening method for fragment-based drug discovery. NMR methods are capable of robustly detecting the binding of fragments to macromolecular targets over an extraordinarily broad affinity range (from covalent to millimolar). This chapter provides a stepwise process for creating an NMR-based fragment screening program. The construction of fragment libraries is described, including compound selection, plating of stocks, and preparation of mixtures. Guidance is given for designing fragment screens, such as choosing the appropriate NMR screening format and method, and optimizing the sample conditions and experimental parameters. The identification and validation of screening hits is described, and

Structural Biology, Vertex Pharmaceuticals, Incorporated, Cambridge, Massachusetts, USA

Methods in Enzymology, Volume 493
ISSN 0076-6879, DOI: 10.1016/B978-0-12-381274-2.00009-1

a number of potential pitfalls are discussed. Rather than detailing one specific screening protocol, this chapter outlines the available options and provides information to enable users to design their own customized fragment screening programs.

1. Introduction

Since its inception with the work of Fesik and coworkers in 1996 (Shuker *et al.*, 1996), fragment-based drug discovery (FBDD) has gradually gained acceptance throughout the pharmaceutical industry and within academia. Once regarded as a niche method, FBDD is now widely used in combination with traditional compound screening for lead discovery, and this has begun to impact mainstream medicinal chemistry programs, with a number of FBDD-derived drugs now in clinical trials (Chessari and Woodhead, 2009; Congreve *et al.*, 2008). The growth of fragment-based screening has been driven largely by its flexibility, practicality, and ease of implementation. It is relatively straightforward to start a FBDD program, since inexpensive fragments are readily available, the experimental methods are simple, and suitable instruments already exist within most pharmaceutical and academic laboratories. The growth of fragment-based approaches has also been fueled by advances in methodology, particularly the successful integration of fragment screening with structure-based drug design (SBDD). The combination of FBDD and SBDD is employed as a principal lead discovery method at a number of small companies and is used in conjunction with biochemical high-throughput screening (HTS) at many others, including virtually all major pharmaceutical companies.

NMR spectroscopy was the first experimental method used for screening fragments. Although other techniques, such as X-ray crystallography, surface plasmon resonance, mass spectrometry, isothermal titration calorimetry, and biochemical screening, have also been applied, NMR remains the predominant method for primary fragment screening. NMR has a distinct advantage over other methods in that it requires almost no assay development: all that is needed is a soluble target. There is no need to crystallize, immobilize, or tag the target. However, NMR consumes relatively large amounts of reagents and has lower throughput than HTS. Because of these disadvantages, other methods may eventually supplant NMR for primary fragment screening (particularly as technology advances), but because of its flexibility and robustness, NMR will likely continue to be widely used, especially as a secondary assay.

While it is relatively easy to get started screening fragments, the ultimate success of a FBDD program depends upon thoughtfully implementing a complete process, including constructing the library, selecting targets, choosing the experimental methods, validating, and following up the hits. There are many reviews available that cover the general principles of FBDD (Leach *et al.*, 2006; Wyss and Eaton, 2007), library design (Hubbard *et al.*, 2007a), experiments, and applications (Alex and Flocco, 2007; Congreve *et al.*, 2008; Erlanson, 2006; Lepre *et al.*, 2004), and in this volume, there are several chapters that discuss strategies for converting fragment hits into leads, so those topics will not be discussed in detail here. This chapter instead provides practical information for setting up and running NMR-based fragment screens while avoiding problems that can be pitfalls for the novice or the unwary.

2. Constructing the Fragment Library

The first consideration when building a fragment screening library is its intended purpose: whether it will be used for a one-time screen against a specific target, a limited number of screens against a class of related targets, or as a universal screening library. The intended purpose determines the library design strategy and whether to invest the time to pretest solubilities and design mixtures, which would be justified for a large, broad-use library but not for a small library to be used only once. The characteristics of the target (molecular size, expression system, and abundance) influence the choice of methods used for primary screening and following up hits, and how many compounds can be reasonably screened.

2.1. Library design

The most common fragment libraries, intended for screening against a wide variety of targets, are diverse sets of compounds selected according to specific criteria, such as maximal diversity or physicochemical properties, and then filtered to remove those containing functional groups associated with chemical reactivity, toxicity, and false positives (Baell and Holloway, 2010; Hann *et al.*, 1999; Rishton, 1997; Walters and Murcko, 2002). We and others have reported screening diverse libraries of 1000–3000 fragments (Albert *et al.*, 2007; Baurin *et al.*, 2004; Hubbard *et al.*, 2007b; Lepre, 2007; Schuffenhauer *et al.*, 2005), but library sizes continue to grow in size as screening technology advances. Researchers at Abbott, for example, report screening ca.10,000 compounds (Petros *et al.*, 2006).

2.1.1. Property-based selection of fragments

The physicochemical properties most frequently used for fragment selection (Table 9.1) are molecular weight, clogP, number of hydrogen bond donors and acceptors, number of rotatable bonds, and, to a lesser extent, polar

Table 9.1 Criteria used to define fragment libraries

Property	Rule of three	Rule of V	Vertex avg[a]
MW (Da)	$\leq$300	150–300	249.7 $\pm$ 49.4
clogP	$\leq$3	$\leq$3	1.5 $\pm$ 1.1
H-bond donors	$\leq$3	$\leq$3	1.0 $\pm$ 0.8
H-bond acceptors	$\leq$3	$\leq$6	3.1 $\pm$ 1.3
Rotatable bonds	$\leq$3	$\leq$6	3.1 $\pm$ 1.8
PSA (Å^2)	$\leq$60	$\leq$80	60.2 $\pm$ 21.5

[a] Average values $\pm$ standard deviation for 1100 compounds in the Vertex diverse fragment libraries.

surface area. The "Rule of three" (Ro3) criteria proposed by Astex (Congreve *et al.*, 2003) have been widely applied to design fragment screening collections (Table 9.1). The Ro3 limits are well suited for screening by X-ray crystallography, since the allowed fragments are usually very soluble (e.g., 25–200 m*M* in aqueous buffer; Hartshorn *et al.*, 2005) and structural information will be available to follow up the hits. For NMR-based fragment screening, however, the Ro3 limits may be too restrictive. The simplicity of Ro3-compliant fragments limits the diversity of the library and can produce hits that are difficult to optimize due to a lack of synthetically accessible functionality. Very small fragments (MW < 150 Da) also have a greater tendency to bind in multiple orientations (Babaoglu and Shoichet, 2006; Siegal *et al.*, 2007). In cases where structural information is unavailable and the hits are being followed up by a directed search strategy, slightly more complex fragments can be preferable because they provide more information to guide the selection of analogs for screening. For targets where hydrogen bonding is important for fragment binding, such as kinases, it is useful to have more than three hydrogen bond acceptors. For these reasons, at Vertex, we use a modified set of criteria, facetiously named the "Rule of V" (Table 9.1).

Beyond property-based selection and functional group-filtering of candidates, a number of strategies have been used to further narrow the selection of fragments for diverse libraries. Clustering and selection from the centroids is often used to pick compounds for which analogs are readily available (Lepre, 2002). Fragments are often chosen to contain scaffolds and side chains commonly found in known drugs (Hartshorn *et al.*, 2005; Johnson *et al.*, 2003; Lepre, 2001, 2002; Siegal and Vieth, 2007), or synthetically accessible functional groups (Erlanson *et al.*, 2004; Schuffenhauer *et al.*, 2005). Other factors are also considered, such as aqueous solubility, novelty, chemical and metabolic stability, availability of analogs, and vendor reliability. Further information about the design of diverse screening libraries is available from many sources (Albert *et al.*, 2007; Baurin *et al.*, 2004;

Blomberg *et al.*, 2009; Chen and Hubbard, 2009; Hann and Oprea, 2004; Hartshorn *et al.*, 2005; Hubbard *et al.*, 2007a,b; Lepre, 2001, 2002; Schuffenhauer *et al.*, 2005). In addition, we and others screen fragment libraries targeted toward certain classes of proteins, such as kinases (Akritopoulou-Zanze and Hajduk, 2008; Aronov and Bemis, 2004), and custom libraries selected for specific targets using virtual screening.

2.1.2. Commercial fragment libraries

Predesigned fragment libraries are increasingly available directly from compound vendors: Table 9.2 lists current commercial suppliers. Commercial libraries can provide well-behaved, diverse collections with which to begin screening, which can then be expanded with more targeted and novel sets of compounds cherry-picked from vendor catalogs.

Predesigned libraries offer a number of advantages over individually selected compounds: the compounds are usually less expensive and in stock, they can often be obtained as DMSO stocks, solubility data may be

Table 9.2 Commercial suppliers of fragment libraries

Vendor	Number of fragments	Comments
ACB blocks	1280	Ro3-compliant
ASDI	1700	Ro3-compliant
Asinex	1800	Solubility > 100 μ*M* (PBS)
	12,000	Modified Ro3
ChemBridge	12,000	Ro3-compliant
ChemDiv	5000	Ro3-compliant
Enamine	1190	Ro3-compliant
	11,717	Modified Ro3
InFarmatik	406	Ro3-compliant
	243	3D fragments
Iota Pharmaceuticals	5500	Ro3-compliant
Key Organics	6356	Ro3-compliant
Life Chemicals	24,823	Modified Ro3
Maybridge (Thermo Fisher)	1500	Ro3-compliant, sol > 1 m*M*
	30,000	Modified Ro3
Otava	3800	Modified Ro3
Prestwick	2800	Ro3-compliant
Pyxis	317	Ro3-compliant
Timtec	3100	Modified Ro3
Zenobia	350	Ro3-compliant

available, and numerous analogs can be purchased for hit follow-up. At the time of this writing, approximately 89,000 nonredundant fragments were available from eight vendors (Asinex, Chembridge, InFarmatik, Key Organics, Life Chemicals, Maybridge, Pyxis, and Zenobia), of which about 53,000 were Ro3-compliant (P. Walters and A. Daltas, unpublished). Some distributors such as eMolecules (www.emolecules.com) maintain user-searchable online databases of millions of compounds and will procure compounds from individual suppliers and prepare DMSO stocks for an additional fee.

2.2. Preparing fragment stocks

Careful library preparation is essential to obtaining good results from fragment screening. Although meticulous selection, testing, and removal of poorly behaved compounds from the collection are tedious and time-consuming, it is well worth the time invested, since a well-crafted library can be used over and over again, and the frustration of pursuing false positives is avoided.

Compounds should be purchased in amounts sufficient to run an ample number of screens, as well as provide enough material to test solubility and stability, titrate hits in secondary assays, and set up crystallizations, while allowing for losses due to transfers and handling. We typically purchase enough solid to make 400 μL of 250 m*M* stock solution (30 mg at MW = 300 Da). While it is tempting to economize by reducing the amount purchased, replenishing compounds is time-consuming and expensive, and compounds rapidly go out of stock: 4 years after constructing a library of diverse fragments from 16 well-known vendors, we reordered 1000 compounds and found that only 63% were still commercially available. It may be cost-effective to order small amounts of candidate compounds to test solubility before investing in large quantities. Care should be taken to avoid liquids or oils if these would complicate the preparation of stock solutions. We tend to avoid salts, although they are usually more soluble in water than the corresponding free acids or bases, because they have a higher propensity to precipitate from DMSO.

Preparing stock solutions is labor-intensive, particularly if the compounds need to be removed from the vendor-provided container and weighed. The latter step can be avoided if the vendor provides a data file with the exact amount of compound in each vial or can dispense solids into pre-tared vials provided by the customer. The correct volume of deuterated DMSO can then be added directly and racks of vials sonicated and/or shaken in an incubator to promote dissolution.

When making DMSO stocks, the target concentration is determined by the desired concentrations of fragments and DMSO in the screening samples and the number of compounds in the planned mixtures. For example, screening of mixtures of five compounds, each in buffer at 500 μ*M* with a maximum of 1% residual DMSO, requires mixture stocks of five compounds in DMSO at 50 m*M* and individual compounds in DMSO at 250 m*M*. Freshly made stocks are checked for undissolved solid and then aliquots are (1) dispensed into 96 V-bottom deepwell blocks and buffer added to make NMR reference samples, (2) diluted to 10 m*M* in DMSO for use in secondary assays, and (3) diluted into 1:1 acetonitrile:water for LC–MS confirmation of compound purity and identity. High-concentration DMSO stocks decompose and precipitate readily if subjected to repeated freeze/thaw cycles or exposed to the atmosphere (from which they rapidly absorb moisture). To minimize this problem, our DMSO stocks are stored frozen under dry nitrogen in 1.4 mL polypropylene Matrix 2D bar-coded microtubes (Thermo Fisher Scientific).

For one-time screens of a relatively small number of compounds, libraries can be constructed quickly using DMSO stocks from an existing HTS compound collection, thereby saving time and money. For example, aliquots of 10 m*M* HTS compounds in nondeuterated DMSO can be dispensed into 96-deepwell plates and then dried down in a centrifugal evaporator equipped with a microplate rotor (SpeedVac, Thermo Fisher Scientific). The dried compounds are redissolved in a 1:5 volume of deuterated DMSO to make 50 m*M* stocks, followed by addition of protein in buffer to make the screening samples.

2.3. Collecting reference NMR spectra and constructing mixtures

When ligand-detected methods are to be used for screening, reference NMR spectra of the individual compounds are collected in common biological buffers such as phosphate (pH 6.5–7.5), d_{18}-HEPES (pH 7–8), or d_{11}-TRIS (pH 7–9). In order to maintain the target pH, the buffer concentration should be at least 20-fold higher than the expected total concentration of fragments in the mixtures (e.g., 50 m*M* buffer for mixtures of five fragments, each at 500 μ*M*). Our typical NMR reference samples contain 500 μ*M* compound in 50 m*M* sodium phosphate buffer, pH 7, 90% D_2O/10% H_2O, and 1% d_6-DMSO, with 0.1 m*M* 4,4-dimethyl-4-silapentane-1-sulfonic acid (DSS) added as an internal chemical shift and concentration standard, and 0.1 m*M* sodium azide as an antimicrobial agent. Buffer is added to predispensed aliquots of DMSO stock and mixed in the 96-deepwell block, then transferred to 5-mm NMR tubes arrayed in a custom rack using a Gilson 215 liquid handler equipped with an eight-channel multiprobe (Gilson, Inc., Middletown, WI). ^{1}H NMR spectra are collected at 295 K using a 500 MHz Bruker Avance II NMR spectrometer equipped with an inverse triple resonance cryoprobe and a BACS-120 robotic sample changer (Bruker Biospin, Billerica, MA). The ^{1}H NMR spectra are used to confirm the identity, purity, and concentration of the fragments and examined for line broadening (which would indicate aggregation). Compounds with suspiciously broad resonances are subjected to ^{1}H NOESY experiments (States *et al.*, 1982) to check for NOEs characteristic of high molecular weight species. Typically, about half to three-quarters of Rule of Three-compliant fragments pass the solubility/aggregation test and 1–2% fail the purity/identity test.

Cocktailing, or screening mixtures of fragments, drastically reduces protein consumption and data collection time but requires more library preparation and introduces the potential for interference between compounds. When ligand-detected NMR methods are used for the screen (so there is no need to deconvolute mixtures to identify hits), it is generally worth investing the effort to create mixtures if the library contains more than a few hundred compounds and will be used multiple times.

Mixtures are designed to avoid possible chemical reactions between fragments and minimize the overlap between the NMR resonances of the components. The ^{1}H NMR reference spectra are processed to artificially broaden the resonances (e.g., using a Gaussian window), which simplifies the spectra by concealing multiplet structure, and then peak-picked, omitting peaks with intensities below a minimum threshold and with chemical shifts in ranges occupied by resonances from water, DMSO, buffer, glycerol, and other additives. We then classify all compounds as acid/base/neutral or electrophile/nucleophile/inert in order to determine allowable combinations, as described by Hann *et al.* (1999). Mixtures are designed, according to user-specified parameters for the minimum separation between peaks (typically 0.04 ppm) and the number of compounds per mixture (5–10), using a custom program that randomly assembles a set of chemically compatible mixtures, scores them according to the number of overlapped NMR resonances, and then uses a simulated annealing protocol to design and test new mixtures until the overlap score no longer improves. Typically, at least a few hundred diverse compounds are needed to generate adequately resolved, chemically compatible mixtures of five.

There is an obvious trade-off between the efficiency of screening larger mixtures and increased difficulty in identifying hits and coping with interactions. Most laboratories screen mixtures of 15 or fewer compounds because larger mixtures are problematic: there is more resonance overlap, compounds are more prone to precipitate and destabilize the protein, and interactions between components are more common (Hann *et al.*, 1999; Lepre, 2001).

The d_6-DMSO stocks of individual compounds are combined to make the mixture stocks, which are then diluted into buffer to collect reference spectra. The ^{1}H NMR spectra of the mixtures are compared to the sums of the individual reference spectra in order to identify intensity changes, new resonances, or shifts in peak positions arising from precipitation or interactions between the components. Because compound aggregation can be present even in optically clear solutions, ^{1}H NOESY spectra are used to check for aggregation in all mixtures. Usually, only a few mixtures show problems, and these can usually be remedied by removing a single offending compound. To avoid repeated freeze/thaw cycles and atmospheric exposure of the mixture stocks, single-use plates are prepared for screening: aliquots of mixtures are dispensed into 96-deepwell blocks, sealed with microplate foils, and stored at −70 °C. At the time of the screen, a 100:1 volume of protein in buffer is added directly to the blocks, mixed, and then transferred to NMR tubes using a liquid handler as described above.

3. Developing the Screen

3.1. Finding sample conditions

Once a target has been selected, it is necessary to find an appropriate construct and sample conditions. Given a choice, it is generally preferable to use a full-length protein rather than an isolated domain. The functional viability of the protein should be confirmed, either by showing activity in a biochemical assay or by binding a known ligand with expected affinity. The nature and relevance of posttranslational modifications (e.g., phosphorylation, glycosylation, acetylation, etc.) should be known, as should the need for cofactors. Terminal tags (such as poly-His tags) may promote aggregation and should be treated cautiously. There is generally no problem with screening in the presence of cofactors (such as saturating ATP, NAD, NMN, Mg^{+2}), even if they give rise to a binding signal, provided there is no direct interference with the fragments or components of the buffer (e.g., Mg^{+2} precipitating with phosphate).

The protein should be checked for aggregation and activity under the same conditions as the NMR screen (buffer, pH, temperature, DMSO concentration, and time course). Size exclusion chromatography and dynamic light scattering can be used to check molecular size and monodispersity, and ^{1}H NMR spectra can be used to evaluate the folding and aggregation state (Page *et al.*, 2005; Rossi *et al.*, 2010). When buffer and stabilizer conditions need to be optimized, small quantities of protein can be screened using solubility (Lepre and Moore, 1998) or thermal shift (Nettleship *et al.*, 2008) as read-outs. High salt concentrations (e.g., >100 m*M* NaCl) should be avoided if possible when using a cryogenically cooled NMR probe, since high sample conductivity leads to loss of sensitivity and sample heating (Hautbergue and Golovanov, 2008; Kelly *et al.*, 2002). If known ligands or cofactors with suitable affinities are available, they should be used to confirm that the protein is competent to bind ligands and determine the protein and ligand concentrations necessary to produce an unambiguous binding response in the NMR screen.

3.2. Choosing a format

Users often begin screening with whatever instrumentation is already available, but some may have a choice of formats, that is, screening in tubes versus a flow probe, or screening in 5-mm versus smaller diameter tubes. Flow systems avoid the need for NMR tubes and there is little or no need for shimming (Stockman *et al.*, 2001), but are vulnerable to clogging due to sample precipitation and potential carryover between samples. Prefiltering samples is helpful in preventing clogs but is not a panacea, since precipitation can be gradual. Primarily because of these problems, flow systems are used less frequently for screening than are tubes.

When screening in NMR tubes, the choice of tube size is dictated by the conflicting demands of high throughput versus low reagent consumption. Standard 5-mm tubes, which require 500–600 μL of sample, have the highest fill factor and hence the highest signal-to-noise at a given sample concentration. Microprobes offer higher signal-to-noise *per unit mass of solute*, such that a 1.7-mm micro cryoprobe has sensitivity comparable to a 5-mm room temperature probe while requiring only about 35 μL of sample (Rossi *et al.*, 2010). If the sample concentration can be increased by three- to fourfold without encountering solubility problems, then a 1.7-mm cryoprobe should have similar throughput to a 5-mm cryoprobe but significantly lower reagent consumption. Because of the fragility and small size of 1.7-mm tubes, they are usually filled using a liquid handler, and robotic systems are currently available that can track and load micro-NMR tubes into the magnet from 96 arrays (e.g., SampleJet, Bruker Instruments, Billerica, MA; 7600-AS, Varian Instruments, Palo Alto, CA).

3.3. Choosing an NMR experiment

The NMR methods used for fragment screening are typically classified as ligand- or receptor-based methods, according to whether ligand binding is detected via observation of ligand or receptor resonances. The details of these methods, along with their pros and cons, have been discussed extensively in the literature (e.g., Lepre *et al.*, 2004) and will be summarized here.

Receptor-based methods observe chemical shift perturbations that occur upon ligand binding in HSQC or TROSY spectra of ^{15}N (Shuker *et al.*, 1996), ^{13}C (Hajduk *et al.*, 2000), or site selectively labeled (Wiegelt *et al.*, 2002) receptors. Although this approach can provide direct information about the ligand-binding site, there are a number of drawbacks. Receptor-based methods are limited to targets that can be expressed and isotopically labeled at high levels (usually in bacteria), and for which the resonances can be resolved (and, preferably, assigned to specific residues), which limits their use to targets of no more than ca. 30–40 kDa in size. Furthermore, these methods do not identify which component of a mixture is binding, so additional deconvolution experiments (testing components individually) are required. Because most of our therapeutic targets are larger than 30 kDa, we use ligand-based methods for primary screening, and even for smaller targets prefer to screen by ligand-based methods and then use TROSY to map the binding sites of the hits.

Ligand-based methods detect changes in the characteristics of the ligand spectrum (such as relaxation rates, NOE, and magnetization transfer) that occur upon binding to the receptor. These methods require much lower receptor concentrations than receptor-based methods, no isotopic labeling or resonance assignments are needed, and the receptor can be of unlimited size. Most ligand-based methods, however, provide no information about the ligand-binding site, which must then be obtained from additional experiments.

Currently, the most widely used ligand-based methods are saturation transfer difference (STD) (Mayer and Meyer, 1999, 2001) and Water-LOGSY (Dalvit *et al.*, 2000, 2001), which rely on transfer of magnetization from the receptor to bound ligands. These methods consume the smallest amounts of receptor, which is present at sub-stoichiometric concentrations (ligand:protein ratio > 20), and are capable of directly detecting ligands with affinities ranging from 10^{-8} to 10^{-3} *M*. Both experiments are susceptible to false positives arising from fragments that aggregate in solution, which can be detected by running additional experiments (e.g., competition or mapping studies) or, preferably, avoided by weeding out such molecules during library construction. STD is often preferred over WaterLOGSY because it is less prone to interference due to magnetization transfer via exchangeable ligand protons, but the latter can be superior for receptors with low proton densities, such as nucleic acids (Lepre *et al.*, 2002). The original STD experiment is often modified to use excitation sculpting or WATERGATE for improved water suppression (Lepre *et al.*, 2004; Peng *et al.*, 2001). A recently developed polarization-optimized version of WaterLOGSY has been reported with higher sensitivity that can increase screening throughput by three- to fivefold over the conventional version (Gossert *et al.*, 2009).

When water-soluble ligands with known affinities for the receptor are available, competitive displacement experiments can be run using ligand-based methods. Competition experiments are commonly performed to confirm the binding site of screening hits and, from the magnitude of the displacement effect, estimate binding affinity (Mayer and Meyer, 2001). Alternatively, a known ligand with relatively weak affinity ($K_D > 10$ μM) can be used as a reporter or spy molecule: screening is performed with the known ligand present in all samples, and a decrease in binding of this reporter indicates competition by a fragment in the mixture (Dalvit *et al.*,

2002; Jahnke *et al.*, 2002). This method can be run at relatively low fragment concentrations, can detect very avid or even covalent binders (which would otherwise give false negatives in STD or WaterLOGSY experiments), and is specific for ligands that bind to the site of interest; however, it is limited by the requirement for suitable reporter molecules, which are often unavailable.

One-dimensional ^{1}H relaxation experiments (Hajduk *et al.*, 1997) are not as popular as STD and WaterLOGSY because they require higher protein concentrations (ligand:protein ratio < 20), and control spectra must be collected separately. Nonetheless, some groups use relaxation-filtered spectra for primary screening or in parallel with other ligand-detected methods (Hubbard *et al.*, 2007b) to confirm ligand binding.

Fluorine-19 relaxation can also be used for screening: ^{19}F spectra are much simpler than ^{1}H, and, owing to its large chemical shift range, ^{19}F often exhibits larger line broadening effects upon binding to a receptor. ^{19}F-labeled reporter molecules are highly sensitive when used in competition experiments such as the FAXS method (Dalvit *et al.*, 2003, 2004). At Novartis, a library of 1152 fluorinated fragments is screened directly using a ^{19}F R_2 filter (Vulpetti *et al.*, 2009). The large ^{19}F chemical shift dispersion and low fragment concentration (18 μM for CF_3-containing fragments) enable screening of large mixtures (36 fragments per sample). This technique offers high throughput (potentially thousands of compounds per day) and low fragment concentration (thus avoiding solubility problems). Unfortunately, the usefulness of ^{19}F-based methods is currently limited by the low number and diversity of purchasable fluorinated fragments. Out of ca. 12 million commercially available compounds, we found 424,000 fragments that conform to modified Rule of Three criteria (Schuffenhauer *et al.*, 2005) and pass our functional group filters (Walters and Murcko, 2002), of which only 43,700 (10.3%) contained one CF or CF_3 group; these compounds are considerably less diverse than the full fragment set (P. Walters, unpublished results).

Alternatively, several groups describe the use of ^{13}C-labeled compounds with inverse-detected line broadening or STD experiments to detect ligand binding, either directly or in competition mode (Feher *et al.*, 2008; Rauber and Berger, 2010; Swann *et al.*, 2010). Although these methods could potentially circumvent the resonance overlap problems of STD/WaterLOGSY and the diversity problems of fluorinated fragments, they are less sensitive than ^{1}H and ^{19}F methods and, more importantly, few ^{13}C-labeled fragments are commercially available.

3.4. Setting experimental conditions

Setting the experimental conditions for the screen requires a compromise between the desire for high throughput and the ability to detect low-affinity binders, which favor the use of high protein and fragment concentrations, and the need to limit protein consumption and avoid solubility problems, which favor the opposite. To test the performance of the NMR experiments under various conditions, we make samples of human serum albumin with salicylic acid (a known ligand with $K_D \sim 140\ \mu M$; Rich *et al.*, 2001) and sucrose (a negative control). Stocks of 100 μ*M* human serum albumin, 10 m*M* salicylic acid, and 10 m*M* sucrose are first prepared in 50 m*M* pH 7 phosphate buffer. Various concentrations of protein and ligands can then be tested by diluting the stocks (into fully deuterated buffer for STD and 10% D_2O/90% H_2O buffer for WaterLOGSY).

Figure 9.1A shows the ^1H NMR spectrum of a test sample of 500 μ*M* salicylic acid and 500 μ*M* sucrose with 5 μ*M* HSA: the salicylic acid and

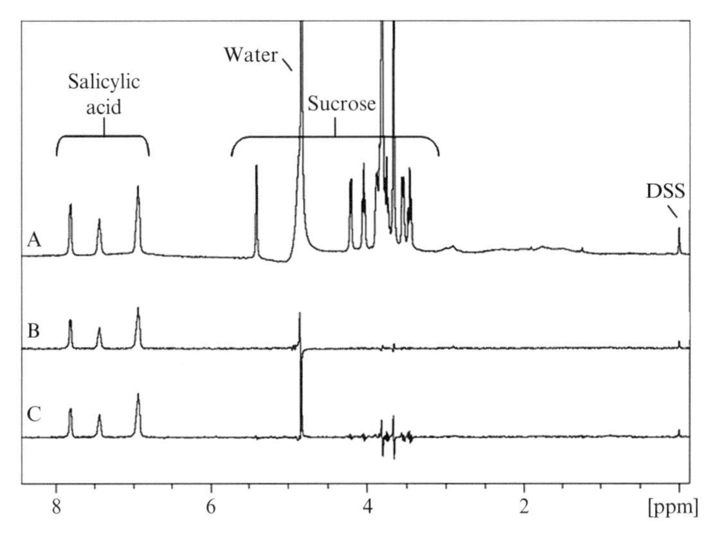

Figure 9.1 Examples of STD spectra collected using HSA/salicylic acid/sucrose standard sample. (A) 500 MHz ^1H NMR spectrum with WATERGATE water suppression (Piotto *et al.*, 1992) of 500 μ*M* salicylic acid, 500 μ*M* sucrose, and 5 μ*M* human serum albumin in 50 m*M* sodium phosphate, pH 7 (90% D_2O, 10% H_2O), 0.1 m*M* sodium azide, 0.1 m*M* 4,4-dimethyl-4-silapentane-1-sulfonic acid (DSS). (B) Saturation transfer difference spectrum (Mayer and Meyer, 1999, 2001) of sample from (A), collected using a 3-s saturation train of 50 ms Gaussian G3 cascade pulses (Emsley and Bodenhausen, 1990) to selectively saturate the HSA methyl resonances at 0.4 ppm. WATERGATE was used for water suppression and a 25-ms spin lock relaxation filter to suppress the protein resonances. Four sets of eight interleaved on and off-resonance (at 20 ppm) scans were collected, for a total experiment time of 4.5 min. (C) STD spectrum collected as in (B), but with a 2 degree variation in sample temperature during the acquisition to illustrate the appearance of subtraction artifacts (most prominent at 3.5–4.0 ppm). All spectra were collected at 298 K using a Bruker Avance II 500 MHz NMR spectrometer equipped with a cryogenically cooled, 5 mm $^1H\{^{13}C, ^{15}N\}$-triple resonance probe. Chemical shifts are referenced to DSS at 0 ppm. Spectra were processed with 4 Hz exponential line broadening using Topspin 2.1 (Bruker Instruments).

sucrose resonances are well separated, and very weak HSA resonances are visible, particularly between 2.5 and 3.0 ppm. Figure 9.1B shows the saturation difference spectrum: salicylic acid produces a strong signal in the difference spectrum while sucrose, which does not bind, produces none. The negative control is useful for identifying artifacts due to instrument instability: Fig. 9.1C demonstrates incomplete subtraction of the sucrose resonances arising from a 2 °C variation in sample temperature during the acquisition.

The sensitivity of the NMR experiments depends upon the concentrations of ligand and protein and the ligand:protein ratio (Peng *et al.*, 2001). For ligand-detected methods, the detected signal also increases with molecular correlation time, so the ligand:protein ratio can be lower for high MW targets; ratios up to 10,000:1 have been used for STD experiments of large targets. For moderately sized proteins (25–70 kDa), we typically screen at a ligand:protein ratio of 500 or 1000:1 (0.5–1 μ*M* protein and 500 μ*M* fragments); screening 1000 fragments in mixtures of five thus requires ca. 5 mg of protein and 30 min per STD spectrum. Throughput can be increased by raising the protein concentration: for example, STD spectra can be collected in 4–5 min using 5 μ*M* protein (Fig. 9.1B).

4. After the Screen

4.1. Identifying and validating fragment hits

Hits are identified by comparing the STD or WaterLOGSY spectra of the mixtures with the individual ^{1}H NMR reference spectra. Although this step lends itself to automation (peaks in screening spectra can be automatically picked and then compared to reference peak lists), the spectrum of every hit should be manually examined. Some apparent hits may be subtraction artifacts due to instrument instability, appearing as narrow resonances that are out of phase with regard to the rest of the spectrum (see Fig. 9.1C). Others may be due to spillover from irradiation of the protein or water, such as the DSS resonances of Fig. 9.1B and C. When some fragment resonances fall under the protein irradiation frequency and are inadvertently saturated, artifacts are typically seen for all other resonances of that fragment. This problem can be diagnosed by using an alternative irradiation frequency (e.g., by saturating protein aromatics at 7.3 ppm).

Even when the fragment mixtures are carefully checked beforehand, some may aggregate when screened, particularly if the buffer conditions differ from those of the reference spectra. All samples should be checked for cloudiness or precipitate at the end of the screen. Spectral indicators of aggregation include apparent binding from all components of a mixture, an unusually broad-based water resonance, and broadening or chemical shift changes of fragment resonances. The latter evidence is obtained from ^{1}H NMR reference spectra, which are collected along with the screening data.

The minimum binding signal used to designate a compound as a hit is based upon the noise level of the data. Typically, the minimum signal is at least five times the root-mean-squares noise, and a value of ten times the noise is used if the samples will be spiked with a known ligand to check for competition (since at least a 50% decrease in binding signal should be observed). The reliability of the hits can be improved by combining results from different experiments, each with different sources of artifacts. For example, the group at Vernalis compares STD, WaterLOGSY, and R_2 relaxation-filtered spectra with and without added competitor. The most robust hits appear in all the three experiments, while those that bind but show no change upon addition of competitor are designated as "binding non-competitively" (Hubbard *et al.*, 2007b).

We find that a small number of otherwise well-behaved fragments appear as hits in nearly every screen, against a wide variety of targets. These frequent hitters are subjected to additional scrutiny during follow-up; some examples are shown in Fig. 9.2. Curiously, a number contains scaffolds that have previously been identified as "privileged" or "preferred" motifs for protein binding (Barelier *et al.*, 2010; Hajduk *et al.*, 2000).

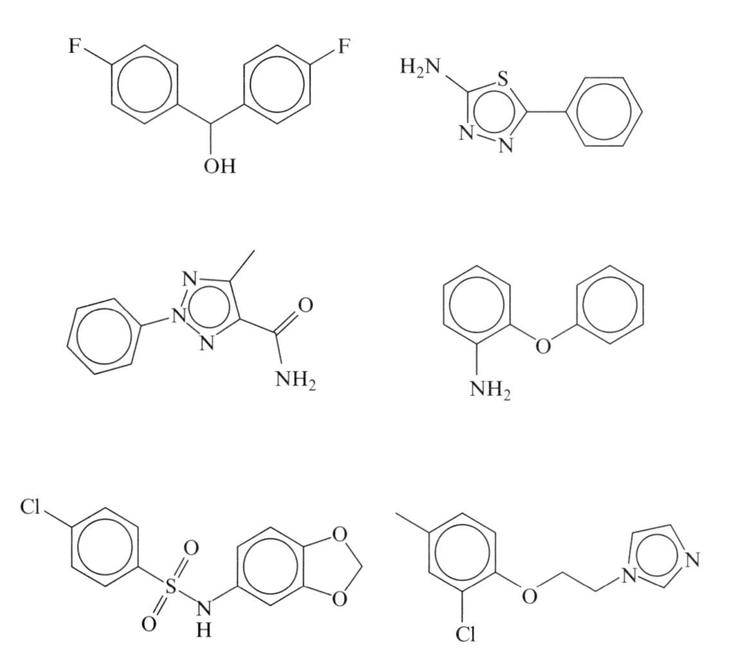

Figure 9.2 Examples of frequent hitters from Vertex fragment screens.

Hits must be validated to confirm that they are binding to the site of interest and determine their binding affinities or biological activities. When a receptor-based screening method is used, binding site information is available from the primary screen, and affinities can be determined (for $K_D \geq$ [receptor]) by titration of the hits (Dalvit, 2009). When ligand-detected methods are used, binding affinities can be measured directly using initial STD rates (Angulo *et al.*, 2010) or by titration of linewidths or WaterLOGSY/STD signal (Fielding *et al.*, 2005), and indirectly via changes in the binding signal from a competing ligand of known affinity (Dalvit *et al.*, 2001).

It is highly desirable to validate hits using "orthogonal" methods (i.e., based on different read-outs and prone to different sources of interference), such as a combination of NMR methods (competition and chemical shift mapping) with biophysical methods (SPR, ITC, fluorescence, etc.) or biochemical methods (enzyme assay, cell surface receptor assay, etc.). Quality control checks (NMR, LC–MS, testing of new lots, etc.) should also be run to check for compound decomposition or contamination.

4.2. Following up hits: Potential pitfalls

The success of FBDD hinges upon converting fragment hits into leads, which is most effective when hits are followed up using both directed searches (testing analogs of hits) and structure-based drug discovery (using X-ray and NMR structures of bound fragments). Many successful examples have been described (Albert *et al.*, 2007; Alex and Flocco, 2007; Congreve *et al.*, 2008; Erlanson, 2006; Hajduk and Greer, 2007; Hubbard *et al.*, 2007a, b; Lepre, 2002, 2007; Lepre *et al.*, 2004; Rees *et al.*, 2004; Wyss and Eaton, 2007), including in this volume, so this discussion will focus on problems that may arise during the follow-up process.

Ideally, as activity and structural data are obtained for the fragment hits and analogs, a coherent picture will emerge. The same relative binding affinities are observed using multiple methods. Multiple actives are found within each scaffold class, and clear structure–activity relationships are evident. A consistent pattern of ligand:protein contacts is observed for the bound fragments, and the binding orientations of the fragments are maintained as new functional groups are added or other fragments are linked.

Sometimes, a clear picture cannot be obtained because the primary screen finds too few or too many hits. A lack of hits may indicate that the target has a low propensity to bind ligands and is not druggable; however, protein that is aggregated, misfolded, or otherwise incompetent to bind ligands can also produce few or no hits. On the contrary, protein that is aggregated, misfolded, or has exposed hydrophobic surfaces can also produce very high hit rates. In such cases, competition experiments may be unable to identify genuine hits, since a high background level of nonspecific binding will occur for all hits.

Membrane proteins are particularly prone to problems with high background binding when ligand-based screening methods are used, making it difficult to identify specific binders. For example, an STD experiment can detect compounds that bind to membrane proteins or the surrounding detergent micelles (even when the protein is selectively irradiated, since magnetization is transferred from protein to micelle). Control experiments using micelles without protein are not entirely reliable for identifying micelle-binders, since some compounds apparently bind to micelles only when protein is present (P. Connolly, unpublished results).

A special challenge is posed by binding sites with shallow binding energy topologies. Typical enzyme targets possess well-defined conformations and binding sites that make specific, energetically favorable interactions with ligands in order to position active site residues in a specific geometry to carry out catalysis. Fragments will bind consistently in the deep energy well of the active site, making similar patterns of protein contacts. In contrast, the binding sites of noncatalytic targets (such as receptors or protein–protein interactions) may be conformationally flexible and bind ligands via multiple, low-affinity interactions spread out over a large area. These binding sites possess many shallow potential energy minima with similar energies. Such sites bind fragments unpredictably: different binding modes and protein contacts are observed for structurally similar fragments (with comparable affinities), and the bound orientations change dramatically upon modification of the scaffold, sometimes accompanied by changes in local protein conformation.

Lead discovery and optimization can be very challenging for binding sites with shallow energy topologies. In order to construct potent leads, it may be necessary to contact several widely separated, low-affinity binding sites, and the use of structural information to guide this process can be complicated by reorientation as the scaffold is modified. This may result in leads with high molecular weights and relatively low binding efficiencies, making them poor starting points for medicinal chemistry optimization.

5. Conclusions

FBDD is very adaptable and readily tailored to the needs and available resources of each user. There is no single best way to implement fragment screening by NMR, so this chapter has outlined a general process for method development, discussed the choices available, and provided information to enable users to select the appropriate experiments and procedures for their applications while avoiding common pitfalls.

In our laboratory, we customize the FBDD approach to each project. For example, fragment screens have been used to assess the druggability of new targets and construct libraries of promising scaffolds prior to HTS screening, drive lead discovery when an HTS assay was unavailable, search for new starting points after HTS failed to find viable leads, find fragments that bind in a nearby pocket in order to extend an existing medicinal chemistry lead, and identify novel scaffolds for second-generation programs or to replace portions of a potential clinical candidate that encountered toxicity or IP problems. As the use of FBDD continues to grow and the technology advances, many more success stories and applications to challenging targets (such as membrane-bound receptors and protein–protein interactions) are expected to appear in the scientific literature.

ACKNOWLEDGMENTS

The author gratefully acknowledges the assistance of Pat Walters and Andrew Daltas with analysis of commercial libraries, Peter Connolly for experimental expertise and helpful discussions, and Jonathan Moore for a critical reading of the chapter.

REFERENCES

Akritopoulou-Zanze, I., and Hajduk, P. J. (2008). Kinase-targeted libraries: The design and synthesis of novel, potent and selective kinase inhibitors. *Drug Discov. Today* **14,** 291–297.

Albert, J. S., Blomberg, N., Breeze, A. L., Brown, A. J. H., Burrows, J. N., Edwards, P. D., Folmer, R. H. A., Geschwindner, S., Griffen, E. J., Kenny, P. W., Nowak, T., Olsson, L., *et al.* (2007). An integrated approach to fragment-based lead generation: Philosophy, strategy and case studies from AstraZeneca's drug discovery programmes. *Curr. Top. Med. Chem.* **7,** 1600–1629.

Alex, A. A., and Flocco, M. M. (2007). Fragment-based drug discovery: What has it achieved so far? *Curr. Top. Med. Chem.* **7,** 1544–1567.

Angulo, J., Enriques-Navas, P. M., and Nieto, P. M. (2010). Ligand-receptor binding affinities from saturation transfer difference (STD) NMR spectroscopy: The binding isotherm of STD initial growth rates. *Chemistry* **16,** 7803–7812.

Aronov, A. M., and Bemis, G. W. (2004). A minimalist approach to fragment-based ligand design using common rings and linkers: Application to kinase inhibitors. *Proteins* **57,** 36–50.

Babaoglu, K., and Shoichet, B. K. (2006). Deconstructing fragment-based inhibitor discovery. *Nat. Chem. Biol.* **2,** 720–723.

Baell, J. B., and Holloway, G. A. (2010). New substructure filters for the removal of pan assay interference compounds (PAINS) from screening libraries and for their exclusion in bioassays. *J. Med. Chem.* **53,** 2719–2740.

Barelier, S., Pons, J., Gehring, K., Lancelin, J.-M., and Krimm, I. (2010). Ligand specificity in fragment-based drug design. *J. Med. Chem.* **53**(14), 5256–5266.

Baurin, N., Aboul-Ela, F., Barril, X., Davis, B., Drysdale, M., Dymock, B., Finch, H., Fromont, C., Richardson, C., Simmonite, H., and Hubbard, R. E. (2004). Design and characterization of libraries of molecular fragments for use in NMR screening against protein targets. *J. Chem. Inf. Comput. Sci.* **44,** 2157–2166.

Blomberg, N., Cosgreve, D. A., Kenny, P. W., and Kolmodin, K. (2009). Design of libraries for fragment screening. *J. Comput. Aided Mol. Des.* **23**(8), 513–525.

Chen, I.-J., and Hubbard, R. E. (2009). Lessons for fragment library design: Analysis of output from multiple screening programs. *J. Comput. Aided Mol. Des.* **23**(8), 603–620.

Chessari, G., and Woodhead, A. J. (2009). From fragment to clinical candidate—A historical perspective. *Drug Discov. Today* **14,** 668–675.

Congreve, M., Carr, R., Murray, C., and Jhoti, H. (2003). A 'rule of three' for fragment-based lead discovery? *Drug Discov. Today* **8,** 876–877.

Congreve, M., Chessari, G., Tisi, D., and Woodhead, A. J. (2008). Recent developments in fragment-based drug discovery. *J. Med. Chem.* **51,** 3661–3680.

Dalvit, C. (2009). NMR methods in fragment screening: Theory and a comparison with other biophysical techniques. *Drug Discov. Today* **14,** 1051–1057.

Dalvit, C., Pevarello, P., Tato, M., Veronesi, M., Vulpetti, A., and Sundstrom, M. (2000). Identification of compounds with binding affinity to proteins via magnetization transfer from bulk water. *J. Biomol. NMR* **18,** 65–68.

Dalvit, C., Fogliatto, G., Stewart, A., Veronesi, M., and Stockman, B. (2001). Water-LOGSY as a method for primary NMR screening: Practical aspects and range of applicability. *J. Biomol. NMR* **21,** 349–359.

Dalvit, C., Flocco, M., Knapp, S., Mostardini, M., Perego, R., Stockman, B. J., Veronesi, M., and Varasi, M. (2002). High-throughput NMR-based screening with competition binding experiments. *J. Am. Chem. Soc.* **124,** 7702–7709.

Dalvit, C., Fagerness, P. E., Hadden, D. T., Sarver, R. W., and Stockman, B. J. (2003). Fluorine-NMR experiments for high-throughput screening: Theoretical aspects, practical considerations, and range of applicability. *J. Am. Chem. Soc.* **125,** 7696–7703.

Dalvit, C., Aridini, E., Fogliatto, G. P., Mongelli, N., and Veronesi, M. (2004). Reliable high-throughput functional screening with 3-FABS. *Drug Discov. Today* **9,** 595–602.

Emsley, L., and Bodenhausen, G. (1990). Gaussian pulse cascades: New analytical functions for rectangular selective inversion and in-phase excitation in NMR. *Chem. Phys. Lett.* **165,** 469–476.

Erlanson, D. A. (2006). Fragment-based lead discovery: A chemical update. *Curr. Opin. Biotechnol.* **17,** 643–652.

Erlanson, D. A., Wells, J. A., and Braisted, A. C. (2004). Tethering: Fragment-based drug discovery. *Annu. Rev. Biophys. Biomol. Struct.* **33,** 199–223.

Feher, K., Groves, P., Batta, G., Jimenez-Barbero, J., Muhle-Goll, C., and Kover, K. E. (2008). Competition saturation transfer difference experiments improved with isotope editing and filtering schemes in NMR-based screening. *J. Am. Chem. Soc.* **130,** 17148–17153.

Fielding, L., Rutherford, S., and Fletcher, D. (2005). Determination of protein-ligand binding affinity by NMR: Observations from serum albumin model systems. *Magn. Reson. Chem.* **43,** 463–470.

Gossert, A. D., Henry, C., Blommers, M. J., Jahnke, W., and Fernandez, C. (2009). Time efficient detection of protein-ligand interactions with the polarization optimized PO-WaterLOGSY NMR experiment. *J. Biomol. NMR* **43,** 211–217.

Hajduk, P. J., Olejniczak, E. T., and Fesik, S. W. (1997) One-dimensional relaxation and diffusion-edited NMR methods for screening compounds that bind to macromolecules. *J. Am. Chem. Soc.* **119,** 12257–12261.

Hajduk, P. J., and Greer, J. (2007). A decade of fragment-based drug design: Strategic advances and lessons learned. *Nat. Rev. Drug Discov.* **6,** 211–219.

Hajduk, P. J., Bures, M., Praestgard, J., and Fesik, S. W. (2000). Privileged molecules for protein binding identified from NMR-based screening. *J. Med. Chem.* **43,** 3443–3447.

Hann, M. M., and Oprea, T. I. (2004). Pursuing the leadlikeness concept in pharmaceutical research. *Curr. Opin. Chem. Biol.* **8,** 255–263.

Hann, M., Hudson, B., Lewell, X., Lifely, R., Miller, L., and Ramsden, N. (1999). Strategic pooling of compounds for high-throughput screening. *J. Chem. Inf. Comput. Sci.* **39,** 897–902.

Hartshorn, M. J., Murray, C. W., Cleasby, A., Frederickson, M., Tickle, I. J., and Jhoti, H. (2005). Fragment-based lead discovery using X-ray crystallography. *J. Med. Chem.* **48,** 403–413.

Hautbergue, G. M., and Golovanov, A. P. (2008). Increasing the sensitivity of cryoprobe protein NMR experiments by using the sole low-conductivity arginine glutamate salt. *J. Magn. Reson.* **191,** 335–339.

Hubbard, R. E., Chen, I., and Davis, B. (2007a). Informatics and modeling challenges in fragment-based drug discovery. *Curr. Opin. Drug Discov. Dev.* **10,** 289–297.

Hubbard, R. E., Davis, B., Chen, I., and Drysdale, M. J. (2007b). The SeeDs approach: Integrating fragments into drug discovery. *Curr. Top. Med. Chem.* **7,** 1568–1581.

Jahnke, W., Floersheim, P., Ostermeier, C., Zhang, X., Hemmig, R., Hurth, K., and Uzunov, D. P. (2002). NMR reporter screening for the detection of high affinity ligands. *Angew. Chem. Int. Ed Engl.* **41,** 3420–3423.

Johnson, E. C., Feher, V. A., Peng, J. W., Moore, J. M., and Williamson, J. R. (2003). Application of NMR SHAPES screening to an RNA target. *J. Am. Chem. Soc.* **125,** 15724–15725.

Kelly, A. E., Ou, H. D., Withers, R., and Doetsch, V. (2002). Low-conductivity buffers for high-sensitivity NMR measurements. *J. Am. Chem. Soc.* **124,** 12013–12019.

Leach, A. R., Hann, M. M., Burrows, J. N., and Griffen, E. J. (2006). Fragment screening: An introduction. *Mol. Biosyst.* **2,** 429–446.

Lepre, C. A. (2001). Library design for NMR-based screening. *Drug Discov. Today* **6,** 133–140.

Lepre, C. A. (2002). Strategies for NMR screening and library design. *In* "BioNMR Techniques in Drug Research," (O. Zerbe, ed.), pp. 1349–1364. Wiley-VCH, Weinheim.

Lepre, C. A. (2007). Fragment-based drug discovery using the SHAPES method. *Expert Opin. Drug Discov.* **2,** 1555–1566.

Lepre, C. A., and Moore, J. M. (1998). Microdrop screening: A rapid method to optimize solvent conditions for NMR spectroscopy of proteins. *J. Biomol. NMR* **12,** 493–499.

Lepre, C. A., Peng, J., Fezjo, J., Abdul-Manan, N., Pocas, J., Jacobs, M., Xie, X., and Moore, J. M. (2002). Applications of SHAPES Screening in Drug Discovery. *Comb. Chem. High Throughput Screening* **5,** 583–590.

Lepre, C. A., Moore, J. M., and Peng, J. W. (2004). Theory and applications of NMR-based screening in pharmaceutical research. *Chem. Rev.* **104,** 3641–3676.

Mayer, M., and Meyer, B. (1999). Characterization of ligand binding by saturation transfer difference NMR spectroscopy. *Angew. Chem. Int. Ed.* **38**(12), 1784–1788.

Mayer, M., and Meyer, B. (2001). Group epitope mapping by saturation difference NMR to identify segments of a ligand in direct contact with a protein receptor. *J. Am. Chem. Soc.* **23,** 6108–6117.

Nettleship, J. E., Brown, J., Groves, M. R., and Geerlof, A. (2008). Methods for protein characterization by mass spectrometry, thermal shift (ThermoFluor) assay, and multiangle or static light scattering. *Methods Mol. Biol.* **426,** 299–318.

Page, R., Peti, W., Wilson, I. A., Stevens, R. C., and Wuthrich, K. (2005). NMR screening and crystal quality of bacterially expressed prokaryotic and eukaryotic proteins in a structural genomics pipeline. *Proc. Natl. Acad. Sci. USA* **102,** 1901–1905.

Peng, J., Lepre, C., Fejzo, J., Abdul-Manan, N., and Moore, J. (2001). Magnetic resonance-based approaches for lead generation in drug discovery. *Methods Enzymol.* **338,** 202–230.

Petros, A. M., Dinges, J., Augeri, D. J., Baumeister, S. A., Betebenner, D. A., Bures, M. G., Elmore, S. W., Hajduk, P. J., Joseph, M. K., Landis, S. K., Nettesheim, D. G., Rosenberg, S. H., *et al.* (2006). Discovery of a potent inhibitor of the antiapoptotic protein Bcl-xL from NMR and parallel synthesis. *J. Med. Chem.* **49,** 656–663.

Piotto, M., Saudek, V., and Sklenar, V. (1992). Gradient-tailored excitation for single-quantum NMR spectroscopy of aqueous solutions. *J. Biomol. NMR* **2,** 661–665.

Rauber, C., and Berger, S. (2010). 13C-NMR detection of STD spectra. *Magn. Reson. Chem.* **48,** 91–93.

Rees, D. C., Congreve, M., Murray, C. W., and Carr, R. (2004). Fragment-based lead discovery. *Nat. Rev. Drug Discovery* **3,** 660–672.

Rich, R. L., Day, Y., Morton, T. A., and Myszke, D. G. (2001). High-resolution and high-throughput protocols for measuring drug/human serum albumin interactions using Biacore. *Anal. Biochem.* **296,** 197–207.

Rishton, G. M. (1997). Reactive compounds and in vitro false positives in HTS. *Drug Discov. Today* **2,** 382–384.

Rossi, P., Swapna, G. V. T., Huang, Y. J., Aramini, J. M., Anklin, C., Conover, K., Hamilton, K., Xiao, R., Acton, T. B., Ertekin, A., Everett, J. K., and Montelione, G. T. (2010). A microscale protein NMR sample screening pipeline. *J. Biomol. NMR* **46,** 11–22.

Schuffenhauer, A., Ruedisser, S., Marzinzik, A. L., Jahnke, W., Blommers, M., Selzer, P., and Jacoby, E. (2005). Library design for fragment based screening. *Curr. Top. Med. Chem.* **5,** 751–762.

Shuker, S. B., Hajduk, P. J., Meadows, R. P., and Fesik, S. W. (1996). Discovering high-affinity ligands for proteins: SAR by NMR. *Science* **274,** 1531–1534.

Siegal, M. G., and Vieth, M. (2007). Drugs in other drugs: A new look at drugs as fragments. *Drug Discov. Today* **12,** 71–79.

Siegal, G., Eiso, A. B., and Schultz, J. (2007). Integration of fragment screening and library design. *Drug Discov. Today* **12,** 1032–1039.

States, D. J., Haberkorn, R. A., and Ruben, D. J. (1982). A two-dimensional nuclear Overhauser experiment with pure absorption phase in four quadrants. *J. Magn. Reson.* **48,** 286–292.

Stockman, B. J., Farley, K. A., and Angwin, D. T. (2001). Screening of compound libraries for protein binding using flow-injection nuclear magnetic resonance spectroscopy. *Methods Enzymol.* **338,** 230–246.

Swann, S. L., Song, D., Sun, C., and Hajduk, P. J. (2010) Labeled ligand displacement: Extending NMR-based screening of protein targets. *ACS Med. Chem. Lett.* **1**(6), 295–299.

Vulpetti, A., Hommel, U., Landrum, G., Lewis, R., and Dalvit, C. (2009) Design and NMR-based screening of LEF, a library of chemical fragments with different local environment of fluorine. *J. Am. Chem. Soc.* **131**(6), 12949–12959.

Walters, W. P., and Murcko, M. A. (2002). Prediction of 'drug-likeness'. *Adv. Drug Deliv. Rev.* **54,** 255–271.

Wiegelt, J., van Dongen, M., Uppenberg, G. J., Schultz, J., and Wikstrom, M. (2002). Site-selective screening by NMR spectroscopy with labeled amino acid pairs. *J. Am. Chem. Soc.* **124**(11), 2446–2447.

Wyss, D. F., and Eaton, H. L. (2007). Fragment-based approaches to lead discovery. *Front. Drug Des. Discovery* **3,** 171–202.

CHAPTER TEN

Binding Site Identification and Structure Determination of Protein–Ligand Complexes by NMR: A Semiautomated Approach

Joshua J. Ziarek, Francis C. Peterson, Betsy L. Lytle, *and* Brian F. Volkman

Contents

Abstract

Over the last 15 years, the role of NMR spectroscopy in the lead identification and optimization stages of pharmaceutical drug discovery has steadily

Department of Biochemistry, Medical College of Wisconsin, Milwaukee, Wisconsin, USA

Methods in Enzymology, Volume 493

ISSN 0076-6879, DOI: 10.1016/B978-0-12-381274-2.00010-8

increased. NMR occupies a unique niche in the biophysical analysis of drug-like compounds because of its ability to identify binding sites, affinities, and ligand poses at the level of individual amino acids without necessarily solving the structure of the protein–ligand complex. However, it can also provide structures of flexible proteins and low-affinity ($K_d > 10^{-6}$ M) complexes, which often fail to crystallize. This chapter emphasizes a throughput-focused protocol that aims to identify practical aspects of binding site characterization, automated and semiautomated NMR assignment methods, and structure determination of protein–ligand complexes by NMR.

Abbreviations

HMQC	heteronuclear multiple-quantum correlation
HSQC	heteronuclear single-quantum correlation
NMR	nuclear magnetic resonance spectroscopy
NOE	nuclear Overhauser effect
NOESY	nuclear Overhauser effect spectroscopy
NUS	nonuniform sampling
RMSD	root mean square deviation
ROESY	rotating frame Overhauser effect spectroscopy
SAR	structure–activity relationship
TOCSY	total correlation spectroscopy
TROSY	transverse relaxation optimized spectroscopy

1. Introduction

The incorporation of structure-based knowledge in drug design has led to a threefold increase in identification of high-potency compounds ($IC_{50} < 100$ nM) from fragment leads (Hajduk and Greer, 2007). Nuclear magnetic resonance spectroscopy (NMR) is an invaluable tool for drug discovery, particularly with the growth of fragment-based screening in the pharmaceutical industry. The SAR (structure–activity relationships) by NMR technique (Shuker *et al.*, 1996) refocused the niche of NMR in drug design from structure determination to characterization and optimization of lead compounds (Peng *et al.*, 2001). With steady progress in production of isotope-labeled proteins (Frederick *et al.*, 2007; Jeon *et al.*, 2005; Tyler *et al.*, 2005a,b; Zhao *et al.*, 2004), instrument sensitivity (Hajduk *et al.*, 1999; Jensen *et al.*, 2010), and methods for rapid data collection and analysis (Frydman *et al.*, 2002; Rovnyak *et al.*, 2004a,b;

Schanda and Brutscher, 2005), NMR is being utilized early in the process of identifying lead molecules because of its ability to quickly recognize and characterize ligand binding regardless of target protein function (Hajduk and Greer, 2007).

This chapter presents a workflow for spectroscopists interested in examining protein–ligand interactions with an emphasis on identifying the binding site and solving the structure of the complex. We first present an efficient process for spectral acquisition, processing, and assignment of a target protein or complex. We then describe an approach to protein-based compound screening that includes experimental validation of *in silico* docking results and quantitative affinity measurements to prioritize further studies. Finally, we detail how to solve the structure of protein–ligand complexes. Various approaches toward completely automating NMR structure determination have been recently reviewed and discussed and are beyond the scope of this chapter (Baran *et al.*, 2004; Güntert, 2009; Markley *et al.*, 2009a; Williamson and Craven, 2009). Instead, we focus on the semiautomatic approach used in our laboratory to determine protein structures and on its application to solving protein–ligand complexes.

2. Automated and Semiautomated Chemical Shift Assignment

2.1. Acquisition and processing of NMR spectra

Before NMR data collection begins, ^{15}N-labeled proteins must first be screened by 1D and 2D NMR to assess folding, aggregation state, stability, ligand binding, and to detect unstructured regions (see Chapter 2 for information on NMR sample preparation methods). If necessary, individual protein domains can be isolated from other regions of the protein to increase solubility and reduce aggregation (Peterson *et al.*, 2007; Waltner *et al.*, 2005). Proteins that are judged amenable to NMR structure determination are then produced in [U-^{15}N,^{13}C]-labeled form and put into a general workflow for NMR-based structure determination, as outlined in Fig. 10.1.

A standardized scheme for acquisition of 2D and 3D spectra sufficient for NMR structure determination is outlined in Table 10.1. On a 600 MHz spectrometer equipped with a cryogenic probe, these experiments can be acquired in 7–10 days for proteins under 25 kDa at concentrations > 0.5 m*M*. This time is significantly shortened (to as little as 30 h) if the protein concentration is high enough that data can be recorded without signal averaging or using nonlinear sampling methods. Continuous data collection on a single sample reduces spectral variations, resulting in improved peak picking and automated chemical shift assignment later on. In the case of protein–ligand complexes, several additional experiments, outlined in Section 2.5, are necessary

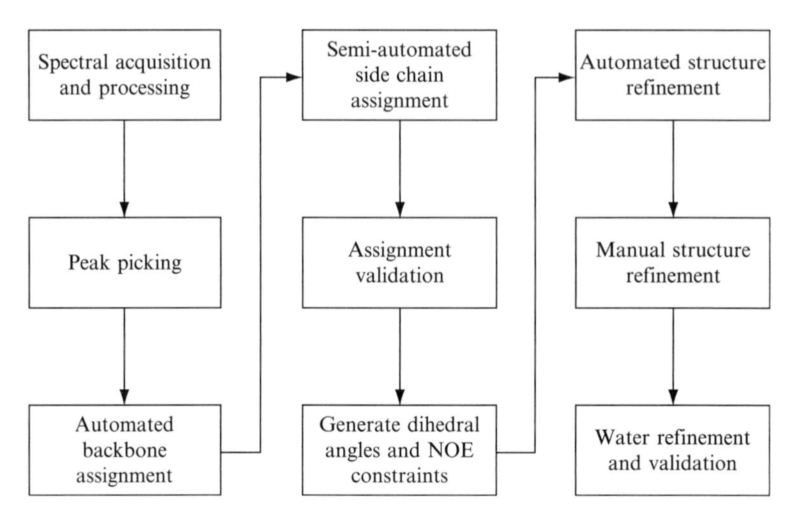

Figure 10.1 Flow chart describing the collection, assignment, and structure determination of a protein–ligand complex.

for ligand assignment. Additionally, intermolecular distance constraints for protein–ligand complexes are determined from one additional experiment, the 3D F1-^{13}C/^{15}N-filtered/F3-^{13}C-edited NOESY-HSQC. This experiment selects for (edits) protons attached to ^{13}C in the F3 dimension, and selects against (filters) protons bound to ^{13}C and ^{15}N in the F1 dimension. Thus, if unlabeled ligand is bound to labeled protein, this experiment will yield only intermolecular NOEs between protein and ligand protons.

As outlined in Table 10.1, all data for the assignment and structure analysis are acquired in three blocks, starting with the NOESY spectra, followed by backbone triple-resonance experiments and then TOCSY-based experiments that provide side chain assignment information. 1D ^{1}H and 2D ^{15}N–^{1}H HSQC spectra are collected before and after every 3D experiment to provide a record of sample and instrument stability over the data collection period. Standard data acquisition parameters, including up-to-date pulse calibrations and spectral parameters, are stored in a common repository on each NMR instrument, so that data acquisition for each new protein can be started rapidly after only a few routine steps (VT adjustment, ^{2}H lock, ^{1}H/^{15}N/^{13}C tuning, shimming, ^{1}H 90° pulse calibration, etc.). Accurate chemical shift referencing in the direct and indirect dimensions using 4,4-dimethyl-4-silapentane-1-sulfonic acid (DSS) is important for consistency between spectra (Markley *et al.*, 1998). Standard 2D and 3D pulse sequences are provided in pulse sequence libraries from the NMR spectrometer manufacturers and are also available from the BioMagResBank (BMRB) Web site (http://www.bmrb.wisc.edu/tools/choose_pulse_info.php).

Table 10.1 Modular scheme for NMR data collection for protein structure analysis

NOESY	Ligand complexes	Backbone	Side chain
3D ^{15}N-NOESY-HSQC	2D F1/F2-^{13}C/^{15}N-doubly-filtered NOESY	3D HNCO	3D C(CO)NH
3D ^{13}C-NOESY-HSQC	3D F1-^{13}C/^{15}N-filtered, F3-^{13}C-edited NOESY	3D HNCA	3D HCCH-TOCSY
2D ^{13}C-HSQC-aromatic		3D HN(CO)CA	2D ^{13}C-HSQC
3D ^{13}C-NOESY-HSQC-aromatic		3D HN(CA)CO	3D H(CCO)NH
		3D HNCACB	3D HBHA(CO)NH

After a target protein or complex has been validated for NMR-based structure determination, the NOESY experiments are collected in a continuous block. The NOESY experiments are used to define distance constraints and therefore are the most sensitive to changes in sample stability. When solving a protein–ligand complex, the unlabeled ligand resonances and intermolecular contacts are detected using isotope filtered from NOESY experiments collected in the next block. Finally, a series of heteronuclear correlation spectra for use in the backbone and side chain chemical shift assignment stages are acquired.

As with data collection, we use a standardized set of scripts to convert and process the time-domain NMR data using NMRpipe (Delaglio *et al.*, 1995). Conversion of the Bruker-formatted data to NMRpipe format is accomplished with the bruk2pipe script, supplied with NMRpipe. For processing, we typically zero-fill the direct ^{1}H dimension to 1024 complex points and apply both forward–backward linear prediction and zero-filling in each indirect dimension to 128 or 256 complex points. In depth analysis of spectral processing is beyond the scope of this text and interested readers are directed to Hoch and Stern's treatise (Hoch and Stern, 1996). After transformation, the resulting frequency-domain spectra are converted to XEASY format (Bartels *et al.*, 1995); however, NMRpipe is compatible with any of the other commonly used spectral analysis programs such as Sparky (Goddard and Kneller, 2001), CARA (Keller, 2004/2005), or NMRView (Johnson and Blevins, 1994).

Reduced dimensionality methods for data collection encode multiple indirect frequencies simultaneously rather than independently, enabling the acquisition of four- or higher-dimensional spectra in a short time and reducing the number of necessary experiments for structure determination. Examples include the G-matrix Fourier transform (GFT) method (Kim and Szyperski, 2003; Shen *et al.*, 2005) and HIFI-NMR (Eghbalnia *et al.*, 2005). Nonuniform sampling (NUS), in which data is collected for only a subset of all incremented evolution periods at nonfixed intervals, is another approach to decreased data collection times that is especially promising for larger proteins (Barna *et al.*, 1987; Rovnyak *et al.*, 2004a,b). The recently released TopSpin 3.0 Bruker software package incorporates NUS as part of a routine workflow, including optimized sampling schemes and data processing via Multi-Dimensional Decomposition (Orekhov *et al.*, 2003). Alternatively, processing via maximum entropy reconstruction (Mobli *et al.*, 2007) has been made more accessible to the general protein NMR community by use of the Rowland NMR Toolkit Script Generator, available as a web server at http://sbtools.uchc.edu/nmr/nmr_toolkit/.

2.2. Peak picking

Once all 2D and 3D data have been processed, the position and intensity of peaks in each experiment must be recorded for submission to an automated chemical shift assignment protocol. Peak picking is one step for which there are relatively few reliable automated programs. As has been discussed by others, this is due to issues with noise, artifacts, and strong peak overlap (Altieri and Byrd, 2004; Williamson and Craven, 2009). Because accurate peak picking is crucial for attaining good results with automated assignment programs, most labs traditionally use manual or semiautomated methods assisted by software such as NMRView (Johnson and Blevins, 1994), XEASY (Bartels *et al.*, 1995), or Sparky (Goddard and Kneller, 2001). AUTOPSY (Koradi *et al.*, 1998) is one of the few commonly used automated peak picking programs; however, even with this sophisticated algorithm, peak lists may still require further editing, especially in regions of overlapped peaks (Malmodin *et al.*, 2003). It remains to be seen whether recent developments such as the PICKY automated peak picking program (Alipanahi *et al.*, 2009b) or more error-tolerant backbone assignment programs such as MARS (Jung and Zweckstetter, 2004) and IPASS (Alipanahi *et al.*, 2009a) will have an effect on currently established peak picking methods.

Our method for picking peaks of the 3D heteronuclear spectra is a semiautomated approach that uses the automated peak picker of XEASY. The basic steps of our peak picking protocol are outlined below:

1. Pick the ^{15}N-HSQC peaks using the automatic in-phase peak picker of XEASY.
2. After making any necessary manual adjustments to the ^{15}N-HSQC peak list, use this list as the basis for picking the HNCO peaks. The HNCO spectrum enables greater resolution of overlapped ^{15}N-HSQC peaks and thus, the ^{1}H and ^{15}N dimensions of its peaks are used as the 2D root for picking peaks in all 3D heteronuclear experiments and the 3D ^{15}N-edited NOESY.
3. Manually inspect the HNCO peak list in XEASY, checking for any missing or extra peaks.
4. Automatically pick the peaks in the HNCA, HNCACB, HN(CO)CA, HN(CA)CO, C(CO)NH, H(CCO)NH, and HBHA(CO)NH spectra. Manually inspect the spectra for erroneous peaks.
5. The ^{13}C-edited aromatic NOESY is picked separately using the ^{13}C aromatic HSQC peak list as the 2D root. For the ^{15}N-edited NOESY and ^{13}C-edited aliphatic NOESY, we wait to pick the peaks until fully assigned HNCO and HCCH-TOCSY peak lists (see Sections 2.3 and 2.4) are available for use as a starting point for manual peak picking.

2.3. Backbone chemical shift assignments

With the advent of numerous semiautomated and automated assignment protocols over the last decade, the backbone resonance assignment process, utilizing standard triple-resonance NMR data, has become a relatively routine process in many laboratories. What once took weeks can now be completed in a day or less for proteins under 25 kDa. Automated resonance assignment programs include GARANT (Bartels *et al.*, 1996, 1997), AutoAssign (Zimmerman *et al.*, 1994, 1997), MARS (Jung and Zweckstetter, 2004), ABACUS (Grishaev *et al.*, 2005b), MATCH (Volk *et al.*, 2008), and PINE (Bahrami *et al.*, 2009), as well as many others which have been previously reviewed (Altieri and Byrd, 2004; Baran *et al.*, 2004; Grishaev *et al.*, 2005a). The programs, PINE and AutoAssign, have the additional advantage of being readily available as web-based servers: http://pine.nmrfam.wisc.edu/ and http://nmr.cabm.rutgers.edu/autoassign/, respectively. Furthermore, the

AutoAssign web server has the option to submit jobs simultaneously to PINE, the results of which can be shown in a comparison chart where the user can easily identify resonance assignment differences between the two programs.

Much of our past work has utilized GARANT for backbone assignments, but PINE is a convenient, new alternative that generates equally good results. PINE accepts peak lists from up to thirteen 3D heteronuclear experiments in either XEASY or Sparky format. The output of PINE is a probabilistic assignment for every residue. This means that it is possible for some residues to have more than one candidate spin system for their chemical shifts. In these cases, an associated probability is assigned to each candidate spin system assigned to a residue. Assigned XEASY peak lists are obtained using PINE as follows:

1. The input files are the sequence file (∗.seq), empty chemical shift list (∗.prot), and peak lists from HNCA, HNCACB, HN(CO)CA, HN (CA)CO, C(CO)NH, H(CCO)NH, and HBHA(CO)NH experiments, all in XEASY format. The empty ∗.prot file is required in order to obtain assigned, XEASY-formatted peak lists from the PINE server.
2. In cases where prior partial assignments are available, such as backbone assignments that are unchanged between the apo and bound forms of the protein, a preassignment file in BMRB format can be uploaded.
3. The files are uploaded to the server: http://miranda.nmrfam.wisc.edu/PINE/. Depending on the size of the protein, the quality of input data, and the number of current jobs, run times vary from 1 h to 1 day.
4. The assignment results, which include secondary structure identification and assigned peak lists, are returned via email.
5. The peak lists are then loaded into XEASY for inspection and verification of backbone assignments, and any missing assignments are completed manually. Alternatively, the PINE-SPARKY extension (Lee *et al.*, 2009; available at http://pine.nmrfam.wisc.edu/PINE-SPARKY/) can be used to complete the assignments interactively. This interface allows the user to simultaneously visualize the spectra and the identity of probable peak assignments.

2.4. Side chain chemical shift assignments

In contrast to the backbone assignments, side chain assignments typically require greater manual intervention due to incomplete data and degeneracy of aliphatic chemical shifts. With ideal C(CO)NH, H(CCO)NH, and HCCH-TOCSY peak lists, both PINE and GARANT are capable of generating automatic side chain assignments. In reality, however, at least 90% completeness of ordered regions is required for automated structure calculations and, therefore, nearly complete assignments require manual analysis (Jee and Güntert, 2003). First, the partial aliphatic side chain assignments obtained from PINE are verified. From these assignments, the side chain ^{1}H–^{13}C cross-peaks in the patterns expected for an HCCH-TOCSY spectrum can be generated using GARANT. This peak list forms the starting point for manual completion of all aliphatic side chain assignments; depending on the completeness of HCCH-TOCSY assignments, the manual processing can be completed within 1 day. Finally, aromatic side chain assignments are completed manually using the ^{13}C aromatic NOESY and the previously determined aliphatic side chain assignments.

1. Verify the assignments for the HSQC, HNCO, HNCACB, HN(CA) CO, C(CO)NH, HBHA(CO)NH, and H(CCO)NH peak lists. Add any missing assignments and save the resulting *.prot file.
2. Generate the HCCH-TOCSY peak list with the genPeaks command of GARANT or an equivalent utility: genPeaks HCCH24 *protein*.seq *protein*.prot hcch.peaks
3. Examine the HCCH-TOCSY peak list in XEASY or another spectral assignment program, picking peaks in any unassigned spin systems. Correct any incorrect or missing assignments; only the diagonal peaks need to be assigned. The methionine H^{ε} peaks can typically be recognized by their characteristic ^{1}H and ^{13}C shifts and the lack of any other correlations. These and other side chain chemical shifts that cannot be assigned due to overlapped or missing resonances are completed later during analysis of the ^{13}C-edited NOESY.
4. Assign NH_2 chemical shifts in the ^{15}N-HSQC on the basis of H^{β}–NH_2 and H^{γ}–NH_2/H^{β}–NH_2 NOE correlations in the ^{15}N-edited NOESY for asparagine and glutamine, respectively.
5. Assign the $C^{\delta}H^{\delta}$ groups of the ^{13}C aromatic NOESY peak list (diagonal peaks only) according to NOE correlations with intraresidual H^{N}, H^{α}, and H^{β} resonances (Lin *et al.*, 2006). Once the chemical shifts of the $C^{\delta}H^{\delta}$ groups are known, their neighboring $C^{x}H^{x}$ or $N^{x}H^{x}$ groups can be assigned.
6. Create a final *.prot file by incorporating assigned chemical shifts from the following experiments in this order:
   - HCCH-TOCSY
   - ^{13}C aromatic NOESY
   - HNCO
   - HSQC
7. Proline ^{13}C chemical shifts can identify *cis* Xaa–Pro peptide bonds (Stanczyk *et al.*, 1989). Checking for the *cis* versus *trans* proline conformation can be done automatically by using the CYANA macro 'cisprocheck' or a program such as POP (Schubert *et al.*, 2002). Prolines in the *cis* conformation should be renamed "cPRO" in the XEASY/CYANA format sequence file. Likewise, any cysteine side chains known to participate in a disulfide bridge should be renamed "CYSS."

2.5. Ligand assignments

The assignment strategy for small molecules is complicated by the lack of isotopic enrichment of ^{15}N and ^{13}C. In the free state, assignment of all protons in the small molecule can be completed using a set of 2D homonuclear spectra that include 2D ^{1}H-TOCSY, 2D ^{1}H-NOESY, and/or 2D ^{1}H-ROESY. However, ligand binding to its target protein induces chemical shift changes in both the protein and ligand with varying magnitudes that are dependent on the local chemical environment. To assign an unlabeled ligand in the bound state, one can use a 2D F1/F2-^{13}C/^{15}N-doubly-filtered NOESY spectrum (Ikura and Bax, 1992) as previously described (Alam *et al.*, 1998). In the doubly filtered NOESY spectrum, only NOE cross-peaks involving ^{12}C-attached protons as the source and destination are present in the final spectrum. Since ^{12}C-attached protons only occur within the small molecule, the doubly filtered NOESY spectrum will only show intramolecular NOEs for the ligand. These NOEs can then be used to sequentially assign the proton shifts for the small molecule.

2.6. Validation

Because the completeness and accuracy of the chemical shift assignments is crucial for the robustness of the subsequent steps of structure determination, the assignments should be checked for errors before proceeding with structure determination. TALOS and CYANA, programs described in Sections 4.2.1 and 4.2.2, automatically check for chemical shifts that are significantly outside the typical range for the particular atom/residue and also report statistics on the completeness of the assignments. A variety of other programs available for validating chemical shift referencing and/or assignments have recently been evaluated (Wang *et al.*, 2010).

3. Identification of Ligand Binding Sites by Chemical Shift Mapping

3.1. HSQC titrations with fragments, compounds, and mixtures

Fragment-based drug design is most effective for identifying relatively weak ($K_d > 10^{-6}$ *M*) compounds with the potential for development into high-affinity molecules. NMR is an optimal technique to test the quality of these fragment leads because it can provide atom-specific information that is most sensitive to low-affinity interactions. A basic outline of how NMR can contribute to the rational drug development process is outlined in Fig. 10.2. Once the chemical shift assignments of the target protein are known

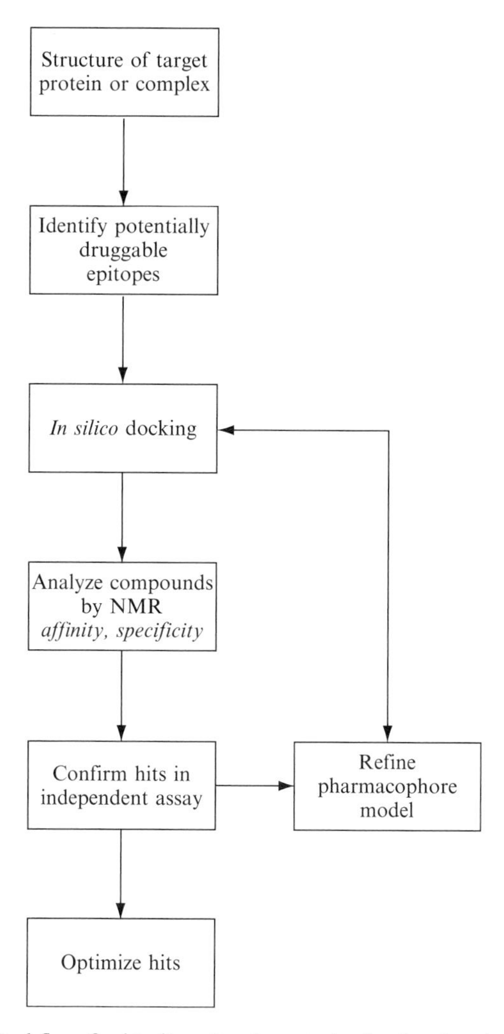

Figure 10.2 Workflow for binding site characterization by chemical shift mapping.

(see Section 2), [U-^{15}N,^{13}C]-enriched protein can be titrated with a putative ligand or binding partner and the resulting changes in chemical shift monitored by 2D correlation spectroscopy. This procedure is called *chemical shift mapping* and can be used to determine binding constants, stoichiometry, and ligand position. 2D ^{1}H–^{15}N HSQC spectroscopy is the most common method for mapping backbone chemical shifts, because every residue type (except proline) contains a highly sensitive probe to changes in local magnetic environment. HMQC (Mandal and Majumdar, 2004) and TROSY (Pervushin *et al.*, 1997) are alternative pulse schemes that provide advantages in certain cases and supply equivalent spectral information. As most protein–ligand interactions are side chain-mediated, 2D ^{1}H–^{13}C HSQCs can provide additional information, but the side chain signals suffer from overlap in large target proteins. Thus, selective ^{13}C-labeling of Ile, Leu, and Val residues can significantly improve resolution without compromising sensitivity (Meyer and Peters, 2003).

Titrations are typically performed in two ways: (1) add known ligand volumes to a target–protein sample, or (2) prepare multiple samples with a constant protein concentration and increasing ligand concentrations. The former method consumes substantially less protein, but suffers from less accurate binding constants due to changing protein concentrations. Using cryoprobe technology without a sample-handling robot, three unique compounds can typically be analyzed per day using this scheme:

1. Collect initial HSQC spectra on 50 μ*M* protein sample.
2. Collect additional HSQC spectra with ligand additions of 25, 50, and 800 μ*M*; comparison of four data points facilitates exchange regime determination and binding affinity quantitation (Fielding, 2003).
3. Overlay spectra and qualitatively examine the magnitude of chemical shift perturbation, specificity, and exchange regime (Fig. 10.3A).
4. Measure chemical shift values of each assigned resonance throughout titration, determine adjusted chemical shift (Eq. (10.5)), and plot as a function of residue number (Fig. 10.3B). Use chemical shift perturbations to quantify binding affinity (Eq. (10.6); Fig. 10.3C). If the target protein structure is known (see Section 4 for structure determination), residues with the most significant perturbations can be highlighted (Fig. 10.3D); this is particularly valuable for identifying ligand specificity when compound docking models have been calculated.

NMR validation can be the most time-consuming aspect of screening large numbers of compounds. To improve throughput, compounds can be grouped together and simultaneously screened by NMR. The most promising mixtures are then analyzed individually. Alternatively, "chromatographic" NMR experiments have been described to use 2D diffusion information to identify high-affinity small molecules from mixtures (Gonnella *et al.*, 1998; Gounarides *et al.*, 1999; Hajduk *et al.*, 1997a,b). The gain in efficiency of mixture analysis strategies is difficult to assess; Ross and colleagues argue that the high binding success rate (~1:3) of compounds suggests that the extra effort spent to deconvolute spectra should be redirected to initially preparing more reagents for individual analysis (Ross *et al.*, 2000). Automated sample handling and rapid 2D NMR acquisition schemes, such as SOFAST-HMQC (Schanda and Brutscher, 2006; Schanda *et al.*, 2005) and single-scan spatial-encoding (Frydman *et al.*, 2002; Gal *et al.*, 2006), can be expected to increase the throughput of NMR-based ligand screening in the near future.

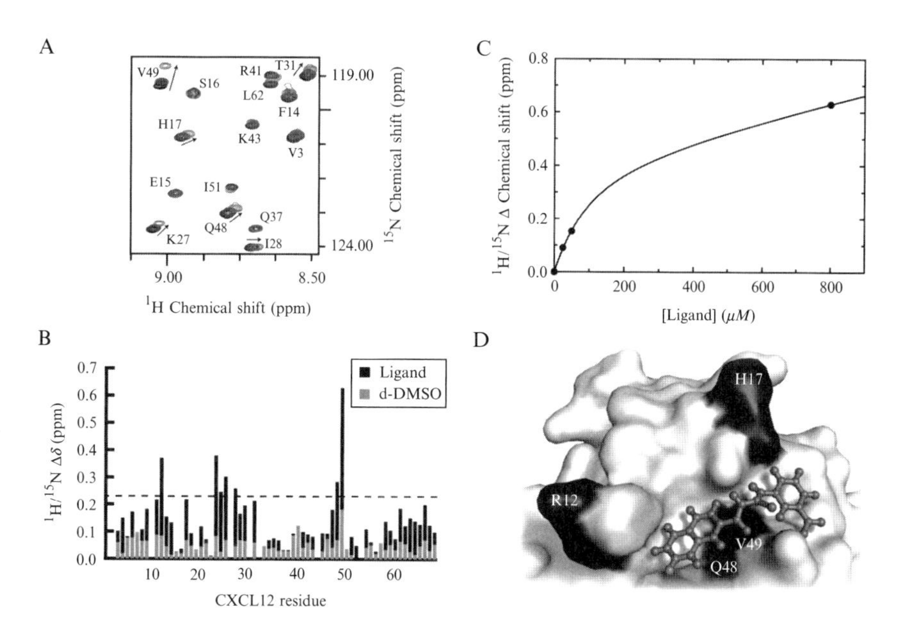

Figure 10.3 An example of binding site identification by chemical shift mapping. (A) Overlays of HSQC spectra monitoring the chemokine CXCL12 being titrated with 0, 25, 50, and 800 μ*M* (colored from black to gray) of 3-(napthylene-2-carbonylthiocarbamoylamino)-benzoic acid (ligand). (B) CXCL12 adjusted chemical shift change (Eq. (10.5)) for each residue as a function of 800 μ*M* ligand (black) or d-DMSO control (gray). The dotted line represents residues with the most significant chemical shift perturbations. (C) Quantification of V49 affinity (K_d = 64 μ*M*) by fitting of chemical shift perturbations to Eq. (10.6). (D) Comparison of significant chemical shift perturbations (black) with predicted docking of the ligand to CXCL12.

3.2. Discrimination between specific and nonspecific binding

As many small molecules are hydrophobic, the solubility of ligands is often improved by reconstitution in an organic solvent; thus, the solvent itself may interact with the target protein. These interactions may be nonsaturable but can produce perturbations of substantial magnitude, and may even reveal the locations of druggable sites in the protein (Byerly *et al.*, 2002; Dalvit *et al.*, 1999; Liepinsh and Otting, 1997). Solvent perturbations are relatively simple to detect by performing a solvent-only titration for comparison with spectra reporting on candidate ligands (Fig. 10.3B).

Another facet of nonspecific binding is the orientation or mode that a compound adopts when associated with a target protein. Lead compounds that exhibit multiple-mode binding should be modified to maintain a single-mode prior to optimization. Analysis of titration line shapes can quickly identify binding modes (Reibarkh *et al.*, 2006); single-mode ligands with micromolar affinity should produce line-broadening at substoichiometric concentrations that sharpen upon saturation. Protein resonances that do not sharpen generally reflect multiple-mode ligand binding. However, this does not preclude the possibility that ligands can modulate the oligomeric state (Veldkamp *et al.*, 2005) or conformation of the protein (Vajpai *et al.*, 2008).

As chemical shift perturbations can result from both ligand-binding and ligand-induced conformational changes, discrimination between direct and indirect effects is crucial to effective chemical shift mapping. Residues with the largest changes in chemical shift are presumed to be specifically involved in the binding interaction, as ligand-induced conformation changes are rare in low-affinity interactions (Cavanagh, 2007). However, the binding site of ligands interacting with target proteins with large conformation changes can be identified by the use of differential chemical shifts. Fesik and colleagues outline a protocol to compare chemical shift perturbations of highly similar compounds (Medek *et al.*, 2000). The chemical shift changes resulting from conformation change should be similar, while ligand-induced perturbations will differ by chemical uniqueness. We performed a similar analysis of differential chemical shifts between sulfated and unsulfated peptides to identify the sulfotyrosine recognition site of CXCL12 in the absence of explicit structural data (Veldkamp *et al.*, 2006).

3.3. Quantitation of binding constants by NMR

The determination of binding constants during compound screening is crucial to ranking compounds for optimization and lead selection. The ability to quantitate affinity by NMR is highly dependent on the binding kinetics. The following is a general outline of the theoretical principles for determining binding reactions on the NMR time scale (for detailed texts on NMR kinetics, see Cavanagh, 2007; Levitt, 2008). Assuming a simple two-state equilibrium, for a protein (P) binding to a ligand (L) to form a complex (PL), with second-order binding kinetics,

$$\mathrm{P} + \mathrm{L} \underset{k_{\mathrm{off}}}{\overset{k_{\mathrm{on}}}{\rightleftarrows}} \mathrm{PL} \tag{10.1}$$

where k_{on} and k_{off} are the rate constants of association and dissociation, respectively. The rate constant for exchange (k_{ex}), or the chemical exchange lifetime ($\tau_{\mathrm{ex}} = 1/k_{\mathrm{ex}}$), is calculated as

$$k_{\mathrm{ex}} = k_{\mathrm{on}}[\mathrm{L}] + k_{\mathrm{off}} \tag{10.2}$$

Comparison of the exchange rate constant (k_{ex}) with the difference in angular frequency ($\Delta\omega$) between an unbound (ω_{P}) and bound (ω_{PL}) resonance,

$$\Delta\omega = \omega_{\mathrm{P}} - \omega_{\mathrm{PL}} \tag{10.3}$$

categorizes the binding interaction into three basic exchange regimes:

$$\begin{aligned} \text{Fast exchange} &: k_{\mathrm{ex}} \gg |\Delta\omega| \\ \text{Intermediate exchange} &: k_{\mathrm{ex}} \sim |\Delta\omega| \\ \text{Slow exchange} &: k_{\mathrm{ex}} = |\Delta\omega| \end{aligned} \tag{10.4}$$

The particular exchange regime a binding reaction belongs to can be readily identified by observation of each resonance throughout a titration; it is important to note that each atom can exhibit its own rate of exchange for a particular binding reaction.

Protein resonances in fast exchange do not change significantly in intensity and continuously exhibit chemical shift changes through the titration. Fast-intermediate exchange is similar to fast exchange, with the exception of diminished intensity and linewidth as the chemical shift changes from unbound and bound. Resonances that significantly diminish, possibly below detection, and then reappear at a different chemical shift with increasing ligand concentrations are in intermediate exchange. In practice, resonances in intermediate exchange may not reappear, because it is not possible to get high enough ligand concentrations depending on solubility and protein concentration. Slow exchange is characterized by a resonance that decreases in intensity but does not change chemical shift with greater ligand concentration; at the same time, a resonance with a unique chemical shift will begin to increase in intensity. Visual inspection of binding kinetics provides a qualitative interpretation of binding affinity as fast exchange is weak ($K_d > 10^{-6}$ M), intermediate exchange is generally between 10 and 100 μM, and slow exchange is strong ($K_d < 10^{-6}$ M).

To facilitate high-throughput quantitative analysis, we will only consider binding reactions in fast exchange. We presume this is valid because most compounds screened will have affinities weaker than 10^{-6} M; however, to minimize error, the need to ensure fast-exchange binding kinetics cannot be understated. Interested readers are directed toward excellent reviews on quantitating slow and intermediate exchange kinetics (Feeney *et al.*, 1979; Fielding, 2003). This is especially important during lead optimization, as compounds are expected to possess high affinities. HSQC chemical shift changes are quantified for each titration point using an adjusted chemical shift value (Δ) of the form,

$$\Delta = \sqrt{(\Delta\delta_{\mathrm{H}})^2 + (c\Delta\delta_{\mathrm{N}})^2} \tag{10.5}$$

in which $\Delta\delta_H$ and $\Delta\delta_N$ are the chemical shift changes of amide proton and nitrogen, respectively. Many values for the scaling factor, c, have been proposed ranging from 0.1 to 0.25 (Geyer *et al.*, 1997; Mulder *et al.*, 1999); we typically employ a value of 0.2 to balance the contributions of ^{1}H and ^{15}N to the magnitude of the perturbation. More complex formulations that use scaling factors specific to each amino acid have also been proposed (Geyer *et al.*, 1997; Schumann *et al.*, 2007). A similar range of scaling factors has been proposed for the measurement of combined ^{1}H–^{13}C shift perturbations (Schumann *et al.*, 2007). The increase in adjusted chemical shift perturbation as a function of ligand concentration is then fitted by nonlinear regression using an equation that accounts for ligand depletion, single-site binding, and nonspecific binding:

$$\Delta = (\Delta_{max}/2[P_T]) \times \left\{ K_d + [P_T] + [L_T] - \sqrt{(K_d + [P_T] + [L_T])^2 - 4[P_T][L_T]} \right\} + NS[L_T] \quad (10.6)$$

in which Δ is the adjusted chemical shift change, Δ_{max} the maximum adjusted chemical shift change at saturation, $[P_T]$ total protein concentration, K_d the equilibrium dissociation constant, $[L_T]$ the total ligand concentration, and NS is nonspecific binding.

3.4. Pharmacophore modeling from comparative shift perturbation analysis

The chemical features of a protein–ligand interface that are responsible for a given biological function define a pharmacophore. They include the hydrogen bond vectors, hydrophobic volumes, and steric properties of active ligands that identify potential inhibitors. It is important to note that pharmacophores do not correspond to common functional groups, and therefore occupy a unique niche in structure-based drug design. A pharmacophore model is often used as a prefilter for *in silico* screening because it can be evaluated more rapidly than explicit docking and may be less prone to susceptible to false positive hits. In a ligand-based computational process, the pharmacophore definition can be enhanced by NMR titration and structure data. Specifically, using the method for comparative chemical shift perturbation analysis (Medek *et al.*, 2000) outlined in Section 3.2, one can identify the binding site of a series of related lead compounds. Once the binding modes of several compounds have been determined by comparative analysis, the pharmacophore elucidation can be performed by any number of programs (as reviewed by Leach *et al.*, 2010).

Structural knowledge of a protein provides invaluable information of exclusion regions (Greenidge *et al.*, 1998; Rella *et al.*, 2006; Tintori *et al.*, 2008); knowledge of binding site volumes accelerates compound docking considerably as only molecules that will fit are examined.

3.5. Epitope definition for docking and *in silico* screening

Both traditionally druggable proteins (enzymes), with distinct binding pockets, and protein–protein interfaces, with relatively flat surfaces, are being successfully targeted for small-molecule inhibitors (Veldkamp *et al.*, 2010; Wells and McClendon, 2007). An integral part of rational drug design is identification of biologically relevant binding surfaces or epitopes. When target protein structures are known, knowledge of epitopes is invaluable for directed compound docking and reducing the computational load of screening chemical libraries. Traditionally, epitope definition utilized structures solved by crystallography; however, NMR is an established alternative that is especially suited for solving weak complexes that are frequently not crystallizable. When full structure determination of the complex is not feasible, NMR can still yield epitope information from HSQC titrations. In the absence of high-resolution structural data, ligand-binding epitopes can be discerned through site-directed mutagenesis or antibody inhibition.

A variety of docking programs have been described for computational screening of chemical libraries, including AutoDock (Morris *et al.*, 1998), DOCK (Ewing *et al.*, 2001), eHiTs (Zsoldos *et al.*, 2006), FlexX (Rarey *et al.*, 1996), and Fred (McGann *et al.*, 2003). In collaboration with other experts, we have successfully implemented epitope-directed compound screening against cytokine-receptor signaling systems using both the DOCK and AutoDock packages. These programs can be used for screening of both private and public chemical libraries. For example, we recently used the NMR structure of the chemokine CXCL12 in complex with a 38-residue peptide of the CXCR4 receptor (Veldkamp *et al.*, 2008) to identify a small-molecule ligand for CXCL12 that inhibits CXCR4-mediated calcium flux (Veldkamp *et al.*, 2010). Specifically, we used the CXCL12-binding site of two CXCR4 residues to define an epitope. Using DOCK, we targeted 1.4 million compounds from the ZINC database (Irwin and Shoichet, 2005), a free virtual library of commercially available small molecules, at this site. In an NMR-based assay of the top-ranked molecules, four of the five tested compounds bound specifically to the targeted epitope, including one with a K_d value of 64 μM. The success of this study and others (Arkin and Wells, 2004; Berg, 2008; Fuller *et al.*, 2009) demonstrates the value of NMR binding data to identify epitopes for virtual screening within the rational drug development process.

4. 3D Structure Determination by NMR

4.1. NMR-derived structural constraints

Solution NMR structure calculations rely mainly on two types of geometric constraints: (1) distance constraints based on analysis of 3D ^{15}N- and ^{13}C-edited NOESY spectra and (2) dihedral angle constraints predicted from chemical shifts. When disulfide bonds, bound metal ions, or other cofactors of known geometry are present, additional constraints are typically included in the form of generic upper and lower distance limits to enforce the covalent cross-links or coordinating bonds. The use of other restraints, such as residual dipolar couplings (RDCs) and pseudocontact shifts, can be directly incorporated into CYANA 3.0, an updated version of the CYANA structure determination program (Herrmann *et al.*, 2002). However, in our experience, these constraint types are not essential to obtain high-quality structural models for most diamagnetic globular domains under ~25 kDa.

4.1.1. Dihedral angles inferred from chemical shifts

After the chemical shift assignments have been completed, constraints for the backbone φ and ψ dihedral angles are generated using the program TALOS+ (Shen *et al.*, 2009). Alternatively, ^{3}J-scalar coupling measurements can be used to estimate dihedral angles from empirically defined relationships (Karplus, 1959); however, our streamlined approach employs chemical shift-derived dihedral angle constraints as they can be obtained without the need for additional spectral acquisition. TALOS+, like the original program TALOS (Cornilescu *et al.*, 1999), compares the H^N, N, C′, H^α, C^α, and C^β chemical shifts from the protein against a database of chemical shifts and backbone geometry in tripeptide segments from known protein structures. The TALOS+ database currently contains 200 proteins. As with the original TALOS program, the ten best database matches for each residue (based on up to 18 individual chemical shifts: six atoms each from three amino acids) are compared to see if they cluster in a consistent region of the Ramachandran (φ vs. ψ angle) plot. TALOS+ utilizes two additional parameters for making predictions. The first is a neural network component whose output is used as an additional empirical term in the conventional TALOS database search. The second is an estimated backbone order parameter S^2 derived from the chemical shifts using the random coil index method (RCI; Berjanskii and Wishart, 2005), which biases against the prediction of backbone dihedral angles in flexible regions. An interface for inspecting and modifying the results is provided, after which the acceptable predictions are converted into a dihedral angle constraint file for use in structure calculations (Cornilescu *et al.*, 1999). TALOS+ also provides an additional feature to precheck chemical shift referencing and possible chemical shift errors. Tabular chemical shift data in the TALOS input format is easily generated either by using the CYANA macro 'taloslist' for chemical shifts in the CYANA/XEASY(.prot) format or with the bmrb2talos script supplied with the TALOS+ software. We have further simplified the data conversion and TALOS+ calculation process with an in-house script (available upon request).

4.1.2. NOE distance constraints

Distance restraints derived from proton–proton nuclear Overhauser effects (NOEs) are the primary NMR data used to define the secondary and tertiary structure of a protein. NOEs develop due to through-space dipole–dipole interactions rather than through-bond interactions. Thus, they contain information on the distances between hydrogen atoms separated by 5 Å or less in space even though the residues may be distal in primary sequence. The intensity of an NOE, that is, the volume of the corresponding cross-peak in a NOESY spectrum, is proportional to the inverse of the sixth power of the distance between the two nuclei due to averaging caused by rotational motion. However, because the NOE is not always a precise reflection of the internuclear distance, structure calculation programs generally translate NOESY cross-peak intensities into upper bounds on interatomic distances rather than fixed distance restraints.

With the availability of a variety of automated assignment programs, the generation of NOE distance constraints from NOESY spectra is a relatively robust and automatic process, at least for the calculation of initial structures. These structures can then be used, if necessary, to expand and correct the preliminary list of NOE assignments for input into another round of structure calculations. The process continues in an iterative fashion, making adjustments to the input parameters and experimental constraints until an acceptable final ensemble of structures is obtained. Programs used for automated assignment include NOAH (Mumenthaler and Braun, 1995), ARIA (Linge *et al.*, 2003a; Nilges, 1995; Rieping *et al.*, 2007), CYANA (Güntert, 2003, 2004; Herrmann *et al.*, 2002), PASD (Kuszewski *et al.*, 2004, 2008), and AutoStructure (Huang *et al.*, 2005, 2006), all of which have been reviewed previously (Güntert, 2003; Williamson and Craven, 2009). Although the automated NOE assignment methods require NOESY peak lists that are accurately referenced relative to the chemical shift assignments, some programs can deal robustly with noisy peak lists. It has been demonstrated that the NOEASSIGN module of CYANA, version 2.1, can yield reliable NOE distance constraints even when one-half to two-thirds of the cross-peaks in the initial NOESY peak lists are noisy or artifactual (Lopez-Mendez and Güntert, 2006), while PASD can tolerate up to 80% incorrectly picked peaks (Kuszewski *et al.*, 2004).

The NOEASSIGN module of CYANA is our preferred tool for automated NOESY peak assignments. Interproton NOE distance constraints generated by NOEASSIGN are derived from three different 3D NOESY spectra: ^{15}N-edited NOESY-HSQC, ^{13}C-aliphatic-edited NOESY-HSQC, and ^{13}C-aromatic-edited NOESY-HSQC. An additional experiment, the 3D F1-^{13}C/^{15}N-filtered/F3-^{13}C-edited NOESY, is added for protein–ligand structures. Because the NOEASSIGN module can cope with a high number of artifacts in the peak lists, only minor manual editing should be necessary at this stage as long as the NOESY spectral referencing matches that of the ^{15}N-HSQC/HNCO and HCCH-TOSCY spectra used to obtain the chemical shift values. Because automated NOE assignment is inseparable from the structure calculation process, the details are described in Section 4.2.

4.2. Automated NMR structure refinement with CYANA

4.2.1. Solving the structure of the free protein

Over the past 6 years, we have successfully used NOEASSIGN-generated distance constraints, in combination with dihedral angle constraints, to solve more than 20 unique structures with CYANA (de la Cruz *et al.*, 2007; Lytle *et al.*, 2004, 2006; Peterson *et al.*, 2005, 2006, 2007; Tuinstra *et al.*, 2007; Vinarov *et al.*, 2004; Waltner *et al.*, 2005). This list includes not only single domain proteins but also symmetrical homodimers (Peterson *et al.*, 2006, 2010) and protein–ligand complexes (Veldkamp *et al.*, 2008). The NOEASSIGN algorithm introduced in CYANA 2.0 is an improved version of the original CANDID algorithm (Herrmann *et al.*, 2002) that is based on a probabilistic treatment of the NOE assignment process. An overall probability of the correctness of possible NOE assignments is calculated as the product of the probabilities for each of the criteria for assignment: agreement between the chemical shift values and the peak positions, compatibility with a preliminary 3D structure, and network-anchoring, that is, the extent of embedding in the network of all other NOE assignments. Automated NOE assignment and structure calculation by the CYANA torsion angle dynamics (TAD) algorithm (Güntert *et al.*, 1997) are combined in an iterative process that comprises seven cycles of automated NOE assignment, followed by a final structure calculation using only unambiguously assigned distance constraints. The structural result of a given cycle is used to guide the NOE assignments in the following cycle, with the precision of the structure normally improving with each subsequent cycle. In the first two cycles, "constraint combination" is applied to medium- and long-range distance restraints in order to reduce the impact of erroneous distance restraints on the resulting structure. In all seven cycles, ambiguous distance constraints are used to generate conformational restraints from NOESY cross-peaks with multiple possible assignments.

In our experience, the NOEASSIGN routine, given the proper input, consistently results in a converged set of initial structures for monomeric proteins. In favorable cases, the result is a well-defined structural ensemble, with backbone RMSD < 1 Å and target function < 5 Å^2. Structures can typically be obtained in less than an hour and often under 30 min, depending on the size of the protein and available computing power. For example, for a 160-residue protein, computation time is 50 min using an 8-core Apple Mac Pro or about 15 min with a 16 node/32 processor Linux cluster.

An example script for running a CYANA calculation with the automated NOE assignment module can be found on the CYANA homepage (http://www.cyana.org/wiki/index.php/Main_Page). The order of the tolerance parameters used by NOEASSIGN is (1) direct ^{1}H, (2) indirect ^{1}H, and (3) indirect ^{15}N/^{13}C. We typically use values of 0.03 ppm for both ^{1}H dimensions and 0.3 ppm for the ^{15}N and ^{13}C dimensions. By leaving the calibration parameter line blank, these parameters are determined automatically such that the median of the upper distance limits for each peak list equals the value of the reference distance variable *dref*, which has a default value of 4.0 Å. In each cycle, and in the final structure calculation, 100 conformers are calculated using the standard simulated annealing schedule with 10,000 torsion angle dynamics steps per conformer. The 20 conformers with the lowest final target function values are analyzed using a variety of statistics, including overall target function, restraint totals, and ensemble coordinate precision (atomic RMSD). In our protocol, NOEASSIGN is initially run in triplicate with identical input files (sequence file, prot file, aco file, and peak lists), changing only the value of "randomseed" for each run. The results of the three NOEASSIGN calculations are then evaluated. While each final structure should converge to fairly similar RMSD and target function values, as reported in the *final.ovw* file, it is more important to assess whether all the three structural ensembles display the same overall tertiary fold by using a structure visualization program such as PyMol or Molmol (Koradi *et al.*, 1996). The presence of any significant deviations is indicative of misassigned NOEs, which may reflect inaccuracies in the chemical shift assignment list. These misassignments can have a significant impact on the structure that may go undetected as refinement proceeds; therefore, it is necessary that they be corrected at the outset. Constraint combination applied during the first two NOEASSIGN cycles helps to minimize these potential structural distortions, but it is not a panacea. It is also important to inspect the statistics of the initial structural ensemble using the "cyanatable" script, because a poorly defined structure ensemble from cycle 1 may ultimately converge in later cycles on a precise but inaccurate final ensemble. NOEASSIGN output can be further evaluated based on the following two criteria for a successful calculation: (1) Less than 25% of the long-range NOEs have been discarded by the automated NOESY assignment routine for the final structure calculation and (2) the backbone RMSD to the mean for the structure bundle of cycle 1 is < 3 Å (Güntert, 2004).

When a satisfactory, consistent ensemble has been obtained using NOEASSIGN, it is typically necessary to inspect and edit the NOESY peak assignments in XEASY (starting from the peak lists generated in cycle 7), focusing on the largest consistent distance and angle violations listed in the *.ovw file as well as any disordered regions or inconsistent conformations between replicate ensembles. Erroneous or missing chemical shift assignments should also be corrected by looking for spin systems (NOESY strips) that remain largely unassigned. NOEASSIGN calculations may then be repeated to satisfy the two criteria defined above. After correcting and verifying all peak lists and chemical shift assignments, a lack of convergence in cycle 1 could indicate the presence of a homodimeric assembly that was not apparent from earlier analysis of the spectra or hydrodynamic measurements of the oligomeric state. After obtaining a satisfactory ensemble of structures by NOEASSIGN, manual structure refinement is performed in order to prepare the ensemble from torsion angle dynamics for a final Cartesian-space refinement calculation in explicit water and subsequent validation steps.

Manual structure refinement proceeds iteratively, using CYANA to calculate an ensemble of structures based on the assigned peak lists (an example script can be found at the CYANA homepage). At the outset of manual refinement, the calibration of NOE intensity/distance conversions should be optimized to decrease both the RMSD and target function. Initial runs are performed with automated calibration in which the parameters for the backbone, side chain, and methyl calibration functions are calculated automatically by the CYANA macro 'calibration.' The calculated parameters found in the output file are then used as a starting point for fine-tuning the calibration. This is accomplished by testing a grid of calibration values (at least three for each peak list, nine calculations altogether) to determine the combination that results in the lowest target function and RMSD. The calibration should also be checked by comparing known interproton distances within secondary structures from well-resolved X-ray structures (Hur and Karplus, 2005) to the corresponding distances in the upper limit distance constraint (.upl) file.

Iterative rounds of CYANA calculations are then performed until no further improvement can be made in the structure. Between each CYANA calculation, the NOESY peak lists are inspected and peak assignments are added, removed, or corrected based on the results of the structure calculations. The consistently violated constraints should be corrected by reassigning cross-peaks, referring to the structure as necessary when there are multiple possibilities. PyMol or Molmol can be used to obtain a list of all atoms within 5.0 Å of an atom of interest. Special attention should be given to verifying the aromatic side chain assignments, since these residues often occupy central locations in the hydrophobic core. In addition to adding and correcting NOE assignments, any consistently violated dihedral angle constraints should be reevaluated or eliminated. Since TALOS constraints rely on chemical shift-based predictions with a finite error rate, a conservative approach preserves only those dihedral angle constraints that correspond to regions of recognizable secondary structure.

4.2.2. Defining CYANA library entries for ligands

Automated NOE structure refinement of protein–ligand complexes using CYANA requires the creation of a molecular topology description for each component of the complex that is not found in the standard CYANA library (e.g., amino acids and nucleotides). The library definition or topology file for a small-molecule ligand contains a description of atom types and nomenclature, covalent connectivities, dihedral angle definitions, and standard geometry of the small molecule. The requirement for such topology files in solving the structures of protein–ligand complexes has increased in parallel with structure-based drug design and was initially addressed by the PRODRG server (Schuttelkopf and van Aalten, 2004) (http://davapc1.bioch.dundee.ac.uk/prodrg/). The PRODRG server takes a PDB file or simple text drawings and generates topology files for GROMOS, GROMACS, WHAT IF, REFMAC5, CNS, O, SHELX, HEX, and MOL2. However, these formats are not suitable for a CYANA topology file. The pdb2reslib module of Olivia (http://fermi.pharm.hokudai.ac.jp/olivia/) addresses this need and allows the user to generate a CYANA topology file from a PDB file. In this section, we describe the process of generating the CYANA topology description for an arbitrary ligand molecule.

Ligand library entries can be generated using a PDB coordinate file as a convenient, widely available description of a molecular structure. If a PDB file is not readily available, one can be created by importing a simplified molecular input line entry specification (SMILES) string (Weininger, 1988, 1989) for the molecule into a graphical chemical editor such as Marvin Sketch (http://www.chemaxon.com/products/marvin/marvinsketch/). The ligand model can then be exported in standard Protein Data Bank format and converted to the CYANA library format using the pdb2reslib module of Olivia. After conversion, the new ligand definition should be inspected and edited to add missing dihedral angles (or remove unnecessary angles), adjust atom nomenclature and numbering, and adjust dihedral angle naming prior to use in CYANA (Fig. 10.4). The edited entry can then be appended to the standard CYANA library file with a unique residue identifier (e.g., DRG). Finally, because molecular dynamics calculations in torsion angle space requires that all components of the system be linked by covalent or virtual bonds in a single chain, the ligand must be incorporated into the protein complex by appending linker residues (e.g., PL, LL, LL2, etc.) and the ligand residue to the sequence (.seq) file, which CYANA will automatically interpret and use to link the ligand to the polypeptide chain.

A

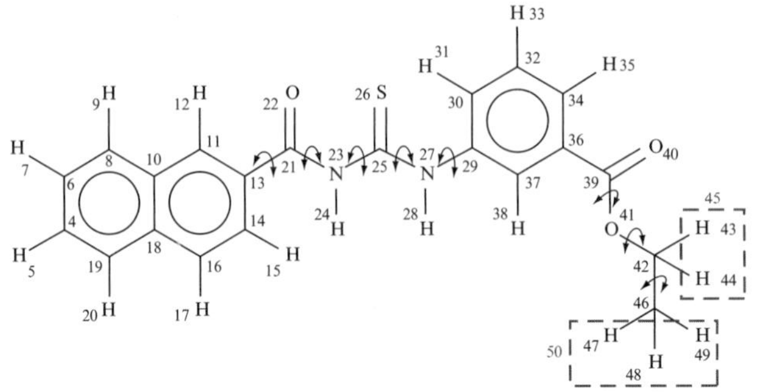

B

	RESIDUE		DRG	8	53	4	50								
	1	ANG1	0	0	0.0000	11 13	21 23	0	4						
	2	ANG2	0	0	0.0000	13 21	23 25	0							
	3	ANG3	0	0	0.0000	21 23	25 27	0							
	4	ANG4	0	0	0.0000	23 25	27 29	0							
	6	ANG5	0	0	0.0000	25 37	29 30	0							
	6	ANG61	0	0	0.0000	36 39	41 42	50	3						
	7	ANG62	0	0	0.0000	39 41	42 46	50							
	8	ANG63	0	0	0.0000	41 42	46 47	50							
	1	Q1	DUMMY	0	0.0000	0.0000	0.0000	0.0000	0	0	0	0	0		
2	2	Q2	DUMMY	0	0.0000	1.4142	0.0000	−1.4120	0	0	0	0	0		
	3	Q3	DUMMY	0	0.0000	2.8284	0.0000	0.0000	0	0	0	0	0		
	4	C1	C_ARO	0	0.0000	−13.1100	−2.4900	−0.4300	5	6	19	0	0		
	6	H1	H_ARO	0	0.0000	−13.9420	−3.0300	−0.5590	4	0	0	0	0		
	6	C2	C_ARO	0	0.0000	−13.2100	−1.0900	−0.2300	4	7	8	0	0		
	7	H2	H_ARO	0	0.0000	−14.1100	−0.6550	−0.2170	6	0	0	0	0		
	8	C3	C_ARO	0	0.0000	−12.0600	−0.3100	−0.0500	6	9	10	0	0		
	9	H3	H_ARO	0	0.0000	−12.1320	0.6770	0.0900	8	0	0	0	0		
	10	C4	C_VIN	0	0.0000	−10.8000	−0.9400	−0.0700	8	11	18	0	0		
	11	C5	C_ARO	0	0.0000	−9.6300	−0.1700	0.1000	10	12	13	0	0		
	12	H5	H_ARO	0	0.0000	−9.6950	0.8190	0.2340	11	0	0	0	0		
	13	C6	C_VIN	0	0.0000	−8.3700	−0.8000	0.0800	11	14	21	0	0		
2	14	C7	H_ARO	0	0.0000	−8.2800	−2.2000	−0.1200	13	15	16	0	0		
	15	H7	H_ARO	0	0.0000	−7.3800	−2.6360	−0.1330	14	0	0	0	0		
	16	C8	C_ARO	0	0.0000	−9.4300	−2.9900	−0.3000	14	17	18	0	0		
	17	H8	H_ARO	0	0.0000	−9.3580	−3.9770	−0.4450	16	0	0	0	0		
	......														
	41	O3	O_EST	0	0.0000	1.0500	−2.2700	−0.0800	39	42	0	0	0		
	42	C14	C_ALI	0	0.0000	2.2500	−3.0900	−0.1400	41	43	44	46	0		
	43	H141	H_ALI	0	0.0000	2.7390	−2.9030	−0.9920	42	0	0	0	45		
	44	H142	H_ALI	0	0.0000	2.8400	−2.8740	0.6380	42	0	0	0	45		
	45	Q14	PSEUD	0	0.0000	2.7895	−2.8885	−0.1770	0	0	0	0	0		
1	46	C15	C_ALI	0	0.0000	1.8600	−4.5700	−0.0900	42	47	48	49	0		
	47	H151	H_ALI	0	0.0000	2.6850	−5.1340	−0.1310	46	0	0	0	50		
	48	H152	H_ALI	0	0.0000	1.3710	−4.7580	0.7620	46	0	0	0	50		
	49	H153	H_ALI	0	0.0000	1.2700	−4.7860	−0.8680	46	0	0	0	50		
	50	Q15	PSEUD	0	0.0000	1.7753	−4.8927	−0.0790	0	0	0	0	0		
	51	Q1	DUMMY	0	0.0000	1.2306	2.3170	−1.1792	0	0	0	0	0		
	52	Q2	DUMMY	0	0.0000	−0.1836	2.3170	−2.5934	0	0	0	0	0		
	53	Q3	DUMMY	0	0.0000	1.2306	2.3170	−4.0076	0	0	0	0	0		

Figure 10.4 Example of small-molecule ligand CYANA library entry. (A) 2D representation of ethyl 3-{[(naphthalen-2-ylformamido)methanethioyl]amino}benzoate with atom numbers indicated. Multiple protons bound to a single heavy atom have the potential for chemical shifts degeneracy (indicated by—gray box) and are given a pseudoatom identifier (colored gray). Rotatable bonds are indicated by arrows. (B) CYANA library entry for the molecule described in panel A after conversion of the PDB file by the pdb2reslib module of Olivia and user editing. Additions to the original PDB file and user changes are highlighted by gray shading. Conversion of a PDB file by

4.2.3. Structure determination of a protein–ligand complex

The structure of a protein–ligand complex is highly valuable in the early stages of structure-based drug design, where lead compounds that interact weakly with their targets may not readily crystallize. In general, the NMR structure of a protein–ligand complex is solved in the same way as the protein itself, as described in the previous sections. The availability of chemical shift and NOE assignments for the ligand-free protein, along with ligand-induced chemical shift perturbation data, simplifies the assignment process for the protein complex. However, structure determination of a protein–ligand complex presents a challenge similar to what we have previously described for NMR structure determination of symmetric protein homodimers (Markley *et al.*, 2009b). As with a protein homodimer, the identification and assignment of intermolecular NOEs between protein and ligand is the principal hurdle, as they are indistinguishable from intramolecular protein NOEs in standard isotope-edited NOESY spectra. Chemical shift degeneracy and their low relative abundance add to the difficulty of detecting NOE contacts that define protein–ligand interface. To overcome these problems, we acquire a 3D F1-^{13}C/^{15}N-filtered, F3-^{13}C-edited NOESY spectrum (Stuart *et al.*, 1999) on a sample of ^{15}N/^{13}C-labeled protein saturated with unlabeled ligand (Fig. 10.5). This experiment detects only NOEs arising between protons directly bound to ^{13}C (protein) and protons bound to any non-^{15}N or ^{13}C atom (ligand) while suppressing all other cross-peaks by isotope filtering and editing. Constraints generated from the filtered NOESY spectrum can therefore be unambiguously assigned to protons on the small-molecule ligand. Additionally, CYANA 3.0 now allows the user to specify NOEs as intermolecular by setting the "color" value of the peak list (column 5) to 9. If ^{13}C/^{15}N-labeled ligand is available, a 3D F1-^{13}C/^{15}N-filtered, F3-^{13}C-edited NOESY spectrum can be collected on a sample containing unlabeled protein and labeled ligand, thereby providing a complementary dataset that further disambiguates atom assignments for intermolecular NOEs. We employed this approach to solve the structures of two protein–peptide complexes, for which [*U*-^{15}N/^{13}C]

---

pdb2reslib adds the necessary pseudoatom definitions (indicated by 1 and gray shading), the dummy atoms required to connect the small molecule to the protein backbone (indicated by 2 and gray shading), and definitions for the dihedral angels (indicated by 3 and gray shading). However, pdb2reslib may not define all potential dihedral angles and the user may choose to add additional dihedral angle definitions (indicated by 4 and gray shading). In this example, the user has chosen to rename the atoms and renumber the molecule sequentially from left to right as shown in panel (A) and rename the angles identified by pdb2reslib (indicated by 3 and gray shading) to reflect the tree structure used in CYANA structure calculations. A detailed explanation for the file format can be found on the CYANAwiki (http://www.cyana.org/wiki/index.php/Residue_library_file).

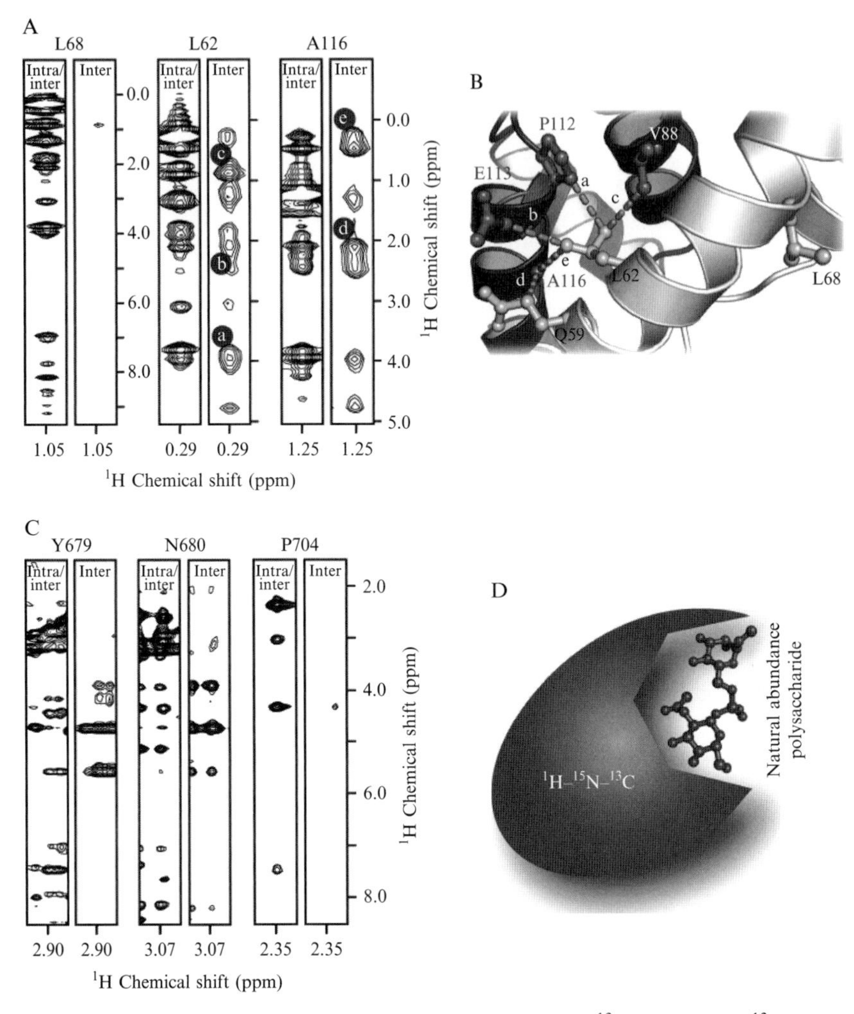

Figure 10.5 Detection of intermolecular NOEs in a 3D F1-^{13}C-filtered, F3-^{13}C-edited NOESY spectrum. (A) Strips extracted from a 3D ^{13}C-edited NOESY spectrum (left strip) contain both intra- and intermolecular NOEs, while strips from the 3D F1-^{13}C-filtered, F3-^{13}C-edited NOESY spectrum (right strip) show only intermolecular NOEs. Strip pairs are shown for MZF1 residues L68, L62, and A116. Leucine 68 is outside the dimer interface and therefore does not have NOEs in the F1-^{13}C-filtered, F3-^{13}C-edited NOESY spectrum. Lower-case letters correspond to NOEs shown in panel (B). (B) Intermolecular NOEs involving Q59, L62, P112, E113, and A116. Lower-case letters correspond to labeled peaks in (A). One monomer is colored in white and the other in gray. (C) Strips extracted from 3D ^{13}C-edited NOESY (left strip) and 3D F1-^{13}C-filtered, F3-^{13}C-edited NOESY (right strip) spectra as in panel (A). Comparison of the strip pairs reveals intermolecular NOEs between a polysaccharide and domain-5 of cation independent-mannose-6 phosphate receptor (CI-MPR). Tyrosine 679 and N680 are in contact with the sugar, while P704 lies just outside the binding pocket. (D) Schematic showing the labeling scheme for the polysaccharide–domain-5 complex.

peptide ligands were produced using standard methods for expression and labeling in *Escherichia coli* (Tyler *et al.*, 2010; Veldkamp *et al.*, 2008).

4.3. Structure refinement in explicit water and validation

The final stages of refinement are reached when the desired goals for precision (e.g., backbone RMSD below ~0.6 Å), agreement with experimental constraints (none violated by more than 0.5 or 5 Å), and good torsion angle geometry (no residues in disallowed regions of ϕ/ψ space) have been achieved. Typically, this requires manual inspection and editing of the NOE assignments that result in consistently violated constraints, as we have previously described (Markley *et al.*, 2009b). NMR structures defined by >15–20 nontrivial constraints/residue typically exhibit high coordinate precision, with RMSD values of 0.5–0.7 Å for backbone atoms and 0.8–1.2 Å for all nonhydrogen atoms. However, it is important to assess the local mobility of the polypeptide backbone using ^{15}N relaxation measurements such as ^{15}N–^{1}H heteronuclear NOE, since flexible regions will often be devoid of long-range NOEs. Consequently, the local NOE constraint density in flexible regions is expected to be low and no extra effort should be made to identify long-range NOE constraints for those residues. Once a consistent set of experimental constraints has been obtained, and the resulting structural ensemble meets the basic acceptance criteria, a final refinement calculation is performed in explicit water with physical force field parameters.

To simplify the water refinement routine, we have developed a script that converts the output of a CYANA calculation into input for water refinement and initiates the water refinement routine, as previously described (Markley *et al.*, 2009b). Our protocol utilizes a method for Cartesian-space molecular dynamics in explicit water using XPLOR-NIH that was shown previously to improve the stereochemical quality of NMR structural ensembles generated with simple force fields (Linge *et al.*, 2003b). Improvements in a variety of validation criteria were observed upon water refinement of NMR ensembles, particularly for the percentage of residues in the most favored region of the Ramachandran (ϕ vs. ψ angle) plot. While this routine is not fully integrated into CYANA calculations, the user can initiate a water refinement run with a single script once a consistent set of experimental constraints has been achieved. The initial structural model and the model refined using restrained molecular dynamics in explicit water are validated against the distance and dihedral angle constraints using WHAT IF and PROCHECK-NMR. Comparison of the input and refinement statistics shows that the Ramachandran statistics of the TAD ensemble from CYANA and the WHAT IF RMS Z scores are systematically improved by refinement in explicit water. Additional information, protocols, and scripts for final structural refinement in explicit water can be obtained from the Northeast Structural Genomics wiki.

(http://www.nmr2.buffalo.edu/nesg.wiki/Structure_Refinement_Using_CNS_Energy_Minimization_With_Explicit_Water)

5. Conclusion

Solution NMR is a powerful tool for the characterization and structural determination of protein–ligand complexes. As a result of structural genomics efforts aimed toward automation of the principal bottlenecks in protein structure determination by NMR, structures of proteins and domains under ~30 kDa can now be solved in a few weeks. Consequently, NMR spectroscopy is a feasible alternative to X-ray structure determination for drug discovery applications, particularly for flexible proteins or membrane proteins that are not amenable to crystallization. The efficiency of our assignment and structure determination protocol is greatly enhanced by maintaining the essential tools: NMR pulse programs and instrument calibrations, spectral conversion and processing scripts, peak picking scripts, and standardized scripts for running programs, such as TALOS+, CYANA, and NIH-XPLOR, all of which are available from the authors upon request. In addition to its capacity to define the structures of protein complexes, NMR remains a robust method to screen and characterize protein–ligand complexes at the atomic level. Each of these applications will be useful as new targets are validated and enter the pipelines for structure-based drug discovery.

REFERENCES

Alam, S. L., Volkman, B. F., Markley, J. L., and Satterlee, J. D. (1998). Detailed NMR analysis of the heme-protein interactions in component IV Glycera dibranchiata monomeric hemoglobin-CO. *J. Biomol. NMR* **11,** 119–133.

Alipanahi, B., Gao, X., Karakoc, E., Balbach, F., Donaldson, L., Arrowsmith, C., and Li, M. (2009a). IPASS: Error tolerant NMR backbone resonance assignment by linear programming Technical Report, No. CS-2009-16. University of Waterloo, http://www.cs.uwaterloo.ca/research/tr/2009.

Alipanahi, B., Gao, X., Karakoc, E., Donaldson, L., and Li, M. (2009b). PICKY: A novel SVD-based NMR spectra peak picking method. *Bioinformatics* **25,** i268–i275.

Altieri, A. S., and Byrd, R. A. (2004). Automation of NMR structure determination of proteins. *Curr. Opin. Struct. Biol.* **14,** 547–553.

Arkin, M. R., and Wells, J. A. (2004). Small-molecule inhibitors of protein-protein interactions: Progressing towards the dream. *Nat. Rev. Drug Discov.* **3,** 301–317.

Bahrami, A., Assadi, A. H., Markley, J. L., and Eghbalnia, H. R. (2009). Probabilistic interaction network of evidence algorithm and its application to complete labeling of peak lists from protein NMR spectroscopy. *PLoS Comput. Biol.* **5,** e1000307.

Baran, M. C., Huang, Y. J., Moseley, H. N., and Montelione, G. T. (2004). Automated analysis of protein NMR assignments and structures. *Chem. Rev.* **104,** 3541–3556.

Barna, J. C. J., Laue, E. D., Mayger, M. R., Skilling, J., and Worrall, S. J. P. (1987). Exponential sampling: An alternative method for sampling in two dimensional NMR experiments. *J. Magn. Reson.* **73,** 69–77.

Bartels, C., Xia, T., Billeter, M., Güntert, P., and Wüthrich, K. (1995). The program XEASY for computer-supported NMR spectral-analysis of biological macromolecules. *J. Biomol. NMR* **6,** 1–10.

Bartels, C., Billeter, M., Güntert, P., and Wüthrich, K. (1996). Automated sequence-specific assignment of homologous proteins using the program GARANT. *J. Biomol. NMR* **7,** 207–213.

Bartels, C., Güntert, P., Billeter, M., and Wüthrich, K. (1997). GARANT—A general algorithm for resonance assignment of multidimensional nuclear magnetic resonance spectra. *J. Comp. Chem.* **18,** 139–149.

Berg, T. (2008). Small-molecule inhibitors of protein-protein interactions. *Curr. Opin. Drug Discov. Devel.* **11,** 666–674.

Berjanskii, M. V., and Wishart, D. S. (2005). A simple method to predict protein flexibility using secondary chemical shifts. *J. Am. Chem. Soc.* **127,** 14970–14971.

Byerly, D. W., McElroy, C. A., and Foster, M. P. (2002). Mapping the surface of *Escherichia coli* peptide deformylase by NMR with organic solvents. *Protein Sci.* **11,** 1850–1853.

Cavanagh, J. (2007). Protein NMR Spectroscopy: Principles and Practice, 2nd edn. Academic Press, Amsterdam, Boston.

Cornilescu, G., Delaglio, F., and Bax, A. (1999). Protein backbone angle restraints from searching a database for chemical shift and sequence homology. *J. Biomol. NMR* **13,** 289–302.

Dalvit, C., Floersheim, P., Zurini, M., and Widmer, A. (1999). Use of organic solvents and small molecules for locating binding sites on proteins in solutions. *J. Biomol. NMR* **14,** 23–32.

de la Cruz, N. B., Peterson, F. C., Lytle, B. L., and Volkman, B. F. (2007). Solution structure of a membrane-anchored ubiquitin-fold (MUB) protein from *Homo sapiens*. *Protein Sci.* **16,** 1479–1484.

Delaglio, F., Grzesiek, S., Vuister, G. W., Zhu, G., Pfeifer, J., and Bax, A. (1995). NMRPipe: A multidimensional spectral processing system based on UNIX pipes. *J. Biomol. NMR* **6,** 277–293.

Eghbalnia, H. R., Bahrami, A., Tonelli, M., Hallenga, K., and Markley, J. L. (2005). High-resolution iterative frequency identification for NMR as a general strategy for multidimensional data collection. *J. Am. Chem. Soc.* **127,** 12528–12536.

Ewing, T. J. A., Makino, S., Skillman, A. G., and Kuntz, I. D. (2001). DOCK 4.0: Search strategies for automated molecular docking of flexible molecule databases. *J. Comput. Aided Mol. Des.* **15,** 411–428.

Feeney, J., Batchelor, J. G., Albrand, J. P., and Roberts, G. C. K. (1979). The effects of intermediate exchange processes on the estimation of equilibrium constants by NMR. *J. Magn. Reson.* **33,** 519–529(1969).

Fielding, L. (2003). NMR methods for the determination of protein-ligand dissociation constants. *Curr. Top. Med. Chem.* **3,** 39–53.

Frederick, R. O., Bergeman, L., Blommel, P. G., Bailey, L. J., McCoy, J. G., Song, J., Meske, L., Bingman, C. A., Riters, M., Dillon, N. A., Kunert, J., Yoon, J. W., *et al.* (2007). Small-scale, semi-automated purification of eukaryotic proteins for structure determination. *J. Struct. Funct. Genomics* **8,** 153–166.

Frydman, L., Scherf, T., and Lupulescu, A. (2002). The acquisition of multidimensional NMR spectra within a single scan. *Proc. Natl. Acad. Sci. USA* **99,** 15858–15862.

Fuller, J. C., Burgoyne, N. J., and Jackson, R. M. (2009). Predicting druggable binding sites at the protein-protein interface. *Drug Discov. Today* **14,** 155–161.

Gal, M., Mishkovsky, M., and Frydman, L. (2006). Real-time monitoring of chemical transformations by ultrafast 2D NMR spectroscopy. *J. Am. Chem. Soc.* **128,** 951–956.

Geyer, M., Herrmann, C., Wohlgemuth, S., Wittinghofer, A., and Kalbitzer, H. R. (1997). Structure of the Ras-binding domain of RalGEF and implications for Ras binding and signalling. *Nat. Struct. Biol.* **4,** 694–699.

Goddard, T. D., and Kneller, D. G. (2001). Sparky 3. University of California, San Francisco.

Gonnella, N., Lin, M., Shapiro, M. J., Wareing, J. R., and Zhang, X. (1998). Isotope-filtered affinity NMR. *J. Magn. Reson.* **131,** 336–338.

Gounarides, J. S., Chen, A., and Shapiro, M. J. (1999). Nuclear magnetic resonance chromatography: Applications of pulse field gradient diffusion NMR to mixture analysis and ligand-receptor interactions. *J. Chromatogr. B Biomed. Sci. Appl.* **725,** 79–90.

Greenidge, P. A., Carlsson, B., Bladh, L. G., and Gillner, M. (1998). Pharmacophores incorporating numerous excluded volumes defined by X-ray crystallographic structure in three-dimensional database searching: Application to the thyroid hormone receptor. *J. Med. Chem.* **41,** 2503–2512.

Grishaev, A., Llin•s, M., and Thomas, L. J. (2005a). Protein structure elucidation from minimal NMR data: The CLOUDS approach. *Methods Enzymol.* **394,** 261–295, Academic Press.

Grishaev, A., Steren, C. A., Wu, B., Pineda-Lucena, A., Arrowsmith, C., and Llinas, M. (2005b). ABACUS, a direct method for protein NMR structure computation via assembly of fragments. *Proteins* **61,** 36–43.

Güntert, P. (2003). Automated NMR protein structure calculation. *Prog. Nucl. Magn. Reson. Spectrosc.* **43,** 105–125.

Güntert, P. (2004). Automated NMR structure calculation with CYANA. *Methods Mol. Biol.* **278,** 353–378.

Güntert, P. (2009). Automated structure determination from NMR spectra. *Eur. Biophys. J.* **38,** 129–143.

Güntert, P., Mumenthaler, C., and Wuthrich, K. (1997). Torsion angle dynamics for NMR structure calculation with the new program DYANA. *J. Mol. Biol.* **273,** 283–298.

Hajduk, P. J., and Greer, J. (2007). A decade of fragment-based drug design: Strategic advances and lessons learned. *Nat. Rev. Drug Discov.* **6,** 211–219.

Hajduk, P. J., Dinges, J., Miknis, G. F., Merlock, M., Middleton, T., Kempf, D. J., Egan, D. A., Walter, K. A., Robins, T. S., Shuker, S. B., Holzman, T. F., and Fesik, S. W. (1997a). NMR-based discovery of lead inhibitors that block DNA binding of the human papillomavirus E2 protein. *J. Med. Chem.* **40,** 3144–3150.

Hajduk, P. J., Meadows, R. P., and Fesik, S. W. (1997b). Discovering high-affinity ligands for proteins. *Science* **278**(497), 499.

Hajduk, P. J., Gerfin, T., Boehlen, J. M., Haberli, M., Marek, D., and Fesik, S. W. (1999). High-throughput nuclear magnetic resonance-based screening. *J. Med. Chem.* **42,** 2315–2317.

Herrmann, T., Güntert, P., and Wuthrich, K. (2002). Protein NMR structure determination with automated NOE assignment using the new software CANDID and the torsion angle dynamics algorithm DYANA. *J. Mol. Biol.* **319,** 209–227.

Hoch, J. C., and Stern, A. S. (1996). NMR data processing. Wiley-Liss, New York.

Huang, Y. J., Moseley, H. N., Baran, M. C., Arrowsmith, C., Powers, R., Tejero, R., Szyperski, T., and Montelione, G. T. (2005). An integrated platform for automated analysis of protein NMR structures. *Methods Enzymol.* **394,** 111–141.

Huang, Y. J., Tejero, R., Powers, R., and Montelione, G. T. (2006). A topology-constrained distance network algorithm for protein structure determination from NOESY data. *Proteins: Struct. Funct. Bioinform.* **62,** 587–603.

Hur, O., and Karplus, K. (2005). Methods of translating NMR proton distances into their corresponding heavy atom distances for protein structure prediction with limited experimental data. *Protein Eng. Des. Sel.* **18,** 597–605.

Ikura, M., and Bax, A. (1992). Isotope-filtered 2D NMR of a protein-peptide complex: Study of a skeletal muscle myosin light chain kinase fragment bound to calmodulin. *J. Am. Chem. Soc.* **114,** 2433–2440.

Irwin, J. J., and Shoichet, B. K. (2005). ZINC—A free database of commercially available compounds for virtual screening. *J. Chem. Inf. Model.* **45,** 177–182.

Jee, J., and Güntert, P. (2003). Influence of the completeness of chemical shift assignments on NMR structures obtained with automated NOE assignment. *J. Struct. Funct. Genomics* **4,** 179–189.

Jensen, D. R., Woytovich, C., Li, M., Duvnjak, P., Cassidy, M. S., Frederick, R. O., Bergeman, L. F., Peterson, F. C., and Volkman, B. F. (2010). Rapid, robotic, small-scale protein production for NMR screening and structure determination. *Protein Sci.* **19,** 570–578.

Jeon, W. B., Aceti, D. J., Bingman, C. A., Vojtik, F. C., Olson, A. C., Ellefson, J. M., McCombs, J. E., Sreenath, H. K., Blommel, P. G., Seder, K. D., Burns, B. T., Geetha, H. V., *et al.* (2005). High-throughput purification and quality assurance of Arabidopsis thaliana proteins for eukaryotic structural genomics. *J. Struct. Funct. Genomics* **6,** 143–147.

Johnson, B., and Blevins, R. (1994). NMRView: A computer program for the visualization and analysis of NMR data. *J. Biomol. NMR* **4,** 603–614.

Jung, Y. S., and Zweckstetter, M. (2004). Mars—Robust automatic backbone assignment of proteins. *J. Biomol. NMR* **30,** 11–23.

Karplus, M. (1959). Contact electron-spin coupling of nuclear magnetic moments. *J. Chem. Phys.* **30,** 11–15.

Keller, R. (2004). Optimizing the Process of Nuclear Magnetic Resonance Spectrum Analysis and Computer Aided Resonance Assignment. A dissertation submitted to the Swiss Federal Institute of Technology Zurich (ETH Zürich) for the degree of Doctor of Natural Sciences. Diss. ETH Nr. 15947.

Kim, S., and Szyperski, T. (2003). GFT NMR, a new approach to rapidly obtain precise high-dimensional NMR spectral information. *J. Am. Chem. Soc.* **125,** 1385–1393.

Koradi, R., Billeter, M., and Wuthrich, K. (1996). MOLMOL: A program for display and analysis of macromolecular structures. *J. Mol. Graph.* **14**(51–55), 29–32.

Koradi, R., Billeter, M., Engeli, M., Güntert, P., and Wuthrich, K. (1998). Automated peak picking and peak integration in macromolecular NMR spectra using AUTOPSY. *J. Magn. Reson.* **135,** 288–297.

Kuszewski, J., Schwieters, C. D., Garrett, D. S., Byrd, R. A., Tjandra, N., and Clore, G. M. (2004). Completely automated, highly error-tolerant macromolecular structure determination from multidimensional nuclear overhauser enhancement spectra and chemical shift assignments. *J. Am. Chem. Soc.* **126,** 6258–6273.

Kuszewski, J. J., Thottungal, R. A., Clore, G. M., and Schwieters, C. D. (2008). Automated error-tolerant macromolecular structure determination from multidimensional nuclear Overhauser enhancement spectra and chemical shift assignments: Improved robustness and performance of the PASD algorithm. *J. Biomol. NMR* **41,** 221–239.

Leach, A. R., Gillet, V. J., Lewis, R. A., and Taylor, R. (2010). Three-dimensional pharmacophore methods in drug discovery. *J. Med. Chem.* **53,** 539–558.

Lee, W., Westler, W. M., Bahrami, A., Eghbalnia, H. R., and Markley, J. L. (2009). PINE-SPARKY: Graphical interface for evaluating automated probabilistic peak assignments in protein NMR spectroscopy. *Bioinformatics* **25,** 2085–2087.

Levitt, M. H. (2008). Spin Dynamics: Basics of Nuclear Magnetic Resonance. 2nd edn. John Wiley & Sons, Chichester, England; Hoboken, NJ.

Liepinsh, E., and Otting, G. (1997). Organic solvents identify specific ligand binding sites on protein surfaces. *Nat. Biotechnol.* **15,** 264–268.

Lin, Z., Xu, Y. Q., Yang, S., and Yang, D. W. (2006). Sequence-specific assignment of aromatic resonances of uniformly 13C, 15N-labeled proteins by using 13C- and 15N-edited NOESY spectra. *Angew. Chem. Int. Ed.* **45,** 1960–1963.

Linge, J. P., Habeck, M., Rieping, W., and Nilges, M. (2003a). ARIA: Automated NOE assignment and NMR structure calculation. *Bioinformatics* **19,** 315–316.

Linge, J. P., Williams, M. A., Spronk, C. A., Bonvin, A. M., and Nilges, M. (2003b). Refinement of protein structures in explicit solvent. *Proteins* **50,** 496–506.

Lopez-Mendez, B., and Güntert, P. (2006). Automated protein structure determination from NMR spectra. *J. Am. Chem. Soc.* **128,** 13112–13122.

Lytle, B. L., Peterson, F. C., Qiu, S. H., Luo, M., Zhao, Q., Markley, J. L., and Volkman, B. F. (2004). Solution structure of a ubiquitin-like domain from tubulin-binding cofactor B. *J. Biol. Chem.* **279,** 46787–46793.

Lytle, B. L., Peterson, F. C., Tyler, E. M., Newman, C. L., Vinarov, D. A., Markley, J. L., and Volkman, B. F. (2006). Solution structure of Arabidopsis thaliana protein At5g39720.1, a member of the AIG2-like protein family. *Acta Crystallogr. F Struct. Biol. Cryst. Commun.* **62,** 490–493.

Malmodin, D., Papavoine, C. H., and Billeter, M. (2003). Fully automated sequence-specific resonance assignments of hetero-nuclear protein spectra. *J. Biomol. NMR* **27,** 69–79.

Mandal, P. K., and Majumdar, A. (2004). A comprehensive discussion of HSQC and HMQC pulse sequences. *Concepts Magn. Reson. Part A* **20A,** 1–23.

Markley, J. L., Bax, A., Arata, Y., Hilbers, C. W., Kaptein, R., Sykes, B. D., Wright, P. E., and Wüthrich, K. (1998). Recommendations for the presentation of NMR structures of proteins and nucleic acids. *Pure Appl. Chem.* **70,** 117–142.

Markley, J. L., Aceti, D. J., Bingman, C. A., Fox, B. G., Frederick, R. O., Makino, S., Nichols, K. W., Phillips, G. N., Jr., Primm, J. G., Sahu, S. C., Vojtik, F. C., Volkman, B. F., *et al.* (2009a). The center for eukaryotic structural genomics. *J. Struct. Funct. Genomics* **10,** 165–179.

Markley, J. L., Bahrami, A., Eghbalnia, H. R., Peterson, F. C., Ulrich, E. L., Westler, W. M., and Volkman, B. F. (2009b). Macromolecular structure determination by NMR spectroscopy. *In* "Structural Bioinformatics," (J. Gu and P. E. Bourne, eds.), 2nd edn. pp. 93–142. Wiley-Blackwell, Hoboken, NJ.

McGann, M. R., Almond, H. R., Nicholls, A., Grant, J. A., and Brown, F. K. (2003). Gaussian docking functions. *Biopolymers* **68,** 76–90.

Medek, A., Hajduk, P. J., Mack, J., and Fesik, S. W. (2000). The use of differential chemical shifts for determining the binding site location and orientation of protein-bound ligands. *J. Am. Chem. Soc.* **122,** 1241–1242.

Meyer, B., and Peters, T. (2003). NMR spectroscopy techniques for screening and identifying ligand binding to protein receptors. *Angew. Chem. Int. Ed. Engl.* **42,** 864–890.

Mobli, M., Maciejewski, M. W., Gryk, M. R., and Hoch, J. C. (2007). An automated tool for maximum entropy reconstruction of biomolecular NMR spectra. *Nat. Methods* **4,** 467–468.

Morris, G. M., Goodsell, D. S., Halliday, R. S., Huey, R., Hart, W. E., Belew, R. K., and Olson, A. J. (1998). Automated docking using a Lamarckian genetic algorithm and an empirical binding free energy function. *J. Comput. Chem.* **19,** 1639–1662.

Mulder, F. A., Schipper, D., Bott, R., and Boelens, R. (1999). Altered flexibility in the substrate-binding site of related native and engineered high-alkaline *Bacillus subtilisins*. *J. Mol. Biol.* **292,** 111–123.

Mumenthaler, C., and Braun, W. (1995). Automated assignment of simulated and experimental NOESY spectra of proteins by feedback filtering and self-correcting distance geometry. *J. Mol. Biol.* **254,** 465–480.

Nilges, M. (1995). Calculation of protein structures with ambiguous distance restraints. Automated assignment of ambiguous NOE crosspeaks and disulphide connectivities. *J. Mol. Biol.* **245,** 645–660.

Orekhov, V. Y., Ibraghimov, I., and Billeter, M. (2003). Optimizing resolution in multidimensional NMR by three-way decomposition. *J. Biomol. NMR* **27,** 165–173.
Peng, J. W., Lepre, C. A., Fejzo, J., Abdul-Manan, N., and Moore, J. M. (2001). Nuclear magnetic resonance-based approaches for lead generation in drug discovery. *Methods Enzymol.* **338,** 202–230.
Pervushin, K., Riek, R., Wider, G., and Wuthrich, K. (1997). Attenuated T2 relaxation by mutual cancellation of dipole-dipole coupling and chemical shift anisotropy indicates an avenue to NMR structures of very large biological macromolecules in solution. *Proc. Natl. Acad. Sci. USA* **94,** 12366–12371.
Peterson, F. C., Lytle, B. L., Sampath, S., Vinarov, D., Tyler, E., Shahan, M., Markley, J. L., and Volkman, B. F. (2005). Solution structure of thioredoxin h1 from *Arabidopsis thaliana*. *Protein Sci.* **14,** 2195–2200.
Peterson, F. C., Hayes, P. L., Waltner, J. K., Heisner, A. K., Jensen, D. R., Sander, T. L., and Volkman, B. F. (2006). Structure of the SCAN domain from the tumor suppressor protein MZF1. *J. Mol. Biol.* **363,** 137–147.
Peterson, F. C., Deng, Q., Zettl, M., Prehoda, K. E., Lim, W. A., Way, M., and Volkman, B. F. (2007). Multiple WASP-interacting protein recognition motifs are required for a functional interaction with N-WASP. *J. Biol. Chem.* **282,** 8446–8453.
Peterson, F. C., Baden, E. M., Owen, B. A., Volkman, B. F., and Ramirez-Alvarado, M. (2010). A single mutation promotes amyloidogenicity through a highly promiscuous dimer interface. *Structure* **18,** 563–570.
Rarey, M., Kramer, B., Lengauer, T., and Klebe, G. (1996). A fast flexible docking method using an incremental construction algorithm. *J. Mol. Biol.* **261,** 470–489.
Reibarkh, M., Malia, T. J., and Wagner, G. (2006). NMR distinction of single- and multiple-mode binding of small-molecule protein ligands. *J. Am. Chem. Soc.* **128,** 2160–2161.
Rella, M., Rushworth, C. A., Guy, J. L., Turner, A. J., Langer, T., and Jackson, R. M. (2006). Structure-based pharmacophore design and virtual screening for novel angiotensin converting enzyme 2 inhibitors. *J. Chem. Inf. Model.* **46,** 708–716.
Rieping, W., Habeck, M., Bardiaux, B., Bernard, A., Malliavin, T. E., and Nilges, M. (2007). ARIA2: Automated NOE assignment and data integration in NMR structure calculation. *Bioinformatics* **23,** 381–382.
Ross, A., Schlotterbeck, G., Klaus, W., and Senn, H. (2000). Automation of NMR measurements and data evaluation for systematically screening interactions of small molecules with target proteins. *J. Biomol. NMR* **16,** 139–146.
Rovnyak, D., Frueh, D. P., Sastry, M., Sun, Z. Y., Stern, A. S., Hoch, J. C., and Wagner, G. (2004a). Accelerated acquisition of high resolution triple-resonance spectra using non-uniform sampling and maximum entropy reconstruction. *J. Magn. Reson.* **170,** 15–21.
Rovnyak, D., Hoch, J. C., Stern, A. S., and Wagner, G. (2004b). Resolution and sensitivity of high field nuclear magnetic resonance spectroscopy. *J. Biomol. NMR* **30,** 1–10.
Schanda, P., and Brutscher, B. (2005). Very fast two-dimensional NMR spectroscopy for real-time investigation of dynamic events in proteins on the time scale of seconds. *J. Am. Chem. Soc.* **127,** 8014–8015.
Schanda, P., and Brutscher, B. (2006). Hadamard frequency-encoded SOFAST-HMQC for ultrafast two-dimensional protein NMR. *J. Magn. Reson.* **178,** 334–339.
Schanda, P., Kupce, E., and Brutscher, B. (2005). SOFAST-HMQC experiments for recording two-dimensional heteronuclear correlation spectra of proteins within a few seconds. *J. Biomol. NMR* **33,** 199–211.
Schubert, M., Labudde, D., Oschkinat, H., and Schmieder, P. (2002). A software tool for the prediction of Xaa-Pro peptide bond conformations in proteins based on 13C chemical shift statistics. *J. Biomol. NMR* **24,** 149–154.

Schumann, F. H., Riepl, H., Maurer, T., Gronwald, W., Neidig, K. P., and Kalbitzer, H. R. (2007). Combined chemical shift changes and amino acid specific chemical shift mapping of protein-protein interactions. *J. Biomol. NMR* **39,** 275–289.

Schuttelkopf, A. W., and van Aalten, D. M. (2004). PRODRG: A tool for high-throughput crystallography of protein-ligand complexes. *Acta Crystallogr. D Biol. Crystallogr.* **60,** 1355–1363.

Shen, Y., Atreya, H. S., Liu, G., and Szyperski, T. (2005). G-matrix Fourier transform NOESY-based protocol for high-quality protein structure determination. *J. Am. Chem. Soc.* **127,** 9085–9099.

Shen, Y., Delaglio, F., Cornilescu, G., and Bax, A. (2009). TALOS+: A hybrid method for predicting protein backbone torsion angles from NMR chemical shifts. *J. Biomol. NMR* **44,** 213–223.

Shuker, S. B., Hajduk, P. J., Meadows, R. P., and Fesik, S. W. (1996). Discovering high-affinity ligands for proteins: SAR by NMR. *Science* **274,** 1531–1534.

Stanczyk, S. M., Bolton, P. H., Dell'Acqua, M., and Gerlt, J. A. (1989). Observation and sequence assignment of a cis prolyl peptide bond in unliganded staphylococcal nuclease. *J. Am. Chem. Soc.* **111,** 8317–8318.

Stuart, A. C., Borzilleri, K. A., Withka, J. M., and Palmer, A. G. (1999). Compensating for variations in H-1-C-13 scalar coupling constants in isotope-filtered NMR experiments. *J. Am. Chem. Soc.* **121,** 5346–5347.

Tintori, C., Corradi, V., Magnani, M., Manetti, F., and Botta, M. (2008). Targets looking for drugs: A multistep computational protocol for the development of structure-based pharmacophores and their applications for hit discovery. *J. Chem. Inf. Model.* **48,** 2166–2179.

Tuinstra, R. L., Peterson, F. C., Elgin, E. S., Pelzek, A. J., and Volkman, B. F. (2007). An engineered second disulfide bond restricts lymphotactin/XCL1 to a chemokine-like conformation with XCR1 agonist activity. *Biochemistry (Mosc.)* **46,** 2564–2573.

Tyler, R. C., Aceti, D. J., Bingman, C. A., Cornilescu, C. C., Fox, B. G., Frederick, R. O., Jeon, W. B., Lee, M. S., Newman, C. S., Peterson, F. C., Phillips, G. N., Jr., Shahan, M. N., *et al.* (2005a). Comparison of cell-based and cell-free protocols for producing target proteins from the *Arabidopsis thaliana* genome for structural studies. *Proteins* **59,** 633–643.

Tyler, R. C., Sreenath, H. K., Singh, S., Aceti, D. J., Bingman, C. A., Markley, J. L., and Fox, B. G. (2005b). Auto-induction medium for the production of [U-15N]- and [U-13C, U-15N]-labeled proteins for NMR screening and structure determination. *Protein Expr. Purif.* **40,** 268–278.

Tyler, R. C., Peterson, F. C., and Volkman, B. F. (2010). Distal interactions within the par3-VE-cadherin complex. *Biochemistry (Mosc.)* **49,** 951–957.

Vajpai, N., Strauss, A., Fendrich, G., Cowan-Jacob, S. W., Manley, P. W., Grzesiek, S., and Jahnke, W. (2008). Solution conformations and dynamics of ABL kinase-inhibitor complexes determined by NMR substantiate the different binding modes of imatinib/nilotinib and dasatinib. *J. Biol. Chem.* **283,** 18292–18302.

Veldkamp, C. T., Peterson, F. C., Pelzek, A. J., and Volkman, B. F. (2005). The monomer-dimer equilibrium of stromal cell-derived factor-1 (CXCL 12) is altered by pH, phosphate, sulfate, and heparin. *Protein Sci.* **14,** 1071–1081.

Veldkamp, C. T., Seibert, C., Peterson, F. C., Sakmar, T. P., and Volkman, B. F. (2006). Recognition of a CXCR4 sulfotyrosine by the chemokine stromal cell-derived factor-1alpha (SDF-1alpha/CXCL12). *J. Mol. Biol.* **359,** 1400–1409.

Veldkamp, C. T., Seibert, C., Peterson, F. C., De la Cruz, N. B., Haugner, J. C., Basnet, H., III, Sakmar, T. P., and Volkman, B. F. (2008). Structural basis of CXCR4 sulfotyrosine recognition by the chemokine SDF-1/CXCL12. *Sci. Signal.* **1,** ra4.

Veldkamp, C. T., Ziarek, J. J., Peterson, F. C., Chen, Y., and Volkman, B. F. (2010). Targeting SDF-1/CXCL12 with a ligand that prevents activation of CXCR4 through structure-based drug design. *J. Am. Chem. Soc.* **132,** 7242–7243.

Vinarov, D. A., Lytle, B. L., Peterson, F. C., Tyler, E. M., Volkman, B. F., and Markley, J. L. (2004). Cell-free protein production and labeling protocol for NMR-based structural proteomics. *Nat. Methods* **1,** 149–153.

Volk, J., Herrmann, T., and Wuthrich, K. (2008). Automated sequence-specific protein NMR assignment using the memetic algorithm MATCH. *J. Biomol. NMR* **41,** 127–138.

Waltner, J. K., Peterson, F. C., Lytle, B. L., and Volkman, B. F. (2005). Structure of the B3 domain from *Arabidopsis thaliana* protein At1g16640. *Protein Sci.* **14,** 2478–2483.

Wang, B., Wang, Y., and Wishart, D. S. (2010). A probabilistic approach for validating protein NMR chemical shift assignments. *J. Biomol. NMR* **47,** 85–99.

Weininger, D. (1988). SMILES, a chemical language and information system. 1. Introduction to methodology and encoding rules. *J. Chem. Inf. Comput. Sci.* **28,** 31–36.

Weininger, D., Weininger, A., and Weininger, J. L. (1989). SMILES. 2. Algorithm for generation of unique SMILES notation. *J. Chem. Inf. Comput. Sci.* **29,** 97–101.

Wells, J. A., and McClendon, C. L. (2007). Reaching for high-hanging fruit in drug discovery at protein-protein interfaces. *Nature* **450,** 1001–1009.

Williamson, M. P., and Craven, C. J. (2009). Automated protein structure calculation from NMR data. *J. Biomol. NMR* **43,** 131–143.

Zhao, Q., Frederick, R., Seder, K., Thao, S., Sreenath, H., Peterson, F., Volkman, B. F., Markley, J. L., and Fox, B. G. (2004). Production in two-liter beverage bottles of proteins for NMR structure determination labeled with either ^{15}N- or ^{13}C-^{15}N. *J. Struct. Funct. Genomics* **5,** 87–93.

Zimmerman, D., Kulikowski, C., Wang, L., Lyons, B., and Montelione, G. T. (1994). Automated sequencing of amino acid spin systems in proteins using multidimensional HCC(CO)NH-TOCSY spectroscopy and constraint propagation methods from artificial intelligence. *J. Biomol. NMR* **4,** 241–256.

Zimmerman, D. E., Kulikowski, C. A., Huang, Y., Feng, W., Tashiro, M., Shimotakahara, S., Chien, C., Powers, R., and Montelione, G. T. (1997). Automated analysis of protein NMR assignments using methods from artificial intelligence. *J. Mol. Biol.* **269,** 592–610.

Zsoldos, Z., Reid, D., Simon, A., Sadjad, B. S., and Johnson, A. P. (2006). eHITS: An innovative approach to the docking and scoring function problems. *Curr. Protein Pept. Sci.* **7,** 421–435.

CHAPTER ELEVEN

Protein Thermal Shifts to Identify Low Molecular Weight Fragments

James K. Kranz* *and* Celine Schalk-Hihi†

Contents

Abstract

Measuring the strength of binding of low molecular weight ligands to a target protein is a significant challenge to fragment-based drug discovery that must be solved. Thermal shift assays are uniquely suited for this purpose, due to the thermodynamic effects of a ligand on protein thermal stability. We show here how to implement a thermal shift assay, describing the basic features and analysis of the protein unfolding data. We then describe in detail the effects of a ligand on the observed stability of the protein to produce a shift in stability. The anatomy of ligand-induced thermal shift data is discussed in detail. We describe the unique aspects of concentration–response curves, the effect of protein unfolding energetics, and the stoichiometry of the interaction.

* Biopharmaceutical Technologies, GlaxoSmithKline Biopharmaceutical Research and Development, Upper Merion, Pennsylvania, USA

† Structural Biology, Johnson & Johnson Pharmaceuticals Research and Development, LLC, Spring House, Pennsylvania, USA

Methods in Enzymology, Volume 493

ISSN 0076-6879, DOI: 10.1016/B978-0-12-381274-2.00011-X

We outline a typical assay development strategy for optimizing dye type and concentration, protein concentration, and buffer conditions. Guidelines are presented to demonstrate the limits of detection for weak-binding ligands, as applied to sulfonamide-based inhibitors of carbonic anhydrase II and applied to nucleotide binding to the death-associated protein kinase 1 catalytic domain.

Abbreviations

1,8-ANS	1-anilinonaphthalene-8-sulfonate
2,6-TNS	2-(*p*-toluidinyl)naphthalene-6-sulfonic acid
DMSO	dimethylsulfoxide

1. Introduction

Fragment-based drug discovery is a viable approach to the assembly of molecules with drug-like properties out of smaller drug fragments. The need to accurately estimate the affinity of fragments to the target protein is of paramount importance in ranking which fragments are most likely to succeed as constituents of the complete molecule. However, there are few general techniques that lend themselves to do so because the ability to measure the binding of fragments is challenged by their relatively small size and corresponding weak affinities.

No single experimental approach is amenable for every fragment-based screen. For proteins that lack an enzymatic activity, few techniques are useful in establishing an experimental estimate of ligand-binding affinity. Biophysical-binding assays represent a powerful approach in assaying ligand binding independent of enzymatic function. One such approach is a thermal shift assay, wherein the intrinsic stability of the target protein is measured by thermally induced unfolding, with a subsequent increase in protein melting temperature due to the stabilizing energy upon binding of a ligand and from which a binding affinity can be measured. This aspect of protein folding has long been appreciated. The concept that proteins undergo conformational changes, with a particular configuration stabilized or destabilized through the binding of a specific low molecular weight ligands, was introduced by Koshland (1958) and Linderstrøm-Lang and Schellman (1959). From different patterns of binding events, thermal shift assays are commonly employed in studies of ligands to better understand mechanistic protein-binding information. Recent technological advances have allowed us to exploit in medium- to high-throughput manner the sensitivity of a protein unfolding event to ligand binding through assay miniaturization and parallelization in plate-based methodologies.

2. Thermal Shift Assays

2.1. General description

Many experimental techniques lend themselves toward use in a thermal shift assay. In differential scanning calorimetry, the differential change in system heat capacity comparing sample to reference cell is monitored as a function of scanning temperature; distinct signatures of either native or denatured forms of the test protein are observed, in addition to a distinctive transition associated with the enthalpic heat evolved upon unfolding. Spectroscopic approaches are also amenable to thermal shift assays, provided they can distinguish behavior of the folded protein, the unfolded protein, and can quantify the transition between the two. When unified with Peltier-based temperature control, thermal shift assays can be performed by applying various spectrophotometric methods: circular dichroism which detects protein secondary structure (Taniuchi and Bohnert, 1975); absorbance spectroscopy (Hermans and Scheraga, 1961); and fluorescence spectroscopy (Eftink, 1997). For fluorescence spectroscopy, either intrinsic tryptophan fluorescence (Epstein *et al.*, 1971) or dye-based approaches (Daniel and Weber, 1966), or a combination of the two (Ramsay and Eftink, 1994) can be applied. Each of these methods has been utilized in performing a thermal shift assay, wherein the characteristic midpoint of unfolding changes in response to a titration of ligand binding is regarded as an apparent binding affinity constant.

While differential scanning calorimetry has the advantage of being a truly label-free technique, being a largely serial technique, its current implementation is severely limited by throughput. Plate-based spectrophotometric instrumentation often lacks appropriate temperature control needed for routine use in ligand binding. *ThermoFluor*® is a system for rapid thermal shift assay (Matulis *et al.*, 2005; Pantoliano *et al.*, 2001). *ThermoFluor* employs a dye whose fluorescence is sensitive to its local environment to monitor the equilibrium between a protein in its folded and unfolded states, thus utilizing dye fluorescence as a measure of protein unfolding. The instrument combines camera-based charge-coupled device detection of dye fluorescence and PCR-type Peltier device for uniform temperature control; it is a plate-based apparatus to assay protein stability and ligand-induced thermal shifts. Other instruments that contain similar components, notably RT-PCR machines, have been adapted also for use in thermal shift assays in measurements of ligand-binding affinities (Lo *et al.*, 2004; Niesen *et al.*, 2007; Zhang and Monsma, 2010). Common among these different instruments is the attribute that each well comprises a distinct protein unfolding assay in a 384-well assay plate, with data collected simultaneously across the entire plate.

A fluorescent-based thermal shift assay measures protein unfolding via a change in the fluorescent properties of bound dye molecules upon protein denaturation. Suitable dyes, such as 1-anilinonaphthalene-8-sulfonate (1,8-ANS), 2-(*p*-toluidinyl) naphthalene-6-sulfonic acid (2,6-TNS), DAPOXYL® derivatives, or dansyl derivatives, possess two key features: their intrinsic fluorescence is strongly quenched by water, and they bind nonspecifically to hydrophobic surfaces such as a protein thereby altering their fluorescent properties. For each target protein, the appropriate concentration of dye is assessed through estimation of the dye-binding affinity to both the folded and unfolded forms of the protein, from plots of either initial or differential fluorescence as a function of dye concentration. Ultimately, a dye concentration is chosen at which the affinity difference for bound dye to the folded and unfolded forms is most disparate.

2.2. A typical assay

(1) *Materials*: A fluorimeter equipped with Peltier-based temperature control (e.g., a *ThermoFluor* instrument) or similar instrumentation (RT-PCR machines) may be employed provided the hardware components (light source, CCD camera, and interference filters) are appropriate for the fluorophore employed to detect protein unfolding; suitable fluorescent dye, generally 1,8-ANS (Invitrogen, A47, Carlsbad, CA); a suitable assay plate such as the black Thermofast 384-well PCR plate from Abgene (Thermo-Fisher, New York, NY).
(2) *Compound solutions*: Test ligands are prepared in DMSO at a 50- to 100-fold concentrated solution, generally in the 10–100 m*M* range. Within a source 384-well plate, compounds are diluted in DMSO serially. For titration, a typical experimental protocol employs a set of 12 wells, comprising 11 different concentrations of a test compound with a single negative control well (DMSO alone).
(3) *Protein solution*: The protein is diluted from a concentrated stock to a working concentration of ~0.5–5 μ*M* protein with ~20–200 μ*M* of 1,8-ANS into a suitable assay buffer. The exact concentrations of protein and dye are defined by experimental assay development studies.
(4) *Ligand sample dispense*: Compounds are dispensed as concentrated (50- to 100-fold) solutions in 100% DMSO using a nanoliter liquid handling system (such as the Echo from Labcyte, Sunnyvale, CA, or the Cartesian Hummingbird, now sold by Zinsser Analytic, Frankfurt, Germany) where common compound dispensed volumes range from 5 to 100 nL. Alternatively, compounds can be diluted to concentrations closer to the final assay concentration, and dispensed in larger volumes (e.g., 2–4 μL of twofold concentration), using multichannel pipettes (Impact pipettes, Matrix Technologies Corp, Hudson, NH) or a plate-based pipetting system (Biomek FX, or equivalent).
(5) *Protein sample dispense*: Typically, 2–4 μL of the protein and dye solution is dispensed from a source 384-well plate or reservoir as suggested for

dilute compounds, at which point the solution is at the final assay concentration with respect to all components.

(6) *Centrifugation and oil dispense*: Finally, following a brief centrifugation ($\sim$1000 $\times$ *g*-force, 1 min) of the assay plate to mix compounds into the protein solution, 1–2 μL silicone oil (DC 200, Sigma-Aldrich) is overlaid onto the solution, followed by an additional centrifugation step ($\sim$1000 $\times$ *g*-force, 1 min). The silicone oil prevents evaporation of sample during heating.

To assess protein thermal stability, the experimental temperature is steadily increased at such a rate as to allow thermal equilibrium to be maintained throughout the experiment. Typical temperature ramp rates range from 0.1 to 10 °C/min but are generally in the range of 1 °C/min. The fluorescence in each well is measured at regular intervals, 0.2–1 °C/image, over a temperature range spanning the typical protein unfolding temperatures of 25–95 °C. At some point during the experiment, a temperature is reached above which the protein native structure becomes thermodynamically unstable and the protein unfolds. In general, there is exposure of hydrophobic surface upon unfolding of a protein. Thus, dyes bind both the native and denatured states of a protein but with slightly different affinities and extent of fluorescent quenching, giving rise to a change in fluorescence upon thermal denaturation. Plotting sample fluorescence as a function of increasing temperature, Fig. 11.1 shows a typical protein melting curve obtained from a thermal shift assay in a 384-well plate. In this example, fluorescence was captured via a *ThermoFluor* instrument with a temperature ramp rate of 1 °C/min and with fluorescence scanning at an interval of 1 °C/image.

2.3. Equations and data analysis

Equations relevant to analysis of thermal shift assays have been described previously (Matulis *et al.*, 2005; Zhang and Monsma, 2010; Zubriene *et al.*, 2009). A thermal stability assay measures the temperature-induced change in equilibrium between the folded, or native state (N), and the denatured, or unfolded state (U). A thermal shift assay measures the additional stabilization of the native state induced by a ligand (L) binding to the native protein as set forth in Eqs. (11.1)–(11.4).

$$U + L_f \underset{\text{unfolding}}{\overset{K_U}{\rightleftharpoons}} N + L_f \underset{\text{binding}}{\overset{K_b}{\rightleftharpoons}} NL_b \tag{11.1}$$

$$\Delta G(T) = \Delta G_U(T) + \Delta G_b(T) \tag{11.2}$$

$$K_U = \frac{[U]}{[N]} = e^{-(\Delta G_U(T)/RT)} \tag{11.3}$$

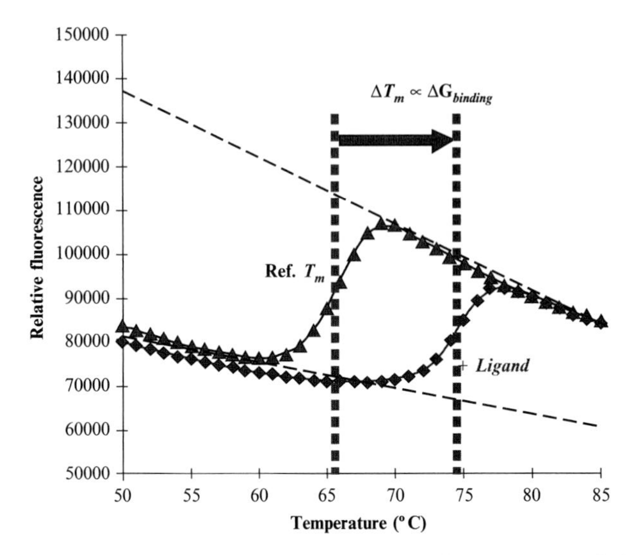

Figure 11.1 Protein unfolding data from a pair of wells in a 384-well thermal shift assay, showing unfolding of a typical protein (bovine carbonic anhydrase II at 2 μ*M*) alone or in the presence of a fixed concentration of a tight-binding ligand (25 μ*M* acetazolamide). The shift in melting temperature from $T_m = 66$ to 75 °C occurs as a result of the added binding energy from the addition of the ligand.

$$K_b = \frac{[NL_b]}{[N][L_f]} = e^{-(\Delta G_b(T))/RT} \tag{11.4}$$

L_f is the concentration of free ligand, L_b is the concentration of bound ligand, K_b is ligand-binding constant [which is equivalent to K_A the ligand association constant or $K_A = (1/K_D)$, where K_D is the ligand dissociation constant], NL_b is the complex of native protein and bound ligand, $\Delta G(T)$ is the total stability of the ligand–protein complex as defined by the temperature-dependent free energy [which is a sum of the protein stability $\Delta G_U(T)$, plus the binding energy $(\Delta G_b(T)]$, K_U is the equilibrium constant for the protein unfolding, T is the temperature (in Kelvin) at any point over the range of the experiment, and R is the universal gas constant in cal/mol K.

3. Binding Affinity in Thermal Shifts

The temperature midpoint, T_m, of a thermal denaturation curve is the characteristic temperature at which the concentration of N and U are equivalent. Ligand binding induces a shift in protein thermal stability; the difference between the T_m in the absence of ligand and the presence is thus a ΔT_m. Since the contribution of binding energy to protein stability is determined by both the binding affinity and the ligand concentration, different ligand-binding events may be compared based on the magnitude of ΔT_m at a fixed concentration of ligand. Alternatively, a series of T_m measurements gathered at widely varying ligand concentrations may be used to determine a K_D of the test ligand.

3.1. Equations for concentration–response curves

To measure the strength of ligand binding from a thermal shift assay, the data are fit in two stages. These are described by below xEqs. (11.5)–(11.9).

$$y(T) = y_F(T) + \frac{y_U(T) - y_F(T)}{1 + e^{\Delta G_U(T_m)/RT}} = y_U(T) + \frac{y_F(T) - y_U(T)}{1 + e^{-\Delta G_U(T_m)/RT}} \quad (11.5)$$

$$y_F(T) = y_{F,T_m} + m_F(T - T_m) \quad (11.6)$$

$$y_U(T) = y_{U,T_m} + m_U(T - T_m) \quad (11.7)$$

$$\Delta G_U(T) = \Delta H_U(T) - T\Delta S_U(T) \quad (11.8a)$$

$$\Delta G_U(T) = \Delta H_{U,T_r} + \Delta C_{p,U}(T - T_r) - T\left(\Delta S_{U,T_r} + \Delta C_{p,U}\ln(T/T_r)\right) \quad (11.8b)$$

$$y(T) = y_{F,T_m} + m_F(T - T_m) + \frac{y_{U,T_m} - y_{F,T_m} + m_U(T - T_m)}{1 + e^{\Delta H_{U,T_r} + \Delta C_{p,U}(T - T_r) - T\left(\Delta S_{U,T_r} + \Delta C_{p,U}\ln(T/T_r)/RT\right)}} \quad (11.9)$$

First, the fluorescence in the thermal shift assay, $y(T)$, is used to determine the protein thermal stability, T_m, at any given concentration of ligand. Two terms in Eq. (11.5) define the fluorescence of 1,8-ANS in the presence of the fully folded, $y_F(T)$, and the fully unfolded protein, $y_U(T)$. Both of these fluorescent baselines are linear functions of temperature, T, as defined in Eqs. (11.6) and (11.7), where Y_{F,T_m} and Y_{U,T_m} describe the temperature-dependent fluorescence of 1,8-ANS in the presence of native and nonnative protein (referenced to the melting temperature T_m), and m_F and m_U are the slopes associated with linear temperature-dependent fluorescence for the folded and unfolded protein, respectively. The protein stability term $\Delta G_U(T)$ in Eq. (11.5) is the same as defined in Eq. (11.2), it may be replaced by the Gibbs–Helmholtz relationships in Eqs. (11.8a) and (11.8b) (Privalov, 1979). Equation (11.8a) shows that the temperature-dependent Gibbs free energy of protein unfolding, $\Delta G_U(T)$, comprises a temperature-dependent enthalpy, $\Delta H_U(T)$, and a temperature-dependent entropy, $\Delta S_U(T)$. These two parameters in turn have their temperature dependence Eq. (11.8b) described by a constant enthalpy, $\Delta H_{U,T_r}$, and entropy, $\Delta S_{U,T_r}$, at the reference temperature, T_r, set to the value of the unfolding T_m of protein in the absence of a ligand. In addition, a heat capacity for protein unfolding, $\Delta C_{p,U}$, defines the temperature dependence of $\Delta H_{U,T_r}$ and $\Delta S_{U,T_r}$. Equations (11.5) and (11.8b) are combined into Eq. (11.9), which is used in a nonlinear least squares minimization algorithm to estimate parameters of Y_{F,T_m} and m_F, Y_{U,T_m} and m_U, $\Delta H_{U,T_r}$, and T_m for each individual sample. The value of $\Delta C_{p,U}$ in Eq. (11.9) is highly correlated with $\Delta H_{U,T_r}$ which is typically an order of magnitude larger than $\Delta C_{p,U}$ and is thus difficult to determine independently from $\Delta H_{U,T_r}$ with nonlinear least squares regression analysis. Instead, the value of $\Delta C_{p,U}$ is held fixed near its value measured by other techniques, generally estimated from differential scanning calorimetry (discussed in Section 4.3), or is estimated based on protein composition (Murphy and Gill, 1991).

In a second stage of data analysis, the T_m value of the protein at each concentration of test ligand is related to the expected effect for a ligand with a given binding affinity, K_b or K_D, which is defined by Eqs. (11.10a) and (11.10b). The relationship between total ligand concentration, total protein concentration, and the two equilibrium constants for protein unfolding and ligand binding is defined by Eq. (11.10a).

$$L_t = (1 - K_U)\left(\frac{P_t}{2} + \frac{1}{K_U K_b}\right) \tag{11.10a}$$

$$\begin{aligned} L_t = {} & \left(1 - e^{-\left(\Delta H_{U,T_r} + \Delta C_{p,U}(T_m - T_r) - T_m\left(\Delta S_{U,T_r} + \Delta C_{p,U} \ln(T_m/T_r)\right)/RT_m\right)}\right) \\ & \times \left(\frac{P_t}{2} + 1/e^{-\left(\Delta H_{U,T_r} + \Delta C_{p,U}(T_m - T_r) - T_m\left(\Delta S_{U,T_r} + \Delta C_{p,U} \ln(T_m/T_r)\right)\right)/RT_m} \right. \\ & \left. \times e^{-\left(\Delta H_b(T_0) + \Delta C_{p,b}(T - T_0) - T\left(\Delta S_b(T_0) + \Delta C_{p,b} \ln(T/T_0)\right)\right)/RT}\right) \end{aligned} \tag{11.10b}$$

P_t is the total protein concentration which is the sum of the concentrations of N, U, and NL_b. The L_t is the total ligand concentration which is the sum of concentrations of L_f and NL_b in Eq. (11.1). Additionally, K_U is described in terms of the temperature-dependent enthalpy, entropy, and heat capacity of protein unfolding in Eq. (11.8b) and K_b is described in terms of an enthalpy, $\Delta H_b(T_0)$, an entropy, $\Delta S_b(T_0)$, and heat capacity, $\Delta C_{p,b}$, of ligand binding in Eq. (11.10b).

3.2. Anatomy of concentration–response curves

Theoretical curves are generated to demonstrate how the dependence of ΔT_m on ligand concentration is affected by several important parameters in the thermal shift assay. The theoretical behavior of weak- to tight-binding ligands based on variations in theoretical affinity is shown (Fig. 11.2A). First, the intrinsic binding affinity, K_D, greatly affects the magnitude of ΔT_m at a given concentration of added ligand. This attribute is particularly important in determining the appropriate test concentration necessary for a robust, quantifiable binding affinity in applications of thermal shift assays to fragment screening where affinities are often expected to be quite weak. For the

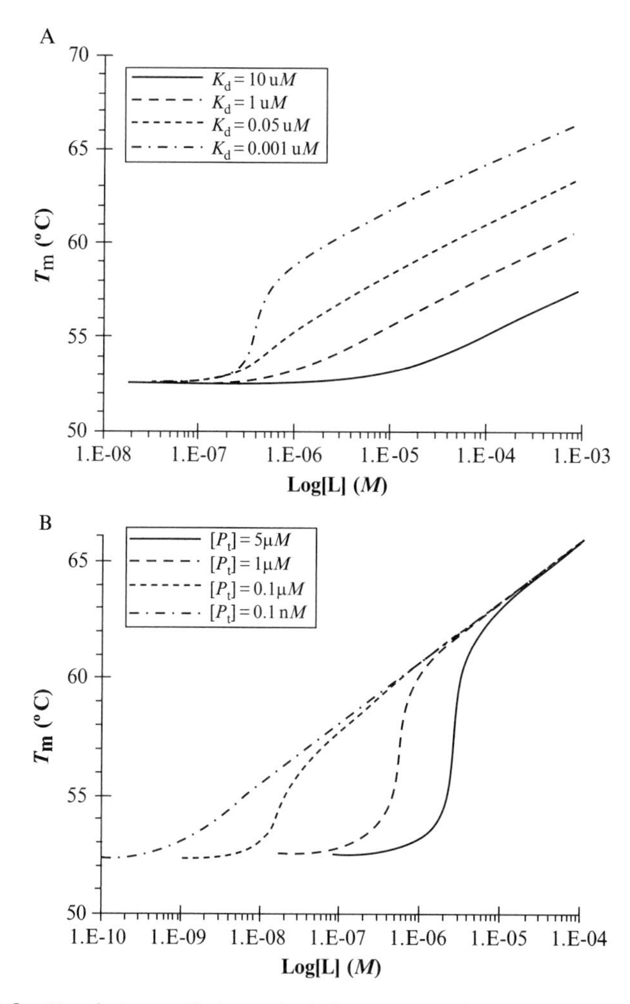

Figure 11.2 Simulations of thermal shift concentration–response curves using Eq. (11.10b). The thermodynamic parameters describing protein stability were held constant for all simulations. (A) Effect of differing ligand-binding affinities ranging from $K_D = 0.001$ to 10 μM, with a constant protein concentration, $P_t = 1$ μM. (B) Effect of varying protein concentrations from $P_t = 0.0001$ to 10 μM, with a constant ligand-binding affinity, $K_D = 0.001$ μM.

two weakest simulated binding curves, K_D is set at 1 and 10 μM (Fig. 11.2A). There is no influence of ligand on protein T_m at ligand concentrations below their respective affinities; such weak-binding inhibitors must be tested at several concentrations above the estimated affinity (e.g., at ligand $> 20\ \mu M$ where $K_D = 10\ \mu M$). In general, it is recommended that affinities be determined from data sets where at least the top three concentrations of ligand result in a ΔT_m above the standard deviation in T_m for the particular test protein (detailed in Section 5).

3.2.1. Compound concentration and solubility

Simulated curves for tight-binding inhibitors in Fig. 11.2A highlight the influence of the protein concentration on the shape of the dose–response curve. In these simulations, a sigmoidal dependence of T_m on ligand concentration is observed just below the protein concentration, consistent with a saturation-binding experiment. Once the protein is saturated by ligand, the influence of additional increasing concentration of ligand on the observed T_m shows a continual increase. This aspect of thermal shift assays is a unique consequence of the Gibbs–Helmholtz equation (Brandts and Lin, 1990; Shrake and Ross, 1992), the result of which is a continuing shift in the equilibrium between native and denatured protein as more and more ligand is added. It is often observed that a saturation point is reached where the T_m becomes independent on ligand concentration, reaching a plateau value of ΔT_m at high ligand concentration. This behavior often correlates with the solubility limit of the relevant ligand, based on independent solubility measurements. Experimental estimates of ligand solubility are necessary for providing a rational justification for omitting data points above this solubility threshold from use in K_D estimations.

3.2.2. Protein concentration

Simulations in Fig. 11.2B further explore the effect of protein concentration on the shape of the dose–response curve, assuming a constant high affinity of $K_D = 1\ nM$. Simulations varied the protein concentration from $[P_t] = 0.1$, 1, or 10 μM as well as the theoretical limit where the protein concentration is infinitely dilute ($[P_t] \ll K_D$). In each of the examples where the protein concentration exceeds the K_D, a sigmoidal dependence of T_m on ligand concentration is observed until protein saturation is achieved. This phenomenon also gives insight into the binding stoichiometry between ligand and protein, but only when the experimental conditions are met such that the K_D is significantly lower ($\sim$50-fold) than the experimental protein concentration.

3.2.3. Protein thermodynamics and stoichiometry

The concentration–response curves in Fig. 11.2 were each simulated utilizing a common set of thermodynamic parameters governing the temperature dependence of the protein folding/unfolding equilibrium, specifically the parameters that comprise $\Delta G_U(T)$ in Eq. (11.9), $\Delta H_{U,T_r}$, $\Delta S_{U,T_r}$, and $\Delta C_{p,U}$. The result is a common, limiting behavior for the dependence of T_m on ligand concentration at concentrations in excess of either the K_D or P_t. The precise values of estimated K_D depend on experimental estimates of protein thermodynamic parameters. So long as a common protein condition is used in comparing a number of different ligands, the rank order of ligand-binding strengths can be ascertained using estimates of protein unfolding enthalpy and heat capacity based on protein composition (Murphy and Gill, 1991).

The actual ligand concentration is generally assumed to be precisely known, though this is rarely true concerning small molecule ligands and fragments used in drug discovery. The actual concentration of the active ingredient can be much higher or more often much lower than is intended, due to a number of reasons including the technical challenge of precisely weighing small quantities of material for solubilization, imprecise knowledge of the correct salt form of the source compound, the loss of solubilized material due to freeze-thaw cycles of source materials, or through the slow decomposition of material through chemo- or photoreactivity over several years' time as is often the case with solubilized libraries of compounds. Compound concentration inaccuracies have an unpredictable effect on the apparent affinity from which decisions are made in structure–activity relationships as fragments are selected for further chemical elaboration.

Figure 11.3 shows a data set taken from a library of nucleotide-competitive compounds that bind a kinase catalytic domain, where it is subsequently learned that the active concentration of a tight-binding ligand is 10-fold lower than originally intended. Attempts to fit the primary data (Fig. 11.3A) show that no scenario of variable affinity or system enthalpy (dominated by the enthalpy of unfolding, $\Delta H_{U,T_r}$) could describe the experimental data. As an example, curve fits in Fig. 11.3A employ an assumed affinity of $K_D = 1\ \mu M$, with values of the enthalpy ranging from $\Delta H_{U,T_r} = 75{,}000$–$250{,}000$ kcal/mol. Each of these curves passes through some of the data points but fails to fit the experimental data. Thus, an affinity value cannot be determined from this data set despite clear evidence of a binding event. However, the apparent sigmoidal behavior of this ligand is a clue that the actual affinity must be tight relative to the experimental protein concentration ($P_t = 4.3\ \mu M$), which could only be the case if the actual ligand concentration was much lower than originally suggested from the expected stock concentration.

A fresh, neat aliquot of this same compound is obtained in sufficient quantities to accurately measure an amount for solubilization and is retested to obtain an affinity of $K_D = 0.015\ \mu M$. Returning to the original data set, we employed an additional fit parameter to allow the stock ligand concentration to be varied and held the affinity constant at $K_D = 0.015\ \mu M$. The results are shown in Fig. 11.3B. The original assumed ligand concentrations (large symbols) are poorly described by a fit of $K_D = 1\ \mu M$. In contrast,

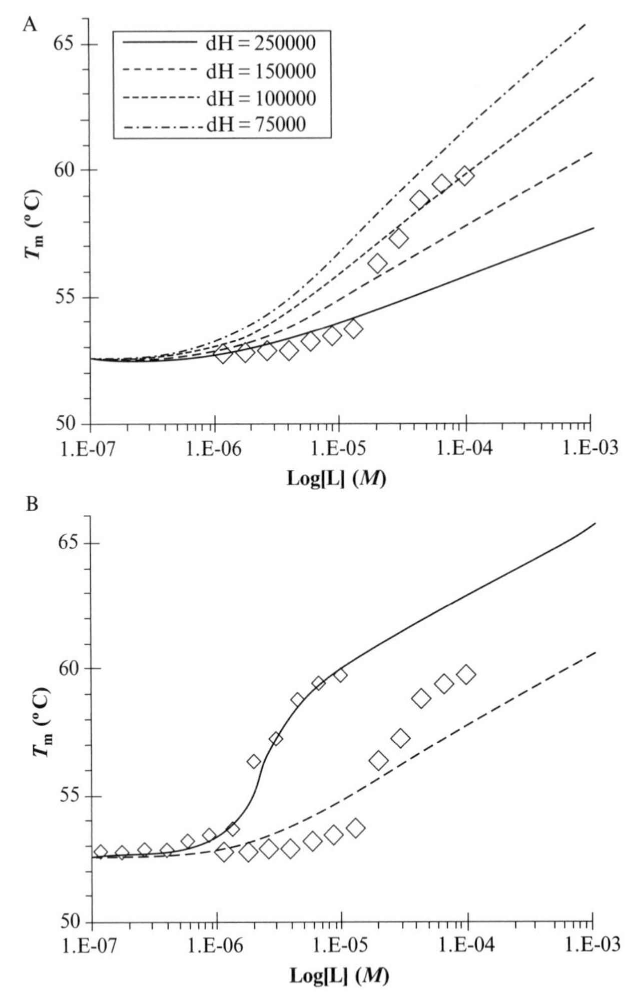

Figure 11.3 (A) Concentration–response data for a nucleotide-site ligand binding to a kinase catalytic domain, fit to Eq.(11.10b) with differing values of the protein folding enthalpy, $\Delta H_{\mathrm{U},T_\mathrm{r}} = 75{,}000–250{,}000$ kcal/mol. For each of these cases, the fits poorly describe the data assuming a $K_\mathrm{D} = 1\ \mu M$. (B) The ligand concentration values associated with original data (large diamonds) in panel (A) are minimized (small diamonds) using a fixed $K_\mathrm{D} = 0.015\ \mu M$ based on independent sample testing. The experimental ligand concentrations for each resulting data point following minimization suggest the intended concentration is ~10-fold higher than the actual concentration.

when each of the ligand concentration associated with each measured T_m is reduced ~10-fold and then fitted with a $K_\mathrm{D} = 0.015\ \mu M$ the resulting curve describes the data excellently. (Similar situations might arise if a racemic mixture of compounds exists with different affinities for the relative forms.) This example highlights the importance of an independent experimental estimate of the actual ligand concentration for samples used in a thermal shift titration experiment.

4. Typical Thermal Shift Assay Development

Various applications of protein thermal stability assays have been published on several different systems (Cimmperman *et al.*, 2008; Grasberger *et al.*, 2005; Kervinen *et al.*, 2006; Matulis *et al.*, 2005; Mezzasalma *et al.*, 2007; Pantoliano *et al.*, 2001; Zubriene *et al.*, 2009). As with any general experimental approach applied to a variety of systems, there are commonalities as well as system-specific aspects of assay development. Here, we briefly describe a common framework that forms the basis for *ThermoFluor* assay development, with some insights into protein specific behavior that must be explored for development of a robust assay. Being a 384-well plate-based format, *ThermoFluor* assays provide an opportunity to assay many binding events in parallel, in low assay volumes (2–4 μL/well) using limited amounts of protein (~100–200 μg/plate). For applications of fragment-based screening or high-throughput screening, potentially thousands of assay plates may be run which requires significant assay optimization in order to minimize protein consumption.

4.1. Appropriate dye and protein concentrations

The basis of *ThermoFluor* is to monitor protein thermal stability with use of a dye that displays a different fluorescent signature when bound to the folded and unfolded protein in monitoring protein thermal stability. The useful fluorescent signal results from nonspecific binding of the dye to the target protein, folded or unfolded, in addition to the signal of the dye that is free in solution. The fluorescence associated with unbound dye and dye bound to the folded protein comprise collectively the initial, background fluorescence. The fluorescent associated with unbound dye and dye bound to the unfolded protein provides collectively the source of signal to monitor temperature-induced unfolding.

The two experimental parameters that must first be optimized are the concentration of dye, such as 1,8-ANS, and the concentration of protein to afford an acceptable signal window. The relative dissociation constants of dye to the unfolded protein, $K_{D,unfolded}$, and to the folded protein, $K_{D,\ folded}$, are determined experimentally from a series of thermal-induced protein denaturating curves obtained over a range of dye concentrations, at a fixed protein concentration. A plot of y_{F,T_m} as a function of variable dye concentration from different samples can be fit to a standard binding equation to estimate $K_{D,\ folded}$, while a plot of the difference in unfolding signal between folded and unfolded forms, $\Delta y = (y_{U,T_m} - y_{F,T_m})$ as a function of variable dye concentration can be fit to a standard binding equation to estimate $K_{D,\ unfolded}$ (Cimmperman *et al.*, 2008). In most situations, the dye has a higher affinity for the unfolded protein, that is, $K_{D,unfolded} < K_{D,\ folded}$. Also, unfolded proteins present more hydrophobic surfaces that form potential dye-binding sites, many of which are sequestered from solvent in the folded form. For this reason, the strength of the fluorescence signal usually depends on the change in exposure of hydrophobic surface. However, prominent examples exist in which 1,8-ANS binds a folded protein tighter than to the unfolded protein (Daniel and Weber, 1966; McClure and Edelman, 1966; Slavik, 1982; Stryer, 1965), that is, $K_{D,\ folded} < K_{D,\ unfolded}$ and consequently the temperature dependence of the observed fluorescence appears inverted from that of a typical dye-based thermal shift assay.

The need to optimize the dye concentration arises from a necessity of maximizing the signal associated with protein unfolding over the contribution of background fluorescence to the observed signal. As the concentration of dye exceeds $K_{D,unfolded}$ for dye-binding to unfolded protein, additional increases in dye concentration only elevate the background fluorescence without affecting the signal difference between folded and unfolded forms. Empirically, the most suitable dye concentration is in the range where binding to unfolded protein is mostly saturated while binding of dye to the folded form is minimized. Typical ranges are 25–200 μM for 1,8-ANS and 10–100 μM for dapoxyl sulfonate, though instrumental configuration (fluorescence wavelength and band-pass settings) also affect the outcome from dye optimization.

The need to optimize protein concentration arises from the practical requirements to produce sufficient quantities of protein for the relevant application. For use in fragment-based screening and other high-throughput screening applications, researchers are also generally concerned with the lower limit of protein concentration that will provide a large enough signal window for fluorescence change upon protein unfolding to meet statistical criteria for assay reproducibility. For any protein, there is a linear dependence of signal due to unfolding on protein concentration. The lower limit of useful protein concentration depends greatly on the specific protein of interest. Many small proteins (e.g., ubiquitin) do not have a substantially different fraction of nonpolar surface area between the folded and unfolded forms and show similar affinities of 1,8-ANS to both forms ($K_{D,unfolded} \sim K_{D,\ folded}$ for 1,8-ANS binding to ubiquitin, data not shown). This behavior restricts the adaptability of certain proteins to a thermal shift binding assay due to issues related to insufficient signal amplitude without going to unusually high protein concentration. Conversely, many larger proteins expose a great deal of surface area upon unfolding which effectively presents many more binding sites for the dye. This phenomenon extends the lower limit of protein concentration needed to carry out a titration experiment. For the majority of systems, the final protein concentration is limited to the range of >25 μg/mL due to detection sensitivity limits and <250 μg/mL. The standard deviations $\sigma(T_m)$ for the controls is in the range of ~0.15–0.20 °C for typical applications. When protein conservation is unnecessary, as in biopharmaceutical formulation development, much higher protein concentrations (1–10 mg/mL) may be preferred.

4.2. Buffer condition profiling

Protein stability is commonly modulated through variations in solution conditions, a strategy that has relevance in applications to protein over-expression on a small scale, in applications to ligand binding and in applications to commercial formulation in the development of therapeutic proteins. A plate-based matrix of solution conditions can be utilized to optimize the final buffer condition for fragment-based screening, to discover what conditions to avoid (such as excessive $NiCl_2$ in the presence of a histidine-tagged protein that is destabilized by Ni^{2+}), in probing the interactions of buffer components with the test protein, or to facilitate the exploration of crystallization condition screening.

High-throughput condition profiling or formulation profiling readily provides ample data on protein stability. However, the assembly of unique condition matrices is somewhat laborious, and importantly qualitative interpretation of the data can be challenging. An extremely efficient implementation of condition-based profiling (1) utilizes a common set of reagents arranged in a matrix of conditions with nonrandom variations and (2) is used repetitively across multiple systems allowing for protein–protein comparisons of stability data.

One strategy to optimize protein stability via optimization of buffer condition has been described for the kinase catalytic domains of Akt-3 and cFMS (Mezzasalma *et al.*, 2007). As part of the study, the protein stability landscape associated with variable protein T_m is measured as a function of NaCl concentration and pH to provide a basis for selecting solution ionic strength and buffer reagent. Ionic strength and pH are generally interdependent in their effects on protein stability. The particular implementation of condition matrix for salt and pH in Mezzasalma *et al.* (2007) is assembled as two 96-well plates (assayed together in pair-wise duplicates on a single 384-well plate). The first of these groupings employs only three buffers (PIPES, HEPES, or $NaPO_4$) at a constant 25 m*M* concentration, with and without 100 m*M* NaCl, with and without 5 m*M* $MgCl_2$, with 5–12 different pH values per buffer/NaCl/$MgCl_2$ combination. This type of layout explores a sufficiently large range of pH within the buffering capacity of different buffer types with similar p*K* values as a means of discerning both a pH dependence of protein stability and any buffer effects between the three buffer types. A single concentration of NaCl and/or $MgCl_2$ is included to provide insights into the ionic strength dependence of differing pH effects. Variations in ionic strength are explored further in the second 96-well plate which employs a common range of NaCl concentrations (50, 100, 200, 300, 400, and 500 m*M*), over an array of pH from 4 to 9, using 12 different combinations of buffer type and pH, where all buffers are present at 25 m*M*.

The combination of these two pH/salt plates employs the concept of a simple set of related conditions from which decisions can be made regarding appropriate buffer type, pH, and ionic strength. First, a change in any individual variable is consistently employed across a change in a second variable; for example, a common array of pH values would be employed in exploration of buffer type or ionic strength. Second, the range of any one variable is sufficiently wide to elicit a measurable effect on protein stability but falls within a relevant range for the intended application; for example, a narrow pH range may be sufficient for biopharmaceutical formulation development of a protein while a wider pH range may be needed to study how protonation affects binding. Third, the matrix of conditions is sampled with sufficient granularity to observe gradual, incremental changes in protein stability; for pH, a change in pH by every 0.2 unit may be required, while for ionic strength, increments of NaCl by 50 m*M* may be necessary. Finally, condition profiling is tested with an eye on biological relevance; intracellular concentration of Ca^{2+} ranges from 0.1 to 10 μ*M*, while extracellular Ca^{2+} is around 500 μ*M* to 1 m*M*, thus the use of >10 μ*M* $CaCl_2$ has no bearing on the functional properties of an intracellular protein but may be relevant to an excreted one.

For applications to biopharmaceutical protein development, successful formulations are generally complex based on a need to balance ideal protein storage conditions against the need for administration by injection. Formulation development might begin with a matrix profiling of pH and salt as described above, from which a particular buffer and pH (and perhaps ionic strength) would be identified; this starting condition could then be used in an exploration of excipients that are relevant for protein formulation. Similarly, for fragment-based drug discovery assays, the desired solution condition would be one that affords minimal loss of dilute protein during robotic manipulation (incorporation of nonionic detergents), maintains protein function (relevant counterions), and approximates the physiologically relevant solution conditions. Since fragments may be delivered from DMSO stocks, the denaturing effect of DMSO must be tested for a given protein target and mitigated in order to enable the screening campaign.

The matrix of buffer conditions is sometime transferable from one protein to another. From the broad profiling, individual components are identified as being relevant for a particular target of interest and are further explored in a more focused manner to ascertain the optimum buffer condition for the protein.

4.3. Additional protein characterization

Proper quantification of K_D values requires quantification of the protein unfolding enthalpy, $\Delta H_{U,T_r}$. To define the unfolding enthalpy, differential scanning calorimetry is the method of choice as described by Klinger *et al.* (2006) and Matulis *et al.* (2005). When the value of $\Delta H_{U,T_r}$ is precisely determined, there is excellent correlation between thermal shift assay derived binding constants and those obtained with other techniques.

5. Dynamic Range of Thermal Shift Assays and Guidelines For a "Significant" Binding Event

The real advantage of thermal shift assays in fragment screening comes from the fact that the observed signal comes from protein unfolding which is independent of ligand size or affinity and is constant across a panel of ligands. Thermal shift sensitivity for extremely tight-binding ligands is exceptionally good in thermal shift titration; an increase in affinity translates to an increase in ΔT_m. Binding affinities in the subpicomolar range are readily determined in thermal shift assays (Brandts and Lin, 1990; Shrake and Ross, 1992). The sensitivity limit in a thermal shift assay for weak-binding ligands is more relevant for fragment-based screening. We show here two examples, one with strong-binding ligands and another with weak-binding ligands.

Sulfonamide-based inhibitors of erythrocytic human carbonic anhydrase II (hCAII; EC 4.2.1.1) are commonly employed as test systems of thermal shift assays (Matulis *et al.*, 2005), and other biophysical binding assays (Cannon *et al.*, 2004; Myszka, 2004). This protein provides a useful test system of fragment-based screening owing to their generally small size and range in affinities. The enzyme contains an active site zinc ion that is known to commonly interact with sulfonamide-based inhibitors, with the sulfonamide of the inhibitor providing a fourth nonprotein ligand to the active site Zn^{2+} (Behnke *et al.*, 2010). A panel of 16 sulfonamides, with molecular weight in the range of 95–370 g/mol, is tested in thermal shift assays using a *ThermoFluor* (Table 11.1). Measured affinities range from very tight binding for ethoxzolamide, $K_D = 2 \times 10^{-10}$ *M*, to undetectable binding for cyclopropanesulfonamide, methanesulfonamide, and S(−)sulpiride when tested up to 200 μ*M* in compound concentration. In all cases, the measured affinities match those determined by other techniques, yet the sensitivity issues often associated with measuring effects of low molecular weight ligands are absent.

Table 11.1 *ThermoFluor* affinities of sulfonamides binding to hCAII

Compound name	K_D (μM)[a]	Mol. wt. (g/mol)
Acetazolamide	0.011 (±0.004)	222.3
Benzenesulfonamide	1.80 (±0.31)	157.2
4-Carboxybenzenesulfonamide	1.39 (±0.30)	201.2
Chloromethoxy-*N*[sulfamoylphenyl-ethyl]benzamide	0.20 (±0.06)	368.8
Chlorothiazide	0.50 (±0.07)	295.7
Cyclopropanesulfonamide	>95	121.2
Dansyl amide	1.91 (±0.48)	250.3
Dichlorophenamide	0.002 (±0.001)	305.2
Ethoxzolamide	0.0002 (±0.0001)	258.3
Furosemide	0.45 (±0.12)	330.7
Methanesulfonamide	>95	95.1
Methazolamide	0.017 (±0.009)	236.3
4-Aminomethylbenzenesulfonamide	29.8 (±5.8)	222.7
Sulfanilamide	15.8 (±2.9)	172.2
S(−)Sulpiride	>95	341.4
Trifluoromethanesulfonamide	0.0036 (±0.0024)	149.0

[a] Binding affinities, K_D, obtained from *ThermoFluor* thermal shift assays were fit assuming a constant $\Delta H = 150{,}000$ cal/mol and $\Delta C_p = 2500$ cal/mol K for hCAII unfolding, and $\Delta H = (-5000)$ cal/mol and $\Delta C_p = (-200)$ cal/mol K for ligand binding. The buffer contained 150 m*M* NaCl, 25 m*M* PIPES, pH 7.0, 0.5 m*M* EDTA, and 0.001% Tween-20. hCAII, sulfonamide inhibitors, and buffer components were all obtained from Sigma-Aldrich.

Two sulfonamides are illustrative of the lower limits of detection. Concentration–response curves are shown for the binding of 4-aminomethylbenzenesulfonamide (222.7 g/mol) and S(−)sulpiride (341.4 g/mol) to carbonic anhydrase (Fig. 11.4). These compounds are tested up to a final high concentration of 200 μM ligand. For each, the majority of samples produced no change in the thermal stability of the enzyme. However, the highest three concentrations of the sulfonamide produce a ΔT_m exceeding twice the standard deviation of the reference wells ($T_m = 54.46 \pm 0.17$ °C), from which an affinity is estimated at $K_D = 35$ μM. For the sulpiride, none of the samples shows a statistically significant shift in T_m; thus a $K_D > 50$ μM.

All ATP-utilizing enzymes, including kinases, bind ATP as a complex with the divalent cation Mg^{2+} (Yan and Tsai, 1999). Removing the divalent cation significantly reduces affinities of the nucleotide. As an example to demonstrate the sensitivity of thermal shift assays to detect weak binders, the catalytic kinase domain of the death-associated protein kinase, or DAPK-1 (Bialik and Kimchi, 2006), is employed in a *ThermoFluor*-based fragment screening assay. A panel of tool compounds is assayed for ligand-binding affinities in the absence of a divalent salt, in the presence

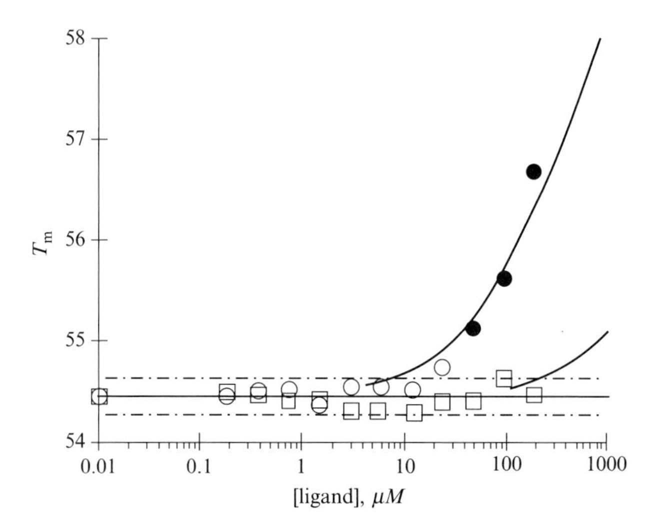

Figure 11.4 4-Aminomethylbenzenesulfonamide (circles) or S(−)sulpiride (squares) binding to human carbonic anhydrase II. The reference protein stability was $T_m = 54.46 \pm 0.17$ °C, depicted by horizontal solid and dotted lines for the mean and ±1 standard deviation from the mean. Each ligand was tested up to 200 μ*M*; only the top three test conditions of 4-aminomethylbenzenesulfonamide (closed circles) gave a shift in T_m that exceeded one standard deviation within error of each measurement. S(−) Sulpiride (open circles) showed no effect on the stability of the enzyme, giving no estimable affinity for this range of ligand concentrations.

of either 5 m*M* $MgCl_2$, or in the presence of 0.5 m*M* $MnCl_2$. The experiment utilizes a ligand concentration range of 0.04–2500 μ*M* (threefold dilutions over 11 wells, plus one negative control well). As shown in Table 11.2, in the presence of either $MgCl_2$ or $MnCl_2$, the affinities show the expected preference of DAPK-1 for ATP and its analogs over other nucleotides or nucleic acids. With ATP, a $K_D = 2.9$ μ*M* (in the presence of Mg^{2+}) is in line with a $K_M \sim 2$–5 μ*M* for ATP turnover in a peptide substrate model system (Velentza *et al.*, 2001). Affinities in the absence of a divalent metal ion are quite weak, the binding affinity for ATP and its analogs drops to a $K_D \sim 500$ μ*M*. For other nucleotides, the binding is either weak ($K_D = 200$–400 μ*M*) or not detectable ($K_D > 1000$ μ*M*). Nucleic acids are weak binders with or without a divalent metal ion, with a K_D in the 100–500 μ*M* range. A nonspecific tight-binding inhibitor of kinases, is used here as a positive control. These binding data highlight the range of weak affinity detectable in a thermal shift assay for compounds with a molecular weight in the 100–550 g/mol range.

In general, at least three test samples in a concentration–response curve should produce a statistically significant change in measured T_m to determine a ligand-binding affinity. Several of the examples presented in Table 11.1 for DapK-1 approach the limits for what constitutes a

Table 11.2 Effect of buffer conditions on nucleotide analog binding affinities to DapK-1

Nucleotide/ analog	K_D (μM) (1×) PBS buffer[a]	K_D (μM) PBS + 5 mM $MgCl_2$	K_D (μM) PBS + 500 μM $MnCl_2$	Mol. wt. (g/mol)
ATP	500 (±200)	2.9 (±0.4)	0.40 (±0.08)	551.1
ATPγS	500 (±200)	2.5 (±0.5)	0.41 (±0.15)	547.0
AMP-PNP	500 (±200)	3.1 (±0.7)	0.45 (±0.0)	529.9
ADP	667 (±250)	3.6 (±0.5)	0.4 (±0.1)	448.1
AMP	500 (±200)	500 (±200)	80 (±40)	391.1
dAMP	>1000	>1000	167 (±83)	354.2
GDP	>1000	400 (±200)	200 (±65)	487.1
GMP	>1000	>1000	>1000	407.1
CMP	>1000	>1000	>1000	323.2
UMP	>1000	>1000	>1000	368.1
cAMP	>1000	>1000	250 (±166)	450.3
Adenosine	250 (±85)	100 (±36)	100 (±36)	267.2
Adenine	500 (±200)	500 (±200)	500 (±200)	135.1
Staurosporine	0.025 (±0.004)	0.025 (±0.006)	0.025 (±0.005)	466.5

[a] Binding affinities, K_D, obtained from *ThermoFluor* thermal shift assays were fit assuming a constant $\Delta H = 110{,}000$ cal/mol, $\Delta S = 374$ cal/mol K, and $\Delta C_p = 2500$ cal/mol K for DapK-1 unfolding, and $\Delta H = (-5000)$ cal/mol and $\Delta C_p = (-200)$ cal/mol K for ligand binding. The concentration of nucleotides used was 0.04–2500 μM. The concentration of protein was 0.6 μM.

"significant," or measurable, binding event for weak ligands. The only practical limitation in conducting a thermal shift assay is the solubility limit of the test ligand. Thus, for fragment-based screening, thermal shift assays are highly useful as a screening and follow-up tool when highly soluble fragments are utilized, allowing estimates of binding affinity in the range of $K_D = 100$ μM to 1 M. No other technique has such a vast range of sensitivity for measuring ligand binding to soluble proteins.

ACKNOWLEDGMENT

We thank the volume editor, Dr. Lawrence Kuo, for his help in the preparation of this chapter.

REFERENCES

Behnke, C. A., *et al.* (2010). Atomic resolution studies of carbonic anhydrase II. *Acta. Crystallogr. D Biol. Crystallogr.* **66,** 616–627.

Bialik, S., and Kimchi, A. (2006). The death-associated protein kinases: Structure, function, and beyond. *Annu. Rev. Biochem.* **75,** 189–210.

Brandts, J. F., and Lin, L. N. (1990). Study of strong to ultratight protein interactions using differential scanning calorimetry. *Biochemistry* **29,** 6927–6940.

Cannon, M. J., *et al.* (2004). Comparative analyses of a small molecule/enzyme interaction by multiple users of Biacore technology. *Anal. Biochem.* **330,** 98–113.

Cimmperman, P., *et al.* (2008). A quantitative model of thermal stabilization and destabilization of proteins by ligands. *Biophys. J.* **95,** 3222–3231.

Daniel, E., and Weber, G. (1966). Cooperative effects in binding by bovine serum albumin. I. The binding of 1-anilino-8-naphthalenesulfonate. Fluorimetric titrations. *Biochemistry* **5,** 1893–1900.

Eftink, M. R. (1997). Fluorescence methods for studying equilibrium macromolecule-ligand interactions. *Methods Enzymol.* **278,** 221–257.

Epstein, H. F., *et al.* (1971). Folding of staphylococcal nuclease: Magnetic resonance and fluorescence studies of individual residues. *Proc. Natl. Acad. Sci. USA* **68,** 2042–2046.

Grasberger, B. L., *et al.* (2005). Discovery and cocrystal structure of benzodiazepinedione HDM2 antagonists that activate p53 in cells. *J. Med. Chem.* **48,** 909–912.

Hermans, J. J., and Scheraga, H. A. (1961). Structural Studies of Ribonuclease. V. Reversible Change Configuration. *J. Am. Chem. Soc.* **83,** 3283–3292.

Kervinen, J., *et al.* (2006). Effect of construct design on MAPKAP kinase-2 activity, thermodynamic stability and ligand-binding affinity. *Arch. Biochem. Biophys.* **449,** 47–56.

Klinger, A. L., *et al.* (2006). Inhibition of carbonic anhydrase-II by sulfamate and sulfamide groups: An investigation involving direct thermodynamic binding measurements. *J. Med. Chem.* **49,** 3496–3500.

Koshland, D. E. (1958). Application of a theory of enzyme specificity to protein synthesis. *Proc. Natl. Acad. Sci. USA* **44,** 98–104.

Linderstrøm-Lang, K., and Schellman, J. A. (1959). Protein structure and enzyme activity. *In* "The Enzymes", Vol. 1 (P. D. Boyer, H. Lardy, and K. Myrbäk, eds.), 2nd Edn. pp. 443–510. Academic Press, New York.

Lo, M. C., *et al.* (2004). Evaluation of fluorescence-based thermal shift assays for hit identification in drug discovery. *Anal. Biochem.* **332,** 153–159.

Matulis, D., *et al.* (2005). Thermodynamic stability of carbonic anhydrase: Measurements of binding affinity and stoichiometry using ThermoFluor. *Biochemistry* **44,** 5258–5266.

McClure, W. O., and Edelman, G. M. (1966). Fluorescent probes for conformational states of proteins. I. Mechanism of fluorescence of 2-p-toluidinylnaphthalene-6-sulfonate, a hydrophobic probe. *Biochemistry* **5,** 1908–1919.

Mezzasalma, T. M., *et al.* (2007). Enhancing recombinant protein quality and yield by protein stability profiling. *J. Biomol. Screen.* **12,** 418–428.

Murphy, K. P., and Gill, S. J. (1991). Solid model compounds and the thermodynamics of protein unfolding. *J. Mol. Biol.* **222,** 699–709.

Myszka, D. G. (2004). Analysis of small-molecule interactions using Biacore S51 technology. *Anal. Biochem.* **329,** 316–323.

Niesen, F. H., *et al.* (2007). The use of differential scanning fluorimetry to detect ligand interactions that promote protein stability. *Nat. Protoc.* **2,** 2212–2221.

Pantoliano, M. W., *et al.* (2001). High-density miniaturized thermal shift assays as a general strategy for drug discovery. *J. Biomol. Screen.* **6,** 429–440.

Privalov, P. L. (1979). Stability of proteins: Small globular proteins. *Adv. Protein Chem.* **33,** 167–241.

Ramsay, G., and Eftink, M. R. (1994). A multidimensional spectrophotometer for monitoring thermal unfolding transitions of macromolecules. *Biophys. J.* **66,** 516–523.

Shrake, A., and Ross, P. D. (1992). Origins and consequences of ligand-induced multiphasic thermal protein denaturation. *Biopolymers* **32,** 925–940.

Slavik, J. (1982). Anilinonaphthalene sulfonate as a probe of membrane composition and function. *Biochim. Biophys. Acta* **694,** 1–25.

Stryer, L. (1965). The interaction of a naphthalene dye with apomyoglobin and apohemoglobin. A fluorescent probe non-polar binding sites. *J. Mol. Biol.* **13,** 482–495.
Taniuchi, H., and Bohnert, J. L. (1975). The mechanism of stabilization of the structure of nuclease-T by binding of ligands. *J. Biol. Chem.* **250,** 2388–2394.
Velentza, A. V., *et al.* (2001). A protein kinase associated with apoptosis and tumor suppression: Structure, activity, and discovery of peptide substrates. *J. Biol. Chem.* **276,** 38956–38965.
Yan, H., and Tsai, M. D. (1999). Nucleoside monophosphate kinases: Structure, mechanism, and substrate specificity. *Adv. Enzymol. Relat. Areas Mol. Biol.* **73,** 103–134, x.
Zhang, R., and Monsma, F. (2010). Fluorescence-based thermal shift assays. *Curr. Opin. Drug Discov. Devel.* **13,** 389–402.
Zubriene, A., *et al.* (2009). Measurement of nanomolar dissociation constants by titration calorimetry and thermal shift assay—Radicicol binding to Hsp90 and ethoxzolamide binding to CAII. *Int. J. Mol. Sci.* **10,** 2662–2680.

CHAPTER TWELVE

HTS Reporter Displacement Assay for Fragment Screening and Fragment Evolution Toward Leads with Optimized Binding Kinetics, Binding Selectivity, and Thermodynamic Signature

Lars Neumann, Konstanze von König, *and* Dirk Ullmann

Contents

Abstract

Parameters such as residence time, kinetic selectivity, and thermodynamic signature are more and more under debate as optimization objectives within fragment-based lead discovery. However, broad implementation of these parameters is hampered by the lack of technologies that give rapid access to binding kinetics and thermodynamic information for large amounts of compound–target interactions. Here, the authors describe a technology—the reporter displacement assay—that is capable of opening this bottleneck and of supporting data-driven design of lead compounds with tailor-made residence time, kinetic selectivity, and thermodynamic signature.

Proteros Biostructures GmbH, Am Klopferspitz 19, Martinsried, Germany

Methods in Enzymology, Volume 493
ISSN 0076-6879, DOI: 10.1016/B978-0-12-381274-2.00012-1

1. Introduction

In the past decades, binding affinity was the main optimization parameter in lead discovery. Medicinal chemistry gathered vast knowledge of how to improve binding affinity of early compound or fragment hits in a short time. Understanding the interdependence of compound structures and binding affinity, the so-called structure activity relationship, represents the core of most lead optimization programs. Many valuable drugs were designed following this path. However, attrition rates are extremely high. A significant disconnect between biochemical affinity data and the affinity measured in cell cultures is more the rule than the exception. Even more seldom a strict correlation has been seen between biochemical affinity data and *in vivo* efficacy. In consequence, for about 5 years an enhanced debate is ongoing within the drug discovery community whether biochemical affinity values might be too simplistic a view on the action of drugs *in vivo*. A biochemical test system represents a closed system with an unaltered compound concentration throughout the whole experiment. In contrast, an organism is an open system characterized by fluctuating compound concentrations. After compound administration, a peak of compound plasma concentration is reached. Thereafter the compound concentration declines due to metabolic processes. Hence once a compound is bound to its target, it is the dissociation rate of the compound–target complex that dictates how long the target is blocked. In consequence, as long as the K_d value of a compound is below the maximal concentration within the organisms, the dissociation rate of the compound–target complex is more diagnostic for the *in vivo* efficacy of a compound than its binding affinity. For this reason Copeland and others used the term *residence time* to describe how long a compound resides on its target and suggested to implement residence time as a crucial lead optimization parameter. Retrospective analysis has shown that among successful drugs there is a significant accumulation of compounds with prolonged residence times, although nearly none of these drugs were initially designed to have a long residence time. Compounds with prolonged residence times appear to have a greater chance to pass all *in vivo* and clinical trials. In addition, examples are reported that show a direct correlation between residence time and *in vivo* efficacy. (Copeland *et al.*, 2006; Inada *et al.*, 1999; Lu and Tonge, 2010; Swinney, 2004, 2006; Tummino and Copeland, 2008; Zhang and Monsma, 2009). Despite this knowledge, in many lead optimization programs residence time is still not considered as optimization parameter. One reason for neglecting residence time is the lack of technologies that allow high-throughput measurements of binding kinetics to the target and eventually also to potential off targets. In this chapter, the authors describe a technology, the reporter displacement assay, which enables quantification of large volumes of binding kinetics. Further it is described how the reporter displacement assay can be used to identify fragments that bind to the target at the site of interest and how to evolve these fragments to leads with the desired residence time.

Another discussion that was sparked a few years ago is the use of calorimetric data to support fragment evolution and lead optimization. More and more reports are accumulating indicating that the natures of most lead optimization cycles result in the production of lead compounds that are binding due to entropic rather than due to enthalpic contributions. Associated problems are lead compounds with low solubility, low selectivity, low metabolic stability, and low oral availability. Multiple reports describe that typically first-in-class drugs are binding due to entropic changes while the best-in-class drugs are binding due to enthalpic changes. It is suggested to use calorimetric methods to identify fragment or compound hits that bind due to an enthalpic contribution and then to control throughout the whole lead optimization process that the initial enthalpic binding properties are not turned into entropic binding properties (Chaires, 2008; Freire, 2008, 2009; Ladbury *et al.*, 2010; Torres *et al.*, 2010). The application of these suggestions is hampered by the low throughput and the extremely high protein consumption of the calorimetric standard technology of isothermal titration calorimetry. This chapter describes how the reporter displacement assay can be used to measure calorimetric data in a high-throughput mode and how calorimetric data can be incorporated into fragment-based lead discovery at very early stages for fragment hit selection.

2. The Reporter Displacement Assay

In order to meet the need for a robust binding assay that is suitable for high-throughput measurement of binding affinities, binding kinetics including residence time and thermodynamic signatures, the reporter displacement assay has been developed. The assay principle has been described previously (Müller *et al.*, 2010; Neumann *et al.*, 2009). In brief, the reporter displacement assay is based on a reporter probe that is distinctively designed to bind to the site of interest of the chosen target enzyme or receptor. The proximity between reporter and protein results in the emission of an optical signal. Fragments and compounds that bind to the same binding site displace the probe to cause a signal loss (Fig. 12.1). Kinetics of signal loss can be monitored. The reporter displacement assay is a homogeneous method that is performed in 384 well plates allowing the measurement of 384 kinetic traces in parallel.

For K_d determination, the amount of signal loss is quantified at different compound concentrations once the system has reached equilibrium (Fig. 12.1). Since the K_d value of the reporter probe and the reporter concentration are known, the compound K_d can be calculated from the

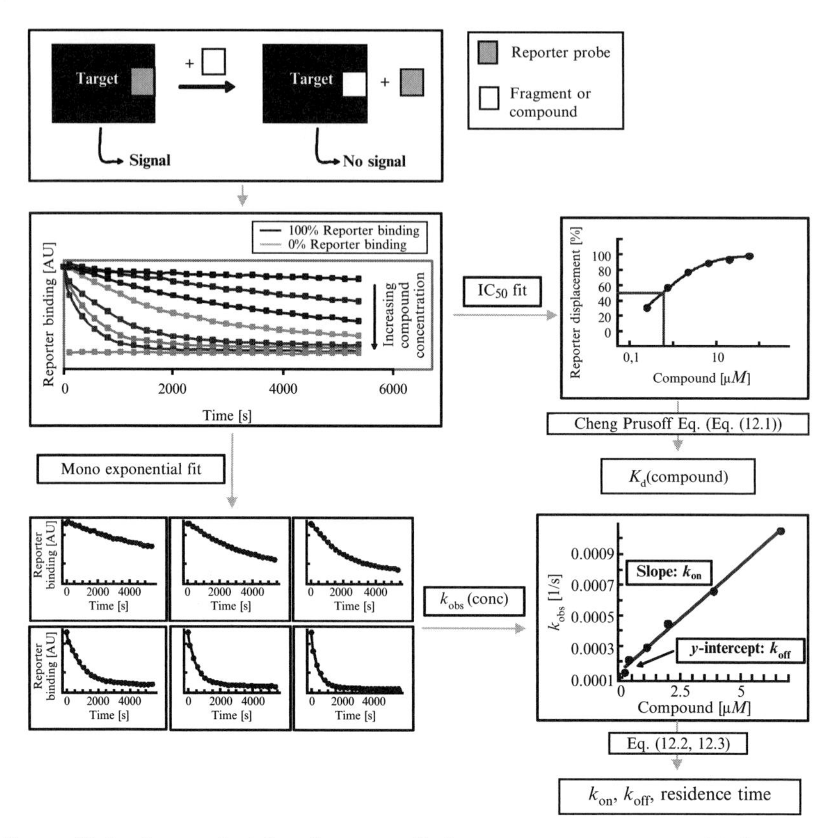

Figure 12.1 Assay principle of reporter displacement assay. Reporter binding to its target causes emission of a specific optical signal. Displacement of reporter by a competing compound of interest results in signal loss that can be monitored over time. Here, the reporter displacement assay is applied to the Sorafenib–bRAF interaction. A quantity of 2 n*M* bRAF was mixed with 12 n*M* ATP site-specific reporter probe in 8 μl 20 m*M* MOPS, pH 7.0, 1 m*M* DTT, 0.01% Tween 20 within a 384-well low volume plate. After 60 min incubation, Sorafenib was added at the desired concentrations. Displacement of the reporter probe was measured continuously over time. The signal for 100% probe binding was measured in the absence of Sorafenib and the signal for 0% reporter binding was quantified in the absence of bRAF. The Sorafenib IC_{50} value was quantified by regular IC_{50} fitting using the percent reporter displacement values after the system had reached equilibrium. The K_d values were calculated using the Cheng–Prusoff equation (Eq. (12.1)). Binding kinetics was analyzed by applying a monoexponential decay equation to the reporter displacement traces. The extracted exponent yielded the observed association rate k_{obs} for each Sorafenib concentration. By replotting the k_{obs}values against the associated Sorafenib concentrations and by fitting to linear Eq. (12.2), k_{on} and k_{off} values were obtained from the slope and the *y*-intercept, respectively. Residence time was calculated by Eq. (12.3).

measured IC_{50} according to the Cheng–Prusoff equation (Eq. (12.1); Cheng and Prusoff, 1973):

$$K_d(\text{compound}) = \frac{IC_{50}}{[1 + [\text{reporter}]/K_d(\text{reporter})]} \quad (12.1)$$

where K_d(compound) is the dissociation constant of the investigated compound, K_d(reporter) is the dissociation constant of the reporter, [reporter] is the concentration of reporter, and IC_{50} is the compound concentration that inhibits 50% of reporter binding.

$$k_{obs} = k_{off} + k_{on}[\text{compound}] \quad (12.2)$$

where k_{obs} is the observed association rate, k_{off} is the dissociation rate, k_{on} is the association rate, and [compound] is the compound concentration.

$$\text{Residence time} = \frac{1}{k_{off}} \quad (12.3)$$

where k_{off} is the dissociation rate.

The kinetic parameters and the residence time are determined by fitting the time trace of each compound concentration to a monoexponential equation (Fig. 12.1). The resulting k_{obs} value (i.e., the observed association rate) is plotted against the compound concentration and a linear fit yields k_{on} as the slope and k_{off} as the y-intercept (Eq. (12.2); Fig. 12.1). The residence time is calculated as the reciprocal of the k_{off} value (Eq. (12.3)).

This setup of the assay opens up flexible choices. For instance, the site of interest on the target protein can be its active site, an allosteric site, or even a binding interface. Accordingly, the reporter probe can be made from any kind of interacting molecule, such as a previously known inhibitor, a cofactor, a protein binding partner, or a peptide. In this way, there do not exist any assay limitations due to the size of the target protein or the compound.

Prerequisite for this type of analysis is: (1) the knowledge of the K_d value for the reporter, and (2) that the reporter dissociation rate k_{off} (reporter) is fast compared to the k_{obs}(compound) in order to ensure that the reporter displacement kinetics reflects the compound binding kinetics and not the reporter dissociation kinetics (see Fig. 12.2).

Thus, in a first step, the K_d values for the reporter are measured by a regular Scatchard analysis. Increasing concentrations of reporter are incubated with the target and reporter binding is measured. The K_d value for the probe is quantified by applying Eq. (12.4) as demonstrated in Fig. 12.3.

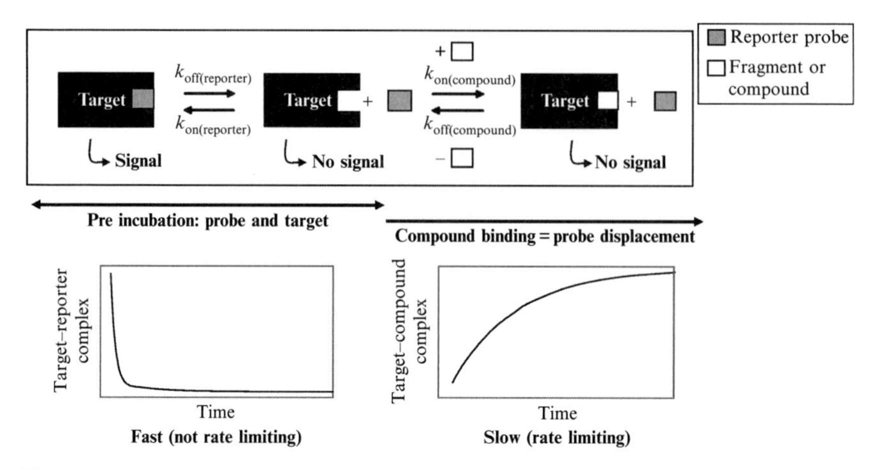

Figure 12.2 Kinetic interplay between reporter dissociation and compound binding. In order to analyze the compound–target interaction kinetics, compound is added to the preformed target–reporter complex. The reporter has to be designed to have fast dissociation kinetics compared to the binding kinetics of the compound to ensure that compound binding and not dissociation of the target–reporter complex is the rate-limiting step.

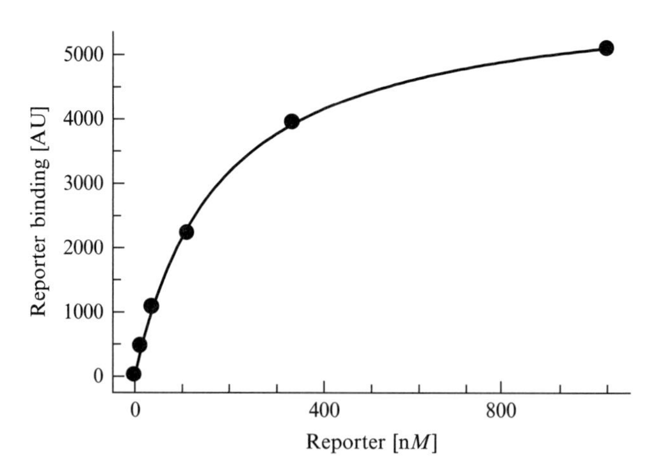

Figure 12.3 Determination of K_d value exemplified for the HDAC1–reporter interaction. Increasing concentrations of HDAC1-specific reporter were incubated with 20 nM HDAC1. Reporter binding was quantified for each reporter concentration and the K_d value was calculated to be 179 nM by applying Eq. (12.4).

$$\text{Signal(reporter binding)} = \frac{\text{BL}_{\text{max}}[\text{reporter}]}{K_d(\text{reporter}) + [\text{reporter}]} \tag{12.4}$$

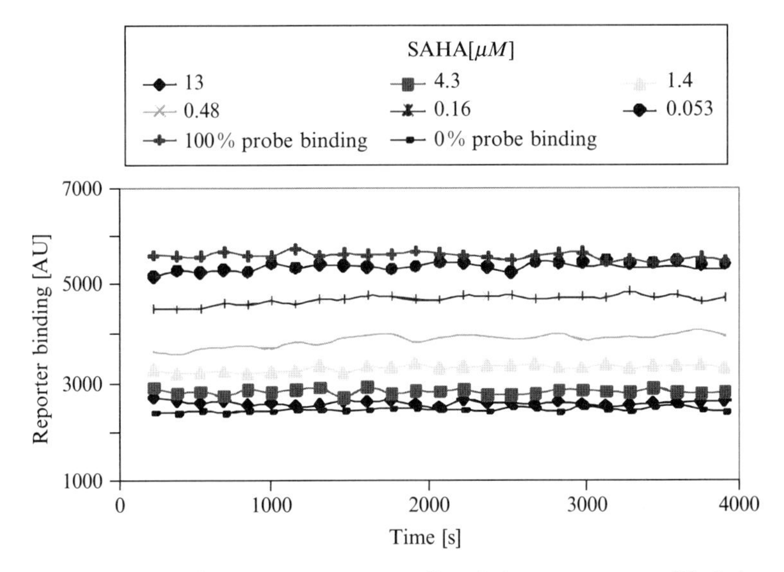

Figure 12.4 Controlling reporter–target dissociation rate exemplified by the HDAC1–SAHA interaction. HDAC1 and reporter were incubated to form the reporter–target complex. Thereafter increasing concentrations of the fast binding reference inhibitor SAHA were added and reporter displacement was measured. Reporter displacement was completely finished within the 40 s that were required to transport the assay plate after compound addition to the plate reader, as indicated by the parallel run of the lines. Thus, reporter dissociation is very rapid and not the rate-limiting step in the measurements of compound with slow binding kinetics. (See Color Insert.)

where BL_{max} is the binding signal that represents occupation of 100% of the binding sites, K_d(reporter) is the dissociation constant of reporter, [reporter] is the concentration of reporter.

In a second step, it is ascertained that the dissociation of the reporter–target complex is not the rate-limiting step. A fast binding compound is added to the reporter–target complex. In case probe displacement occurs instantaneously, within the time of compound addition and first measurement, it can be concluded that reporter dissociation is fast and not rate limiting (Fig. 12.4).

3. Residence Time and Kinetic Selectivity in Fragment Evolution

Residence time becomes a more and more appreciated optimization parameter in conventional and fragment-based lead discovery. The reporter displacement assay represents an ideal technology to support an entire fragment-based lead discovery program that is aiming at generating lead compounds with defined binding affinities and residence times.

Usually, the first step of a fragment lead optimization program is to screen a fragment library for fragments that bind to the target, here exemplified for a fragment screen against p38α. In this screen, 1023 fragments were screened at a concentration of 2 mM against p38α using the reporter displacement technology. A reporter was used that addresses the ATP binding site and the back pocket region of p38α ensuring that only fragments were identified that bound to the binding site of interest, while fragments that bound nonspecifically to the protein surfaces did not contaminate the hit list. In all, 69 fragments were found to block reporter binding by more than 50% (Table 12.1). Subsequent IC_{50} determination verified 81% of the initial hit fragments. IC_{50} values varied between 248 and 5490 μM (Neumann *et al.*, 2009).

In the second phase, structural complexity is added to selected fragment hits in order to improve binding affinity and residence time. Fig. 12.5 illustrates how the reporter displacement assay is used to monitor optimization of binding affinity and residence time in parallel by simultaneously studying structure activity and structure kinetic relationship. Compound 1 is taken from the primary screen described above. Compound 2 and 3 exemplify fragments with increasing complexity and have been previously described as p38α binding fragments (Paragellis *et al.*, 2002; Regan *et al.*, 2003; Zaman *et al.*, 2006). BIRB 796 serves as an example for a lead-like compound that contains parts of compounds 1–3 within its structure. Compound 1 shows a binding affinity of $K_d = 1165$ μM. Values for binding kinetics cannot be quantified because binding is so fast to fall beyond the detection limit of the assay. Binding equilibrium is reached in the 50 s between compound addition and the first measurement, resulting in parallel time traces for the reporter probe binding signal in Fig. 12.5. Compound 2 shows a dramatically increased binding affinity of 800 nM as compared to compound 1. However, binding kinetics is still too fast to be quantified. Adding further compound complexity results in compound 3 which possesses a strongly improved binding affinity of 32 nM. In addition,

Table 12.1 Screen of 1023 fragments against p38α

Reporter displacement (%)	Number of fragments	Hit rate (%)
>90	2	0.2
>80	8	0.8
>70	16	1.6
>60	32	3.1
>50	69	6.7

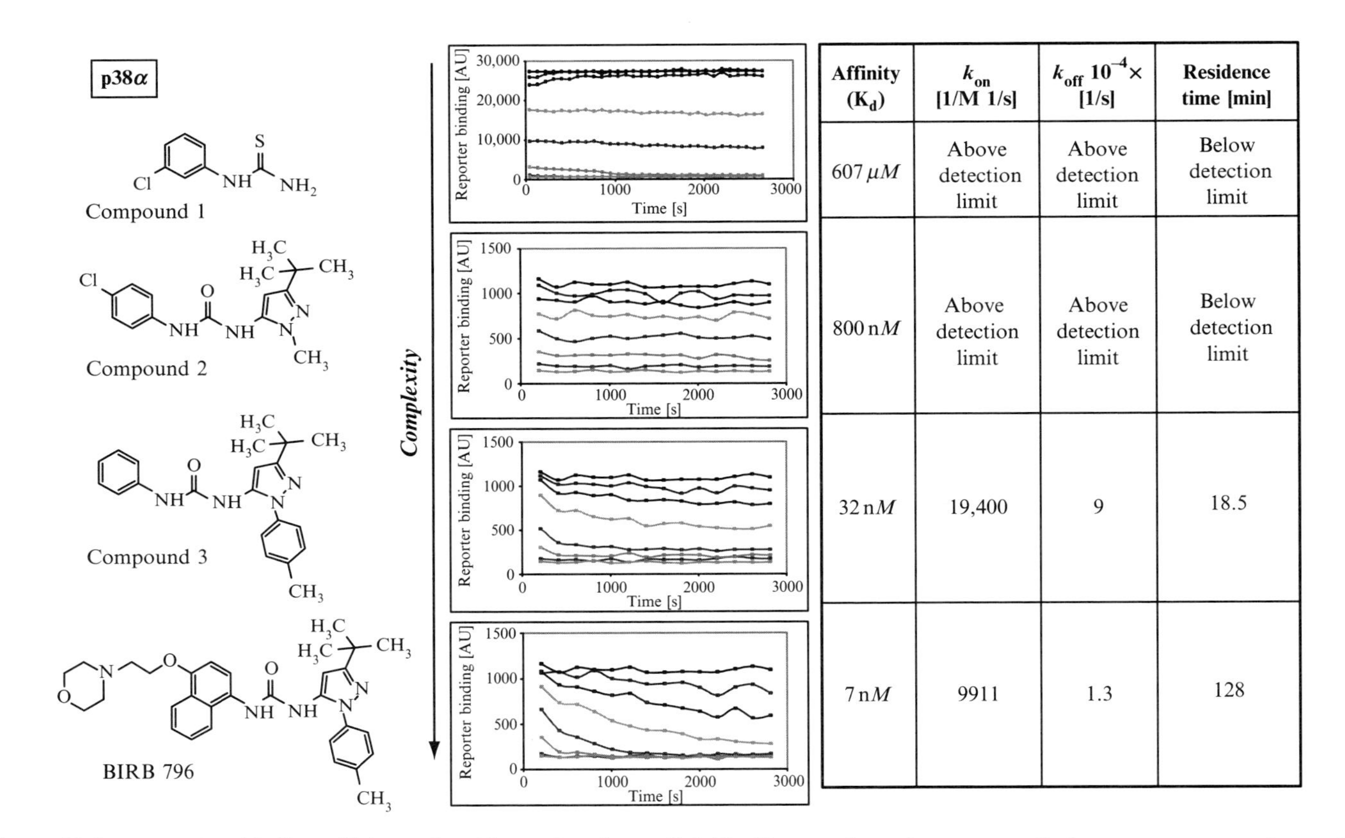

Affinity (K_d)	k_{on} [1/M 1/s]	k_{off} $10^{-4}\times$ [1/s]	Residence time [min]
607 μM	Above detection limit	Above detection limit	Below detection limit
800 nM	Above detection limit	Above detection limit	Below detection limit
32 nM	19,400	9	18.5
7 nM	9911	1.3	128

Figure 12.5 Improving binding affinity and residence time in parallel. To illustrate how the reporter displacement assay can be used to evolve fragments to lead compounds with optimized binding affinity and residence time, compounds with increasing complexity are characterized using the reporter displacement assay.

detectable binding kinetics is reached. An association rate constant k_{on} and a dissociation rate constant k_{off} for the complex of compound 3 binding to p38α of 19,400 M^{-1} s^{-1} and 0.0009 s^{-1}, respectively, are found. These values are in good agreement with the k_{on} and k_{off} values of 73,000 M^{-1} s^{-1} and 0.0016 s^{-1}, respectively, that have been reported (Regan *et al.*, 2003). Applying Eq. (12.3), the k_{off} value of 0.0009 s^{-1} translates into a residence time of 18.5 min. Finally, the lead-like compound BIRB 796 shows an affinity of 7 n*M* and a residence time of more than 2 h. Thus, due to the capability of the reporter displacement assay to quantify binding kinetics and binding affinity in parallel for large numbers of compounds in very short time frames, the reporter displacement assay is ideally suited to be incorporated into medicinal chemistry cycles with a set up time that can be as short as 1–2 weeks.

Besides binding affinity and residence time, selectivity is a crucial factor for drug efficacy *in vivo*. In the past decades, selectivity has been monitored by measuring binding affinities of lead compounds against a number of potential targets. Considering the observation that residence time is at least equally important as affinity for *in vivo* efficacy, it becomes obvious that selectivity profiling solely based on affinity values will be misleading. Figure 12.6 uses Sorafenib to demonstrate the wide variety of residence times that can be found for a single compound. Among the 15 examined Sorafenib targets, the longest residence time was found for cKIT to be 811 min followed by CDK8/CycC and bRAF with 576 and 568 min, respectively. Besides these, very long residence time values, medium residence times of 24 and 45 min were found for targets such as DDR1 and DDR2. For other targets such as TAOK3 and TIE2, even residence times below 2 min were detected.

In order to illustrate the effect of residence time, the *in vivo* performance of Sorafenib against four targets is simulated in Fig. 12.7. For the calculations, a maximal plasma concentration (C_{max}) of 3 μ*M* and a plasma half-life of 0.5 h are assumed. Both values reflect typical values that are found for lead-like and also drug-like compounds. Further, K_d and residence time values quantified by the reporter displacement assay (Figs. 12.6 and 12.7) are considered for the calculations. The four displayed targets are selected in order to cover all four possible combinations between affinity and residence time. CDK8/CyC and DDR1 are both high-affinity targets for Sorafenib with K_d values in the double digit nanomolar range. However, Sorafenib has a very long residence time with CDK8/CycC of 576 min but only a medium residence time with DDR1 of 24 min. cKIT is a low-affinity target with a micromolar K_d value but has a very long residence time of 811 min. Finally, TAOK3 is a low-affinity target with a short residence time. Since the C_{max} is even above the K_d value of both low-affinity targets, all four targets are at least partially occupied and inhibited by the compound after drug administration. After having reached C_{max}, the compound concentration declines due to metabolic actions. In consequence, targets that have a

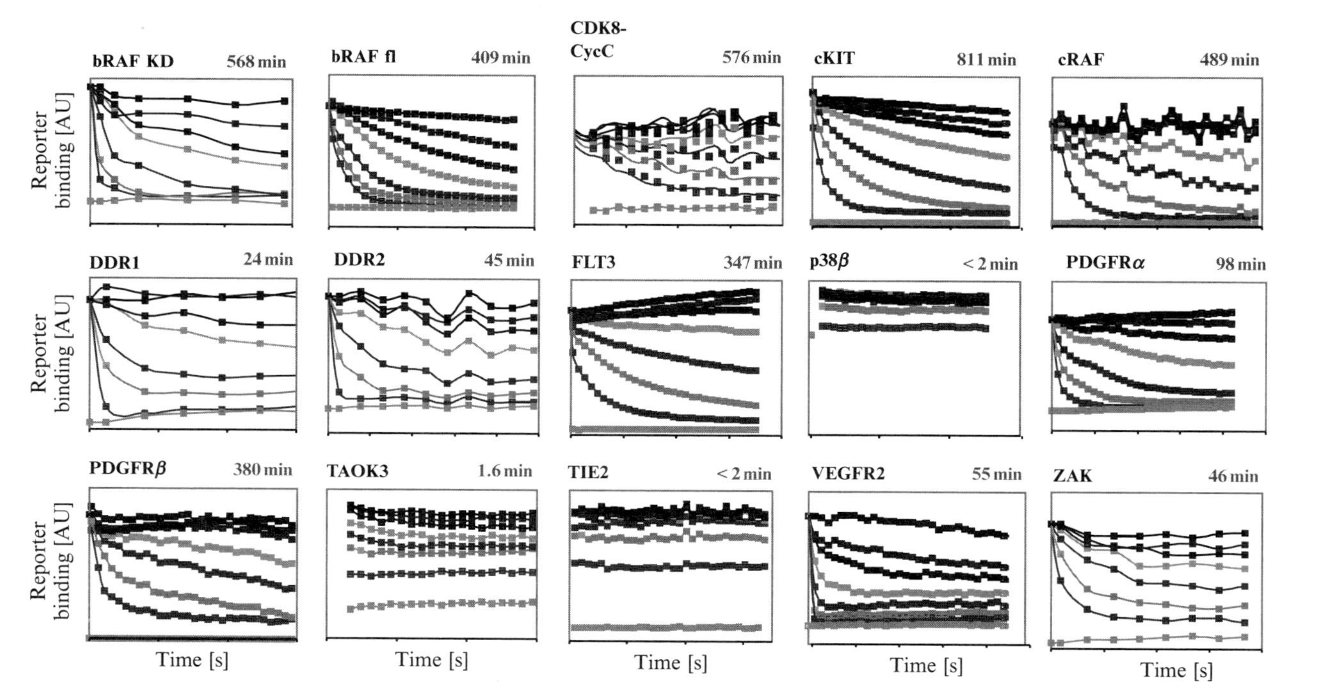

Figure 12.6 Kinetic selectivity profiling of Sorafenib. Residence times of Sorafenib on 15 targets are measured using the reporter displacement assay. For each target reporter displacement is given over time in the presence of increasing Sorafenib concentrations. The color coding of the traces reflects the Sorafenib concentration (Sorafenib concentration: red > orange > green > light blue > dark blue > purple). Black traces represent reporter binding signal over time in the absence of Sorafenib and gray traces represent the signal in absence of target, indicating the signals for 100% and 0% reporter binding, respectively. Target names are given in black and residence times in blue. For illustration purposes, time scales differ between the different targets and are not shown. (See Color Insert.)

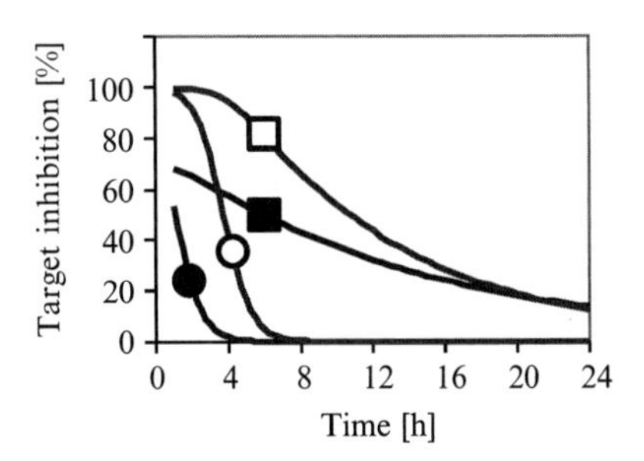

Target	K_d [μM]	Residence time [min]	Affinity/residence time
□ CDK8-CycC	0.030	576	High/long
○ DDR1	0.072	24	High/short
■ cKIT	2.3	811	Low/long
● TAOK3	2.2	1.6	Low/short

Figure 12.7 Simulation of *in vivo* efficacy against four targets. Target inhibition over time was calculated for four selected Sorafenib targets. Calculations were performed assuming a maximal plasma concentration (C_{max}) of 3 μ*M* and a plasma half-life of 0.5 h. Inhibition for each target was projected by calculating target occupancy using the indicated K_d and residence time values. The K_d and residence time values were extracted from Fig. 12.6.

long residence time will be occupied longer by the compound than targets with a short residence time and are therefore inhibited more effectively over time. For example, although CDK8/CycC and DDR1 have nearly the same affinity for the compound, DDR1 is blocked by 50% after 5 h, while CDK8/CycC is still blocked by 100%. After 7 h DDR1 is not blocked anymore at all while CDK8/CycC activity is still inhibited by 90%. The role of residence time becomes even clearer when comparing the inhibition of DDR1 and cKit. Although DDR1 is a high-affinity target and cKit a low-affinity target, DDR1 and cKIT are inhibited to an equal extent already after 4 h post C_{max}. After 7 h, DDR1 is not inhibited anymore at all but the low-affinity target cKit is still blocked by 50%. Most notably, also the inhibition of CDK8/CycC and cKit draw level after 18 h. Obviously, TAOK3 as a low-affinity target with a short residence time is not efficiently inhibited. Thus as long as the K_d value of the target–compound interaction is below the *in vivo* compound concentration, residence time and not binding affinity represents the main biochemical indicator for *in vivo* efficacy. In consequence, medicinal resources may be better invested in increasing the residence time of a lead series instead of trying to increase the binding affinity, for example, from a K_d value of 100 to 10 n*M*. In addition, Fig. 12.7 makes obvious that kinetic selectivity profiling identifies other targets as potential off-targets than the classical selectivity profiling that is based on affinity comparisons. In the presented example, kinetic selectivity profiling identified cKit as a potential target for the compound in addition to CDK8/CycC. Omitting the time dimension of an *in vivo* situation, a conventional affinity-based selectivity profiling would have identified DDR1 and not cKit as a potential off-target.

4. High-Throughput Thermodynamics in Fragment Evolution

For several years, it has been under debate whether lead optimization would benefit from adding calorimetric data to decision making within medicinal chemistry (Chaires, 2008; Ferenczy and Keserü, 2010; Freire, 2008, 2009; Ladbury *et al.*, 2010; Torres *et al.*, 2010). Inhibitors that bind to their targets due to enthalpic rather than due to entropic changes are thought to be less prone to problems such as low solubility, low selectivity, low oral availability, and low metabolic stability. For targets like HIV protease and 3-hydroxy-3-methlyglutaryl coenzyme A (HMG-CoA), it has been shown that binding of early drugs is driven by entropy, while later marketed drugs with more advantageous clinical properties bind due to changes in enthalpy. Due to their superior properties, the more advanced enthalpic drugs have replaced the earlier entropic drugs (Freire, 2008).

Like for any other chemical or physical process, a decrease of the Gibbs free energy G is prerequisite for compound–protein binding. The change of free energy (ΔG) has to be negative for the binding process. High-affinity binding interactions are characterized by larger negative changes of free energy than low-affinity interactions as given in Eq. (12.5). According to Eq. (12.6), ΔG can be negative due to either a very large negative change of enthalpy (ΔH) or a very large positive change of entropy (ΔS), or both.

$$\Delta G = -RT\ln\left(\frac{1}{K_\mathrm{d}}\right) \tag{12.5}$$

where ΔG is the change of free energy, R is the gas constant, T is the temperature, and K_d is the dissociation constant.

$$\Delta G = \Delta H - T\Delta S \tag{12.6}$$

where ΔG is the change of free energy, ΔH is the change of enthalpy, T is the temperature, and ΔS is the change of entropy.

ΔH is a measure for the net changes of the strength of all involved noncovalent interactions, for example, hydrogen bonds and polar interactions.

Thus, in order to increase binding affinity, hydrogen bonds between compound and target have to be positioned in perfect distance and angle to assure that the enthalpic gain overcompensates the enthalpic penalty that is paid for breaking the hydrogen bonds between compound and water molecules during desolvation. Similar considerations apply to polar interactions. Thus, due to the high distance and angle dependence of hydrogen bonds, favorable enthalpic drug–target interactions are difficult to engineer into compound structures. The main reason for entropically favored drug–target binding are hydrophobic interactions. Highly ordered water molecules that are coordinated to hydrophobic moieties are released into the bulk water upon compound binding. Since hydrophobic interactions are less distance- and not angle-dependent, affinity increases are more straightforward to realize by adding hydrophobic moieties to the inhibitors than, on the contrary, by constructing drugs with a precisely oriented network of hydrogen bonds and/or polar interactions within the target binding pocket. Thus, target binding of most drugs is driven by entropy. Interestingly, binding of most natural ligands is driven by enthalpy (Olsson *et al.*, 2008).

In order to assure that fragment-based lead discovery campaigns yield leads with an enthalpically favored binding, two measures are required. First, since enthalpic compounds are more challenging to construct than entropic compounds, it is beneficial to use enthalpic fragments as starting point for fragment evolution. Second, binding thermodynamics should be controlled during the entire fragment evolution and lead optimization process in order to ensure that an initially enthalpically favored binding is not turned into an entropically favored binding, while seeking to increase binding affinities.

Isothermal calorimetry represents the standard technology for thermodynamic measurements. However, isothermal calorimetry is hampered by its low throughput and its extremely high target protein consumption. These drawbacks prevent one from measuring the thermodynamic properties of the entire hit lists from screening campaigns in order to identify fragments with optimal thermodynamic signatures for fragment evolution. For the same reason, the thermodynamic signatures of compounds during the optimization cycles are not controlled.

The reporter displacement assay represents a high-throughput binding assay with low target protein consumption and can be used in combination with van't Hoff analysis to overcome this bottleneck in the determination of large numbers of thermodynamic signatures. The van't Hoff analysis requires the measurement of K_d values at several temperatures, for example, 4–35 °C. Subsequently, the change of free energy (ΔG) is calculated (see Eq. (12.5)) and plotted against the associated temperature (see Fig. 12.8). According to Eq. (12.6), the change of enthalpy (ΔH) is given by the y-intercept while the change of entropy (ΔS) is given by the slope. Although van't Hoff analysis may yield misleading results, for example, if large heat capacity changes are involved in the compound–target binding process,

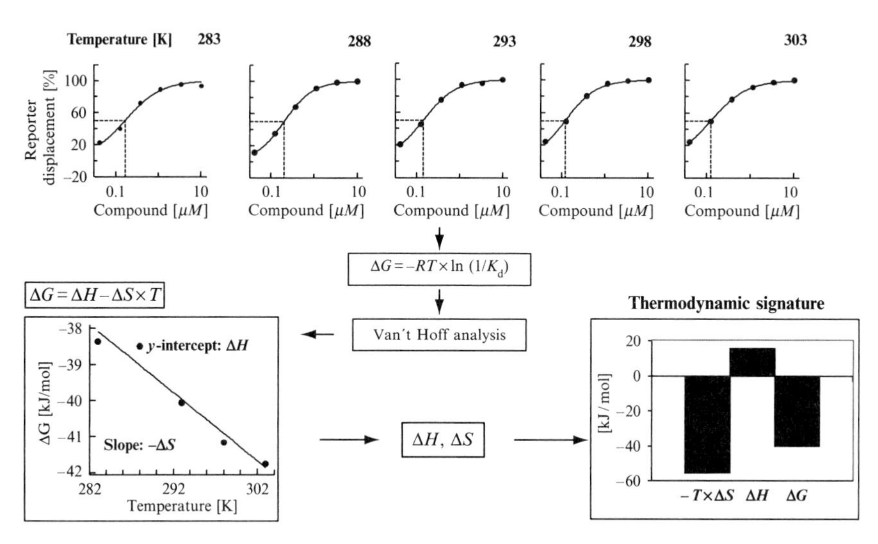

Figure 12.8 Van't Hoff analysis for reporter displacement assay data exemplified for the Sorafenib–bRAF interaction. The IC_{50} values (50% probe displacement) were measured at various temperatures and K_d values determined using Eq. (12.3). The ΔG values were calculated (see Eq. (12.5)) and plotted against temperature. ΔH and $-\Delta S$ can be quantified by fitting the data to Eq. (12.6). Data are visualized by a thermodynamic signature plot. Binding of Sorafenib to bRAF is driven by entropy and disfavored by enthalpy.

van't Hoff analysis has proven to be a reliable method to determine changes of entropy and enthalpy in numerous cases (Chakraborty *et al.*, 2010; Deinum *et al.*, 2002; Hedge *et al.*, 2002; Lohman *et al.*, 1996; Navratilova *et al.*, 2007; Neumann *et al.*, 2002; Papalia *et al.*, 2008; Walsh, 2010).

The reporter displacement assay delivers K_d values by measuring IC_{50} values at a known reporter concentration using a reporter with the predetermined value K_d(reporter). By applying the Cheng–Prusoff equation (Eq. (12.1)), the value for K_d(compound) of the investigated compound is calculated from the measured IC_{50} value.

Since the value for K_d(reporter) depends on the temperature, the determination of K_d(reporter) for all investigated temperatures is a prerequisite to use the reporter displacement assay for van't Hoff analysis. Figure 12.9 exemplifies the K_d(reporter) determination for the interaction with p38α from 10 to 35 °C. Reporter binding was quantified in the presence of increasing reporter concentrations at various temperatures and fitted to Eq. (12.4) in order to calculate the values for K_d(reporter) (Table 12.2).

Subsequently, ΔG values were calculated for each temperature using Eq. (12.5), as shown in Table 12.1, and plotted against the associated temperatures given in Kelvin (Fig. 12.10). The van't Hoff analysis was performed for the interaction between the reporter and p38α. The data

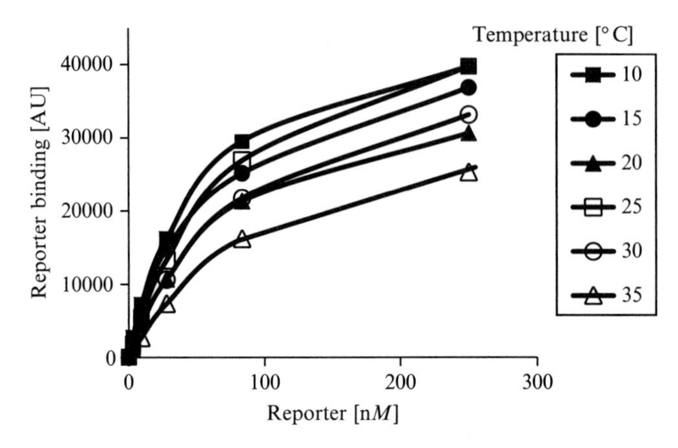

Figure 12.9 Determination of K_d values for the interaction between reporter and p38α from 10 to 35 °C. Increasing reporter concentrations were incubated with p38α at various temperatures, and reporter binding was quantified. The K_d values were determined by fitting the data to Eq. (12.4), and ΔG values were calculated using Eq. (12.5). The values are summarized in Table 12.1.

Table 12.2 K_d and ΔG values for reporter–p38a interaction at various temperatures

T [°C]	**10**	**15**	**20**	**25**	**30**	**35**
T [K]	283	288	293	298	303	308
K_d [nM]	54	61	72	81	89	105
ΔG [J mol^{-1}]	−39,360	−39,753	−40,042	−40,444	−40,877	−41,125

were fitted to Eq. (12.6) to quantify ΔH and ΔS values. ΔH and $-\Delta S$ values were determined to be -19 kJ mol^{-1} and -72 mol J^{-1} K^{-1}, respectively. Thus, reporter binding to p38α is driven by both enthalpy and entropy. The thermodynamic signature is visualized in Fig. 12.11.

Having determined the K_d(reporter) values for all investigated temperatures, the reporter displacement assay in combination with van't Hoff analysis represents a powerful tool that allows access to thermodynamic information of large numbers of compound–target interactions. Thus, an entire hit sets from regular and fragment screening campaigns can be analyzed for their thermodynamic signatures. Fragments can be identified that bind due to a change in enthalpy rather than due to a change in entropy and serve as starting points for fragment evolution. Figure 12.12 exemplifies the thermodynamic signatures of all fragment hits from a screening campaign. Interestingly, the binding of all fragments have a favorable enthalpic contribution. In contrast, entropy changes occur in favor of the binding of some fragments and in disfavor of other fragments. However, more fragments have a favorable entropic binding contribution. For most of the

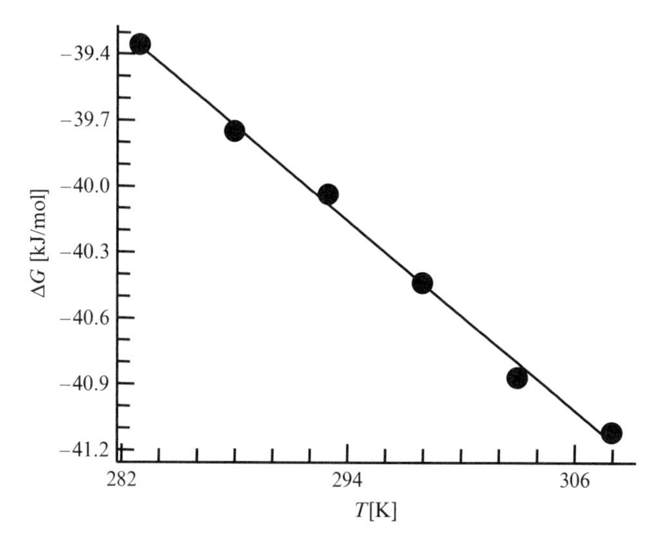

Figure 12.10 Van't Hoff analysis for reporter–p38a interaction. The ΔG values were plotted against temperature in Kelvin (Table 12.1) and fitted to Eq. (12.6). ΔH values were extracted from the y-intercept and $-\Delta S$ from the slope. ΔH and $-\Delta S$ were determined to be -19 kJ mol^{-1} and -72 mol J^{-1}K^{-1}, respectively.

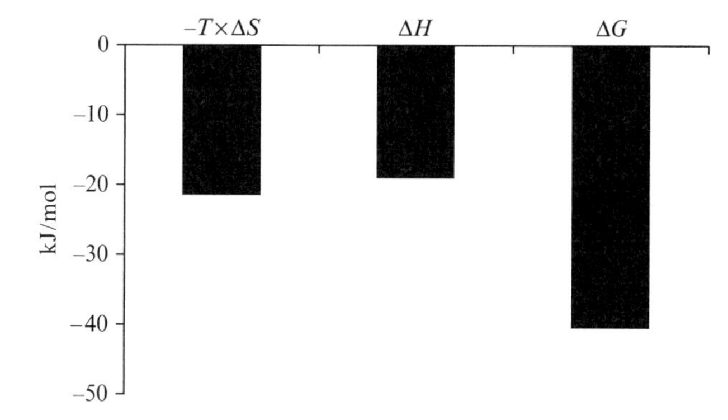

Figure 12.11 Thermodynamic signature of the reporter–p38α interaction. The ΔG, ΔH, and $-T\Delta S$ ($T = 298$ K) values were plotted. The ΔG value was taken from Table 12.1 and ΔH and ΔS were retrieved from Fig. 12.10 and visualized in a thermodynamic signature plot.

fragments that have an entropically favorable binding component, the entropy changes contribute less to the binding affinity than the enthalpy changes. The authors have observed similar thermodynamic characteristics for hits from additional fragment screening campaigns, suggesting that fragments typically are enthalpic binders and can be identified as starting point if thermodynamic analysis is applied to entire fragment hit selections.

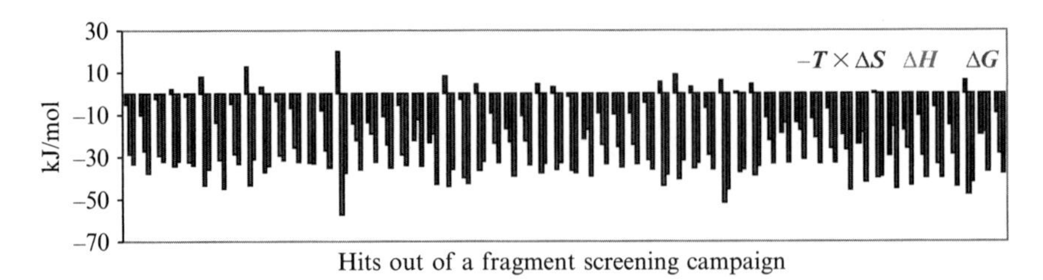

Figure 12.12 Thermodynamic signatures of all hits from an entire fragment screening campaign. The ΔH and ΔS values were determined for all hits from a fragment screening campaign using the reporter displacement assay in combination with van't Hoff analysis. The thermodynamic signatures were plotted for 298 K.

In contrast to fragments, most of the investigated lead-like compounds and even marketed drugs do not bind due to a favorable enthalpic contribution. Larger compounds often solely bind due to large changes of entropy as exemplified for the Sorfenib–bRAF interaction in Fig. 12.8. One explanation for this striking enrichment of entropic compounds among leads and drugs might be the nature by which most lead optimizations cycles are performed (Fig. 12.13). HTS or fragment screening campaigns deliver hits as starting points. Chemically nearest neighbors of these hits are synthesized and submitted for IC_{50} or K_d determination. Those compounds that show the most significant increase in affinity serve as starting points for the next compound synthesis cycle. This optimization cycle is maintained until the desired affinity is achieved. As discussed above, it is a challenging task to increase binding affinity by optimizing the enthalpic interactions due to their strict dependence on angle and distance. Entropic interactions are less angle- and distance-dependent and are therefore more straightforward to realize. In consequence the likelihood to increase affinity by alterations of entropy is largely higher than the likelihood to increase affinity by alteration of enthalpy. Lead optimization processes that do not control whether the increase in binding affinity is caused by enthalpic or by entropic optimization are therefore designed to produce entropic lead compounds.

Figure 12.14 illustrates a typical switch of the thermodynamic signature of an enthalpically binding fragment to an entropically binding lead-like compound within a lead optimization cycle that is not thermodynamically controlled. Thermodynamic signatures are measured for the binding between p38α and (1) a fragment screening hit (compound 1), (2) two further evolved fragments (compounds 2 and 3), and (3) BIRB 796 as an example for a lead-like compound. All four compounds have been described in the literature before as p38α binding compounds. (Neumann *et al.*, 2009; Paragellis *et al.*, 2002; Regan *et al.*, 2003; Zaman *et al.*, 2006). The fragment (compound 1) binds due to a change of enthalpy while the entropic change disfavors the binding. Compound 1 serves as an ideal enthalpic starting point for fragment

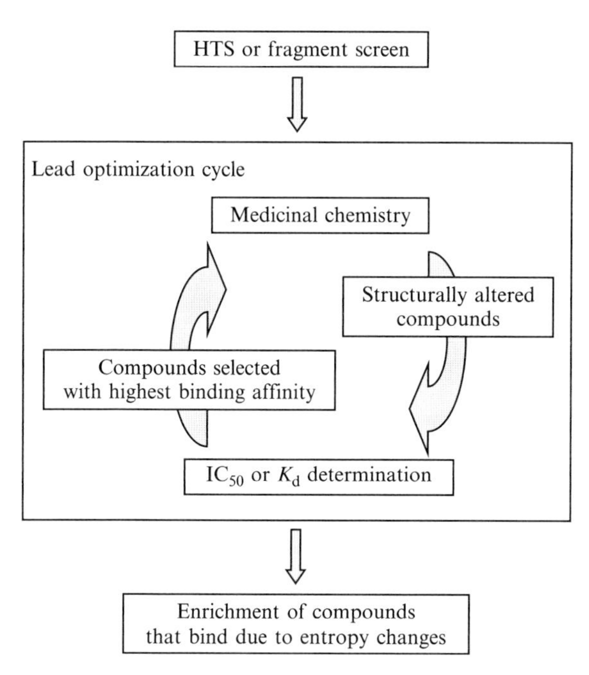

Figure 12.13 Typical lead optimization cycles enrich compounds that bind due to entropy.

evolution. Adding the pyrazol ring (compound 2) increases the affinity dramatically without altering the thermodynamic signature significantly and serves as an example that affinity can be improved without relying on entropic binding contribution by addition of hydrophobic moieties. In contrast, compound 3 exemplifies the widely applied method to increase affinity by adding hydrophobic moieties to the fragments. Compound 3 is evolved from compound 2 by adding a hydrophobic phenyl ring to the pyrazol ring and shows a 25-fold increased affinity. The thermodynamic signature reveals that the increase in affinity is realized due to the switch from a disfavorable entropic term to a favorable one. Enthalpy still does favor the binding but to a lesser extent than for compound 2. BIRB 796 represents an example for a lead-like compound that may have been evolved from compound 3 by adding another hydrophobic moiety to further increase affinity. BIRB 796 carries another hydrophobic phenyl group and shows a 4.6-fold increased affinity. The thermodynamic signature reflects typical thermodynamic properties of lead-like compounds that are produced following the described lead optimization procedure. BIRB 796 binding is solely supported by a large change in entropy. In contrast, enthalpy changes are highly in disfavor of binding, thereby reducing binding affinity.

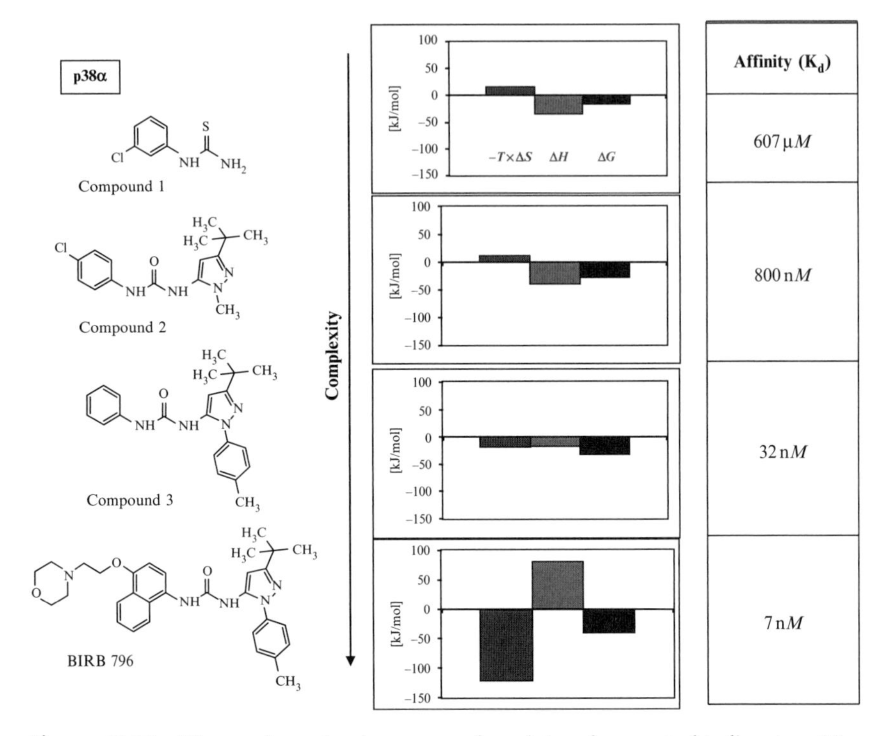

Figure 12.14 Thermodynamic signatures of evolving fragments binding to p38α. Thermodynamic signatures for the binding between p38α and compounds with increasing complexity are measured using the reporter displacement assay. Binding of the smallest fragment (compound 1) is solely driven by enthalpy and disfavored by entropy. With increasing fragment/compound complexity enthalpic contribution becomes less favorable and entropic contribution more favorable. Binding of the high complexity compound BIRB 796 is completely driven by entropy and disfavored by enthalpy and therefore shows the inverse thermodynamic signature of compound 1. Binding affinities were measured using the reporter displacement assay (Neumann *et al.*, 2009).

Figure 12.14 exemplifies how an enthalpic fragment is turned into an entropic lead-like compound by applying standard lead optimization cycles seeking rapid affinity improvement without thermodynamic control. Like many other lead-like compounds, BIRB 796 shares the resulting common problems such as low solubility, low selectivity, low metabolic stability, and low oral availability. In the past, isothermal calorimetry has not been capable of measuring the thermodynamic signature of every compound synthesized in the course of a lead optimization program in order to prevent the production of entropic leads. However, high-throughput thermodynamic methods, like the reporter displacement assay, open the door for thermodynamic characterization in early drug discovery.

5. Conclusion

The reporter displacement assay represents a high-throughput binding assay that allows access to unprecedented data volumes describing binding affinity (K_d), binding kinetics (k_{on}, k_{off}, residence time), and thermodynamic signatures (ΔH, ΔS), in this way opening new avenues in fragment-based lead discovery. Co-optimization of binding affinity and binding kinetics is possible. Lead compounds with tailor-made residence times can be produced on a rational base. Instead of static affinity-based selectivity, kinetic selectivity can be used to ensure high *in vivo* efficacy and low toxicity and calorimetric decision making can be incorporated into early lead discovery phases.

Acknowledgments

We thank Doris Hafenbradl and Marc Nicola Sommer who were heavily involved in the conceptional development of the reporter displacement assay. We further thank Allegra Ritscher, Birgit Flicke, Elisabeth Schneider, Stefanie Bauer, and Stefanie Gspurning for their never ending support at the bench and we are grateful for indispensable discussions with Gerhard Müller, Tim Woodcock, and Peter Sennhenn.

References

Chaires, J. B. (2008). Calorimetry and thermodynamics in drug design. *Annu. Rev. Biophys.* **37,** 135–151.

Chakraborty, M., Sengupta, A., Bhattacharya, D., Banerjee, S., and Chakrabarti, A. (2010). DNA binding domain of RFX5: Interactions with X-box DNA and RFXANK. *Biochim. Biophys. Acta* **1804,** 2016–2024.

Cheng, Y., and Prusoff, W. H. (1973). Relationship between the inhibition constant (KI) and the concentration of inhibitor which causes 50 per cent inhibition (I50) of an enzymatic reaction. *Biochem. Pharmacol.* **22,** 3099–3108.

Copeland, R. A., Pompliano, D. L., and Meek, T. D. (2006). Drug–target residence time and its implications for lead optimization. *Nat. Rev. Drug Discov.* **5,** 730–739.

Deinum, J., Gustavsson, L., Gyzander, E., Kullman-Magnusson, M., Edström, A., and Karlsson, R. (2002). A thermodynamic characterization of the binding of thrombin inhibitors to human thrombin, combining biosensor technology, stopped-flow spectrophotometry, and microcalorimetry. *Anal. Biochem.* **300,** 152–162.

Ferenczy, G. G., and Keserü, G. M. (2010). Thermodynamics guided lead discovery and optimization. *Drug Discovery Today* **15,** 919–932.

Freire, E. (2008). Do enthalpy and entropy distinguish first in class from best in class? *Drug Discov. Today* **13,** 869–874.

Freire, E. (2009). A thermodynamic approach to the affinity optimization of drug candidates. *Chem. Biol. Drug Des.* **74,** 468–472.

Hedge, S. S., Dam, T. K., Brewer, F., and Blanchard, J. S. (2002). Thermodynamics of aminoglycoside and acyl-coenzyme A binding to the salmonella enterica AAC(6′)-Iy aminoglycoside N-acetyltransferase. *Biochemistry* **41,** 7519–7527.

Inada, Y., Ojima, M., Kanagawa, R., Misumi, Y., Nishikawa, K., and Naka, T. (1999). Pharmacologic properties of candesartan cilexetil-possible mechanisms of long-acting antihypertensive action. *J. Hum. Hypertens.* **13,** 75–80.

Ladbury, J. E., Klebe, G., and Freire, E. (2010). Adding calorimetric data to decision making in lead discovery: A hot tip. *Nat. Drug Discov.* **9,** 23–27.

Lohman, T. M., Overman, L. B., Ferrari, M. E., and Kozlov, A. G. (1996). Highly salt-dependent enthalpy change for *Escherichia coli* SSB protein-nucleic acid binding due to ion-protein interactions. *Biochemistry* **35,** 5272–5279.

Lu, H., and Tonge, P. J. (2010). Drug–target residence time: Critical information for lead optimization. *Curr. Opin. Chem. Biol.* **14,** 467–474.

Müller, G., Sennhenn, P. C., Woodcock, T., and Neumann, L. (2010). The retro-design concept for novel kinase inhibitors. *IDrugs* **13,** 457–466.

Navratilova, I., Papalia, G. A., Rich, R. L., Bedinger, D., Brophy, S., Condon, B., Deng, T., Emerick, A. W., Guan, H. W., Hayden, T., Heutmekers, T., Hoorelbeke, B., *et al.* (2007). Thermodynamic benchmark study using Biacore technology. *Anal. Biochem.* **364,** 67–77.

Neumann, L., Abele, R., and Tampé, R. (2002). Thermodynamics of peptide binding to the transporter associated with antigen processing (TAP). *J. Mol. Biol.* **324,** 965–973.

Neumann, L., Ritscher, A., Müller, G., and Hafenbradl, D. (2009). Fragment based lead generation: Identification of seed fragments by a highly efficient fragment screening technology. *J. Comput. Aided Mol. Des.* **23,** 501–511.

Olsson, T. S. G., Williams, M. A., Pitt, W. R., and Ladbury, J. E. (2008). The thermodynamics of protein-ligand interactions and solvation: Insights for ligand design. *J. Mol. Biol.* **384,** 1002–1017.

Papalia, G. A., Giannetti, A. M., Arora, N., and Myszka, D. G. (2008). Thermodynamic characterization of pyrazole and azaindole derivatives binding to p38 mitogen-activated protein kinase using Biacore T100 technology and van't Hoff analysis. *Anal. Biochem.* **383,** 255–264.

Paragellis, C., Tong, L., Churchill, L., Cirillo, P. F., Gilmore, T., and Graham, A. G. (2002). Inhibition of p38 MAP kinase by utilizing a novel allosteric binding site. *Nat. Struct. Biol.* **9,** 268–272.

Regan, J., Paragellis, C. A., Cirillo, P. F., Gilmore, T., Hickey, E. R., Peet, G. W., Proto, A., Swinamer, A., and Moss, N. (2003). The kinetics of binding to p38MAP kinase by analogues of BIRB 796. *Bioorg. Med. Chem. Lett.* **13,** 3101–3104.

Swinney, D. (2004). Biochemical mechanisms of drug action: What does it take for success? *Nat. Rev. Drug Discov.* **3,** 801–808.

Swinney, D. (2006). Opportunities to minimise risk in drug discovery and development. *Expert Opin. Drug Discov.* **1,** 627–633.

Torres, F. E., Recht, M. I., Coyle, J. E., Bruce, R. H., and Williams, G. (2010). Higher throughput calorimetry: Opportunities, approaches and challenges. *Curr. Opin. Struct. Biol.* **20,** 1–8.

Tummino, P. J., and Copeland, R. A. (2008). Residence time of receptor–ligand complexes and its effect on biological function. *Biochemistry* **47,** 5481–5492.

Walsh, S. T. R. (2010). A biosensor study indicating that entropy, electrostatics, and receptor glycosylation drive the binding interaction between interleukin-7 and its receptor. *Biochemistry* **49,** 8766–8778.

Zaman, G. J., van der Lee, M. M., Kok, J. J., Nelissen, R. L., and Loomans, E. E. (2006). Enzyme fragment complementation binding assay for p38alpha mitogen-activated protein kinase to study the binding kinetics of enzyme inhibitors. *Assay Drug Dev. Technol.* **4,** 411–420.

Zhang, R., and Monsma, F. (2009). The importance of drug-target residence time. *Curr. Opin. Drug Discov. Dev.* **12,** 488–496.

CHAPTER THIRTEEN

Fragment Screening Purely with Protein Crystallography

John C. Spurlino

Contents

Abstract

We screen for fragments using X-ray crystallography as the primary screen. There are several unique features in our screening methodology. As a result of using X-ray diffraction as our primary screen, we do not use affinity data to bias our data collection or design in progressing hits toward a lead. Another difference in our methodology is that we choose to group our compounds as shape-similar groups. We also screen in a first pass mode without recollecting failed diffraction experiments. This method of screening results in an average loss of 5–10% of the data sets for the primary screen. The remaining data sets offer enough information to successfully advance three to five scaffolds into the secondary library design. We do not deconvolute the wells which show evidence of fragment binding by repeating the soaks with single compounds. Instead, evaluation of the possible fragments is done by refinement and examination of

Structural Biology, Johnson & Johnson Pharmaceutical Research and Development, LLC, Pennsylvania, USA

Methods in Enzymology, Volume 493
ISSN 0076-6879, DOI: 10.1016/B978-0-12-381274-2.00013-3

the resulting electron density difference maps. These methods allow us to complete the initial screen of a primary library of fragments in less than 3 months. A secondary library of fragments is designed using the base structures with electron density envelopes from the successful fragment hits of the primary library. Chemistry is chosen to probe interactions with the target and push the observed binding pocket limits in order to more clearly define the plasticity and range of possible extensions to the scaffolds chosen. The secondary library compounds are also screened in shape-similar groupings of five that are chosen without the knowledge of binding affinity. Our approach is a completely orthogonal one from traditional high-throughput screening in finding novel compounds.

1. Introduction

The concept of fragment-based drug discovery (FBDD) was pioneered at Abbott and first broached in the SAR by NMR paper of Fesik (Hajduk *et al.*, 1997; Shuker *et al.*, 1996) and quickly followed by extension to crystallography by Muchmore *et al.* (2000) and Nienaber *et al.* (2000). An advantage of a fragment screening approach derives from the small size of fragments allowing a much higher hit rate than a typical high-throughput screening (HTS; Schuffenhauer *et al.*, 2005). The use of small fragments, however, comes with the requirement for a more sensitive screening methodology because the binding affinities tend to be in the high micromolar to low millimolar versus the nanomolar to low micromolar for typical HTS methods. Additionally, the projections of groups that interact with the protein target are limited and therefore can be sampled efficiently by a smaller number of compounds than in HTS screens (Hann *et al.*, 2001). The relatively shallow energy gradients that can be sampled by the small fragments allow for the discovery of starting points that offer opportunity to increase the affinity and specificity along with introduction of good pharmaceutical properties. The high information content gained with the initial starting scaffold make structure an important aspect of the fragment method for evolving the starting hits while maintaining the good pharmaceutical properties that are present from the beginning.

The adaption of fragment screening with X-ray crystallography was faster than the availability of robust hardware and software solutions to meet the need for collection of large numbers of data sets. Many groups turned to prescreening the binding of fragments to targets with affinity or other biophysical measurements of binding in order to reduce the number of crystallographic experiments required (Murray and Blundell, 2010). The results of this prescreening were then used to select the fragments at some arbitrary affinity level to use in structure determination. The structures were either determined singularly or in groups to further reduce the number of experiments required.

The structural genomics effort has resulted in a rapid increase in the automation of X-ray diffraction data collection and structure determination in the past decade (Adams *et al.*, 2002, 2010; Holton and Alber, 2004; Service, 2001). At the same time, X-ray diffraction detectors have undergone rapid improvement from increased sensitivity and data-transfer speeds to the use of a shutterless detector. The improvements in both equipment and software that accompanied this structural revolution have made the use of fragment-based screening more tractable by purely X-ray crystallography.

1.1. Basic operations

We use two sources of X-rays to collect diffraction data. A Rigaku MicroMax-007HF Microfocus rotating anode with a Saturn 944 CCD area detector and an ACTOR® robot is our home source (see Fig. 13.1). This system also includes a LN40 liquid nitrogen generator to supply liquid nitrogen to the holding dewar on a continuous basis allowing operation over long weekends without the need to worry about the crystals cryogenic status.

Data collection is handled by JDirector and D*trek software (Pflugrath, 1999, 2004) using autocentering. The beam is large enough for the loop-based autocentering to be effective. The data collection strategy is determined by the space group making sure that a complete data set is collected.

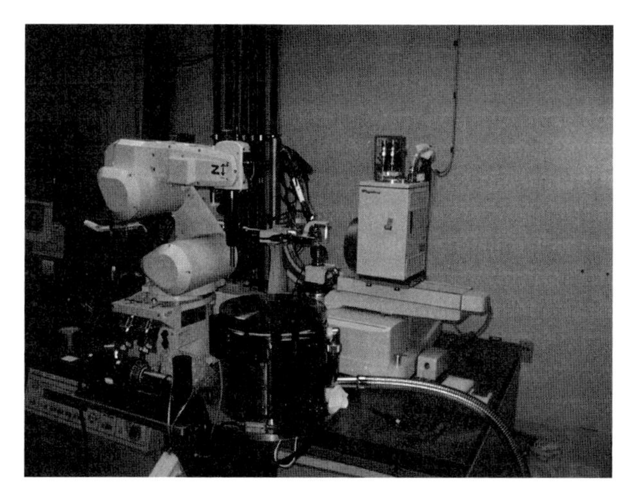

Figure 13.1 Shown above is our home setup with the Saturn detector, ACTOR® robotic system and the 60-crystal cryo dewar on top of the Rigaku 007 rotating anode, the LN40 liquid nitrogen generator is located off screen.

The typical data set requires 2 hours to collect. This system allows 24/7 data collection with a dewar capacity of 60 crystals for approximtely120 h of unattended operation. We also routinely collect data at synchrotrons. The beam lines of choice offer the Dectris Pilatus 6M pixel array detector (Marchal and Wagner, 2010). This detector allows data sets to be routinely collected in less than 5 minutes including approximately 3 minutes to mount and center the crystal. The short data collection time makes it more efficient to collect 180° of data, which guarantees a complete data set, rather than calculate a strategy to collect a complete data set in fewer degrees. We primarily collect synchrotron data at either PXI or PXII at the Swiss Light Source (PXI) or IMCA at the APS. Average times for data collection are shown in Table 13.1.

Data processing is normally a routine matter. We use a number of programs to accomplish this—HKL2000 (Otwinowski and Minor, 1997) and D*trek and XDS (Kabsch, 1993). For routine data sets, they all do a good job and the results are indistinguishable (Table 13.2). The use of APRV (Kroemer *et al.*, 2004) and XDS to process Pilatus data allows us to script the processing of the data to run automatically for the integration of the large number of data sets that can be collected in 8 hours.

Structure refinement is conducted using the PHENIX system (Adams *et al.*, 2010) and the model is fit to the electron density using Coot (Emsley and Cowtan, 2004). Most data sets can be processed with the starting apo model using one round of rigid body refinement followed by three rounds of positional refinement. Occasionally, the refinement will fail resulting in *R*-values over 40% and requiring an initial molecular replacement run to be performed in order to obtain a suitable starting model. After the initial refinement, the model is inspected and geometry, rotamer, and bad fits are

Table 13.1 Routine statistics for data collection

X-ray source	Detector	Average crystal mounting and centering time (min)	Average data set collection time (min)	Shipping crystals to synchrotron (days)	Data disk return (days)
Rigaku 007	Saturn 944	4.0 (robotic, auto)	120	0	0
IMCA	Pilatus 6M	4.0 (robotic, manual)	2.5	2	2
SLS	Pilatus 6M	2.5 (manual, manual)	2.5	2	2

The time delays involved in the transport of crystals to the synchrotron and return of the data disk must be accounted for in the general logistics of collecting data. These delays are the same regardless of whether 1 or 100 data sets are collected.

Table 13.2 Comparison of a data set processed by the different crystallographic packages

Program	Number of reflections	Number of unique reflections	*R*-factor (last shell)	I/σ (last shell)	Completeness (last shell)	Refined structure RMSD
HKL2000	84,263	17,879	5.1 (12.0)	41.0 (13.4)	98.6 (94.0)	0.163 versus D*trek
D*trek	84,718	17,886	5.7 (12.8)	19.3 (6.5)	98.6 (96.7)	0.166 versus XDS
APRV/XDS	84,568	17,882	5.1 (11.7)	29.3 (12.8)	98.6 (96.0)	0.164 versus HKL2000

RMSD, root mean square distance.

The overall statistics are close for all three programs. The ultimate comparison is the RMSD between atoms of the refined structure from each processed data set. The starting structure was the same for all refinements which were done by PHENIX. The geometry, Ramachandran plot, Rotamer plot, and final *R*-values are all comparable.

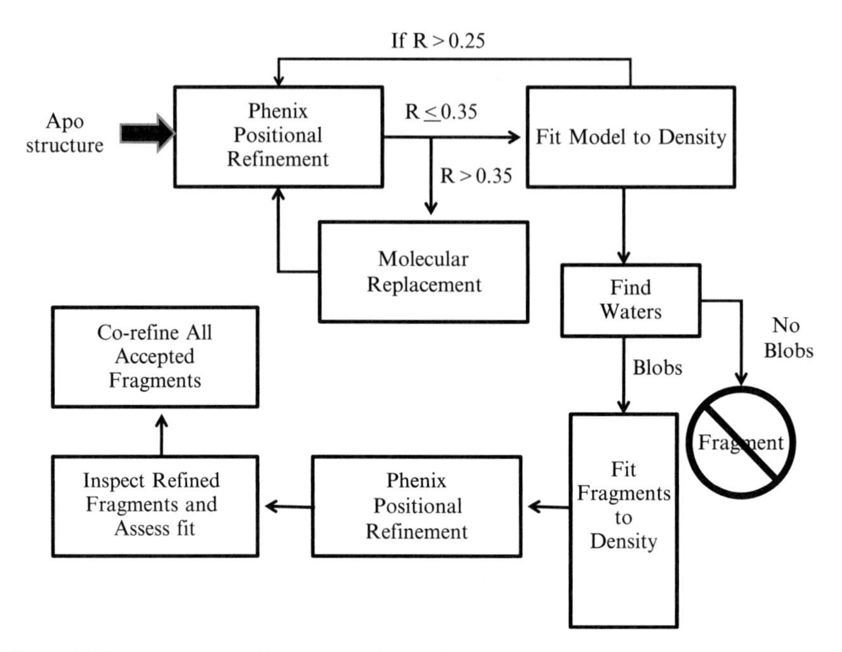

Figure 13.2 Processing scheme for refining fragment data and evaluating fragments in soaks. The individual fragments identified are refined separately. The electron-density difference maps are examined, and the final set of fragments that appear valid is then corefined and approximate occupancies determined.

corrected. The quality of the refined structure is checked with Molprobity (Chen *et al.*, 2010). Water molecules are then added to unfilled electron density using Coot in the $2F_o - F_c$ maps at a level that are above 1.8 σ and between 2.4 and 3.2 Å from the protein. Coot flags any large volumes of unfilled density at the same time as water searches are done for easy identification of potential fragments. If there are unfilled volumes (or blobs) of electron density, they are inspected and evaluated for the best fit with the potential fragments (see Fig. 13.2).

1.2. The Johnson & Johnson Pharmaceutical Research and Development Paradigm[1]

The paradigm at Johnson & Johnson has several unique aspects to it that we feel offers an advantage in both speed and information content. Of course before FBDD screening can begin there must be a robust crystal system that is ready to be soaked. The crystals must be able to be generated in large

[1] Lawrence Kuo, John Spurlino, Brett Tounge, Frank Lewandowski, and Zhihua Sui are codevelopers of the Johnson &Johnson FBDD paradigm.

numbers. Over 500 crystals can be used in the course of an FBDD screen (180–200 in the primary library screen, 100–150 in both the secondary and tertiary library screen).

The overall time to complete a FBDD program from crystal to lead averages 13 months (see Fig. 13.3). Because many of the steps can be accelerated for important programs, there is the potential to shorten this timeline. The time for the synthesis of the secondary and tertiary libraries is unpredictable, but the majority of the compounds are delivered within the allocated time in the flowchart. The design of the next library can begin before the screen has been completed if sufficient information from a large number of hits occur early on in the screening.

1.2.1. Electron density generated evolution of structure

We use the technique of electron density generated evolution of structures (EDGES) to drive the design of a drug lead from an initial fragment hit. We depend on the information from several rounds of X-ray screening and directed chemical synthesis to drive the evolution of the small fragments into lead-like compounds before introducing affinity measurement into the final selection design process for lead declaration. The electron density maps generated from the refinement of the apo or empty structure of the protein against the X-ray data collected from a crystal that was soaked with a mixture of fragments represent the average shape of the fragments that were bound to the target protein. Sometimes the resulting density is due

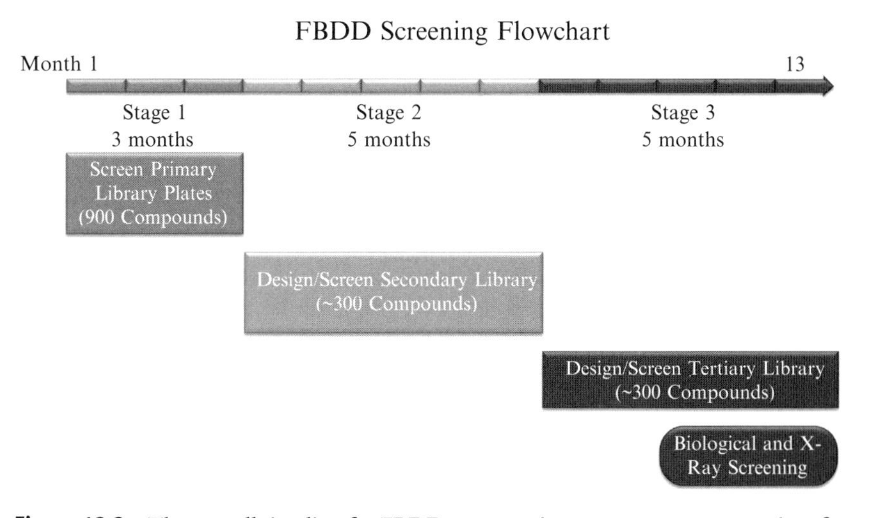

Figure 13.3 The overall timeline for FBDD representing an average response time for the progression of lead generation from the time a FBDD-ready crystal system is obtained. Stages 2 and 3 include the time for chemical synthesis of the fragment libraries.

to more than one fragment binding in the crystal. Instead of treating this as a problem to be solved by redoing the experiment with all the individual compounds that make up the soak, we prefer to use the resulting electron density maps as an envelope for possible fragment evolution. The outlines of these electron density maps present a volume that represents the binding hot spot interaction zone. We will apply computational methods to deconvolute the possible fragments and get a general idea of the makeup of the potential fragments that bind. A recent review on shape and medicinal chemistry (Nicholls *et al.*, 2010) highlights the growing importance of shape in drug discovery.

The second round of screening expands and refines the interactions discovered in the initial screen. The purpose of the second library is to push the edges of the fragments discovered in the primary screen. Additions are made to the initial fragment in order to fill space and add interactions with the binding pocket. The hit rate is expected to be higher than with the primary library. In order to sufficiently understand the limits of the binding pocket the design should push the expected boundaries of the binding pocket. We continue to group members of the library in shape-similar cohorts of five fragments. We do not employ any affinity measurements of the members of this library to drive any decisions on what to screen or in the next round of fragment library design.

The third round polishes the information obtained in the second round and adds affinity measurements to the selection criteria for final screening. The third round is essentially a structure-based drug design exercise and should result in a lead or lead-like compound with or without further chemical syntheses.[2]

1.2.2. No affinity measurements in the beginning

HTS campaigns start with an affinity cutoff and are driven entirely by a structure–activity relationship developed on affinity and sometimes these campaigns are dominated by nuisance hits with poor drug-like properties (Shoichet, 2006). Many FBDD screens also start with a prescreening that is affinity based. We have chosen to approach lead discovery from a different angle. By screening entirely based on X-ray crystallography, we do rely on information gleaned from the three-dimensional, atomic-resolution structure of the bound protein–fragment complex to drive compound development.

Three things make FBDD screening purely with X-ray crystallography possible: (1) A smaller number of fragments is required to effectively sample

[2] We have declared leads straight out of compounds in the tertiary library without added chemical synthesis efforts. We also have experience where further structure-guided chemical synthesis is needed to evolve a compound to qualify to be a lead.

chemical space. (2) There is an increased speed and automation of X-ray data collection, processing, and refinement. (3) Our goal is to augment, not replace, traditional HTS screening. With the current technology, we can efficiently screen 1000 fragments by grouping in sets of five using structure alone.

FBDD has been seen as an alternative to HTS to avoid chasing potency (Hajduk and Greer, 2007), so it is somewhat ironic that many FBDD screens start by triage with an affinity measurement. The Johnson & Johnson paradigm follows the original principle of FBDD and is only guided by three-dimensional structure at atomic resolution.

2. The Primary Library Screen

The fragment evolution starts with a primary library of fragments. The constituents of the primary library are defined as having more than 5 and less than 15 nonhydrogen atoms, a simple pharmacophore, and drug-like properties. We use an agnostic screen for the primary library and do not tailor it to the target we are screening. Because we use the same library for all targets, it is easy and rapid to deploy. Although it is possible to enrich the number of hits from the primary screen by biasing it toward privileged scaffolds for the target of interest, we feel this is counterproductive (see Chapter 7) (Kuo, 2011). By looking for fragments that are known to bind, no new or novel information will be gained.

Typical data collection times for the primary library are 15 days if done at home with a rotating anode and ACTOR® robot. If the data is collected on a Dectris shutterless pixel array detector at a synchrotron, the entire library can be collected in less than 18 h (average 5 min per data set). Since the primary library is preplated and ready to deploy, there is no lag time for synthesis or plating of target-specific libraries. The time to complete the initial screening of the primary library can be compressed to less than 10 weeks (see Fig. 13.4).

This library is screened using 180–190 crystals that have been soaked with groups of approximately five shape-similar compounds. As a general rule of thumb, data sets are not recollected due to destruction of crystals in the soak, nondiffracting crystals, or poor-quality data sets. If there is less than a 3% response, then this restriction is lifted and the missing data sets will be tried in a second round of primary library soaks. Conditions are adjusted for this round to improve chances of a successful data collection. Deconvolution of the wells is not done for any of the primary data sets; instead, an assignment of the most probable compound(s) is done based on X-ray refinement. In fact, the composite electron density map that results from the binding of multiple fragments contains valuable information.

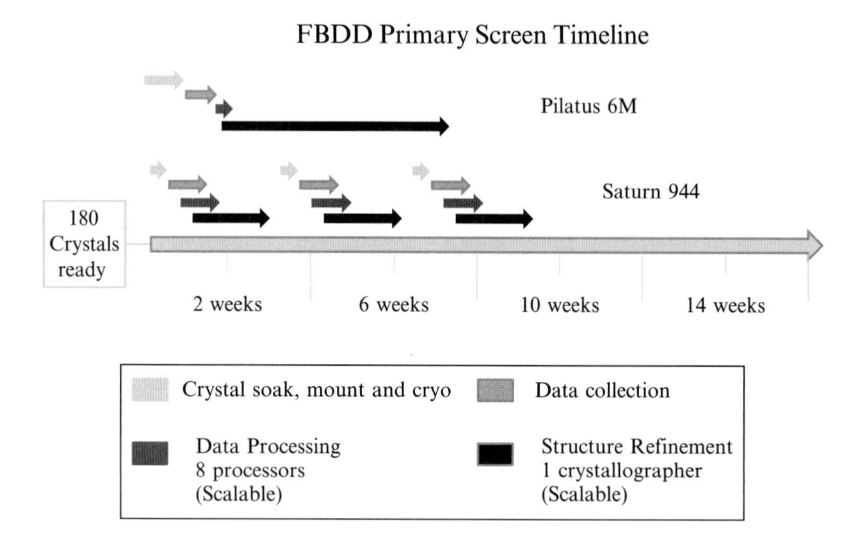

Figure 13.4 The timeline for X-ray data collection of the primary library can be compressed given sufficient resources. The major time consuming event is the refinement of the costructure, which is scalable by adding more crystallographers. The timeline for data collection at home allows for the data to be processed as soon as X-ray data collection is finished and then passed onto structure refinement. The synchrotron offers quicker data collection times with an overall time savings of about 2 weeks, but there is a penalty in that the crystals and data must be shipped so that the immediate accessibility of the data is delayed.

2.1. The primary library fragments

The first choice made for the fragment screen is the design of a library of fragments. The collection typically has an average number of about 1000 compounds for direct screening and 10,000–40,000 for prescreening modes. Many use the Astex rule of three (Congreve *et al.*, 2003) in the design and selection of their fragments. The "rule of three" states that the molecular weight (MW) $\leq$ 300 Da, the CLogP be $\leq$3, the number of H-bond donors be $\leq$3, the number of rotatable bonds be $\leq$3, and with the additional limit that the polar surface area is restricted to $\leq$60 Å^2. This rule can be seen as an extension of Lipinski's "rule of five" for good pharmacokinetic properties (Lipinski *et al.*, 2001) for fragments. Everyone starts with a large diverse selection designed to sample chemical space with small, 100–250 MW, compounds. These libraries are further filtered by chemists to remove groups and moieties that are known to be liabilities. The bias is to eliminate fragments that have bad pharmacokinetic properties or known synthetic or stability problems. The superset is then selected to obtain the best coverage of chemical space from approximately 1000 compounds. The result is a biased, diverse set of compounds.

We have designed a 900 compound library of fragments. We use the same library for all targets to look for initial hits. The design and selection of these libraries for Johnson & Johnson are covered in depth in Chapter 1 (Tounge and Parker, 2011). A further advantage of using this target agnostic primary library is an increase in the speed of deployment of an FBDD campaign, since its library is preplated and ready to soak crystals immediately.

2.2. Grouping by shape similarity

Screening of fragments in groups is a common practice used in X-ray screening to reduce the number of data sets collected and is typically done with three to eight compounds per soak (Blundell *et al.*, 2006; Davies *et al.*, 2009). Even with modern automation and equipment improvements, individually screening 1000 compounds by X-ray diffraction is not practical. Due to the logistics of soaking times, it is not possible to set up the soaks for more than 180 crystals in a day. The preparation of 1000 crystals would therefore span at least a week. The structural refinement for 1000 crystals would take 80 days by itself, and while much of it can be automated, there is a significant portion that still requires human interpretation. The grouping of fragments for soaks to reduce the number of structures needed therefore makes sense. Table 13.3 shows the times associated with screening crystals by X-ray diffraction.

There are, however, disadvantages to grouping compounds. If more than one fragment can bind, there is the possibility that one fragment will dominate and another weaker fragment will not be seen at all in the experiment. As in all screening exercises, false negatives are inevitable. The use of fragment grouping has enough advantages in the gain of speed to offset any potential loss of information.

Our primary library is subdivided into three 60-well with each well consisting of mixtures of five compounds. The grouping of compounds can be done in different ways. A random grouping with no consideration done

Table 13.3 Time course for various events in an FBDD screen

Crystals	Place crystal in soaking solution (h)	Mount crystals and freeze (h)	Data collection on a Pilatus detector (h)	Data processing (scalable) (h)	Data refinement (scalable) (h)
60	1.5	2	8	30	120
180	4.5	6	24	90	360
1000	25	34	130	500	2000

The time for data refinement is based on a 5% hit rate, where 95% of the data does not require additional refinement after the first round.

for the shape of the constituents of the group is possible, but almost never used. The groupings are done to give some advantage in the resolution of bound fragment structures. Most groupings have tended to be done to maximize the difference between the compounds making up the group in order to make identifying the bound compound from the electron density easier. The premise being that different-shaped compounds would not bind in the same site and the resulting electron density will unambiguously allow the placement of the fragment (Hartshorn *et al.*, 2005). The use of halide atoms (Cl, Br, I) in unique binding projections has also been used by different companies (SGX, now part of Lilly; Zenobia) to help in this discrimination. These atoms are also often good leaving groups allowing for the ease in creating additional extensions to the library (Antonysamy *et al.*, 2008). The halide atom must allow the extension of the fragment from the location at which they are bound to be of value in synthetic extension of the fragment. However, these halide atoms can often be bound in a closed cavity that does not allow expansion. We have included these atoms more for their chemical properties than as either synthesis handles or locator atoms.

A premise that is made in selecting different-shaped compounds is that the difference in shape should preclude more than one from binding in the same site and that the different shape will allow easy differentiation of the fragments. This premise is not always true. There are cases where the resulting electron density from a soak cannot be resolved to identify the bound fragment(s) and the information is delayed while individual crystal soaks of the fragments in the mixture are prepared, data collected, and individual structures determined. Another problem arises from the fact that when dissimilar compounds can bind at the same location the resulting density may represent only one of the fragments present in the soak.

The choice to group compounds by shape-similar criteria has a number of advantages. The similarity of the shape can reenforce the density at the binding site where the atoms of the compounds overlap when more than one compound binds at the same site. The affinities of compounds that have a similar shape are more likely to allow partial occupancy by all the groups that can bind. The discrimination between the shape-similar compounds can often be accomplished through multiple X-ray refinements using each of the potential candidates as a starting point and then comparing the resulting electron density maps. Many times the co-refinement of occupancy can even result in a rough rank ordering of the compounds present. The structural information is a very powerful and informative tool that is used in the design of subsequent libraries to tease out the full EDGES relationship. Even in the case where the exact nature of the individual compound cannot be exactly determined by experimental methods the basic group and orientation is apparent. The information provided allows the chemists to design further compounds based on the general group immediately without the need for deconvolution. Furthermore, the subtleties of the binding mode are refined in the next stage by the appropriate design of a secondary library (e.g., the position of heteroatoms in a ring or preferred substituent on a ring system).

Several examples of primary library hits are presented as examples of the various modes of hits and separation that can be encountered in a first screen where the hit rate is typically 3%, but can be as high as 25%. In all cases, the initial refinement of the data is done without any water molecules present. The resulting electron density maps and refined structure is then examined. The protein is corrected for geometry, rotamer, and positional errors. The maps are also examined for density that is too large to be a simple water molecule (blobs). If a blob of density is seen, then the experiment is marked as a possible fragment bound success and the fragments that were present in the soak are examined for possible candidates to refine. Water is also added at this point and placed at high difference peaks (1.8 σ in $2F_o - F_c$ maps and 3 σ in $F_o - F_c$ maps[3]) within hydrogen bonding distance of another molecule. If there is a potential fragment identified, then placed water molecules are edited to remove any water molecules in close proximity to the potential fragment (i.e., large contiguous blobs of density). These examples serve to demonstrate typical outcomes from soaking a mixture of compounds and the information that is gained in each.

2.3. Examples of primary library screening

As an illustration of different outcomes in a primary library screen several experiments with the fatty acid binding protein 4 (FABP4) is shown. FABP4 is a carrier protein for fatty acids that has been indicated in a number of disease areas (Furuhashi and Hotamisligil, 2008) including diabetes (Furuhashi *et al.*, 2007; Mohler *et al.*, 2009), heart disease (Rhee *et al.*, 2009), obesity (Yun *et al.*, 2009), and asthma (Shum *et al.*, 2006). The FABP4 structure is a beta-barrel with an interior cavity that holds the fatty acids. The entrance to the interior cavity is sealed by a phenylalanine residue. In the case of FABP4, the most robust crystals are produced from protein that had carried over a bound fatty acid through purification. The analysis of the purified protein solution shows that a number of different fatty acids are present in the solution (data not shown). Palmitic acid is a

[3] The σ is defined as the standard deviation of the X-ray density map. The two maps typically used are the $2F_o - F_c$ and $F_o - F_c$. The $2F_o - F_c$ map shows where the model of the fitted structure is in agreement with the collected data and is reenforced where missing atoms should be placed. A well-fit map will cover most atoms at a 1.5 σ level. The $F_o - F_c$ map shows where the model of the fitted structure does not agree with the data. The positive contour levels indicate where there should be atoms that are not in the model and the negative contour levels indicate where atoms should not be placed. A contour level of 3 σ is sufficient to show regions of concern.

major component and is consistent with the observed electron density seen in the "apo" structure (see Fig. 13.5).

The following examples represent a case where the fragment must either displace a bound molecule or cohabitate with the molecule that is already present. The high concentration (~5 m*M*) of each fragment in the soak should be able to displace the prebound ligand by mass action, since the concentration of the bound ligand is low. During the soak a crystal

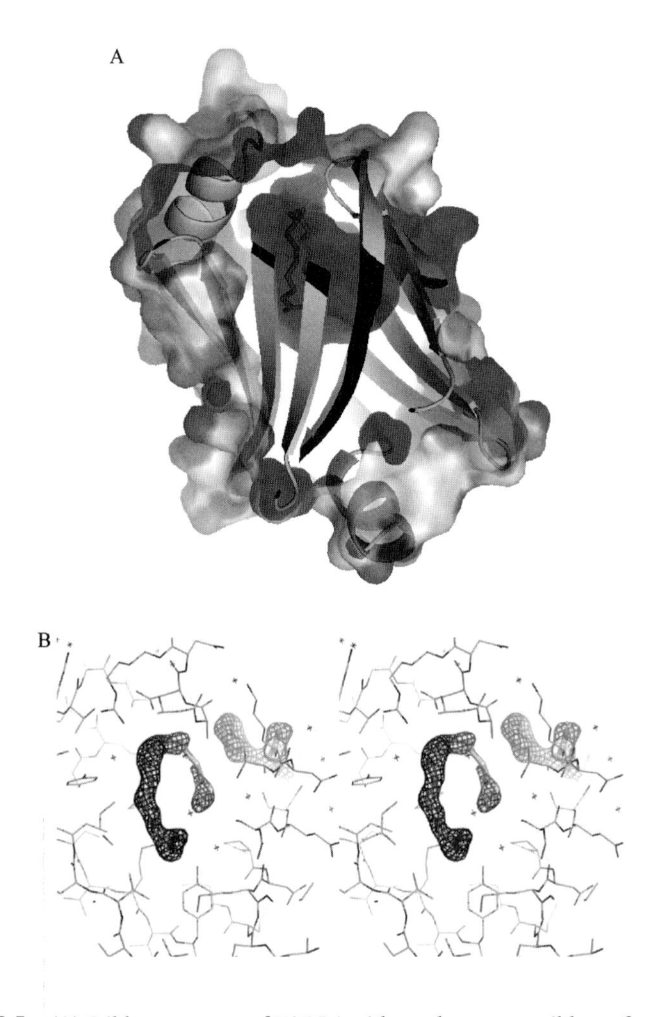

Figure 13.5 (A) Ribbon cartoon of FABP4 with a solvent accessible surface showing the bound palmitic acid in the interior cavity displayed using PyMOL® (DeLano, 2008). (B) A stereoview of the refined $2F_o - F_c$ electron density map contoured at 1.5 σ showing the palmitic acid and the Phe57 residue (right side) that forms the cap of the binding cavity.

approximately 64,000 μm^3 is placed in a 1 mL drop with the 5 mM fragments. The concentration of the bound ligand in the drop would be in the nanomolar range based on an approximately 1:1 ratio of ligand to protein and a 50% solvent content in the crystal.[4]

There are some additional factors to take into consideration for crystals with bound ligands. Factors such as high local concentration of protein in the crystal and local recapture may slow the equilibrium. The 24-h soak we employ is usually sufficient to allow the displacement of any endogenous ligand and the binding of fragments. The affinity of the fragment required to displace a bound ligand is greater than that required to substitute in an empty protein. Crystal packing forces (e.g., where protein conformation is constrained but required for binding) and the lattice of protein molecules can also interfere with the simple mass action forces. When dealing with crystals that have a ligand bound we do an evaluation for the suitability of the crystal for an FBDD campaign. An initial screen to test the FABP4 system using a known low-affinity fragment like molecule showed that fragments were indeed able to displace the bound fatty acid and the full screen was started.

The starting structure for the FBDD campaign is the fatty acid bound form of FABP4. A typical structure is shown with the bound fatty acid and the closed form of the protein with Phe57 closing the entry portal to the interior of the protein. The starting structure model used to refine against the X-ray data collected from the fragment soaks did not include either the palmitic acid or water molecules, even though this structure was refined with a bound palmitic acid in the interior cavity. The resulting $F_o - F_c$ difference maps typically showed density consistent with the palmitic acid when no substitution took place. Equally obvious were times when the fatty acid is completely expelled from the binding cavity and the fragment density is well defined. Intermediate cases required a more careful examination. Cohabitation of the binding pocket is usually straightforward with density for both the fatty acid and fragment being visible with no contiguous density between them. All cases where a fragment and the palmitic acid are both observed there is some displacement of at least a part of the palmitic acid from its binding location as seen in the "apo" structure. Many times the displacement of the palmitic acid results in the movement of the Phe 57 residue which typically closes the binding cavity in the fatty acid bound

[4] A standard crystal is 100 × 100 × 100 μm which equals 1.0×10^{-9} L. The unit cell of this protein contains four molecules and has a volume of 5.85×10^{-22} L (65.445 × 88.940 × 99.658 Å). The concentration of the bound ligand in the crystal is approximately equal to the number of molecules in the unit cell of the protein divided by Avagadro's number divided by the volume of the crystal which equals 11.4 mM [(4 molecules/6.022×10^{23} molecules/mole)/5.85×10^{-22} L]. The final concentration of the bound ligand in the soaking solution is equal to the molarity of the ligand in the crystal times the volume of the crystal divided by the total volume of the experiment (the volume of the soaking solution plus the volume of the crystal) which equals 11.4 nM [11.4 mM × 1.0×10^{-9} L/(1.0×10^{-3} L + 1.0×10^{-9} L]. So the concentration of the fragment is approximately 44,000 times that of the bound ligand.

form of FABP4. Some cases present a binding site with partial density for what could be the fragment and little or no density for the fatty acid. The decision on whether to designate the resulting experiment a substitution is made based on refinement statistics and the amount of coverage of the fragment at lower density thresholds.

The soak for the first example shows electron density in the $F_o - F_c$ map that is consistent with three of the fragments in the mixture (Fig. 13.6). The three fused ring systems (fragments *1a*, *1b*, and *1c*) are all possible fits to the electron density. The other two fragments do not fit the observed density and were ruled out as possible candidates for substitution. There is also density which shows the partially displaced fatty acid. The acid head group remains in the same position as seen in the "apo" structure; however, the tail of the fatty acid has been displaced by a bound fragment. The phenylalanine side chain that normally closes the portal for FABP4 shows no density for the position that it occupies in the refined "apo" structure and additionally shows some negative density from the $F_o - F_c$ map. The Phe57 residue does not appear to be well localized in this structure, although there are several weak areas of electron density that have reasonable geometry. The most likely is chosen as the base case for further refinements for this soak.

All three fragments are docked to the density along with a palmitic acid molecule and refined. The refined maps for two of the complexes are shown in Fig. 13.6. The refinement of fragment *1a* shows some residual density (Fig. 13.6C), but it has the best overall fit to the $2F_o - F_c$ map. The fused seven-membered ring (fragment *1b*) shows good refinement with very little extra density seen in the resulting $F_o - F_c$ map; however, the coverage in the $2F_o - F_c$ map is not as complete as for the six-membered ring. The third possibility, fragment *1c*, showed extra density in both the $2F_o - F_c$ map and $F_o - F_c$ map. When the three fragments are simultaneously refined, the resulting occupancy suggested a majority of the substitution is for fragment *1a* and the other two only contributed approximately 5% each. Three fragments from this experiment are selected to be used in the further design of the secondary library. The consensus piece of these fragments is the tetrazole ring fused to another ring. In this example, the position of all heteroatoms is overlapping in approximately the same orientation for all possible substituents, but the size of the fused ring is not. An advantage is that the different sized rings allow a different projection angle to be used to make interactions with the binding pocket surface and multiple probes for binding can be explored without the need for further deconvolution.

Another example from the FABP4 system results in the complete displacement of the bound fatty acid. In this case, the palmitic acid residue does not refine well and shows a large area of negative density that interrupts the path of the palmitic acid as seen in the "apo" structure. Two fragments are reasonable models for the observed electron density maps (see Fig. 13.7).

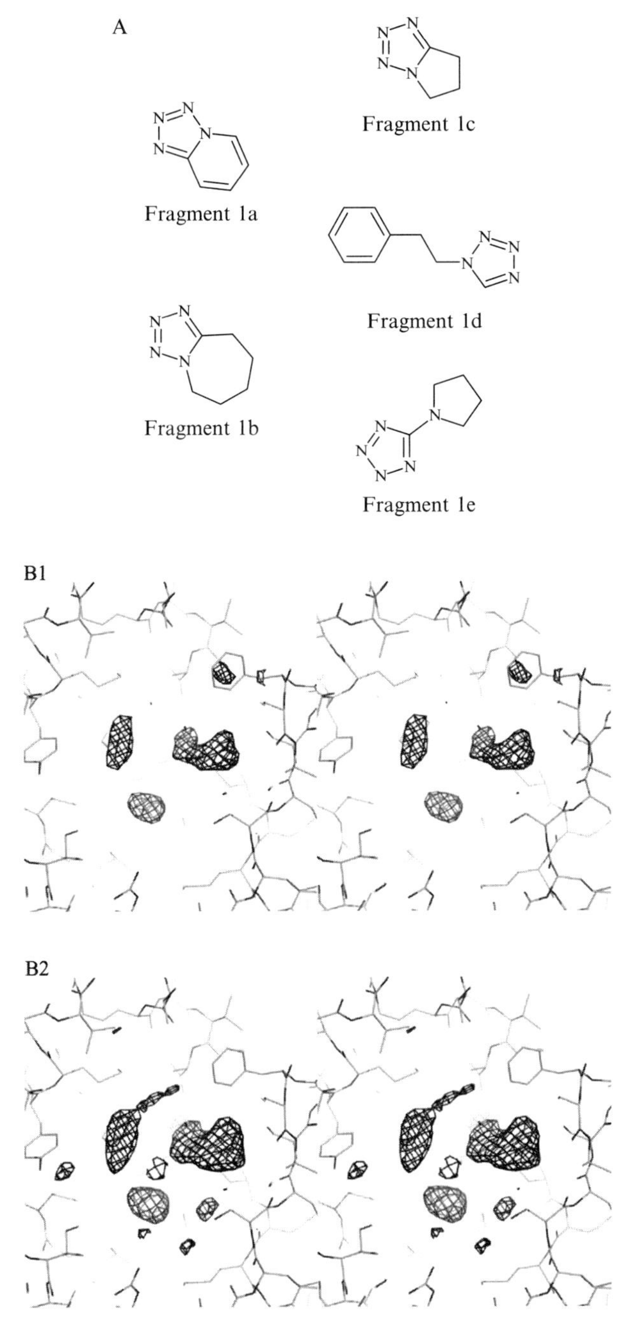

Figure 13.6 (Continued)

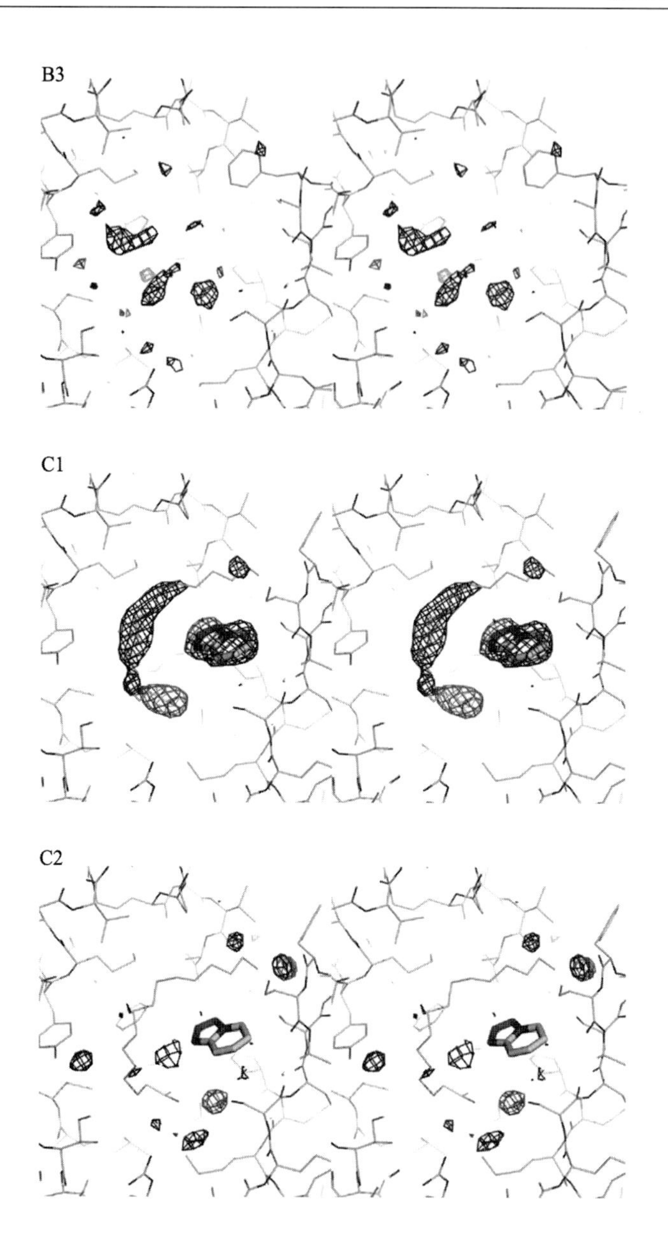
B3
C1
C2

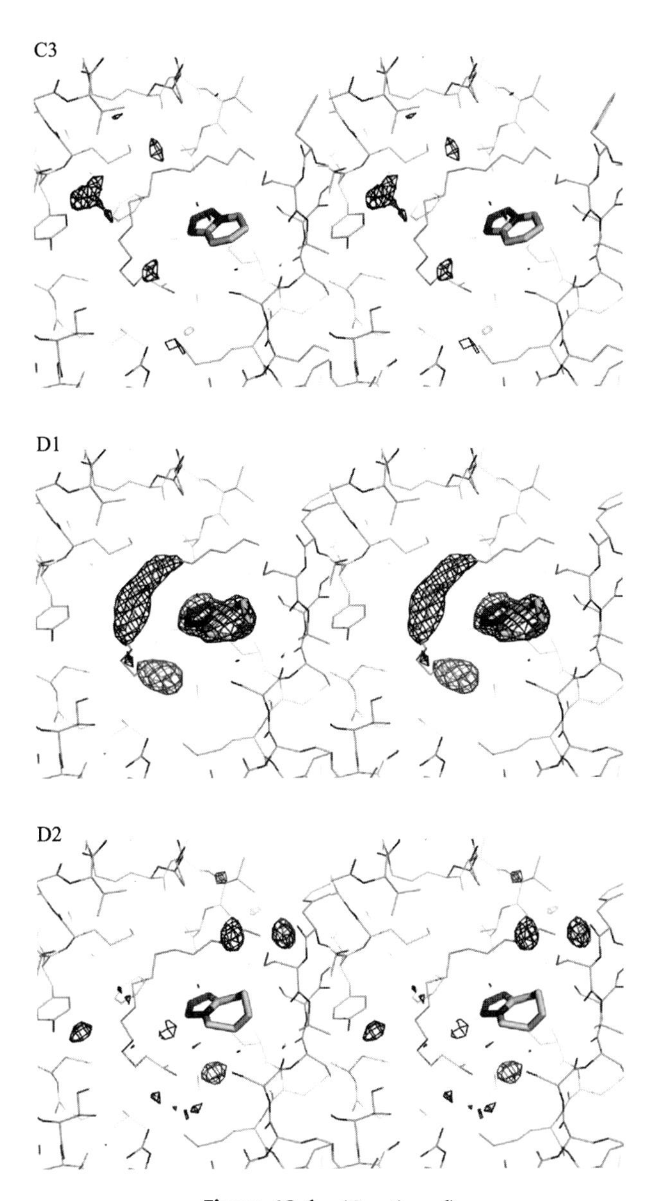

Figure 13.6 (Continued)

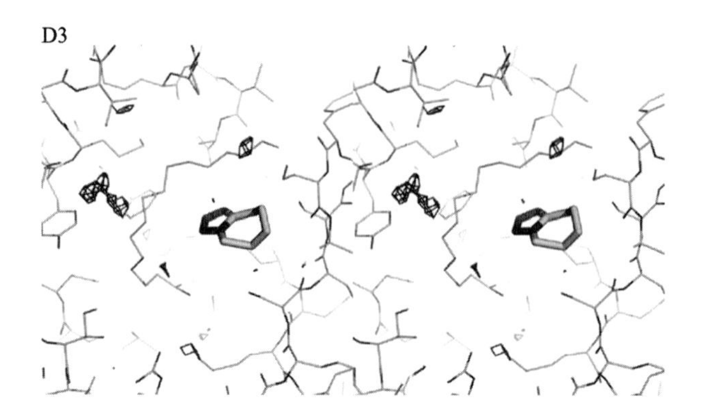

Figure 13.6 (A) The five fragments that made up the soak. (B) Stereoviews of the initial electron density maps that show extra electron density where the fragment binds. The $2F_o - F_c$ map (B1) is contoured at 1.5 σ. The stereoviews for the $F_o - F_c$ maps for 3.0 σ (B2) and -3.0 σ (B3) are shown. (C) Stereoview of the electron density maps resulting from the refinement of fragment *1a* [$2F_o - F_c$ map at 1.5 σ (C1) and $F_o - F_c$ maps for 3.0 σ (C2) and -3.0 σ (C3)] are shown. (D) Stereoviews of the electron density maps for fragment *1b* [$2F_o - F_c$ map (D1) and $F_o - -F_c$ maps for 3.0 σ (D2) and -3.0 σ (D3)] with the same contour levels are also shown.

The refinement of the two fragments indicated that the fluorine substituted fragment (fragment *2e*) is a superior choice to the chlorine substituted one (fragment *2b*). The electron difference maps are of similar shape and covered approximately the same area on the fragment, although fragment *2e* fits somewhat better due to a relative shift in fragment position in the refinement caused by the heavier chlorine atom for fragment *2b*. The fluorine atom shows *B*-factors[5] that are consistent with the rest of the fragment, whereas the chlorine atom shows elevated *B*-factors. The phenyl ring shows an interrupted electron density map even when contoured at 1.2 σ in the refinement of both fragments. To further validate fragment substitution, an additional round of refinement is done replacing the fragments with water molecules. The resulting map is shown in Fig. 13.7E. The lack of convincing density for the palmitic acid residue in this soak is positive evidence for the presence of a fragment, as failed soaks still show the presence of a palmitic acid group after initial refinement. Although the occupancy for

[5] *B*-factors or temperature factors indicate the precision of the atom positions fit to the electron density maps. Atom positions can be uncertain because of general thermal motion. The B-factor also accounts for the general disorder in the crystal from which the structure was determined. In a high-quality model, *B*-factors reflect the mobility or flexibility of various parts of the molecule. High *B*-factors mean greater uncertainty about the actual atom position. Values of 60 or greater may imply disorder (e.g., free movement of a side chain or alternative side-chain conformations). Values of 80, 20, and 5 are equivalent to uncertainties of 1.0, 0.5, and 0.25 Å, respectively.

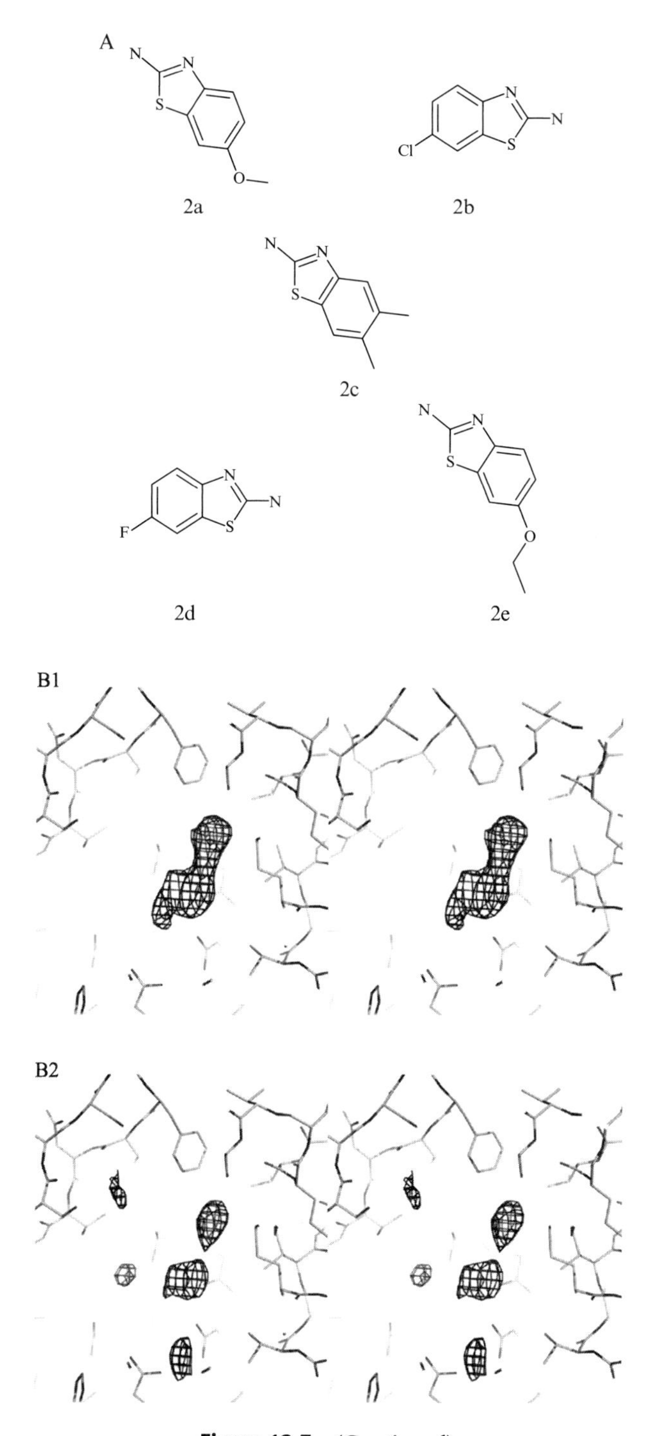

Figure 13.7 (Continued)

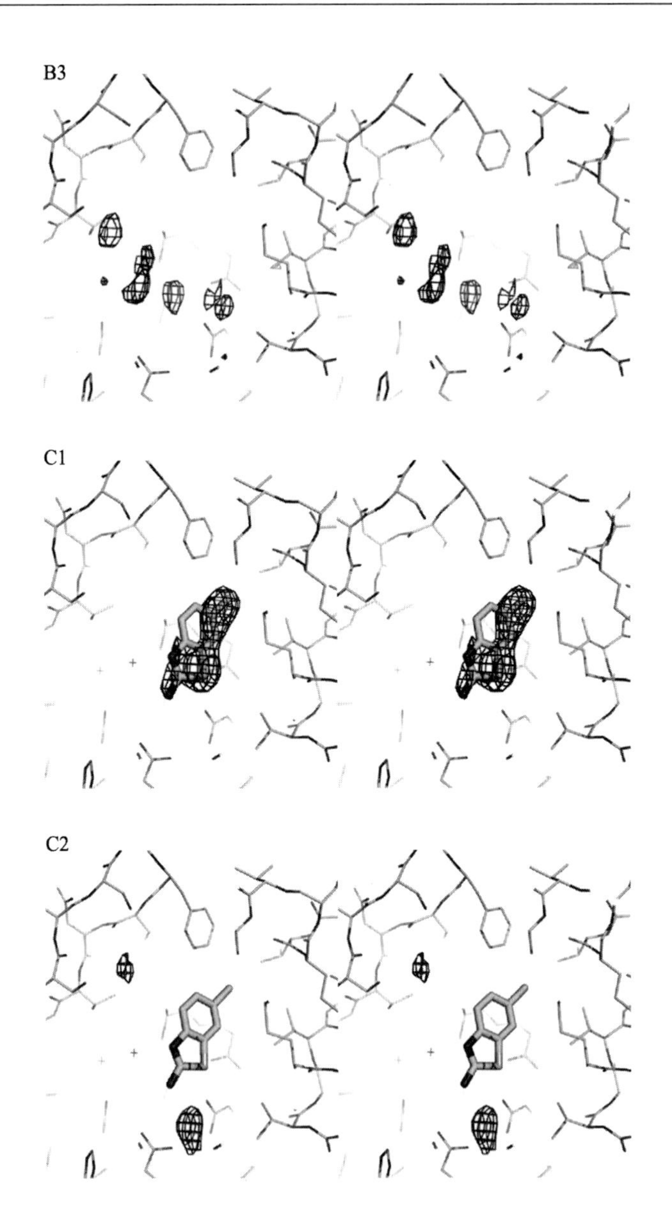
B3
C1
C2

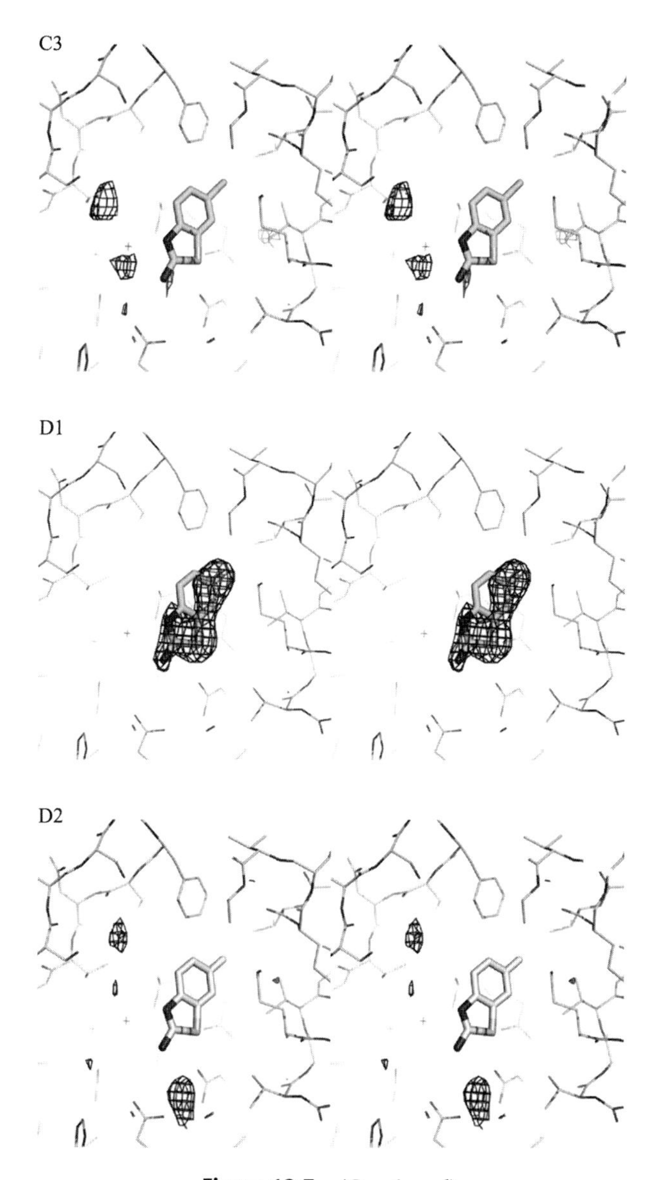

Figure 13.7 (Continued)

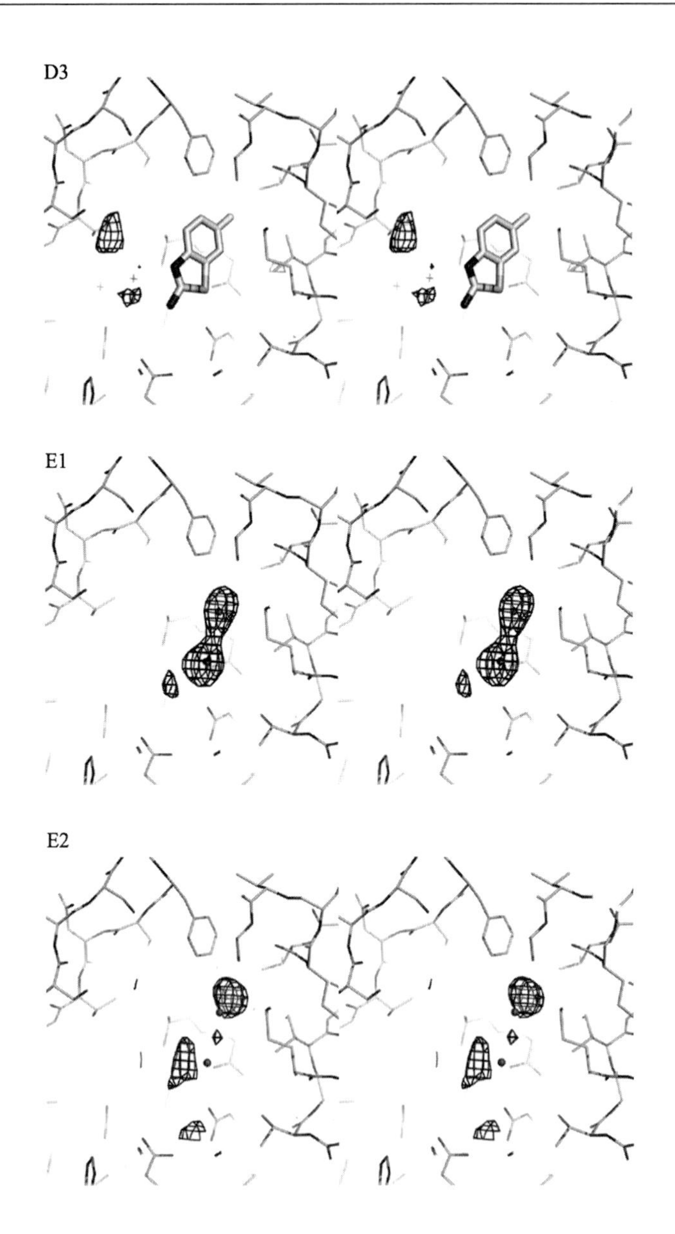
D3
E1
E2

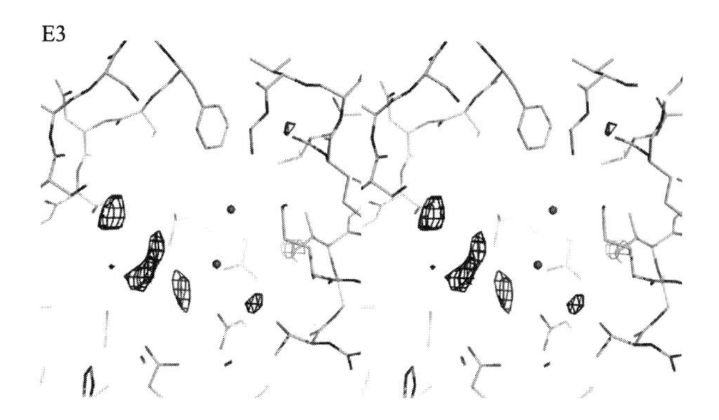

Figure 13.7 (A) The five fragments that made up the soak. (B). Stereoviews of the initial electron density maps that show extra electron density where the fragment binds. The $2F_o - F_c$ map (B1) is contoured at 1.5 σ. The stereoviews for the $F_o - F_c$ maps for 3.0 σ (B2) and -3.0 σ (B3) are shown. (C) Stereoviews of the electron density maps resulting from the refinement of fragment *2b* [$2F_o - Fc$ map at 1.5 σ (C1) and $F_o - F_c$ maps for 3.0 σ (C2) and -3.0 σ (C3)] are shown. (D) Stereoviews of the electron density maps for fragment *2d* [$2F_o - F_c$ map (D1) and $F_o - F_c$ maps for 3.0 σ (D2) and -3.0 σ (D3)]. (E) Stereoviews of the electron density maps for refinement of water at the site [$2F_o - F_c$ map (E1) and $F_o - F_c$ maps for 3.0 σ (E2) and -3.0 σ (E3)] are shown.

this fragment soak is less than complete it is convincing and the fragments are accepted into the design process for the secondary library. One feature of the two previous examples is the lack of direct hydrogen bonding with the protein from any of the fragments. The evolution of these fragments reaches out from the core binding groups to try and form hydrogen bonds with nearby residues (an arginine, aspartic acid, or glutamine).

The third example from the FABP4 screen shows only a single possible fragment that accounts for the observed electron density. This time the head group of the fatty acid has been displaced from its "apo" position, but the fatty acid has not been completely ejected from the FABP4 binding pocket. Even with the need to displace the palmitic acid and the cohabitation of the binding cavity, the electron density clearly shows the binding mode (see Fig. 13.8). The need to displace a bound compound is not ideal requiring more potent fragments than binding to an empty cavity, but these examples prove that FBDD methodology can even be applied in such a case when suitable empty crystals cannot be obtained.

The last example for primary library soaks demonstrates the possibility that multiple fragments may bind at different sites within the crystal. The protein has a more open and extended binding pocket which allows more sites for substitution. Multiple substitution is often seen in these cases and the binding can be by either the same fragment or a different fragment. With a small fragment the adjacent fragments can be joined or one fragment

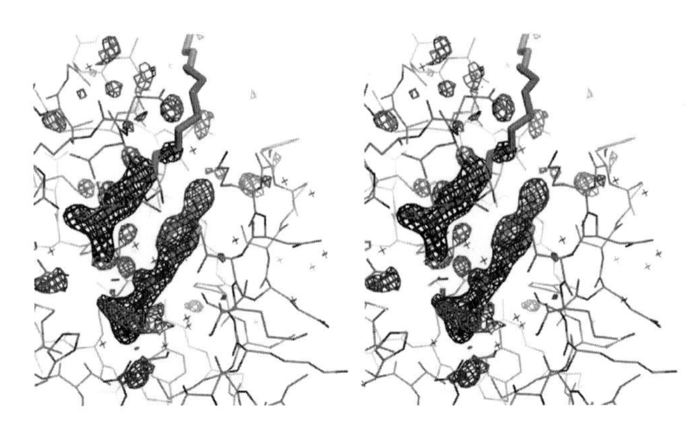

Figure 13.8 A Stereo picture of electron density maps resulting from the refinement of fragment 3c with the $2F_o$–F_c map (blue) contoured at 1.5s. The $2F_o$–F_c maps are colored green (3.0s) and red (−3.0s). This view clearly shows the displacement of the palmitic acid by the fragment. (See Color Insert.)

can be chosen as a base and substituents added to build into volume occupied by the second fragment.

Additionally, there have been cases where the binding of the second fragment is dependent on the binding of the first. In this case, the initial fragment soak shows two fragments bound near the active site and a rearrangement of a portion of the protein. While deconvolution is typically not done for primary library fragments, the nature of this event justifies further experiments. The two fragments are very similar in shape, but distinguishable from each other and the remaining fragments in the mixture in the original soak. The two fragments identified as being bound are soaked into crystals both individually and together in duplicate for a total of six separate experiments. The resulting structures reaffirm that when soaked together both are bound and the protein rearrangement occurred. The fragment that bound near the active site residue is also observed in the same location when soaked individually; however, no protein rearrangement occurs. The second fragment that bound in close proximity to the first fragment and proximal to the site of protein rearrangement does not bind when soaked individually. This dual fragment propagated that structural rearrangement is a serendipitous event and would not have been caught in single fragment structure determination from prescreening using biophysical or biochemical solution activity as the detection tool.

2.4. Primary library screen conclusions

These examples demonstrate the efficacy of using mixtures of like shaped fragments in an X-ray only screening with no deconvolution. Even though there are some possibilities of missing fragments that could bind, the depth of coverage and the expansion of the found fragments in the secondary library design should alleviate any major holes in the chemical–structural space coverage. The exclusion of a fragment by a much tighter binding fragment is not as detrimental to the process when the compounds are grouped by shape similarity. The base pharmacophore should still be present in the prefered binder and the evolution in the secondary library should recover most, if not all of the missing information.

The primary library is screened in a first-pass high-throughput mode. The first subset of 60 crystals enables us to evaluate the soaking protocol for the crystal system. The experimental process is not perfect and some crystals do not produce a good data set. There are a number of reasons for the lack of a data set. The crystal may be destroyed by the soak. The diffraction data may not be able to be processed due to multiple spots from a cracked crystal, poor resolution (worse than 3.2 Å), or other deformations of the unit cell. If the loss of crystals is too great, adjustments are made to soaking conditions or an alternative crystal form is evaluated. We do not recollect missing data sets. This strategy is chosen as an efficiency enhancing step since the fine tuning of soaking conditions to obtain every missing data set could add weeks to the process. There is typically a 3–5% dropout rate for the primary library. The coverage of chemical space from the remaining crystals is sufficient to enable the design of a secondary library. In some cases where the hit rate is high there is enough information to move forward with the secondary library after collecting only two thirds of the primary fragment library.

In general, we find that sufficient information is present to allow the design team progress three to five distinct scaffolds into designing fragments for the secondary library screen. The design team consists of a member from each of the crystallography, computational chemistry, and chemistry teams. The differing view point of each member of the team results in a better design cycle. A more detailed examination of the design of compounds for secondary and tertiary libraries is covered in Chapter 16 (Lanter *et al.*, 2011).

We find that the advantages of speed of progression into the secondary library by not deconvoluting the similar shaped fragments or recollecting failed soaks does not limit the information obtained. The chemists input into the design phase for compounds of the secondary library tests many of the ambiguities that can be present due to our methodology. For example, the position of heteroatoms in a ring system are sufficiently sampled in the secondary library compound design (and grouping) to define any ambiguities remaining from those in the primary screen. The general flow of data for the primary screen is represented by Fig. 13.4. Even with shipping times the speed of data collection at the synchrotron can allow the entire primary library of fragment to be collected in less than a week. The processing and refinement of 180 data sets, however, takes a full-time person month. The data collection rate of 15 days to collect the same data sets at home on a rotating anode with the ACTOR® robot is still in line with the ability to process and refine the them. The preferred choice is to collect crystals at home when possible and to use the synchrotron when required to meet demand or when the resolution obtained at home is not sufficient. The synchrotron also offers an advantage of several weeks if a rush project is encountered.

3. The Secondary Library Screen

3.1. Secondary library fragment design principles

Fragments in the secondary library is designed by expansion and linking of the hits found from the primary library fragments. Growing the fragment is a more prudent method as the variations in binding orientation are more easily accommodated by the protein. The expansion of selected primary library hits is based on the structural information gained in the primary library screen. Similar scaffolds from primary hits can be used to develop a fused system when they only partially overlap. Projections from common cores are used as a basis to explore further interactions with the binding site. In all cases, the structural landscape of the binding site from all bound structures is used as a ground state for the size and direction for adding atoms to the core atoms of the fragment expansion set for the secondary library. A more complete treatment of the design of fragments in secondary libraries can be found in Chapter 16. (Lanter *et al.*, 2011)

The design is meant to push the limits of the binding pocket and therefore the number of hits. As an indication of the increase in number of hits and in some cases the number of multiple occupancies Table 13.4 shows statistics for several programs.

The similarity of fragments grouped in the secondary library is greater than that seen in the more diverse primary libraries. One advantage of

Table 13.4 A comparison of the hit rates of primary and secondary libraries for several screens

Program	Primary data sets (attempted)	Primary hits (fragments)	Secondary data sets (attempted)	Secondary hits (fragments)
Synthetase	99 (120)	37 (49)	49 (49)	12 (15)
Kinase	94 (180)	45 (57)	41 (52)	19 (25)
Peptidase	244 (309)	14 (24)	52 (56)	15 (23)
ATPase	135 (180)	7 (13)	62 (71)	32 (53)

There are five compounds in the soaking mixture for each experiment. The number of data sets represents a completed structure determination. Hits represent a successful fragment soak. A hit can contain more than one fragment.

grouping by similarity is particularly evident at this stage in that a rough structure activity relationship can often be assigned based on the co-refinement of grouped fragments. Another advantage comes in terms of speed. Fragments are designed from primary hits and tend to have a larger number of similar cores. The synthesis of these fragments arrive in groups that come from the same synthetic pathway and are easily grouped into shape-similar sets. The early grouping of fragments allows the screening of the secondary library to begin before the entire library is made, effectively shortening the time for progression into tertiary library design.

3.2. Example of a secondary library screen

The following subset of fragments are designed to probe the pocket present meta to the phenol group found in the primary library screen. Four of the five fragments soaked in this example can be fit into the resulting density (see Figs. 13.9 and 13.10). The initial refinements suggest that all four might be partially present. Compound *4a* is the only one that cannot fit the density. The inspection of the resulting electron density maps from the individual refinements indicates that no one fragment can account for all the binding. The density maps from the refinement of fragment *4b* (Fig. 13.11) show some positive density around the Cl atom suggestive of a tetrahedral substitution geometry (either the CF_3 or the tertbutyl group). There is also some negative density near the methyl substituent of the morpholine group, this compound is the only one with such a substitution. The refinement of fragment *C* like the fragment *4b* refinement shows density around the flourine suggestive of either fragment *4d* or *4e* (Fig. 13.12). There is also a pattern of negative and positive density around the phenol group, this pattern suggests that the ring is shifted in refinement to normalize the lack of electrons provided by the flourine substituent. The refinement of fragment *4d* is the best single fragment fit; however, there is some negative density around the tertbutyl group and a small amount of positive density where the methyl substituent from fragment *4b* would be (see Fig. 13.13). The final fragment refined shows a large area of negative

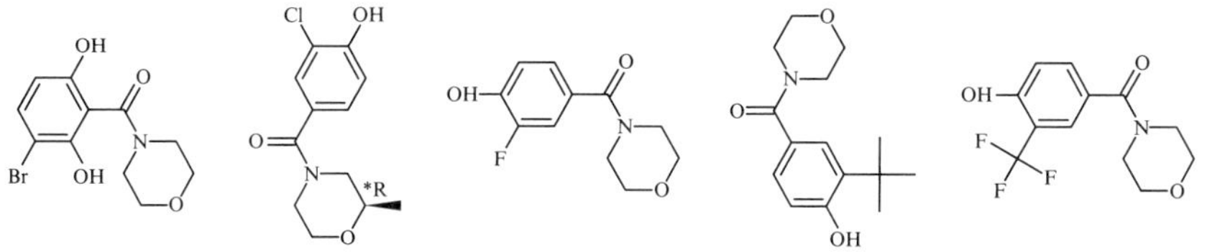

4a 4b 4c 4d 4e

Figure 13.9 The five compounds that make up the soak.

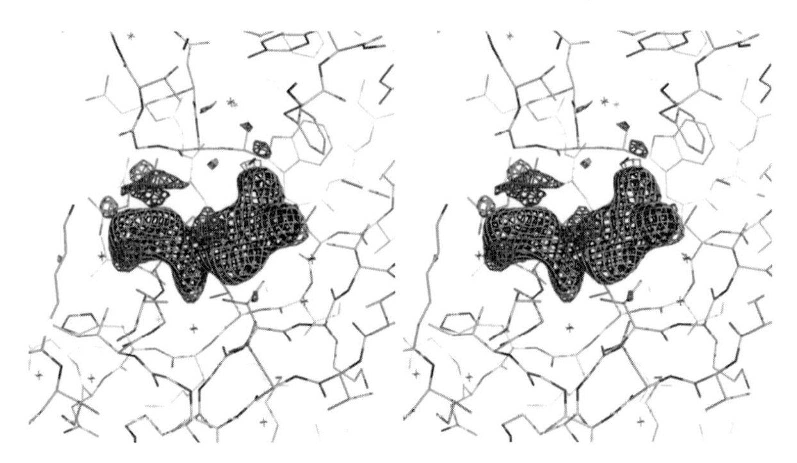

Figure 13.10 Stereoviews of the initial electron density maps showing the region for the fragment binding. The $2F_o - F_c$ map (A) is contoured at 1.5 σ. The stereoviews for the $F_o - F_c$ maps for 3.0 σ (B) and -3.0 σ (C) are shown.

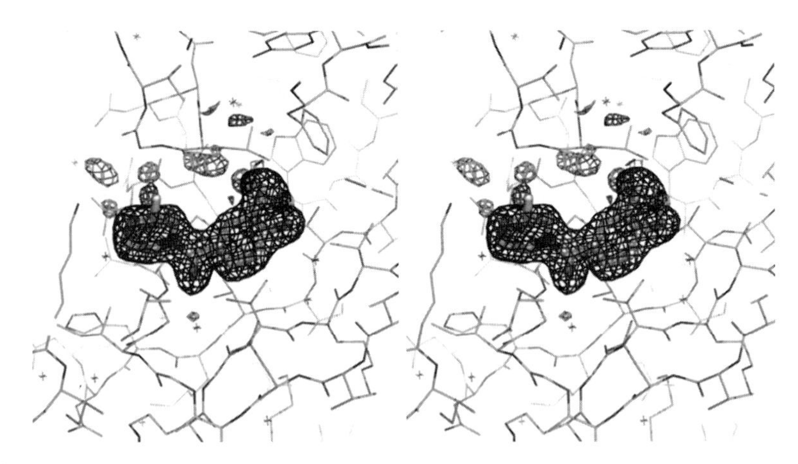

Figure 13.11 Stereoviews of the refined electron density maps for fragment *4b* are shown. The $2F_o - F_c$ map (A) is contoured at 1.5 σ. The stereoviews for the $F_o - F_c$ maps for 3.0 σ (B) and -3.0 σ (C) are shown. There is some extra density around the chlorine atom (B) that is not accounted for by this fragment. The methyl group on the morpholino also protrudes from the density in the $2F_o - F_c$ map (A) and is in close proximity to some negative density (C).

density in the $F_o - F_c$ map around the flourine atoms (see Fig. 13.14) suggesting only partial occupancy or the tertbutyl group as a preferential fragment. All four fragments were then prepared for co-refinement and the resulting density (Fig. 13.15) shows a good fit. The resulting percent

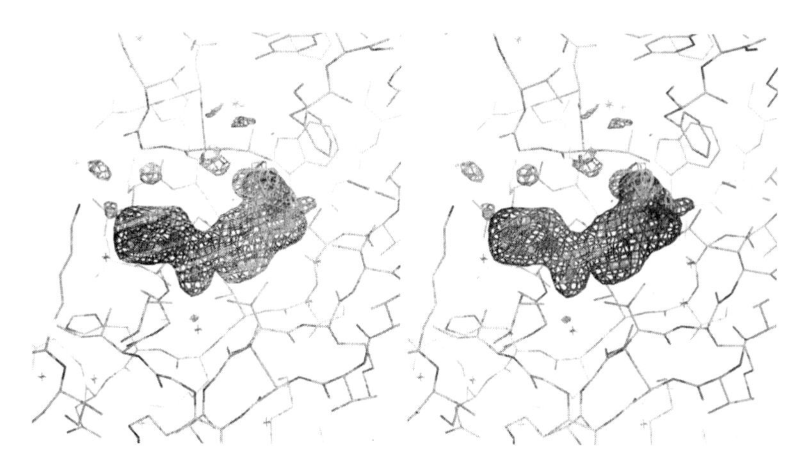

Figure 13.12 Stereoviews of the refined electron density maps for fragment *4c* are shown. The $2F_o - F_c$ map (12A) is contoured at 1.5 σ. The stereoviews for the $F_o - F_c$ maps for 3.0 σ (12B) and -3.0 σ (12C) are shown. The positive difference density that appears around the fluorine atom suggests that fragment *4d* or *4e* would be a better fit.

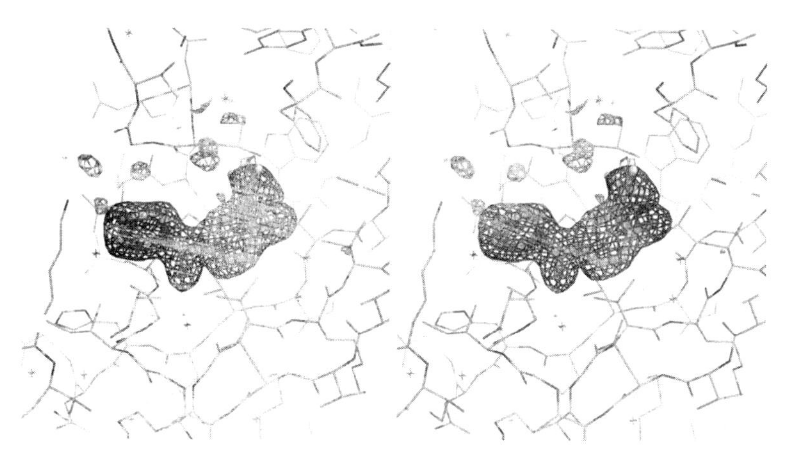

Figure 13.13 Stereoviews of the refined electron density maps for fragment *4d* are shown. The $2F_o - F_c$ map (A) is contoured at 1.5 σ. The stereoviews for the $F_o - F_c$ maps for 3.0 σ (B) and -3.0 σ (C) are shown.

occupancy of the four fragments was as follows: fragment *4b* 26%, fragment *4c* 24%, fragment *4d* 22%, and fragment *4e* 30%. The simultaneous refinement of fragments works best when there are only two fragments possible, but even with four fragments, it does give a rough estimate and an indication of possibilities for refinement in the tertiary library for exploring the projection into the pocket.

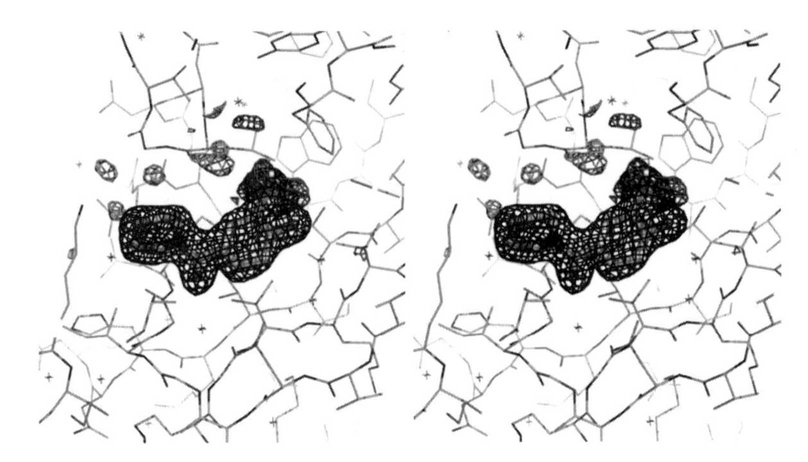

Figure 13.14 Stereoviews of the refined electron density maps for fragment *4e* are shown. The $2F_o - F_c$ map (A) is contoured at 1.5 σ. The stereoviews for the $F_o - F_c$ maps for 3.0 σ (B) and -3.0 σ (C) are shown. The negative density around the trifluoro atoms suggests that they only partially occupy the site or that the *t*-butyl substituent would be better. (See Color Insert.)

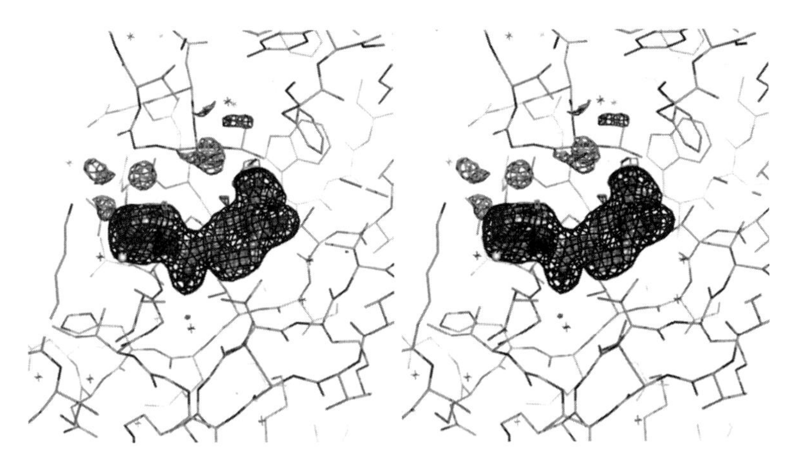

Figure 13.15 Stereoviews of the refined electron density maps for co-refinement of all four fragments are shown. The $2F_o - F_c$ map (A) is contoured at 1.5 σ. The stereoviews for the $F_o - F_c$ maps for 3.0 σ (B) and -3.0 σ (C) are shown. (See Color Insert.)

3.3. Secondary library screen conclusions

The use of shape-similar grouping has a more profound effect on secondary libraries than on the primary library where there is more diversity. Since fragments in secondary libraries are derived from a limited set of cores their diversity is concentrated on the projections from the base scaffold found from fragments in the primary screen. The groupings therefore have more similarity and tend to result in more cases where multiple fragments can co-occupy the binding within the crystal matrix. A majority of the cases are two fragments that potentially occupy the same area. The co-refinement of these fragments gives a good relative assessment of the relative affinity. The occupancy will approach zero when the fragment is not really present and is a good indication for a bad choice. When there are three or four possible fragments the relative values of the occupancy are not as reliable; however, a fragment that is not present will still be flagged.

4. Why Screening Purely with X-Ray Structures Works

Many of the procedures and processes employed at Johnson & Johnson are used to shorten the time and increase the efficiency of producing a lead-like compound from a fragment-based screen in 12–18 months. Our goal is to produce one to three lead-like compounds with different scaffolds. We see the use of X-ray crystallography-based FBDD as a complementary method to HTS screening and other biophysical screening methodologies and not as a replacement for them.

One strategy to shorten the time needed to complete an X-ray structure-based screen is that we do not recollect missed data sets from soaks. Our initial testing of the crystal system prior to embarking on an FBDD screen confirms the ability to reproduce the crystals and the ability to soak fragment like compounds into them. We expect for a number of crystals to not produce usable data sets for a variety of reasons. The crystal may not survive the soak. The crystal may not diffract to a sufficient resolution. Either of these conditions may occur because the fragments affect the crystal packing. Some crystals are lost in the mounting, cryoprotecting, and freezing. The percentage of crystals that are lost from the primary library soaks varies from 15% to 50% (see Table 13.4). The number of different fragments that are found, however, is sufficient to advance five to eight different chemical synthetic schemes to prepare for fragments in the secondary library. The secondary libraries typically offer better crystal survival rates (0–25% crystal loss).

Additional time is saved by not physically deconvoluting the soaks for wells that have a fragment hit. The hit rate varies from 3% to 30% for wells with a bound fragment versus soaks attempted for the primary library. The well hit rate translates to 1–8% hit rate for the fragments themselves (see Table 13.4). This hit rate increases if only the wells that produced a usable X-ray diffraction data set are used to calculate the percentage of hits. If physical deconvolution were done then an additional 35–225 crystal soaks and data set collections would have to be done. In some cases that would result in a greater number of crystals needed than for the primary screen itself, which would translate to a doubling of the time required to obtain an answer. The use of a computational deconvolution is much faster and when coupled to grouping by shape similarity offers enough information to proceed to the design of the next fragment library.

One advantage of screening solely with X-ray diffraction and not prescreening with affinity is that at the end of the screen the structural data is immediately available for use in the design of the next library. There is additional time saved by not prescreening; however, prescreening will typically reduce the number of fragments to soak for structure determination thereby saving some time. There are additional logistical considerations for prescreening including the need to plate and configure the soaks for the primary library fragments that are found to bind, the decision to group, or determine the structures individually that add time to any prescreening method. Obviously, as for any screening methods our approach is not suitable for all protein targets; hardy and enduring protein single crystals being a key determinant of success.

The overlap between crystallographic hits, biophysical screening hits, and solution-based activity hits is a variable.[6] In other words, all methods uncover hits not found by the other. So while there is no single best method for pursuing FBDD and all of the commonly used methods have some advantages over others, we believe that our use of purely X-ray protein crystallography to guide hit-to-lead progression offers advantages in providing an orthogonal approach to augment other screening approaches.

ACKNOWLEDGMENTS

The author is grateful to Richard Alexander, Alan Gibbs, Marta Abad, Carsten Schubert, Frank Lewandowski, Cindy Milligan, Brett Tounge, Zhihua Sui, and Lawrence Kuo for their collaboration.

REFERENCES

Adams, P. D., *et al.* (2002). PHENIX: Building new software for automated crystallographic structure determination. *Acta Crystallogr. D* **58,** 1948–1954.

Adams, P. D., *et al.* (2010). PHENIX: A comprehensive Python-based system for macromolecular structure solution. *Acta Crystallogr. D* **66,** 213–221.

Antonysamy, S. S., *et al.* (2008). Fragment-based discovery of hepatitis C virus NS5b RNA polymerase inhibitors. *Bioorg. Med. Chem. Lett.* **18,** 2990–2995.

Blundell, T. L., *et al.* (2006). Structural biology and bioinformatics in drug design: Opportunities and challenges for target identification and lead discovery. *Philos. Trans. R. Soc. B Biol. Sci.* **361,** 413–423.

[6] Unpublished in-house observations.

Chen, V. B., *et al.* (2010). MolProbity: All-atom structure validation for macromolecular crystallography. *Acta Crystallogr. D* **66,** 12–21.

Congreve, M., *et al.* (2003). A [']Rule of Three' for fragment-based lead discovery? *Drug Discov. Today* **8,** 876–877.

Davies, D. R., *et al.* (2009). Discovery of leukotriene A4 hydrolase Inhibitors using metabolomics biased fragment crystallography. *J. Med. Chem.* **52,** 4694–4715.

DeLano, W. L. (2008). The PyMOL Molecular Graphics System. DeLano Scientific LLC, Palo Alto, CA, USA.

Emsley, P., and Cowtan, K. (2004). Coot: Model-building tools for molecular graphics. *Acta Crystallogr. D* **60,** 2126–2132.

Furuhashi, M., and Hotamisligil, G. S. (2008). Fatty acid-binding proteins: Role in metabolic diseases and potential as drug targets. *Nat. Rev. Drug Discov.* **7,** 489–503.

Furuhashi, M., *et al.* (2007). Treatment of diabetes and atherosclerosis by inhibiting fatty-acid-binding protein aP2. *Nature* **447,** 959–965.

Hajduk, P. J., and Greer, J. (2007). A decade of fragment-based drug design: Strategic advances and lessons learned. *Nat. Rev. Drug Discov.* **6,** 211–219.

Hajduk, P. J., *et al.* (1997). Discovery of potent nonpeptide inhibitors of stromelysin using SAR by NMR. *J. Am. Chem. Soc.* **119,** 5818–5827.

Hann, M. M., *et al.* (2001). Molecular complexity and its impact on the probability of finding leads for drug discovery. *J. Chem. Inf. Comput. Sci.* **41,** 856–864.

Hartshorn, M. J., *et al.* (2005). Fragment-based lead discovery using X-ray crystallography. *J. Med. Chem.* **48,** 403–413.

Holton, J., and Alber, T. (2004). Automated protein crystal structure determination using elves. *Proc. Natl. Acad. Sci. USA* **101,** 1537–1542.

IMCA. https://www.imca.aps.anl.gov/tiki-index.php?page=IMCA. Accessed on 12/11/10.

Kabsch, W. (1993). Automatic processing of rotation diffraction data from crystals of initially unknown symmetry and cell constants. *J. Appl. Crystallogr.* **26,** 795–800.

Kroemer, M., *et al.* (2004). APRV—A program for automated data processing, refinement and visualization. *Acta Crystallogr. D* **60,** 1679–1682.

Kuo, L. C. (2011). How to Avoid Rediscovering the Known. *Methods Enzymol.* **493,** 159–168.

Lanter, J., Zhang, X., and Sui, Z. (2011). Medicinal Chemistry-Inspired Fragment-Based Drug Discovery. *Methods Enzymol.* **493,** 423–448.

Lipinski, C. A., *et al.* (2001). Experimental and computational approaches to estimate solubility and permeability in drug discovery and development settings. *Adv. Drug Deliv. Rev.* **46,** 3–26.

Marchal, J., and Wagner, A. (2010). Performance of PILATUS detector technology for long-wavelength macromolecular crystallography. *Nucl. Instrum. Methods Phys. Res. A* doi:10.1016/j.nima.2010.06.142.

Mohler, M. L., *et al.* (2009). Recent and emerging anti-diabetes targets. *Med. Res. Rev.* **29,** 125–195.

Muchmore, S. W., *et al.* (2000). Automated crystal mounting and data collection for protein crystallography. *Structure* **8,** R243–R246.

Murray, C. W., and Blundell, T. L. (2010). Structural biology in fragment-based drug design. *Curr. Opin. Struct. Biol.* **20,** 497–507.

Nicholls, A., *et al.* (2010). Molecular shape and medicinal chemistry: A perspective. *J. Med. Chem.* **53,** 3862–3886.

Nienaber, V. L., *et al.* (2000). Discovering novel ligands for macromolecules using X-ray crystallographic screening. *Nat. Biotechnol.* **18,** 1105–1108.

Otwinowski, Z., and Minor, W. (1997). Processing of X-ray diffraction data collected in oscillation mode. *In* "Methods in Enzymology," (C. W. Carter, Jr. and R. M. Sweet, eds.), Vol. 276, pp. 307–326. Academic Press, San Diego, CA.

Pflugrath, J. W. (1999). The finer things in X-ray diffraction data collection. *Acta Crystallogr. D Biol. Crystallogr.* **55,** 1718–1725.

Pflugrath, J. W. (2004). Macromolecular cryocrystallography—Methods for cooling and mounting protein crystals at cryogenic temperatures. *Methods* **34,** 415–423.

PXI. http://sls.web.psi.ch/view.php/beamlines/px/index.html. Accessed on 12/11/10.

Rhee, E. J., *et al.* (2009). The association of serum adipocyte fatty acid-binding protein with coronary artery disease in Korean adults. *Eur. J. Endocrinol.* **160,** 165–172.

Schuffenhauer, A., *et al.* (2005). Library design for fragment based screening. *Curr. Top. Med. Chem.* **5,** 751–762.

Service, R. F. (2001). Robots enter the race to analyze proteins. *Science* **292,** 187–188.

SGX. http://www.sgxpharma.com/. now part of Lilly.

Shoichet, B. K. (2006). Screening in a spirit haunted world. *Drug Discov. Today* **11,** 607–615.

Shuker, S. B., *et al.* (1996). Discovering high-affinity ligands for proteins: SAR by NMR. *Science* **274,** 1531–1534.

Shum, B. O. V., *et al.* (2006). The adipocyte fatty acid-binding protein aP2 is required in allergic airway inflammation. *J. Clin. Invest.* **116,** 2183–2192.

Tounge, B. A. and Parker, M. H. (2011). Designing a Diverse High-Quality Library for Crystallography-Based FBDD Screening. *Methods Enzymol.* **493,** 3–20.

Yun, K. E., *et al.* (2009). Association between adipocyte fatty acid-binding protein levels and childhood obesity in Korean children. *Metab. Clin. Exp.* **58,** 798–802.

Zenobia. http://www.zenobiatherapeutics.com/home_page.htm. Accessed on 12/11/10.

CHAPTER FOURTEEN

Computational Approach to *De Novo* Discovery of Fragment Binding for Novel Protein States

Zenon D. Konteatis,* Anthony E. Klon,* Jinming Zou,* *and* Siavash Meshkat†

Contents

* Department of Design, Ansaris, Four Valley Square, Blue Bell, Pennsylvania, USA
† Discovery Technologies Department, Ansaris, Four Valley Square, Blue Bell, Pennsylvania, USA

Methods in Enzymology, Volume 493
ISSN 0076-6879, DOI: 10.1016/B978-0-12-381274-2.00014-5

Abstract

In silico fragment-based drug discovery has become an integral component of the new fragment-based approach that has evolved over the past decade. Protein structure of high quality is essential in carrying out computational designs, and protein flexibility has been shown to impact prospective designs or docking experiments. Here we introduce methodology to calculate protein normal modes and protein molecular dynamics in torsion space which enable the development of multiple protein states to address the natural flexibility of proteins. We also present two fragment-based sampling methods, grand canonical Monte Carlo and systematic sampling, which are used to study protein–fragment interactions by generating fragment ensembles and we discuss the process by which these ensembles are linked to design ligands.

1. Introduction

Fragment-based drug discovery (FBDD) approaches have joined high-throughput screening as widely accepted methods for identifying hit and lead compounds for drug development projects (Hajduk and Greer, 2007). *In silico* FBDD has emerged as a new contributor to the quest for more diverse, novel, and efficient molecules that can become oral drugs (Chen and Shoichet, 2009; Pellecchia, 2009). A previous chapter in this issue (DesJarlais, Chapter 6) covers various computational techniques used in FBDD, so this chapter will focus on two types of technologies and approaches recently developed to study protein flexibility and to calculate fragment binding free energies at Ansaris.

A protein model of sufficient quality is necessary for computational fragment simulations to be successfully used in drug design. Such a model usually becomes available through X-ray crystallography, NMR studies, or by homology modeling. For better small molecule design results, accounting for protein flexibility is usually required (Beier and Zacharias, 2010). Here we discuss simulation algorithms in torsion space that enable protein modeling approaches including molecular dynamics, normal modes, and modal Monte Carlo simulations. These tools are used to produce unique or consensus protein states that can account for established or even unknown protein–ligand binding interactions. When applied to protein–protein interaction problems, these methods parallel the natural protein motions and may uncover new pockets on the protein surface that static crystallographic poses do not capture.

The fragment simulation technologies are comprised of the grand canonical Monte Carlo and the systematic sampling methods which explore fragment–protein interactions and provide estimates of free energy of binding of each fragment with the target protein. These methods are used to identify binding sites on the protein and to characterize them for druggability. Fragment interactions in several regions of the protein can point to pockets near previously established binding sites or to allosteric sites that can provide new molecule designs. By joining adjacent fragments in the binding site we computationally assemble whole molecules, easily avoiding the linking problem faced in other FBDD approaches.

2. Protein Modeling

Accurate protein structures are crucial for structure-based drug discovery and in particular for fragment-based drug design. Experimental methods such as X-ray crystallography and nuclear magnetic resonance are widely used to obtain the atomic positions. These methods have some limitations. X-ray crystallography can only be applied to proteins that crystallize well. NMR can currently be applied to smaller protein molecules. As a result of these limitations, computational protein modeling is often carried out starting with known initial experimental structures. Protein structure computations often employ one of the following methods:

1. Energy minimization
2. Molecular dynamics
3. Normal modes analysis
4. Monte Carlo simulation

Energy minimization is the most basic method for dealing with molecules. Most methods are local. They start with a conformation of a molecule from an X-ray position and find the nearest local energy minimum, that is, a "nearby" conformation that has lower energy and whose change in energy gradient falls below a certain value. Energy minimization employs techniques of mathematical optimization. The most widely used mathematical tools are the conjugate gradient and steepest descent methods (Flannery *et al.*, 1988).

2.1. Protein molecular dynamics simulations in torsion space

Molecular dynamics is the simulation of evolution of atomic positions in time, being acted on by the relevant interatomic forces. When using an empirical force field, the equations of motion are essentially Newton's second law:

$$Mq = F \tag{14.1}$$

Here, q is a vector of the generalized coordinates (i.e., degrees of freedom), M is the mass matrix of the system, and F is the vector of forces acting on the generalized coordinates. Conservative forces are computed as the gradient of the energy of the system. Molecular dynamics solves these equations by time integration. This exercise amounts to computing the acting forces and mass distribution at each time step and then solving the acceleration and velocities of the moving bodies. The degrees of freedom in the equations of motion are typically the position of all atoms involved in the simulation. For each molecule, these degrees of freedom can be alternatively formulated as internal coordinates, plus the rigid transformation of the molecule. The rigid transformation includes rotation and translation components. The internal coordinates include bond stretching, angle bending, and torsional degrees of freedom. Given the fact that bond stretching and angle bending do not significantly contribute to the overall motion of a molecule, the motion can be captured by torsional degrees of freedom and a rigid transformation. For large molecules, like proteins, one can gain a major advantage in computational efficiency by considering only the torsional degrees of freedom.

Molecular dynamics has many applications in protein modeling. Primarily it is used to generate alternative states of a protein–ligand complex. Near a local minimum, molecular dynamics can explore states near equilibrium. However, given a sufficiently high prescribed temperature, molecular dynamics can overcome energy barriers that are proportional to that temperature. Hence, this can be used to overcome energy barriers and sample other low-energy states of the system. For a protein–ligand complex molecular dynamics can be used to study the strength of interactions between the two molecules. It can also be used as a validation tool for the protein structure by assessing the protein molecule's stability over the simulation time.

2.2. Protein normal mode analysis in torsion space

Normal mode analysis is another powerful tool to study the intrinsic motions of molecules. It captures the coordinated motion of the atoms in terms of vibration modes of the structure at various frequencies. At any local energy minimum, the actual motion of a molecule can be computed by a weighted sum of the normal mode displacements. To obtain normal modes, we start with the undamped, second-order ordinary differential equation of motion:

$$Mq + Kq = 0 \tag{14.2}$$

Here, q is a vector of the generalized coordinates (i.e., degrees of freedom), M is the mass matrix of the system (captures the mass distribution of the system), and K is the stiffness matrix (describes the flexibilities of the system). This ordinary differential equation can be solved as a generalized eigenvalue problem that satisfies:

$$Kv = \omega^2 M\, v \tag{14.3}$$

Here, ω is the modal frequency (the square root of the eigenvalue) and ν is mode shape (the eigenvector). For large complex molecules such as proteins, the first few normal modes can often capture their most fundamental motion. This motion is frequently related to the opening of the binding site. Normal modes can hence be used to capture interesting protein states which are relevant for drug design.

Multiple relevant protein states can be obtained using molecular dynamics and normal modes. Each such protein state can be treated as rigid and computationally sampled with various fragments to probe the binding site as discussed in Section 3. In theory, a large number of protein states have to be sampled to obtain the full statistical–mechanical behavior of the system. In practice, with a judicious choice of a few protein states, one should be able to capture the preferred binding mode of a series of related compounds. Figure 14.1 shows two states of p38 MAP kinase generated by molecular dynamics. The protein was subjected to energy minimization in Cartesian space, and then in torsion space. The torsion space computation was carried out by the Imagiro system (Carnevali *et al.*, 2003). The 300 K thermostat was achieved using Langevin dynamics.

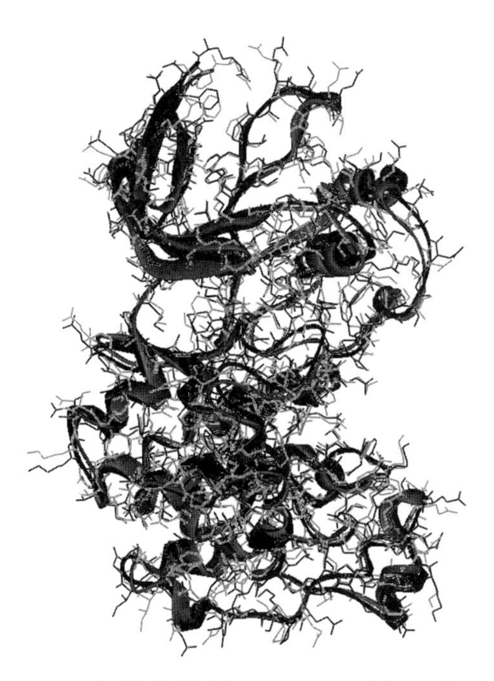

Figure 14.1 Two states of p38 MAP kinase generated by molecular dynamics. The two states are taken from the same MD simulation trajectory. The state at 200 ps is shown in blue. The state at 500 ps is shown in red. (For interpretation of the references to color in this figure legend, the reader is referred to the Web version of this chapter.)

Figure 14.2 demonstrates how normal modes can be used to generate a protein state of p38 Map kinase, starting from the crystallographic state. The computation was carried out by the Imagiro system (Carnevali *et al.*, 2003), in torsion space. The protein was subjected to energy minimization in Cartesian space, and then in torsion space. The stiffness and mass matrices were computed analytically. Modal amplitudes were scaled to reproduce the motion at 300 K.

2.3. Protein Monte Carlo simulations

Finally, Monte Carlo simulations can be used to explore the conformations of molecules. The basic principle of the Monte Carlo method is stochastic sampling of the degrees of freedom, with probabilities designed to capture the likelihood that a given conformation will occur based on its energy. Hence, lower-energy conformations are heavily sampled in Monte Carlo simulations, and higher energy conformations, while accessible, are sampled less frequently. For protein simulations, the degrees of freedom can be atomic coordinates, torsion angles, or even the normal modes of the protein. This is an example of a Canonical Monte Carlo method, where the number of atoms in the system is fixed during the simulation (Carnevali *et al.*, 2003). In Section 3.1, the grand canonical Monte Carlo method is applied to sample the binding of rigid fragments to a rigid protein. In this variant, the number of particles in the system is also a degree of freedom.

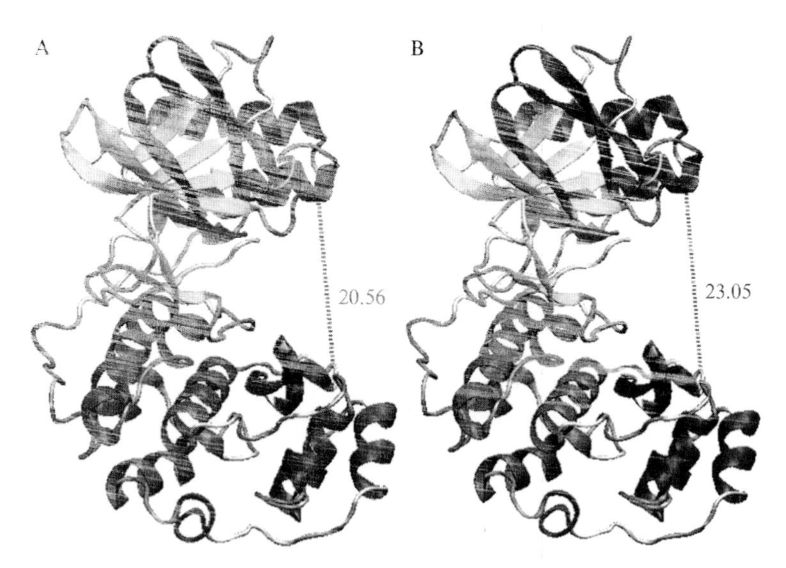

Figure 14.2 Two states of p38 MAP Kinase. (A) p38 at crystallographic conformation. (B) Opening of the binding site due to its fundamental normal mode.

2.4. Protein homology modeling by the TICRA method

Several methods have been developed to compute protein structures (Nayeem *et al.*, 2006; Zhang, 2008). TICRA is a method that can be used to compute the structure of protein–ligand complexes. Briefly, this method involves a sequence of transplant-insert-constrain-relax-assemble iterations to compute a protein or protein–ligand structure. The TICRA iterations used to compute protein or protein–ligand structure are:

Transplant: If necessary, transplant a desired feature from one protein structure into the protein model. The feature is typically a group of related atoms in a loop or subdomain, or explicit water molecules.

Insert: Insert known active molecules into the current protein model. The new molecule may be a whole ligand or a fragment of it. If the approximate position of a fragment is not known, systematic sampling of that fragment is performed in the presence of the current conformation of the protein, with or without additional constraints, to identify low-energy poses of the fragment, as described in Section 3.2.

Constrain: If necessary, add constraints/restraints between positions of the atoms in the protein and other molecules, such as a ligand or explicit water molecules.

Relax: Perform a sequence of short-duration molecular dynamics and energy minimization steps on the modified protein in the presence of new molecules and constraints.

Assemble: Systematically search the relaxed protein for fragments of the expected molecule, and assemble the molecule from its fragment distributions. These iterations are performed one fragment at a time. At the conclusion of the iterations, the protein and ligand complex is computed.

The TICRA method has been used to obtain several models for computational drug discovery programs. For example, the p38 DFG-out state was computed from two known crystal structures: p38 DFG-in state (Wang *et al.*, 1998), abl DFG-out state (Lee *et al.*, 2008), and the binding mode of a known inhibitor (Gill *et al.*, 2005).

3. Fragment Binding in Protein Model: Methods for Free Energy Calculation

Once a protein model has been selected for design, fragments are simulated against it using two different methods which calculate the free energy of binding: (a) grand canonical Monte Carlo and (b) systematic sampling. These free energy methods are theoretically more accurate as they include both enthalpy and entropy in the calculations. Detailed descriptions of these methods follow.

3.1. Grand canonical simulations of fragment binding

The grand canonical Monte Carlo simulation of fragment binding to a target protein takes place through the annealing of the chemical potential (Fig. 14.3). The protein model is placed in a simulation box with repeating boundary conditions.

The model is then solvated by the fragments and a predetermined number of Monte Carlo steps are taken to explore the rotational and translational motions of the fragment in the box (Guarnieri and Mezei, 1996). After each step, the Metropolis criterion (Metropolis *et al.*, 1953) is used to determine whether the moves are kept or rejected. The probability of keeping a move is taken from the grand canonical ensemble. A calculated parameter, the *B*-value, is used during the annealing process and is related to the excess chemical potential (μ), relative to the chemical potential of a reference state (μ_{ideal}), by Eq. (14.4):

$$B = (\mu - \mu_{\text{ideal}})/kT + \ln(\langle N \rangle) \tag{14.4}$$

where N is the number of ligands in the system and T is the temperature of the system. The overall protocol for carrying out the grand canonical simulations is briefly described as follows.

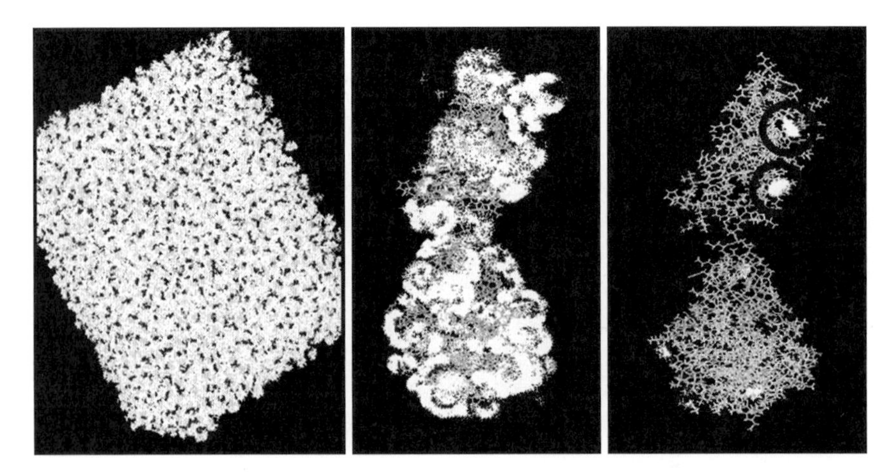

Figure 14.3 Graphical representation of the grand canonical ensemble method for simulating fragments. A protein model is placed in a box flooded with virtual fragments (left panel). Weakly bound fragments evaporate off the protein during simulated annealing at defined energy levels (center panel). Fragment clusters forming the most favorable interactions with the protein are revealed in the final stages of the simulation (right panel).

1. The protein model prepared as described in Section 2 is placed in a unit cell with periodic boundary conditions.
2. For each fragment whose binding energy is being calculated against the protein model, the simulation box is filled such that the protein model is bathed in a solution of that particular fragment. For each fragment being studied, a separate simulation is carried out. For flexible fragments, multiple conformers are generated in advance and separate simulations are run for each conformer.
3. The initial and final values for the *B*-value as well as the step size may be varied by the user. The initial *B*-value factor is typically set at 10 and is reduced by values of 1.0 or 0.5 until either a predetermined minimum *B*-value or a convergence criterion is reached. The convergence criterion (Metropolis *et al.*, 1953) is an adjustable parameter for the average number of fragments in the unit cell for a given *B*-value.
4. Each time the *B*-value is reduced, fragments are inserted and deleted. Insertions and rejections are retained or discarded according to the Metropolis Monte Carlo criterion using the grand canonical ensemble probability function until the fragment density equilibrates at the desired chemical potential. The probability of accepting an insertion or deletion is given by Eqs. (14.5) and (14.6) (Clark *et al.*, 2006):

$$\rho_{\text{insert}} = \min\left[1, \frac{\exp(-\Delta E/KT + B)V}{N+1}\right] \tag{14.5}$$

$$\rho_{\text{insert}} = \min\left[1, \frac{\exp(-\Delta E/KT - B)N}{V}\right] \tag{14.6}$$

   where V is the volume of the system and ΔE is the change in potential energy for the move being attempted.
5. Monte Carlo rotational and translational moves are applied to the fragments in the simulation box. The nonbonded interactions between the fragments and the protein are calculated after each step, but the fragments do not interact with each other (Clark *et al.*, 2009a,b). At each *B*-value, a user-defined number of Monte Carlo steps are taken. We typically use 2,000,000 steps at each *B*-value during the simulation. The probability of accepting the rotational or translational movement of a given fragment is given by Eq. (14.7) (Clark *et al.*, 2006):

$$\rho_{\text{move}} = \min[1, \exp(-\Delta E/kT)] \tag{14.7}$$

6. If the minimum specified *B*-value or convergence criterion is reached, the simulation ends. Otherwise, steps 3–6 are repeated as necessary.

3.2. Systematic sampling of fragment binding

Systematic sampling is a method where binding pose and interaction energy of a fragment in a protein is computed at all favorable positions that cover the region of interest (Clark *et al.*, 2009a,b). To make this approach sufficiently fast to be practical three computational aids are employed:

1. The space of fragment translations is sampled on a 3-D grid of a specified spacing (*translational sampling grid resolution*)
2. The space of fragment rotations is sampled up to a specified effective distance (*rotational sampling resolution*)
3. The energy field of the protein is sampled on a 3-D grid of a specified spacing (*energy grid resolution*)

3.2.1. Translational sampling

The sampling of fragment translations is straightforward: we successively set the translation vector **t** to points of a uniform 3D rectangular grid, consisting of the vectors

$$\mathbf{t}_{ijk} = i\Delta_x\hat{\mathbf{x}} + j\Delta_y\hat{\mathbf{y}} + k\Delta_z\hat{\mathbf{z}} \tag{14.8}$$

where i, j, and k are integers, $\hat{\mathbf{x}}$, $\hat{\mathbf{y}}$, and $\hat{\mathbf{z}}$ are unit vectors in the coordinate directions, and Δ_x, Δ_y, and Δ_z are translational resolutions in the three coordinate directions.

3.2.2. Rotational sampling

Like translations, rotations also form a 3D space. However the rotation space is not Euclidean. This means that no set of three rotational degrees of freedom can be isolated separately to provide a "uniform" coverage of rotation space. However, rotations do form a metric space, so a consistent notion of distance can be defined. It is preferable to get a uniform coverage of the atomic positions that have resulted from the application of the rotations. For example, it is desirable to sample more densely rotations around a short axis of the fragment, since those rotations generate larger atomic displacements than rotations around a long axis of the fragment. To address this, we define the "distance" between two fragment rotations as the atomic root mean square displacement generated when the fragment is moved from the first rotation to the second one. With this metric, the distance between two rotations does not simply depend on the angle between the two rotations, and it instead takes into account the ligand shape. To ensure uniform coverage of the rotation space, the process of selecting fragment rotations is started with a large set S_0 consisting of N_R randomly selected fragment rotations. We select from S_0 the fragment rotations to be used in the sampling, and we call S_1 (a subset of S_0) the set of n_R chosen fragment rotations.

3.2.3. Protein energy field grid

Once a fragment pose has been selected, we need to compute the interaction energy between the fragment and the protein for that pose. The procedure described below works whenever the interaction between the fragment and the protein is a sum of pair potential terms. Here we describe the interaction energy E as a sum of van der Waals energies plus electrostatic interaction using a distance-dependent dielectric constant:

$$E = \sum_{ab} \left[\frac{kq_a q_b}{r_{ab}^2} + \varepsilon_{ab} \left(\frac{\sigma_{ab}^{12}}{r_{ab}^{12}} - 2\frac{\sigma_{ab}^6}{r_{ab}^6} \right) \right] \tag{14.9}$$

where index a runs over fragment atoms, index b runs over atoms of the protein, q_a and q_b are atomic charges, ε_{ab} and σ_{ab} are van der Waals parameters for the atom pair (a, b), r_{ab} is distance between atoms a and b, and k is the electrostatic constant. The $1/r_{ab}{}^2$ dependence of the electrostatic term is due to the usage of AMBER (Cornell *et al.*, 1995) distance-dependent dielectric constant. In what follows, $\mathbf{r}_a$ and $\mathbf{r}_b$ represent the position vectors of atoms a and b, respectively. This can be rewritten as:

$$E = \sum_{a} [q_a \phi(\mathbf{r}_a) + \psi_a(\mathbf{r}_a)] \tag{14.10}$$

where $\phi(\mathbf{r})$ and $\psi_a(\mathbf{r})$ are potential scalar fields that only depend on the protein, independent of the positions of the fragment atoms:

$$\phi(\mathbf{r}) = \sum_{b} \frac{kq_b}{(\mathbf{r} - \mathbf{r}_b)^2} \tag{14.11}$$

$$\psi_a(\mathbf{r}) = \sum_{b} \varepsilon_{ab} \left[\frac{\sigma_{ab}^{12}}{(\mathbf{r} - \mathbf{r}_b)^{12}} - 2\frac{\sigma_{ab}^6}{(\mathbf{r} - \mathbf{r}_b)^6} \right]. \qquad 14.12)$$

The number of distinct $\psi_a(\mathbf{r})$ fields equals the number of distinct atom types in the fragment.

3.2.4. Practical considerations for resolutions

To avoid excessive computation, a higher value of grid resolution is preferred. However, the grid must provide sufficient accuracy to represent protein binding sites. It has been empirically determined that an energy grid resolution of 0.25 Å is sufficient to approximate typical potential wells in protein binding sites. Likewise, a sampling resolution of 0.25 Å is able to capture binding modes of fragments in most binding sites.

3.2.5. Application examples of systematic sampling

Figure 14.4 shows a few poses of the urea fragment sampled against p38 MAP kinase protein. The inhibitor which contains a urea fragment is shown in transparent mode. One of the sampled poses is very close to the crystallographic pose of urea in the inhibitor. The protein was prepared in the Imagiro system (Carnevali *et al.*, 2003). This process involved assigning protonation states to ionizable residues, choosing the appropriate histidine tautomers, addition of missing atoms, and minimizing the energy of the system in the presence of the crystallographic ligand and water molecules.

Figure 14.5 shows computed versus experimental binding free energies for rigid fragments benzene, benzofuran, indole, indene, toluene, thianaphthene, *o*-xylene, *m*-xylene, and *p*-xylene, with T4 lysozyme (Clark *et al.*, 2009a,b).

4. Protein Binding Site Characterization Via Fragment Simulations

The results of the fragment simulations carried out with either the grand canonical Monte Carlo or systematic sampling algorithms can be used to identify and analyze binding sites on the protein of interest. An unprecedented binding site for an unknown protein or a novel site on a known protein can be identified based on the resulting distributions of a set of probe

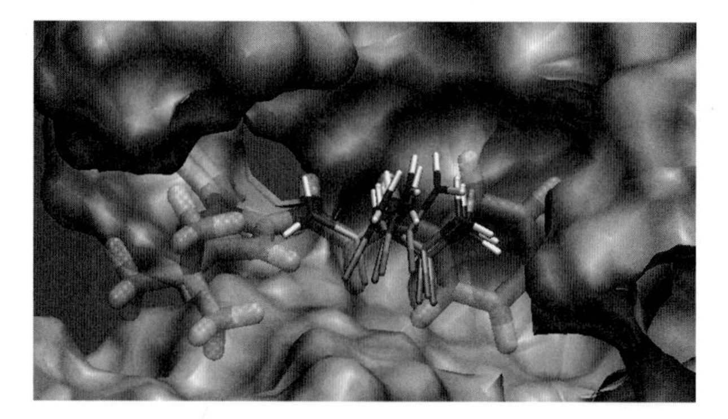

Figure 14.4 Representative systematic sampling poses of the urea fragment in the p38 MAP Kinase allosteric pocket. A few poses of the urea fragment generated by systematic sampling are shown in atom colors. The known inhibitor which contains urea is shown in green in transparent mode. (For interpretation of the references to color in this figure legend, the reader is referred to the Web version of this chapter.)

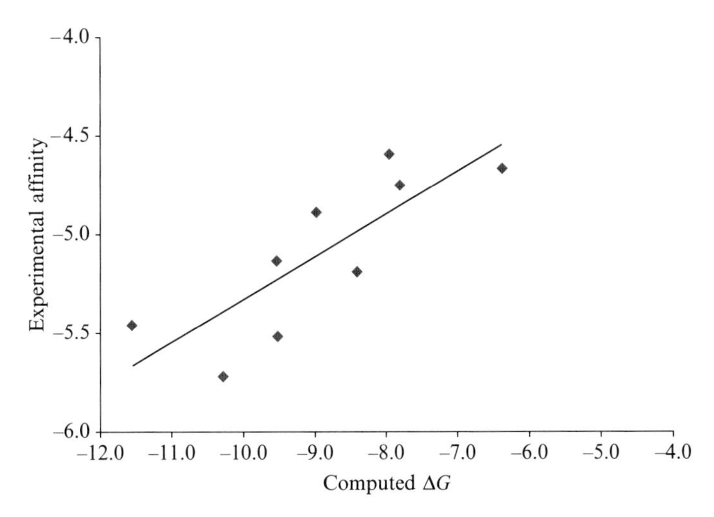

Figure 14.5 Computed versus experimental binding free energies (kcal/mol) for rigid compounds benzene, benzofuran, indole, indene, toluene, thianaphthene, *o*-xylene, *m*-xylene, and *p*-xylene, with T4 lysozyme.

fragments at low *B*-values. A thorough analysis of the site reveals the types of intermolecular interactions that are preferred by the protein. This information is used to guide the design of small molecule binders. Once key interacting residues are identified, fragment maps for each residue are generated from the full simulation, which typically includes thousands of fragments. Fragment maps allow for a richer selection of fragments with bonding geometries, chemistries, and interaction types for a given residue, adding to the diversity of chemical space that can be explored when designing new molecules.

4.1. Binding site identification

Novel binding sites for a protein model are identified through evaluation of the results of grand canonical Monte Carlo simulations for a small set of ~60 probe fragments (Table 14.1). These probe fragments fall into one of five categories based on their chemical nature and are classified as hydrogen bond acceptors, hydrogen bond donors, aliphatic hydrophobic, aromatic, or mixed chemistry fragments containing two or more of the first four.

Five criteria are broadly used to identify potentially novel binding sites.

1. The binding site should have regions where simulated small molecule fragments bind with high affinity. The calculated *B*-value for a favorable site will vary from protein to protein, but as seen in Fig. 14.6, the

Table 14.1 Sample probe set used for binding site identification in Monte Carlo simulations

Aromatic	Hydrocarbon	Hydrogen bond donors	Hydrogen bond acceptors	Mixed chemistry
1,2-Dichlorobenzene	*cis*-Butene	Ethanol	Acetaldehyde	2-Aminopyrimidine
1,3,5-Trifluorobenzene	Cyclopentane	Guanidine	Acetone	2-*tert*-Butylphenol
1,3-Tichlorobenzene	Cyclopropane	Indole	Acetonitrile	3-Aminopyrazole
Benzene	Ethane	Methanol	Dimethylsulfoxide	4-Aminopyridine
Chlorobenzene	Isobutene	Methylamine	Ester	5-Methyltetrazole
Flurenyl	Methane	Phenol	Ether	Succinimide
Indane	Neopentane	Pyrrole	Pyridine	Acetamide
Naphthalene	Norbornane	Thiol	Pyrimidine	Benzimidazole
N-methylindole	Propane		Tetrahydrofuran	Catechol
Propyne	Propene		Thioether	Diaminopyrimidine
Thiophene	*trans*-Butene		Trimethylamine	Urea
				Imidazole
				Imidazolindin-2-one
				Methanesulfonamide

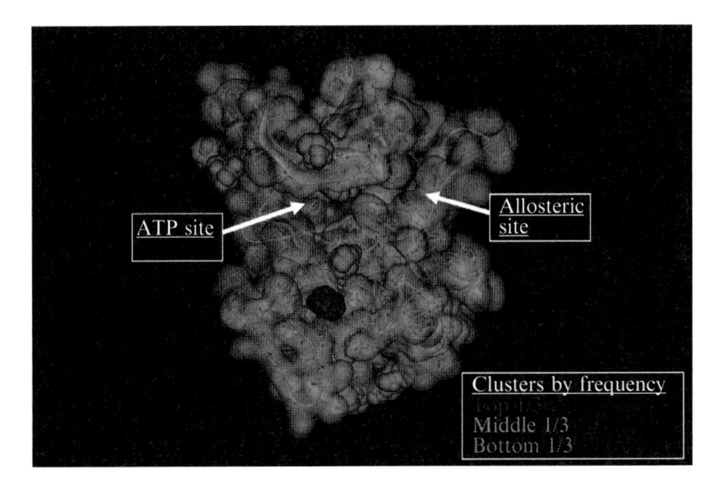

Figure 14.6 Distribution of clusters from Table 14.2 on the surface of a protein kinase. Note the presence of two clusters in the ATP site and three clusters in the allosteric site. Frequency refers to the times a specific cluster is present as the cluster radius and number of fragments change: the more persistent a cluster location is across fragments and cluster radii, the higher its frequency is.

fragments with the best calculated binding affinities will be concentrated in a few sites at the lower *B*-values.

2. These high-affinity sites preferably should not be flat surfaces. Cavities are generally preferred as drug targets because the desolvation of hydrophobic portions of a small molecule drives the initial association of the protein–ligand complex. For certain classes of targets, such as protein–protein interactions, it may not be necessary to satisfy this criterion. However, even in these instances, the emphasis should be on the identification of subpockets on the protein–protein interfaces that may not be exploited by the native protein complex.
3. There should be no tightly bound waters in the active site that would prevent fragment binding. Note that this does not necessarily mean that there should be no waters present in the active site. Indeed, key water molecules may be crucial for ligand binding in certain instances. Tightly bound waters that do not preclude fragment binding may be included in the protein model and used for another round of fragment simulations.
4. The site should be relatively compact, preferably not more than $\sim$15 Å across at its widest point. This helps to prevent the design of inordinately large molecules that may have other undesirable drug-like characteristics due to their size.
5. The putative site must accommodate clusters of fragments of differing chemistry (aromatic, hydrophobic, donors, acceptors, and mixed) in close proximity such that they may be linked together via covalent

bonding. This step helps to ensure that specificity for the target may be achieved while addressing the problem of how to link together fragments which bind to neighboring high-affinity sites (Fig. 14.7).

It should be stressed that the sites targeted for design may not necessarily meet all of the five criteria mentioned. Rather, these five criteria are general guidelines that may not all be applicable to all proteins. In cases where there is prior knowledge about the binding site of interest, the information gained from a careful analysis of the binding sites is used to guide the design of small molecule binders. In cases where there are no known sites or a novel site is desired for a target protein, the analysis of pockets using these criteria may be used to prioritize which pocket to target for design.

4.2. Fragment maps

Fragment mapping is typically carried out by examining the intermolecular interactions between key residues in the binding site and the fragment distributions resulting from the grand canonical Monte Carlo or systematic sampling simulations. This step is carried out to identify the types and character of fragments near key residues that have favorable *B*-values. A distance dependant filter is used to search for nearby fragments. If desired, additional filters may be applied for criteria such as favorable geometries for intermolecular hydrogen bonds, and various filters may be combined to generate new filters. The result is a fragment distribution of favorable poses

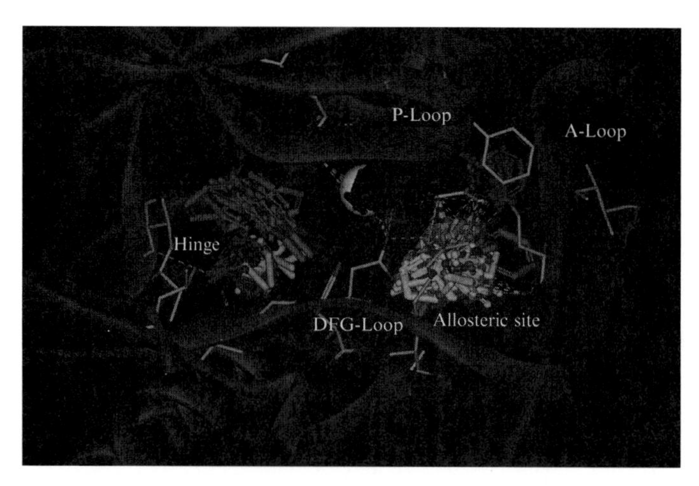

Figure 14.7 A close-up view of the ATP and allosteric sites for the protein kinase in Fig. 14.6 showing an example of a druggable site. Note the different types of fragments within covalent bonding distance in the binding site and the presence of one high-affinity water molecule (cyan sphere). (See color insert.)

for each fragment fitting the specified criteria. This approach may be used as a starting point for design, such as the identification of hinge-binding fragments with hydrogen bond accepting groups in the ATP site of protein kinases. It may also be used to strategically target nearby residues when selecting new fragments to link to an existing built molecule. In this way, a molecule may be sequentially built up fragment by fragment while maximizing favorable interactions with the protein. It is for this reason that criterion 5 from Section 4.1 is highly desired when identifying new binding sites.

4.3. Water maps

Because water molecules are treated as separate molecular fragments in the grand canonical Monte Carlo simulations, the binding of high-affinity waters to the binding site of interest may be studied in detail. The analysis of water maps uses essentially the same techniques from binding site identification and fragment mapping. As implied by step 3 in Section 4.1, the *B*-values for high-affinity waters bound to key amino acid residues can be evaluated. At this point, a decision must be made about how to treat a particular water molecule. If there are a sufficient number of fragments that occupy the same volume with lower *B*-values, then designs to this region should use these fragments which are predicted to bind to the protein more favorably. On the other hand, if there are few fragments with sufficiently low *B*-values to compete with water binding, a new protein model may be generated including the tightly bound waters identified through either the grand canonical Monte Carlo or systematic sampling simulations. The fragment simulations may then be repeated, considering the tightly bound water molecule as part of the protein model.

4.4. Application examples and druggability of protein sites

We briefly compare the cases of a favorable and an unfavorable site for drug design using the analysis of the fragment maps discussed in Sections 4.1–4.3. Figure 14.6 shows the structure of a kinase against which a small set of probe fragments (Table 14.1) has been run. The set of probe fragments is clustered at each location on the protein surface by varying the cluster radius and observing the number of unique fragments present in each cluster. Table 14.2 shows the minimum number of different fragments observed in each cluster for the corresponding cluster radius. This cluster analysis is not dependent on *B*-values, but rather on the inherent distributions of the fragments themselves.

The resulting clusters may then be evaluated interactively, as shown in Fig. 14.6. For this particular kinase, the highest concentration of five clusters of fragments, including three high-affinity clusters, is observed in

Table 14.2 Cluster count as a function of cluster radius (in Å) and minimum number of fragments

	Cluster radius (Å)									
Minimum number of fragments	2.0	1.9	1.8	1.7	1.6	1.5	1.4	1.3	1.2	1.1
7	26	26	26	26	26	25	24	24	24	20
8	16	15	14	14	13	12	12	12	12	9
9	8	8	8	8	8	8	7	7	7	6
10	4	4	4	4	4	4	4	4	3	3
11	2	2	2	2	2	2	2	2	2	2
12	0	0	0	0	0	0	0	0	0	0

The larger the number of fragments and the smaller the radius, the more robust the cluster is.

the interface between the N-terminal and C-terminal lobes. Upon closer examination (see Fig. 14.7), two clusters are observed in the ATP pocket, consisting of one dominated by probe fragments with an aromatic character, while the other contains both hydrogen bond acceptor and hydrocarbon probe fragments in the hinge region. The cluster occupying the hinge region is one of the high-affinity clusters. As expected in the case of a protein kinase, the fragment maps imply that the best fragments to begin design within the ATP site are those with aromatic character and hydrogen bond acceptors. The allosteric pocket is occupied by three clusters of probe fragments. One lower-affinity cluster with hydrogen bond donating character occupies a site in competition with a tightly bound water molecule. One of the two high-affinity clusters in the allosteric site is hydrogen bond accepting in character, while the other is occupied by hydrogen bond acceptor, aromatic, and mixed chemistry probe fragments. Most of the fragments are within covalent bonding distance of one another and varied character is observed throughout the binding site. In addition, there is only one high-affinity water molecule to compete against for ligand binding.

By contrast, Fig. 14.8 shows the fragment map for an unfavorable binding site using the set of probe fragments. This putative pocket was identified as a low-affinity site on the basis of the cluster analysis. Although the three clusters shown are close enough to make covalent linkages, the interactions are dominated by hydrophobic character, as seen by the presence of the aromatic and hydrophobic probe fragments, with only one site for possible hydrogen-bond donor probes. The site is also very flat and solvent exposed, rendering it an undesirable target for the design of small molecule binders.

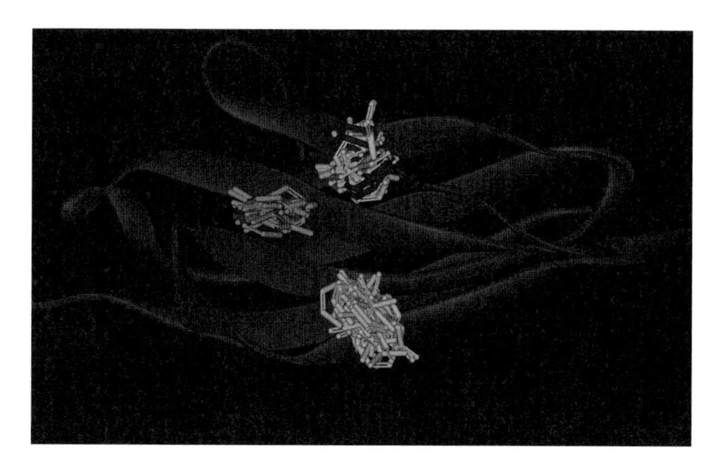

Figure 14.8 View of a site that is not ideal for the design of small molecule inhibitors. Note that there are fewer clusters than in Fig. 14.7, with less diversity of interaction types. The site is also shallow and solvent exposed.

5. Fragment-Based Design

The fragment–protein interaction results obtained by either the grand canonical or systematic sampling simulation are used to design new molecules in the identified binding site. The fragment ensembles are the result of interrogating each protein region by each fragment and as such the *B*-value rank ordering represents the protein preference for fragments. In order to prospectively design whole molecules from the fragment ensembles, several steps are necessary and are outlined below.

5.1. Identification of key fragments

High-affinity sites on the protein surface are identified by clusters of fragments as described in Section 4.1 These correspond to "hot spots" (Wells and McClendon, 2007) or anchoring locations on the protein, where fragments of differing character interact very productively with specific protein residues or protein subpockets. Fragment maps (see Section 4.2) in these subpockets are used to identify subsets of the sampled fragments that have optimum interactions with the protein and have available positions from which to join neighboring fragments to build molecules. These fragments are selected as "key fragments" from which to initiate molecular designs. Alternatively, if the binding site is well known and designs are desired for that site, then key residues on the protein are selected (e.g., Met109 on p38 kinase) and a filter is applied to select fragments that interact specifically with each of those residues.

5.2. Linking fragments to generate *de novo* molecules

Each selected key fragment is used to build a two fragment molecule by directly connecting from a specified atom of the key fragment to the nearest fragments available within bonding distance and angle tolerances. Each fragment is not a single pose but rather an ensemble of poses representing the natural motion of the fragment within the region of the protein. Fig. 14.9A shows four such fragment ensembles or distributions that occupy a region near the DFG-out state of the p38 kinase. Fig. 14.9B displays the linked fragment subdistributions to form a p38 allosteric inhibitor and Fig. 14.9C shows the final molecule moving within the boundaries of the protein site. This capability to generate whole molecule distributions is an advance in free energy methods and promises to give more realistic small molecule–protein dynamic interactions.

5.3. Processing molecules and final design selections

Once a set of designed molecules is assembled representative poses are minimized energetically using a molecular mechanics minimizer such as the one in Macromodel (Mohamadi *et al.*, 1990). Minimized poses are then checked against the original poses to ensure that the correct interactions with the protein have been maintained. Those molecules that upon minimization lose their original interactions, mainly due to internal molecular strain, are not selected for further analysis. The correctly designed molecules are then considered for synthetic accessibility by synthetic chemists and only molecules deemed synthesizable within realistic timeframes are selected for actual synthesis and testing in biochemical assays. Molecular descriptors such as molecular weight, logP, and polar surface area are calculated and final selection of molecules to be synthesized is made. In all 5–15 molecules per class are synthesized and tested and the results fed back into the design cycle to improve designs

6. Future Directions

The technologies and design approach presented here have been successfully employed in multiple drug discovery projects, some of which have progressed to lead optimization and preclinical testing. Technology enhancements currently at the research stage hold great promise to make fragment-based drug design even more accurate and efficient.

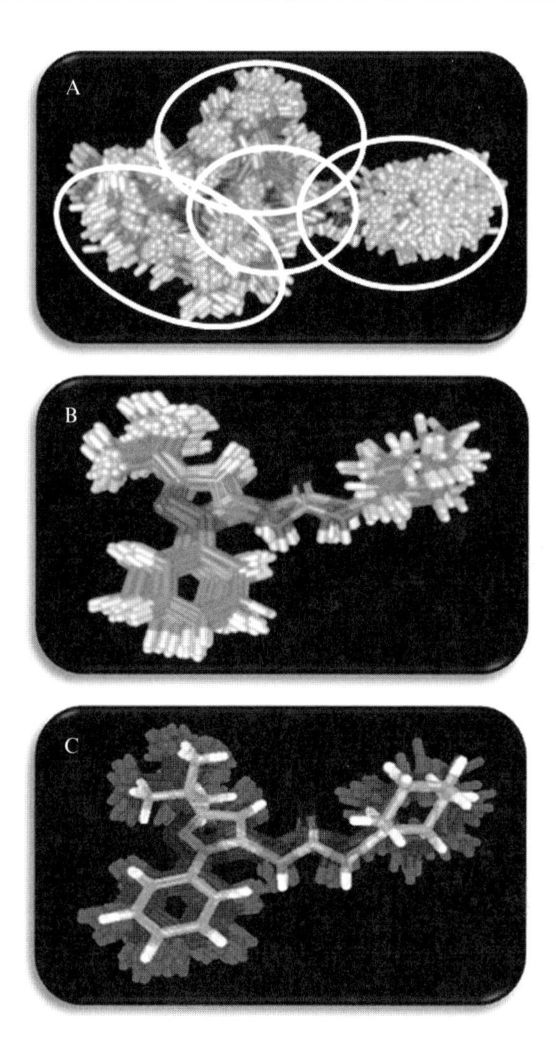

Figure 14.9 Building molecules from fragment distributions. (A) Four different fragments are shown with their total distributions in a particular region of the p38 protein (not shown for clarity). (B) Starting from the urea distribution (middle of structure) and joining nearby fragments the whole molecule is build with full distribution. (C) The center pose of the molecule is displayed along with the other acceptable binding poses in the binding site.

6.1. Multiple protein states studied simultaneously

Accounting for protein flexibility is a must for improved designs and computational affinity prediction. Normal modes and molecular dynamics have been used to produce multiple protein states that are energetically accessible from a starting X-ray crystal structure. In projects where multiple co-crystal structures are available, they can also form a set of protein states that can be used in *in silico* FBDD. Traditionally our fragment simulations have been carried out sequentially in each of these protein states. Though extremely laborious, this process promises to improve affinity prediction and binding pose identification (Beier and Zacharias, 2010). The challenge is to select enough relevant protein states that correctly represent the accessible conformational space of the protein without increasing the false positive poses for designed ligands. The torsion based normal modes and molecular dynamics simulations appear very promising in this selection process as they study the natural motions of proteins more rigorously.

6.2. Fragment simulation ensembles in multiple protein states

A "consensus" protein state can be generated by accounting for multiple existing ligands that bind to the same binding site. Fragment simulations on such a consensus state have already proven an advancement in designs in our experience. Expanding this consensus protein state to include other states where new pockets are present or where existing pockets rearrange will provide avenues for novel fragment-based designs. This type of protein change has been shown to exist on systems involved in protein–protein interactions (Wells and McClendon, 2007). Fragment ensembles generated by either of our methods allow for direct comparisons between protein states. Processes to take advantage of this capability are being developed so ligands differentiating between states can be designed more efficiently. Such an approach will enable better ligand designs for a target protein with multiple states, and also address selectivity issues between protein targets with close structural similarity (e.g., kinases of the same family).

6.3. Areas of research to improve binding affinity prediction

The accurate prediction of ligand–protein binding affinity has been the desired outcome of all computational design algorithms to date, albeit without anyone achieving the universal solution. Limitations in properly accounting for protein flexibility, solvation, force-field accuracy, conformational entropy of the ligand and protein upon complexation, and charge representation have all contributed to this inaccuracy. Beside the improved protein flexibility approach discussed above, a solvation treatment is being researched using our unique water simulations where protein–water interactions are identified and quantified. Water distributions allow the understanding of differential hydration tendencies of various protein subpockets and permit the development of strategies to deal with these differences in an explicit manner. New methods for improved charge representation such as quantum mechanical treatments also hold high promise to help achieve higher accuracy in affinity prediction. Efforts to incorporate these methods in our computational methodology are being considered for future implementation.

ACKNOWLEDGMENTS

The authors express their gratitude to the following colleagues who have contributed to the development and application of the technologies discussed: F. Guarnieri, M. Sherman, P. Carnevali, M. Clark, F. Hollinger, J. Wiseman, G. Toth, T. Fujimoto, J. Ludington, M. Kelly, E. Springman, G. Talbot, and K. Milligan.

REFERENCES

Beier, C., and Zacharias, M. (2010). Tackling the challenges posed by target flexibility in drug design. *Expert Opin. Drug Discov.* **5**(4), 347–359.

Carnevali, P., Toth, G., Toubassi, G., and Meshkat, S. (2003). Fast protein structure prediction using Monte Carlo simulations with modal moves. *J. Am. Chem. Soc.* **125,** 14244–14245.

Chen, Y., and Shoichet, B. K. (2009). Molecular docking and ligand specificity in fragment-based inhibitor discovery. *Nat. Chem. Biol.* **5**(5), 358–364.

Clark, M., Guarnieri, F., Shkurko, I., and Wiseman, J. S. (2006). Grand canonical Monte Carlo simulation of ligand-protein binding. *J. Chem. Inf. Model.* **46,** 231–242.

Clark, A. M., Meshkat, S., and Wiseman, J. S. (2009a). Grand canonical free-energy calculation of protein-ligand binding. *J. Chem. Inf. Model.* **49,** 934–943.

Clark, M., Meshkat, S., Talbot, G. T., Carnevali, P., and Wiseman, J. S. (2009b). Fragment-based computation of binding free energy by systematic sampling. *J. Chem. Inf. Model.* **49,** 1901–1913.

Cornell, W. D., Cieplak, P., Bayly, C. I., Gould, I. R., Merz, K. M., Ferguson, D. M., Spellmeyer, D. C., Fox, T., Caldwell, J. W., and Kollman, P. A. (1995). A second generation force field for the simulation of proteins, nucleic acids, and organic molecules. *J. Am. Chem. Soc.* **117**(19), 5179–5197.

Flannery, B. P., Teukolsky, S. A., and Vettering, W. T. (1988). Numerical Recipes in C: The Art of Scientific Computing. W.H. Press, Cambridge University Press, 317–324.

Gill, A., Frederickson, M., Cleasby, A., Woodhead, S., and Carr, M. (2005). Identification of novel p38 MAP kinase inhibitors using fragment-based lead generation. *J. Med. Chem.* **48,** 414–426.

Guarnieri, F., and Mezei, M. (1996). Simulated annealing of chemical potential: A general procedure for locating bound waters. Application to the study of the differential hydration propensities of the major and minor grooves of DNA. *J. Am. Chem. Soc.* **118,** 8493–8494.

Hajduk, P. J., and Greer, J. (2007). A decade of fragment-based drug design: Strategic advances and lessons learned. *Nat. Rev. Drug Discov.* **6**(3), 211–219.

Lee, T., Ma, W., Zhang, X., Giles, F., Cortes, J., Kantarjian, H., and Albitar, M. (2008). BCR-ABL alternative splicing as a common mechanism for imitanib resistance: Evidence from molecular dynamics simulations. *Mol. Cancer Ther.* **12,** 3834–3841.

Metropolis, N., Rosenbluth, A. W., Rosenbluth, M. N., Teller, A. H., and Teller, E. (1953). Equation of state calculations by fast computing machines. *J. Chem. Phys.* **21,** 1087–1092.

Mohamadi, F., Richard, N. G. J., Guida, W. C., Liskamp, R., Lipton, M., Caufield, C., Chang, G., Hendrickson, T., and Still, W. C. (1990). Macromodel - an integrated software system for modeling organic and bioorganic molecules using molecular mechanics. *J. Comput. Chem.* **11**(4), 440–467.

Nayeem, A., Sitkoff, D., and Krystek, S., Jr. (2006). A comparative study of available software for high-accuracy homology modeling: From sequence alignments to structural models. *Protein Sci.* **15,** 808–824.

Pellecchia, M. (2009). Fragment-based drug discovery takes a virtual turn. *Nat. Chem. Biol.* **5** (5), 274–275.

Wang, Z., Canagarajah, B. J., Boehm, J. C., Kassisa, S., Cobb, M. H., Young, P. R., Abdel-Meguid, S., Adams, J. L., and Goldsmith, E. J. (1998). Structural basis of inhibitor selectivity in MAP kinases. *Structure* **6,** 1117–1128.

Wells, J. A., and McClendon, C. L. (2007). Reaching for high-hanging fruit in drug discovery at protein-protein interfaces. *Nature* **450,** 1001–1009.

Zhang, Y. (2008). Progress and challenges in protein structure prediction. *Curr. Opin. Struct. Biol.* **18,** 342–348.

SECTION THREE

EXAMPLES

CHAPTER FIFTEEN

Lead Generation and Examples: Opinion Regarding How to Follow Up Hits

Masaya Orita,* Kazuki Ohno,* Masaichi Warizaya,†
Yasushi Amano,† *and* Tatsuya Niimi*

Contents

Abstract

In fragment-based drug discovery (FBDD), not only identifying the starting fragment hit to be developed but also generating a drug lead from that starting fragment hit is important. Converting fragment hits to leads is generally similar to a high-throughput screening (HTS) hits-to-leads approach in that properties associated with activity for a target protein, such as selectivity against other targets and absorption, distribution, metabolism, excretion, and toxicity (ADME/Tox), and physicochemical properties should be taken into account. However, enhancing the potency of the fragment hit is a key requirement in FBDD, unlike HTS, because initial fragment hits are generally weak. This enhancement is presently achieved by adding additional chemical groups which bind to additional parts of the target protein or by joining or combining two or more hit fragments; however, strategies for effecting greater

* Chemistry for Leads, Chemistry Research Labs, Drug Discovery Research, Astellas Pharma Inc., Tsukuba, Ibaraki, Japan
† Advanced Genomics, Molecular Medicine Research Labs, Drug Discovery Research, Astellas Pharma Inc., Tsukuba, Ibaraki, Japan

Methods in Enzymology, Volume 493
ISSN 0076-6879, DOI: 10.1016/B978-0-12-381274-2.00015-7

improvements in effective activity are needed. X-ray analysis is a key technology attractive for converting fragments to drug leads. This method makes it clear whether a fragment hit can act as an anchor and provides insight regarding introduction of functional groups to improve fragment activity. Data on follow-up chemical synthesis of fragment hits has allowed for the differentiation of four different strategies: fragment optimization, fragment linking, fragment self-assembly, and fragment evolution. Here, we discuss our opinion regarding how to follow up on fragment hits, with a focus on the importance of fragment hits as an anchor moiety to so-called hot spots in the target protein using crystallographic data.

1. Introduction

Although potency is the most important factor in validation, selection, and further optimization of a lead compound, many variables related to absorption, distribution, metabolism, excretion, and toxicity (ADME/Tox) should be considered for successful drug discovery. Of the many physicochemical parameters relevant to drug discovery, molecular size—represented by molecular weight or the number of nonhydrogen heavy atoms (HA)—is known to be related to a number of ADME/Tox properties. As a general rule, potency within the same chemical series is often strongly correlated with molecular size. However, increasing molecular size generally worsens ADME/Tox parameters such as solubility, bioavailability, and metabolic stability. Indeed, despite increasing molecular size of drugs in the clinical candidate stages, mean size declines in every subsequent phase to market. In addition to molecular size, lipophilicity and polar surface area (PSA) have also been reported to be associated with ADME/Tox parameters. Comparison of molecular properties of compounds at various stages of development reveals that lipophilicity of compounds currently being synthesized by leading drug discovery companies is significantly greater than that of recently discovered oral drugs and compounds in the clinical phase (Leeson and Springthorpe, 2007). This observed increase in lipophilicity tends to elicit nonspecific binding to many proteins, including those engaged in metabolism and toxicity, resulting in attrition in drug development. The importance of PSA to intestinal permeability and oral bioavailability has also been described previously (Ertl *et al.*, 2000; Palm *et al.*, 1997). The balance between potency and ADME/Tox parameters, which can be defined as a drug's "quality," is becoming increasingly important in drug discovery.

The typical path from fragment hits to leads and from high-throughput screening (HTS) hits to leads are described in Fig. 15.1. Typical HTS hits share almost the same molecular size, lipophilicity, and PSA values as lead

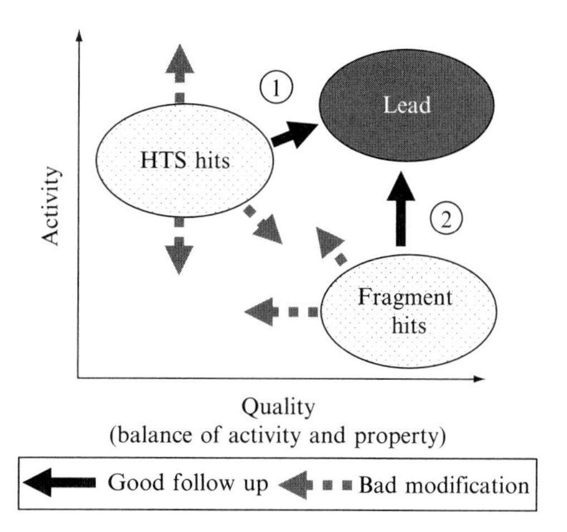

Figure 15.1 Schematic representation of typical paths from HTS hits and fragment hits as starting compounds for lead generation. Typical HTS hits share almost the same molecular size, lipophilicity, and PSA values as lead compounds but show lower activity, indicating lower quality (balance of activity and molecular properties) in HTS hits than in leads; quality of these hits must therefore be improved to become useful leads. In contrast, while typical fragment hits also show low activity, they typically exhibit comparable quality to lead compounds and need no improvement in quality during the fragment hit-to-lead process.

compounds but show lower activity, indicating lower quality in HTS hits than in leads. To become useful leads, potency of HTS hits must be increased while retaining these values, or reducing them if need be. In contrast, typical fragment hits show much lower activity and proportionally lower values in molecular size, lipophilicity, and PSA when compared with lead compounds. Fragment hits have sufficient room to grow large in size and achieve enhanced potency. Further in contrast to HTS hits, fragment hits exhibit comparable quality to lead compounds and need no improvement in quality during the fragment hit-to-lead process. Recent advances in structure biology, including experimental technology (X-ray crystallography and nuclear magnetic resonance) and computational chemistry, have facilitated the rational design of hit compounds by introducing appropriate chemical groups to enhance compound activity. However, increasing activity while retaining a constant molecular size is still difficult. While HTS and fragment-based drug discovery (FBDD) each have pros and cons as lead discovery technologies, FBDD retains the advantage with regard to ease in controlling quality, making FBDD an attractive approach for its efficiency and efficacy.

One major concern with regard to FBDD is whether or not a fragment hit can serve as an adequate starting point for construction of a lead compound. Fragment hits typically have a molecular weight of approximately 250 kDa and activity in the order of 10^{-3} to 10^{-5} *M*; these hits can only occupy a limited region of a target protein's binding pocket, contributing only ~50% of the total binding energy of a typical lead compound with 10^{-8} to 10^{-9} *M*-order activity (Akritopoulou-Zanze and Hajduk, 2009). On this point, we believe that an optimized fragment hit constitutes the compound most appropriate for use in construction of a lead compound, for two reasons: First, it is widely recognized that relatively small regions of the binding surface contribute to a large part of binding energy. Although this "hot spot" concept has not been studied extensively, a recent study reasoned why a small region can function as a molecular recognition motif for several proteins through the analysis of desolvation effects (Young *et al.*, 2007). Second, the translational and rotational entropic loss experienced by a compound on binding to a protein is estimated to be 15–20 kJ/mol for 10^{-3} *M* order affinity (Borsi *et al.*, 2010; Murray and Verdonk, 2002), that a fragment hit with an apparent low affinity of 10^{-4} *M* order intrinsically contributes to interaction corresponding to 10^{-7} *M* affinity to overcome this entropic barrier. We hypothesize that an optimized fragment hit can act as an anchor at the hot spot of a target protein and become an ideal lead compound. However, one of the most important key to success in FBDD is how to find and generate such an optimized fragment hit.

Here, we first describe ligand efficiency, which is a metric used to assess quality of fragment hits and related compounds in FBDD. We then provide successful examples of FBDD and examine them with respect to the interaction of anchor moieties with hot spots, detailing how this interaction functions in fragment hit identification and fragment hit-to-lead processing.

2. Ligand Efficiency

Ligand efficiency (LE) is the binding energy per HA (Hopkins *et al.*, 2004) and is interpreted as a measurement of goodness of interaction between a given compound and its target protein. In addition, because it represents the balance of potency and molecular size, which is related to many ADME/Tox parameters as mentioned above, LE can also be used to assess the quality of hit compounds. Although theoretically the intrinsic binding energy (ΔH) should be used for calculating LE, actual binding energy (ΔG) is usually used with the following equation for practical use:

$$\mathrm{LE} = \Delta G/\mathrm{HA} \quad (15.1)$$

where $\Delta G = -RT\ln K_\mathrm{d}$, $-RT\ln K_\mathrm{i}$, or $-RT\ln(\mathrm{IC}_{50})$.

LE was originally proposed as a useful parameter for selecting and optimizing lead compounds; more recently, however, LE is being applied in FBDD, in the prioritization of fragment hits and conversion of fragment hits to leads. The Abbott group carried out a retrospective analysis of 18 highly optimized inhibitors, demonstrating the remarkably linear relationship between potency and molecular mass during ideal optimization and thereby indicating that LE values stay almost constant during ideal fragment-to-lead processing (Hajduk, 2006). A search through the literature turns up 30 other unique examples in the field of FBDD in which similar tendencies were observed (Orita *et al.*, 2009a).

Various modified LE metrics have been proposed to date (summarized in Table 15.1). The Abbott group proposed the percentage efficiency index (PEI) and binding efficiency index (BEI), in which molecular weight is used as the index of molecular size because it reflects the contribution of different types of heteroatoms (e.g., fluorine vs. iodine; Abad-Zapatero and Metz, 2005). As mentioned above, lipophilicity and PSA are known to be related to ADME/Tox parameters. In the proposed indices ligand-lipophilicity efficiency (LLE; Leeson and Springthorpe, 2007) and logP/ligand efficiency (LELP; Keseru and Makara, 2009), lipophilicity is directly introduced into functions to assess the quality of compounds. The Abbott group also proposed the surface-binding efficiency index (SEI), which is based on PSA (Abad-Zapatero and Metz, 2005). These ligand efficiency indices are interrelated and may be used either alone or in combination.

LE values have recently been reported to be intrinsically related to molecular size, with smaller molecules having inherently greater LE values than larger ones. The Johnson & Johnson group first proposed an empirical "fit quality" metric to compensate for this dependency by calculating the maximum LE for a certain HA, using affinity data derived from the Binding Database (Reynolds *et al.*, 2007, 2008), and the AstraZeneca group proposed the measurement of size-independent ligand efficiency (SILE; Nissink, 2009). The concept of size dependency is simply explained as follows: when a small ligand binds to a protein "hot spots," LE is high. In contrast, large ligands bind not only to "hot spots" but also to other regions, causing a reduction in LE. Size-dependency is thus related to the anisotropic environment of the protein pocket. Friesner's group developed a novel, computationally efficient method of analyzing environments for solvating water in active site pockets, with results showing that the active sites of proteins provide extremely diverse environments for solvating water in this manner (Abel *et al.*, 2008; Thijs *et al.*, 2009; Young *et al.*, 2007). In addition, Reynolds *et al.* presented two physical interpretations of the size-dependency: (1) Hydrophobic interaction, which is a major driving force for protein ligand binding, is related to the hydrophobic accessible surface area of ligands. Large ligands can only tolerate limited increases in hydrophobic accessible surface area. (2) Large complex ligands have many points of contact with an active

Table 15.1 Ligand efficiency indices

Name	Definition	Comment	Reference
LE	$-RT\ln(K_d)/(\mathrm{HA}) \approx -RT\ln(\mathrm{p}K_i)/(\mathrm{HA})$	Original definition of ligand efficiency	Hopkins *et al.* (2004)
BEI	$\mathrm{p}K_i$ (or $\mathrm{p}K_d$)/MW	Corrected by MW instead of HA	Abad-Zapatero and Metz (2005)
SEI	$\mathrm{p}K_i$ (or $\mathrm{p}K_d$)/PSA	Corrected by PSA instead of HA	Abad-Zapatero and Metz (2005)
LLE	$\mathrm{p}K_i$ − clogP (or logD)	Metric of acceptable lipophilicity per unit of *in vitro* potency	Leeson and Springthorpe (2007)
FQ	$\mathrm{LE}/\ (-\ 0.064 + 0.873 \times e^{(-\ 0.026\ \times\ \mathrm{HA})})$	Empirical score for correction of size-dependency	Reynolds *et al.* (2007)
FQ	$\mathrm{LE}/(0.0715 + 7.5328/(\mathrm{HA}) + 25.7079/(\mathrm{HA})^2 + -361.47222/(\mathrm{HA})^3)$	Empirical score for correction of size-dependency	Reynolds *et al.* (2008)
%LE	$\mathrm{LE}/\ (\varphi^{\log_2(10/\mathrm{HA})}) \times 100$ $\varphi = 1.618$ (golden ratio)	Empirical score for correction of size-dependency	Orita *et al.* (2009a)
LELP	logP/LE	Metric of contribution of lipophilic component to LE	Keseru and Makara (2009)
SILE	$-RT\ln(\mathrm{p}K_i)/(\mathrm{HA})^{0.3}$	Empirical score for correction of size-dependency	Nissink (2009)

LE, ligand efficiency; BEI, binding efficiency index; SEI, surface-binding efficiency index; LLE, ligand-lipophilicity efficiency; FQ, fit quality; LELP; ligand-efficiency-dependent lipophilicity, SILE; size-independent ligand efficiency; HA, number of nonhydrogen heavy atoms; MW, molecular weight.

site pocket. Satisfying multiple molecular recognition sites with a single ligand often leads to structural constraints, which makes ligand affinity weak (Reynolds *et al.*, 2008).

Our group also introduced %LE derived using Kuntz's data (Kuntz *et al.*, 1999), which include the strongest-binding ligands those authors found capable of removing nontypical drug molecules such as heavy metals and carbon monoxide (Orita *et al.*, 2009a). The relationship between LE and HA of maximal affinity ligands is described in Fig. 15.2, from which we made the following observations: when HA is 10, LE is almost 1; when HA increases by twofold, the maximal LE decreases by $1/\varphi$-fold (φ = golden ratio, 1.618). Based on these findings, we estimated the maximal LE by curve-fitting Kuntz's data as follows:

$$\text{Maximal LE} = \varphi^{\log_2(10/\text{HA})} \tag{15.2}$$

$$\varphi^2 = \varphi + 1 \tag{15.3}$$

$$1/\varphi = \varphi - 1 \tag{15.4}$$

$$\%\text{LE} = \text{LE}/\max \text{LE} \times 100 \tag{15.5}$$

Lines were subsequently added to Fig. 15.2 using the above equations. Using the simple formulae in Eqs. (15.3) and (15.4), this maximal LE is extremely easy to memorize. When the HA increases by twofold, the maximal LE increases by $1/\varphi$ ($=\varphi - 1$, 0.618). When HA decreases by twofold, the maximal LE increases by φ (=1.618). When HA increases by

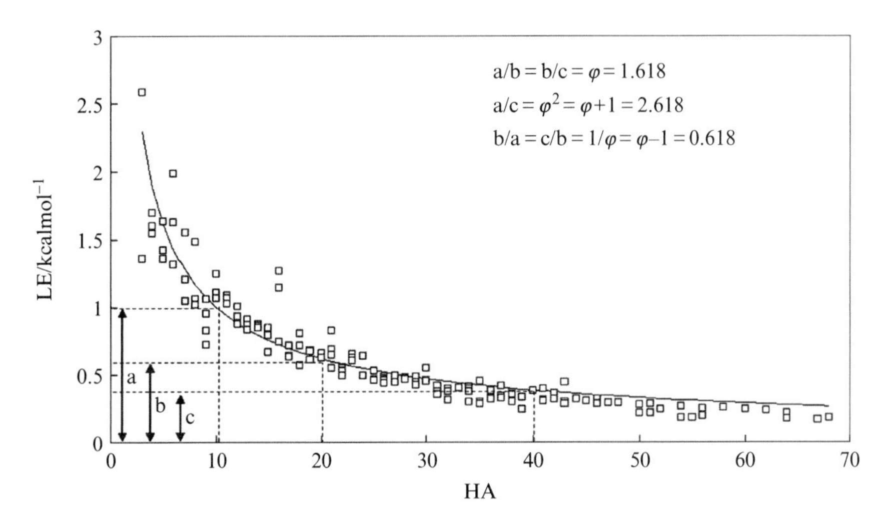

Figure 15.2 Size-dependency of activity of the strongest binder.

fourfold, the maximal LE decreases by φ^2 ($=\varphi + 1$, 2.618). %LE can be a useful index, describing the ratio of the LE of a compound to that of the strongest-binding ligand with the same HA. The concept of maximal LE has been confirmed using data from clinical candidates and launched drugs which were obtained from the databases provided by GVK BioSciences (Hyderabad, India). As expected, in almost all cases (5368/5813, 92%), the LE value of a compound was smaller than the maximal LE, and thus the % LE value was under 100% (average %LE for clinical candidates and launched drugs: 76%). These data confirm that the %LE index is useful not only for the prioritization of fragment hits but also during fragment-to-lead processing.

In our previous study, we investigated the change in LE and change in % LE during conversion of fragments to leads (Orita *et al.*, 2009a). The relationship between LE values for fragments and leads is shown in Fig. 15.3A. The average LE value for fragments (0.40) was almost as same as that for leads (0.38). The relationship between %LE values for fragments and leads is shown in Fig. 15.3B. In almost all cases, %LE for leads was larger (average: 76%) than that for fragments (average: 55%), indicating that this index increases when fragments are successfully optimized to leads. Of particular note is the fact that the average %LE value for leads was almost the same as the average for clinical candidates and launched drugs. However, in 28 out of 30 cases (93%), the %LE value exceeded 40%. To get a rough idea, %LE target level for leads and criteria for fragments should be set to 76% and 40%, respectively. In Section 3, %LE will be used to clarify our strategy about fragment to lead process.

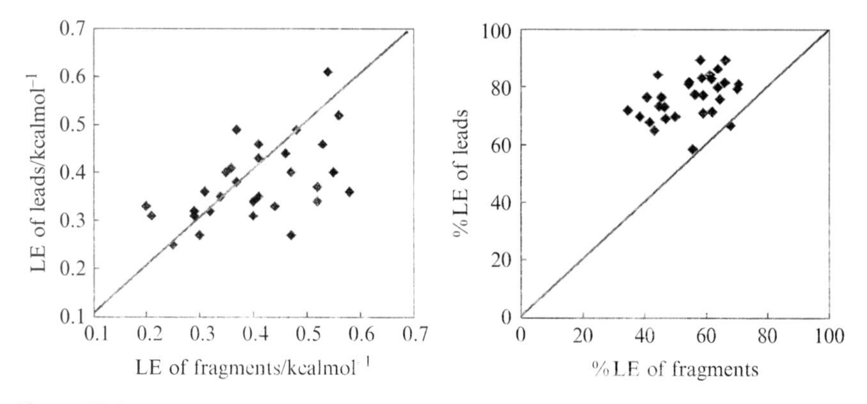

Figure 15.3 Comparison of LE (A) and %LE values (B) between fragments and leads.

3. Four Different Approaches for Converting Fragment Hits to Leads

Approaches for converting fragment hits to leads fall into one of four categories: fragment optimization, fragment linking, fragment self-assembly, and fragment evolution (Alex and Flocco, 2007; Erlanson, 2006; Keseru and Makara, 2006; Rees *et al.*, 2004). Selecting the appropriate approach for a target is naturally important in effectively converting fragments to drug leads. Below, we describe each of these four classifications and outline the practical use of our proposed fragment-to-lead chart.

3.1. Fragment linking

Fragment linking was first reported as an FBDD approach by the Abbott group, a pioneer in this field, and involves two or more fragments which bind to different parts of the binding pocket being linked to obtain a molecule with a higher affinity (Shuker *et al.*, 1996). Substantial improvement in activity can be achieved if both the length and conformation of the linker are optimal for linking fragments to interact with their respective optimal binding site and if the linker part contacts to the binding pocket well (Shuker *et al.*, 1996). In such a case, the affinity (free energy of binding) of the linked compound is expected to be approximately equal to at least the sum of the affinities of the fragments involved; in many cases, however, the observed affinity of the linked compound may be greater than the sum of its parts due to the linking effect (Borsi *et al.*, 2010; Jencks, 1981; Shuker *et al.*, 1996). This additional increase in affinity can be explained by the translational and rotational entropy lost by each fragment (Murray and Verdonk, 2002). Based on the theoretical consideration of the entropy lost from a fragment when it complexes with a target protein, Page and Jencks estimated a potential increase in affinity of approximately eightfold (Page and Jencks, 1971). Although the experimentally observed range of increase in activity generally ranged from ~ 0 to 3 orders (Borsi *et al.*, 2010; Murray and Verdonk, 2002), enhancement of activity via the linking effect is an attractive key point in this approach.

Our proposed fragment-to-lead-chart demonstrating the activity enhancement achievable via the fragment linking approach is shown in Fig. 15.4A. Molecular size, the number of heavy atoms is shown on the x-axis, and the $-\log K_i$ value (pK_i) is shown on the y-axis. Points A and B indicate hit fragments A and B, respectively. If the linker region has no effect on the size or potential of the linked compound, the linked compound is expected to be plotted at point C, which corresponds to the sum of the two hit fragment vectors (for A and B). However, because of the linking

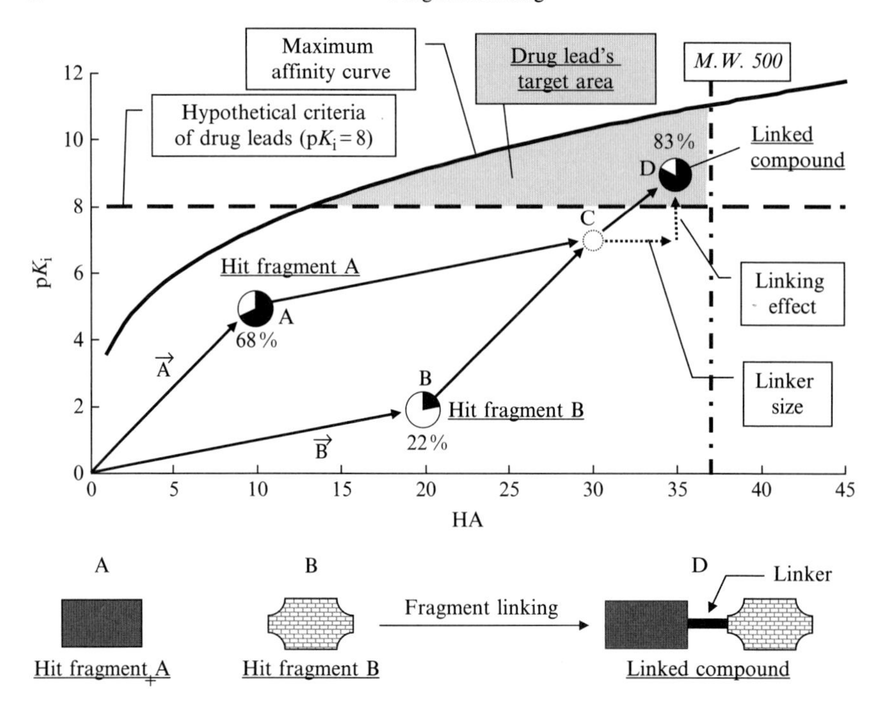

A
Fragment linking
Maximum affinity curve
Drug lead's target area
M.W. 500
Hypothetical criteria of drug leads (pK_i = 8)
83%
D
Linked compound
C
Hit fragment A
A
68%
Linking effect
Linker size
B
Hit fragment B
22%
pK_i
HA
0
2
4
6
8
10
12
0
5
10
15
20
25
30
35
40
45
A
B
D
Linker
Fragment linking
Hit fragment A
Hit fragment B
Linked compound

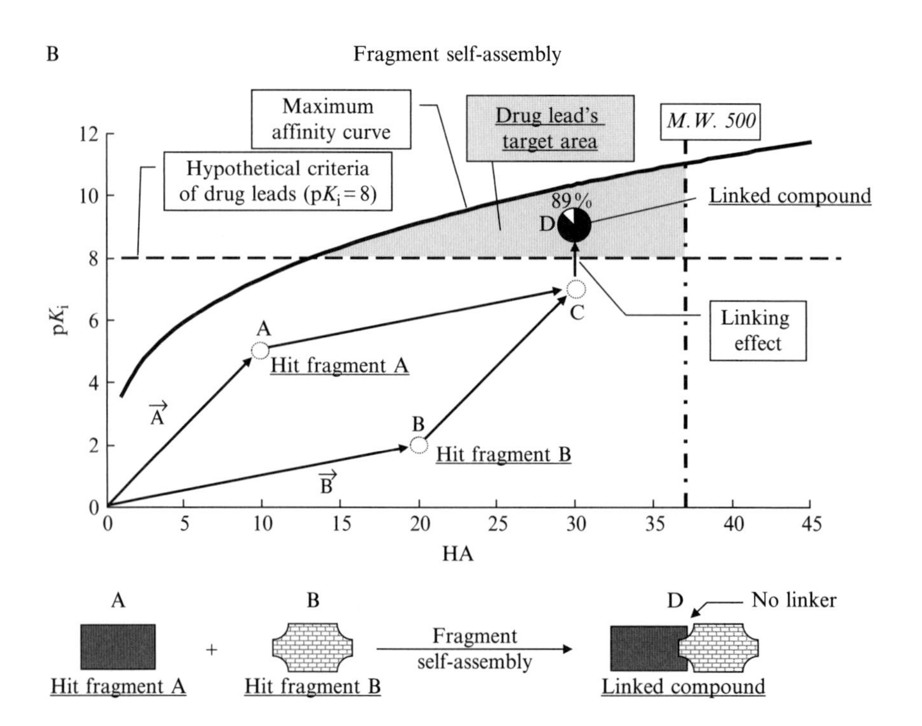

B
Fragment self-assembly
Maximum affinity curve
Drug lead's target area
M.W. 500
Hypothetical criteria of drug leads (pK_i = 8)
89%
D
Linked compound
C
Linking effect
A
Hit fragment A
B
Hit fragment B
pK_i
HA
0
2
4
6
8
10
12
0
5
10
15
20
25
30
35
40
45
A
B
D
No linker
+
Fragment self-assembly
Hit fragment A
Hit fragment B
Linked compound

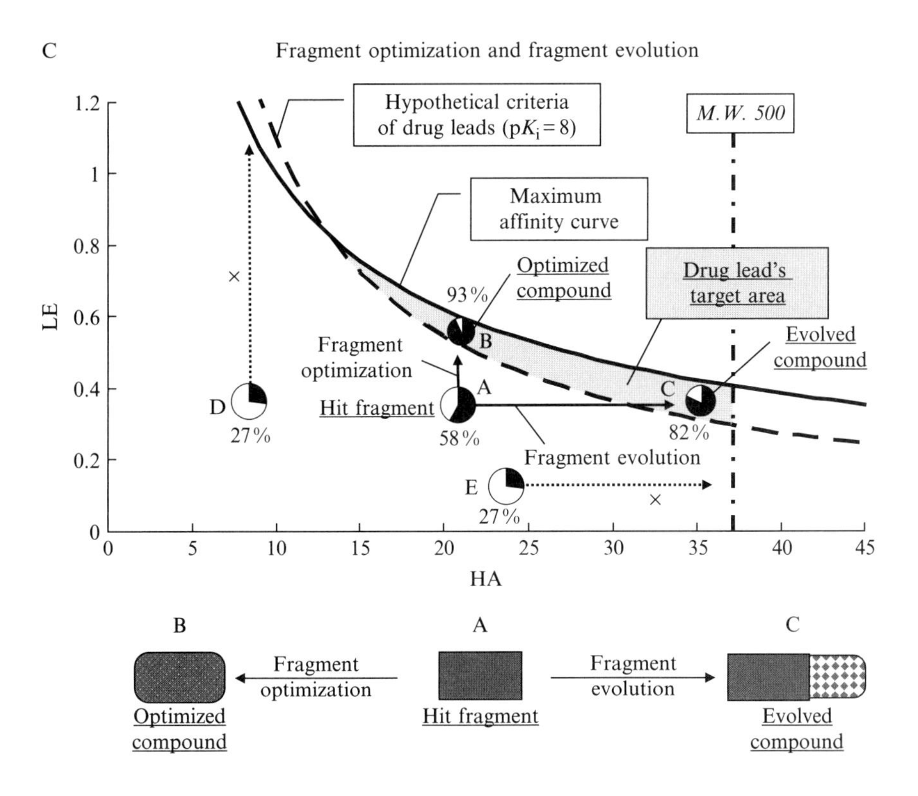

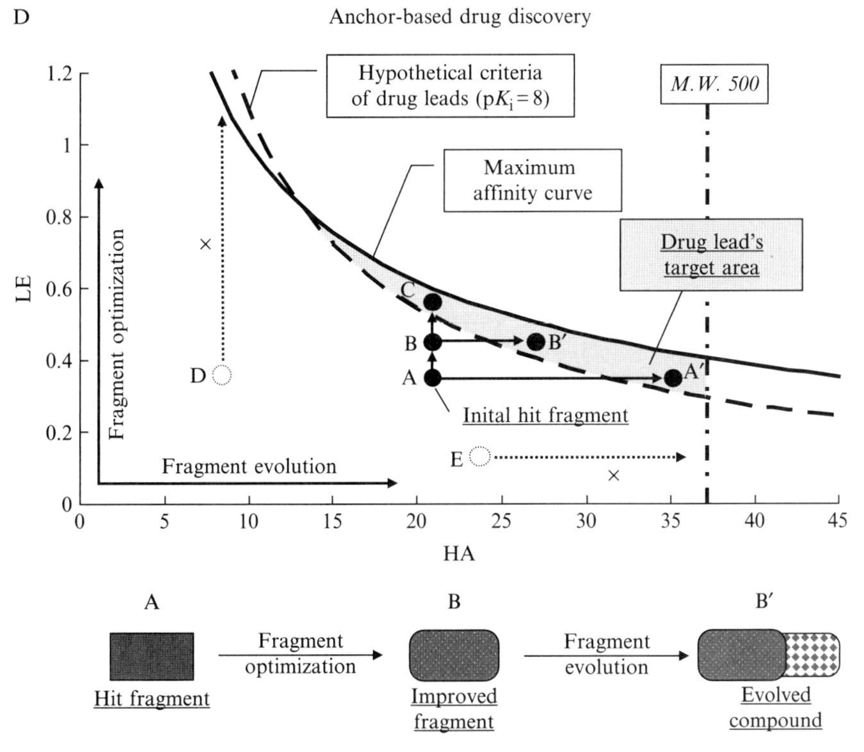

Figure 15.4 (Continued)

effect on affinity and the increase in size on addition of a linker part, the linked compound is generally plotted approximately at point D, which is shifted slightly in the top right direction from point C. Here, we can see that by linking two fragments whose pK_i values are 5 (hit fragment A: $K_i = 10$ μM, %LE = 68%) and 2 (hit fragment B: $K_i = 10$ mM, %LE = 22%), the strong linked compound has a pK_i value of 9 (%LE = 83%). In this case, the linking effect is "2," calculated by subtracting the pK_i values of the hit fragments from that of the linked compound (linking effect = (pK_i of linked compound) − [(pK_i of hit fragment A) + (pK_i of hit fragment B)] = 9 − (5 + 2) = 2).

Lipinski proposed the rule-of-five filter for drug-likeness, which states that molecular weight is <500 g/mol, the number of hydrogen-bond donors ≤5, the number of hydrogen-bond acceptors ≤10, and the clogP ≤ 5 (Lipinski *et al.*, 1997). The perpendicular dashed line in Fig. 15.4A indicates a molecular weight of ~500 g/mol and is plotted together with a horizontal dashed line showing a pK_i value of 8 ($K_i = 10$ nM), which are typical hypothetical criteria for affinity in drug leads. The maximum affinity curve, which indicates the affinity of Kuntz's strongest binding ligands and is converted from our estimated equation of maximal LE (Orita *et al.*, 2009a), is also plotted in Fig. 15.4A.

As described above, it is a reasonable assumption that compounds plotted above the maximum affinity curve are impossible to obtain in general. As such, the gray region in Fig. 15.4A, which is bounded on the left by the perpendicular dashed line describing molecular weight as ~500 g/mol, above by the horizontal dashed line describing pK_i value as 8 and below by the curve illustrating the affinity of Kuntz's strongest binding ligands, can be defined as the target area for obtaining drug leads during fragment-to-lead processing. To efficiently obtain drug leads via the fragment linking approach, we must carefully plan and consider a strategy for converting fragments to leads, as shown in Fig. 15.4A.

Figure 15.4 Different approaches to converting fragment hits to leads. (A) Plot of HA versus activity value (pK_i) in fragment linking. The perpendicular dashed line indicates a molecular weight of ~500 g/mol. The horizontal dashed line shows that pK_i value is 8 ($K_i = 10$ nM). The curving solid line is the maximum affinity curve, which indicates the affinity of Kuntz's strongest binding ligands. The gray region is the drug lead's target area. The %LE value is shown via circle graph. (B) Plot of HA versus activity value (pK_i) in fragment self-assembly. The figure legend is the same as that for (A). (C) Plot of HA versus LE in fragment optimization and evolution. The perpendicular dashed line indicates a molecular weight of ~500 g/mol. The dashed curve shows that pK_i value is 8 ($K_i = 10$ nM). The curving solid line is the maximum LE curve, which indicates the affinity of Kuntz's strongest binding ligands. The gray region is the drug lead's target area. The %LE value is shown via circle graph. (D) Plot of HA versus LE in anchor-based drug discovery. Figure legend is the same as that for (C).

The fragment linking approach to processing fragments into leads uses structural information for fragment binding modes, such as the crystal structures of the hit fragment–target protein complexes or intermolecular nuclear Overhauser effect between two hit fragments for a single target protein. These data are generally important for achieving effective molecular design of the linked compound. Also important is the synthesis method determined by the chemist, as designing a synthesizable linked compound can be particularly difficult. However, despite the classic attractiveness and merits of fragment linking, this method is not particularly easy to carry out, due to difficulties in finding more than two fragments which bind to different parts of a binding pocket and in designing compounds without disrupting the binding modes of hit fragments. Given such difficulties, relatively little success has been achieved with the fragment linking approach in terms of the potential and drug- or lead-likeness of the obtained linked compound.

3.2. Fragment self-assembly

Fragment self-assembly is an attractive concept in the fields of both chemistry and FBDD, involving reacting fragments together in order to create more active inhibitors in the presence of a target protein. Under this approach, the target site serves as a template for chemical synthesis, and two powerful methods—dynamic combinatorial chemistry (Huc and Lehn, 1997; Lehn, 1999; Lehn and Eliseev, 2001; Ramstrom and Lehn, 2002) and *in situ* click chemistry (Kolb and Sharpless, 2003; Kolb *et al.*, 2001; Lewis *et al.*, 2002)—have been developed and are widely employed. Dynamic combinatorial chemistry involves linking the reactive fragments (building blocks) together via a reversible reaction, thereby allowing a constant interchange of building blocks and resulting in a mixture of library members that is in equilibrium, thermodynamically selecting combinations of fragments that act as inhibitors. *In situ* click chemistry is a strategy which relies on irreversible, target-guided synthesis and employs a set of highly reliable reactions without side reactions or destruction of the target protein. The defining difference between these two complementary methods is that dynamic combinatorial chemistry is reversible chemistry, while *in situ* click chemistry is irreversible. Unlike traditional iterative cycles of synthesis and screening to generate leads, these methodologies allow for generation of libraries based on spontaneous assembly of the reactive fragments.

Our proposed fragment-to-lead-chart demonstrating the activity enhancement achievable via the fragment self-assembly approach is shown in Fig. 15.4A. The defined target area in this figure is the same as that described in the fragment linking approach. Just as with that approach, the affinity of the assembled compound in the self-assembly approach is expected to be approximately equal to at least the sum of the affinities of the fragments used as the reagents for synthesizing the assembled compound, and enhancement of activity via the linking effect can also be expected. Because the fragment self-assembly approach involves direct linking without using a linker, HA of the assembled compound is generally equal to the sum of the fragment HA, or sometimes slightly lower. Taking these points into account, the assembled compound is expected to be plotted approximately at point C in the fragment-to-lead chart, shifting upwards from the point marking the sum of the vectors of the two original fragments (reagents).

Because a target protein is used to select the reagents and synthesis of the inhibitor, fragment self-assembly does not require information on the affinities of the fragments involved, and thus information regarding fragment hit activities (points A, B, and C in Fig. 15.4B) are not needed when using approach; however, library design for the reactive fragments and optimization of reaction conditions to avoid degradation of the target protein are extremely important in this approach, and structural information is useful for selecting reagents when designing the fragment library. However, as with the fragment linking approach, using the fragment self-assembly approach presents the difficulty of finding more than two fragments which bind to different parts of the binding pocket.

3.3. Fragment optimization

Fragment optimization is similar to the common hit-to-lead approach, maintaining an almost constant molecule size of the initial hit compound. In this approach, a fragment's activity and drug-like properties are improved by implementing minor structural changes, an effect similar to that observed in traditional optimization research of lead compounds conducted by medicinal chemists.

Our proposed fragment-to-lead-chart demonstrating the activity enhancement achievable via the fragment optimization and fragment evolution approaches is shown in Fig. 15.4C. Although the x-axis of this plot shows molecular size, just as in Fig 15.4A and B, the y-axis shows LE value, unlike the previous figures. While the pK_i value is useful in the fragment linking and self-assembly approaches because it facilitates determining additive properties of activities of two or more fragment hits, LE value is more useful in the fragment optimization and evolution approaches because it facilitates determining activity enhancement of single fragment hit.

Under the fragment optimization approach, hit fragments are optimized to drug leads while maintaining an almost constant molecule size; thus, the process vector for converting a fragment to a drug lead tends to adopt a perpendicular orientation when plotted. The perpendicular dashed line in Fig. 15.4C indicates a molecular weight of $\sim$500 g/mol and is plotted together with a dashed curve showing a pK_i value of 8 (K_i = 10 nM) and the curve that shows the affinity of Kuntz's strongest binding ligands (Orita *et al.*, 2009a). The target area for obtaining drug leads during fragment-to-lead processing are plotted here in a manner similar to that in Fig 15.4A and B. To efficiently obtain drug leads via the fragment optimization approach, we must consider the perpendicular route from point A ($pK_i = 5.4$, LE = 0.35, %LE = 58%) to point B ($pK_i = 8.65$, LE = 0.56, %LE = 93%) shown in Fig. 15.4C.

Given that compounds with low molecular weight generally show good drug-like parameters, the fragment optimization approach is quite useful for obtaining potential drug-leads. However, improving the activity value without increasing molecular weight of a hit fragment can be extremely difficult. Further, the likelihood of successfully completing fragment optimization difficulties reduced with decreasing fragment size; if fragment optimization is performed taking point D ($pK_i = 1.8$, LE = 0.35, %LE = 27%) as the starting point, satisfying the hypothetical criteria for drug leads ($pK_i = 8$) is difficult, as the hypothetical criteria curve is plotted higher than the maximum LE curve when HA is smaller than 13. Under these conditions, it is a reasonable assumption that compounds plotted above the maximum affinity curve are impossible to obtain in general (Orita *et al.*, 2009a).

3.4. Fragment evolution

Fragment evolution involves adding to a compound functional groups that bind to additional parts of the target protein. Of the four methods for converting fragment hits to leads, this approach is the most straightforward procedure and is the most widely and successfully applied. Unlike the fragment linking and fragment self-assembly approaches, fragment evolution does not require two or more hit fragments which bind to different parts of the pocket, and unlike fragment optimization, it does not require enchancement of a hit's LE value. Structure-based drug design facilitates quick, efficient conversion of fragments to drug-leads. Further, practical application of parallel synthesis or combinatorial chemistry in the evolution step is also possible, and the approach is often quite convenient because related compounds containing the hit fragment as a partial structure are already stocked in-house or are commercially available.

As mentioned above, our proposed fragment-to-lead-chart demonstrating the activity enhancement achievable via the fragment evolution approach is shown in Fig. 15.4C. Because hit fragments are typically optimized to drug leads in this approach by adding functional groups that bind to adjacent regions of the binding pocket, HA is generally increased during conversion of fragments to leads in fragment evolution. In contrast, constancy is evident in the LE value for a given series of molecules progressing from fragment hits to drug leads, as described in Section 3.4. Therefore, in the fragment evolution approach, hit fragments are generally optimized to drug leads while maintaining an almost constant LE value, thereby contributing to a horizontal tendency from point A (pK_i = 5.4, LE = 0.35, %LE = 58%) to point C (pK_i = 9.0, LE = 0.35, %LE = 82%) in Fig. 15.4C.

If the hit fragment selected for conversion occurs at point D (pK_i = 2.6, LE = 0.15, %LE = 27%) in Fig. 15.4C, an evolved compound cannot typically be obtained within the target area (gray region), due to the LE value's constancy during hit fragment evolution. When postulating constancy, fragments with LE values below 0.2 must have greater than 55 HA to achieve a pK_i above 8, meaning that the molecular weight of the evolved compound will exceed 600 g/mol, a value too large for a drug lead. Therefore, before advancing to the evolution step in fragment processing, the LE value of the fragment should be increased as much as possible, using techniques such as analogue search.

Effectively obtaining drug leads via this approach will require follow-up synthesis of hit fragments to avoid obtaining low LE values, and accomplishing this will require structural information, such as data regarding crystal structures of hit fragment–target protein complexes. However, more and more successes using this simple fragment evolution technique are being reported, as shown in Table 15.2, whose details are described in Section 4.

4. Following Up on Hits: Anchor-Based Drug Discovery

Target sites of proteins typically contain so-called hot spots: smaller sub-sites of the ligand-binding regions of proteins which contribute a major part of the free energy of ligand binding (DeLano, 2002; Hajduk *et al.*, 2005; Kortemme and Baker, 2002). Hot spots are likely to bind small core drug-like compounds that serve as molecular anchors with higher affinity than other regions of the binding site. However, despite the generally weak affinity of fragment hits identified via FBDD, the much smaller and simple structures of these fragments allow it to exert favorable contact, including key hydrogen bonding, with the target protein, enabling the fragments to bind to target sites with high LE values. As such, the concept that fragment hits preferentially fill hot spots in the ligand binding site is widely accepted (Arkin and Wells, 2004; Ciulli *et al.*, 2006; Thanos *et al.*, 2003). Fesik *et al.* in the Abbott group reported that small molecules bind almost exclusively to well-defined, localized regions of proteins, independent of their affinity (Hajduk *et al.*, 2005), and Murray and Verdonk in the Astex group

Table 15.2 Successful examples of fragment-based drug design

Number / *Target* / Stage	Fragment: Structure interaction	Fragment: Information	Drug/lead: Structure interaction	Drug/lead: Information
Ex. 1 *B-Raf(V600E)* Phase3 Clinical	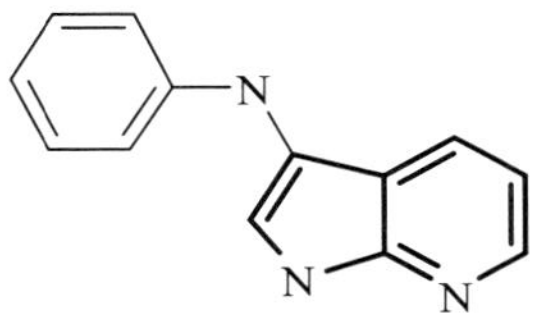	PDB: 3C4E MW: 209 HA: 16 IC_{50}: ~100 μ*M* clogP: 3.38 LE: ~0.34 %LE: ~47% LLE: ~0.62	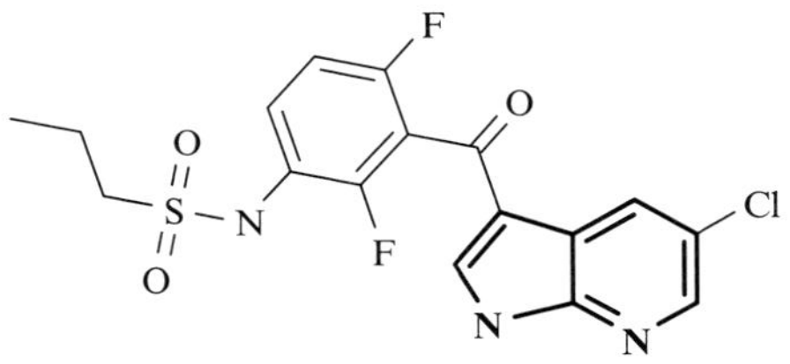	PDB: 3C4C MW: 414 HA: 27 IC_{50}: 0.013 μ*M* clogP: 2.30 LE: 0.40 %LE: 79% LLE: 5.59
RMSD 0.68 Å	#cHB[a]: 1 #non-cHB[b]: 0 #contact[c]: 9		#cHB: 1 #non-cHB: 4 #contact: 12	
Ex. 2 *Aurora Kinase* Phase2 Clinical	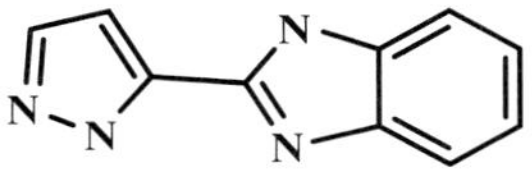	PDB: 2W1D MW: 184 HA: 14 IC_{50}: 910 n*M* clogP: 2.18 LE: 0.59 %LE: 74% LLE: 3.86	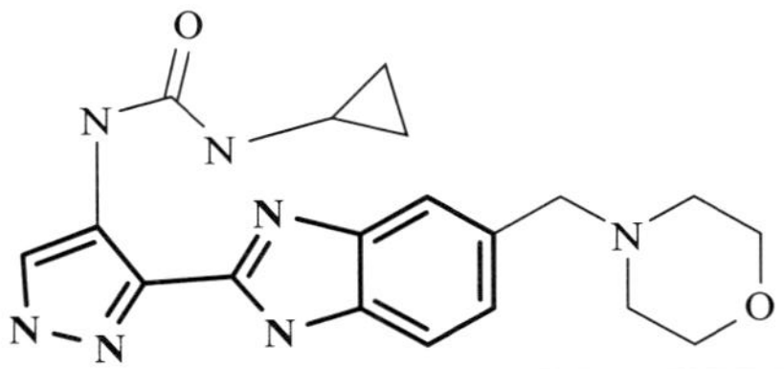	PDB: 2W1G MW: 382 HA: 28 IC_{50}: 52% at 3 n*M* [d] clogP: 2.41 LE: 0.42 %LE: 85% LLE: 6.15

(continued)

Table 15.2 (*continued*)

Number *Target* Stage	Fragment		Drug/lead	
	Structure interaction	Information	Structure interaction	Information
RMSD 0.46 Å	#cHB: 3 #non-cHB: 0 #contact: 5		#cHB: 3 #non-cHB: 0 #contact: 11	
Ex. 3 *CDK2* Phase2 Clinical	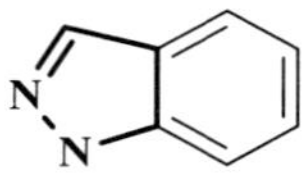	PDB: 2VTA MW: 118 HA: 9 IC_{50}: 64% at 1 mM [e] clogP: 1.63 LE: 0.64 %LE: 60 LLE: 2.62	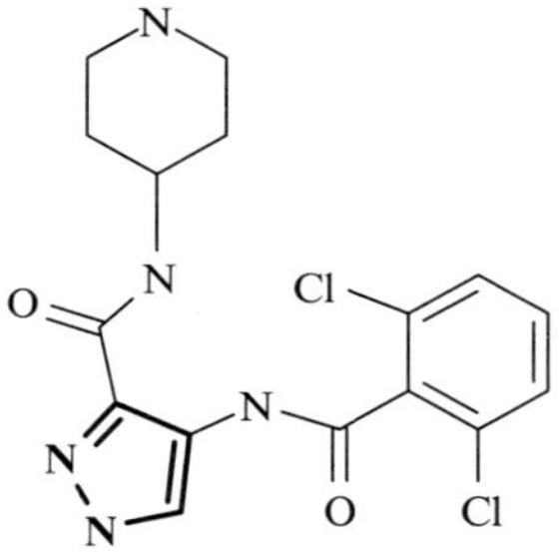	PDB: 2VU3 MW: 382 HA: 25 IC_{50}: 0.047 μ*M* clogP: 0.33 LE: 0.40 %LE: 75% LLE: 7.00
RMSD 0.67 Å	#cHB: 2 #non-cHB: 0 #contact: 5		#cHB: 2 #non-cHB: 2 #contact: 12	

Ex. 4 *HSP90* Phase2 Clinical	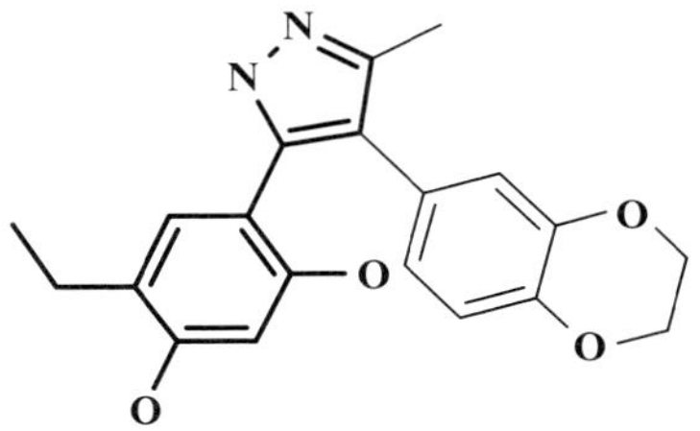	PDB: 2BT0 MW: 352 HA: 26 IC_{50}: 0.28 μM clogP: 3.01 LE: 0.34 %LE: 67% LLE: 3.54	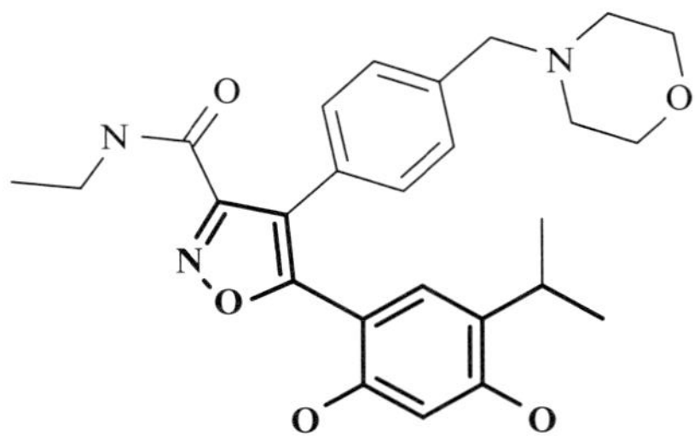	PDB: 2VCI MW: 466 HA: 34 IC_{50}: 0.021 μM clogP: 2.81 LE: 0.31 %LE: 72% LLE: 4.87
RMSD 0.33 Å	#cHB: 2 #non-cHB: 0 #contact: 11		#cHB: 2 #non-cHB: 2 #contact: 12	
Ex. 5 *PPARs* Phase2 Clinical	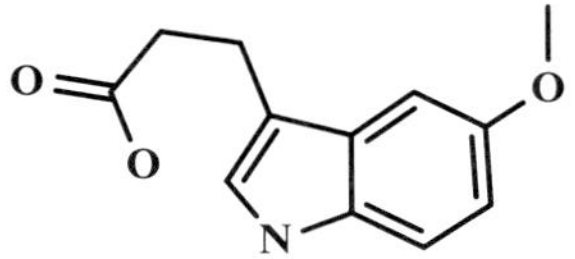	PDB: 3ET0 MW: 219 HA: 16 EC_{50}: ~150 μM clogP: 1.91 LE: ~0.33 %LE: ~45% LLE: ~1.91	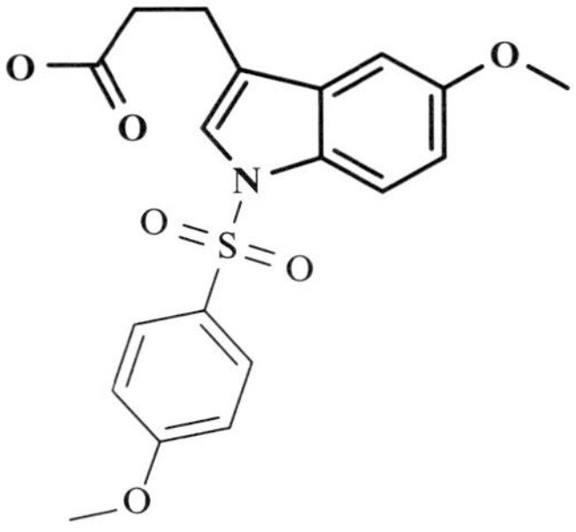	PDB: 3ET3 MW: 389 HA: 27 EC_{50}: 0.37 μM clogP: 3.63 LE: 0.32 %LE: 65% LLE: 2.8
RMSD 0.62 Å	#cHB: 4 #non-cHB: 0 #contact: 9		#cHB: 4 #non-cHB: 0 #contact: 13	

(continued)

Table 15.2 (*continued*)

Number *Target* Stage	Fragment		Drug/lead	
	Structure interaction	Information	Structure interaction	Information
Ex. 6 *LTA4H* Phase2 Clinical (discontinued)	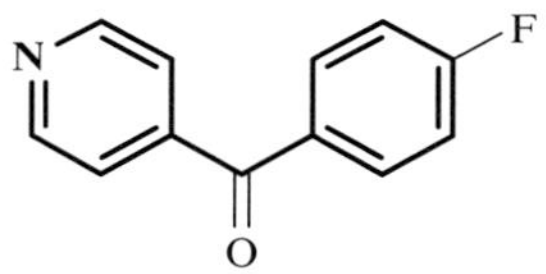	PDB: 3FU0 MW: 201 HA: 15 IC_{50}: 5.3 m*M* clogP: 2.25 LE: 0.21 %LE: 27% LLE: 0.03	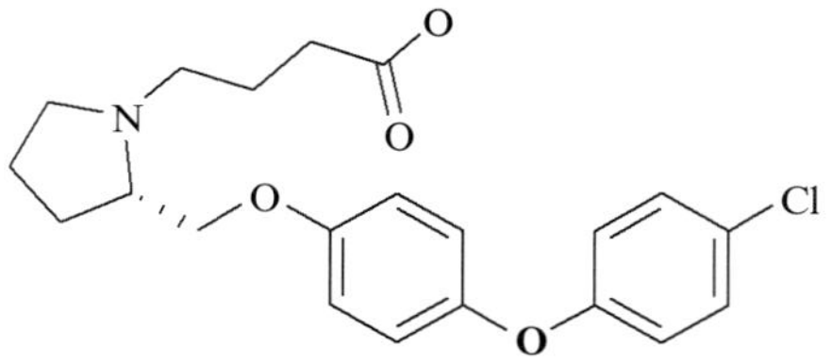	PDB: 3FH7 MW: 389 HA: 32 IC_{50}: 4.2 μ*M* clogP: 4.31 LE: 0.23 %LE: 51% LLE: 1.07
RMSD 1.24 Å	#cHB: 0 [e] #non-cHB: 0 #contact: 6		#cHB: 0 #non-cHB: 4 #contact: 17	
Ex. 7 *CDK2* Discovery	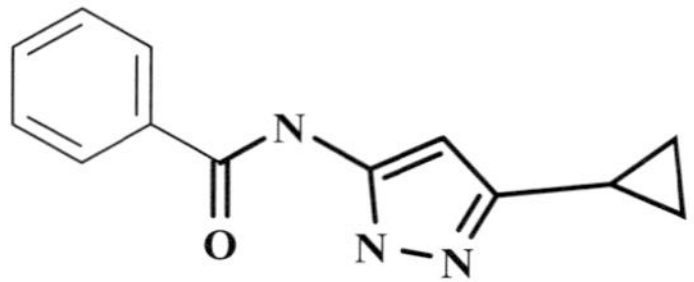	PDB: 1VYZ MW: 227 HA: 17 IC_{50}: 0.29 μ*M* clogP: 2.24 LE: 0.52 %LE: 76% LLE: 4.30	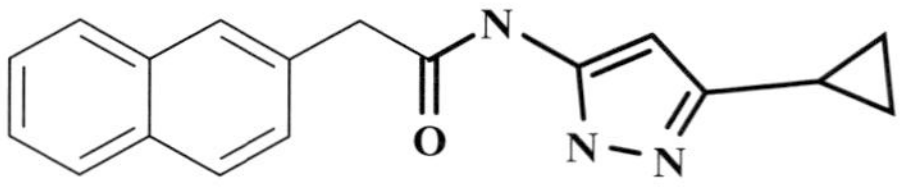	PDB: 1VYW MW: 291 HA: 22 IC_{50}: 0.037 μ*M* clogP: 3.47 LE: 0.46 %LE: 80% LLE: 3.96
RMSD 0.77 Å	#cHB: 3 #non-cHB: 0 #contact: 10 [f]		#cHB: 3 #non-cHB: 0 #contact: 8	

Ex. 8 *p38 α* Discovery	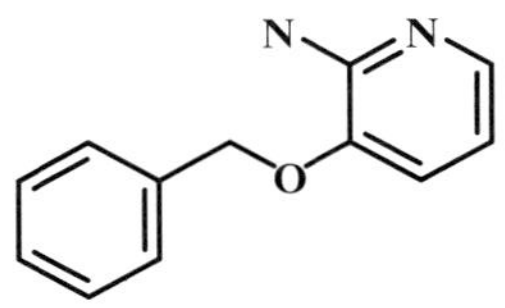	PDB: 1W7H MW: 200 HA: 15 IC_{50}: 1300 μ*M* clogP: 2.71 LE: 0.26 %LE: 35% LLE: 0.18	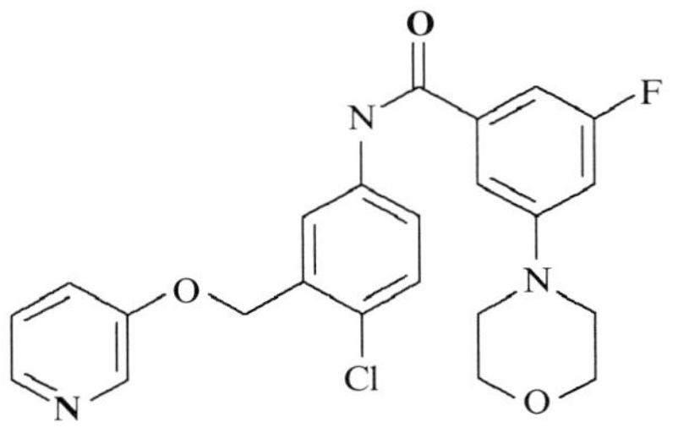	PDB: 1W83 MW: 446 HA: 31 IC_{50}: 0.065 μ*M* clogP: 3.03 LE: 0.32 %LE: 69% LLE: 4.16
RMSD 1.49 Å	#cHB: 1 #non-cHB: 1 #contact: 9		#cHB: 1 #non-cHB: 2 #contact: 11	
Ex. 9 *p38 α* Discovery	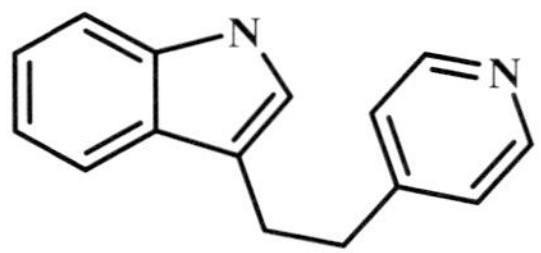	PDB: 1W84 MW: 222 HA: 17 IC_{50}: 35 μ*M* clogP: 3.08 LE: 0.36 %LE: 52% LLE: 1.38	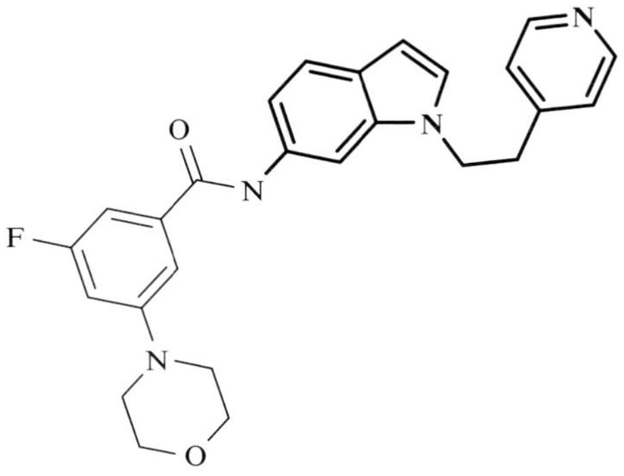	PDB: 1WBT MW: 445 HA: 33 IC_{50}: 0.34 μ*M* clogP: 3.80 LE: 0.27 %LE: 61% LLE: 2.67
RMSD 0.94 Å	#cHB: 1 #non-cHB: 1 #contact: 8		#cHB: 1 #non-cHB: 2 #contact: 12	

(*continued*)

Table 15.2 *(continued)*

Number *Target* Stage	Fragment		Drug/lead	
	Structure interaction	Information	Structure interaction	Information
Ex. 10 *Protein Kinase B* Discovery	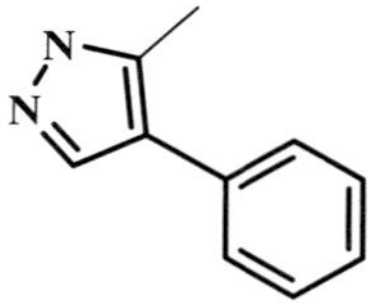	PDB: 2UW3 MW: 158 HA: 12 IC_{50}: 80 μ*M* clogP: 2.10 LE: 0.47 %LE: 53% LLE: 2.00	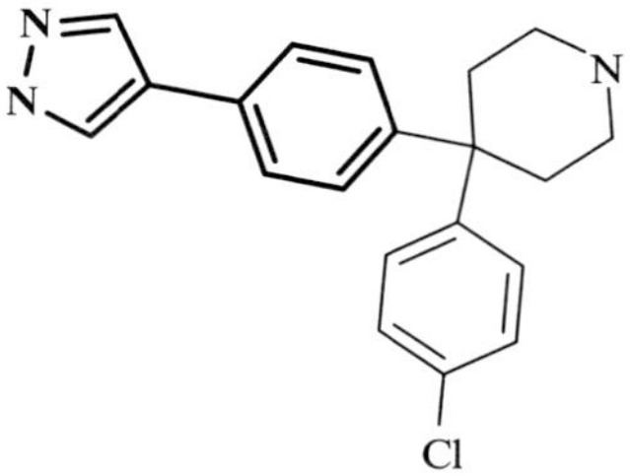	PDB: 2UW7 MW: 338 HA: 24 IC_{50}: 0.018 μ*M* clogP: 4.60 LE: 0.44 %LE: 81% LLE: 3.14
RMSD 0.53 Å	#cHB: 2 #non-cHB: 0 #contact: 10		#cHB: 2 #non-cHB: 1 #contact: 11	
Ex. 11 *Protein Kinase B* Discovery	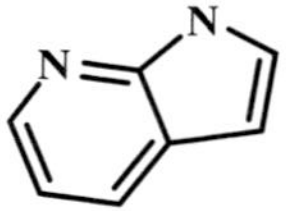	PDB: 2UVX MW: 118 HA: 9 IC_{50}: $>$100 μ*M* clogP: 1.18 LE: $<$0.61 %LE: $<$56% LLE: $<$2.82		PDB: 2VO7 MW: 343 HA: 24 IC_{50}: 0.006 μ*M* clogP: 2.80 LE: 0.47 %LE: 86% LLE: 5.42
RMSD 0.54 Å	#cHB: 2 #non-cHB: 0 #contact: 10		#cHB: 2 #non-cHB: 0 #contact: 15	

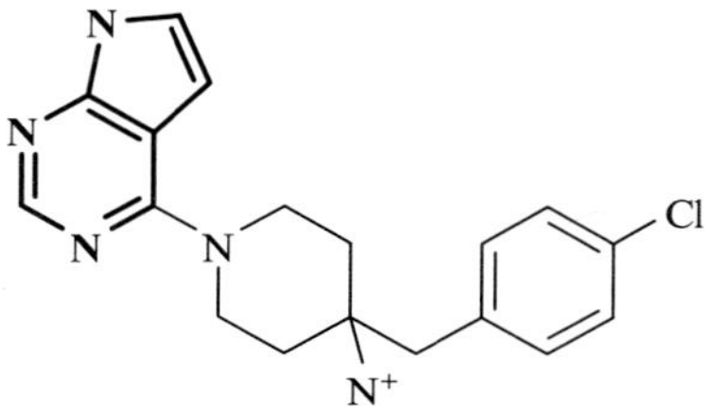

Ex. 12 *ERK2* Discovery	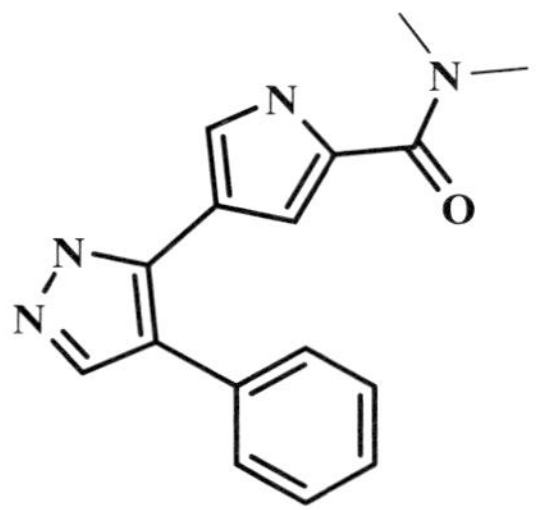	PDB: 2OJG MW: 280 HA: 21 IC_{50}: 2.3 μ*M* clogP: 1.19 LE: 0.37 %LE: 61% LLE: 4.45	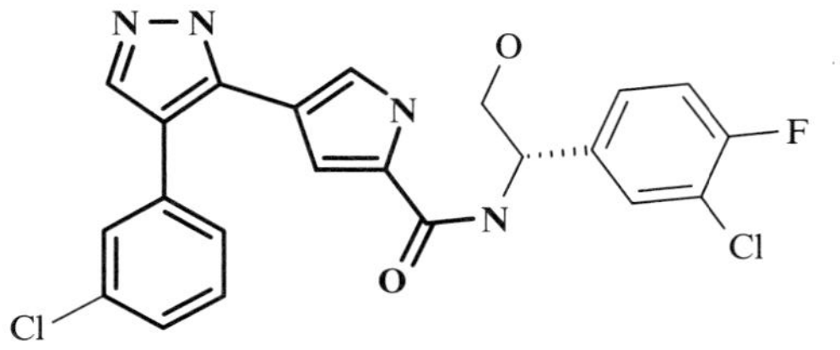	PDB: 2OJJ MW: 459 HA: 31 IC_{50}: 0.002 μ*M* clogP: 4.32 LE: 0.38 %LE: 84% LLE: 4.38
RMSD 0.31 Å	#cHB: 4 #non-cHB: 0 #contact: 12[f]		#cHB: 4 #non-cHB: 0 #contact: 11	
Ex. 13 *JAK2* Discovery	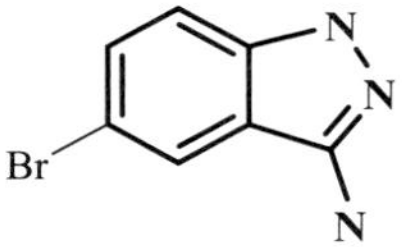	PDB: 3E62 MW: 212 HA: 11 IC_{50}: 40.9 μ*M* clogP: 2.24 LE: 0.54 %LE: 58% LLE: 2.15	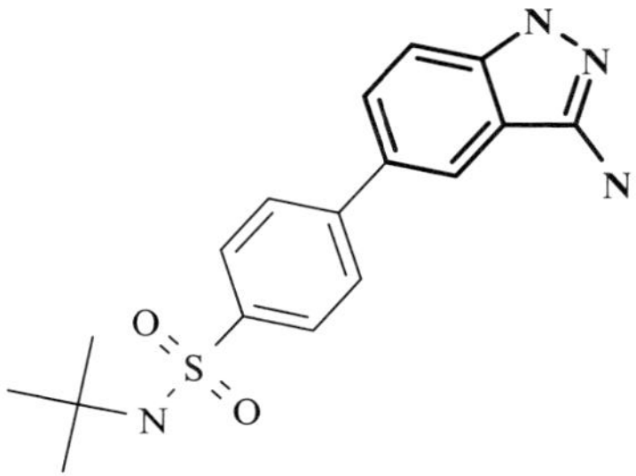	PDB: 3E64 MW: 344 HA: 24 IC_{50}: 0.078 μ*M* clogP: 3.15 LE: 0.40 %LE: 74% LLE: 3.96
RMSD 0.21 Å	#cHB: 3 #non-cHB: 0 #contact: 6		#cHB: 3 #non-cHB: 0 #contact: 12	

(continued)

Table 15.2 *(continued)*

Number *Target* Stage	Fragment Structure interaction	Fragment Information	Drug/lead Structure interaction	Drug/lead Information
Ex. 14 *BACE1* Discovery	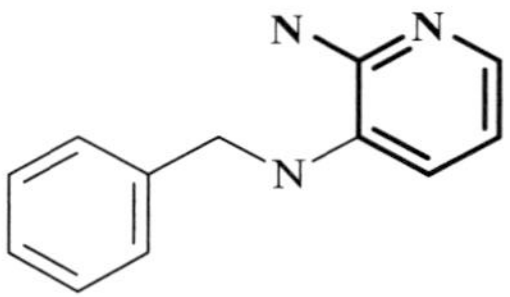	PDB: 2OHM MW: 199 HA: 15 IC_{50}: 310 μ*M* clogP: 2.24 LE: 0.32 %LE: 42% LLE: 1.27	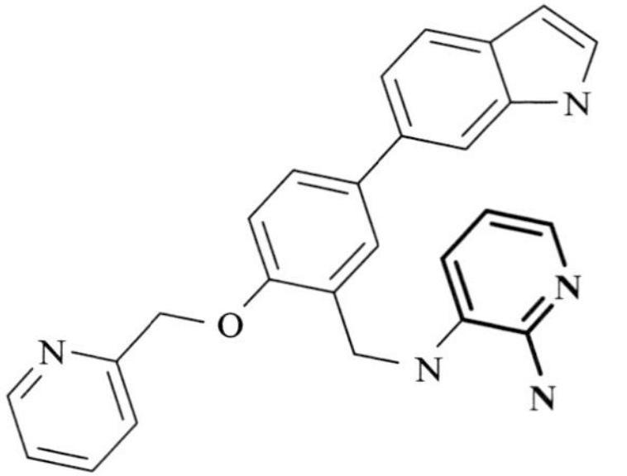	PDB: 2OHU MW: 422 HA: 32 IC_{50}: 4.2 μ*M* clogP: 4.31 LE: 0.23 %LE: 51% LLE: 1.07
RMSD 1.71 Å	#cHB: 4 #non-cHB: 0 #contact: 6		#cHB: 4 #non-cHB: 2 #contact: 11	
Ex. 15 *BACE1* Discovery	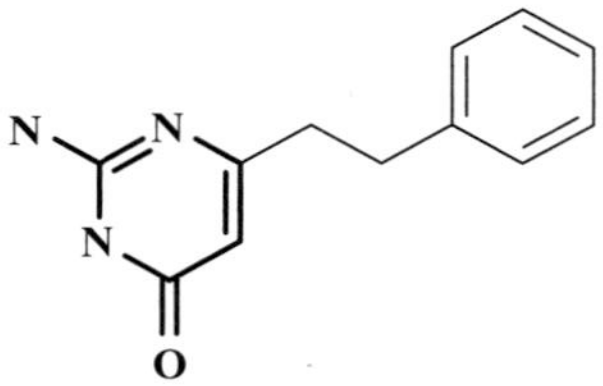	PDB: 2V00 [g] MW: 215 HA: 16 IC_{50}: 660 μ*M* clogP: 1.01 LE: 0.27 %LE: 38% LLE: 2.17	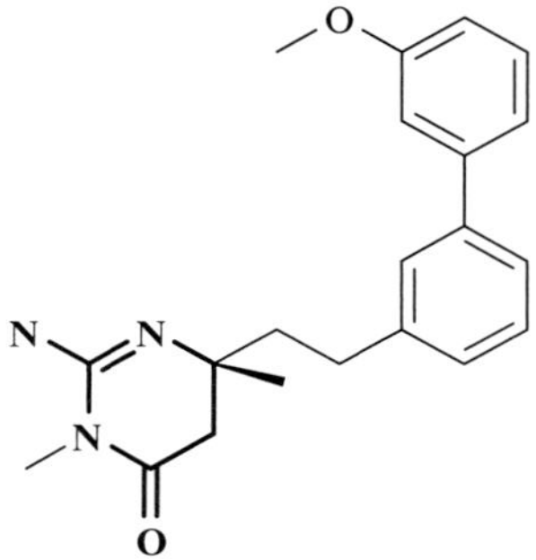	PDB: 2VA7 MW: 351 HA: 26 IC_{50}: 0.08 μ*M* clogP: 3.42 LE: 0.37 %LE: 72% LLE: 3.68

RMSD 2.35 Å	#cHB: 4 #non-cHB: 1 #contact: 8		#cHB: 4 #non-cHB: 0 #contact: 12	
Ex. 16 *Thrombin* Discovery	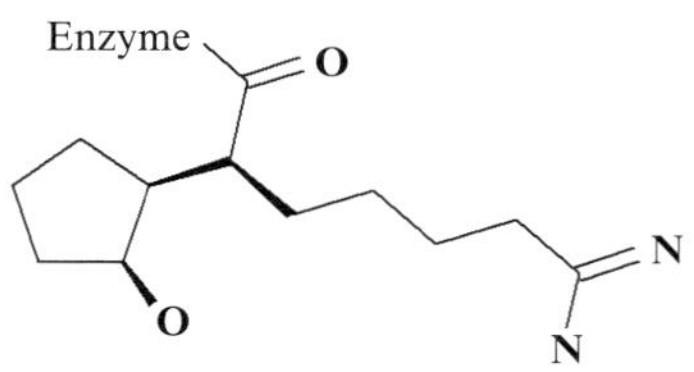	PDB: 1QHR MW: 226 HA: 16 IC_{50}: 0.13 μ*M* clogP: 0.00 LE: 0.59 %LE: 81% LLE: 6.89	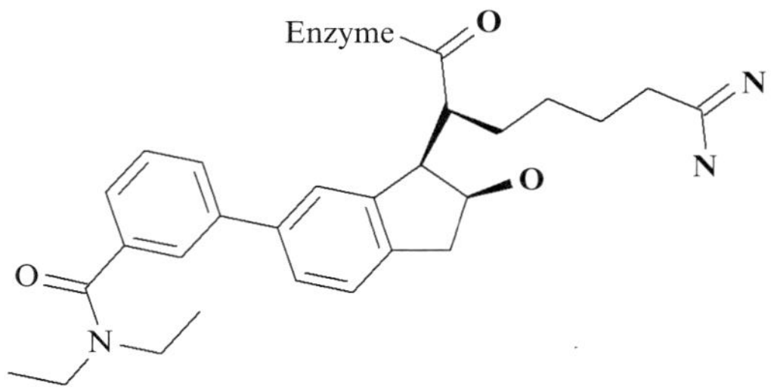	PDB: 1QJ7 MW: 466 HA: 33 IC_{50}:0.008 μ*M* clogP: −0.55 LE: 0.33 %LE: 77% LLE: 8.65
RMSD 0.78 Å	#cHB: 6 #non-cHB: 0 #contact: 12		#cHB: 6 #non-cHB: 1 #contact: 14	
Ex. 17 *uPA* Discovery	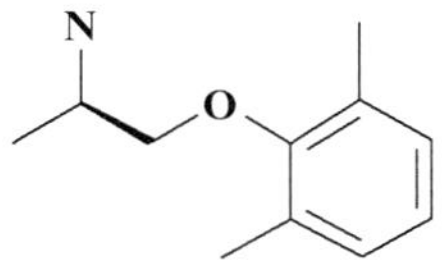	PDB: 2VIN MW: 179 HA: 13 IC_{50}: >1000 μ*M* clogP: 2.57 LE: <0.31 %LE: <38% LLE: <0.43	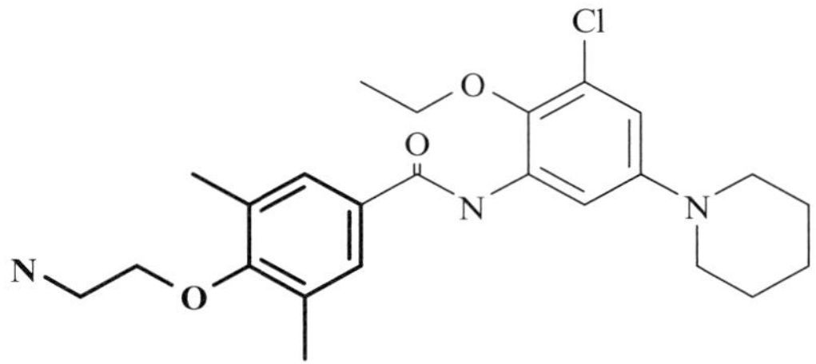	PDB: 2VIW MW: 446 HA: 31 IC_{50}: 0.072 μ*M* clogP: 4.51 LE: 0.31 %LE: 69% LLE: 2.63
RMSD 0.45 Å	#cHB: 3 #non-cHB: 0 #contact: 9		#cHB: 3 #non-cHB: 1 #contact: 10	

(continued)

Table 15.2 *(continued)*

Number *Target* Stage	Fragment Structure interaction	 Information	Drug/lead Structure interaction	 Information
Ex. 18 HSP90 Discovery	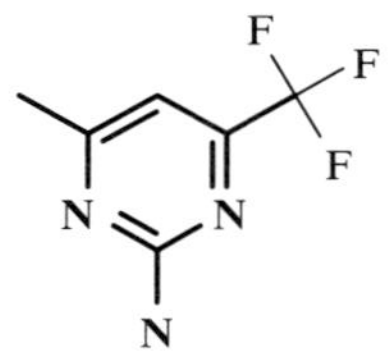	PDB: 2QFO MW: 177 HA: 12 IC_{50}: 18 μ*M* clogP: 1.14 LE: 0.54 %LE: 61% LLE: 3.60	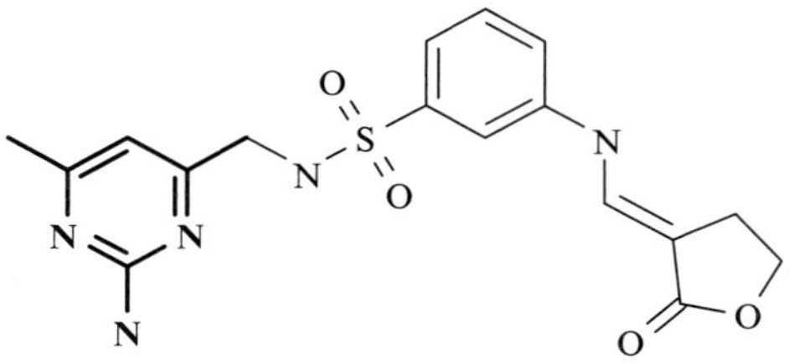	PDB: 2QG0 MW: 389 HA: 27 IC_{50}: 1.9 μ*M* clogP: 0.76 LE: 0.29 %LE: 58% LLE: 4.96
RMSD 0.16 Å	#cHB: 1 #non-cHB: 0 #contact: 5		#cHB: 1 #non-cHB: 3 #contact: 12	
Ex. 19 *HSP90* Discovery	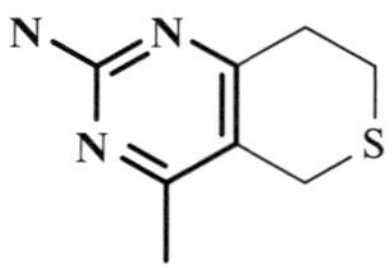	PDB: 3FT5 MW: 181 HA: 12 IC_{50}: 15 μ*M* clogP: 0.70 LE: 0.55 %LE: 62% LLE: 4.12	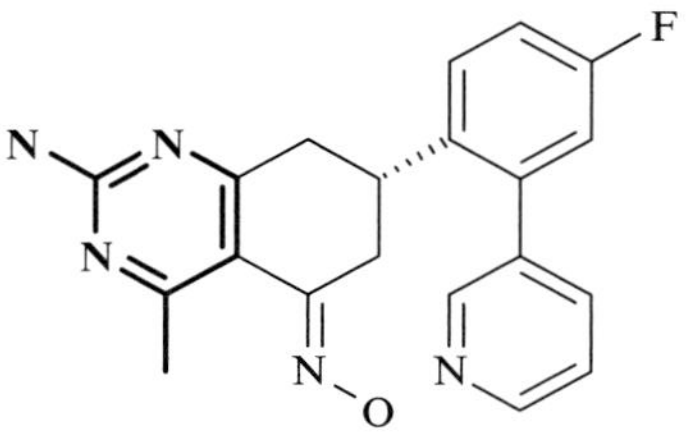	PDB: 3FT8 MW: 363 HA: 27 IC_{50}: 0.03 μ*M* clogP: 3.72 LE: 0.38 %LE: 76% LLE: 3.80
RMSD 0.35 Å	#cHB: 1 #non-cHB: 0 #contact: 5		#cHB: 1 #non-cHB: 0 #contact: 10	

	Fragment		Lead	
Ex. 20 *HSP90* Discovery	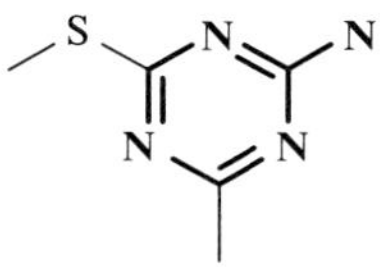	PDB: 2WI2 MW: 156 HA: 10 IC_{50}: 350 μ*M* clogP: 0.57 LE: 0.47 %LE: 47% LLE: 2.89	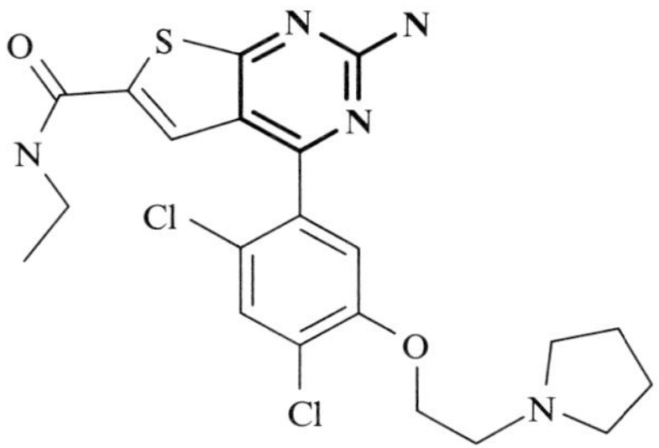	PDB: 2WI7 MW: 480 HA: 31 IC_{50}: 0.058 μ*M* clogP: 4.76 LE: 0.32 %LE: 70% LLE: 2.48
RMSD 0.50 Å	#cHB: 1 #non-cHB: 0 #contact: 3		#cHB: 1 #non-cHB: 1 #contact: 7	
EX. 21 *PDE4* Discovery	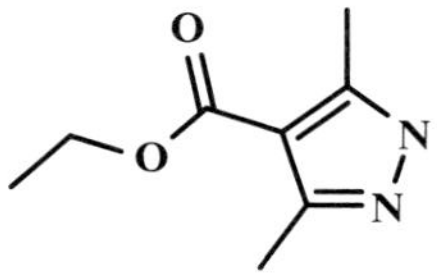	PDB: 1Y2B MW: 168 HA: 12 IC_{50}: 82 μ*M* clogP: 1.69 LE: 0.46 %LE: 53% LLE: 2.40	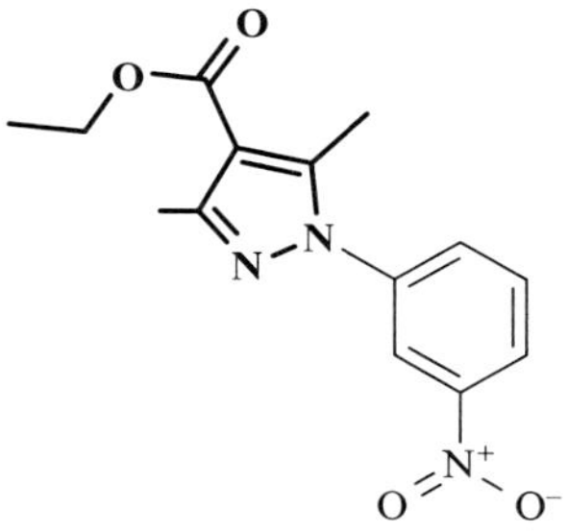	PDB: 1Y2K MW: 289 HA: 21 IC_{50}: 0.021 μ*M* clogP: 3.22 LE: 0.50 %LE: 83% LLE: 4.46
RMSD 0.30 Å	#cHB: 1 #non-cHB: 0 #contact: 6		#cHB: 1 #non-cHB: 0 #contact: 9	

(*continued*)

Table 15.2 *(continued)*

Number *Target* Stage	Fragment		Drug/lead	
	Structure interaction	Information	Structure interaction	Information
Ex. 22 DHNA Discovery	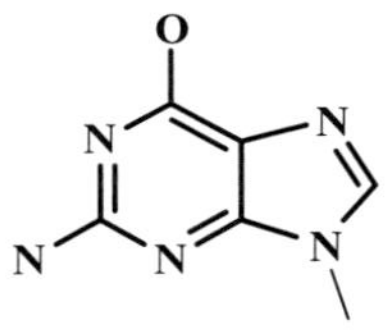 O, N, N, N, N, N	PDB: 1RRW MW: 165 HA: 12 IC_{50}: 28 μ*M* clogP: 0.31 LE: 0.52 %LE: 59% LLE: 4.24	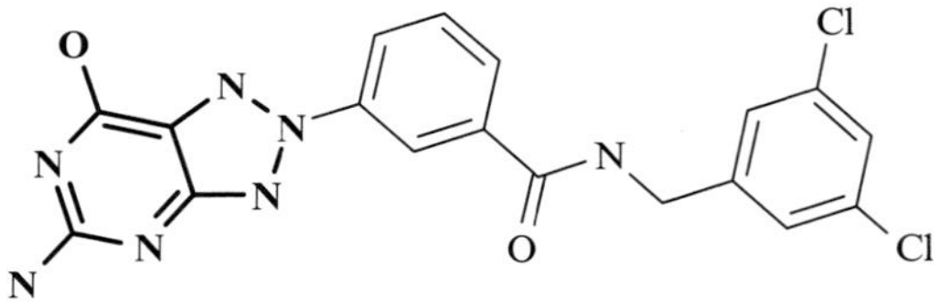 O, N, N, N, N, N, N, N, O, Cl, Cl	PDB: 1RS4 MW: 432 HA: 29 IC_{50}: 0.068 μ*M* clogP: 2.98 LE: 0.34% LE: 71% LLE: 4.19
RMSD 0.81 Å	#cHB: 3 #non-cHB: 0 #contact: 3		#cHB: 3 #non-cHB: 1 #contact: 6	
Ex. 23 *NS5B-RNA polymerase (HCV)* Discovery	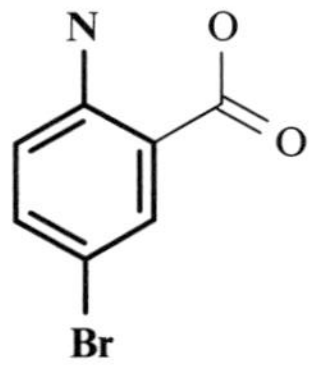 N, O, O, Br	PDB: 3CIZ MW: 216 HA: 11 IC_{50}: 200–500 μ*M* clogP: 2.20 LE: 0.46–0.41 %LE: 49–44% LLE: 1.50–1.10	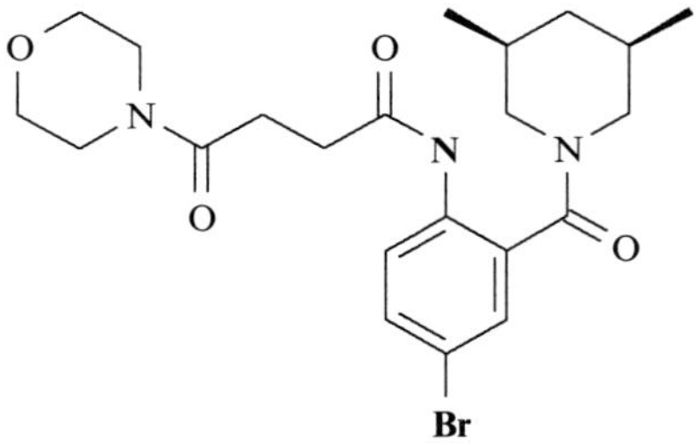 O, N, O, O, N, N, O, Br	PDB: 3CJ5 MW: 480 HA: 30 IC_{50}: 0.25 μ*M* clogP: 3.12 LE: 0.30 %LE: 64% LLE: 3.48
RMSD 0.96 Å	#cHB: 0 #non-cHB: 1 #contact: 3		#cHB: 0 #non-cHB: 3 #contact: 9	

Example	Fragment	Fragment data	Lead	Lead data
Ex. 24 *Mycobacterium tuberculosis Pantothenate Synthetase* Discovery	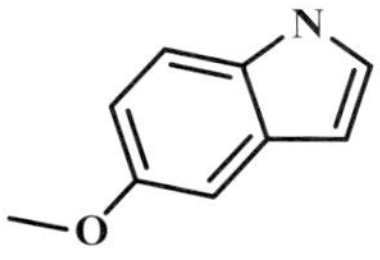	PDB: 3IMG MW: 147 HA: 11 IC_{50}: 1100 μ*M* clogP: 2.15 LE: 0.37 %LE: 39% LLE: 0.81	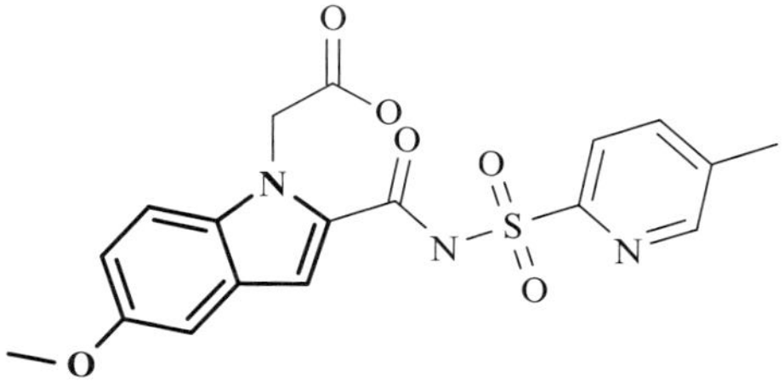	PDB: 3IUE MW: 403 HA: 28 IC_{50}: 1.5 μ*M* clogP: 3.02 LE: 0.28 %LE: 58% LLE: 2.80
RMSD 0.29 Å	#cHB: 1 #non-cHB: 1 #contact: 8		#cHB: 1 #non-cHB: 5 #contact: 13	
Ex. 25 MetAP2 Discovery	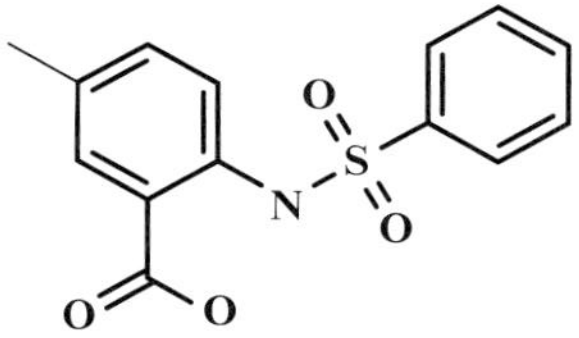	PDB: 1YW7 MW: 291 HA: 20 IC_{50}: 1.4 μ*M* clogP: 3.68 LE: 0.40 %LE: 65% LLE: 2.17	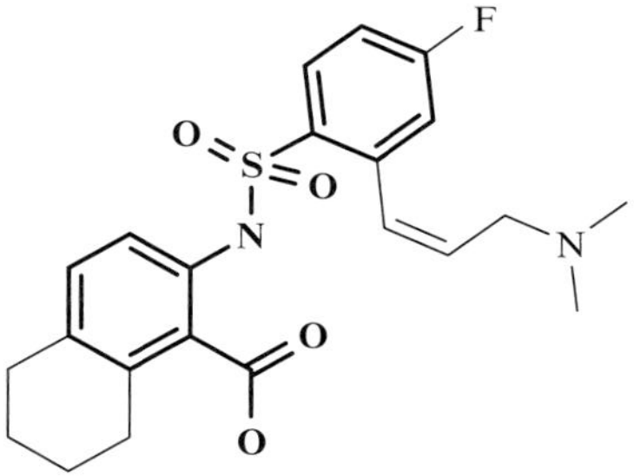	PDB: 1YW9 MW: 433 HA: 30 IC_{50}: 0.019 μ*M* clogP: 3.03 LE: 0.35 %LE: 75% LLE: 4.69
RMSD 0.28 Å	#cHB: 3 #non-cHB: 0 #contact: 11		#cHB: 3 #non-cHB: 1 #contact: 12	

[a] #cHB: number of conserved hydrogen bonds between fragment and lead. The distance between the hydrogen bond donor and acceptor should be ≤3.3 Å.
[b] #non-cHB: number of nonconserved hydrogen bonds between a fragment and lead.
[c] #contact: number of residues within 4.0 Å of a fragment or lead. #cHB, #non-cHB, and #contact were calculated using Ligand Explorer 3.8.RMSD values are calculated between the coordinates of core structures in the fragment hit and the lead compound indicated in bold. Calculations are made based on the pair of superimposed target proteins complexed with the fragment and lead, in which all calcium atoms are used in superposition.
[d] A logit function was used to estimate the pIC_{50} value.
[e] One indirect hydrogen bond is mediated by a water molecule.
[f] Due to conformational changes caused by the lead compound, the number of the contact residues decreases despite an increase in molecular weight.
[g] Endothiapepsin was used as a surrogate for BACE-1.

suggested that the molecular anchor is a fragment that will bind to the enzyme with low affinity (Murray and Verdonk, 2002). Because effective improvement of affinity, including selectivity, can also be expected with the addition of interactions with neighboring regions of the binding pocket to a fragment anchor, identification of hot spots and fragment anchors are important steps in the FBDD approach (Landon *et al.*, 2007, 2009).

Table 15.2 lists examples of compounds developed via FBDD approaches. Examples which satisfied the following two criteria are listed in the table: the crystal structures of the target protein (or surrogate) complexed with both fragment and drug lead were deposited in the Protein Data Bank, and the binding affinity (as IC_{50}, EC_{50}, %inhibition, or K_d) was reported for both the fragment and drug lead. Using these criteria, we identified 25 fragment/drug lead pairs. A total of 12 companies or academic institutions had conducted research involving these examples, with target proteins of 17 enzymes and 1 receptor. Of the 25 examples, six drugs have advanced into the clinical study stage, with the highest developing stage being Phase 3 (PLX-4720; Example 1 in Table 15.2). Key points characterizing these successful examples are as follows:

(a) Generally, at least one hydrogen bond exists between each fragment hit and the target protein.
(b) Generally, hydrogen bonds to the target protein are conserved between the fragment hit and drug lead, although the hydrogen bond may disappear on the chemical modification of the fragment hit.
(c) Generally, the binding modes of the core structure of fragment hit are conserved in the drug lead. The average root-mean-square distance (RMSD) of a core structure between fragment hit and drug lead is 0.71 Å.
(d) During processing of fragment hits to drug leads, the average pIC_{50} and number of contact residues increases from 4.5 to 7.3 and from 7.4 to 10.5, respectively. The improvement in activity is believed to be due to increased interactions.
(e) During processing of fragment hits to drug leads, the average value of clogP increases from 1.89 to 3.10, indicating that lipophilicity of hit fragment increases in follow-up synthesis.
(f) The improved activity of hit fragments from points (d) and (e) is believed to come mainly from hydrophobic interactions between additional lipophilic portions of drug leads and the secondary site of the target pocket.
(g) Generally, LE, LLE, and %LE values for starting fragments exceed 0.30, 1.0, and 40%, respectively.

Rejto and Verkhivker in the Agouron group suggested that an effective molecular anchor must meet both the thermodynamic requirement of relative energetic stability of a single binding mode and have consistent kinetic accessibility (Rejto and Verkhivker, 1996). We suggest that almost every fragment hit listed in Table 15.2 can function as a fragment anchor, because crystallographic studies have proven that the binding mode of a core fragment structure is the same between the initial fragment hit and final drug lead. For example, as an example of a compound obtained via FBDD in Table 15.2, ERK2 inhibitor (Aronov *et al.*, 2007) is discussed (Example 12 in Table 15.2). Figure 15.5 shows a superimposition of the crystal structures of pyrazolylpyrrole derivatives complexed with ERK, including initial fragment hits and final drug leads, which was reported in a paper released by the Vertex group. As is evident in the figure, the binding modes of the core pyrazolylpyrrole scaffold, including hydrogen bonding with the hinge region of the kinase, are completely conserved (RMSD of a core structure between fragment hit and drug lead is very small, 0.31 Å), and the pyrazolylpyrrole scaffold also has accessible affinity (K_i = 2.3 μM) and high ligand efficiency (LE = 0.37). Taken together, these findings indicate that pyrazolylpyrrole scaffold has relative energetic stability for a single binding mode, supporting the notion that pyrazolylpyrrole scaffold can function as an effective molecular anchor. Such anchoring of the fragment can be further confirmed in many other examples in Table 15.2.

The above points (a), (b), and (c) indicate that the examples shown in Table 15.2 can prove the importance of fragment anchors for successful

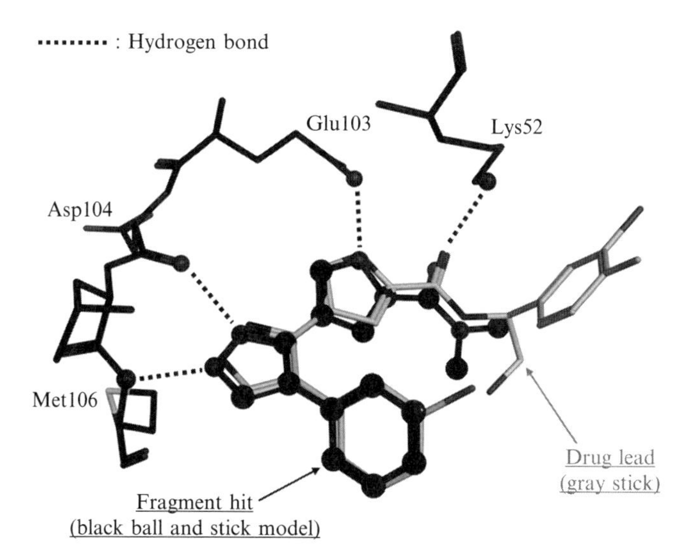

Figure 15.5 Overlay of binding modes of initial fragment hits (black ball and stick model) and final drug lead (gray stick) of pyrazolylpyrrole derivatives complexed with ERK. The dashed lines indicate hydrogen bonds. The binding modes of the core pyrazolylpyrrole scaffold and four hydrogen bonds are completely conserved.

FBDD, and as such the modern FBDD approach can be redubbed "anchor-based drug discovery" (ABDD). Further, points (d), (e), and (f) indicate that the second interaction site of the target protein is mainly hydrophobic, and thus addition of a hydrophobic group to a fragment hit is useful in achieving greater interaction. Point (g) clarifies that starting fragments should have good LE, LLE, and %LE values, a finding consistent with our previous study (Orita *et al.*, 2009a). Taken together, these points all indicate that optimization of fragment anchors is important for successful FBDD.

Figure 15.6 represents our ideal image combining ABDD with the fragment-to-lead chart in Fig. 15.4D.

(1) In this case, the protein target sites are believed to be mainly two sites: a hot spot and the secondary site (another region of the binding site), and fragment anchors bind to the hot spot.
(2) We believe that key requirement of the molecular anchor is hydrogen bonding to the hot spot, which fixes the binding mode, as observed with pyrazolylpyrrole ERK2 inhibitor (Example 12 in Table 15.2). As such, we depicted the hot spot to be a hydrophobic and accessible region which includes a donor or acceptor site for a hydrogen bond. This is a reasonable postulation in many cases (Hajduk *et al.*, 2005), and hydrogen bonds are known to strengthen in a hydrophobic environment (Young *et al.*, 2007).
(3) In ABDD, the fragment library should be designed such that hydrogen bonging donors or acceptors can interact efficiently with hot spots. Since fragment screening is performed in high concentrations in order to detect weak interactions between fragments and a target site, such a design would be particularly useful, as it would predictably cause the

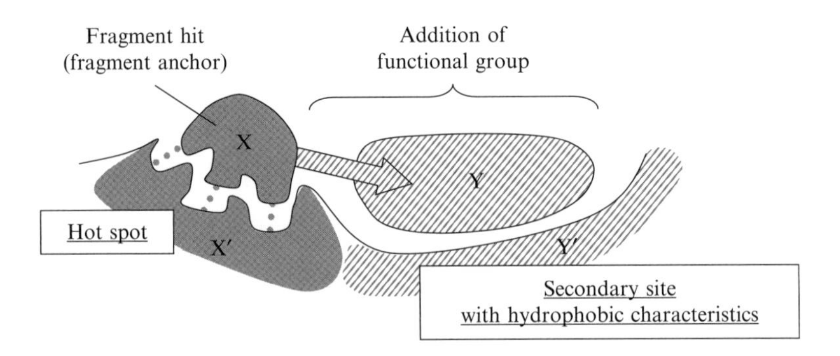

Figure 15.6 Schematic for anchor-based drug discovery. Regions X′ and Y′ indicate hot spots and secondary sites in the binding pocket, respectively. X is the fragment hit (fragment anchor) and Y is the additional part which is introduced during follow-up synthesis.

fragments in the library to become more soluble. Solubility is also preferred in drugs.

(4) ABDD initially stresses identification of fragment anchors with high LE values. From the fragment-to-lead chart (Fig. 15.4D), we hypothesize that initial hits in fragment screening will be plotted at point A. As described previously, fragment-to-lead processing using fragment optimization and fragment evolution approaches result in increases in x- and y-axis values, respectively (directions described with arrows in Fig. 15.4D). After obtaining an initial fragment hit, an analogue search maintaining an almost constant molecule size should be performed. This process can be expressed in the conversion of point A to B or C in Fig. 15.4D and is the same as in the fragment optimization approach.

(5) If the analogue search in (4) leads to point C, this indicates that the drug lead can only be obtained via fragment optimization. However, such an approach is difficult in many cases, particularly for compounds with fewer than 13 HA, as described previously. More often than not, only slightly active compounds (plotted at point B) are obtained, or no compound stronger than that at point A is found. In such cases, the fragment anchor must next be identified and extended.

(6) The validation of the fragment anchor which should be performed follow-up synthesis toward drug lead should be determined using hydrogen bond anchoring, as well as the activity such as LE and %LE which are metrics to assess the quality of fragment hits, and the chemical tractability for addition of functional groups that bind to secondary site of the target protein. Crystallographic study therefore is important for not only evidencing specific binding between fragment anchors and target proteins, but also in obtaining relevant and detailed descriptions of three-dimensional interaction.

(7) We have depicted the secondary site as being primarily hydrophobic, as noted in points (d), (e), and (f) above. We therefore believe that adding lipophilic parts to fragment anchors is an effective way of improving potency of a fragment hit. Fortunately, chemical addition of a lipophilic moiety is simpler and easier to accomplish than adding a hydrophilic moiety, particularly in the case of parallel or combinatorial synthesis; further, a large number of compounds can be synthesized simultaneously, as reactions using lipophilic reagents generally do not require protecting groups.

(8) During fragment-to-lead processing, the lipophilicity of a compound generally increases with the addition of lipophilic moieties. Activity of fragment hits must be improved using structural information while carefully monitoring ADME/Tox parameters. In the fragment-to-lead chart (Fig. 15.4D), addition of a lipophilic moiety corresponds to routes from A to A′ or B to B′ and is the same as that using the fragment evolution approach.

(9) As mentioned above, during fragment-to-lead processing, an analogue search is conducted to identify hit fragments, and the optimized fragment is then evolved. However, some researchers first evolve an identified hit fragment, and then conduct optimization of the fragment subsequently. In our experience, we have found that the former method is "better" from a productivity point of view, for the following reasons: (A) The affinity of a starting fragment largely affects that of the lead. In other words, weak activity of a hit fragment or its derivatives presents an unattractive scaffold (Fig. 15.4C, case E). (B) Evolving from a fragment with weak activity sometimes causes a change in a fragment's binding mode, resulting in a confusing SAR pattern. To obtain a clear understanding of SAR, a strong fragment should be adopted as a starting point. (C) Fragment derivatives sometimes induce conformational changes in the target protein pocket, opening up novel subpockets. Such derivatives are thought to be promising, because strong inhibitors sometimes cause conformational changes of the target protein. However, these interesting fragments may be missed when using latter approach.

We strongly advise identification of fragment anchors with relative energetic stability in a single binding mode when using an ABDD approach with fragment optimization or evolution. While a strong, active lead may indeed be formed from a fragment showing multiple binding modes with no hydrogen bonds to the target site, there is risk of complicating the structure-based fragment-to-hit conversion, as pointed out by Rejto and Verkhivker (1996). The fragment anchor is the only molecule which provides a direct route to structure-guided parallel or combinatorial synthesis when exploring new inhibitors. We believe that using ABDD approach described in this section can give the greatest chance of success in FBDD and accelerate the lead discovery process.

5. Conclusions

Successful application of FBDD requires not only accurate measurement of the weak affinities of fragments but also enhancement of fragments to drug leads. In the present chapter, we described our opinion regarding how best to follow up the fragment hits. A researcher must select an FBDD approach taking into account the target protein, contents of the fragment library, and the throughput of potential structural methods, such as X-ray crystallography and NMR, and biophysical techniques, such as mass spectrometry, surface plasmon resonance, isothermal titration calorimetry, and high-concentration screening. In some cases, combined use of the four FBDD approaches may be necessary (Niimi *et al.*, 2001; Orita *et al.*,

2009b), and we emphasized in particular the importance of the ABDD approach. Over the past 10 years, FBDD has gained popularity among pharmaceutical companies and members of academia as a simple, quick, and productive approach to identifying drug leads, and more than 10 compounds have already entered clinical development using this technique (Orita *et al.*, 2009b). FBDD approaches will likely continue to attract further attention and lead to development of even more compounds.

REFERENCES

Abad-Zapatero, C., and Metz, J. T. (2005). Ligand efficiency indices as guideposts for drug discovery. *Drug Discov. Today* **10,** 464–469.

Abel, R., Young, T., Farid, R., Berne, B. J., and Friesner, R. A. (2008). Role of the active-site solvent in the thermodynamics of factor Xa ligand binding. *J. Am. Chem. Soc.* **130,** 2817–2831.

Akritopoulou-Zanze, I., and Hajduk, P. J. (2009). Kinase-targeted libraries: The design and synthesis of novel, potent, and selective kinase inhibitors. *Drug Discov. Today* **14,** 291–297.

Alex, A. A., and Flocco, M. M. (2007). Fragment-based drug discovery: What has it achieved so fair? *Curr. Top. Med. Chem.* **7,** 1544–1567.

Arkin, M. R., and Wells, J. A. (2004). Small-molecule inhibitors of protein-protein interactions: Progressing towards the dream. *Nat. Rev. Drug Discov.* **3,** 301–317.

Aronov, A. M., Baker, C., Bemis, G. W., Cao, J. R., Chen, G. J., Ford, P. J., Germann, U. A., Green, J., Hale, M. R., Jacobs, M., *et al.* (2007). Flipped out: Structure-guided design of selective pyrazolylpyrrole ERK inhibitors. *J. Med. Chem.* **50,** 1280–1287.

Borsi, V., Calderone, V., Fragai, M., Luchinat, C., and Sarti, N. (2010). Entropic contribution to the linking coefficient in fragment based drug design: A case study. *J. Med. Chem.* **53,** 4285–4289.

Ciulli, A., Williams, G., Smith, A. G., Blundell, T. L., and Abell, C. (2006). Probing hot spots at protein-ligand binding sites: A fragment-based approach using biophysical methods. *J. Med. Chem.* **49,** 4992–5000.

DeLano, W. L. (2002). Unraveling hot spots in binding interfaces: Progress and challenges. *Curr. Opin. Struct. Biol.* **12,** 14–20.

Erlanson, D. A. (2006). Fragment-based lead discovery: A chemical update. *Curr. Opin. Biotechnol.* **17,** 643–652.

Ertl, P., Rohde, B., and Selzer, P. (2000). Fast calculation of molecular polar surface area as a sum of fragment-based contributions and its application to the prediction of drug transport properties. *J. Med. Chem.* **43,** 3714–3717.

Hajduk, P. J. (2006). Fragment-based drug design: How big is too big? *J. Med. Chem.* **49,** 6972–6976.

Hajduk, P. J., Huth, J. R., and Fesik, S. W. (2005). Druggability indices for protein targets derived from NMR-based screening data. *J. Med. Chem.* **48,** 2518–2525.

Hopkins, A. L., Groom, C. R., and Alex, A. (2004). Ligand efficiency: A useful metric for lead selection. *Drug Discov. Today* **9,** 430–431.

Huc, I., and Lehn, J. M. (1997). Virtual combinatorial libraries: Dynamic generation of molecular and supramolecular diversity by self-assembly. *Proc. Natl. Acad. Sci. USA* **94,** 2106–2110.

Jencks, W. P. (1981). On the attribution and additivity of binding energies. *Proc. Natl. Acad. Sci. USA* **78,** 4046–4050.

Keseru, G. M., and Makara, G. M. (2006). Hit discovery and hit-to-lead approaches. *Drug Discov. Today* **11,** 741–748.

Keseru, G. M., and Makara, G. M. (2009). The influence of lead discovery strategies on the properties of drug candidates. *Nat. Rev. Drug Discov.* **8,** 203–212.

Kolb, H. C., and Sharpless, K. B. (2003). The growing impact of click chemistry on drug discovery. *Drug Discov. Today* **8,** 1128–1137.

Kolb, H. C., Finn, M. G., and Sharpless, K. B. (2001). Click chemistry: Diverse chemical function from a few good reactions. *Angew. Chem. Int. Ed.* **40,** 2004–2021.

Kortemme, T., and Baker, D. (2002). A simple physical model for binding energy hot spots in protein-protein complexes. *Proc. Natl. Acad. Sci. USA* **99,** 14116–14121.

Kuntz, I. D., Chen, K., Sharp, K. A., and Kollman, P. A. (1999). The maximal affinity of ligands. *Proc. Natl. Acad. Sci. USA* **96,** 9997–10002.

Landon, M. R., Lancia, D. R., Yu, J., Thiel, S. C., and Vajda, S. (2007). Identification of hot spots within druggable binding regions by computational solvent mapping of proteins. *J. Med. Chem.* **50,** 1231–1240.

Landon, M. R., Lieberman, R. L., Hoang, Q. Q., Ju, S. L., Caaveiro, J. M. M., Orwig, S. D., Kozakov, D., Brenke, R., Chuang, G. Y., Beglov, D., *et al.* (2009). Detection of ligand binding hot spots on protein surfaces via fragment-based methods: Application to DJ-1 and glucocerebrosidase. *J. Comput.-Aided Mol. Des.* **23,** 491–500.

Leeson, P. D., and Springthorpe, B. (2007). The influence of drug-like concepts on decision-making in medicinal chemistry. *Nat. Rev. Drug Discov.* **6,** 881–890.

Lehn, J. M. (1999). Dynamic combinatorial chemistry and virtual combinatorial libraries. *Chem. Eur. J.* **5,** 2455–2463.

Lehn, J. M., and Eliseev, A. V. (2001). Chemistry—Dynamic combinatorial chemistry. *Science* **291,** 2331–2332.

Lewis, W. G., Green, L. G., Grynszpan, F., Radic, Z., Carlier, P. R., Taylor, P., Finn, M. G., and Sharpless, K. B. (2002). Click chemistry in situ: Acetylcholinesterase as a reaction vessel for the selective assembly of a femtomolar inhibitor from an array of building blocks. *Angew. Chem. Int. Ed.* **41,** 1053–1057.

Lipinski, C. A., Lombardo, F., Dominy, B. W., and Feeney, P. J. (1997). Experimental and computational approaches to estimate solubility and permeability in drug discovery and development settings. *Adv. Drug Deliv. Rev.* **23,** 3–25.

Murray, C. W., and Verdonk, M. L. (2002). The consequences of translational and rotational entropy lost by small molecules on binding to proteins. *J. Comput.-Aided Mol. Des.* **16,** 741–753.

Niimi, T., Orita, M., Okazawa-Igarashi, M., Sakashita, H., Kikuchi, K., Ball, E., Ichikawa, A., Yamagiwa, Y., Sakamoto, S., Tanaka, A., *et al.* (2001). Design and synthesis of non-peptidic inhibitors for the Syk C-terminal SH2 domain based on structure-based in-silico screening. *J. Med. Chem.* **44,** 4737–4740.

Nissink, J. W. M. (2009). Simple size-independent measure of ligand efficiency. *J. Chem. Inf. Model.* **49,** 1617–1622.

Orita, M., Ohno, K., and Niimi, T. (2009a). Two 'Golden Ratio' indices in fragment-based drug discovery. *Drug Discov. Today* **14,** 321–328.

Orita, M., Warizaya, M., Amano, Y., Ohno, K., and Niimi, T. (2009b). Advances in fragment-based drug discovery platforms. *Expert Opin. Drug Discov.* **4,** 1125–1144.

Page, M. I., and Jencks, W. P. (1971). Entropic contributions to rate accelerations in enzymic and intramolecular reactions and the chelate effect. *Proc. Natl. Acad. Sci. USA* **68,** 1678–1683.

Palm, K., Stenberg, P., Luthman, K., and Artursson, P. (1997). Polar molecular surface properties predict the intestinal absorption of drugs in humans. *Pharm. Res.* **14,** 568–571.

Ramstrom, O., and Lehn, J. M. (2002). Drug discovery by dynamic combinatorial libraries. *Nat. Rev. Drug Discov.* **1,** 26–36.
Rees, D. C., Congreve, M., Murray, C. W., and Carr, R. (2004). Fragment-based lead discovery. *Nat. Rev. Drug Discov.* **3,** 660–672.
Rejto, P. A., and Verkhivker, G. M. (1996). Unraveling principles of lead discovery: From unfrustrated energy landscapes to novel molecular anchors. *Proc. Natl. Acad. Sci. USA* **93,** 8945–8950.
Reynolds, C. H., Bembenek, S. D., and Tounge, B. A. (2007). The role of molecular size in ligand efficiency. *Bioorg. Med. Chem. Lett.* **17,** 4258–4261.
Reynolds, C. H., Tounge, B. A., and Bembenek, S. D. (2008). Ligand binding efficiency: Trends, physical basis, and implications. *J. Med. Chem.* **51,** 2432–2438.
Shuker, S. B., Hajduk, P. J., Meadows, R. P., and Fesik, S. W. (1996). Discovering high-affinity ligands for proteins: SAR by NMR. *Science* **274,** 1531–1534.
Thanos, C. D., Randal, M., and Wells, J. A. (2003). Potent small-molecule binding to a dynamic hot spot on IL-2. *J. Am. Chem. Soc.* **125,** 15280–15281.
Thijs, B., Ramy, F., and Woody, S. (2009). High-energy water sites determine peptide binding affinity and specificty of PDZ domains. *Protein Sci.* **18,** 1609–1619.
Young, T., Abel, R., Kim, B., Berne, B. J., and Friesner, R. A. (2007). Motifs for molecular recognition exploiting hydrophobic enclosure in protein-ligand binding. *Proc. Natl. Acad. Sci. USA* **104,** 808–813.

CHAPTER SIXTEEN

Medicinal Chemistry Inspired Fragment-Based Drug Discovery

James Lanter, Xuqing Zhang, *and* Zhihua Sui

Contents

Abstract

Lead generation can be a very challenging phase of the drug discovery process. The two principal methods for this stage of research are blind screening and rational design. Among the rational or semirational design approaches, fragment-based drug discovery (FBDD) has emerged as a useful tool for the generation of lead structures. It is particularly powerful as a complement to high-throughput screening approaches when the latter failed to yield viable hits for further development. Engagement of medicinal chemists early in the process can accelerate the progression of FBDD efforts by incorporating drug-friendly properties in the earliest stages of the design process. Medium-chain acyl-CoA synthetase 2b and ketohexokinase are chosen as examples to illustrate the importance of close collaboration of medicinal chemists, crystallography, and modeling.

Medicinal Chemistry, Johnson & Johnson Pharmaceutical Research and Development, L.L.C., Spring House, Pennsylvania, USA

Methods in Enzymology, Volume 493
ISSN 0076-6879, DOI: 10.1016/B978-0-12-381274-2.00016-9

1. Introduction

1.1. Lead generation in medicinal chemistry

Before getting into the details of the involvement of medicinal chemistry in the fragment-based drug discovery (FBDD), it is useful to give a brief overview of the means by which lead molecules have historically been identified. Blind screening was the first approach to lead discovery and remains a major component of modern drug discovery efforts. Originally, this screening was done with a heavy pharmacology component which has obvious drawbacks with regard to the consumption of screening materials, throughput, test subject handling, and interpretation of the results.

Over the years, biological research emphasis shifted from pharmacological to biochemical and finally, the realm of molecular biology. With these changes, the discovery paradigm shifted to locating compounds that acted at only one molecular target via a specific pathway. As both technology and scientific understanding progressed, modern high-throughput screening (HTS) techniques and equipment emerged, allowing the fairly rapid screening of hundreds of thousands of molecules against the molecular target of interest. Concurrently, advances in both chemistry and robotics yielded combinatorial and parallel synthesis approaches to build vast libraries of compounds specifically for the molecular target. Involvement of molecular modelers was frequently engaged when enough information (i.e., crystal structures or homology models) was available to provide a more rational basis, but frequently, libraries were simply prepared on the basis of availability of starting materials and a tractable and/or patentable scaffold. The automation was also applied to biology as HTS enabled the rapid testing of compound libraries prepared by the medicinal chemists. Depending on the target and the properties of the library being screened, HTS campaigns sometimes do not yield viable leads for further investigation.

For almost as long as the blind screening method, rational approaches to lead generation have served as a complementary approach. A useful tool for the utilization of all rational drug design strategies is quantitative structure–activity relationship (QSAR) analysis. This methodology began to take hold after the seminal paper by Hansch *et al.* (1962). As processor power increased, the methods and capabilities of the computational approach grew in complexity and relevance. Evolution of QSAR with synthetic and medicinal chemistry principles as well as crystallography and molecular modeling techniques falls under the structure-based drug design (SBDD) approach to drug discovery. This advance became further refined as awesome computing power became widespread and molecular modeling more mainstream.

One method of rational design is to chemically modify a natural product of known pharmacological properties to separate desirable and undesirable features. Alteration of the well-known coca leaf extract cocaine in the late nineteenth century is a good example of this method (Fig. 16.1). Pioneering work by Albert Einhorn led to the discovery of procaine which maintained the local anesthesia effects while removing the euphoric, addictive, and some of the toxic properties. Procaine became a lead template for further research in this class of molecules, yielding compounds with improved properties like tetracaine that are still in wide use today. The field of natural products isolation and synthesis is an important adjunct to these efforts, as the testing of intermediates on the way to the final product can yield similar results, as was found in the case of taxotere (Guenard *et al.*, 1993). In general, the natural products-based approach led to enormous parallel advances in synthetic organic chemistry that continue into the present day. These advances not only allowed the production of natural products in the quantities needed for further evaluation but also enabled the synthesis of "unnatural" analogs which often had even better properties than their congeners. In fact, many medicinal chemists come from an organic synthesis background, having received their medicinal training "on the job."

A second means of lead generation involves exploitation of detailed biochemical knowledge of the fate of endogenous chemicals. For example, it is well documented that the circulating androgen, testosterone, is transformed into its more active metabolite 5α-dihydrotestosterone through the action of 5α-reductases in various tissues. Using the structural information of the substrate and product of this biochemical process, inhibitors such as finasteride (Bakshi *et al.*, 1995) were designed to inhibit this reaction (Fig. 16.2). This example is also illustrative of the principles of SBDD which has been frequently used in these instances. Many drugs have been discovered through this approach, including contraceptives, prostaglandins, and steroidal anti-inflammatory agents.

Related to biochemical knowledge-based strategies are those based on observations of the metabolism of administered drugs or clinical outcomes not related to the primary study endpoint. The antibacterial agent, Prontosil, is an example of the former phenomenon. This compound was discovered during blind pharmacological screening of azo dyes for antibacterial

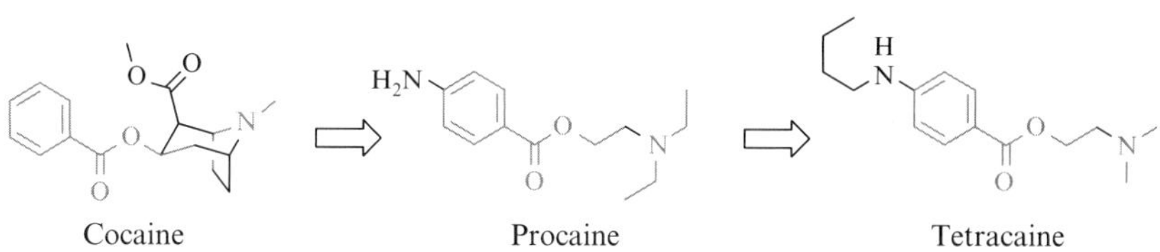

Cocaine
(local anestetic, euphoric
addictive and toxic)

Procaine
(local anestetic,
PABA metabolite)

Tetracaine
(local anestetic,
no PABA metabolite)

Figure 16.1 Cocaine modification.

properties. Its lack of *in vitro* antibacterial activity was puzzling until it was proposed that the active *para*-aminobenzenesulfonamide component was actually released *in vivo* by enzymatic action. This material was subsequently synthesized, tested, and found to have good activity both *in vitro* and *in vivo*, forming the basis for the "sulfa-drug" antibiotic class (Fig. 16.3). This class of drugs then led to the discovery of Tolbutamide from the antibacterial candidate Carbutamide. An astute medical professional was able to attribute side effects observations related to stimulation of insulin secretion during a clinical trial administration of the latter molecule. Subsequent SAR work removed antibacterial properties while maintaining the blood sugar-modulating effects. The discovery of the erectile dysfunction mitigation properties of the PDE-5 inhibitor Sildenafil (Boolell *et al.*, 1996) during a cardiovascular clinical trial (Kling, 1998) is a more recent example of "clinical lead generation," yielding two second-generation drugs for this indication (Ravipati *et al.*, 2007).

Finally, modification of known pharmacophores, sometimes recently referred to pejoratively as the "me too" approach, has been a very fertile

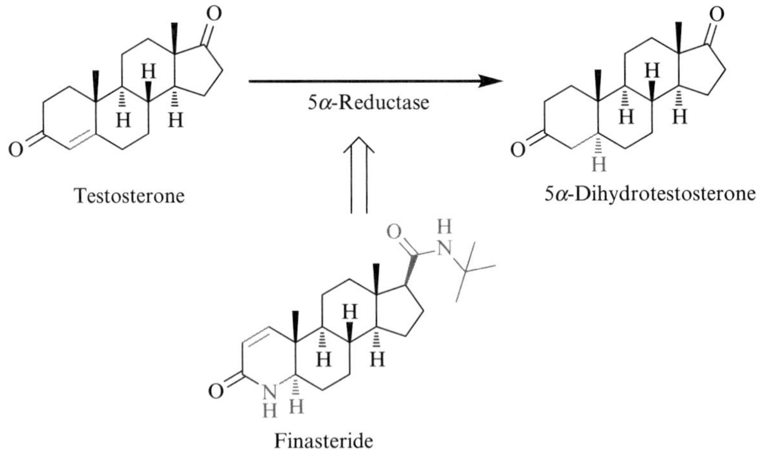

Testosterone

5α-Reductase

5α-Dihydrotestosterone

Finasteride

Figure 16.2 Finasteride mechanism of action.

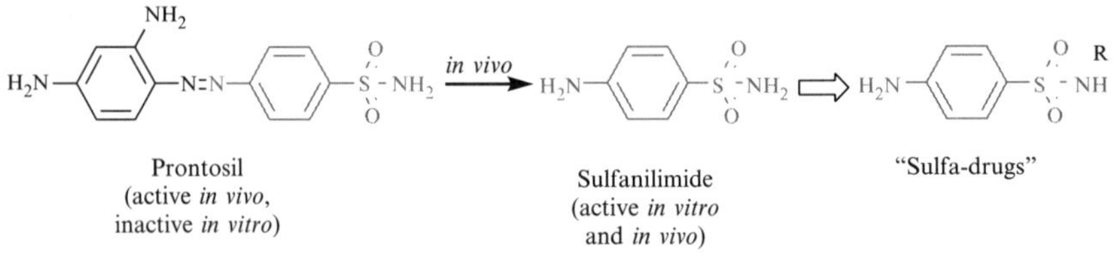

Prontosil (active *in vivo*, inactive *in vitro*)

in vivo

Sulfanilimide (active *in vitro* and *in vivo*)

"Sulfa-drugs"

Figure 16.3 The genesis of sulfa drugs.

means of lead generation. The entire class of benzodiazepine psychoactive agents can trace their origin to the tranquilizer Librium, itself discovered by blind screening (Fig. 16.4). Likewise, the serendipitous discovery of the penicillins launched the β-lactam class of antibiotics. Though over 70 years old, this class of drugs is still very actively researched (me-50^{+}) today and has spun off other research into ancillary agents such as β-lactamase inhibitors (Drawz and Bonomo, 2010).

Given the back and forth trajectory of drug discovery efforts and the success stories from each method in the field, it is apparent that no single methodology is a solution to every problem. It is also apparent that medicinal chemistry is a consistent factor throughout this continuum. For success, even in the early stages of the FBDD process, knowledge of medicinal chemistry is essential to provide a potentially more druggable starting point than HTS approaches. Likewise, a strong background in synthetic organic chemistry yields more efficient preparation of the target libraries.

1.2. Fragment-based drug design

The basic premise of FBDD (Murray and Rees, 2009) is to start with the most atom-efficient leads possible through the screening of much smaller molecules (fragments) than are typically screened. This has several advantages, including minimizing possible sites of metabolism, improving the prospects of oral bioavailability and intellectual property (IP) novelty, and, after several iterations of design, possibly better selectivity than "random" HTS hits, all resulting in enhanced ligand efficiency (Bembenek *et al.*, 2009). Given their small size, fragments would not typically be discovered by HTS screening, as most such systems are run in the micromolar range, while fragment affinities are usually several orders of magnitude lower (millimolar range). Several sensitive biophysical techniques have been utilized to address this issue, including isothermal calorimetry (ITC; Orita *et al.*, 2009), mass spectroscopy (MS; Hannah *et al.*, 2010), surface plasmon resonance (SPR; Neumann *et al.*, 2007), nuclear magnetic resonance

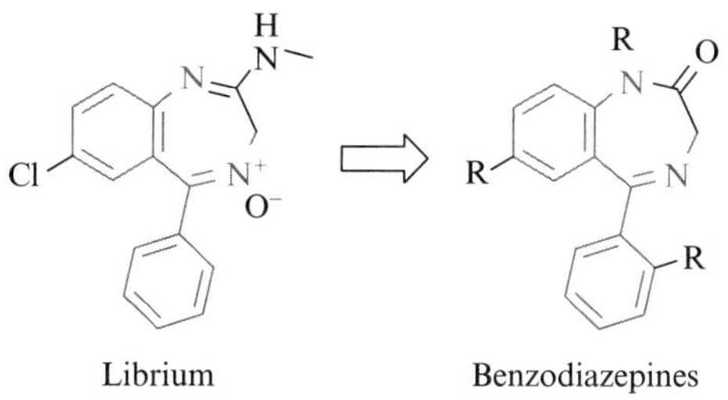

Librium Benzodiazepines

Figure 16.4 From librium to benzodiazepines.

spectroscopy (NMR; Zartler and Mo, 2007), fluorescence correlation spectroscopy (FCS+; Hesterkamp *et al.*, 2007), and X-ray diffraction (Hartshorn, *et al.*, 2005). In contrast to HTS approaches, use of these techniques generally requires large amounts of the purified protein of interest and solubility of fragments at a high concentration in the screening medium. All techniques have their unique capabilities and drawbacks; as our experience has been limited to X-ray crystallographic methods, we will focus on those for this discussion.

X-ray crystallography obviously requires crystal growth and therefore is limited to targets that can be crystallized; thus, membrane-bound targets (i.e., GPCRs, ion channels, etc.) are not suitable for this approach. Among suitable targets, there are many selection criteria which are discussed in other chapters. Likewise, once the target is selected, there are numerous technical details, including methods of crystallization, inclusion auxiliaries such as ATP in the crystallization process, methods of screening (i.e., cocrystallization vs. soaking), and data collection that can be found in this volume. Medicinal chemistry engagement in X-ray-driven FBDD begins at the screening library stage. While the population of such libraries is typically the province of molecular groups, some input from medicinal chemists is useful to ensure that members of the library all contain stable, nonreactive functional groups and have reasonable physical properties. In our protocol, members of the screening library (itself containing $\sim$1000 members) must meet the following criteria: small (MW $<$ 250, containing 6–15 heavy atoms), fewer than four hydrogen bond acceptors, fewer than four hydrogen bond donors, two or fewer rings, and no unspecified chiral centers. The library is then clustered into groups of five similar compounds for screening.

The process of FBDD, regardless of the screening technique, begins with the design and population of a screening or primary library. Among the many selection criteria for this library are low molecular weight, limited molecular complexity, the ability to build off the scaffold, and the inclusion of a limited number of heteroatoms. The fragments that show affinity for the target are then used as the basis for the design of the secondary library design. This involves building off the various positions of the fragment (sprouting) to explore hydrophobic pockets, hydrophilic pockets, metal chelates, and other features of the three-dimensional environment. Alternatively, two fragments (either the same or different) found to be in close proximity in the protein can be joined (merging) by a linker suitable to the intervening space. The results of secondary library screening are then evaluated and used as the basis for tertiary library design which is essentially a more refined version of the secondary library design process. Hits from the tertiary design are then typically potent enough to be screened in more conventional *in vitro* assays. Any compound showing promising potency is then evaluated for their ADME properties, pharmacokinetic parameters, and potential off-target effects. Reasonable outcomes from these secondary assays then lead to the molecules being progressed into lead optimization and ultimately, preclinical candidacy.

2. Medicinal Chemistry Engagement in Fragment-Based Drug Design

Design of the primary library is ideally a close collaboration between the computational, medicinal, and synthetic organic chemistry disciplines. Even when restricting the fragment molecular weight to compounds below 250 Da, there are more possible combinations of heavy atoms than there is ability or manpower to synthesize them. Knowledge of chemical synthesis is critical to cast the widest possible net in terms of structural diversity while keeping in mind the feasibility of their preparation and chemical stability. Medicinal chemistry principles can then be used to narrow the scope of the library in the context of features such as druggable functionality, improved physical characteristics, and avoidance of known metabolic liabilities. Computational techniques can be used to narrow the field to around 1000 members while covering the widest possible chemical space. At this stage of the design, "extreme" functional groups that are suboptimal in final drug molecules such as primary alcohols, carboxylic acids, and very basic amines can be helpful in teasing out strong interactions with protein residues.

More emphasis on the principles of medicinal chemistry occurs after the results of the primary library crystallographic screening emerge. At this stage, bioisosteric replacement of functional groups such as carboxylic acids with suitable heterocycles or sulfonamides can be explored to "front load" physiochemical characteristics. Likewise, replacements of carbinol moieties with amides can offer the possibility of bridging a polar pinchpoint between two hydrophobic pockets.

The majority of secondary library design involves sprouting of functionality off the primary library hits to fill hydrophilic or hydrophobic pockets. To maintain the principle of atom efficiency, these functional groups should not be more than 6–8 additional heavy atoms; of these, no more than 2–3 hydrogen bond donor or acceptors should be included. Placement of the sprouts, although heavily informed by the three-dimensional environment observed in the crystal structure, should be somewhat flexible on the scaffold. Therefore, synthetic tractability is a key selection factor for secondary library design. For example, a fragment that contains coordinatively saturated heteroatoms in all the positions where one wishes to conduct hydrophobic sprouting might lower its priority in the design queue versus a fragment hit where all such positions are "open."

For exploration of hydrophobic pockets, both alkyl chains and aryl rings should be used. The former allow much greater flexibility but come with an entropy cost. In addition to branched and straight chains, it is also useful to incorporate oxygen atoms where synthetically feasible as they afford minor improvements in aqueous solubility. Although substituted phenyl rings are less flexible, they allow easy tuning of both sterics and electronics. In particular, we have found that *ortho–meta–para* phenol–pyridine–chlorine scanning provides a quick and usually easy-to-synthesize means of exploring these factors.

Hydrophilic pocket exploration largely depends on the size and nature of the protein functionality. Traditionally heterocycles are used to explore such loci but alkyl chains containing polar moieties like alcohols, amines sulfones and acids should also be employed to maximize exploration of the region. The latter groups also introduce a greater degree of flexibility in terms of rotatable bonds than their relatively flat and rigid heteroaromatic counterparts. For interaction with strongly basic groups such as arginine and strongly acidic residues like glutamic acid, it can be useful to include weakly complementary functionality such as phenols and pyridine residues, respectively. As with hydrophobic pockets, it is advisable to include some flexibility in the design.

During this process, it is important to note the degree of resolution/refinement of the crystal structures. For example, it can be difficult to differentiate between a nitrogen and a carbon atom on an aromatic ring due to their similar electron density in some levels of refinement. Therefore, the orientation of a ligand in the molecular target should be viewed with some flexibility. Also, it is not uncommon to see either multiple copies and/or orientations of library members in the solved structures. It is our position that these hits represent particularly attractive candidates for secondary library design since they have more than one "shot on goal" with the caveat that each orientation must be assessed independently when designing the secondary library. In addition, for cases where either multiple "copies" of a single molecule or two different molecules exist in a solved structure, opportunity for merging the fragments exists. At the secondary library design stage, however, this can be tricky since significant reorientation of the molecules may occur as they are elaborated; as a result, we usually save the majority of such efforts for tertiary library design.

Since the affinity of the fragments at this point is usually low, the binding mode of any fragment could change under different conditions. Although the crystal structures are in high resolution, treating the particular binding mode as a high-resolution interaction often leads to disappointment. In practice, we use a "fuzzy logic" in our design at this stage. In other words, many possible structures are designed for one particular hit. This point will be illustrated in greater detail in Section 3.2.

Although not critical at this stage, it is helpful to include analogs that improve druggability. For example, the replacement of suboptimal functionality such as aromatic nitro groups and halides with spatially equivalent functionality should be included in the library design. After completion of the design and synthesis, the secondary library is progressed into X-ray screening but usually not evaluated in biological assays in our practice to avoid bias toward any given template.

Design of the tertiary library follows largely the same process as that of secondary library. As noted above, there may be more chances of success for merging strategies at this stage as the opportunities and limitations of the molecular target architecture relative to the secondary library hits become apparent. Since the goal of the tertiary library design is to generate a lead, much greater emphasis should be placed on incorporating (or excising) functional groups to enhance druggability. At this stage, library members are evaluated in X-ray screening and relevant biological assays. To accelerate selection of leads for optimization, hits that emerge from this process should be evaluated for ADME characteristics and against any potential off-target effects related to their structural features (i.e., hERG activity for amines or kinase activity for substituted imidazoles).

3. Case Studies—MACS2b and Ketohexokinase

With these processes of X-ray-based lead generation delineated, it is useful to place them in the context of some real-world examples to illustrate the principles outlined above. First, we will use an example of primary library hit selection and the associated secondary library design and outcome for medium-chain Acyl-CoA synthetase 2b (MACS2b). Then, we will briefly cover the same process for ketohexokinase (KHK) with an emphasis on the tertiary library design and lead declaration. In both cases, the same primary library was screened by soaking crystals of the appropriate enzyme with groups of five similarity clustered compounds. X-ray crystallographic screening was then conducted and the resulting solved structures examined for the orientation of the primary fragments in the enzyme structure.

3.1. Background

MACS2 (Kochan *et al.*, 2009) is a member of the acyl-CoA synthetase family of enzymes that initiate the metabolism of fatty acids through the formation of fatty acid-CoA conjugates. The breakdown of the favored substrates for MACS2b, octanoic, and decanoic acid, is used by the body for energy generation. Interruption of this process is postulated to lead to elimination of the substrates through the kidneys with subsequent breakdown of undesirable longer chain fatty acids occurring to furnish energy to the body. In this way, it is hoped that overnutrition can be resolved resulting in lean insulin-sensitive tissues. KHK, a member of the ribokinase superfamily of kinases, catalyzes the conversion of the furanose form of fructose to fructose-1-phosphate (Raushel and Cleland, 1977) in the presence of adenosine triphosphate (ATP) and potassium ion (K^+). KHK, accompanied by aldolase B and triokinase, are the predominant enzymes accounting for the bulk of dietary fructose metabolism. Inhibition of KHK's activity would suppress carbon supply for FA and VLDL synthesis, thereby establishing it as an attractive therapeutic target. Small molecule inhibitors of KHK may be a practical means to achieve a desired KHK modulation.

3.2. MACS2b secondary library design and outcome

Screening of the primary library for *apo*-MACS2b yielded fragments binding to three discrete regions of the enzyme: the coenzyme A (CoA) pocket, the fatty acid-adenosine (FAAD) pocket, and a tertiary (Tert) hinge region between two domains of the protein. The fragments were evaluated and target compounds proposed according to the process noted in Fig. 16.5. For each region of the enzyme, two chemically diverse fragments upon which the design of the secondary library would be based were selected based on diverse orientation in crystal structure, maximum ability to explore space, and (chemical) structural diversity (Fig. 16.6). For each of these fragments, a library of compounds was designed based on their orientation in crystal structure to maximize the exploration of space in the appropriate pocket, taking into account both steric and stereoelectronic factors. Although there were a few possibilities for merging fragments, for the most part, this design

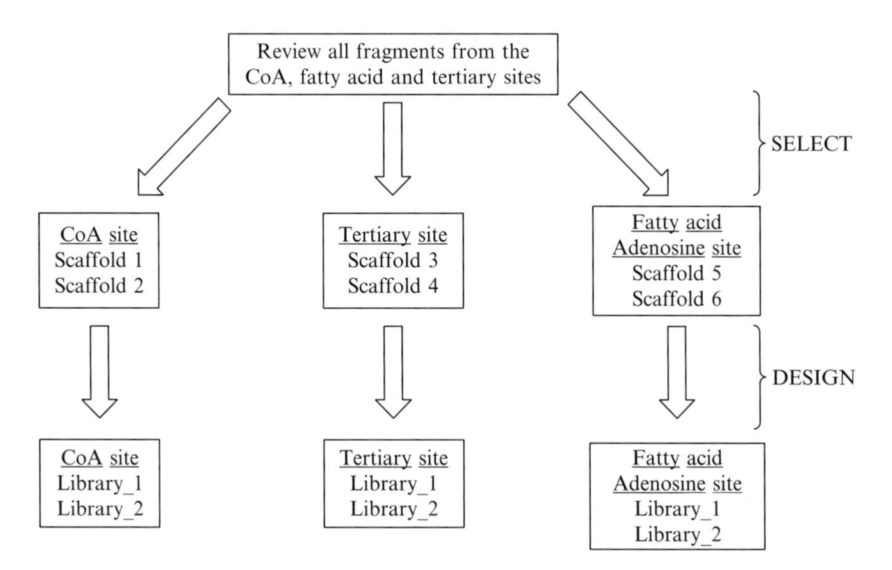

Figure 16.5 *apo*-MACS2b library design process.

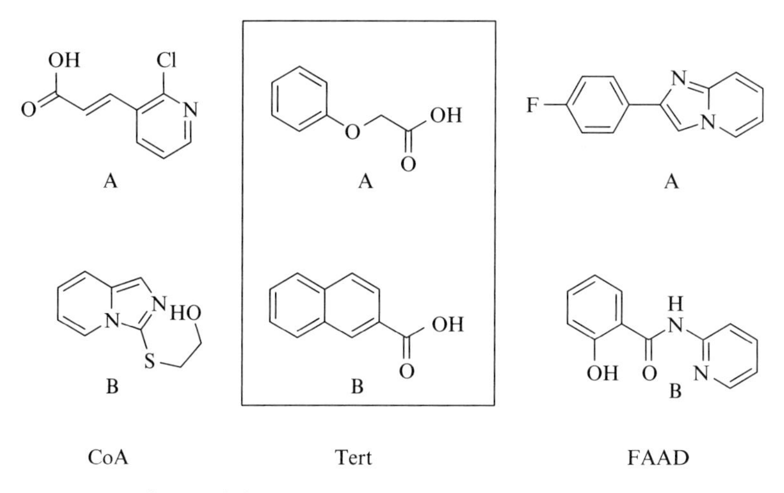

Figure 16.6 Hits selected from the primary screen.

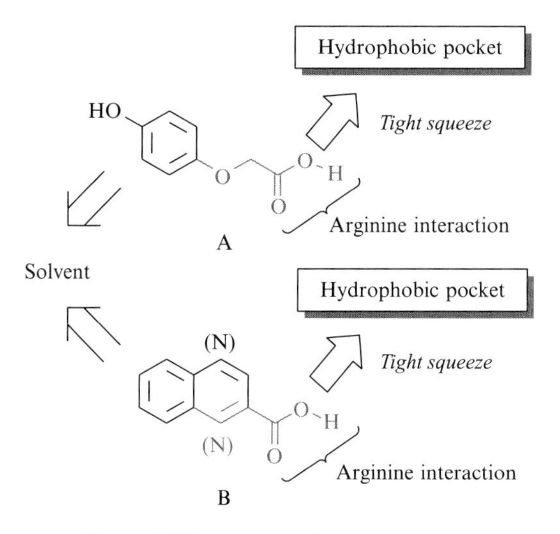

Figure 16.7 Tertiary site environment.

was done by sprouting off suitable areas on the primary hits, taking into account synthetic accessibility.

Considering just the design process for the tertiary site, a rough sketch of the local environment (Fig. 16.7) was used as a guideline during the design. Both fragments bound close to the exterior of the enzyme, with an arginine residue serving as an anchor for the carboxylic acid moiety in both fragments. Just beyond this interaction, there was a narrow pinch point which opened into a fairly large hydrophobic pocket. The design of the fragments therefore incorporated all acidic isosteres to maintain the arginine interaction while including a narrow alkyl linker to bridge the bottleneck to the hydrophobic region. Fragment **A** offered a higher degree of flexibility to possibly allow it to better squeeze into the pocket, while Fragment **B** maintained a more rigid form to avoid unfavorable entropic costs associated with more rotatable bonds. Fragment **A** was also found in the CoA pocket of the enzyme, adding to its potential value by giving it more shots on goal. The design process for these two fragments, noted in Scheme 16.1, is illustrative of the fuzzy logic philosophy mentioned in Section 3.1. For Fragment **A**, although the critical arginine interaction occurred in a constrained steric environment, there was some flexibility in the solvent exposed exterior. Keeping in mind that in a real-world situation, proteins are a dynamic environment, linkers to the acidic moiety were chosen with a variety of lengths relative to the aryl ring. In addition, nonclassical acid bioisosteres such as sulfonamides and hydroxamic acids and 1,3,4-triazoles were included even though their accommodation in the arginine site

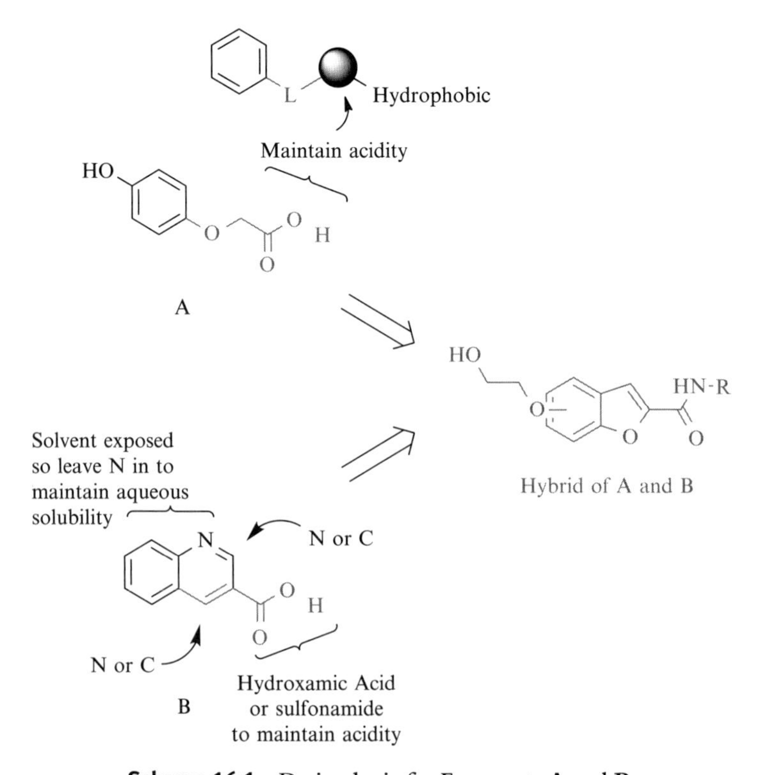

Scheme 16.1 Design logic for Fragments **A** and **B**.

appeared very tight in the crystal structure. Hydroxyl substitution on the solvent-exposed region of the aryl ring was conserved to enhance aqueous solubility in the crystal soaking medium. For this same reason, nitrogen substitution was also maintained on Fragment **B** derivatives, though the position of the heteroatom was varied. Since both positional isomers of the nitrogen atom were consistent with X-ray crystallographic data of the primary fragment, both were incorporated into the secondary library design. Finally, a hybrid structure (C) was also designed by including both positional isomers of the solubilizing glycol chain.

Generalized structures of the final Tert library are shown in Fig. 16.8. As part of the design process, these proposed structures were overlaid on the original hit in the crystal structure to gain confidence in the "fit" of the target compounds (Fig. 16.9). As can be seen in the figure, all designed structures had a reasonable overlap with the original hit and appear to attain the goal of filling the large hydrophobic pocket beyond the critical arginine interaction. One of the more interesting hits to come out of the secondary library design process was the benzofuran (shown in the box). It was interesting because while it did not bind in the Tert region, it was found in the CoA pocket. This illustrates the utility of including fragments binding in multiple sites. A small group of these analogs were tested and found to have activity in an *in vitro* Thermofluor assay, as noted in Table 16.1, further

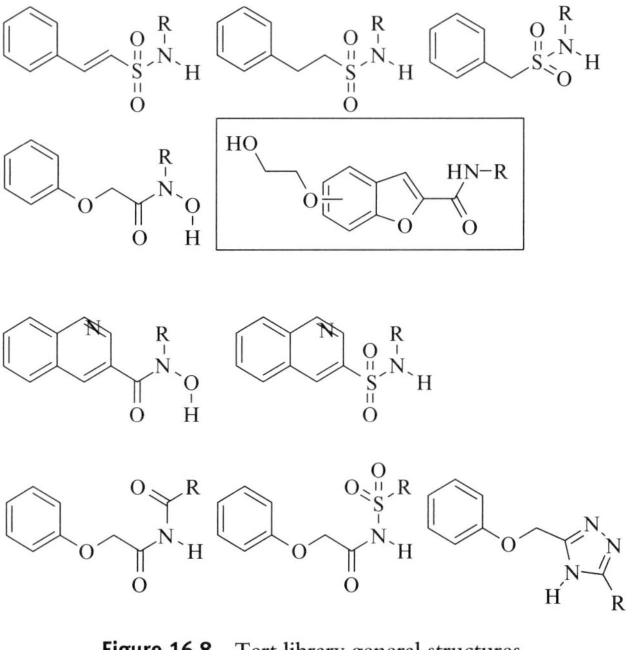

Figure 16.8 Tert library general structures.

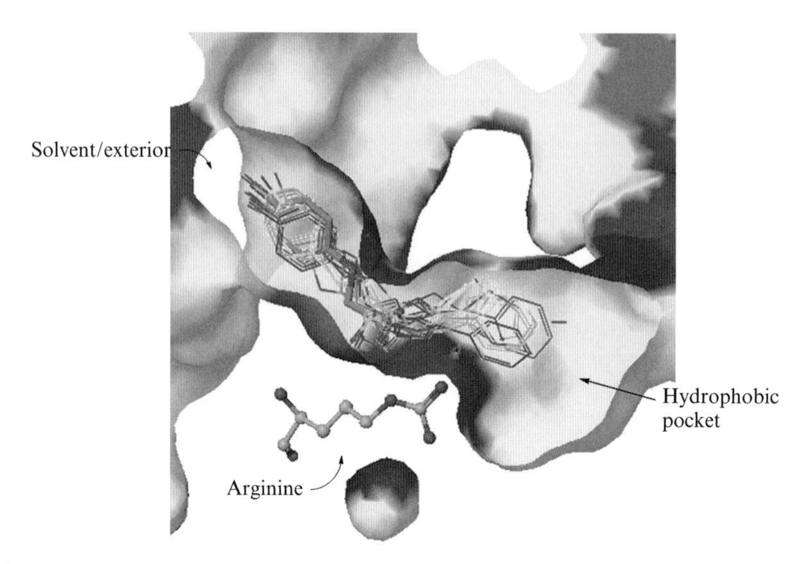

Figure 16.9 A model of overlap between the primary library hit (green) and selected secondary library fragments in the Tert site. (See Color Insert.)

Table 16.1 MACS2b activity of the Tert site hit

HO–CH2CH2–O– (benzofuran) –C(=O)–HN–**R**   **A**   HO–CH2CH2–O– (benzofuran) –C(=O)–HN–**R**   **B**

Substructure	R	TF MACS2b (μ*M*)
A	Bn	6.25
A	2-Cl-Bn	55.56
A	3-Cl-Bn	5.00
A	*4-Cl-Bn*	*0.67*
A	*n*-Bu	>77.0
A	Et	>77.0
B	Bn	7.41
B	2-Cl-Bn	>77.0
B	3-Cl-Bn	2.50
B	*4-Cl-Bn*	*0.67*
B	*n*-Bu	>77.0
B	Et	>77.0

validating the hit. It is also notable that there is already the indication of some nascent SAR about the benzene ring, regardless of the isomer of the glycol ring (2- vs. 3- vs. 4-chloro substitution). Note that the entire process from primary hit evaluation to the design and synthesis of the secondary library culminating in the discovery of a molecule with sub-micromolar *in vitro* potency was accomplished in only 5 months.

3.3. KHK tertiary library design and outcome

The FBDD protocol followed here is initiated by X-ray crystallography screening of a primary library of fragments, followed by iterative design and synthesis of secondary and tertiary libraries. This methodology is guided not by experimental bioaffinity measurements, but by structure–density relationships (SDRs). While the primary library screen focuses on a wide range of hits for a potentially flexible evolution, the goal of the secondary library screen is to confirm the original binding mode observed among the primary library hits and also achieve high diversity on a number of chemotypes for tertiary library progression. The design criteria for secondary library contain more complicated factors such as novelty and diversity, medicinal chemistry tractability, and availability of fragment-derived SDRs. Unlike the secondary library screening, the tertiary library screening emphasizes on lead generation of novel chemotypes. Empirical medicinal chemistry optimization including solution activity test is incorporated into the tertiary library screening.

Fragment evolution has been by far the most successful method of fragment optimization. It is conceptually the most straightforward and particularly when allied with a high degree of structural information such as X-ray crystallography gives the medicinal chemist a valuable advantage in the validation and subsequent optimization of a hit. However, in contrast to the many FBDD programs where fragment elaboration consisted primarily of evolution, more concepts are borrowed from traditional medicinal chemistry into tertiary library design targeting lead-like structures. Pyrazole **1** and **2** was identified as hits binding to the ATP binding site of KHK by X-ray crystallographic screening in the primary and secondary libraries. Despite having a potency that was too weak to be accurately measured, the Fragment **2** was optimized in the tertiary library. Scheme 16.2

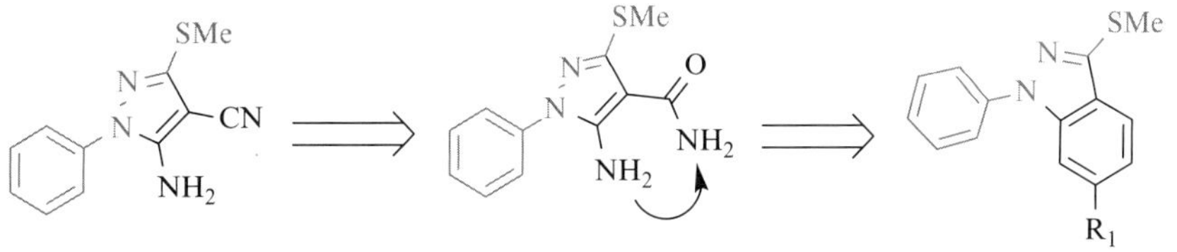

1 Primary hit — **2** Secondary hit — **3** Tertiary desgin

Scheme 16.2 From pyrazole to indazole.

illustrated that a combination of fragment fusion followed by evolution from the primary library hit—pyrazole **1** and secondary library hit—pyrazole **2** yielded indazole **3** in the tertiary library design. The evidence for the incorporation of indazole core in the tertiary library was that an intramolecular H-bonding between the 4-carboxamide and 5-amine functional groups of the secondary library hit **2** is highly likely, as revealed by X-ray crystallography, molecular mechanics, and virtual docking experiments. Furthermore, as indicated by X-ray crystallography, these two functional groups did not engage in obvious H-bonding with the protein. Thus, the [6, 5] fused systems such as indazole became one of the logical designs in the tertiary library provided that indazole can be tolerated in the binding pocket. From the perspective of medicinal chemistry, indazole **3** allowed further substitution on the phenyl ring (R_1) to extend into a new pocket, which could be easily synthesized and quickly derivatized at the 6-position of the indazole core. According to SDRs of the secondary library, minimal tolerance to modification of the thiomethyl group on **2** was observed. However, the thiomethyl group is usually considered as a metabolically disfavored functional group due to its high tendency toward chemical or enzymatic oxidation. To frontload the pharmaceutical properties for lead-like structures, we replaced the thiomethyl group of **3** with the metabolically favored functional isosteres such as methoxy, methyl, or ethyl group of **3** in the tertiary library (Scheme 16.3). Interestingly, in contrast to the observation from the secondary library, some analogs of **3** substituted by methoxy, methyl, or ethyl group were equipotent compared with the corresponding thiomethyl indazole **2**. For example, replacement of the thiomethyl group of **4** (KHK enzyme activity, IC_{50} of 560 n*M*) with the ethyl

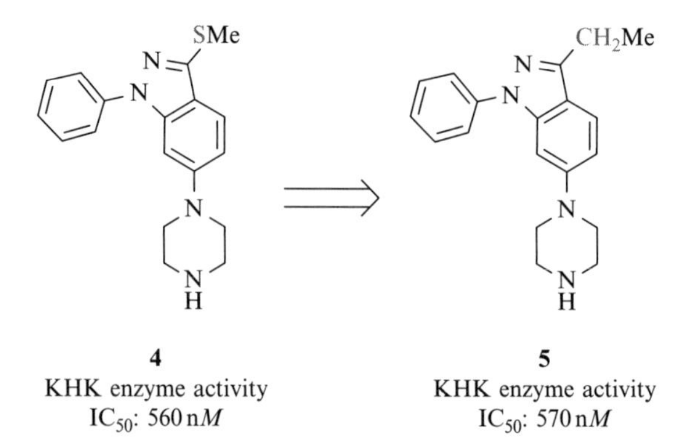

Scheme 16.3 Frontloading pharmaceutical property by replacement of metabolically unstable substitutions.

group gave indazole **5** with KHK enzyme activity of IC_{50} of 570 n*M*, which demonstrated that ethyl was an effective substitution for the thiomethyl group.

The bioisostere approach is commonly utilized in medicinal chemistry as an effective strategy to obtain novel chemical scaffolds with strong intellectual properties (IP) position. It is not common to bring this strategy into early stages of a FBDD program for the purposes of lead generation. However, we realized that early introduction of bioisosteres would greatly expand the scope of the tertiary library. Furthermore, we found that this strategy provided us with fast turnaround time, reasonable synthetic difficulty, and broad chemical diversity, which allowed us to identify lead structures more efficiently. For example, incorporation of the bioisosteres of indazole **6**, such as pyrazolopyrimidine **7**, imidazopyridine **8**, pyrazolopyrimidinone **9**, and isomeric indazole **10**, was conducted in the tertiary library (Scheme 16.4). Although these isosteric heterocyclic [6, 5] core

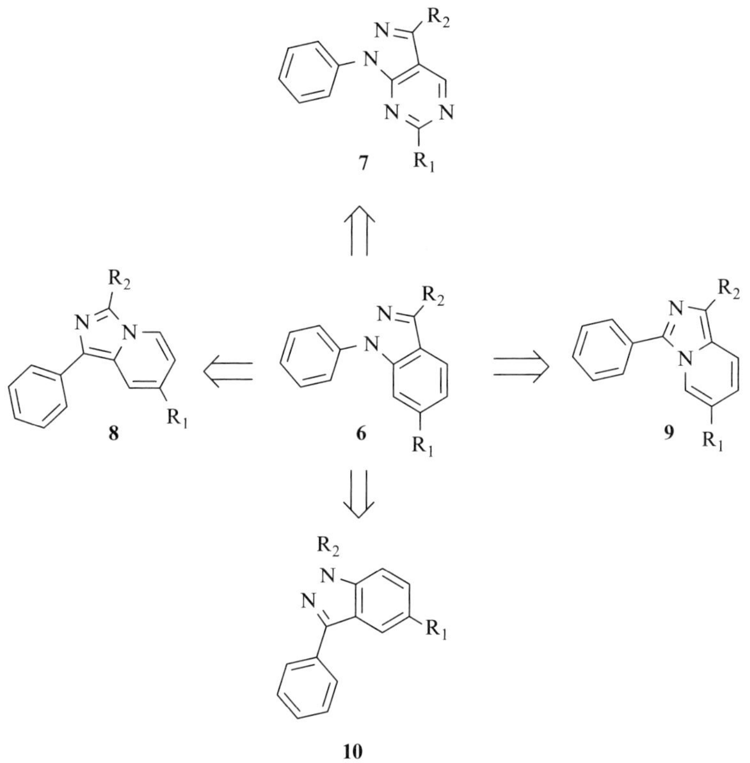

Scheme 16.4 Bioisotere strategy in tertiary library design.

systems showed striking structural similarity for maintaining H-bonding with the conserved water via an appropriately positioned heteroatom within the five-membered heterocycle (e.g., N-2 of the pyrazole), they could provide very different physicochemical properties. Our data indicated that the bioisostere approach was generally well tolerated around the parent indazole scaffold, which highlighted the value of obtaining diverse chemotypes.

Replacement of a pharmaceutically disfavored core or functional group with its bioisostere with a favored pharmaceutical property in the tertiary library stage can facilitate the lead generation process in the entire FBDD program, as well as the follow-up lead optimization driven by traditional medicinal chemistry. Screening small collections of structurally "related" but "friendly" scaffolds against pharmaceutically liable fragment hits can aid the discovery of novel leads with better safety, pharmacology, and bioavailability profiles. One of the secondary library hits in KHK program—benzotriazole **11** contained a pharmaceutically disfavored triazole group which was potentially unstable under certain chemical or enzymatic conditions (Scheme 16.5). The triazole group is also well recognized as a latent troublesome group to induce undesired side effects or serious toxicity. However, the structural data revealed that appropriate H-bonding between the 2-N atom of the Fragment **11** with water was critical for the fragment–protein interaction. Hence, several bioisosteres **12** and **13** with better pharmaceutical properties such as benzisoxazole, indazole, and benzisothiazole were designed into the tertiary library to follow this trend. Not surprising, some of these tertiary library structures with evolution at the 5-position of the core showed low micromolar solution activity. For example, evolution of the secondary library hit benzotriazole **11** at 5-position with a carboxylic group afforded the tertiary library hit **14** with KHK enzyme activity of IC_{50} of 1.70 μM (Scheme 16.6). The strategy of the tertiary library then focused on turning the benzotriazole core into

11 Secondary library hit

12 Tertiary library desgin X = O, S, NMe

13 Tertiary library desgin

Scheme 16.5 Frontloading pharmaceutical properties by replacement of chemically unstable and toxicologically troublesome core.

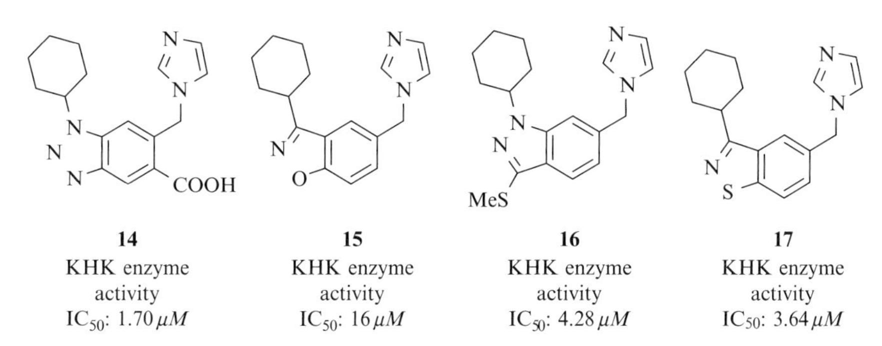

Scheme 16.6 Modification of the core to deliver the lead-like structures.

pharmaceutical friendly cores. While **15** with benzisoxazole core showed only 16 μ*M* KHK activity, benzisothiazole **16** and indazole **17** improved the activities to low micromolar range.

Indazole **3** (Fig. 16.10) was considered a major phenotype for the tertiary library. Docking of indazole **3** with an X-ray crystal structure of pyrazole **1** revealed that the tertiary library structures overlapped closed at the core and leaved ~6–7 Å of the distance toward the carboxylate side-chain of β-clasp residue Asp27B at the closed catalytic site in ATP-site fragment binding. Medicinal chemistry was introduced into the design at the region. A series of close analogs were synthesized and evaluated as KHK inhibitors. Considering the feature of the acidic residue of Asp27B, we designed tertiary library fragments containing basic amino groups targeting potentially favorable spots for ionic ligand–protein interaction. The size, length, rigidity, and geometry of the linkers were modified for favorable interaction between negatively charged Asp27B residues and the protonated fragments. We observed that certain modifications at the R_1 group on the indazole core were well tolerated, which was very advantageous for inhibitor design. The optimization process was then viewed as a medicinal chemistry-driven approach. We focused on identifying lead-like structures in the tertiary library by taking advantage of the flexibility of the ionic interaction between the basic amino group containing side chain and acid residue on the protein. The solution activity was still the main driving force for lead generation, but ADME/PK profiles of the fragments were also considered. Varying the terminal basic amino group at both 3- and 4-positions of the piperidine ring (**18** and **19** in Table 16.2) resulted in equipotent compounds, which indicated that the ionic interaction was strong regardless of the geometry of the terminal amino group located in the binding pocket. The size of the rings was also well tolerated, as illustrated by compound **18**, **20**, and **21** containing corresponding 6-, 5-, and 4-membered rings. We also modified the terminal basic amino pharmacophore by introducing the R_1 group of

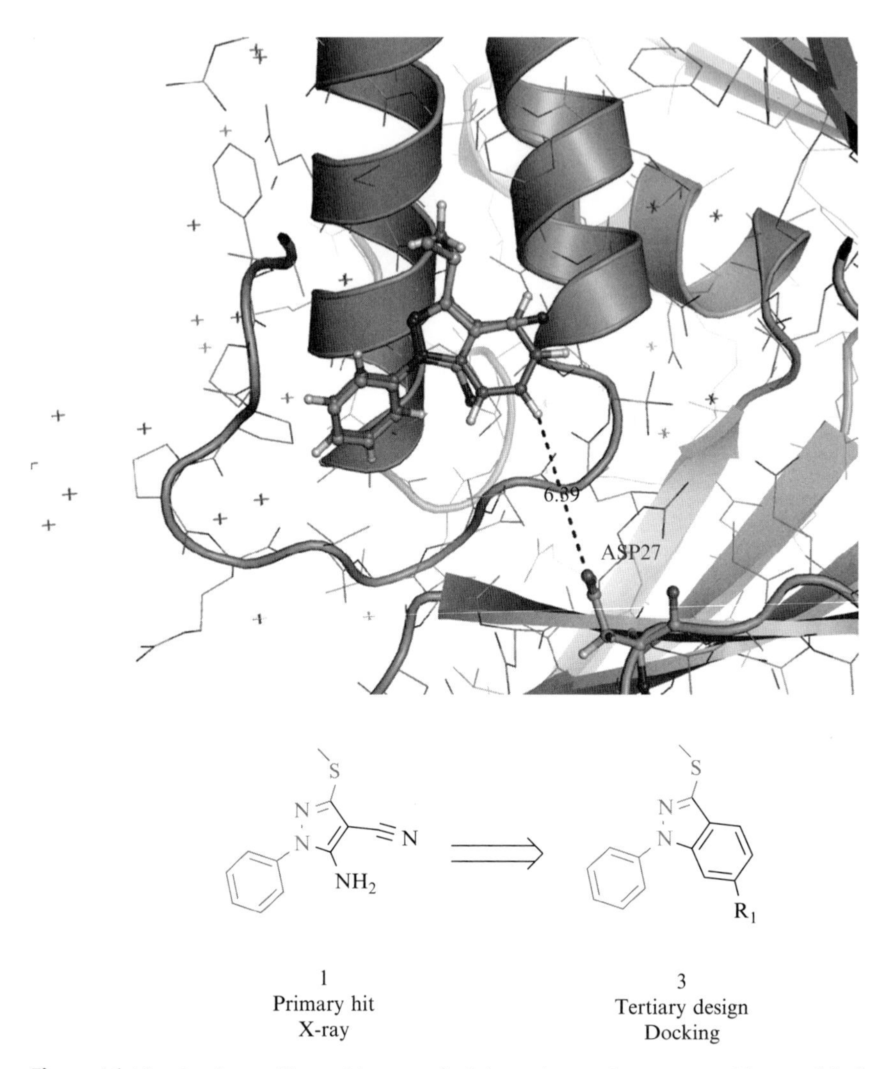

Figure 16.10 A primary library hit pyrazole **1** (green) crystal structure with a modeled tertiary library structure indazole **3** (magenta). (See Color Insert.)

variable length at the 6-position of the indazole core. Compounds **23** and **24** bearing one- and two-atom between the core and the piperidine showed comparable activities in KHK enzyme assay compared with compound **22** with direct linkage. Continuous optimization of the indazole scaffold in the tertiary library was illustrated in Table 16.2. The basicity of the terminal amine played an important role for an efficient interaction with the acidic residue on the protein. For example, ethylation of the secondary amino

Table 16.2 SAR on the R_1 group of the indazole scaffold in tertiary library

ID	R_1	KHK enzyme IC_{50} (μM)
18		0.33
19		0.31
20		0.59
21		0.22
22		0.33
23		0.71
24		0.44

Table 16.3 SAR of the indazole scaffold in tertiary library

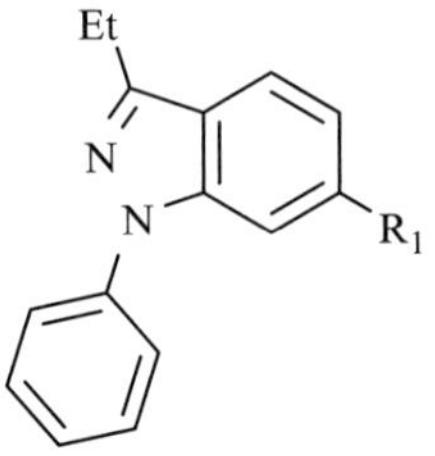

Et N N R_1

ID	R_1	KHK enzyme IC_{50} (μM)
25	 —N NH	0.57
26	 —N N—Et	11.5
27	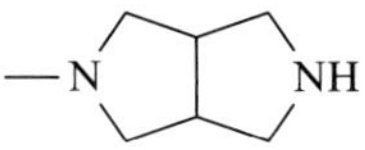 —N NH	0.59
28	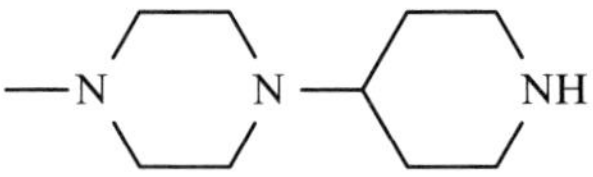 —N N— NH	0.56
29	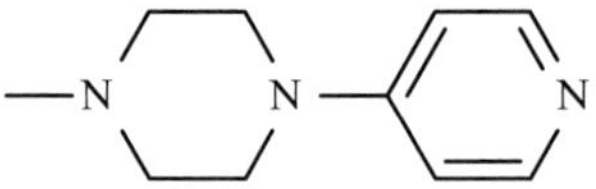 —N N— N	3.52
30	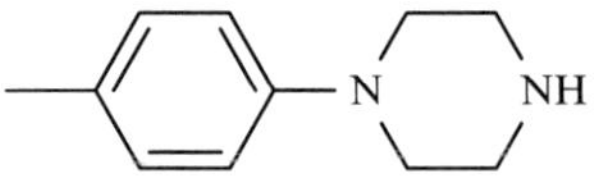 — N NH	9.80

group of compound **25** resulted in a 20-fold drop of the activity in compound **26**. We also looked at the effect of replacing the terminal piperidine ring of compound **28** with a pyridine ring of compound **29**. Compound **29** bearing a relatively weak basic group displayed weaker KHK enzyme activity as compared with compound **28**, although the impact of rigidity on the pyridine ring could not be ruled out in this example. According to SAR findings in Table 16.1, we observed that flexible linkers at 6-position of the indazole scaffold were preferred. This was further investigated by introduction of additional rings as either fused or head to tail-linked forms (compound **27** and **28** in Table 16.3). The modifications in this region clearly showed that the size and length were well tolerated. While the basic amino group was crucial for KHK enzyme activity in the indazole scaffold, a well-positioned linker had a great impact for a favorable ionic interaction. Replacement of the more flexible piperazine ring of **27** with the rigid phenyl ring of **30** dropped the activity almost 20-fold, which indicated the planar feature of the phenyl ring leading the terminal basic amino group to the disfavored distance toward the Asp27B residue. Optimization in the indazole series produced several fragments with low nanomolar KHK enzyme activity. The lead compound was additionally shown to have promising selectivity versus related kinases and a good ADME and pharmacokinetic profile.

One of the key criteria for tertiary library is that the fragments must be synthetically accessible. However, evolution of the fragments from the secondary library to the tertiary library sometimes requires more challenging synthetic routes in order to accomplish the synthesis of the core structures with additional substitutions. For example, evolution of the secondary library hit—quinoxaline **31** would be extremely time-consuming given the relatively high synthetic difficulty on substitution at the 6-position (Scheme 16.7). Based on the structural data and on the results of the docking experiments of quinoxaline and indazole scaffold **3**, we observed that the N atoms of quinoxaline scaffold did not interact directly with the protein or the conserved water. Therefore, the core was replaced by the synthetically

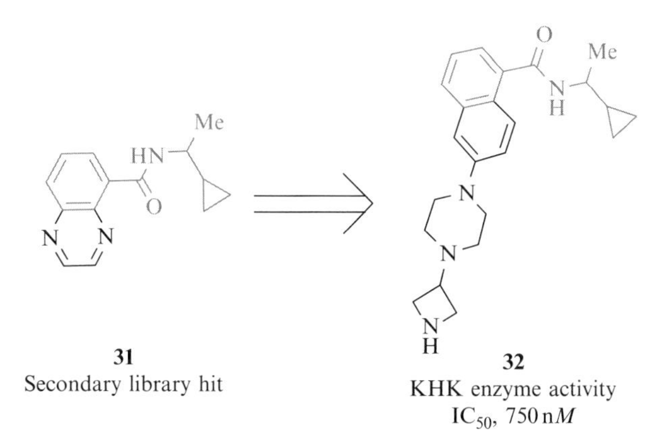

Scheme 16.7 Chemistry-driven tertiary library design.

more feasible naphthalene moiety. Moreover, we applied the similar design principle from indazole scaffold **3** into naphthalene scaffold by evolution at the 6-position of the secondary library fragment according to their close binding modes. Remarkably, this strategy resulted in a novel naphthalene scaffold in the tertiary library with submicromolar KHK enzyme activity (IC_{50}, 750 n*M*), as shown in compound **32**. Hence, extension of the naphthalene core provided not only a reasonable handle for basic amine attachment for interaction with Asp27B but also a handle for easy chemical derivatization.

4. Conclusion

FBDD is a powerful tool for drug discovery and offers medicinal chemists a guided playground for the process ranging from the generation of novel hit scaffolds all the way through the lead declaration stage. Knowledge of medicinal chemistry is essential for FBDD library design to ensure the tractability of the leads generated by the exercise. By incorporating druggable functional groups early in the design phase, the quality of the leads generated by this technique is greatly enhanced. The experience of synthetic organic chemistry is equally important for the design of the library and its rapid preparation. Combined, these disciplines can shorten the overall project time and increase the chances for success. One of the key factors limiting the use of FBDD is the availability of biophysical techniques to measure the weak binding of primary library members. As existing techniques evolve in capability and applicability and new methods of screening are developed, it is likely that FBDD will become an even more widely employed strategy for lead generation. If some of the success stories from its use are any indication, leads so generated should result in a higher success rate for programs employing the FBDD approach than more conventional means such as blind HTS.

Acknowledgment

The authors are grateful to Richard Alexander, Weimei Sun, Alan Gibbs, Marta Abad, John Spurlino, Jian Li, John Geisler, Cailin Chen, Yuanping Wang, and Lawrence Kuo for their collaboration on the MACS2b and KHK projects.

References

Bakshi, R. K., Rasmusson, G. H., Patel, G. F., Mosley, R. T., Chang, B., Ellsworth, K., Harris, G. S., and Tolman, R. L. (1995). 4-Aza-3-oxo-5α -androst-1-ene-17β-Narylcarboxamides as dual inhibitors of human type 1 and type 2 steroid 5α -reductases. dramatic effect of N-Aryl substituents on type 1 and type 2 5α -reductase inhibitory potency. *J. Med. Chem.* **38**(17), 3189–3192.

Bembenek, S. D., Tounge, B. A., and Reynolds, C. H. (2009). Ligand efficiency and fragment-based drug discovery. *Drug Discov. Today* **14**(5/6), 278–283.

Boolell, M., Allen, M. J., Ballard, S. A., Gepi-Attee, S., Muirhead, G. J., Naylor, A. M., Osterloh, I. H., and Gingell, C. (1996). Sildenafil: An orally active type 5 cyclic GMP-specific phosphodiesterase inhibitor for the treatment of penile erectile dysfunction. *Int. J. Impot. Res.* **8**(2), 47–52.

Drawz, S. M., and Bonomo, R. A. (2010). Three decades of β-lactamase inhibitors. *Clin. Microbiol. Rev.* **23**(1), 160–201.

Guenard, D., Gueritte-Voegelein, F., and Potier, P. (1993). Taxol and taxotere : Discovery, chemistry, and structure-activity relationships. *Acc. Chem. Res.* **26**(4), 160–167.

Hannah, V. V., Atmanene, C., Zeyer, D., Van Dorsselaer, A., and Sanglier-Cianferani, S. (2010). Native MS: An 'ESI' way to support structure- and fragment-based drug discovery. *Future Med. Chem.* **2**(1), 35–50.

Hansch, C., Maloney, P. P., Fujita, T., and Muir, R. M. (1962). Correlation of biological activity of phenoxyacetic acids with Hammett substituent constants and partition coefficients. *Nature* **194,** 178–180.

Hartshorn, M. J., Murray, C. W., Cleasby, A., Frederickson, M., Tickle, I. J., and Jhoti, H. (2005). Fragment-based lead discovery using x-ray crystallography. *J. Med. Chem.* **48**(2), 403–413.

Hesterkamp, T., Barker, J., Davenport, A., and Whittaker, M. (2007). Fragment based drug discovery using fluorescence correlation spectroscopy techniques: Challenges and solutions. *Curr. Top. Med. Chem.* **7**(16), 1582–1591.

Kling, J. (1998). From hypertension to angina to Viagra: On its way to becoming Viagra, UK-92, 480 changed from a drug for hypertension to a drug for angina, and then changed again when a 10-day toleration study in Wales turned up an unusual side effect. *Mod. Drug Discov.* **1**(2), 31, 33–34, 36, 38.

Kochan, G., Pilka, E. S., von Delft, F., Oppermann, U., and Yue, W. W. (2009). Structural snapshots for the conformation-dependent catalysis by human medium-chain acyl-coenzyme A synthetase ACSM2A. *J. Mol. Biol.* **388**(5), 997–1008.

Murray, C. W., and Rees, D. C. (2009). The rise of fragment-based drug discovery. *Nat. Chem.* **1**(3), 187–192.

Neumann, T., Junker, H.-D., Schmidt, K., and Sekul, R. (2007). SPR - based fragment screening: Advantages and applications. *Curr. Top. Med. Chem.* **7**(16), 1630–1642.

Orita, M., Warizaya, M., Amano, Y., Ohno, K., and Niimi, T. (2009). Advances in fragment-based drug discovery platforms. *Expert Opin. Drug Discov.* **4**(11), 1125–1144.

Raushel, F. M., and Cleland, W. W. (1977). Bovine liver fructokinase: Purification and kinetic properties. *Biochemistry* **16**(10), 2169–2175.

Ravipati, G., McClung, J. A., Aronow, W. S., Peterson, S. J., and Frishman, W. H. (2007). Type 5 phosphodiesterase inhibitors in the treatment of erectile dysfunction and cardiovascular disease. *Cardiol. Rev.* **15**(2), 76–86.

Zartler, Edward R., and Mo, Huaping (2007). Practical aspects of NMR-based fragment discovery. *Curr. Top. Med. Chem.* **7**(16), 1592–1599.

CHAPTER SEVENTEEN

Effective Progression of Nuclear Magnetic Resonance-Detected Fragment Hits

Hugh L. Eaton *and* Daniel F. Wyss

Contents

Abstract

Fragment-based drug discovery (FBDD) has become increasingly popular over the last decade as an alternate lead generation tool to HTS approaches. Several compounds have now progressed into the clinic which originated from a fragment-based approach, demonstrating the utility of this emerging field. While fragment hit identification has become much more routine and may involve

Global Structural Chemistry, Merck Research Laboratories, Kenilworth, New Jersey, USA

Methods in Enzymology, Volume 493
ISSN 0076-6879, DOI: 10.1016/B978-0-12-381274-2.00017-0

different screening approaches, the efficient progression of fragment hits into quality lead series may still present a major bottleneck for the broadly successful application of FBDD. In our laboratory, we have extensive experience in fragment-based NMR screening (SbN) and the subsequent iterative progression of fragment hits using structure-assisted chemistry. To maximize impact, we have applied this approach strategically to early- and high-priority targets, and those struggling for leads. Its application has yielded a clinical candidate for BACE1 and lead series in about one third of the SbN/FBDD projects. In this chapter, we will give an overview of our strategy and focus our discussion on NMR-based FBDD approaches.

1. Introduction

Versatile nuclear magnetic resonance (NMR) methods are available to study the interaction of a ligand with its drug target, which in this discussion will focus on proteins. Such methods can be used for fragment-based NMR screening, the subsequent progression of fragment hits into leads based on structure–activity relationship (SAR) and structural information, and to support different stages in the lead generation process, ranging from hit characterization early in the process to late stage lead optimization. They can broadly be categorized into target- versus ligand-based NMR methods depending on whether signals from the drug target or the ligand are detected to characterize the intermolecular interaction. Each of these methods has advantages and limitations and can provide information about the ligand–target interaction at various levels of detail, including the determination of ligand affinities and potencies or their binding site and binding mode when bound to the drug target. NMR experiments can be selected to fit the target size and type, the program status, and the resources that are available. Therefore, different NMR screening and follow-up strategies may be selected for different FBDD campaigns.

In this chapter, we first discuss the planning for a NMR-based FBDD campaign, say a few words on how best to prioritize different fragment hits for hit progression, suggest how to progress fragment hits into valid lead series, and finally describe a successful FBDD campaign.

2. How to Plan for a Successful NMR-Based FBDD Campaign?

2.1. Initial considerations

Once a target has been validated and prioritized for drug discovery in a therapy area, it is up to the NMR expert to decide if and how an NMR-based FBDD campaign can make a critical contribution to the project given the available resources and time restraints. Factors that must be considered include but are not limited to knowledge of the mechanism of action of the target and how a proposed drug might effect this mechanism; the existence of biochemical and cell-based assays for the proposed mechanism; the existence of reference compounds; the availability of the target and ease of expression; the size of the target and its oligomeric state in solution; the stability of the target and requirements for cofactors, lipids, or accessory proteins to maintain the structure and/or the activity of the target; availability of chemistry resources to progress potential fragment hits; and ease of generating 3D structural data to assist fragment hit progression. Sometimes, a project team has very specific questions that it would like to have answered, and a focused NMR effort may be better suited than an extensive screening campaign. Therefore, it is good to learn as much as possible about the system in question in order to design an appropriate strategy for NMR-based drug discovery. The highest priority target may not necessarily be the best target for NMR-based FBDD. So, we typically consider various factors, including the amenability of a target to this approach and its probability of success before committing significant resources. Some targeted NMR studies early in the project help make critical decisions as to how to approach a fragment-based NMR campaign.

2.2. NMR screening methods

Target-detected NMR methods (Fig. 17.1A) have distinct advantages that they may reveal structural information about the ligand binding site and its binding mode with the drug target, can detect site-specific ligand binding over a virtually unlimited affinity range, are very reliable, and can be used to derive ligand affinities for weak fragment hits that are in fast exchange on the NMR time scale ($K_D > \sim 10\ \mu M$) or for submicromolar hits when combined in a competition format (Table 17.1). However, they require large amounts of isotope-labeled drug target necessitating expression of the protein in a host that allows high-expression yields ($> \sim 1$ mg/L) and cost-effective isotope labeling (typically, *Escherichia coli*), and knowledge of the 3D structure of the drug target and NMR assignments (or at least a map) of the active site residues to reveal active site binders. Therefore, target-detected NMR approaches are limited to a subset of drug targets (molecular weight $< \sim 40$–60 kDa) that give quality NMR spectra and do not aggregate at relatively high concentrations (~ 25–80 μM) in an aqueous NMR buffer.

Target-detected NMR screens monitor chemical-shift perturbations in the heteronuclear single quantum coherence (HSQC) spectrum of a labeled protein as a small molecule (or a mixture of small molecules) is added (Shuker *et al.*, 1996). The most commonly used labeling scheme is to uniformly label the protein with ^{15}N. The HSQC spectrum of a uniformly

A

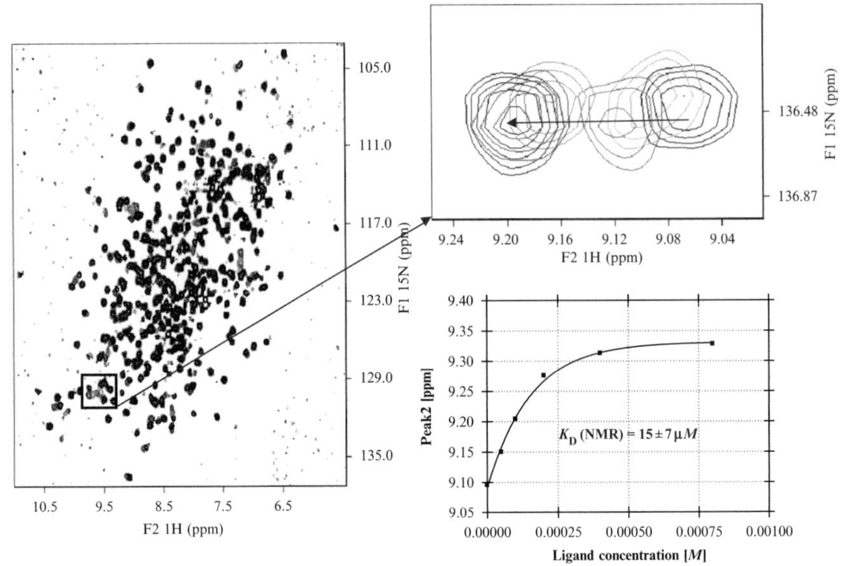
105.0
111.0
117.0
123.0
129.0
135.0
F1 15N (ppm)
10.5
9.5
8.5
7.5
6.5
F2 1H (ppm)
136.48
136.87
F1 15N (ppm)
9.24
9.20
9.16
9.12
9.08
9.04
F2 1H (ppm)
Peak2 [ppm]
K_D (NMR) = 15 ± 7 μM
Ligand concentration [M]

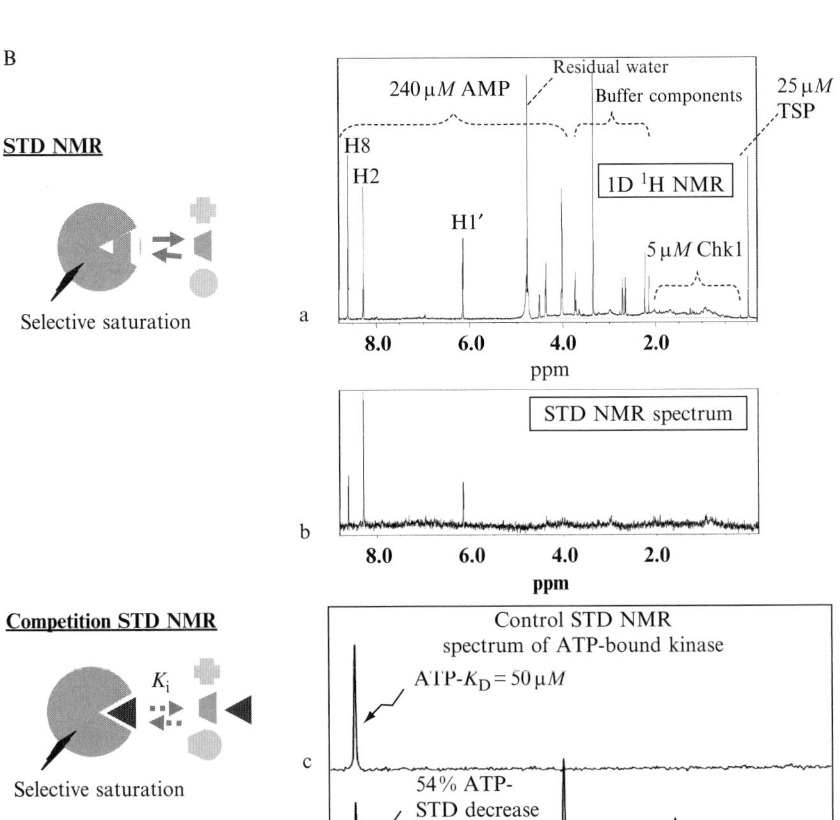
B
STD NMR
Selective saturation
240 μM AMP
Residual water
Buffer components
25 μM TSP
H8
H2
H1′
1D ¹H NMR
5 μM Chk1
a
8.0
6.0
4.0
2.0
ppm
STD NMR spectrum
b
8.0
6.0
4.0
2.0
ppm
Competition STD NMR
K_i
Selective saturation
Weak ligand-marker
High affinity ligand -competitor
Control STD NMR spectrum of ATP-bound kinase
ATP-K_D = 50 μM
c
54% ATP-STD decrease
K_i = 30 μM
d
Binding epitope of competitor

^{15}N-labeled protein contains a resonance for almost every amide N–H pair in the protein, and if these resonances have previously been assigned to the primary sequence of the protein, the binding site of the small molecule can be localized to several residues in the protein. If resonance assignments are not available, but there are reference compounds which are known to bind to the target, these reference compounds can be used to map or "fingerprint" residues in the binding site. If these same residues are perturbed during a screen, it is likely that the screened molecule binds at the same site as that of the reference molecule. Even if no reference compound is available, the pattern of perturbed residues can be used to "bin" small molecules into potentially overlapping binding regions. Finally, for small molecules that are in fast exchange on the chemical-shift timescale, an NMR-K_D can be determined by titrating the protein with the small molecule and monitoring the magnitude of the chemical-shift perturbations as the concentration of the small molecule increases. NMR-K_D determination is particularly useful when functional assays for a target have not yet been developed or are problematic.

HSQC of uniformly ^{15}N-labeled protein can work well up to about 40–60 kDa. For protein targets larger than this, spectral overlap becomes a major problem, and methods that simplify the spectrum and improve signal noise are needed. HSQC of proteins in which methyl groups are labeled with ^{13}C has been used to simplify spectra while still providing good coverage of the target (Hajduk *et al.*, 2000). This labeling scheme also has the advantage of yielding a favorable threefold sensitivity increase. In order to further simplify the HSQC spectrum of a large protein, amino-acid-type-selective labeling can be used with either ^{15}N labeling or ^{13}C labeling of methyl carbons. In amino-acid-type-selective labeling, labeling is confined to either a single amino acid type (i.e., Phe) or a small group of amino acid types (i.e., Ile, Leu, Val). Choice of which amino acid types to label will be based on the presence of an amino acid type in the binding site (if the binding site is known) and/or the distribution of the amino acid in the primary sequence of the protein. Not every protein is amenable to the

Figure 17.1 BioNMR tools to support fragment hit identification and progression, and lead identification and optimization can broadly be categorized into target- versus ligand-detected methods depending on whether signals from the target or the ligand are detected to monitor binding. (A) Target-detected NMR method, in this case, ^{15}N-HSQC, depends on following the movement of cross-peaks as the small molecule is added. If a titration is performed, an NMR-K_D may be extracted. (B) Ligand-detected NMR method, in this case, STD, (a) 1D control spectrum of AMP/kinase; (b) STD spectrum of AMP/kinase. Only resonances of atoms which contact the protein are present in the STD spectrum; (c) STD spectrum of ATP/kinase complex; (d) STD of ATP/kinase/competitor. The STD signal due to ATP is decreased because ATP is partially displaced from the binding site by the competitor. New STD signals for the competitor appear, compared to spectrum (c).

Table 17.1 Advantages and disadvantages of target- and ligand-based NMR screening methods

	Advantages	Disadvantages
Target-detected methods		
	Detect high- and low-affinity ligands	Require large amounts of isotope-labeled protein
	Structure-based	Knowledge of 3D structure of target protein and NMR assignments (or map) of at least active site residues required
	Yield ligand binding site information	
	Detect site-specific binding only	
	Method is very reliable	Limited to smaller protein targets ($<\sim$40–60 kDa)
	SAR for weak ligands ($K_D > \sim 10\ \mu M$)	High protein concentration required ($\sim$25–80 μM)
	SAR for higher affinity ligands by competition	
Ligand-detected methods		
STD	No isotope labeling required	Does not reveal ligand binding site
	No protein size limit	
	Lower protein concentration required ($\sim$1–5 μM)	Signal may be due to binding at multiple sites
	No quality protein NMR spectra required	
Competition-STD	Detects high-affinity ligands	Requires "marker" with known K_D and/or binding site
	Locates binding site and derives K_i relative to "marker" with known binding site and K_D	

labeling schemes required for target-based screens or may not produce quality NMR spectral data. In these cases, ligand-based screens may be employed.

Ligand-detected NMR methods (Fig. 17.1B, Table 17.1) can be applied much more broadly than target-detected NMR methods, since they require ~10–100-fold less drug target, do not require isotope-labeling, and have no upper molecular weight size limitation (in fact, they work better on large proteins). Although some details about the ligand binding epitope may be obtained, ligand-detected NMR methods do not reveal the ligand binding site on the drug target. Ligand-based screens rely on monitoring the change in some NMR parameter of the ligand upon its binding to the protein. One of the most useful of these methods is saturation transfer difference (STD) spectroscopy, and its variant, competition-STD (c-STD) spectroscopy

(Mayer and Meyer, 2001; McCoy *et al.*, 2005; Wang *et al.*, 2004). If spins anywhere in the protein are saturated, the saturation will quickly spread throughout the protein by spin-diffusion and will be transferred to a ligand if it has long-enough residence time in the binding site. If the ligand has fast-enough off-rate, the bound-state saturation will be observed on the free state of the ligand, with its narrow resonances. In practice, the STD experiment works well for the range $0.1\ \mu M < K_D < 1\ mM$, with a protein concentration of 0.5–5.0 μM and ligand present in 50–500 times molar excess.

The presence of signal in the STD spectrum of a ligand/protein complex must be interpreted in the broadest possible sense: there may be relatively tight binding at one binding site, weak binding at multiple sites, or some combination of the two. If there is a reference ligand with a known binding site, c-STD may be used to localize the binding site of a screened molecule. Competition-STD is a two-part experiment. First, the STD spectrum of the reference molecule is obtained. Next, the competitor is added, and the STD spectrum of the ternary mixture (reference molecule, competitor molecule, protein) is obtained. If both molecules are competing for the same binding site, the STD signal of the reference molecule will decrease. The magnitude of the decrease can be used to estimate the affinity of the competitor if the affinity of the reference is known, and the two molecules are strictly competitive with each other for the same binding site (Dalvit, 2008). Since c-STD can help rule out weak, nonspecific binding, it is an extremely valuable addition if well-characterized reference molecules are known for the target.

Finally, substrate-based functional NMR assays can be used to derive percentage inhibition or IC_{50} values (Dalvit *et al.*, 2003). In our experience, functional NMR assays can also reveal valuable details about the mode of action of modulators, since the substrate, the product, and the ligand can be monitored in simple 1D ^1H NMR spectra.

From the previous discussion, it becomes clear that depending on the knowledge and characteristics of a drug target, an appropriate NMR screening method needs to be selected for any given FBDD campaign. Moreover, suitable NMR methods can be selected to derive ligand affinity or potency to assist SAR development (Table 17.2).

Table 17.2 BioNMR methods to determine ligand affinities/potencies

Method	Affinity/Potency Range
Target-based (2D HSQC/c-HSQC)	$K_D > \sim 10\ \mu M$; $K_i < \mu M$
Ligand-based (1D c-STD)	$K_i \sim nM$–mM
Substrate-based functional assays (1D NMR)	% inhibition; IC_{50}

2.3. Library considerations

Fragment-based approaches probe chemical space more efficiently, are less dependent on legacy compound collections, and can provide hits for difficult targets. One of the great advantages of NMR-based methods is the ability to reliably identify weak binders with K_Ds in the low millimolar range while still obtaining useful structural information about their potential binding site(s). With this affinity cut-off, screening a library of 1000–2000 fragments will result in multiple hits for almost any target. The selection of these 1000–2000 compounds for an NMR-based screening library can be critical to the success of the endeavor, and details of this important topic are covered elsewhere (Jacoby *et al.*, 2003; Lepre, 2010). Candidate molecules are filtered to ensure favorable physicochemical properties and a lack of reactive functional groups. Issues of "chemical diversity" versus "drug-likeness" must be balanced. Library members may be synthetic cores for which chemically elaborated back-up libraries are readily available for fast SAR. The chosen fragment screening method may, to some degree, also influence the design of such a fragment library. Nowadays, several companies sell fragment-based screening libraries as part of their business, since FBDD has become increasingly popular over the last decade.

Once candidate library members have been chosen, they must be validated by experiment. For each library member, the chemical structure is verified, the purity of the sample determined, and aqueous solubility measured. DMSO-*d6* stock solutions of the library must be plated and stored in a way which minimizes freeze/thaw cycles and exposure to atmospheric water. In order to facilitate the identification of hits in ligand-detected NMR methods, library members are plated so that each screening cluster is "chemical-shift encoded"; that is, within each cluster, there are no degenerate chemical shifts between cluster members. Target-detected NMR screening methods do not have such requirements, but fragments in an "active" cluster must be deconvoluted to identify hit(s).

2.4. Protein production

Target-detected NMR screening works best if the drug target is relatively small and can be produced cost-effectively in uniformly isotope-labeled form at high yields. This is usually achieved using an *E. coli* expression system because of the advantages of ease of handling, rapid growth, high-level protein production, and low cost for isotope labeling (Lian and Middleton, 2001). However, many eukaryotic proteins are not functionally expressed in *E. coli*, due to problems related to disulfide bond formation, posttranslational modifications, and folding. In such cases, isotope labeling in non-*E. coli* prokaryotic and eukaryotic cells is required to produce proteins suitable for target-detected NMR screening (Takahashi and

Shimada, 2010). Isotope labeling in yeast, insect cells, or mammalian cells can be achieved. However, this usually increases the cost of protein production for NMR screening significantly, and only amino acid-type selective isotope-labeled protein may be feasible. The interested reader is referred to an excellent topical series for further details (Wagner, 2010). In contrast, ligand-detected NMR screening does not require isotope-labeled material, works better for larger targets, requires less material, and does not require high-quality protein NMR spectra; thus, these methods are more broadly applicable, and a broader range of expression hosts can be considered for protein production (Yin *et al.*, 2007).

The choice of NMR screening buffer can be critical to a successful screening campaign. In general, the buffer must be chosen to promote folded, monomeric protein with relatively high solubility and long-term stability. The choice of a good buffer includes consideration of pH, buffer components, added salt and salt concentration, and the addition of stabilizing agents such as glycerol or detergents. The choice of NMR screening buffer may also be somewhat dependent on the NMR screening methods, as ligand-detected methods benefit from D_2O buffers, while ^{15}N-HSQC-based methods require the use of H_2O. Also, finding a suitable buffer for ligand-based methods may be less challenging, as these methods require lower protein concentration ($\sim$1–5 μM) than do target-based methods that require $\sim$25–80 μM protein concentration.

3. How to Prioritize NMR-Detected Fragment Hits for Lead Generation?

3.1. Criteria for selecting fragment hits for follow-up

3.1.1. Ligand efficiency and ligand lipophilicity efficiency

Many hits found through NMR screening will have weak affinity. These weak binders may be a good starting point for lead generation if they exhibit good ligand efficiency (LE) and good ligand lipophilicity efficiency (LLE). Ligand efficiency estimates the efficiency of a binding interaction with respect to the number of nonhydrogen atoms and is a way of normalizing the binding energy by the size of the molecule (Bembenek *et al.*, 2009; Hopkins *et al.*, 2004). Ligand lipophilicity efficiency is a measure of the minimally acceptable lipophilicity per unit of *in vitro* potency: LLE (LS) = pIC50 − c log P (Leeson and Springthorpe, 2007) or a normalized LLE(AT) = 0.111 − (−1.36 × LLE(LS)/number of heavy atoms) can be used for practical reasons for fragment hits, as proposed by the Astex group (Mortenson and Murray, 2009). Chemists will have more freedom to elaborate low molecular weight, high ligand efficiency hits before reaching unacceptable limits of molecular weight and complexity which often lead to compounds that exhibit unacceptable solubility, absorption, and permeability properties. Similarly, fragments with good LLE provide more room to increase lipophilicity during lead optimization without reaching an unfavorable physical profile for the drug-candidate.

3.1.2. Thermodynamic considerations

Affinity, or binding energy, is comprised of two components, enthalpy and entropy. It has recently been proposed that there are advantages to starting with enthalpically driven leads (Freire, 2008; Scott *et al.*, 2009; Ward and Holdgate, 2001), in which binding arises from specific molecular interactions such as hydrogen bonds, salt bridges, and van der Waals interactions. In contrast, entropically driven binding generally arises from nonspecific hydrophobic interactions. Isothermal calorimetry is the tool of choice for determining the relative contributions of entropy and enthalpy to binding affinity. The information from isothermal calorimetry is best interpreted in conjunction with a detailed structural model of the binding interaction (usually from X-ray crystallography) and provides a strong starting point for optimization of a lead series. The relative balance of entropy and enthalpy will, of course, change as optimization progresses, but thermodynamic analysis and detailed structure models can go a long way toward explaining unexpected SAR and providing guidance on where to focus synthetic efforts.

3.1.3. Chemical novelty and tractability

The tractability of a hit for chemical elaboration will be judged by project chemists, who have the expert knowledge needed to assess the possibilities for elaboration of a hit with substituents, or recasting of a chemotype into an isostere. Chemists will also assess the hit for chemical novelty, especially important if the target has been extensively studied by other groups. Close interaction with project chemists is key to the success of a project. In the early stage of a project, a core structure that can easily be derivatized may be advantageous for hit progression.

3.2. Fragment hit validation and initial SAR development

Cross-validation of NMR results with information yielded by other biophysical, biochemical, and cell-based assays can be critical to the progression of a hit. Access to other assay methods is especially important when STD is used as the NMR screening method, since STD reveals no information on the ligand binding site and is more susceptible to unrecognized nonspecific binding. Results from biophysical methods such as surface plasmon resonance, thermal denaturation, and isothermal calorimetry, in addition to X-ray crystallography and structure-based NMR studies, can be used to validate NMR hits. If available, biochemical and cell-biological functional assays are valuable tools to probe the interaction of a hit with its target.

Even before project chemists become actively involved, SAR may be quickly progressed by testing obvious analogs of the initial hit from readily available commercial or internal sources, which may include "expansion" libraries that have been prepared based on members of the screening library. The value of a chemotype or structural motif is more clear if a series of molecules have been studied, and some initial SAR is seen. Based on results from the first round of analoging, project chemists will usually have ideas for further SAR development. It is important that the "iteration time" between submission of new compounds for testing and the reporting of test results back to the project team be as short as possible to maintain project momentum.

3.3. Binding site and binding mode evaluation

Target-based NMR methods can often provide this critical information, especially if site-specific assignments are available from the literature or can be obtained internally and the 3D structure of the drug target is known. The detailed binding mode of a fragment hit by NMR can, however, only be obtained for smaller targets with molecular weights up to about 20 kDa, and requires significant resources. Thus, X-ray crystallography becomes the method of choice for determining the detailed binding mode of a fragment hit.

The preferred binding mode within a chemical series can change even within the same binding site as substituents are changed, thus confusing SAR development. In these cases, knowledge of the detailed binding modes of key members within a lead series may become critical for efficient hit progression.

4. How to Progress NMR-Detected Fragment Hits into Leads?

4.1. Critical differences to typical HTS campaigns

Although fragment hits that are chosen for follow-up tend to be only weak binders initially, their ligand efficiency typically is quite high. This means that the number of atoms involved in the desired interaction with the drug target is usually high for fragment hits. Typical HTS hits, however, tend to be larger and, although having higher potency, contain portions in the molecule that are not directly involved in the desired interaction with the drug target. Therefore, the hit-to-lead optimization process is fundamentally different between fragment hits and typical hits from HTS. Fragment hits need to be extended into nearby binding pockets by increasing their molecular weight to gain potency, whereas the potency of HTS hits often need to be increased without a significant increase of the molecular weight of the initial hit (Rees *et al.*, 2004). Structural information about the binding mode of a fragment hit can be critical for efficient hit-to-lead optimization, as discussed above. Therefore, we prefer to apply this FBDD approach to must-win targets and drug targets for which X-ray or NMR structures can be obtained. Whenever possible with this approach, we like to provide the chemist with low molecular weight, high LE compounds for which we know their binding mode to the drug target, thereby providing chemists with more room for optimizing PK/ADME(T) properties during the lead optimization process. Thus, a structure-focused FBDD approach can produce leads for very challenging targets where other methods may fail (see BACE example below).

4.2. Tailor follow-up approach to circumstances

Follow-up strategies for fragment hits strongly depend on the nature and characteristics of the drug target and the fragment hits. For more challenging targets, structural data is critical for efficient fragment hit-to-lead optimization, whereas for other targets with deep, well-defined active sites, this may not necessarily be the case. In the latter case, high-concentration biochemical screening of libraries that contain "lead-like" compounds (Hann and Oprea, 2004) may be more efficient than a structure-based NMR fragment screen, especially if a robust functional assay can be developed. High-concentration biochemical screens have the distinct advantage that they already provide a functional readout for the fragment hit, and the hit-to-lead process follows a more traditional progression path. However, high-concentration screening of fragment libraries may be prone to larger numbers of "false positives," and orthogonal biophysical methods may become important to prune fragment hit lists.

Although tethering/linking fragments that bind to proximal binding sites can, in principle, yield high-affinity-linked molecules, this approach is often not very practical due to difficulties finding proximal binders, knowing their detailed binding mode, and due to limitations in linker chemistry and optimization (Chung *et al.*, 2009). Thus, expanding/growing initial fragment hits into more potent leads has become much more common for FBDD. FBDD approaches may also become very useful in better understanding the contributions of individual components of an existing lead (Barelier *et al.*, 2010), or for improving an existing lead by "fragment hopping" (Ji *et al.*, 2009).

5. In-house Example of a Successful FBDD Campaign

BACE has been a high-priority therapeutic target for the treatment of Alzheimer's disease (AD) throughout the pharmaceutical industry over the last decade. It is a membrane aspartic acid protease that is localized to the acidic compartments of endosomes/lysosomes in the central nervous system. As a consequence, a BACE inhibitor needs to be able to cross the blood–brain barrier. This makes traditional aspartic protease inhibitors, which typically are large and peptidic, unsuitable as BACE inhibitors. Moreover, the BACE active site is extended, shallow, and hydrophilic. Therefore, the development of potent, selective, orally active, and brain penetrant low molecular weight compounds has been a huge challenge for the entire industry (Durham and Shepherd, 2006; Stachel, 2009).

Much of the early drug discovery efforts had focused on the development of transition state peptidomimetics that were known from the aspartic acid protease field (Ghosh *et al.*, 2000; Maillard *et al.*, 2007). Although this approach yielded highly potent and selective BACE inhibitors, the resulting compounds lacked *in vivo* efficacy likely due to their large molecular weight and suboptimal pharmacokinetic (PK) properties. We have used a highly structure-driven approach consisting of the integrated application of target-detected fragment-based NMR screening, X-ray crystallography, structure-based design, and structure-assisted chemistry together with innovative biology to develop first-in-class clinical candidates that are progressing toward proof-of-concept for the inhibition of BACE1 in AD (Wang *et al.*, 2010; Zhu *et al.*, 2010).

5.1. Fragment hit identification

We developed an efficient protocol for the large-scale production of a fully processed soluble version of [^{15}N]-labeled BACE1 for SbN and X-ray crystallography in which the pre- and pro-sequences are autocatalytically removed within about 3 days at room temperature or 18 days at 4 °C at protein concentrations of ~5–10 mg/ml (Wang *et al.*, 2005). This refolding protocol from inclusion bodies yielded around 40 mg BACE1 per liter cell paste. We could use NMR to monitor structural details of the autocatalytic conversion which revealed a major structural rearrangement in the N-terminal lobe from a partially disordered to a well-folded conformation, suggesting that the pro-sequence may assist the proper folding of the protein. Once the protein was completely folded, we could recycle it multiple times for SbN.

We screened over 10,000 fragments in clusters of 12 to identify active site-directed hits by ^{15}N-HSQC NMR (Wang *et al.*, 2010). Initially, we did not have NMR resonance assignments. In order to not delay SbN, we initially identified peaks of active site residues of BACE1 by binding known peptide inhibitors from the literature and then screened for fragments which showed chemical-shift perturbations of some of those peaks. Eventually, we obtained sequence-specific NMR resonance assignments for BACE1 which then allowed us to study ligand binding in more structural detail (Liu *et al.*, 2004). Overall, we identified nine distinct chemical classes of active site binders to BACE1 in the 30 μ*M*–3 m*M* range, as determined by NMR titration experiments.

Among our initial fragment hits were several amidine-containing chemotypes, including the isothiourea 1 (Fig. 17.2A). We then tested over 200

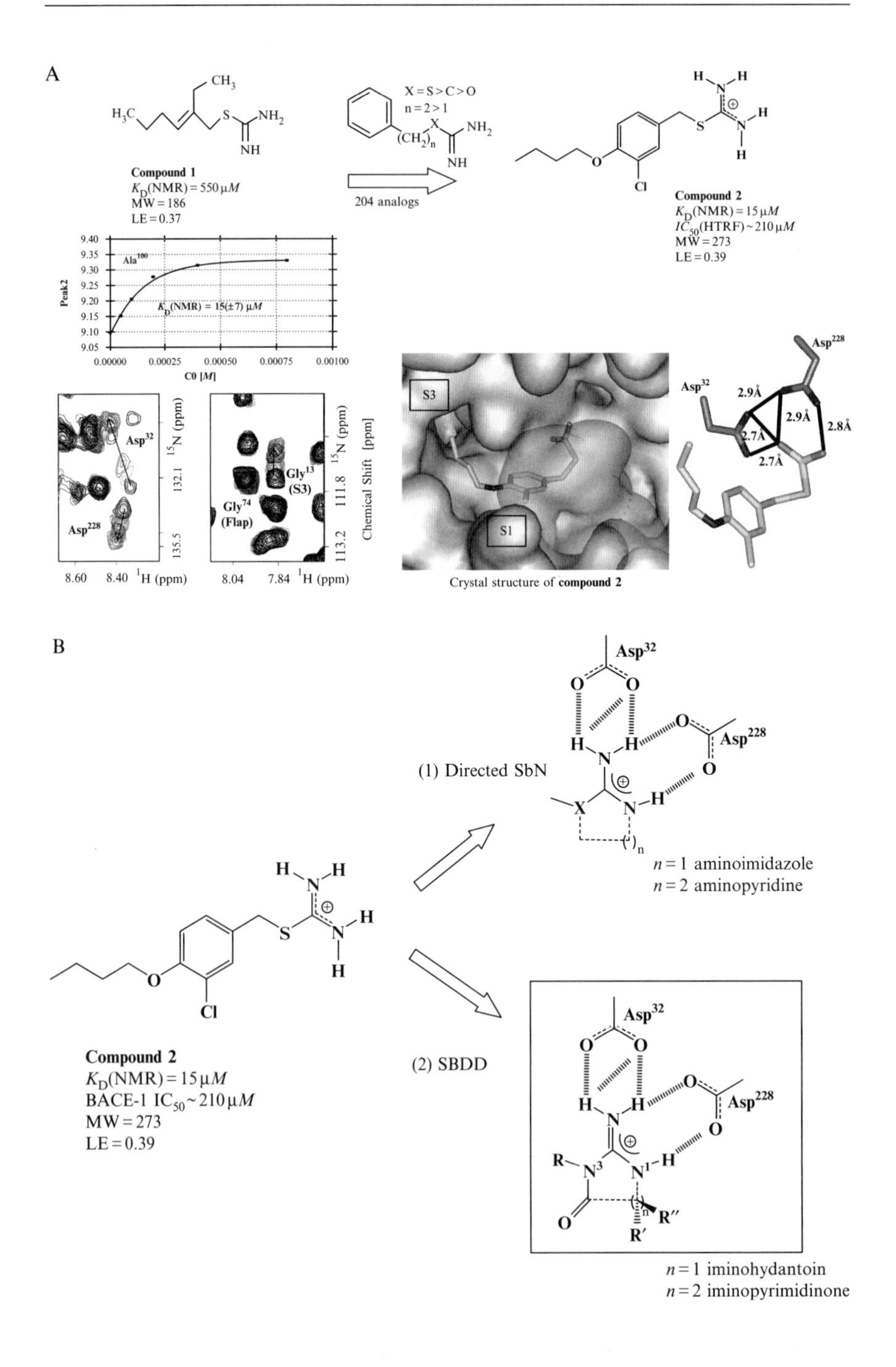
A
Compound 1
K_D(NMR) = 550 μM
MW = 186
LE = 0.37
X = S > C > O
n = 2 > 1
204 analogs
Compound 2
K_D(NMR) = 15 μM
IC_{50}(HTRF) ~ 210 μM
MW = 273
LE = 0.39
Ala[100]
K_D(NMR) = 15(±7) μM
Peak2
C0 [M]
Asp[32]
Asp[228]
Gly[13] (S3)
Gly[74] (Flap)
^{15}N (ppm)
^{1}H (ppm)
Chemical Shift [ppm]
S3
S1
Crystal structure of compound 2
2.9Å
2.8Å
2.7Å
B
(1) Directed SbN
n = 1 aminoimidazole
n = 2 aminopyridine
(2) SBDD
Compound 2
K_D(NMR) = 15 μM
BACE-1 IC_{50} ~ 210 μM
MW = 273
LE = 0.39
n = 1 iminohydantoin
n = 2 iminopyrimidinone

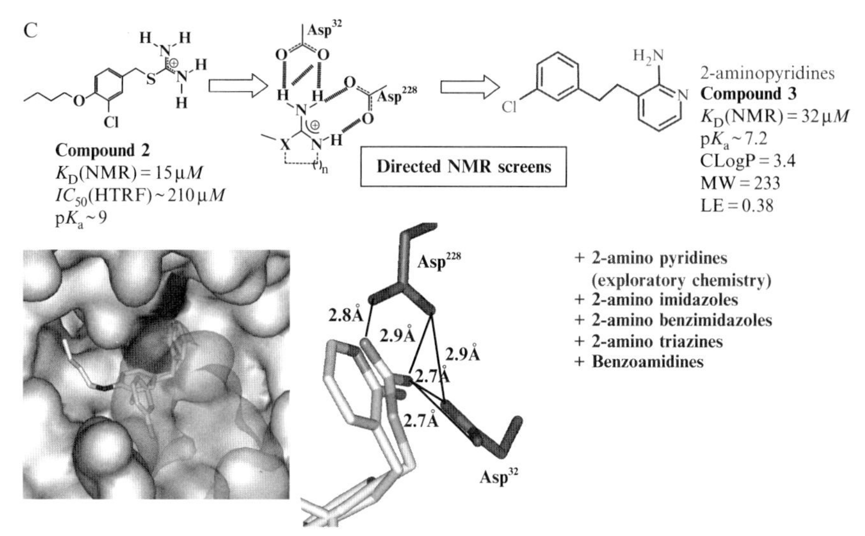

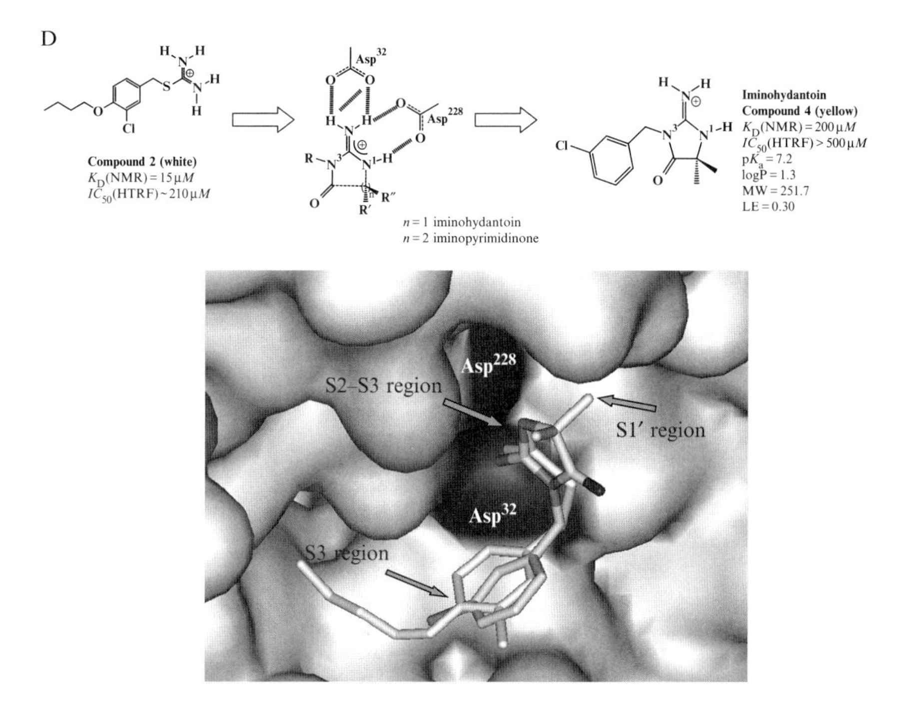

Figure 17.2 Fragment-based NMR screening of BACE-1 yielded isothiourea and related fragment hits as starting points for lead identification. (A) Isothiourea fragment hit identification and optimization by NMR and X-ray crystallography. (B) Search for heterocyclic isothiourea isosteres. (C) 2-Aminopyridines and related heterocyclic isothiourea isosteres were identified through directed SbN. (D) The structure-based design of prototype iminohydantoins yielded attractive starting points for the development of novel low molecular weight BACE-1 inhibitors.

analogs by NMR to derive initial SAR and discover isothiourea 2 which showed an NMR-K_D of 15 μM (LE = 0.39) and weak activity in an enzymatic assay. The NMR chemical-shift perturbation data suggested that compound 2 binds to the two active site aspartates and extends into the S3 pocket while leaving the flap untouched in its "open" apo-conformation. Subsequently, the cocrystal structure of compound 2 with BACE1 revealed details as to how the isothiourea moiety forms an extensive H-bond donor acceptor array with the two active site aspartates and places the chlorophenyl ring into the S1 pocket and extends deep into the shallow S3 pocket through the butyl-ether group. From that point onward, this fragment was used in an X-ray soaking system to solve the X-ray structures of over 1000 BACE1 inhibitors that were followed in this project.

Unfortunately, potential hydrolytic instability of the isothiourea moiety of compound 2 renders it unsuitable for drug development. Therefore, we started an extensive search for heterocyclic isothiourea isosteres that would be chemically attractive with an appropriate basicity (pK_a range ~6–10) to maintain the critical H-bonding network with the two active site aspartates while limiting the number of H-bond donating groups. We pursued two approaches (Fig. 17.2B). In the first approach, we carried out focused NMR screens to identify heterocyclic structures including 2-aminoimidazoles and 2-aminopyridines to bind into the active site of BACE1 (Wang *et al.*, 2010). In the second approach, we designed cyclic acylguanidines, including iminohydantoins and iminopyrimidinones (Zhu *et al.*, 2010).

5.2. Focused search for pharmaceutically attractive isothiourea isosteres

While our general NMR fragment screening was still in progress, we initiated focused-directed NMR screens of heterocyclic isothiourea isosteres that were available from our corporate library. During this process, we identified several heterocyclic cores as active site BACE1 binders, which included 2-aminopyridines, 2-aminoimidazoles, 2-aminobenzimidazoles, 2-aminotriazines, and benzoamidines, whereas other related cores were not identified as hits (Fig. 17.2C). In the 2-aminopyridine series, we discovered compound 3 which bound to the two active site aspartates with an NMR-K_D of 32 μM (LE = 0.38), as judged by the NMR chemical-shift perturbation data. Interestingly, the X-ray crystal structure of this fragment in complex with BACE1 revealed the same H-bonding network as was previously seen for compound 2. Only a few months into the SbN/FBDD campaign, compound 3 provided the first attractive starting point for chemical elaboration. Exploratory chemistry on the 2-aminopyridine series was initiated. Small chemical libraries based on the 2-aminopyridine-phenethyl core were built to explore this chemotype. Several analogs with activities in the micromolar range were identified, and crystal structures for some of these suggested the synthesis of 3,6-disubstituted 2-aminopyridine which yielded first submicromolar inhibitors in this series. However, the planar nature of the 2-aminopyridine core and the difficulties in synthesizing 3,6-disubstituted analogs prevented the easy development of more potent BACE1 inhibitors in this series.

In an alternate approach, novel cyclic acylguanidine active site binding cores such as iminohydantoin and iminopyrimidinone were conceptualized (Fig. 17.2B) in which the critical Asp binding amidino motif, common to SbN hits 2 and 3 and of similar weak basicity, is conserved. Disubstitution at C5 (iminohydantoin) or C5 and C6 (iminopyrimidinone) would simultaneously provide direct access to both prime and nonprime binding sites adjacent to the catalytic aspartate residues, with substitution on the second ring nitrogen providing a further handle for accessing binding pockets adjacent to the active site. To test this hypothesis, the prototype iminohydantoin (compound 4) and its N1-analog were designed and synthesized, in which the 3-chlorobenzyl substituent was predicted to bind in the S1 pocket, in analogy to 2 and 3. We were delighted to find that iminohydantoin 4 bound to BACE1 with an NMR-K_D of 200 μ*M*, whereas no binding was observed for its N1-analog. An X-ray structure of 4 in complex with BACE1 confirmed that 4 bound as predicted (Fig. 17.2D). We then tested several related N3- and N1 analogs. We consistently found by NMR that the N3-, but not the N1-prototype iminohydantoins bound into the active site of BACE1. About a year into the SbN/FBDD approach, we had now discovered a very attractive novel core structure that was chemically stable and provided ample opportunities to extend the molecule into nearby substrate binding pockets using simple hydantoin chemistry.

5.3. Fragment hit-to-lead progression

We quickly identified, however, a second binding mode of the iminohydantoin core in the active site of BACE1 by X-ray crystallography. This is represented by compound 5 in which the ligand-BACE1 H-bonding network is conserved, but the iminohydantoin core is flipped in the active site (Fig. 17.3A). This observation turned out to be highly significant, as this proved to be the preferred binding mode as lead optimization evolved. Simple changes in the substituents could cause the iminohydantoin core to flip or tilt in the binding pocket (Fig. 17.3B). Therefore, structural data was crucial for chemists to understand otherwise confusing SAR.

It was important to demonstrate quickly that we can produce potent iminohydantoin BACE1 inhibitors with submicromolar IC_{50}s in the enzymatic assay. The binding mode of iminohydantoin 7 (Fig. 17.3B) suggested that cyclohexylmethyl and cyclohexylethyl extensions into the respective hydrophobic S1 and S2′ pockets should achieve this goal (Fig. 17.3C). The resulting iminohydantoin 8 was in fact the first submicromolar inhibitor in

A

Asp228

Asp32

NH

Cl

N

NH

O

Compound 4 (orange)
K_D(NMR) = 200 μ*M*
LE = 0.30

Compound 5 (blue)
K_D(NMR) = 120 μ*M*
IC_{50}(HTRF) ~ 300 μ*M*
LE = 0.27

B

Thr232

Mode B

Asp228

S1′

Asp32

S3

Mode A

S1

Lys107 / Phe108

112.2

113.0

Gly21

113.8

8.82 8.73 8.64

Compound 6
IC_{50}(HTRF) = 8 μ*M*
NMR: Mode A

Compound 7
IC_{50}(HTRF) = 73 μ*M*
NMR: Mode B

C

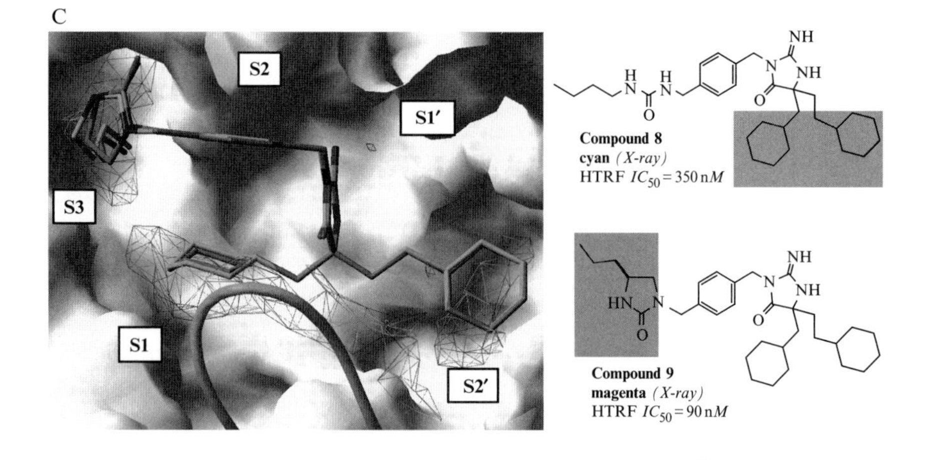

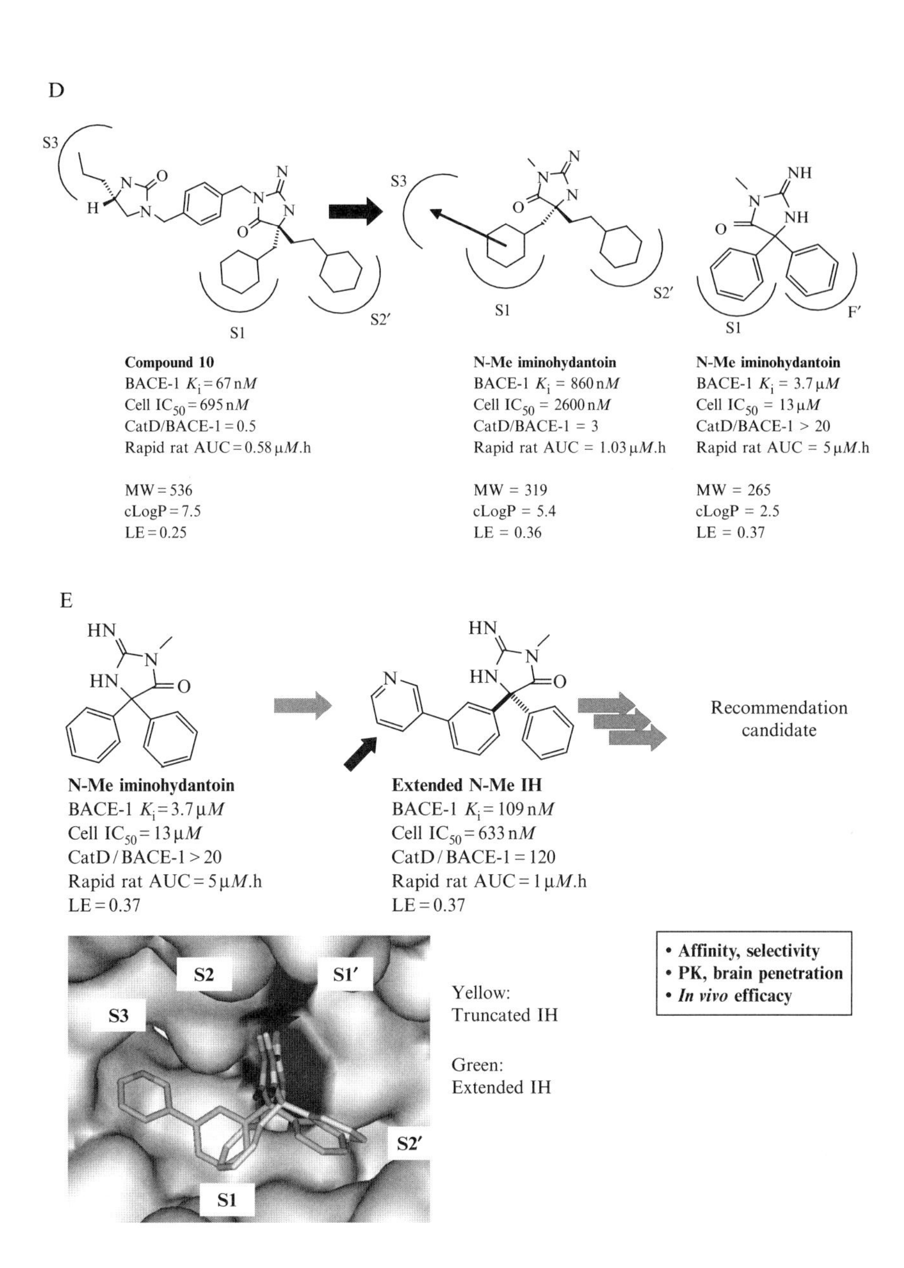

Figure 17.3 Iminohydantoin fragment hit progression. (A) A second binding mode of the iminohydantoin core in the active site of BACE1 was revealed by X-ray crystallography. (B) Simple changes in the substituents could cause the iminohydantoin core to flip or tilt in the active site; thus, structural data simplified SAR development. (C) Structure-based design of the first series of submicromolar iminohydantoin BACE-1 inhibitors. (D) Truncated *N*-methyl iminohydantoins provided a more direct way to build toward S3. (E) Early examples of iminohydantoins and iminopyrimidinons with good overall inhibitor profiles.

this series. Its crystal structure confirmed the underlying structure-based design and suggested that a further increase in potency should be possible by introducing a cyclic urea with the propyl extension in the proper S-configuration. Again, the resulting iminohydantoin 9 bound to BACE1 as expected and showed an increased potency in the enzymatic assay. Isolation of the single stereoisomer with 4(S)/4(R) configuration yielded Compound 10 with a cellular IC_{50} in the submicromolar range (Fig. 17.3D). However, the resulting compounds became non-lead-like with significantly reduced LE and increased $c \log P$. Fortunately, truncated *N*-methyl iminohydantoins showed much higher LE and provided opportunities to build into the S3 pocket more directly without increasing the molecular weight of the iminohydantoins as much. Especially, the 5,5′-diphenyl iminohydantoins showed an overall favorable profile with respect to cellular potency, selectivity, rat PK, brain penetration, and lead-like properties.

The *N*-methyl 5,5′-diphenyl iminohydantoin core offered several opportunities to build into the surrounding S1–S3 and S2′ substrate binding pockets. Fig. 17.3E shows an early example in which a simple pyridine extension toward the S3 pocket from the meta-position of the phenyl in the S1 pocket yielded a highly ligand-efficient BACE1 inhibitor with a submicromolar potency in the cellular assay. Eventually, the team was successful in developing a first-in-class recommendation candidate that has entered clinical trials and is currently progressing toward proof-of-concept for the inhibition of BACE1 in Alzheimer's disease. Ring expansion to the iminopyrimidinone and structure-assisted SAR development were critical to develop BACE1 inhibitors with high affinity and selectivity, excellent PK properties to achieve brain penetration, and ultimately, efficacy *in vivo* (Stamford, 2010).

ACKNOWLEDGMENTS

We thank our current and former colleagues for their invaluable contributions to the work described in this chapter.

REFERENCES

Barelier, S., Pons, J., Marcillat, O., Lancelin, J.-M., and Krimm, I. (2010). Fragment-based deconstruction of Bcl-xL inhibitors. *J. Med. Chem.* **53,** 2577–2588.

Bembenek, S. D., Tounge, B. A., and Reynolds, C. H. (2009). Ligand efficiency and fragment-based drug discovery. *Drug Discov. Today* **15,** 278–283.

Chung, S., Parker, J. B., Bianchet, M., Amzel, L. M., and Stivers, J. T. (2009). Impact of linker strain and flexibility in the design of a fragment-based inhibitor. *Nat. Chem. Biol.* **5,** 407–413.

Dalvit, C. (2008). Theoretical analysis of the competition ligand-based NMR experiments and selected applications to fragment screening and binding constant measurements. *Concepts Magn. Reson. Part A* **32A,** 341–372.
Dalvit, C., Ardini, E., Flocco, M., Fogliatto, G. P., Mongelli, N., and Veronesi, M. (2003). A general NMR method for rapid, efficient, and reliable biochemical screening. *J. Am. Chem. Soc.* **125,** 14620–14625.
Durham, T. B., and Shepherd, T. A. (2006). Progress toward the discovery and development of efficacious BACE inhibitors. *Curr. Opin. Drug Discov. Dev.* **9,** 776–791.
Freire, E. (2008). Do enthalpy and entropy distinguish first in class from best in class? *Drug Discov. Today* **13,** 869–874.
Ghosh, A. K., Shin, D., Downs, D., Koelsch, G., Lin, X., Ermolieff, J., and Tang, J. (2000). Design of potent inhibitors for human brain memapsin 2 (β-Secretase). *J. Am. Chem. Soc.* **122,** 3522–3523.
Hajduk, P. J., Augeri, D. J., Mack, J., Mendoza, R., Yang, J., Betz, S. F., and Fesik, S. W. (2000). NMR-based screening of proteins containing ^{13}C-labeled methyl groups. *J. Am. Chem. Soc.* **122,** 7898–7904.
Hann, M. M., and Oprea, T. I. (2004). Pursuing the leadlikeness concept in pharmaceutical research. *Curr. Opin. Chem. Biol.* **8,** 255–263.
Hopkins, A. L., Groom, C. R., and Alex, A. (2004). Ligand efficiency: A useful metric for lead selection. *Drug Discov. Today* **9,** 430–431.
Jacoby, E., Davies, J., and Blommers, M. J. (2003). Design of small molecule libraries for NMR screening and other applications in drug discovery. *Curr. Top. Med. Chem.* **3,** 11–23.
Ji, H., Li, H., Martasek, P., Roman, L. J., Poulos, T. L., and Silverman, R. B. (2009). Discovery of highly potent and selective inhibitors of neuronal nitric oxide synthase by fragment hopping. *J. Med. Chem.* **52,** 779–797.
Leeson, P. D., and Springthorpe, B. (2007). The influence of drug-like concepts on decision-making in medicinal chemistry. *Nat. Rev. Drug Discov.* **6,** 881–890.
Lepre, C. A. (2011). Practical aspects of NMR-based fragment screening. *Methods Enzymol.* **493,** 219–239.
Lian, L.-Y., and Middleton, D. A. (2001). Labelling approaches for protein structural studies by solution-state and solid-state NMR. *Prog. Nucl. Magn. Reson. Spectrosc.* **39,** 171–190.
Liu, D., Wang, Y.-S., Gesell, J. J., Wilson, E., Beyer, B. M., and Wyss, D. F. (2004). Backbone resonance assignments of the 45.3 kDa catalytic domain of human BACE1. *J. Biomol. NMR* **29,** 425–426.
Maillard, M. C., Hom, R. K., Benson, T. E., Moon, J. B., Mamo, S., Bienkowski, M., Tomasselli, A. G., Woods, D. D., Prince, D. B., Paddock, D. J., Emmons, T. L., Tucker, J. A., *et al.* (2007). Design, synthesis, and crystal structure of hydroxyethyl secondary amine-based peptidomimetic inhibitors of human beta-secretase. *J. Med. Chem.* **50,** 776–781.
Mayer, M., and Meyer, B. (2001). Group epitope mapping by saturation transfer difference NMR to identify segments of a ligand in direct contact with a protein receptor. *J. Am. Chem. Soc.* **123,** 6108–6117.
McCoy, M. A., Senior, M. M., and Wyss, D. F. (2005). Screening of protein kinases by ATP-STD NMR spectroscopy. *J. Am. Chem. Soc.* **127,** 7978–7979.
Mortenson, P. N., and Murray, C. W. (2009). Ligand lipophilicity efficiency—Assessing lipophilicity of fragments and early hits. RSC Fragments 2009, Poster 9, Astra Zeneca Alderley Park, UK, March 4–5, 2009.
Rees, D. C., Congreve, M., Murray, C. W., and Carr, R. (2004). Fragment-based lead discovery. *Nat. Rev. Drug Discov.* **3,** 660–672.
Scott, A. D., Phillips, C., Alex, A., Flocco, M., Bent, A., Randall, A., O'Brien, R., Damian, L., and Jones, L. H. (2009). Thermodynamic optimisation in drug discovery: A case study using carbonic anhydrase inhibitors. *ChemMedChem* **4,** 1985–1989.

Shuker, S. B., Hajduk, P. J., Meadows, R. P., and Fesik, S. W. (1996). Discovering high-affinity ligands for proteins: SAR by NMR. *Science* **274,** 1531–1534.

Stachel, S. J. (2009). Progress toward the development of a viable BACE-1 inhibitor. *Drug Dev. Res.* **70,** 101–110.

Stamford, A. (2010). Discovery of small molecule, orally active and brain penetrant BACE1 inhibitors. 239th ACS National Meeting, San Francisco, CA, March 22, 2010.

Takahashi, H., and Shimada, I. (2010). Production of isotopically labeled heterologous proteins in non-*E. coli* prokaryotic and eukaryotic cells. *J. Biomol. NMR* **46,** 3–10.

Wagner, G. (2010). A topical issue: Production and labeling of biological macromolecules for NMR investigations. *J. Biomol. NMR* **46,** 1–2.

Wang, Y. S., Liu, D., and Wyss, D. F. (2004). Competition STD NMR for the detection of high-affinity ligands and NMR-based screening. *Magn. Reson. Chem.* **42,** 485–499.

Wang, Y.-S., Beyer, B. M., Senior, M. M., and Wyss, D. F. (2005). Characterization of autocatalytic conversion of precursor BACE1 by heteronuclear NMR spectroscopy. *Biochemistry* **44,** 16594–16601.

Wang, Y.-S., Strickland, C., Voigt, J. H., Kennedy, M. E., Beyer, B. M., Senior, M. M., Smith, E. M., Nechuta, T. L., Madison, V. S., Czarniecki, M., McKittrick, B. A., Stamford, A. W., *et al.* (2010). Application of fragment-based NMR screening, X-ray crystallography, structure-based design, and focused chemical library design to identify novel μM leads for the development of nM BACE-1 (β-Site APP Cleaving Enzyme 1) inhibitors. *J. Med. Chem.* **53,** 942–950.

Ward, H. J., and Holdgate, G. A. (2001). Isothermal titration calorimetry in drug discovery. *Prog. Med. Chem.* **38,** 309–376.

Yin, J., Li, G., Ren, X., and Herrler, G. (2007). Select what you need: A comparative evaluation of the advantages and limitations of frequently used expression systems for foreign genes. *J. Biotechnol.* **127,** 335–347.

Zhu, Z., Sun, Z.-Y., Ye, Y., Voigt, J., Strickland, C., Smith, E. M., Cumming, J., Wang, L., Wong, J., Wang, Y.-S., Wyss, D. F., Chen, X., *et al.* (2010). Discovery of cyclic acylguanidines as highly potent and selective β-site amyloid cleaving enzyme (BACE) inhibitors: Part I – Inhibitor design and validation. *J. Med. Chem.* **53,** 951–965.

CHAPTER EIGHTEEN

Advancing Fragment Binders to Lead-Like Compounds Using Ligand and Protein-Based NMR Spectroscopy

Till Maurer

Contents

Abstract

The application of NMR in fragment-based lead discovery (FBLD) has quickly developed from a sensitive method for the identification of low-affinity binders to an important tool in the hit-to-lead process. NMR can play a constructive role in the process from identifying those fragments with the best potential toward a biochemically active compound to developing them into molecules with high affinity and selectivity to a given target protein. NMR hit-to-lead involves revising the lead identification process at the beginning of a fragment-based drug discovery project, the primary screen, and also looking toward protein-detected NMR methods in

Department of Structural Biology, Genentech Inc., South San Francisco, California, USA

Methods in Enzymology, Volume 493
ISSN 0076-6879, DOI: 10.1016/B978-0-12-381274-2.00018-2

advancing compounds from fragment hit into and through fragment hit-to-lead. With the development of higher sensitivity cold NMR probes, ligand-based NMR methods can be successfully applied to a majority of projects found in a pharmaceutical pipeline. Having matured from the original concepts such as SAR by NMR (Shuker, S. B., Hajduk, P. J., Meadows, R. P., Fesik, S. W. (1996) Discovering high-affinity ligands for proteins: SAR by NMR. *Science* **274** (5292), 1531–1534.), projects that base their lead matter on fragment hits are close to or already in the clinic (Woodhead, A. J., Angove, H., Carr, M. G., Chessari, G., Congreve, M., Coyle, J. E., Cosme, J., Graham, B., Day, P. J., Downham, R., Fazal, L., Feltell, R., *et al.* (2010) discovery of (2,4-dihydroxy-5-isopropylphenyl)-[5-(4-methylpiperazin-1-ylmethyl)-1,3-dihydroisoindol-2-yl]methanone (AT13387), a novel inhibitor of the molecular chaperone Hsp90 by fragment based drug design. *J. Med. Chem.* **53**, 5956–5969, Chessari, G., and Woodhead, A. J. (2009). From fragment to clinical candidate: A historical perspective. *Drug Discov. Today* **14** (13–14), 668–675.). Generating new ideas toward new binding modes and mechanisms of action as well as new intellectual property will be the standard by which the success of FBLD will need to be measured. A strategy outlining the various steps involved in NMR hit-to-lead is provided. By means of a specific example, the workflow is described to guide the reader through the experimental setup.

1. Introduction

NMR is an exquisitely sensitive method for probing the surface of a macromolecule for binding interaction with a small molecule or fragment using ligand-based NMR detection methods. These methods include STD (Mayer and Meyer, 2001), water-LOGSY (Dalvit, 2009), t1roh (Hajduk *et al.*, 1997), noe-pumping (Chen and Shapiro, 2000), differential line broadening (LaPlante *et al.*, 1999), and shift perturbation methods (De Marco *et al.*, 1986), for clarity, this chapter will focus on the first two. The high sensitivity for detecting ligand binding can turn out to be a caveat when utilizing NMR to simply detect binding interaction. It is therefore important to consider the hit-to-lead process before a screen is actually started, as the design of a primary screen will need to be optimized for the desired starting points in hit-to-lead. It is advantageous in a structure-based drug design project, and in particular, for fragment-based lead discovery (FBLD) that the place of binding and its architecture are structurally well defined and the enzymatic mechanism is well understood.

Three strategies may be used:

1. The primary NMR screen is directed toward a particular binding site by use of a documented inhibitor of known affinity as a scout or competitor compound.
2. The aim of the screen is to identify a secondary binding site to that of the known inhibitor. Secondary site binding information can easily be

extracted from the primary screen data set to address other mechanisms of action such as allosteric inhibition. A secondary binding site can be particularly attractive when new binding pockets are formed only when the known inhibitor or ligand is already bound.

3. The screen is designed without a specific binding pocket in mind. Here, one would perform the NMR binding experiments with the goal to identify compounds that bind anywhere, an attractive possibility of finding completely novel binding sites, binding modes, and potential mechanisms. The more challenging aspect in this process is to then determine how this primary hit may be optimized into constructive binding that produces activity.

2. Strategies for Defining Hits

So, how do we define a hit? This question must be addressed at the beginning of the process and the answer then will determine which compounds are of interest at the analysis stage. A generic type of binding interaction will be too broad a scope to tackle with a large fragment library. Ligand-based NMR will detect up to molar binding affinity, a good example being dimethyl sulfoxide, which will bind to many different proteins and regularly show up as a hit in a screen simply because of its presence in the ligand solutions. Fragment hit rates of up to 30% of a library can become a challenge in data analysis and follow-up.

NMR is ideal to triage or follow up the hits of any prior primary screen, be it biophysical such as Surface Plasmon Resonance or through a high-throughput activity-based assay. With hits from both work flows, NMR can aid to characterize the binders in terms of binding site, and binding affinity. Characterization of the binding site can be achieved with ligand-based competition experiments. With the availability of isotope-labeled protein that is well behaved in terms of aggregation, stability, and isotropic tumbling the absolute binding affinity can be determined. In addition, known assignments refine the quality of information that can be gained by chemical “shift mapping”—the projection of shift perturbations of a given nucleus to the surface of a protein-binding site. Shift mapping will often lead to an unambiguous binding orientation of the hit. In cases where no assignments for a given labeled protein are available, a small molecule of a known binding site can be used to map changes in chemical shifts to resonance signals belonging to residues involved in binding. Using this information, the binding site of primary hits can be grouped into primary hits binding to similar sites. It is the desired outcome of the filtering process to considerably triage the total number of primary hits into potentially productive binders.

Fragment hit-to-lead will follow the primary screen and is perceived as one or a combination of the following workflows:

1. *Primary hit characterization:* primary hits are analytically characterized using NMR to determine solubility in the screening buffer, which will often be similar to the biochemical assay conditions but can also be expanded to downstream conditions, that is, those used for crystallization. In the same workflow, the integrity of the fragment compound can be confirmed based on the 1D proton spectrum and stability in the buffer conditions checked. The primary hits are then measured using a ligand-detected NMR competition experiment for binding to a specific site and can be grouped into classes according to their preferred binding pocket.
2. *Analoging:* scanning of vendor collections, internal fragment collections, or synthesis will lead to compounds with a similar scaffold as the primary hit. Using ligand-based NMR, one can test for binding to the binding site of the protein targeted by the primary hit and/or active site. In this workflow, compounds can be ranked by affinity through mutual competition and a series of tool compounds identified.
3. *Shift mapping and docking:* using structural information from the target protein, the bound conformation of a primary hit can be modeled using structural information from the target protein, chemical shift perturbation data from ligand-binding studies of isotopically labeled protein, and computational methods. Based on this information, chemical elaboration of the primary hit will confirm the docked structure. Use of NMR titration experiments will allow determination of binding affinity and fast control of the effects of chemical elaboration. Iteration of this process will quickly allow compounds with lead-like affinity to be attained.

While NMR can and should be used for the initial screening of a compound library, the true strength and potential are found in the follow-up of initial hits. NMR offers a unique set of atomic-level observables for the hit-to-lead advancement not found in other techniques. This advantage is outlined in the included example. With these workflows in place, the definition of a primary hit can be kept very broad and NMR used to quickly identify potentially productive starting points for hit-to-lead.

3. The Fragment Library and Protein Production

The details of the fragment library and primary screen are extensively reported elsewhere in this volume, and only a short summary of the use of NMR in fragment library quality control and the primary screen will be presented here.

The evaluation of NMR-binding data in a FBLD project requires the acquisition of a database of 1D NMR spectra of the compounds in the fragment library. Because fragment hits are usually low-affinity binders, they need to be followed up in the hit-to-lead process at high concentrations. Crystallization of fragment hits, for example, will often be done well above its low binding affinity. Compound qualities that will benefit the later stages of the fragment-to-lead process include good aqueous solubility, no effect on solution pH at high ligand concentrations, and a low propensity to aggregate. NMR can determine these in an economical fashion. A dilution series and short 1D proton measurements can quickly be analyzed to determine a kinetic solubility, as NMR is a quantitative method. Addition of an STD experiment at one of the concentrations will additionally show if the compound form aggregates and is in equilibrium with its monomeric form, as this leads to intense signals in the absence of protein. The STD measurement thus also serves the purpose of making sure that potential false positives are not measured in the primary screen. If sample preparation is done manually, this will only be amenable for small fragment libraries. Up to 2000–5000 compounds can be characterized if the lab is equipped with a pipetting system and connected NMR sample handling hardware. Such an automation setup is described in TECAN Journal (2009).

An FBLD project will only make sense if a functionally well-defined and structurally enabled protein target is available. Ideally, this requirement means that a soakable crystallographic system is available, protein production is feasible with milligrams per liter of culture yields, and *Escherichia coli* protein expression will facilitate affordable isotopic labeling. Isotopic labeling in insect cell expression is also becoming a routine procedure but is only feasible if high expression rates can be achieved (Brüggert *et al.*, 2003). NMR can bridge the gap between conditions other biophysical techniques need because more often than not NMR experiments can be performed to match both the cell environment, the conditions in a biological assay and in part also the crystallization tray. An NMR hit confirmation experiment can mean going through different conditions and monitoring tool compound binding under these conditions.

With stable protein in hand, the primary screen will consist of measurements of ligands, or mixtures thereof using ligand-based NMR methods that observe the proton signals of the ligand. A high ligand-to-protein ratio of typically 50–100:1 is necessary for the STD and water-LOGSY experiments, thus the protein will be at a very low concentration of 2.5–10 μM. The primary screen of a library of 2000 fragments measured in mixtures of five will need ca. 5–20 mg of purified protein. A screen of 2000 compounds consisting of 400 experiments taking about 30 min per experiment can be completed with 200 h measurement time.

4. NMR Follow-Up and Fragment Hit-To-Lead

NMR can unfold its strength in the triaging and characterization of primary hits by bridging the gap that often occurs between confirmed binding and structural information gained by X-ray crystallography.

For NMR followup, the source of a hit can be any other biophysical method detecting direct physical interaction, such as SPR, but also compounds from a biochemical assay with an activity readout.

NMR hit-to-lead will start with binding ligand characterization using unlabeled protein, the workflow illustrated in Fig. 18.1. A competition experiment with a binder of a known binding site and affinity will always be the follow-up of choice in a fragment screening project that is focused on a specific drug-able active site. This focus does not rule out that novel binding sites and modes can be targeted. As NMR is fast, a qualitative differentiation of active site binder versus a secondary site such as a cofactor binding site can quickly group hits according to their binding site. In the same workflow, compounds with novel binding sites can be singled out.

On their own the clear differentiation of hit verses nonhit is often not easily achieved, as signal intensity of the ligand observed spectra will vary greatly overall from target to target and from ligand to ligand. The signal intensity can be mostly attributed to the exchange of magnetization due to

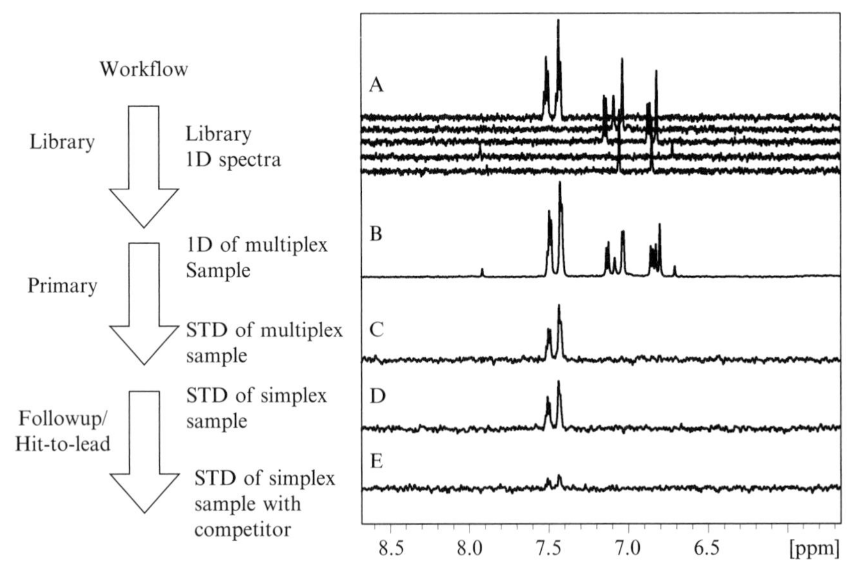

Figure 18.1 Spectra of a fragment screen. (A) The overlay of 1D proton spectra of five compounds dissolved to 250 μ*M* in PBS, 99% ^2H_2O, pH 7.4. (B) The 1D proton spectrum of the mixture of the same five ligands with protein. Signals from the protein are not visible due to the low concentration of 2.5 μ*M*. (C) The STD spectrum of the mixture shown in (B). The signals of one of the compounds in (A) (shown in yellow) gives clear signals indicating binding. (D) Simplex follow-up of the hit from (C). (E) Same as spectrum (D) but with the addition of a high-affinity active site binder. Modulation of the signal indicates competition from active site binding. The low-intensity signal that remains in (E) indicates additional binding at another site. (For interpretation of the references to color in this figure legend, the reader is referred to the Web version of this chapter.)

the kinetics of binding and the dipolar interaction, which is distance dependent and reflects the geometry of the protein-binding site. This property makes qualification of hits based solely on NMR signal intensity difficult. One way to prioritize hits is to use a defined signal intensity cutoff, such as the STD amplification factor (Szczepina *et al.*, 2009), which is 1 minus the ratio of the resonance intensities between on- (I_{sat}) and off-resonance (I_o) irradiation of the enzyme times the molar excess of the ligand can be used to prioritize hits:

$$1 - (I_{sat}/I_o) \times \text{ligand excess}$$

This equation has an arbitrary component as the relative or absolute signal strength is a reflection of multiple factors such as spatial proximity of ligand protons to protein protons in the STD experiment or the degree of water complexation in the water-LOGSY measurement. The strong dependence of water-LOGSY signal intensity on solution conditions and the relative ease of automating measurements in ^2H_2O favor the use of STD.

The confirmed next step of hit-to-lead will be ligand titration experiments using isotope-labeled protein. A fast on–off kinetic is particularly amenable to using NMR as the binding detection method, given that low-affinity binders are the focus of FBLD. In the titration, signals will gradually shift with increasing ligand concentration and as a rule, only signals stemming from residues in the vicinity of the binding site will show these shifts. This phenomenon is in contrast to shift perturbations stemming from high-affinity ligands that will show secondary shift effects (Schumann *et al.*, 2007) and modulation of protein dynamics that make data analysis more time-consuming.

The shifting peaks are easy to pick up and, with a known assignment, can be mapped to the surface of the protein structure. This structure will often have been determined by X-ray crystallography. It is certainly desirable to use the structure calculated based on NMR data acquired under the same conditions to determine the ligand costructure. In reality, it can be difficult to achieve protein assignments and a structure calculation in the rather fast turnover rates for projects in pharmaceutical research. However, within project timelines, it is very often possible to sequentially assign a protein in the time window X-ray crystallography requires for its structure determination.

The chemical shift map is a first step toward obtaining a three-dimensional binding information and will allow identification of ligands that bind in a defined site and similar orientation. Still well away from the resolution X-ray can achieve, this information is very important for crystallographic follow-up, as NMR is the only other method that will directly monitor where a ligand binds. If all primary hits from biophysical methods are selected for follow-up with crystallography without this type of NMR-triage, the large numbers of "false positives" will make a successful crystallographic soaking less likely.

5. Characterizing Binding Modes and Co-Structure Information through Docking

If the high-resolution coordinates of an X-ray costructure are available, advancing ligands with high micromolar to low millimolar binding affinity will quickly be possible. Thus, it should be a goal for a given fragment project to have a soakable crystallographic system set up. NMR chemical shift perturbation data can be used to obtain information on the binding orientation and NMR based ligand optimization can be utilized to reach increased binding affinity to a point where soaking will be successful or a biological assay readout possible. For example, isotope-labeled protein will allow the determination of a shift map to allow identification of the binding site and a binding orientation. The chemical shift map is a projection of experimentally observed changes in the position of signals in the NMR spectrum, often ^{13}C HSQC or ^{15}N HSQC–TROSY spectra but also 2D projections of triple resonance experiments such as 3D HNCO or HN (CO)CA. For the amide proton and nitrogen shifts, this means that the observable atoms are often quite a distance away from where the interaction actually takes place. Shift mapping can become misleading as the chemical shift of amide groups are very sensitive to pH and small changes in the hydrogen bonding patterns of protein backbones. This means some shift changes may not be related to a direct interaction but due to second-sphere effects in hydrogen bonding networks. In contrast, side chains will be closer to the ligand molecule and often allow a better resolved shift map. With that in mind, the shift map can be visualized by color coding the magnitude of the changes scaled to the gyromagnetic ratio of the observed nucleus as a projection into the surface of the protein structure. This usually shows clearly where a ligand binds to the target but is more often not sufficient to calculate a structure of the protein–ligand complex.

Still, there is valuable information in the shift map, and using it in the context of calculated docking modes can be a first step toward utilizing shift information for a protein–ligand costructure.

A visual comparison of the observable chemical shift changes and a set of calculated docking modes can often very easily discriminate these into more or less probable possibilities. Figure 18.3 shows different docked costructures based purely on the protein X-ray coordinates from the Protein Data Base PDB (RCSB Protein Database http://www.pdb.org) and a docking protocol from the program GLIDE (Schrödinger Software, http://www.schrodinger.com/). It becomes clear that some of the orientations are not compatible with the experimental NMR data. Integration of chemical shift changes in a more quantitative sense into the scoring function of a docking run has been demonstrated and can lead to well-defined structural models (González-Ruiz and Gohlke, 2009).

A docking orientation will be analyzed in terms of the chemical environment of the protein active site in order to identify potential interaction partners that can be addressed by chemical modification of the ligand. Modules that will identify potential interaction partners in the target protein active site will bring forth such suggestions. Also, the validity of confirmed docking orientations will be increased by chemical confirmation in the fragment hit-to-lead phase through "SAR by catalog" or directed synthesis by chemists. This procedure initiates an iterative process that, on the one hand, confirms the model, because similar compounds will be expected to show similar NMR chemical shifts. On the other hand, will lead to a gain in binding affinity because more binding interaction partners are addressed.

This important stage in fragment hit-to-lead can quickly lead to an order of magnitude gain in binding affinity. The results of a chemical synthesis can easily and quickly be confirmed in the NMR shift perturbation, with a ^{15}N confirmation measurement taking 1 h. Figure 18.2 shows the ^{15}N shift map of a primary hit in panel A and analogs of the primary hit, in panels B, C, and D. Similar shift perturbations are observed with all analogs. Using the docked conformation shown in Fig. 18.3, the simple chemical modification of the primary hit leads to the perturbations map depicted in Fig. 18.4. It is interesting here to note that the primary hit did not engage residues in the active site as can be seen by the lack of chemical shift changes of the right side of panel B, whereas the spectra of the final derivative shown on the left show clear perturbation. In the example above, an increase of binding affinity of a factor 100 occurs with only one round of chemical modification. This outcome can be particularly fast to achieve if the starting affinity is very low (in the range of high three-digit micromolar), as an additional interaction will contribute to a high degree to the increase in K_d.

Regardless of the origin of a fragment hit, be it a biophysical method such as SPR or an HTS activity-based assay, protein-detected NMR can triage the hit-to-lead process through high numbers of compounds to a point where a ligand is identified that has the affinity needed to be successful in soaking and costructure determination.

6. The Process in an Example

6.1. Sample for setup and parameter test

To test the parameter setup for the NMR experiments, a solution of 20 μM human serum albumin with glucose and benzoic acid at 500 μM in PBS pH 7.4 made in 99.9% ^2H_2O is a good test sample. The pH needs no adjustment

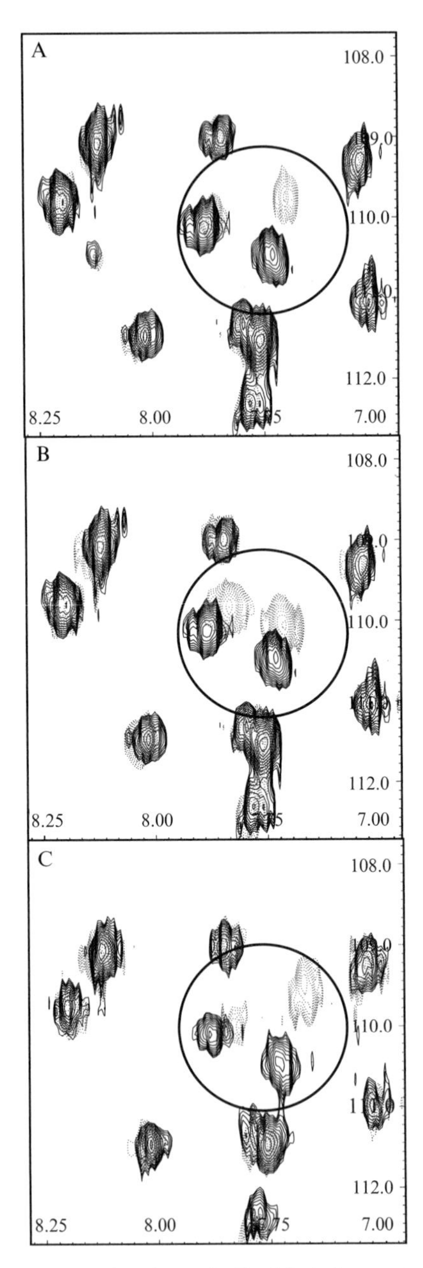

Figure 18.2 Shift mapping using isotopically labeled protein. Shown are the ^{15}N TROSY-HSQC spectra of (A), the primary hit from Fig. 18.1. The protein concentration was 50 μ*M* in 95% H_2O, 5% ^2H_2O, and PBS pH 7.4. Ligand was added from a d6-dimethyl sulfoxide stock solution at 100 m*M* to a concentration of 1 m*M*. Panels (B), (C), and (D) show the shift maps of analogs of the primary hit. All spectra show similar shift changes indicating a similar binding mode.

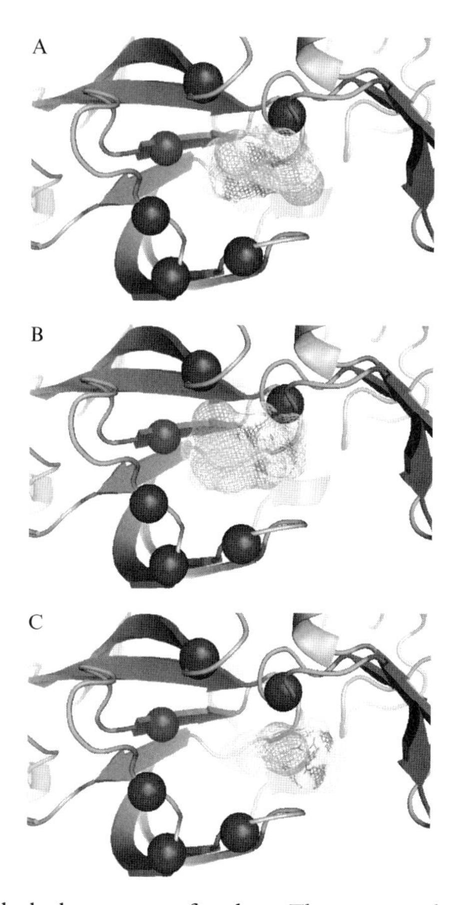

Figure 18.3 The docked structures of analogs. The compounds of the shift maps in 18.2 are shown here. Docking orientation A and B are in line with the shift changes seen in the spectra, indicated by the red spheres. Orientation C is less likely due to the distance to the perturbed amides but also due to lack of shift perturbations at other, closer residues. (For interpretation of the references to color in this figure legend, the reader is referred to the Web version of this chapter.)

to account for deuterium effects if commercial PBS tablets for pH 7.4 are dissolved in pure ^2H_2O. The binding experiment, of which there are examples in the vendor software packages, will show clear binding signals for benzoic acid and no signals for glucose in the STD experiment, whereas the glucose will appear positive and the benzoic acid negative in the water-LOGSY spectrum.

A second sample is used for competition titration. It is a 20- μM solution of human serum albumin and acetaminophen at a concentration of 250 μM. The stepwise addition of ibuprofen starting at 10 μM will lead to a corresponding reduction of the acetaminophen signals to a value close to the noise level.

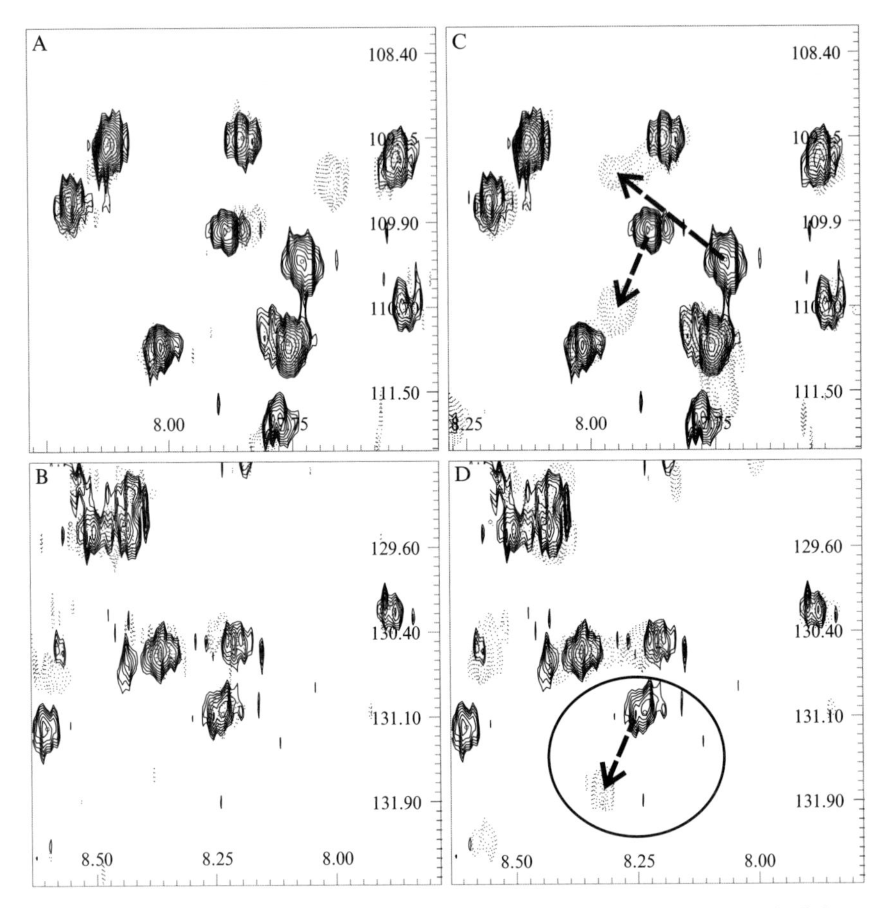

Figure 18.4 The effect of chemical modification of the primary hit. Shown on the left are the shift perturbations of the primary hit. On the right are shown the shifts due to the chemically modified primary hit. The conditions are the same throughout. Far bigger shifts are observed at the original binding site residues, as shown in panel (A), at the same time, new shifts are also observed in another region of the active site, as shown in panel (D).

Water-LOGSY experiments will always be performed in 96% H_2O/4% ^2H_2O. Although STD measurements can also be done in water, greater sensitivity will be gained when the sample is in ^2H_2O.

6.2. Important parameters

The sample temperature is best set to 5–10 °C as it is often accompanied by a still better signal-to-noise when compared to room temperature, presumably due of the modulation of on and off rates. The presaturation power (a pulse train selectively applied to irradiate only protein protons) is an important parameter in the STD experiment. If the power is set to too high a level, the lower pulse selectivity will lead to magnetization of the ligand directly and thus to false positive results. In case of overdriven presaturation power, the acetaminophen signals will not shrink proportional to the amount of ibuprofen that is added as they should because they compete for the same binding site. Water-LOGSY is sensitive to the solution conditions such as pH and ligand-to-protein ratio and needs to be tested for each new series.

6.3. Test of the target protein with a tool compound

The setup for an FBLD project is tested with the protein of interest at a concentration of 2.5–5 μM in proton-free buffer such as phosphate-buffered saline or using deuterium-labeled buffer compounds such as d11-TRIS. The first important test will be for binding detection of a ligand of low affinity and its displacement by a high-affinity binder for the same binding site. This means that a tool compound should be available to test and tweak the setup. Often, a ligand of nanomolar affinity will be available from literature. A low-affinity compound of millimolar affinity may be harder to come by. In this case, a focused screen using a small subset of the final library can be used to identify tool compounds if these are not available *a priori*. The reverse setup, where a low-affinity binder (reporter) is present and the fragment ligands are added and the signals of the reporter are monitored, can have advantages if, for example, the reporter has a simple fluorine label and hence only one signal needs to be monitored. With the advent of cold fluorine NMR probes, this looks to be a promising path forward. An experiment where the stability of the protein is monitored over the screen time period is also important, as protein stability can vary under different solution conditions.

6.4. Automation

Two types of automation are available to date. The first utilizes a robotic pipetting system to prepare the sample and an NMR tube handling system to transfer the sample to and from the magnet. The other has a magnet-mounted sample storage system that accommodates single NMR tubes and racks of 96 tubes of different sizes. With a 20–40 min duration for each experiment in a discontinuous screen, the last sample of a 96 NMR tube containing a microtiter plate will have been in contact with fragments for 30 h prior to being measured. This is certainly not applicable to unstable proteins. Automated "just in time" sample preparation where a robot mixes the sample at a predetermined time before each new measurement is desirable will lead to the same timing of every sample and thus more consistent results. Also, a pipetting system will accommodate automated titration experiments.

6.5. Primary screen data analysis

The deconvolution of mixture NMR spectra is straightforward because each ligand will have set of unique signals, as illustrated in the primary screen workflow in Fig. 18.1A–C. The choice of mixture components can be tuned to minimize overlap and allow unique identification of primary hits from mixtures. Data analysis represents a bottleneck of a fragment screen but there are software solutions available, for example, the drug discovery package of AMIX (Bruker BioSpin, Karlsruhe).

6.6. Follow-up of primary hits using ligand-based NMR

All primary hits recorded in mixtures will be followed up by measurement of the hit alone to confirm binding, this part of the hit-to-lead utilizing the same ligand methods as the primary screen (Fig. 18.1D). At this stage, a ranking may be done based on the signal intensity. This is not exactly quantifiable, as there are many factors determining the signal intensity/integral. Programs that deconvolute the exchange and relaxation effects contributing to the signal amplitude have been developed to calculate quantitative STD signal intensities (Jayalakshmi and Krishna, 2002). Better binning of hits can be accomplished by a competition experiment, as shown in Fig. 18.1E. That will allow the differentiation of active site binders verses secondary site binders. The thus identified primary site hits will move forward to the next step of hit-to-lead. If a secondary binding site is the focus of the screen, compounds whose signals remain with the competitor will indicate nonactive site binding. Such secondary binding sites may be interesting if there is a postulated allosteric or otherwise relevant mechanism by which protein activity can be modulated. Observation of binding exclusively upon addition of the competitor may be an indication of a secondary binding site where the competitor has opened a pocket in the active site now accessible to the fragment or a structural rearrangement has exposed a second site attractive for binding. This finding can be followed up to develop these primary hits into ligands of a completely new mechanism by exploiting active site binding pockets that only become available upon competitor binding. By analysis of the rate by which the competitor reduces the signal of the original hit, a measure of binding site selectivity can be estimated. Competition experiments of similar chemotypes will confirm that a defined binding mode exists and allow structure–affinity relationships to generate the first round of hit-to-lead candidates. Relative binding affinities can be determined by titration experiments, so at this stage, it will be advantageous if labeled protein is available.

6.7. Shift mapping with labeled protein

The triaged hit-to-lead list is now further characterized by titration experiments using isotope-labeled protein and under protein observation methods such as TROSY (Pervushin *et al.*, 1997) or HSQC (Bodenhausen and

Rubin, 1980). Samples of 50 μ*M* labeled protein are mixed with the fragment and the chemical shift changes monitored. With 50 compounds to characterize, a first round of measurements with the ligand at 1 m*M* will be followed by a titration experiment in order to extract the binding affinity. The chemical shift perturbations are fitted against the ligand concentration to a function based on a one-site binding model.

It is here that the true strength of NMR comes to play, that the binding site and affinity can be determined, and at the same time, the overall effect of the ligand on protein dynamics and behavior can be reported. The fast on/off kinetics of low-affinity binding of fragments make data interpretation easier than in cases of high-affinity binders, as here, a simple overlay of spectra at increasing ligand concentrations will show the signals stemming from the residues in the interaction site. This situation can become more complicated with fragments of higher binding affinity, as the exchange rate may lead to line broadening, and this may happen together with line-broadening effects due to modulation of the protein mobility in regions quite far away from the actual binding site (Schumann *et al.*, 2007). Where primary fragment hits are chemically elaborated, the turnaround from compound synthesis to binding affinity and orientation is very fast, with the NMR data acquisition and interpretation taking just a matter of hours. Additional chemical shift perturbations will reflect the increase in ligand size and thus binding interactions with new areas of the protein.

6.8. Docking

Parallel to the shift-based lead design process, fragment docking can be utilized to suggest orientations of a fragment bound in the active site. Analyzed together with NMR data, different docking orientations can be ranked with respect to the observed chemical shift changes. With the example shown in Fig. 18.3, highly scored docking orientations can be triaged by inspection of expected chemical shift changes at residues postulated to be within the binding sphere and their comparison with measured values. Methods to integrate the NMR data into the docking routine are in development (González-Ruiz and Gohlke, 2009) and will make this process more straightforward. By the iterative processes of chemical modification, docking, and experimental shift mapping, an increase in binding affinity can be accomplished. Probing chemical space in the vicinity of the original hit can also be accomplished by ordering similar compounds out of a vendor catalog. To guide this process, docking programs such as MOE offer the option of locating nearby potential interaction partners on the protein side (MOEv, 2007) based on the docked conformation. This iterative process will be the path of choice forward, particularly when X-ray cocrystal structures of primary hit elaborations are not easily available.

7. Summary and Conclusions

In an FBLD project, an NMR primary screen will yield compounds that interact with the target molecule. As the detection window of NMR spans several orders of magnitude, ranging from high nanomolar to high millimolar, up to 30% of a given library can show up as a target binder/ primary hit. It is thus important to separate unproductive and nonoptimizable binders from suitable candidates in the fragment hit-to-lead process. This starts with the design of the primary screen where competition experiments will help focus on a sum of active site binders that can be followed up. Characterization of these in both primary screen follow-up using ligand-based NMR techniques and measurement with isotope-labeled protein will clearly define a fragment set with binding to the target site, at the same time delivering information on binding affinity and orientation. The optimization process is largely driven by costructure information, stemming either from docking using NMR information or preferably by X-ray crystallography with high-resolution coordinates in an acceptable time window most often attained by soaking of ligands into protein crystals. The steps before a ligand goes into a crystallization tray are very important, as the success of the crystallographic process is highest with compounds that bind a preferred site in a defined orientation.

NMR is becoming increasingly important in FBLD due to its scope of information from binding detection to costructure determination. Being the only method that works under the conditions found in biological systems, NMR can act as a triage for hits from other techniques matching the experimental conditions and delivering qualitative binding site information. Fast turnaround allows its application to structure–activity relationship. In all stages of FBLD, NMR can contribute information toward a lead compound of high affinity and potentially novel intellectual property.

Acknowledgment

The author thanks the structural biology and protein engineering departments at Genentech for critical reading of the manuscript.

References

Brüggert, M., Rehm, T., Shanker, S., Georgescu, S., and Holak, T. A. (2003). A novel medium for expression of proteins selectively labeled with 15 N-amino acids in Spodoptera frugiperda (Sf9) insect cells. *J. Biomol. NMR* **25**(4), 335–348.

Bodenhausen, G., and Rubin, D. J. (1980). Natural abundance nitrogen-15 NMR by enhanced heteronuclear spectroscopy. *Chen. Phys. Lett.* **69,** 185–189.

Chen, A., and Shapiro, M. J. (2000). NOE pumping. 2. A high-throughput method to determine compounds with binding affinity to macromolecules by NMR. *J. Am. Chem. Soc.* **122**(2), 414–415.
Chessari, G., and Woodhead, A. J. (2009). From fragment to clinical candidate: A historical perspective. *Drug Discov. Today* **14**(13-14), 668–675.
Dalvit, C. (2009). NMR methods in fragment screening: Theory and a comparison with other biophysical techniques. *Drug Discov. Today* **14**(21-22), 1051–1057.
De Marco, A., Laursen, R. A., and Llinas, M. (1986). 1 H-NMR spectroscopic manifestations of ligand binding to the kringle 4 domain of human plasminogen. *Arch. Biochem. Biophys.* **244,** 727–741.
González-Ruiz, D., and Gohlke, H. (2009). Steering protein-ligand docking with quantitative NMR chemical shift perturbations. *J. Chem. Inf. Model* **49**(10), 2260–2271.
Hajduk, P. J., Olejniczak, E. T., and Fesik, S. W. (1997). One-dimensional relaxation- and diffusion-edited NMR methods for screening compounds that bind to macromolecules. *J. Am. Chem. Soc.* **119,** 12257–12261.
Jayalakshmi, V., and Krishna, N. R. (2002). Complete relaxation and conformational exchange matrix (CORCEMA) analysis of intermolecular saturation transfer effects in reversibly forming ligand-receptor complexes. *J. Magn. Reson.* **155**(1), 106–118.
LaPlante, S. R., Cameron, D. R., Aubry, N., Lefebvre, S., Kukolj, G., Maurice, R., Thibeault, D., Lamarre, D., and Llinàs-Brunet, M. (1999). Solution structure of substrate-based ligands when bound to hepatitis C virus NS3 protease domain. *J. Biol. Chem.* **274**(26), 18618–18624.
Mayer, M., and Meyer, B. (2001). Group epitope mapping by saturation transfer difference NMR to identify segments of a ligand in direct contact with a protein receptor. *J. Am. Chem. Soc.* **123**(25), 6108–6117.
MOEv2007.09. Developed and distributed by Chemical Computing Group. http://www.chemcomp.com.
Pervushin, K., Riek, R., Wider, G., and Wuthrich, K. (1997). Attenuated T2 relaxation by mutual cancellation of dipole-dipole coupling and chemical shift anisotropy indicates an avenue to NMR structures of very large biological macromolecules in solution. *Proc. Natl. Acad. Sci. USA* **94,** 12366–12371.
Schumann, F. H., Riepl, H., Maurer, T., Gronwald, W., Neidig, K. P., and Kalbitzer, H. R. (2007). Combined chemical shift changes and amino acid specific chemical shift mapping of protein-protein interactions. *J. Biomol. NMR* **39**(4), 275–289.
Shuker, S. B., Hajduk, P. J., Meadows, R. P., and Fesik, S. W. (1996). Discovering high-affinity ligands for proteins: SAR by NMR. *Science* **274**(5292), 1531–1534.
Szczepina, M. G., Zheng, R. B., Completo, G. C., Lowary, T. L., and Pinto, B. M. (2009). STD-NMR studies suggest that two acceptor substrates for GlfT2, a bifunctional galactofuranosyltransferase required for the biosynthesis of Mycobacterium tuberculosis arabinogalactan, compete for the same binding site. *Chembiochem* **10**(12), 2052–2059.
TECAN Journal Edition 2, 2009.
Woodhead, A. J., Angove, H., Carr, M. G., Chessari, G., Congreve, M., Coyle, J. E., Cosme, J., Graham, B., Day, P. J., Downham, R., Fazal, L., Feltell, R., *et al.* (2010). Discovery of (2,4-Dihydroxy-5-isopropylphenyl)-[5-(4-methylpiperazin-1-ylmethyl)-1,3-dihydroisoindol-2-yl]methanone (AT13387), a novel inhibitor of the molecular chaperone Hsp90 by fragment based drug design. *J. Med. Chem.* **53,** 5956–5969.

CHAPTER NINETEEN

Electron Density Guided Fragment-Based Drug Design—A Lead Generation Example

Marta C. Abad, Alan C. Gibbs, *and* Xuqing Zhang

Contents

Abstract

We describe here a method using protein crystallography as the sole detection tool for fragment-based lead discovery. The methodology consists of iterative design, synthesis, and X-ray crystallographic screening of three libraries of compounds. Target-specific compound design, by way of active site electron density in the presence of a bound fragment hit and the intentional lack of solution activity bias form the basis of our approach. We provide an example of this alternative fragment-based drug design (FBDD) method, detailing results from a campaign using ketohexokinase to generate a unique lead series with promising drug-like properties.

Structural Biology and Medicinal Chemistry, Johnson & Johnson Pharmaceutical Research and Development, L.L.C., Spring House, Pennsylvania, USA

Methods in Enzymology, Volume 493
ISSN 0076-6879, DOI: 10.1016/B978-0-12-381274-2.00019-4

1. Introduction

Over the last two decades, success stories describing the discovery of leads and clinical candidates using various fragment-based drug design (FBDD) methodologies have become abundant in the primary literature (Congreve *et al.*, 2008; Hajduk and Greer, 2007; Warr, 2009). These methods differ in fragment detection method, screening technology, chemistry strategy, and the protocols used in their application. A wide range of biophysical techniques, including nuclear magnetic resonance (Jhoti *et al.*, 2007), mass spectrometry (Poulsen and Kruppa, 2008), isothermal titration calorimetry (Ciulli *et al.*, 2006), surface plasmon resonance (Miura, 2010; Neumann *et al.*, 2007), and X-ray crystallography (Congreve *et al.*, 2007; Murray *et al.*, 2007; Nienaber *et al.*, 2000), have been applied to the detection of fragments binding to macromolecules of interest. Often, consecutive screens are employed. Several chemistry strategies, including "classical" library design, dynamic combinatorial chemistry (Nestler, 2005), and kinetic target-guided synthesis (Hu and Manetsch, 2010), to mention a few, have been leveraged to effectively cover a chemical space of interest when elaborating fragment hits. The protocols employed in FBDD vary with respect to occurrence and emphasis on screening technologies and chemistry strategies used throughout the FBDD campaign. Although successful FBDD methods differ in screening technologies and chemistry strategies, they are similar in that in most cases, the major emphasis is placed on bioaffinity measurements to guide subsequent compound design.

2. Electron Density Guided FBDD

Our FBDD methodology is guided not by solution activity data, but by high-content X-ray diffraction data. Fragment electron density is used to reveal where and how a fragment hit binds. Comparison and superpositioning of two or more sets of fragment electron density reveal the similarities as well as differences in binding interactions among the fragment hits. The finding is then used to direct the design of subsequent fragment libraries either by sprouting from a fragment of choice or by fusion of nonhydrogen atoms from multiple fragments that occupy different areas in the feature map. Additional information can be derived from the overlap of like scaffolds to elucidate functional group effects within a series. The quality and occurrence of electron density data for a particular chemotype will determine the course of its evolution during the FBDD campaign.

We have described in part our methodology in a recent publication of a lead that binds to the ATP site of ketohexokinase (Gibbs *et al.*, 2010).

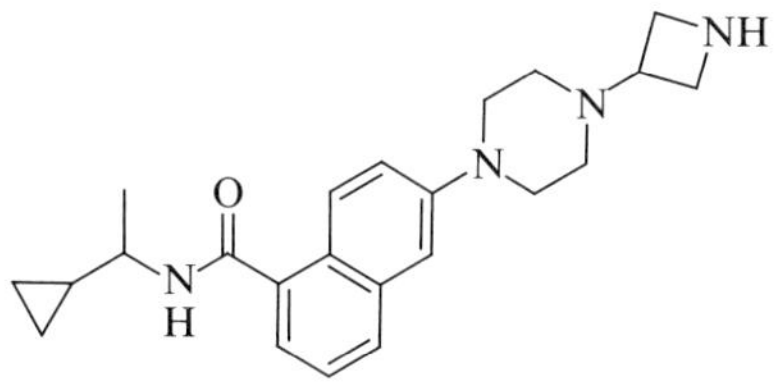

Compound 1

Figure 19.1 Aryl-amide lead compound

Herein, we describe in greater detail our development of an alternative lead molecule, an arylamide ATP competitive inhibitor of ketohexokinase. This lead, compound **1**, is shown in Fig. 19.1 and is derived from a fragment screen using electron density as the primary guide without the knowledge of binding affinity.

Following X-ray screening, discernable fragment electron density is defined as a hit. Electron density described here pertains solely to density in the active site of the protein that is otherwise absent in an *apo* reference structure of the protein. Electron density is of interest if found within a radius of 5 Å of the active site of the enzyme. Data sets containing bound fragment electron density are indexed, processed, and the X-ray structure determined. During refinement, a fragment is placed and optimized by electron density weighing to give its appropriate orientation and binding mode. If the electron density of the fragment is ambiguous, the fragment may be placed and refined in more than one orientation. X-ray crystallographic statistics for select structures described in this chapter are listed in Table 19.1.

The set of electron density characteristics we use to define a bound fragment are size, position, and complementarity. Size is defined by the size of the electron density "visible" in the active site which differentiates between solvent, inorganic ion, and true fragment molecules. Inorganic ions and solvent molecules are smaller and occupy less volume than fragments. Position is defined as the location, juxtaposition, and orientation of the electron density in an environment-dependent manner within the active site. For example, is the electron density near polar or hydrophobic regions of the enzyme active site? Complementarity is defined by the goodness of the fit between the electron density and a fragment. Since the electron density seen could be that of several binding fragments in the cocktail soak (Spurlino, 2011), complementarity is important when determining which fragment in the cocktail fits most reasonably and in what orientation(s).

Figure 19.2 offers an example of size, position, and complementarity of fragments found in the active site of ketohexokinase. In Fig. 19.2A, different fragments are seen to bind together with a sulfate ion which is present in the crystallization solution. In Fig. 19.2B, the fragment bound can be placed

Table 19.1 Crystallographic data and refinement statistics

Compound	AMP-PNP + fructose			
Space group	$P2_12_12_1$	$P2_12_12_1$	$P2_12_12_1$	$P2_12_12_1$
Unit cell (Å)	$a = 82.8$, $b = 86.0$, $c = 136.9$	$a = 82.4$, $b = 85.6$, $c = 135.8$	$a = 83.4$, $b = 85.1$, $c = 137.2$	$a = 83.5$, $b = 86.9$, $c = 139.0$
Resolution (Å)	2.3	2.3	2.5	2.4
Completeness (%)	98.2	99.6	98.8	98.7
R_{merge} (%)	7.1	7.3	4.5	8.5
$\langle I \rangle / \langle \sigma_I \rangle$	18.9	16.4	16.0	9.1
R_{factor} (%)	20.7	19.9	23.3	23.1
R_{free} (%)	23.6	23.5	27.1	26.4
RMSD bonds (Å)	0.008	0.008	0.008	0.008
RMSD angles (°)	1.12	1.054	1.10	1.167

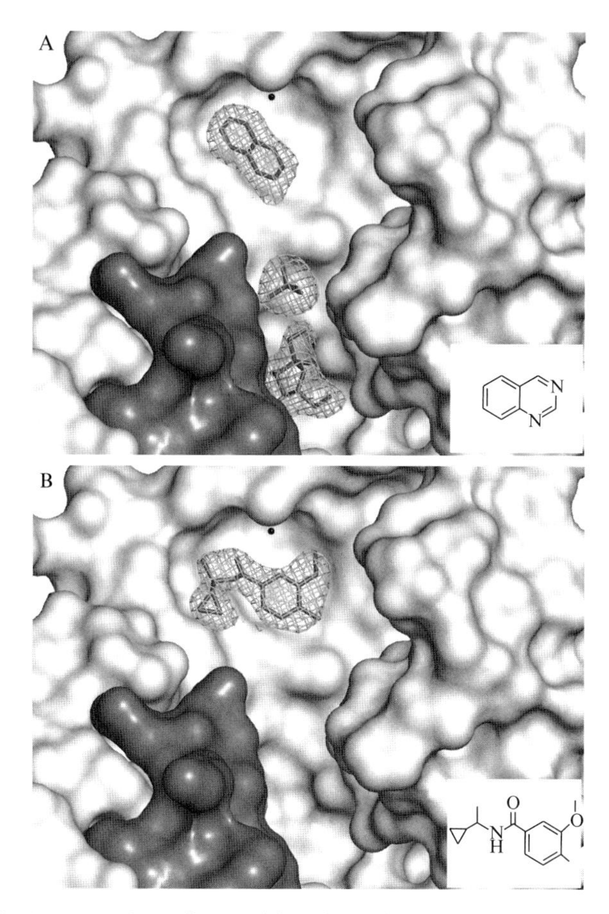

Figure 19.2 Superposition of several hits from the primary library screen showing examples of size, position, and complementarity of fragments in the active site of ketohexokinase. (A) Fragments are sometimes seen to bind together with a sulfate ion which is present in the crystallization solution. (B) The fragment is placed in a single, unambiguous orientation in contrast to the fragment bound at the same location (in A) for which more than one orientation is possible.

in a single, unambiguous orientation. The advantage of screening fragments with protein single crystals and X-ray diffraction is that the number of fragments as well as their locations is immediately known when the resolved electron density map is of high quality.

3. Ketohexokinase

Dietary fructose consumption has increased dramatically in recent years, in large part due to the overuse of high-fructose corn syrup (Gross *et al.*, 2004). High-fructose diets have been shown to promote a variety of metabolic disturbances, including obesity, hyperlipidemia, hypertension, insulin resistance, and in some cases, metabolic syndrome (Basciano *et al.*, 2005).

An important enzyme in fructose metabolism is ketohexokinase (EC 2.7.1.3). Ketohexokinase catalyzes the conversion of D-fructose to fructose-1-phosphate. Fructose-1-phosphate is then converted to triglycerides, fatty acids, and eventually, very low-density lipoprotein (Elliott *et al.*, 2002). Modulation of ketohexokinase may help relieve the aforementioned metabolic disturbances. A desired modulation of ketohexokinase may be accomplished by the use of small-molecule inhibitors.

3.1. Ketohexokinase X-ray structure

Ketohexokinase crystallizes as a dimer in the asymmetric unit with the dimer interface formed through interactions between the two β-sheet domains, forming a β-clasp domain (Gibbs *et al.*, 2010; Trinh *et al.*, 2009). Figure 19.3 depicts the general topology of the ketohexokinase dimer and the active site. The structure reveals that the enzyme possesses two domains—an α/β/α domain and a β-sheet domain. It also shows that each monomer offers a slightly different C_α conformation along the β-clasp domain. This change in backbone conformation is corroborated by molecular dynamics simulations that show that the enzyme readily adopts multiple conformations between open and closed states (Gibbs and Abad, unpublished data). The differences in conformation lead to a smaller binding site in one of the two monomers. Our X-ray data indicate that only the monomer with the smaller binding site binds the substrate. The smaller binding site positions the residues closer, allowing ATP and sugar to be situated in a position for catalysis. An apo X-ray structure of ketohexokinase (unpublished data) confirms that the size difference of the active site between the two monomers is not due to substrate binding. Throughout this chapter, we will refer the "closed" conformation as the active monomer A. We will refer the "open" conformation as the inactive monomer B.

Among the predominant features observed in many of our ketohexokinase X-ray structures are a conserved water molecule and a conserved sulfate ion in the ATP binding site. The conserved water molecule hydrogen bonds to the main chain carbonyl of F245 and also interacts with many bound fragments, ATP, and AMP-PNP (Fig. 19.3). This water molecule has also been observed in previously determined structures of ketohexokinase as well as other ribokinase superfamily member structures (Sigrell *et al.*, 1997, 1998; Trinh *et al.*, 2009). The conserved sulfate ion occupies the same position as the γ-phosphate of ATP and interacts with residues G255, G257, and R108 (Trinh *et al.*, 2009; Gibbs *et al.*, 2010).

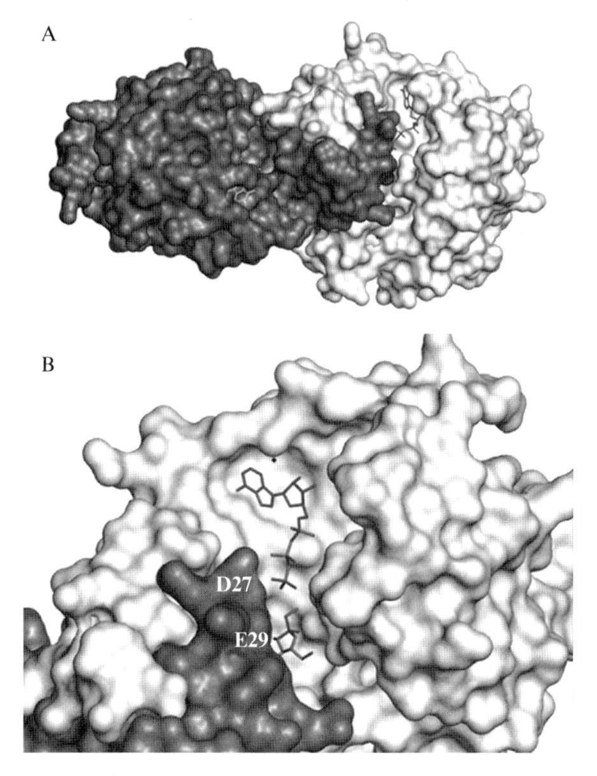

Figure 19.3 Structural features of ketohexokinase. Panel A illustrates the β-clasp holding the two monomers together, monomer A (white) and monomer B (gray). Panel B shows a close-up view of the ketohexokinase catalytic site, ATP, and fructose binding sites. The conserved water (small sphere at 12 o'clock), AMP-PNP, and fructose are visible. AMP-PNP forms hydrogen bonds through N-3 of adenine to the conserved water, and through HO-3′ of ribose to G229. D-Fructose is positioned below the terminal amidophosphate of AMP-PNP (γ-phosphate of ATP). Extension of D27B into the active site is clearly visible at the bottom of the figure. D-Fructose binds in a sterically constrained pocket and is held in place by a bidentate interaction between HO-3′ and HO-4′ with D15, a hydrogen bond between HO-1′ and N45, and a hydrogen bond between HO-6′ and E29B.

4. Ketohexokinase FBDD

We initiated our ketohexokinase FBDD campaign with an X-ray crystallographic screen of a *primary* library of fragments (Tounge, 2011). The *primary* library consists of approximately 900 fragments and is designed as a general-purpose library with no explicit target class biases, that is, no target-privileged functional groups or scaffolds. The compounds were

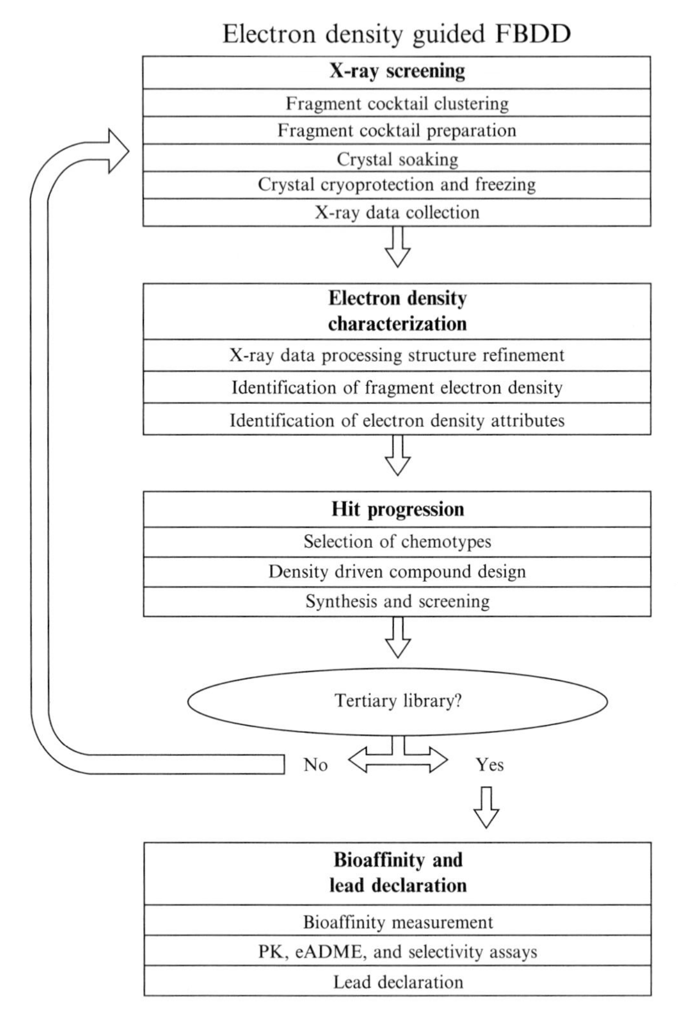

Figure 19.4 Schematic representation of the electron density guided FBDD method.

screened as cocktails of five structurally similar fragments. Fragments are grouped by structural similarity to reinforce the identification of binding of a specific scaffold or molecular core to help minimize scaffold ambiguity when active site density lacks definition and its molecular constitution is not obvious. Grouping fragments by structural similarity offers a binary "yes" or "no" answer as to whether a particular chemotype or scaffold binds the target protein. A flowchart of our FBDD method is illustrated in Fig. 19.4.

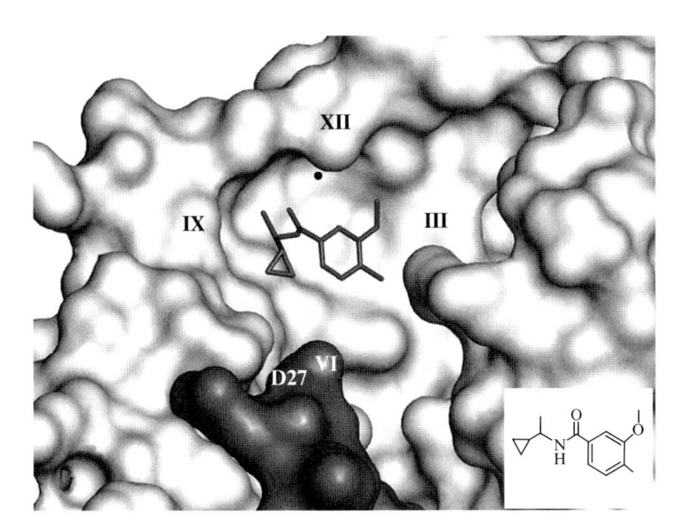

Figure 19.5 The ketohexokinase ATP site feature map. A *primary* library arylamide is shown hydrogen bonding to the conserved water. Water, in the reference frame, is at 12 o'clock.

4.1. Screening for hits with the *primary* fragment library

The *primary* library screen reveals many fragments binding to the ATP site. Most of the hits are found in monomer A. In addition, several fragments are also found to bind the fructose site and a few others bind on the protein surface or in the dimer interface. Following *primary* library screening, approximately 60 data sets are refined. In this study, we use only information derived from fragments bound in the ATP site of monomer A to design compounds to go into the *secondary* and *tertiary* libraries. The data obtained from the primary library hits are used to define regions in the ATP binding site to be pursued in the design of compounds for the second and third rounds of screening. These regions are referenced below to the position on a clock.

Figure 19.5 shows a fragment hit, an arylamide, in the ATP binding site of ketohexokinase. At 12 o'clock is the conserved water molecule observed in most of our X-ray structures and in the X-ray structures of a variety of ribokinase superfamily. The conserved water interacts with the backbone of F245 and the arylamide series carbonyl. At the 2 o'clock position is the region where the ribose of ATP is positioned as defined by the side chains of C282 and F260. This pocket shows a preference for small alkyl, specifically thio-ether, and aryl functional groups. F260 interacts through hydrophobic and π–π stacking interactions with fragment aromatic groups. Although edge-to-face interactions are observed, the majority of π–π stacking involves face-to-face interactions between fragment and F260 side chain, as illustrated in Fig. 19.6. For this reason, we have included in both our

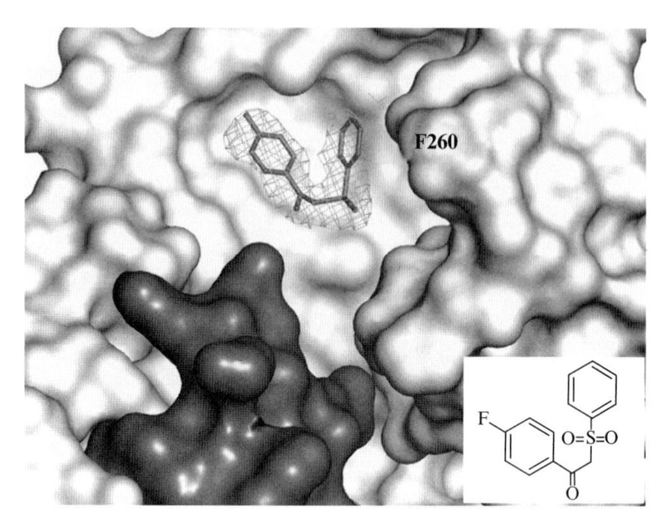

Figure 19.6 Face-to-face π stacking with F260.

secondary and *tertiary* libraries compounds containing π systems that potentially bind either edge-to-face or face-to-face with F260. Several examples of sulfamide-containing fragments are found to bind with the sulfamide group inserted in the small hydrophobic pocket at the 2 o'clock position. This interaction appears to be a combination of hydrogen bonding (between fragment and conserved water) and π–π interaction (between fragment and the side chain of F260). Figure 19.7 highlights electron density associated with one of these fragments. We incorporated sulfamides, sulfonamides, and sulfones in a variety of secondary library compounds to take advantage of this interaction, unfortunately we met with little success.

At the 4 o'clock position is a polar region where the triphosphate of ATP binds. At the 6 o'clock end of the phosphate channel, we often find the aforementioned (Section 3.1) conserved sulfate ion bound; this sulfate superimposes to low RMSD with the γ-phosphate of ATP. At the 5 o'clock position is the small, highly polar fructose site. At the 6 o'clock position is a hydrophobic region where the adenosine group of ATP binds. Many of our fragments with aromatic rings, fused aromatic rings, or bicyclic rings bind in this region. It is assumed that favorable interactions between the sulfhydryl side chain of C289 and the π system of an aromatic fragment (or adenosine of ATP) are the basis for this aromatic "friendly" region (Imai *et al.*, 2007). The sulfhydryl of C289 lies directly underneath the fragment π system. Figure 19.8 displays a *primary* library fragment involved in this interaction.

At the 7 o'clock position is a hot spot formed by the β-clasp residues from monomer B, specifically its D27 (D27B), inserting into the active site of monomer A. Fragment–protein interactions in this region are

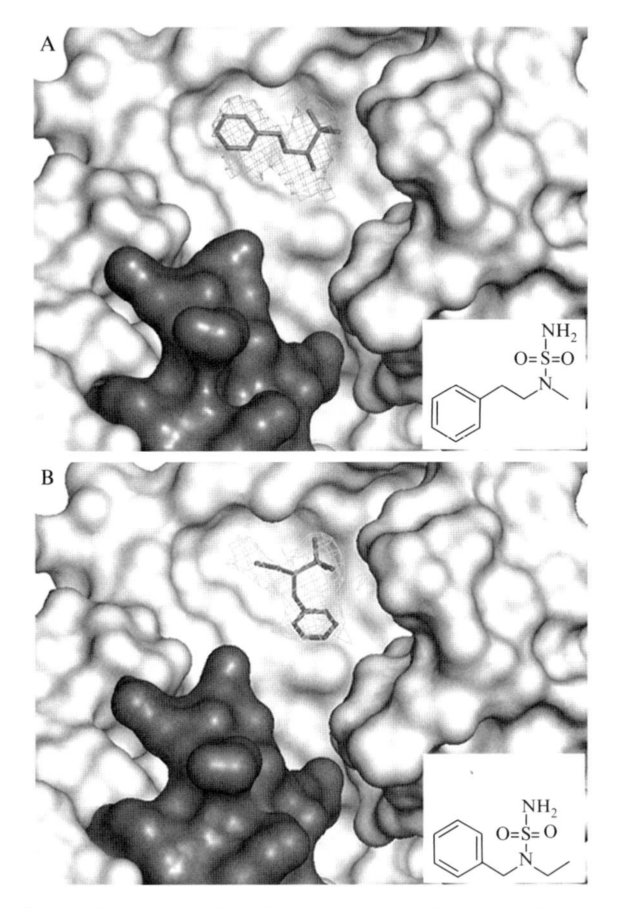

Figure 19.7 Examples of sulfamide from the *primary* library.

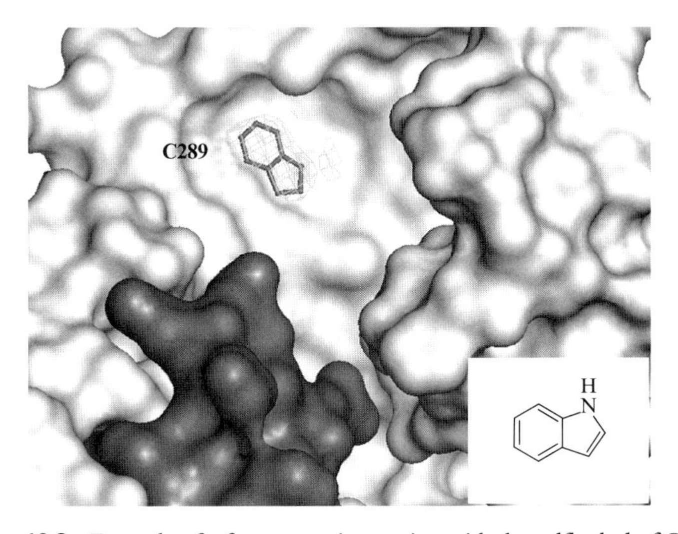

Figure 19.8 Example of a fragment π interaction with the sulfhydryl of C298.

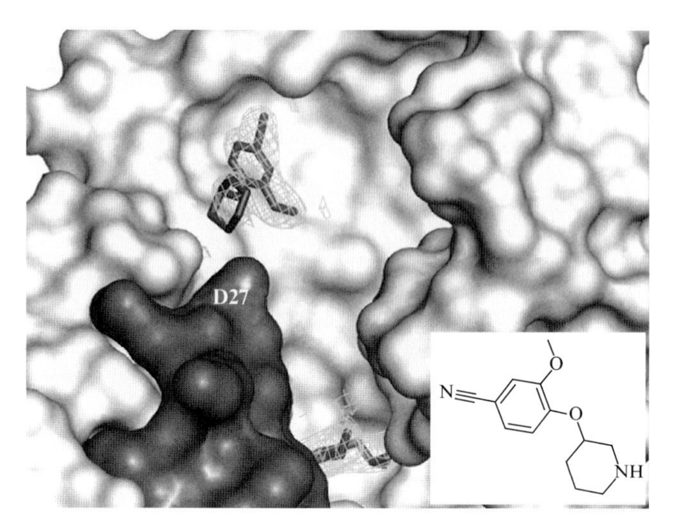

Figure 19.9 Example of interaction between a charged amine (N^+) and D27B.

characterized by attractive charge–charge interactions. The interactions are primarily between the negative charge of carboxylate D27B and the charged amines of fragments. Figure 19.9 depicts a *primary* library fragment with a piperidine ring involved in the charge–charge interaction with D27B. Various *secondary* and *tertiary* library fragments are designed to exploit this interaction. At the 8–10 o'clock positions is the solvent-exposed region that defines the edge of the ATP binding site. A group of proline residues (246–248) form the outermost edge of this pocket. Small aromatic rings or alkyl substituent of fragments commonly bind in this region.

4.2. Fragment orientation and movement

A primary assumption underlying FBDD is that fragments bind in a similar orientation and juxtaposition as a substructure of the eventual, elaborated molecule. Although this is often the case, there are exceptions. In fact, it is reasonable to assume that rotation or translation among fragment members of the same chemotype will occur. This assumption is supported by inherent lower binding affinities and greater degrees of freedom of small fragments relative to larger elaborated molecules.

Fragment rotation and translation may be due to physicochemical properties intrinsic in either the compound, or the protein, or both. Protein conformation may also influence fragment binding. Whatever the cause of fragment rotation or translation during the course of compound elaboration, it is not necessarily detrimental to the goal of discovering a lead. In certain cases, fragment movement may be beneficial if it occurs early in the FBDD campaign as a wider range of substitution and design possibilities become apparent.

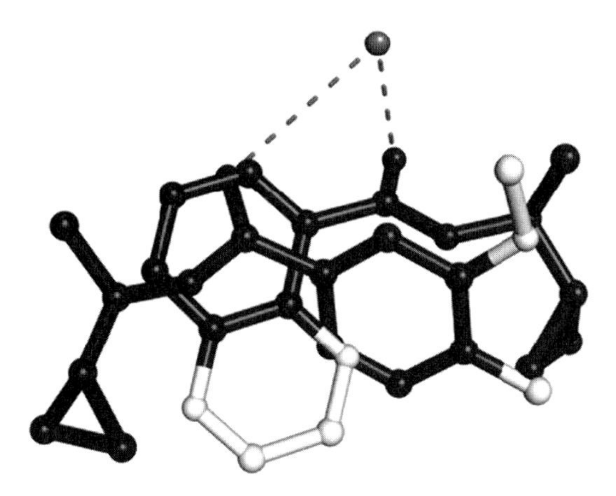

Figure 19.10 Superposition of a phenyl-amide (from the *primary* library) and a quinoxaline-amide (from an arylamide in the *secondary* library). Common cores are represented by black sticks. The compounds have rotated 180°, around the plane of the amide, with respect to each other. Both compounds maintain the hydrogen bond to the conserved water molecule.

Throughout the course of our ketohexokinase FBDD campaign, we observed various movements within chemotypes. Notably, fragment rotation within the lead series of arylamides readily occurs. An example is shown in Fig. 19.10 for two arylamides that rotate 180° from each other with respect to the amide bond. The rotation is taken into consideration in the design of compounds for the *tertiary* library by designing compounds that can be accommodated in the active site regardless of rotation around the amide bond of the scaffolds. Specifically, substitutions at both positions 2 and 3 of the phenyl ring are introduced.

Compound transformations or rotations relative to the "parent" may or may not allow common interactions with the protein between different binding modes. When rotated, the quinoxaline in Fig. 19.10 maintains the key hydrogen bond to the conserved water molecule. In contrast, a pair of tricyclic heterocycles (5-methoxy-tetrahydro-β-carboline and 5-methoxy-tetrahydro-γ-carboline) adopts very different binding modes with apparent lack of common interactions, as shown in Fig. 19.11. One analog accepts a C289-S(H)-π interaction (Fig. 19.11A), whereas the other assumes an interaction with its methoxy in the 2 o'clock hydrophobic pocket of the enzyme (Fig. 19.11B). Structurally, the only difference between the two molecules is the position of one nitrogen atom in the piperidine ring (β vs. γ).

Of the four chemotypes that progressed through the FBDD campaign, three of them (benzotriazoles, pyrazoles, and arylamides) show an alternate rotation around a central axis. Figure 19.12 highlights the 180° rotation

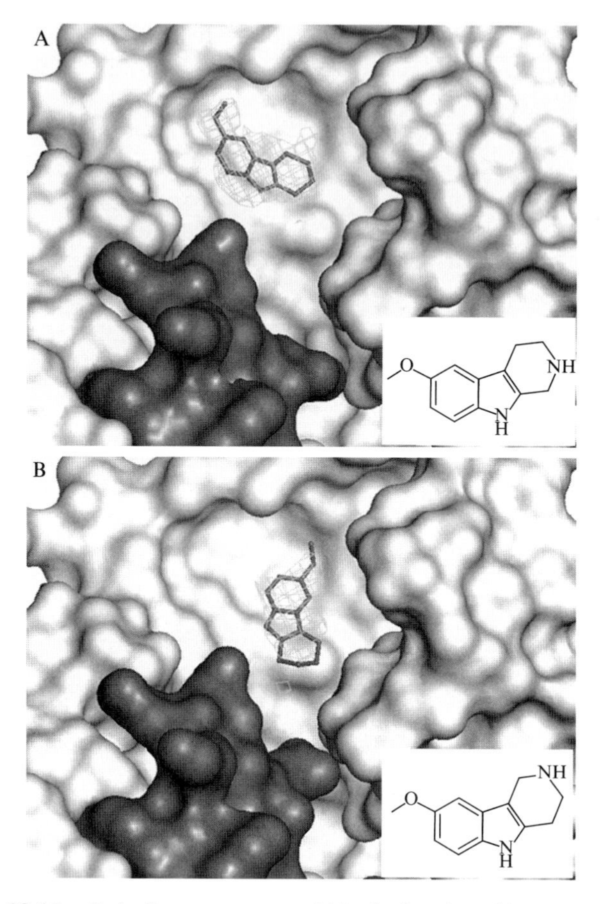

Figure 19.11 Carboline movement within the ketohexokinase active site.

(perpendicular to the benzotriazole ring) of the two benzotriazole binding modes. The hydrogen bond to the conserved water molecule is maintained between the two binding modes albeit in two different ways: through the triazole ring of the core scaffold or through the pendent aromatic heterocycle. The sulfhydryl interaction of C289 is maintained in both binding modes through opposing faces of the aromatic core.

The pyrazole chemotype adopts two binding modes in the ketohexokinase active site, as illustrated in Fig. 19.13. The pyrazole scaffold is the central axis in a 180° rotation parallel to the plane of the pyrazole. This binding mode results in positioning the 1-phenyl at the 2 o'clock region near F260 or at the 10 o'clock position near the three consecutive proline residues (246–248). Our observations suggest that the pyrazoles predominantly bind with the 1-phenyl positioned at the 10 o'clock position, making

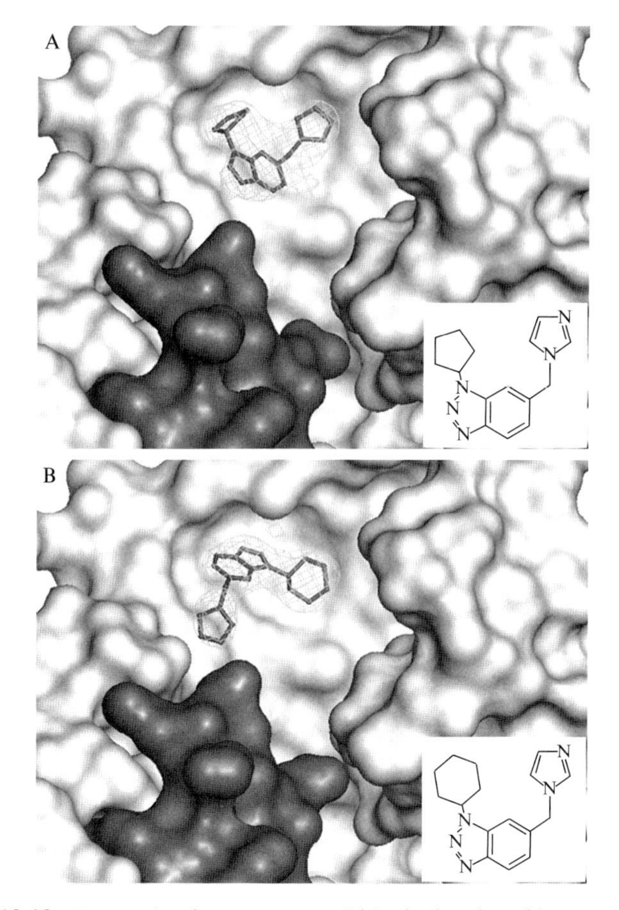

Figure 19.12 Benzotriazole movement within the ketohexokinase active site.

favorable hydrophobic contacts with the proline residues. Both observed binding modes maintain a hydrogen bond to the conserved water molecule through N-2 of the pyrazole.

4.3. Simultaneous fragment binding

On a few occasions, we observe more than one fragment-binding ketohexokinase simultaneously. For example, two fragments bind in the ATP site or one fragment binds in the ATP site and another fragment in the fructose site. Two instances of simultaneous binding involve two identical fragments binding in the ATP site, as shown in Fig. 19.14. The biaryl fragments in Fig. 19.14A contain two aromatic rings (phenyl and thiazole); the fragments in Fig. 19.14B contain one aromatic ring and a saturated heterocycle

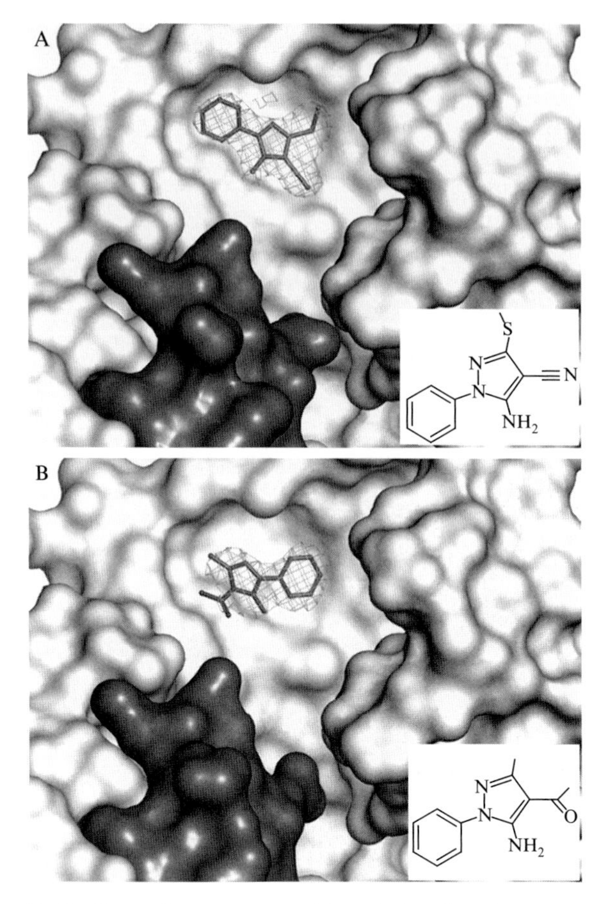

Figure 19.13 Pyrazole movement within the ketohexokinase active site.

(piperizine). Both pairs of fragment bind ketohexokinase in a similar manner. Significant π–π stacking exists between one member of each pair and F260 of ketohexokinase. The fragment that lies parallel to the plane of the page interacts with C289. Incidentally, in Fig. 19.9 is an example of one fragment binding in the ATP site and the other in the fructose site.

4.4. *Secondary* library

Hits that belong to six chemotypes from the *primary* library (pyrazoles, arylamides, benzotriazoles, pyridines, sulfones, and triazolones) are chosen for follow-up (see Fig. 19.15). A total of 350 compounds are prepared for the *secondary* library. Novelty, diversity, synthesis tractability, and

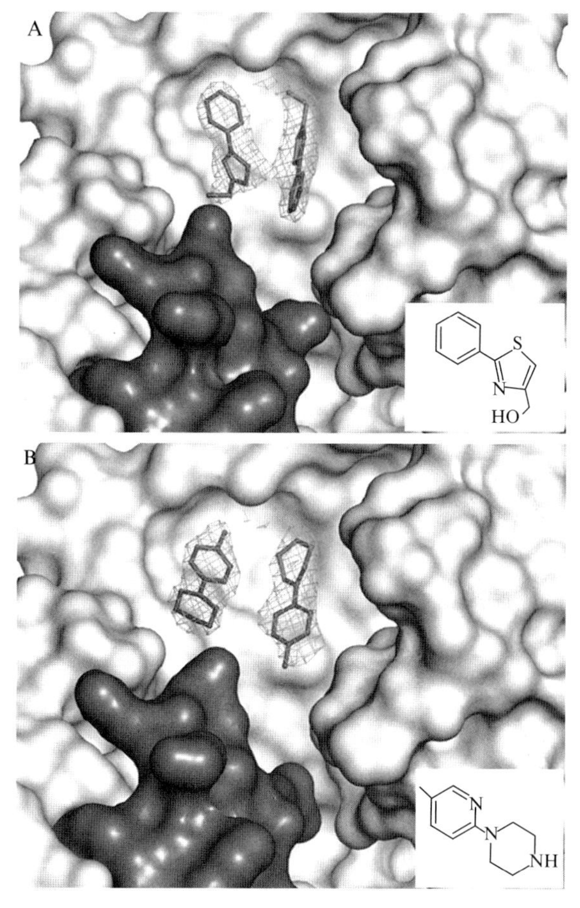

Figure 19.14 Two examples of identical fragments binding simultaneously in the ATP site.

intellectual property space are emphasized in the design of compounds for the second round of screening. Hits are seen for compounds that belong to four of the six chemotypes. No hits are found among compounds that belong to the sulfone and triazolone series.

4.5. *Tertiary* library

The goal of the *primary* library screen is to find a broad range of structurally distinct fragment hits, whereas the goal of the *secondary* library is to narrow the number of progressible compounds. Our preparation of a *tertiary* library places greater emphasis on synthesis of compounds with lead-like features.

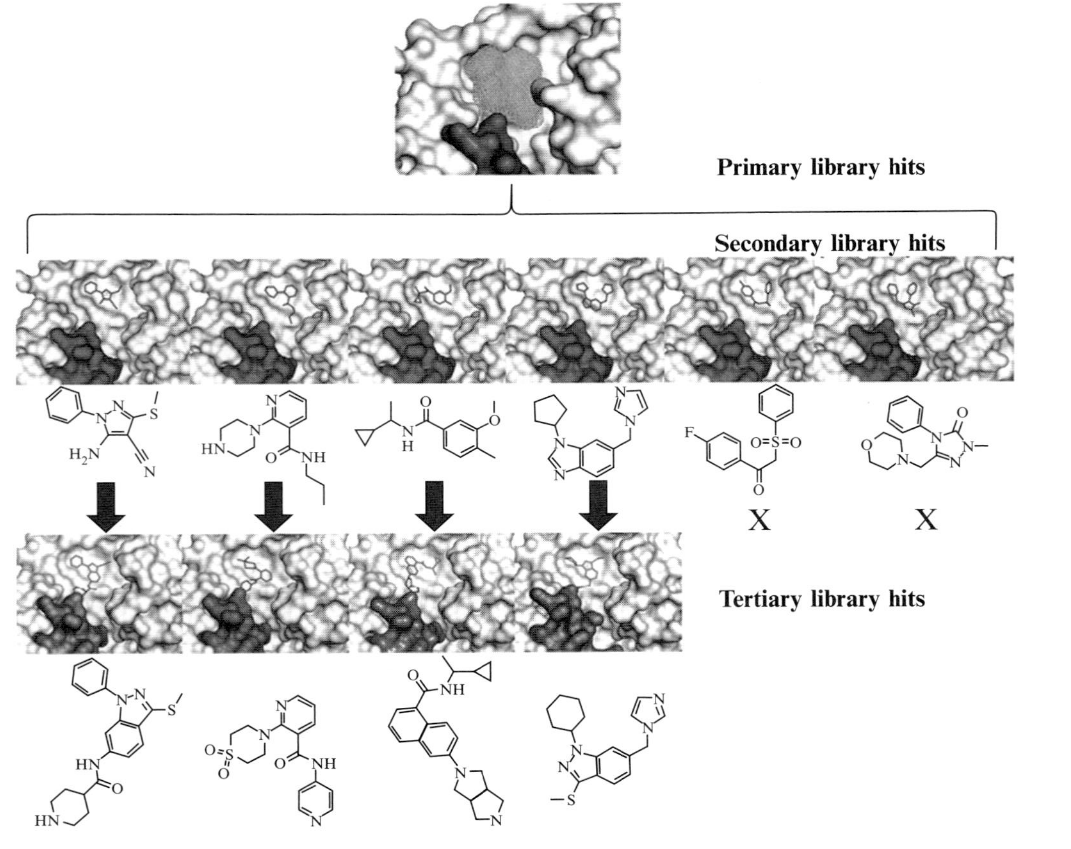

Figure 19.15 Ketohexokinase electron density guided FBDD results. Six hits from the *primary* library were selected as a starting point to design compounds for the *secondary* library. Hits from the *secondary* library guided design of four chemotypes in the *tertiary* library design. Three of the four *tertiary* library series offer hits with affinity greater than 5 μ*M* in the ketohexokinase activity assay.

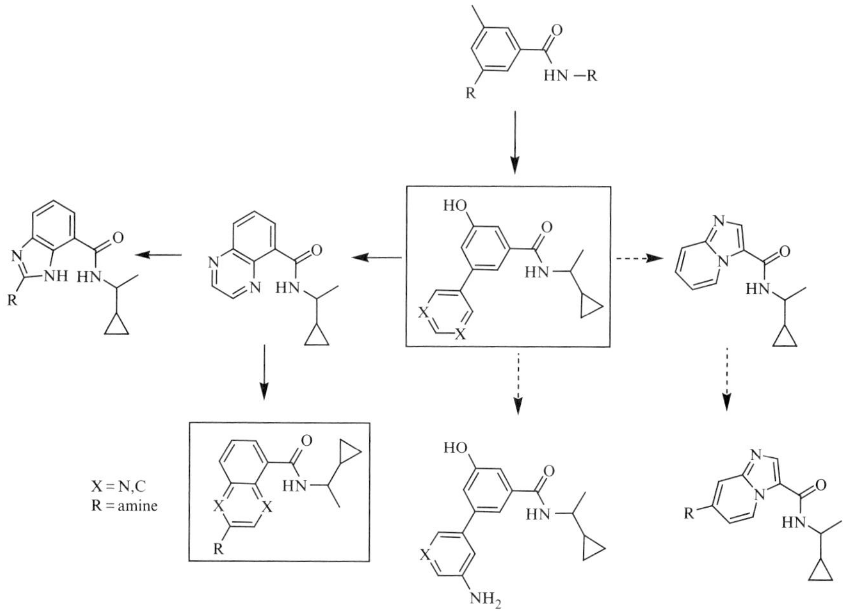

Figure 19.16 A *tertiary* sublibrary design scheme. The carboxamide, important for hydrogen bonding to the conserved water, is maintained in all compounds. Boxed structures indicate scaffolds with at least one *tertiary* library representative testing with affinity greater than 5 μ*M* in the ketohexokinase enzyme assay.

In contrast to the *secondary* library where fragment elaboration consists primarily of "growing," fragment "fusion" is favored in the design of compounds for the *tertiary* library. This is illustrated in Fig. 19.16 for the arylamide series. Electron density comparisons between arylamides and various six-membered aromatic rings seen for hits in the *primary* and *secondary* libraries reveal a nonhydrogen atom overlap that leads to the design of biaryl and fused ring compounds.

As basic amines are seen (Fig. 19.5) around the 7 o'clock region of the active site making favorable interactions with D27B, amines are attached onto the arylamide scaffold to optimize this interaction. The extension toward D27B is accomplished without interfering with amide carbonyl hydrogen bonding to the conserved water. In addition, another aim is that this extension of the arylamide scaffold would provide phosphate channel attractive substituent at the 4 o'clock position of the active site. Unfortunately, this is not realized, as compounds with polar functional groups that are predicted to produce favorable interactions do not result in discernable electron density.

5. First View of the Solution Activity of the Arylamide Lead

All *tertiary* library compounds are tested for solution activity in an enzyme assay. Three of the four *tertiary* library scaffolds (Fig. 19.15) show inhibition with affinity higher than 5 μ*M* in the enzyme assay for at least one of its representatives. Among the arylamide series, compound **1** shows an IC_{50} value of 750 n*M* in the ketohexokinase enzyme assay. Using recombinant proteins, compound **1** has no significant activity (>10 μ*M*) against any of the five major human CYP450 (CYP1A2, CYP3A4, CYP2C9, CYP2C19, CYP2D6). Compound **1** is stable in human and rat liver microsomes (112% remaining at 10 min in HLM, 123% in RLM, and 113% in MLM). The solubility of compound **1** is very good; 35.9 μ*M* in PBS pH 7.4, 40.9 μ*M* in PBS at pH 2.0, 36.5 μ*M* in water, and 40 μ*M* in acetonitrile. The compound is also found to be highly permeable through a monolayer of Caco-2 cells *in vitro*, with an efflux ratio of 3.61 (Papp A–B: 1.32×10^{-6} cm/s, and Papp B–A: 4.75×10^{-6} cm/s). In hERG channel binding assays, compound **1** gives an IC_{50} of 28.0 μ*M*. In single-cell patch clamp studies, inhibitions of the membrane K^+ current Ikr are 8.1% at 0.1 μ*M*, 10% at 0.3 μ*M*, and 18% at 3 μ*M*, compared to 5.9%, 11.3%, and 11.5% with solvent control. In the Cerep selectivity panel screens, compound **1** hits three receptors >70% at the screening concentration of 10 μ*M*; they are A3 (80% inhibition), D2S (58% inhibition), and NK2 (81% inhibition). In the Invitrogen protein kinase panel screens against 200 kinases, compound **1** at 10 μ*M* hits only Aurora kinase A with 60% inhibition, MAP kinase-activated protein kinase 2 with 56%, and MAP kinase-activated protein kinase 3 with 56% in the presence of 10 μ*M* ATP. Selectivity against the ribokinase superfamily has not been tested.

The methodology described above provides an additional route to the discovery of lead compounds by FBDD. Our ketohexokinase FBDD campaign highlights the potential outcomes of the methodology with compounds that are structurally unique, lead-like in pharmacokinetic properties, minimally restrained by intellectual property space, and possess significant binding affinity toward ketohexokinase. All of these properties are materialized in compound **1**.

Acknowledgments

We thank the volume editor, Dr. Lawrence Kuo, for his help in the preparation of this chapter. We also acknowledge the help for work presented here from the Structural Biology, ADME, and Analytical Chemistry departments at Johnson & Johnson Pharmaceutical Research & Development, LLC. The authors also thank the staff at IMCA-CAT for assistance during X-ray data collection. The use of the IMCA-CAT beamline 17-ID and 17-BM at the Advanced Photon Source is supported through a contract with Hauptman-Woodward Medical Research Institute. Use of the Advanced Photon Source is supported by the U.S. Department of Energy, Office of Science, Office of Basic Energy Sciences, under Contract No. W-31-109-Eng-38.

REFERENCES

Basciano, H., Federico, L., and Adeli, K. (2005). Fructose, insulin resistance, and metabolic dyslipidemia. *Nutr. Metab.* **2,** 5.

Ciulli, A., Williams, G., Smith, A. G., Blundell, T. L., and Abell, C. (2006). Probing hot spots at protein-ligand binding sites: A fragment-based approach using biophysical methods. *J. Med. Chem.* **49,** 4992–5000.

Congreve, M., Aharony, D., Albert, J., Callaghan, O., Campbell, J., Carr, R. A. E., Chessari, G., Cowan, S., Edwards, P. D., Frederickson, M., *et al.* (2007). Application of fragment screening by X-ray crystallography to the discovery of aminopyridines as inhibitors of β-secretase. *J. Med. Chem.* **50,** 1124–1132.

Congreve, M., Chessari, G., Tisi, D., and Woodhead, A. J. (2008). Recent developments in fragment-based drug discovery. *J. Med. Chem.* **51,** 3661–3680.

Elliott, S. S., Keim, N. L., Stern, J. S., Teff, K., and Havel, P. J. (2002). Fructose, weight gain, and the insulin resistance syndrome. *Am. J. Clin. Nutr.* **76,** 911–922.

Gibbs, A. C., Abad, M. C., Zhang, X., Tounge, B. A., Lewandowski, F. A., Struble, G. T., Sun, W., Sui, Z., and Kuo, L. C. (2010). Electron Density Guided Fragment-Based Lead Discovery of Ketohexokinase Inhibitors. *J. Med. Chem.* **53,** 7979–7991.

Gross, L. S., Li, L., Ford, E. S., and Liu, S. (2004). Increased consumption of refined carbohydrates and the epidemic of type 2 diabetes in the United States: An ecologic assessment. *Am. J. Clin. Nutr.* **79,** 774–779.

Hajduk, P. J., and Greer, J. (2007). A decade of fragment-based drug design: Strategic advances and lessons learned. *Nat. Rev. Drug Discov.* **6,** 211–219.

Hu, X., and Manetsch, R. (2010). Kinetic target-guided synthesis. *Chem. Soc. Rev.* **39,** 1316–1324.

Imai, Y. N., Inoue, Y., and Yamamoto, Y. (2007). Propensities of polar and aromatic amino acids in noncanonical interactions: Nonbonded contacts analysis of protein – ligand complexes in crystal structures. *J. Med. Chem.* **50,** 1189–1196.

Jhoti, H., Cleasby, A., Verdonk, M., and Williams, G. (2007). Fragment-based screening using X-ray crystallography and NMR spectroscopy. *Curr. Opin. Chem. Biol.* **11,** 485–493.

Miura, T. (2010). Fragment screening using surface plasmon resonance optical biosensor technology. *Yakugaku Zasshi* **130,** 341–348.

Murray, C. W., Callaghan, O., Chessari, G., Cleasby, A., Congreve, M., Frederickson, M., Hartshorn, M. J., McMenamin, R., Patel, S., and Wallis, N. (2007). Application of fragment screening by X-ray crystallography to β-secretase. *J. Med. Chem.* **50,** 1116–1123.

Nestler, H. P. (2005). Combinatorial chemistry and fragment screening–two unlike siblings? *Curr. Drug Discov. Technol.* **2,** 1–12.

Neumann, T., Junker, H. D., Schmidt, K., and Sekul, R. (2007). SPR-based fragment screening: Advantages and applications. *Curr. Top. Med. Chem.* **7,** 1630–1642.

Nienaber, V. L., Richardson, P. L., Klighofer, V., Bouska, J. J., Giranda, V. L., and Greer, J. (2000). Discovering novel ligands for macromolecules using X-ray crystallographic screening. *Nat. Biotechnol.* **18,** 1105–1108.

Poulsen, S.-A., and Kruppa, G. H. (2008). In situ fragment-based medicinal chemistry: Screening by mass spectrometry. *Fragment Based Drug Discov.* 159–198.

Sigrell, J. A., Cameron, A. D., Jones, T. A., and Mowbray, S. L. (1997). Purification, characterization, and crystallization of Escherichia coli ribokinase. *Protein Sci.* **6,** 2474–2476.

Sigrell, J. A., Cameron, A. D., Jones, T. A., and Mowbray, S. L. (1998). Structure of Escherichia coli ribokinase in complex with ribose and dinucleotide determined to 1.8 A resolution: Insights into a new family of kinase structures. *Structure* **6,** 183–193.

Spurlino, J. C. (2011). Fragment screening purely by X-ray crystallography. *Methods Enzymol.* **493,** 321–358.

Tounge, B. A., and Parker, M. H. (2011). Designing a diverse high-quality library for crystallography-based FBDD screening. *Methods Enzymol.* **493,** 3–20.

Trinh, C. H., Asipu, A., Bonthron, D. T., and Phillips, S. E. V. (2009). Structures of alternatively spliced isoforms of human ketohexokinase. *Acta Crystallogr. D Biol. Crystallogr.* **65,** 201–211.

Warr, W. A. (2009). Fragment-based drug discovery. *J. Comput. Aided Mol. Des.* **23,** 453–458.

CHAPTER TWENTY

Experiences in Fragment-Based Lead Discovery

Roderick E. Hubbard*,† *and* James B. Murray*

Contents

Abstract

This chapter summarizes the experience at Vernalis over the past decade in developing and applying fragment-based discovery methods across a range of different targets. The emphasis will be on the practical aspects of the different biophysical techniques (surface plasmon resonance (SPR), differential scanning fluorimetry (DSF), isothermal titration calorimetry, nuclear magnetic resonance, and X-ray crystallography) that can be used to identify

* Vernalis (R&D) Ltd., Granta Park, Cambridge, United Kingdom
† YSBL & HYMS, University of York, Heslington, York, United Kingdom

Methods in Enzymology, Volume 493
ISSN 0076-6879, DOI: 10.1016/B978-0-12-381274-2.00020-0

fragments that bind to targets and a discussion of the criteria and strategies for selecting and evolving fragments to lead compounds.

1. Introduction

The central theme of fragment-based discovery is to screen a small library (500–2000 molecules) of low molecular weight (MW) compounds (typically 110–250 Da) and then to evolve these fragments to generate lead compounds. The methods have two main advantages over conventional screening. Because the fragments are so small, they are less likely to have steric clashes that preclude fitting into a particular binding site. Secondly, a small number of such low MW compounds, if chosen wisely, can represent an extremely large chemical space. For these reasons, there have been substantial developments in the methods over the past 15 years and an accelerating interest in applying them across a broad range of targets within an increasing number of organizations.

There have been a number of recent reviews that summarize the ideas and successes (Fischer and Hubbard, 2009; Schulz and Hubbard, 2009). Most of the examples published to date have been on targets for which crystal structures are available with examples from kinases (Howard *et al.*, 2009), ATPases (Brough *et al.*, 2009), nuclear receptors (Artis *et al.*, 2009), phosphodiesterases (Card *et al.*, 2005), and proteolytic enzymes (Geschwindner *et al.*, 2007). Nuclear magnetic resonance (NMR) methods have also provided structural information, such as for the Bcl-2 family, where NMR methods identified initial fragments which gave inspiration for eventual design of inhibitors which are currently in clinical trials (Oltersdorf *et al.*, 2005).

Establishing a fragment-based discovery approach has four main aspects: library design and maintenance, identifying which fragments bind (screening), characterizing the binding of hit fragments, and chemical evolution of fragments to hit compounds. In this chapter, we summarize some of the experiences at the company Vernalis over the past decade in developing and applying the methods to a series of protein targets. The overall process is summarized in Fig. 20.1 (Hubbard *et al.*, 2007b). A library of some 1200 fragments is screened by ligand-observed NMR methods to identify the fragments which bind. The hit fragments are then characterized by various biophysical methods while attempting to determine the crystal structure of the fragment binding to the protein. The structural information is then used to guide the iterations of evolution of the fragments to larger hit and subsequent optimized lead compounds.

Details of the different components of this process will be discussed in this chapter. This will include both published material and some new analyses of trends and experiences in the various areas.

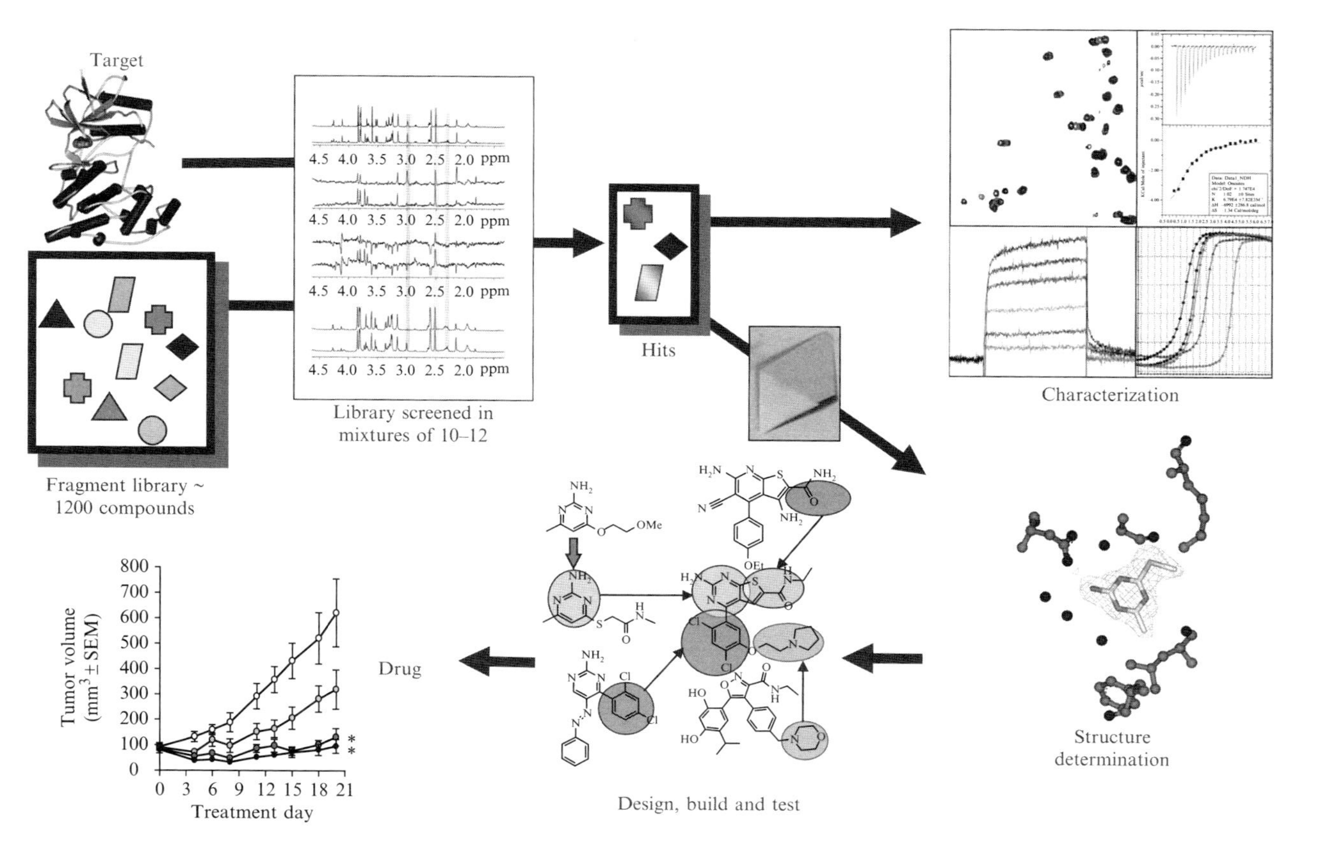

Figure 20.1 Fragment-based ligand discovery at Vernalis, the SeeDs [Structural exploitation of experimental Drug starting points (Hubbard *et al.*, 2007a,b)]. See text for description.

2. Maintaining and Enhancing a Fragment Library

We have published a number of reviews of the physicochemical, cheminformatics, and medicinal chemistry criteria used to select compounds for a fragment library (Chen and Hubbard, 2009; Hubbard *et al.*, 2007a). These are similar to the properties used for the first versions of the Vernalis library (Baurin *et al.*, 2004) where the primary criteria were to select a maximum of 1200 compounds which were of MW less than 250 Da, contained suitable functional groups for chemical synthesis, did not contain known reactive or toxic functionalities, and were predicted to be soluble at high concentrations (>2 m*M* aqueous). Subsequently, the Vernalis library has continuously evolved with fragments removed based on results from screening of a range of targets, the experiences of the medicinal chemists in using the library, and the long-term stability of the compounds. Fragments have then been added from analysis of current commercial libraries and considering synthetic intermediates generated within projects. In addition, fragments are acquired or synthesised based on ideas generated from in-house projects or seen in the literature.

There has been some debate about the optimal number of compounds in a fragment library. In our experience, there are two main experimental constraints: (1) A considerable effort is required to undertake frequent analysis of the library to ensure that compounds remain pure and soluble to allow the high concentrations at which fragments need to be screened. (2) There are practical limits on the number of fragments that can be taken into detailed analysis and validation of fragment binding. An important consideration in deciding the number of compounds in the library is the degree of chemical space that is represented. A recent analysis (Fink and Reymond, 2007) suggests that the size of chemical space accessible by known chemistry increases by an average of over eightfold with each nonhydrogen atom in the molecule. This means that a 1000-member fragment library of average MW 190 (14 nonhydrogen atoms, the Vernalis library) is equivalent to a library of over 10^9 molecules of average MW 280 and 10^{18} compounds with an average MW 450 (32 nonhydrogen atoms). There are many approximations in this extrapolation but it does emphasize that perhaps the effective coverage of chemical space is more sensitive to the size of the compounds than the number in the library.

The Vernalis library is maintained at about 1200 fragments, with average ($\pm$SD) properties as MW 190 $\pm$ 41 Da, *S* log *P* of 0.3 $\pm$ 1.3, 13.5 $\pm$ 2.9 nonhydrogen atoms, 4.5 $\pm$ 1.5 heteroatoms, 2.3 $\pm$ 1.6 rotatable bonds, 3.2 $\pm$ 1.2 hydrogen bond acceptors, and 1.9 $\pm$ 1.3 donors. Figure 20.2 provides more details on the MW and number of rotatable bonds of the

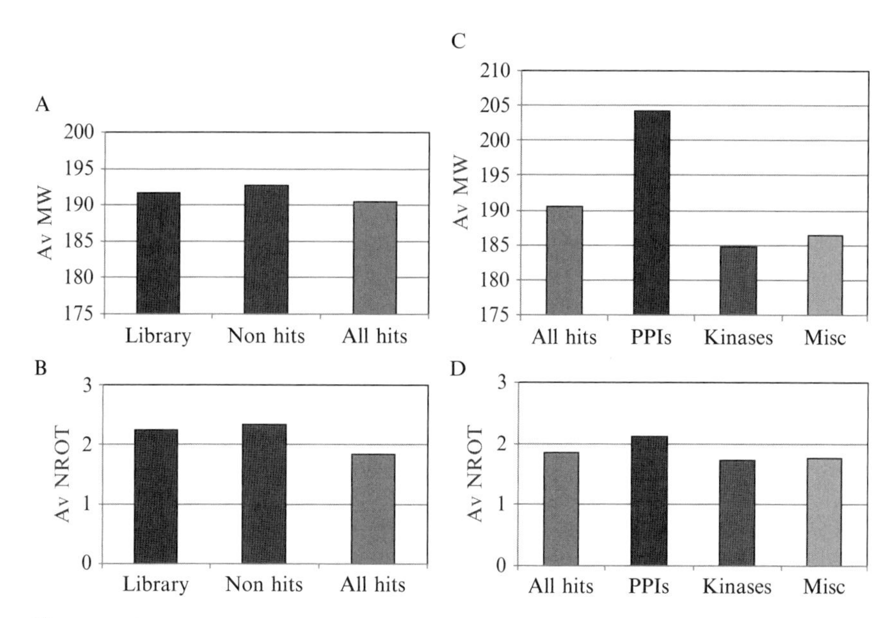

Figure 20.2 The average MW (A) and number of rotatable bonds (B) for the complete Vernalis fragment library and the compounds that are found to be hits or nonhits in screening campaigns. Average MW (C) and average NROT (D) broken down for the different categories of hits. Data taken from Chen and Hubbard (2009).

complete library compared to the properties of hits and nonhits from the library, and also a breakdown of these properties for the different classes of target that have been studied. A more detailed analysis of the hit rates is presented below and in Chen and Hubbard (2009). In summary, the properties of the hits are approximately the same as the nonhits, emphasizing the quality of the overall library. As would be expected, slightly larger and more complex fragments are required to register as hits against the more challenging protein–protein interaction (PPI) targets than the kinases with more well-defined binding sites.

3. Issues with Different Methods for Fragment Screening

Fragments are small and bind to their target with relatively low affinity, with useful fragments often having an affinity between 0.1 μ*M* and 10 m*M*. Reliable detection of low affinity binding has driven and relied on the development of a range of assays, predominantly based on biophysical methods. Figure 20.3 summarizes the range of methods and their effective affinity

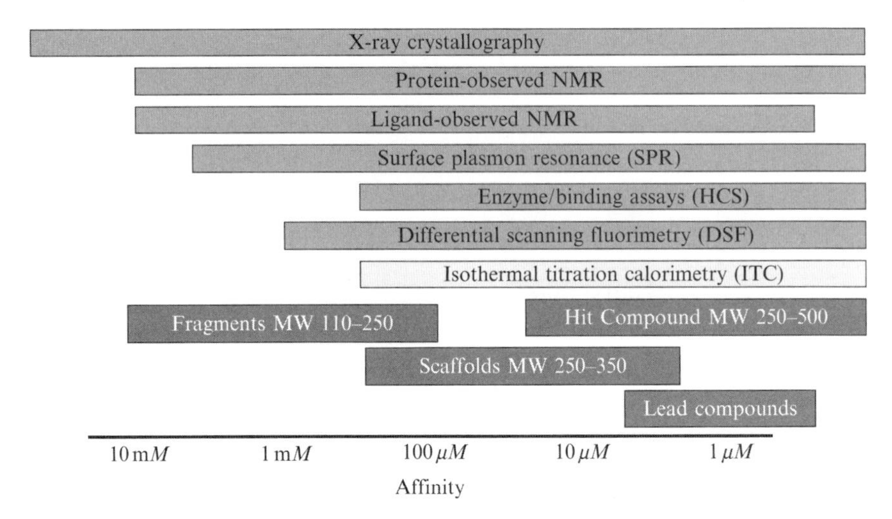

Figure 20.3 Affinity ranges accessible for detecting compound binding to macromolecular target. See text for details.

ranges for robust detection of binding, against some indicative ranges for binding affinities of the different levels of compounds. We and most others regard fragments as compounds with MW < 250, with larger compounds (MW 250–350) termed scaffolds (Card *et al.*, 2005). In the following subsections, we describe some of the features, experimental limitations, and our recent experiences in using this range of screening techniques against a number of targets. The data set is rather sparse as there are only a few targets for which more than one screening method has been used systematically. However, some useful lessons have been learned.

3.1. X-ray crystallography

This is the most information-rich technique. Although it is not possible to measure an affinity, the structure provides details about binding pose and guidance for subsequent fragment evolution. However, there are considerable experimental requirements: the crystal has to survive the high concentrations of ligand required for soaking; the crystal packing has to allow ready access to the binding site; and there must be access to large amounts of synchrotron time to collect the many data sets required and the streamlined protocols and software required for processing and analyzing the structural data. In our experience, even when a suitable crystal form is available, it can sometimes take a number of attempts to obtain a crystal structure of a known hit fragment binding to the protein. The method gives many false negative results, probably because crystal soaking is more a kinetic than a thermodynamically controlled process. It is also an immense investment of resource to deliver the simple yes/no answer required when screening.

3.2. Protein-observed NMR

NMR techniques (such as HSQC) were the first used for fragment screening (Shuker *et al.*, 1996), and NMR can deliver rich information about binding modes and structures for some systems. However, NMR experiments require isotopically labeled protein (difficult if cannot be expressed in bacteria) and there is an effective size limit of about 40 kDa. Nonetheless, NMR experiments such as HSQC are extremely valuable for titrations to determine K_D, as seen later. In our hands, we have found that as little as 20 μ*M* protein (0.5 mL, 15 min data collection) gives sufficient signal-to-noise ratio to reliably determine the K_D in ^{15}N HSQC titrations.

3.3. Differential scanning fluorimetry

In a typical experiment, we use 2 μ*M* protein and 2 m*M* ligand in the presence of SYPRO orange (Invitrogen) in a total volume of 40 μL. The experiment is performed in an RT-PCR machine (after Vedadi *et al.*, 2006). We define a hit as a compound that gives a change in melting temperature of at least two standard deviations of the error in the measurement.

3.4. Surface plasmon resonance

In all the surface plasmon resonance (SPR) experiments reported in this chapter, the protein was attached to a Ni^{2+} NTA chip surface via an N-terminal double-his-tag construct, typically $His_{(6-8)}$ + amino acid spacer (38-7) + $His_{(6-8)}$ + protease cleavage site, followed by the open reading frame of the protein. Experiments were carried out using the same buffer conditions, pH 7.4, 0.005% P20, 10 m*M* HEPES, and 150 m*M* NaCl. We have been able to characterize the binding of fragment hits in the affinity range 0.1–1000 μ*M*. For these direct binding methods with the targets described here, the affinity range that can be reliably determined is limited primarily by the aqueous solubility of the fragment. Our fragment library has a minimum aqueous solubility of 2 m*M* and this is the maximum concentration used in our experiments, thus limiting the reliable determination of affinity to 1 m*M*.

3.5. Ligand-observed NMR

Details of the experimental procedures and protocols for screening fragment libraries by ligand-observed NMR are summarized in Chen and Hubbard (2009). Fragments at 500 μ*M* each are added as mixtures of 8–12 fragments to 10 μ*M* protein in an NMR tube and data from three NMR experiments

(STD, Water-LOGSY, and CPMG) collected in the absence and presence of a competitor ligand. It is not necessary to spend a great deal of time designing the fragment mixtures. A set of fragments are mixed, 1D NMR spectra collected, and a check made that (as usually seen) at least one distinct peak can be seen for each ligand and that the ligands are not interacting (perturbation of spectra). A major advantage of NMR screening is that the state of the ligands and protein can be assessed at each assay point, checking for compound or protein degradation or aggregation. The NMR experiments are monitoring transfer of magnetization to ligands that are in equilibrium with the protein and then probed in solution. So, there is no limit on the size of protein. In many cases, a larger protein is preferable for a more effective transfer of signal. Also, as long as one of the fragments is not binding extremely tightly, more than one compound can be observed binding to the protein in a mixture at the same time.

The spectra are assessed manually, looking for changes in peaks that are consistent with binding which is lost when the competitor ligand is added. Fragments that are hits in all the three NMR experiments (STD, Water-LOGSY, and CPMG) are classified as Class 1, those that are hits in two NMR experiments as Class 2, and those that are hits in one experiment as Class 3. The changes in NMR signals are due to a multitude of not fully understood effects; in our experience, the classification reflects how reliable the binding is as determined by success in crystal soaking, where a suitable crystal form is available. Our general experience is that crystal structures are obtained on the first attempt for about 70% of Class 1 hits, falling to 40% of Class 2 hits, and rarely for Class 3 hits.

3.6. Comparison of SPR, NMR, and DSF

A subset of the fragment library, 80 compounds, was screened against HSP90 by NMR, differential scanning fluorimetry (DSF), and SPR. The NMR screen gave 10 Class 1 competitive hits and 7 noncompetitive hits. SPR generated 14 hits and DSF gave 19 hits. Of these, 8, 7, and 7 compounds gave a crystal structure from NMR, SPR, and DSF hits, respectively on the first attempt at soaking. In all, there were 7 hits common to all the three methods. One of these hits failed to generate a crystal structure either via soaking or cocrystallization. Potentially, this fragment is binding to an alternate site on HSP90. The NMR methods identified two noncompetitive hits that were not found by SPR or DSF, one of which gave a crystal structure in the presence of PU3. SPR methods did not find any unique hits; DSF found 7 unique hits, although these were not pursued further (possibly binding to alternate sites on the protein).

For this example of HSP90, the three methods found similar hits binding to the ATP binding site. However, the competitive step in the NMR experiment has the advantage of identifying true inhibitors of the ATP site directly.

Table 20.1 Validation by DSF of fragment hits from NMR

	Kinase-4	Enzyme-2	Kinase-3
NMR Hits[a]	77	94	37
%DSF Hits[b]	23	11	8
%Structure-NMR[c]	17	10	81
%Structure-DSF[d]	28	92	21

[a] The number of fragments identified as Class 1 hits from a screen of the full 1200 fragment library.
[b] The percentage of the NMR hits that show a shift greater than two standard deviations in the melting temperature of the protein target in a DSF experiment.
[c] The percentage of the NMR hits that gave a crystal structure on the first attempt at soaking into apo-crystals.
[d] The percentage of structures obtained on the first attempt for the NMR hits that were also hits in DSF.

3.7. Confirming hits from NMR by differential scanning fluorimetry

Table 20.1 summarizes the results obtained when hits from an NMR screen were validated by DSF with an analysis of how many of those hits subsequently gave a crystal structure. In all cases and in contrast to HSP90 above, DSF confirmed substantially fewer hits than were seen in NMR, although there was variability in the percentage of those fragments that gave crystal structures. For Enzyme 2 and Kinase 4, DSF was predictive of which fragments gave a structure; however, for kinase 3, many more structures were obtained for fragments that were not hits in DSF. The lack of hit detection for kinase 3 is difficult to rationalize; there does not appear to be a correlation with affinity or with MW of the fragment hits. Despite the controls used, it may relate to compound behavior under these conditions, such as aggregation. We note that no aggregation was detected in the NMR experiments. Our experience suggests that using DSF, as described here, is not suitable as a frontline hit-finding method; however, it does play a useful role in the subsequent characterization of the hits.

3.8. Confirming hits from NMR by SPR

Table 20.2 summarizes some recent experiences in using SPR to characterize hits identified with ligand-observed NMR methods for four kinases and a PPI target. The top line of the table demonstrates that SPR is able to confirm at least 80% of the NMR hits as some sort of a binder. These hits are made up of three categories. About half of the SPR hits bind with 1:1 stoichiometry, give a full dose–response curve with top concentration at least twice that of the K_D, and thus for which a K_D can be determined. The second category is hits that appear to bind with 1:1 stoichiometry,

Table 20.2 Validation by SPR of fragment hits from NMR

	Kinase-1	Kinase-2	Kinase-3	Kinase-4	PPI-1
%hit[a]	100	83	87	84	86
$\%K_D$[b]	100	55	43	35	39
%10–50%[c]	0	13	19	41	29
$\% > 1{:}1$[d]	0	15	25	8	18

[a] Percentage of the fragments identified as binding to the target by NMR that were confirmed by SPR.
[b] Percentage of the NMR hits for which a K_D could be determined.
[c] Percentage of the NMR hits for which SPR gave a clear response, but insufficient to determine K_D.
[d] Percentage of the NMR hits for which SPR showed evidence of more than 1:1 binding stoichiometry.

but where the K_D cannot be determined because they are only 10–50% active at the top concentration which can be reliably used. Finally, there are a significant and surprising number (around 20%) of the hits where the binding stoichiometry clearly exceeds 1:1—this category also includes the few compounds that dissociated rapidly and those that gave abnormal dissociation curves.

These super-stoichiometric binders have been discussed in the SPR literature (Giannetti *et al.*, 2008; Navratilova and Hopkins, 2010) and there is some debate about whether they are valid hits or not. Aggregation has been demonstrated to be a significant problem for high-throughput screening (Coan and Shoichet, 2008). Although no aggregation was observed in the course of the ligand-observed NMR screen, we have detected aggregation in some cases under the SPR experimental conditions. However, we have frequently been able to obtain crystal structures of such hits; for example, we were able to generate useful crystal structures on the first attempt for 60% of the super-stiochiometric hits found for kinase-3. In addition, about 60% of these super-stoichiometric compounds are hits in the DSF assay. It is interesting to note that fragment hits that have inspired lead series in HSP90 (Brough *et al.*, 2009) and PIN1 (Potter *et al.*, 2010) were super-stoichiometric binders, demonstrating that these overbinding compounds can be valid hits. The binding of additional copies is probably due to the information-rich fragments finding other small pockets and clefts on the surface of the protein in addition to the main biologically relevant binding site.

3.9. High concentration screening versus NMR

Although the various biophysical methods such as NMR, SPR, and X-ray can provide a wealth of other information, the primary purpose of a fragment screen is to identify which fragments bind to the target at some concentration. It can be expensive (in time, resource, facilities, and amounts of protein) to screen using biophysical methods and so it would be beneficial if primary assays (either activity or binding) could be used to identify hits. There are many pitfalls in such screening (Macarron, 2006) but for fragments, the principal challenge is configuring the assay to withstand the high concentration of ligand required to identify low-affinity hits in high concentration screening (HCS). The central component for most high-throughput assays is the competitive step, where the putative hit affects the fraction bound of a known labeled ligand that is monitored (directly or indirectly). To achieve such competition, the hit needs to be screened at much higher concentrations. For example, to achieve an IC_{50}, the concentration of the ligand needs to be three times the K_D and 10 times to achieve an IC_{80}. This, combined with the often limiting solubility of medicinally relevant fragments, is frequently a significant challenge for HCS. This solubility versus activity is more pronounced for challenging targets such as PPIs where the nature of the target site is likely to require larger ligands, than say a kinase, to register a response. Figure 20.4 illustrates the problem. For a set of targets, up to 35 validated NMR hits were assayed in an HCS for activity or binding. For some targets (HSP90, kinase-1, PIN1, etc.), a high proportion of the NMR hits registered as hits in the assays; however for others, such as HSP70 and PPI-1, there were few, if any, hits that registered.

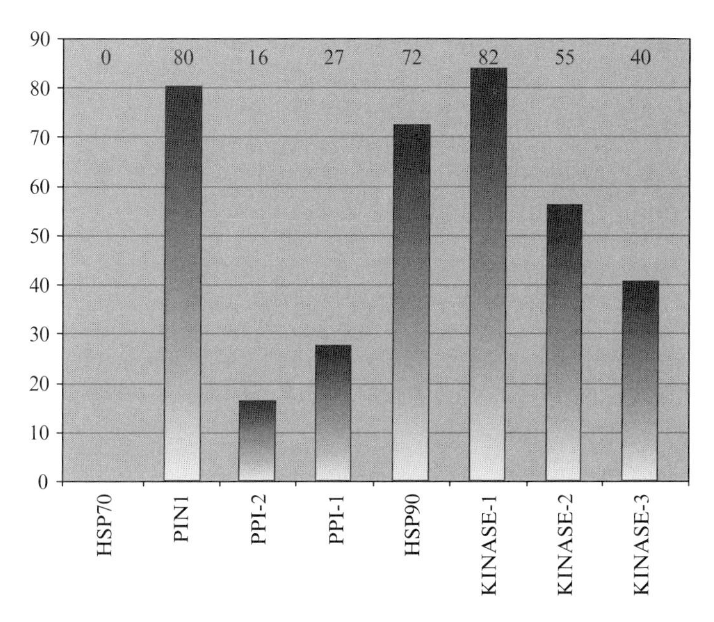

Figure 20.4 A histogram showing the percentage of fragment hits from a ligand-observed NMR screen that are hits in a high concentration binding or enzyme activity assay for a set of targets.

This high false negative rate is not due to the nature of the assays as such—the hit rates for kinase-2 and kinase-3, for example, are much lower than for kinase-1, even though the same format assay is used.

There are also issues with high false positive hit rates in HCS. For HSP90, the full fragment library was screened and for PPI-2, a random 10% of the library was screened, in both cases with a fluorescence polarization assay monitoring displacement of a labeled ligand. More than four times the number of hits was obtained by HCS in both cases compared to NMR and although not pursued in detail, many of the additional hits were not validated by other biophysical methods nor was an X-ray structure obtained. For PIN1, 51 hits were obtained from screening the whole fragment library using a functional assay, of which six could be validated using biophysical methods such as protein-observed NMR and X-ray crystallography.

These experiences confirm that, for the size of compounds in the Vernalis fragment library, ligand-observed NMR is the most robust technique for reliably identifying fragments that bind to a range of proteins. The technique has a dynamic range suitable to all target classes currently being prosecuted, as illustrated in Fig. 20.5. For well-defined binding sites, such as

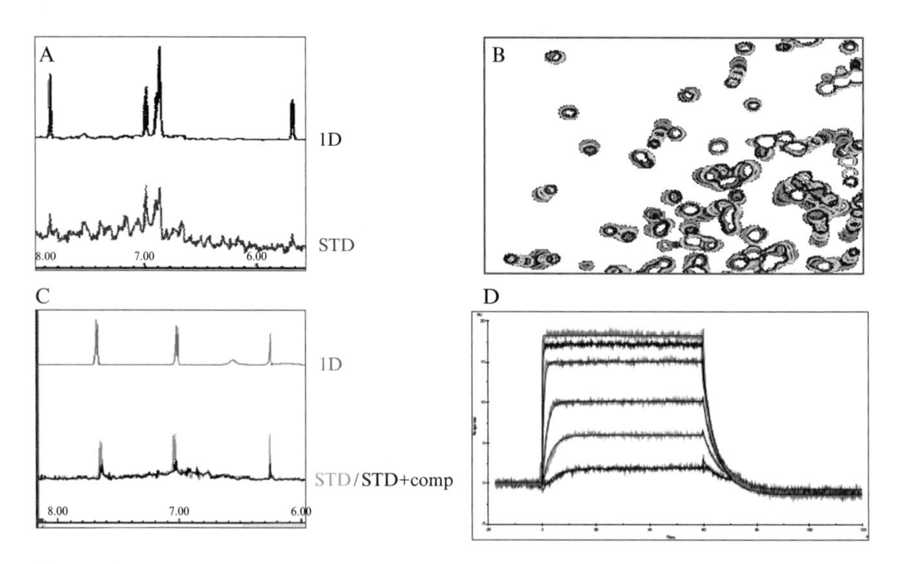

Figure 20.5 The dynamic range of detecting fragment binding by ligand-observed NMR spectroscopy. (A) The 1D NMR spectrum (top) and STD spectrum (bottom) for a fragment binding to a PPI-1 and (B) a titration measured by HSQC NMR spectrum for the same fragment binding to PPI-1, with an estimated K_D of 3.8 m*M*. (C). The 1D NMR spectrum (top) and STD spectrum (bottom, light is compound alone, dark is compound plus competitor) for a fragment binding to kinase-1 and (D) a titration measured by SPR for the same fragment binding to kinase-1 with an estimated K_D of 90 n*M* (cK_i of 120 n*M* from kinase activity measurement)

kinases, small fragments bind with submicromolar affinity; for larger PPI targets, the binding may be millimolar activity. Ligand-observed NMR can detect this full range of affinity. An additional advantage is that NMR can monitor the physical state of the protein and ligand at each assay point, confirming that the protein and fragment are still in solution, whether the protein has unfolded or aggregated, and whether the fragment has decomposed or is aggregating or interacting in the mixture.

We have identified two major issues for the method. First—the amount of protein required can be prohibitive for some targets—a full screen with validation of 1200 fragments typically uses some 20–40 mg of protein. Secondly, it will require additional experiments to identify the fragment hits that are binding to allosteric or cryptic sites on the protein. These hits could be competitive or noncompetitive. Additional experiments to identify allosteric sites can be accomplished in several ways. Firstly, if the protein is amenable to protein-observed methods such as HSQC/TROSY (e.g., up to ~35 *k*Da), then these sites can be readily characterized. A second subscreen can be run where the noncompetitive hits are competed with one another; if the "hit-rate" is high, then there probably exists an additional site that may be druggable (see below). If there are indications of second sites from other experiments, then the screen can be run in the presence of a known central binding site compound. There can clearly be issues if these compounds induce allosteric changes that affect binding, but if binding can be confirmed by other techniques (ITC, SPR, DSF), then more extensive protein-observed NMR experiments may be able to characterize binding.

4. Hit Rates for Different Classes of Target

Recently, we have analyzed the output from our fragment screening campaigns against a set of 11 targets (Chen and Hubbard, 2009). Figure 20.6 shows a plot of the percentage of fragments that were Class 1 hits (see above) for each target, with the *x*-axis as the notional druggability of the binding site calculated using the SiteMap algorithm (Halgren, 2009). This druggability score is empirically derived from the shape and constitution of the binding pockets seen in the structure of the protein. The general trend is that the fragment hit rate reflects the druggability of the target, but there are some discrepancies.

Many fewer hits were obtained for HSP70 (labeled A on Fig. 20.6) than expected from the structure. The active site (Fig. 20.7A) is well formed and contains many of the features expected for a druggable site. However, many of the side chains, and in particular, the solvent, appear to be quite flexible, as seen in the crystal structures overlaid in Fig. 20.7B of HSP70 in complex with various ligands. It is possible that there is an increased entropic penalty for fragments binding to such a flexible active site, thus reducing the observed hit rate.

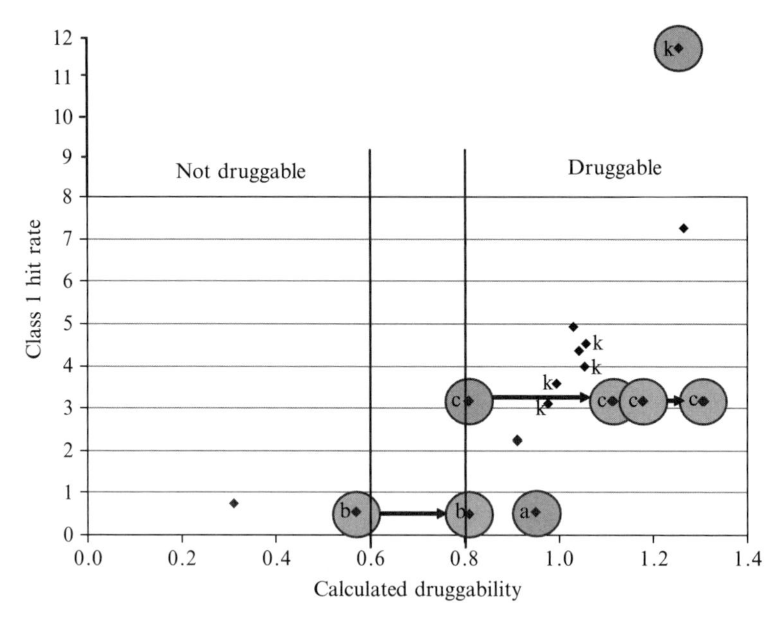

Figure 20.6 The relationship between validated hit rate from a fragment screen and target druggability. k, kinases; a, Hsp70; b, Pin-1; c, a PPI target. The gray circles highlight particular data discussed in the text and the red arrows show that the notional druggability of the target active site alters as different ligands select different conformations of the protein. (See Color Insert.)

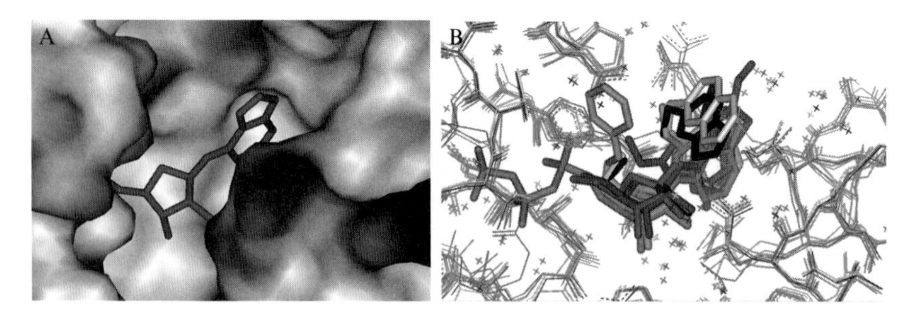

Figure 20.7 Details of crystal structures of Hsp70 bound to various ligands: (A) ADP bound to Hsp70 with electrostatic potential surface; (B) overlay of series of Hsp70 liganded structures showing variability in side chain and solvent positions.

A different consequence of conformational change is observed for PIN1 (labeled B on Fig. 20.6). Only six fragments could be validated as binding to this target, in reasonable agreement with the druggability score calculated from the structure. However, it was possible to progress to more potent hits (see Potter *et al.*, 2010; Williamson *et al.*, 2009); however, these more potent compounds select a different protein conformation, as seen in Fig. 20.8,

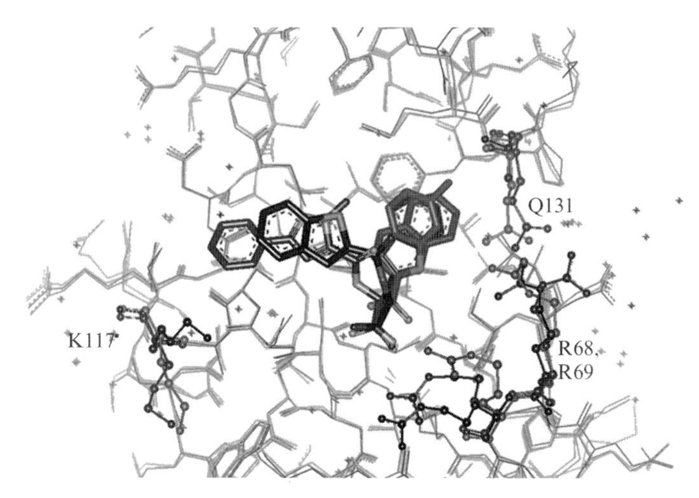

Figure 20.8 Details of crystal structures of the binding site of Pin-1 bound to various ligands. The ligands are drawn in thick stick representation. The displayed protein atoms are in thin stick, colored the same as the corresponding ligand C atoms. The side chains which adopt different conformations in the structures (R68, R69, K117, Q131) are shown in thin ball and stick.

where movements in some large, charged surface residues create a more druggable active pocket. This change in observed protein structure in response to ligand binding was even more striking for a PPI target (labeled C on Fig. 20.6). The fragment hit rate was higher than expected from the initial structure alone; subsequent structures of the target with optimized compounds showed selection of conformations with additional drug-like pockets, as reflected in the druggability score.

Two main lessons can be learned from this analysis. The first is that experimental fragment screening can provide an early assessment of the tractability of a target for a drug discovery campaign, and reflect issues with the target that are not apparent from inspection of the protein structure alone. Secondly, it emphasizes that additional pockets can open up on the binding sites of some targets, and these can be exploited by appropriate ligands to give higher affinity binding. Based on these observations, we have concluded that calculations of druggability from a static structure alone can be misleading.

5. Success Stories in Fragment Evolution

The past 5 years has seen an increasing number of compounds derived from fragments entering clinical trials (reviewed in Chessari and Woodhead, 2009; Congreve *et al.*, 2008; Schulz and Hubbard, 2009). In the first SAR by NMR projects from a group at Abbott (Shuker *et al.*, 1996), fragments were identified and characterized in discrete binding sites, and potent hits generated by linking. There have been other impressive examples of this approach (such as Bcl-2 (Oltersdorf *et al.*, 2005) and HSP90 (Huth *et al.*, 2007) inhibitors). However, most proteins do not have separate, discrete binding cavities, and if any fragments can be found, then it is challenging to design linking chemistry which will retain the position and orientation of the two linked fragments. Most of the successful fragment evolution campaigns have involved either structure-guided growth or merging of fragment information with other compounds. This is illustrated by two published examples from Vernalis work on HSP90.

An important concept to emerge in the past 10 years is that of ligand efficiency (the free energy of binding per ligand nonhydrogen atom; Hajduk, 2006; Hopkins *et al.*, 2004). This not only emphasizes that a small (10 nonhydrogen atoms), weak binding (5 m*M*) fragment is of equal quality to a larger (40 nonhydrogen atoms), high affinity (1 n*M*) binding compound; it is also a valuable metric to ensure that effective interactions are being made as a fragment is evolved.

Figure 20.9 shows the progression from a resorcinol fragment hit to the HSP90 inhibitor, AUY922, which is currently in Phase II clinical trials for cancer. The medicinal chemistry route to this compound has been described in detail elsewhere (Brough *et al.*, 2008); here, we discuss the key features. A ligand-observed NMR screen of an early, 719-member library identified 17 fragments, including a number of resorcinols. The determination of the structure of a resorcinol fragment to HSP90 (Fig. 20.9A) showed that although the fragment binds only weakly (estimated at 2 m*M* affinity), it adopts a unique position, with a network of protein–ligand hydrogen bonds, many mediated by water molecules. The rCAT database of 3.5 *M* commercially available compounds (Baurin *et al.*, 2004) was searched for resorcinols, which were assessed for binding to HSP90 by molecular docking calculations. From this, some 100 compounds were purchased and assayed, one of which (Fig. 20.9B) had submicromolar affinity for HSP90 and showed growth inhibition of HCT116 cells, with PD marker changes consistent with a HSP90 mode of action (data not shown). This so-called SAR by catalog approach is an efficient way of growing fragments to hits and is a particularly powerful way of generating SAR around a fragment through mining a compound collection (commercial or corporate).

Structure-guided medicinal chemistry was then used to optimize the properties of the lead, with the most significant changes as highlighted in Fig. 20.9C. The introduction of the amide functionality (1) on the central ring allows a hydrogen bond network to be established between G97 and K58, with a large increase in affinity. The change from pyrazole to isoxazole in the compound core (2) was suggested from some of the hits identified in the SAR by catalog exercise. This one atom change (O–N) has a dramatic

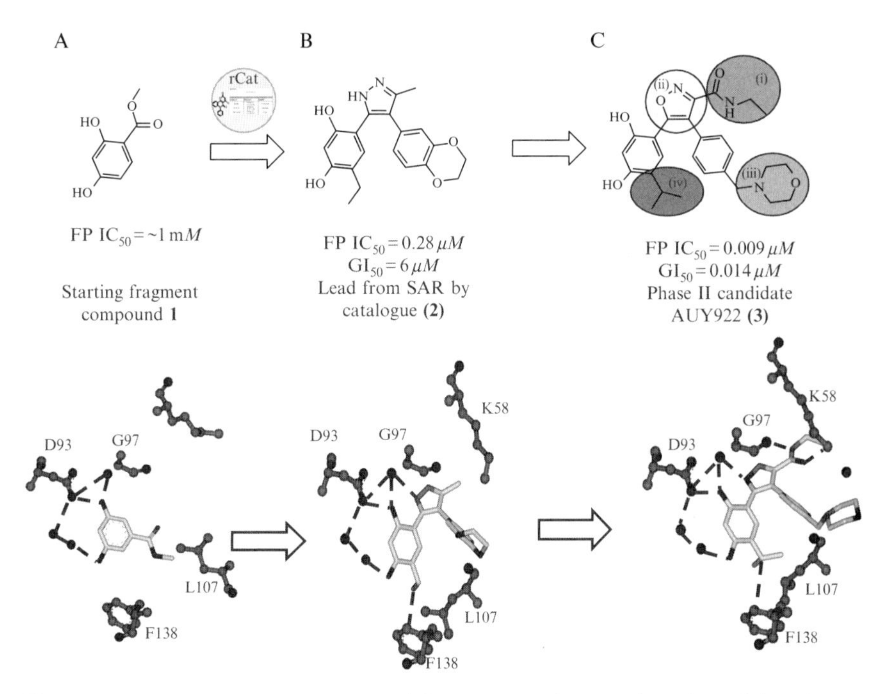

Figure 20.9 Evolution from resorcinol fragment to Phase II clinical candidate (compound 3) for Hsp90. The top panel shows the chemical structures of (A) a fragment, compound 1, (B) the initial lead compound (2), and (C) the optimized clinical candidate (3). The bottom panel shows details from the crystal structure of each of the compounds bound to HSP90, where the blue dashed lines represent key hydrogen bonds. The FP IC_{50} values are for the displacement of a fluorescently labeled probe, the GI_{50} values are growth inhibition of HCT116 cells, and rCAT is the Vernalis corporate virtual library of compounds (Baurin *et al.*, 2004). The highlighted portions of compound 3 (AUY922), labeled (i)–(iv) are discussed in the text. (For interpretation of the references to color in this figure legend, the reader is referred to the Web version of this chapter.)

effect on cellular potency, primarily through an order of magnitude reduction in the off-rate for the compound binding. The mechanism for this is unclear, but it is probably due to the differential activation energy required to make and break the various hydrogen bond networks. This increased target residency is a major contributor to the eventual *in vivo* efficacy of the clinical candidate. The region marked (3) is exposed to solvent and the morpholino group makes additional optimal contacts to the protein. Finally, the replacement of chlorine by isopropyl (4) was found to be the optimum substituent to give an appropriate balance of *in vivo* biological activity. The isopropyl makes an enhanced interaction with F138 but additionally may modulate the glucoronidation of the phenolic groups, thus affecting excretion dynamics.

From a structural and molecular design point of view, the fragment evolution summarized in Fig. 20.10 is particularly attractive. Structures

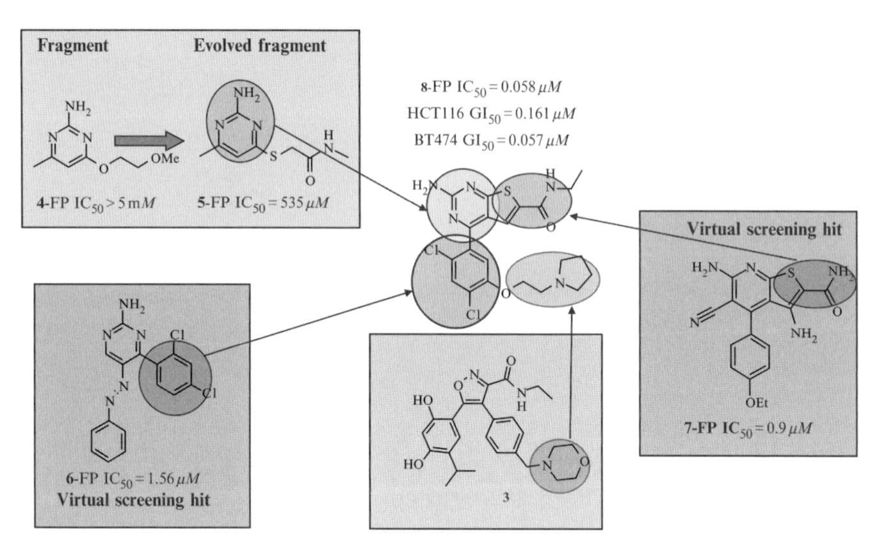

Figure 20.10 Combining features from fragment and virtual screening hits against HSP90 to discover the oral candidate BEP-800 (compound 12). See text for details and Brough *et al.* (2009).

were determined for an SAR by catalog fragment hit (compound **4**), two very different chemical classes identified by virtual screening (**6** and **7**), and the clinical candidate described earlier, AUY 922 (compound **3**). These structures suggested a merging of chemical features to construct the core thienopyrimidine scaffold, which was then decorated with functional groups taken from the different hits. This resulted in BEP-800 (compound **8**), an orally bioavailable and potent HSP90 inhibitor which entered pre-clinical trials. A similar fragment and hit-merging strategy has also been published for the kinase, PDPK1 (Hubbard, 2008).

In both of these examples, the essential details of fragment binding and the protein structure remain constant as the compound is evolved. However, some changes in protein conformation can be observed, as for example, L107 moves in HSP90 (Fig. 20.9).

6. Thoughts on How to Decide Which Fragments to Evolve

In many ways, deciding which fragment(s) to progress is one of the main challenges in fragment-based methods, as the decision will have a defining impact on the direction and chance of success of the program. The fragment screening methods we have described here typically can generate over 50 diverse hits for many targets. In our experience, it is important to bring together as much information as possible before committing the chemistry resources to particular series. Although the details of how these decisions are made will vary, the overall strategy is similar for most of our projects.

The overriding objective is to generate a robust model of how a fragment binds to the target, a structural rationale for the activity of related compounds and a tractable synthetic chemistry strategy for achieving the required affinity and selectivity while retaining ligand efficiency.

Where crystallography is routine, we determine the crystal structure of as many fragments bound to the target as possible. For some systems, it is possible to generate many tens of crystal structures quite routinely, usually by crystal soaking, with occasional validation in cocrystallization.. These structures identify the chemical vectors that are available for elaboration of the fragments, allowing ideas to be generated of how to increase affinity and modulate selectivity. Where crystal structures are not available, we use NMR and/or modeling methods to generate and test models of how the ligands bound (see Section 7).

In parallel with the determination of this structural information, we characterize the binding of the fragments to the target by the most appropriate biophysical method(s)—be that SPR, ITC, DSF, or HSQC NMR. Alongside this biophysical characterization, we explore the chemical space available to the fragments—through database searching and limited library synthesis, as well as a more general appraisal of the options for chemical synthesis. This is a phase we term "hit expansion." For most targets, it is too early at the fragment stage to consider patentability, but for some classes such as kinases, it can be necessary to consider the IP potential of particular fragments at this early stage.

The database searching identifies near neighbor compounds available from commercial suppliers or within the corporate collections which contain substructures of the fragment known to bind (a process often called SAR by catalog). This process can use a set of substructure queries, ligand-based pharmacophores, and fingerprint and similarity searches to find suitable compounds for assessment. If a crystal structure of the fragment bound to the target is available, then focused docking calculations can be used to prioritize acquisition of such available compounds. This hit expansion should explore both core modifications (where MW changes are minimal) and larger compounds more in classical lead-like space (250–350 MW), containing the original fragments as substructures. It should be noted that an overemphasis on "headline potency" can make it tempting to prioritize larger compounds; for this reason, the ligand efficiency needs to be monitored to guard against drifting into a high-throughput screening screen via a fragment detour.

The alternate approach (which we usually conduct alongside database search) is to synthesize small libraries to explore the SAR potential of the most desirable hits, especially on fragments that are not well exemplified in commercial or in-house libraries. This approach can be a way to progress fragments in the absence of structural information, although with such weak binding, it can be difficult to generate a sufficiently directional SAR—too many compounds will just be inactive, not providing much help in generating a model for binding and proposals for optimization.

One final comment on characterizing the binding of compounds during the hit expansion and hits to leads phases. Recently, there has been increased interest in the analysis and consequences of the kinetic profile of drug candidates (see Copeland *et al.*, 2006 for an early perspective, and more recently Lu and Tonge, 2010). Recently, we have begun using the kinetic profile rather than simply the affinity profile to track SAR and make decisions during these early hit expansion stages. The parameter we focus upon is not the observed on-rate but the off-rate (as defined in Copeland *et al.*, 2006). This highlights the primary determinant of compound activity in both target affinity and efficacy as drugs. For example, at the early stages of our fragment evolution, we may have two compounds with the same affinity; however, one has a 10-fold slower off-rate that is compensated for by a 10-fold slower on-rate. This slower off-rate indicates stabilization of a lower energy protein-ligand complex. Either way, the bound conformation is a productive one with respect to complex half-life (compound/drug residence time), and this information can help select the most appropriate compounds for further optimization. The advent of high-throughput real-time SPR equipment, such as the Biacore T100, now allows this kinetic assessment to be made on a large number of compounds rapidly enough to inform the medicinal chemistry cycle.

7. Final Comments

There is a continued need for more rapid and cost-effective methods for detecting and characterizing fragment binding to targets, and the success of fragment-based discovery is stimulating many new developments. The last few years have seen many developments in SPR and ITC methods—currently, other sensor technologies are in development. Perhaps the most striking consequence of the development of fragment-based lead discovery has been the intense application of orthogonal biophysical methods to characterize binding modes and interactions as compounds are optimized. This more questioning approach to the characteristics of ligand binding is having an impact in drug discovery beyond the fragment-based practitioners. This is particularly apparent as the fragment methods are increasingly integrated alongside high-throughput screening in large companies. This is providing the opportunity to use a fragment screen as a "window" into the compound collection (Crisman *et al.*, 2008), or encouraging the triaging of high-throughput screening hits with ligand efficiency rather than absolute affinity as a cut-off.

Fragment-based methods are now firmly established as an effective route to generate novel starting points for drug discovery against a range of targets for which structural information is readily available (see references in Section 1). One of the major challenges is progressing fragments in the absence of crystal structures.

Recently, we have been pursuing a number of challenging PPI targets for which it has proved impossible to generate crystal structures with fragments bound. In many cases, this is because the binding site on the protein prefers to stack with another copy of itself in the crystal, making it difficult to introduce a weak binding fragment. We have found that NMR approaches such as X-filtered NOESY can provide sufficient information about which parts of a fragment are binding to which parts of a target to generate a model of adequate quality to guide chemistry. Once higher affinity hits are obtained, then it has proved possible to induce a different crystal form from which a clear structure emerges, confirming the NMR guided model.

One of the anticipated benefits of fragment screening is the ability to generate tractable hits against novel and challenging targets, such as protein–protein interfaces or the new classes of targets emerging from modern biology—epigenetics, control of protein turnover, multiprotein complexes, and so on. It will be challenging to generate structural information on many of these targets, so the development of methods to progress in the absence of crystal structures (such as NMR-guided models) or in the complete absence of structure (such as off-rate screening) is an important area for development to bring the power of fragment-based methods to this next generation of therapeutic targets.

ACKNOWLEDGMENTS

The authors are extremely grateful to the past and present scientists at Vernalis for providing such a rich store of results from which to construct this chapter.. In particular, we acknowledge the contributions of Ijen Chen for library design and analysis, Ben Davis for all things NMR, Julia Smith for thermal shift assays, Natalia Matassova for SPR, and Paul Brough and James Davidson for medicinal chemistry.

REFERENCES

Artis, D. R., Lin, J. J., Zhang, C., Wang, W., Mehra, U., Perreault, M., Erbe, D., Krupka, H. I., England, B. P., Arnold, J., Plotnikov, A. N., Marimuthu, A., *et al.* (2009). Scaffold-based discovery of indeglitazar, a PPAR pan-active anti-diabetic agent. *Proc. Natl. Acad. Sci. USA* **106,** 262–267.

Baurin, N., Aboul-Ela, F., Barril, X., Davis, B., Drysdale, M., Dymock, B., Finch, H., Fromont, C., Richardson, C., Simmonite, H., and Hubbard, R. E. (2004). Design and characterization of libraries of molecular fragments for use in NMR screening against protein targets. *J. Chem. Inf. Comput. Sci.* **44,** 2157–2166.

Brough, P. A., Aherne, W., Barril, X., Borgognoni, J., Boxall, K., Cansfield, J. E., Cheung, K. M., Collins, I., Davies, N. G., Drysdale, M. J., Dymock, B., Eccles, S. A., *et al.* (2008). 4, 5-diarylisoxazole Hsp90 chaperone inhibitors: Potential therapeutic agents for the treatment of cancer. *J. Med. Chem.* **51,** 196–218.

Brough, P. A., Barril, X., Borgognoni, J., Chene, P., Davies, N. G., Davis, B., Drysdale, M. J., Dymock, B., Eccles, S. A., Garcia-Echeverria, C., Fromont, C., Hayes, A., *et al.* (2009). Combining hit identification strategies: fragment-based and in silico approaches to orally active 2-aminothieno[2, 3-d]pyrimidine inhibitors of the Hsp90 molecular chaperone. *J. Med. Chem.* **52,** 4794–4809.

Card, G. L., Blasdel, L., England, B. P., Zhang, C., Suzuki, Y., Gillette, S., Fong, D., Ibrahim, P. N., Artis, D. R., Bollag, G., Milburn, M. V., Kim, S. H., *et al.* (2005). A family of phosphodiesterase inhibitors discovered by cocrystallography and scaffold-based drug design. *Nat. Biotechnol.* **23,** 201–207.

Chen, I. J., and Hubbard, R. E. (2009). Lessons for fragment library design: Analysis of output from multiple screening campaigns. *J. Comput. Aided Mol. Des* . **23,** 603–620.

Chessari, G., and Woodhead, A. J. (2009). From fragment to clinical candidate–A historical perspective. *Drug Discov. Today* **14,** 668–675.

Coan, K. E., and Shoichet, B. K. (2008). Stoichiometry and physical chemistry of promiscuous aggregate-based inhibitors. *J. Am. Chem. Soc.* **130,** 9606–9612.

Congreve, M., Chessari, G., Tisi, D., and Woodhead, A. J. (2008). Recent developments in fragment-based drug discovery. *J. Med. Chem.* **51,** 3661–3680.

Copeland, R. A., Pompliano, D. L., and Meek, T. D. (2006). Drug-target residence time and its implications for lead optimization. *Nat. Rev. Drug Discov.* **5,** 730–739.

Crisman, T. J., Bender, A., Milik, M., Jenkins, J. L., Scheiber, J., Sukuru, S. C., Fejzo, J., Hommel, U., Davies, J. W., and Glick, M. (2008). "Virtual fragment linking": An approach to identify potent binders from low affinity fragment hits. *J. Med. Chem.* **51,** 2481–2491.

Fink, T., and Reymond, J. L. (2007). Virtual exploration of the chemical universe up to 11 atoms of C, N, O, F: Assembly of 26.4 million structures (110.9 million stereoisomers) and analysis for new ring systems, stereochemistry, physicochemical properties, compound classes, and drug discovery. *J. Chem. Inf. Model.* **47,** 342–353.

Fischer, M., and Hubbard, R. E. (2009). Fragment-based ligand discovery. *Mol. Interv.* **9,** 22–30.

Geschwindner, S., Olsson, L. L., Albert, J. S., Deinum, J., Edwards, P. D., de Beer, T., and Folmer, R. H. (2007). Discovery of a novel warhead against beta-secretase through fragment-based lead generation. *J. Med. Chem.* **50,** 5903–5911.

Giannetti, A. M., Koch, B. D., and Browner, M. F. (2008). Surface plasmon resonance based assay for the detection and characterization of promiscuous inhibitors. *J. Med. Chem.* **51,** 574–580.

Hajduk, P. J. (2006). Fragment-based drug design: How big is too big? *J. Med. Chem.* **49,** 6972–6976.

Halgren, T. A. (2009). Identifying and characterizing binding sites and assessing druggability. *J. Chem. Inf. Model.* **49,** 377–389.

Hopkins, A. L., Groom, C. R., and Alex, A. (2004). Ligand efficiency: A useful metric for lead selection. *Drug Discov. Today* **9,** 430–431.

Howard, S., Berdini, V., Boulstridge, J. A., Carr, M. G., Cross, D. M., Curry, J., Devine, L. A., Early, T. R., Fazal, L., Gill, A. L., Heathcote, M., Maman, S., *et al.* (2009). Fragment-based discovery of the pyrazol-4-yl urea (AT9283), a multitargeted kinase inhibitor with potent aurora kinase activity. *J. Med. Chem.* **52,** 379–388.

Hubbard, R. E. (2008). Fragment approaches in structure-based drug discovery. *J. Synchrotron Radiat.* **15,** 227–230.

Hubbard, R. E., Chen, I., and Davis, B. (2007a). Informatics and modeling challenges in fragment-based drug discovery. *Curr. Opin. Drug Discov. Devel.* **10,** 289–297.

Hubbard, R. E., Davis, B., Chen, I., and Drysdale, M. J. (2007b). The SeeDs approach: Integrating fragments into drug discovery. *Curr. Top. Med. Chem.* **7,** 1568–1581.

Huth, J. R., Park, C., Petros, A. M., Kunzer, A. R., Wendt, M. D., Wang, X., Lynch, C. L., Mack, J. C., Swift, K. M., Judge, R. A., Chen, J., Richardson, P. L., *et al.* (2007). Discovery and design of novel HSP90 inhibitors using multiple fragment-based design strategies. *Chem. Biol. Drug Des.* **70,** 1–12.

Lu, H., and Tonge, P. J. (2010). Drug-target residence time: Critical information for lead optimization. *Curr. Opin. Chem. Biol.* **14,** 467–474.

Macarron, R. (2006). Critical review of the role of HTS in drug discovery. *Drug Discov. Today* **11,** 277–279.

Navratilova, I., and Hopkins, A. L. (2010). Fragment screening by surface plasmon resonance. *ACS Med. Chem. Lett.* **1,** 44–48.

Oltersdorf, T., Elmore, S. W., Shoemaker, A. R., Armstrong, R. C., Augeri, D. J., Belli, B. A., Bruncko, M., Deckwerth, T. L., Dinges, J., Hajduk, P. J., Joseph, M. K., Kitada, S., *et al.* (2005). An inhibitor of Bcl-2 family proteins induces regression of solid tumours. *Nature* **435,** 677–681.

Potter, A. J., Ray, S., Gueritz, L., Nunns, C. L., Bryant, C. J., Scrace, S. F., Matassova, N., Baker, L., Dokurno, P., Robinson, D. A., Surgenor, A. E., Davis, B., *et al.* (2010). Structure-guided design of alpha-amino acid-derived Pin1 inhibitors. *Bioorg. Med. Chem. Lett.* **20,** 586–590.

Schulz, M. N., and Hubbard, R. E. (2009). Recent progress in fragment-based lead discovery. *Curr. Opin. Pharmacol.* **9,** 615–621.

Shuker, S. B., Hajduk, P. J., Meadows, R. P., and Fesik, S. W. (1996). Discovering high-affinity ligands for proteins: SAR by NMR. *Science* **274,** 1531–1534.

Vedadi, M., Niesen, F. H., Allali-Hassani, A., Fedorov, O. Y., Finerty, P. J., Jr., Wasney, G. A., Yeung, R., Arrowsmith, C., Ball, L. J., Berglund, H., Hui, R., Marsden, B. D., *et al.* (2006). Chemical screening methods to identify ligands that promote protein stability, protein crystallization, and structure determination. *Proc. Natl. Acad. Sci. USA* **103,** 15835–15840.

Williamson, D. S., Borgognoni, J., Clay, A., Daniels, Z., Dokurno, P., Drysdale, M. J., Foloppe, N., Francis, G. L., Graham, C. J., Howes, R., Macias, A. T., Murray, J. B., *et al.* (2009). Novel adenosine-derived inhibitors of 70 kDa heat shock protein, discovered through structure-based design. *J. Med. Chem.* **52,** 1510–1513.

CHAPTER TWENTY-ONE

Fragment Screening of Infectious Disease Targets in a Structural Genomics Environment

Darren W. Begley,[*,§] Douglas R. Davies,[*,§] Robert C. Hartley,[*,§] Thomas E. Edwards,[*,§] Bart L. Staker,[*,§] Wesley C. Van Voorhis,[†,§] Peter J. Myler,[‡,§] *and* Lance J. Stewart[*,§]

Contents

[*] Emerald BioStructures, Bainbridge Island, Washington, USA
[†] Department of Medicine, University of Washington, Seattle, Washington, USA
[‡] Seattle Biomedical Research Institute, Seattle, Washington, USA
[§] Seattle Structural Genomics Center for Infectious Disease, Seattle, Washington, USA

Methods in Enzymology, Volume 493
ISSN 0076-6879, DOI: 10.1016/B978-0-12-381274-2.00021-2

Abstract

Structural genomics efforts have traditionally focused on generating single protein structures of unique and diverse targets. However, a lone structure for a given target is often insufficient to firmly assign function or to drive drug discovery. As part of the Seattle Structural Genomics Center for Infectious Disease (SSGCID), we seek to expand the focus of structural genomics by elucidating ensembles of structures that examine small molecule–protein interactions for selected infectious disease targets. In this chapter, we discuss two applications for small molecule libraries in structural genomics: unbiased fragment screening, to provide inspiration for lead development, and targeted, knowledge-based screening, to confirm or correct the functional annotation of a given gene product. This shift in emphasis results in a structural genomics effort that is more engaged with the infectious disease research community, and one that produces structures of greater utility to researchers interested in both protein function and inhibitor development. We also describe specific methods for conducting high-throughput fragment screening in a structural genomics context by X-ray crystallography.

1. Introduction

1.1. Structural genomics of infectious disease proteins

Since its inception in 2000, the US-based Protein Structure Initiative has rapidly advanced high-throughput techniques for cloning, expression, purification, crystallization, and structure determination (Fox *et al.*, 2008). This pipeline approach has led to technological advances that dramatically reduce the per-target time and expense required to generate high-resolution protein structures. A decade later, the Protein Structure Initiative and other structural genomic efforts have generated nearly 5000 high-resolution structures, and annually deposit the most unique structures in the Protein Data Bank (PDB; Dessailly *et al.*, 2009; Nair *et al.*, 2009). Despite this impressive output, the Protein Structure Initiative has fallen short of its original aim to make "3D atomic-level structures of most proteins easily obtainable from knowledge of their corresponding DNA sequences" (Burley *et al.*, 2008), leading external academic reviewers to question its overall goals and scientific value (Moore, 2007; Petsko, 2007; Service, 2008). One criticism often levied against structural genomics is a lack of engagement with the greater research community. High-resolution structure data for a well-characterized target can provide invaluable details on binding surfaces, allosteric, and other biochemical behavior. However, biological function is rarely ascertained from structure alone, and protein structures whose function is uncharacterized can be seen as possessing limited value to biological research. Since the goal of the Protein Structure Initiative is to determine as many unique structures as possible, targets are often pushed through the pipeline long before functional characterization is validated, with putative function assigned based on sequence similarity. The reality is that target-specific functional assays are difficult and time-consuming endeavors which cannot keep pace with current structural genomics efforts (Gerlt, 2007).

The mission of the Seattle Structural Genomics Center for Infectious Disease (SSGCID) is to determine 500 protein structures from NIAID Category A, B, and C organisms (Myler *et al.*, 2009). With this project, we seek to address some of the above-mentioned concerns regarding structural genomics by selecting potential drug targets that would directly benefit the infectious disease research community. To drive this effort, SSGCID target selection is guided by essential, known, or putative function; this process provides a list of gene products which in some way could be the focus of drug design efforts. In pursuit of biologically relevant structures, we solicit requests directly from the scientific community. Most of our collaborators conduct research in academic laboratories, and many share our interest in generating novel leads for underserved infectious diseases. Today, there is an ever greater need for such research, as drug resistance continues to occur in pathogenic bacteria and other organisms, while the pipeline of future antimicrobial drugs dries up (Boucher *et al.*, 2009). Our collaborative work includes a dedicated effort to solve small molecule–protein complexes for which there are proven or experimental inhibitors against homologous targets in other species (Myler *et al.*, 2009; Van Voorhis *et al.*, 2009). The structural information gained from protein–ligand complexes can greatly accelerate functional assignment of targets for which sequence similarity is insufficient for annotation, as well as largely uncharacterized gene products with little available *in vitro* activity data (Hermann *et al.*, 2007; Rakus *et al.*, 2009; Song *et al.*, 2007). Furthermore, structures of infectious disease proteins in complex with inhibitors can provide critical information in reoptimizing current drugs for maximum host–parasite selectivity when a structure of the human homolog is available. Thus, in generating one or more ligand-bound structures for an infectious disease protein, we aim to fulfill our goal of providing a "blueprint for structure-guided drug design" (Myler *et al.*, 2009).

1.2. Ligand-bound structure ensembles

In collaborating with biologists and other external researchers on SSGCID targets, we have found that *apo* structures, although valuable in and of themselves, are often not sufficient to facilitate drug discovery efforts. At a minimum, infectious disease researchers would like information on the native substrate, or a close chemical analog, bound to the target in question. Such information offers scientific support confirming, or correcting, the annotated function of a given target and provides validated ligands that can serve as chemical probes for biological assays. Obtaining one or more inhibitor-bound structures can also guide the initial rounds of structure-based drug design, especially for enzymes and proteins that act upon small molecule metabolites. Like other structural genomics initiatives, the SSGCID is ideally suited to generate *ensembles* of structures for a given target, employing the same structural genomics pipeline put in place to determine novel, *apo* protein structures.

When a new protein structure has been solved, the majority of the work—from sequence validation, expression, purification, and crystallization to structure refinement—has been completed. By comparison, obtaining a series of ligand-bound structures requires little additional effort. Moreover, there is often a significant amount of protein remaining from purification that can be used for ligand binding studies by X-ray crystallography and other biophysical methods. With this asset of protein in hand, we believe that the additional work required to obtain one or more ligand-bound structures is well justified. For example, if there are known inhibitors of the human homolog, structures of those small molecules complexed with one or more microbial targets would be highly desirable to the greater scientific community. Complexes with known inhibitors of homologous targets further support the functional characterization of the specific protein of interest, and can provide a basis for the first stages of drug design and development. This information can be of particular value in identifying point mutations or other structural mechanisms implicated in the development of observed drug resistance in pathogenic organisms (Yuvaniyama *et al.*, 2003).

1.3. Fragment screening and X-ray crystallography

One strategy for generating an ensemble of structures for a single protein is fragment-based screening (Begley *et al.*, manuscript in preparation; Brough *et al.*, 2009; Card *et al.*, 2005; Davies *et al.*, 2009; Hartshorn *et al.*, 2005). Fragment screening by X-ray crystallography with an all-purpose, unfocused compound library allows both the testing of proven and the discovery of new chemical moieties that bind to a protein. The size and reduced complexity of typical fragment molecules ($\leq$300 Da) allow for a diverse sampling of chemical space with a compound library small enough for practical applications in crystallographic screening (Duarte *et al.*, 2007; Hann *et al.*, 2001; Makara, 2007; Rees *et al.*, 2004). Due to their relatively small size, the overall affinity of fragments tends to be weaker, but their per-atom binding efficiency can be much higher than that of larger molecules (Abad-Zapatero and Metz, 2005; Congreve *et al.*, 2003; Hajduk, 2006a; Hopkins *et al.*, 2004). For most small molecules consisting of 15 or more non-hydrogen atoms, contributions to binding free energy increase linearly at best with atomic mass (Hajduk *et al.*, 2005; Kuntz *et al.*, 1999). This energetic correlation suggests that the binding energy of an initial, fragment-sized scaffold greatly determines the potential outcome of further optimization efforts, and that highly efficient fragment binders are therefore valuable starting points in drug design. Furthermore, fragments tend to bind protein "hot spots," localized areas which contribute the majority of binding energy to complex formation, even for large protein–protein interaction surfaces (DeLano, 2002; Hajduk *et al.*, 2005; Wells and McClendon, 2007). Probing such surfaces is a much more challenging prospect with larger compounds, such as those typically employed in high-throughput screening (HTS). The additional bulk and steric effects for such compounds often prevent hit detection by HTS, including those that embed or possess valuable chemical motifs within their scaffolds which can bind as individual fragments (Payne *et al.*, 2007; Rees *et al.*, 2004; Wells and McClendon, 2007). Thus, fragment-based screening can identify the most important regions of larger binding pockets which might be missed by screening larger compounds, and can be used to assess hits in the critical early stages of drug development (Bembenek *et al.*, 2009; Hajduk, 2006a; Reitz *et al.*, 2009).

Protein-based nuclear magnetic resonance (NMR) spectroscopy and X-ray crystallography are two biophysical methods which can be used to conduct screening of macromolecular targets as well as structurally characterize the bound conformations of fragment hits (Nienaber *et al.*, 2000; Shuker *et al.*, 1996). Early indications of success by these methods have driven considerable investment in fragment screening technologies, which include ligand-based NMR (Assadi-Porter *et al.*, 2008; Lepre *et al.*, 2004; Mayer and James, 2004; Pellecchia, 2005), mass spectrometry (Swayze *et al.*, 2002), surface plasmon resonance (Navratilova and Hopkins, 2010; Neumann *et al.*, 2007), differential phase interferometry (Bornhop *et al.*, 2007), fluorescence correlation (Hesterkamp *et al.*, 2007), and calorimetry (Ladbury *et al.*, 2010). Many of these solution-based methods are well suited for the rapid screening and detection of weak-binding interactions between fragments and a broad range of macromolecular targets. However, some ligand-observe methods are prone to generating false positives, while most require additional measures to confirm binding site specificity, and none provide detailed structural information on the complex. Fragment screening by X-ray crystallography has the advantage that hits are simultaneously identified and structurally characterized in their target-bound state. Although not entirely dependent upon structural information (Erlanson *et al.*, 2004), the majority of successful fragment screening campaigns reported in the literature have utilized experimentally determined binding modes to advance weak but efficient fragments into potent inhibitors (Brough *et al.*, 2009; Card *et al.*, 2005; Davies *et al.*, 2009; Hartshorn *et al.*, 2005; Murray and Blundell, 2010; Sanders *et al.*, 2004; Shuker *et al.*, 1996). Recent successes using fragment screening and structure-based methods against infectious diseases have also yielded novel leads for the inhibition of fatty acid (Mochalkin *et al.*, 2009), pantothenate (Ciulli *et al.*, 2008), and folate (Mpamhanga *et al.*, 2009) biosynthesis in pathogenic organisms. It is clear that complex structures obtained by fragment screening can propel novel lead synthesis and development and, therefore, can have a much greater impact on the work of external researchers investigating new ways to combat infectious diseases than *apo* structures alone.

2. Methods

2.1. Fragment library selection, pooling, and storage

Much has previously been published on fragment library selection and the various determinants for suitable candidate molecules, all of which depend on the biophysical technique(s) that are used for screening (Bemis and Murcko, 1996; Duarte *et al.*, 2007; Ertl *et al.*, 2006; Fejzo *et al.*, 1999; Hajduk, 2006b; Hubbard *et al.*, 2007; Muller, 2003; Siegel and Vieth, 2007; Smith and Morowitz, 2004; Zartler and Shapiro, 2008). We employ a diverse, metabolome-inspired fragment library that consists of a structurally diverse array of over 1500 water-soluble metabolites, their bioisosteres, and close chemical analogs (Davies *et al.*, 2009). In striving to establish a compact fragment set, we sought the greatest diversity within this biological context among fragment-sized molecules which already exist in nature, such as substrates, natural products, allosteric regulators, and other associates of metabolic pathways. We took into account ligand efficiency (Abad-Zapatero and Metz, 2005; Andrews *et al.*, 1984; Mochalkin *et al.*, 2009), privileged scaffolds (Bemis and Murcko, 1996; Duarte *et al.*, 2007; Ertl *et al.*, 2006; Muller, 2003), shape diversity (Binkowski and Joachimiak, 2008; Hajduk, 2006b; Lipkus *et al.*, 2008), physicochemical properties, and other parameters in order to maximize coverage of chemical space. Among the fragments returned from these searches, those which would best represent their class while also facilitating various methods of chemical elaboration were prioritized. The result of this cross-parameter search guided by the natural metabolome was the Fragments of Life™ (FOL) small molecule screening collection, our in-house fragment compound library (Davies *et al.*, 2009). In our view, application of this all-purpose library maximizes the likelihood of finding native substrates or substructures bound, while simultaneously probing unpredicted chemistry for binding interactions with our chosen targets.

Compounds selected for inclusion in our FOL library were purchased from various vendors and tested for ≥ 50 m*M* solubility in methanol. This initial level of solubility facilitates the use of ligands at millimolar concentrations needed to perform ligand-based NMR studies and crystal soaks. The use of methanol also avoids precipitation issues associated with long-term storage of test compounds more commonly prepared in dimethyl sulfoxide solution (Waybright *et al.*, 2009). We require compound purity to be $\geq 95\%$ as confirmed by NMR and LCMS, prior to dissolution and preparation for screening. Compounds were rapidly dissolved in methanol and stored at −20 °C to prevent concentration-altering evaporation. We then prepared mixtures of FOL compounds, to test a greater number of fragments for each crystal soaked. Several techniques, including NMR spectroscopy and X-ray crystallography, can utilize mixtures of diverse fragments for screening, which can later be deconvoluted from a single experiment with reasonable confidence. However, we follow up every mixture hit with individual ligand experiments, to confirm fragment identity and binding.

To prepare our pools, we used computational distance metrics to randomly sort our fragments into groups of eight diverse molecules (Davies *et al.*, 2009). The fragments in each pool were then visually inspected to ensure the greatest likelihood of accurate identification of a single molecule from X-ray diffraction data taken on a mixture-soaked crystal. Once pooled, solutions were visually inspected; any signs of precipitation, aggregation, or destabilization upon mixing meant fragments would be recombined into different mixtures to ensure maximum solubility and stable long-term storage. The majority of our fragment pools consist of eight ligands per mixture, dissolved in methanol at 50 m*M* total ligand. This results in a nominal concentration of 6.25 m*M* for each fragment in a pool during the soak. For individual fragment soaks, such as those conducted to confirm identities of binding molecules, we typically use stocks at 25 m*M*.

We intentionally selected an all-purpose, volatile solvent for our fragment stock solutions to provide optimal flexibility when setting up crystallographic experiments. We allow methanol-based fragment stocks to evaporate prior to preparing soaking drops, leaving virtually no solvent residue behind which might dissolve or interfere with the protein crystal. This allows the rapid preparation of crystallization plates with dry films of fragment mixtures that can be stored indefinitely. Methanol is also miscible with many crystallization buffers, allowing the direct addition of fragment stock solutions when necessary. Using an evaporative solvent further facilitates concentration scaling within the confines of the crystallization plate, allowing different fragment-protein ratios for target-specific experiments. However, methanol creates certain liquid handling challenges because of evaporation, and care must be taken to keep solutions in sealed vials as much as possible. In preparing our FOL library, we conducted informal evaporation studies and showed that a borosilicate vial with a Teflon cap containing 2.5 mL of methanol loses less than 14 mg of methanol per day when stored at room temperature. Storage of the same solution in the same vial at −20 °C reduces loss of solvent by evaporation to negligible levels (data not shown). Hence, we have found borosilicate glass vials with plastic screw caps and Teflon inserts to be an acceptable medium for long-term storage of methanol stocks at −20 °C.

2.2. Target selection

Each target nominated for fragment studies within the SSGCID must first be assessed, to ensure prioritization of the most relevant and potentially impactful targets. Targets are nominated by the external scientific community and approved by our NIAID program officer or proposed internally, then reviewed and approved by an external panel of scientific advisors. Approved targets are then evaluated for their feasibility in crystallographic fragment screening studies. In Chapter 4 of this volume, we discuss and analyze characteristics of protein targets which are amenable to fragment screening. To date, the majority of targets selected for fragment screening studies have been evaluated using various experimental criteria to predict the likelihood of success in obtaining ligand-bound structures, as described below:

1. To begin, a protein must first be successfully crystallized and a structure determined, whether in *apo* form or with a targeted ligand bound.
2. Protein crystals were typically selected which diffract to ≤ 1.8 Å in order to reliably obtain ≤ 2.0 Å datasets using our in-house X-ray source. Resolution in this range permits the correct orientation of a bound ligand to be determined from the experimentally observed electron density.
3. Crystals with small unit cells and high symmetry space groups are preferable to minimize data analysis time, but we have established no fixed limit on either criterion to be used as a requirement for fragment screening.
4. If any metals, cofactors, prosthetic groups, or other components are necessary to make relevant binding site(s) available on the protein, they are added at millimolar concentration to the crystallization drop.
5. Protein structures with well-ordered electron density for polyatomic anions or buffer components observed in the predicted or known active site(s) are generally undesirable, as they will compete with fragments for the binding pocket(s). In our experience, citrates, phosphates, sulfates, acetates, and other such charged groups which appear highly ordered in a crystal can be difficult to displace with weak-binding fragments. Crystallization conditions that require polyatomic anions are therefore to be avoided, if possible.
6. Strongly acidic or basic conditions are generally undesirable, as these can have adverse effects on the stability of the target. However, nonneutral pH is sometimes implicated in the physiological relevance of a target, such as with pH sensor proteins. The pH of the crystal soaking drop will affect the protonation state of fragments as well as protein side chains, and should therefore be taken into account when conducting specific pH-range screens.

2.3. Crystal growth plates

Once a candidate protein has been selected, 96-well sitting-drop "farm" trays are plated with the same condition in each well and reservoir. Crystallization conditions are developed from the random sparse matrix screening effort employed to obtain the original crystal and structure, prior to fragment studies. The precipitant chosen for fragment-soaking crystals is usually the same, or based on consensus between several successful conditions which led to crystal growth. Typically, drops consist of 0.4 μL protein solution mixed with 0.4 μL precipitant in the sitting-drop well, with 80–100 μL of precipitant placed in the reservoir. Our standard protein solution contains approximately 20 mg/mL protein with 500 m*M* NaCl, 20 m*M* HEPES, 5% glycerol, and 2 m*M* DTT at a pH = 7.0. Similar buffer conditions have been shown to allow flash-freezing in PCR-sized tubes with minimal effects on protein precipitation (Deng *et al.*, 2004). Wells are sealed with transparent tape, and ideally generate useable, uniform crystals within 24 h to 7 days of sitting. If crystals take longer than a week to grow, or appear nonuniform between wells, further optimization and consideration of additives is done, depending on the target.

2.4. Fragment soak plates

To prepare soak plates, 1.0 μL drops of pooled fragments dissolved in methanol are added to each of several sitting-drop tray wells and allowed to evaporate overnight. A 1.0 μL drop of soak solution is then added to one well per pool to resolubilize the fragments, prior to adding protein crystals. The other wells containing fragment dry films are reserved for cryoprotecting and repeat experiments. Although methanol-based fragment drops can also be plated serially to create soaking mixtures, consistent plating is difficult for volumes < 0.25 μL with highly evaporative solvents, while plating > 1.0 μL runs the risk of fragment material deposition above the level of the soak drop. We typically use Emerald BioSystems Compact Jr. crystallization plates (EBS-XJR) for long-term storage of dry film and fragment pool solutions. Important factors for storage are chemical resistance to degradation, shape of wells for easy harvesting, proper sealing to prevent contamination, and accommodation of drop volumes up to 2.5 μL without wicking to the sealant.

2.5. Crystal soaking

Once the *apo* starting crystals of the target protein are grown, they are transferred to sitting-drop wells for fragment soaking. Typically, 1.0 μL of the crystallization condition is used as the soak solution, and is first added to resolubilize dry film fragment pools. Additional target-specific components can also be added to the soak, such as catalytically important metal ions, or small molecules which facilitate fragment binding to specific sites through cocomplexation. Then two or more protein crystals are transferred to each individual well using cryo loops, and the wells sealed using Crystal Clear Sealing tape (Hampton Research, Aliso Viejo, CA). Prior to sealing, 20 μL of the crystallization solution (without additives) is added to the reservoir to maintain vapor pressure and prevent the drop from drying out completely.

Ideal soaking solutions are those which contain cryoprotectants, and which solubilize the fragments without destroying the protein crystal. If one component of the crystal precipitant has cryoprotecting qualities, we often increase that component to $\geq$25% (v/v) so fragment-soaked crystals can be directly frozen in liquid nitrogen from the soak drop. To maximize fragment solubility, 10% (v/v) or more dimethyl sulfoxide is sometimes added, for crystals previously proven to tolerate this cosolvent concentration without dissolving. Crystals are typically soaked for 24 hrs to 1 week or longer, depending on the target, to ensure maximum fragment binding. We have found that some targets with highly ordered buffer components or coordinated water molecules in the active site require longer soak periods, in order for fragments to effectively displace them within the crystal. We usually perform trial soaks with 24–48 fragment pools, as well as any known substrates or binding ligands found through literature searches, before screening the entire FOL library.

2.6. Cryoprotection

After fragment soaking, one or more protein crystals per pool are harvested by looping and frozen in liquid nitrogen. Unless the soak condition was designed as a cryoprotectant, crystals are dipped into a cryogenic protecting drop immediately prior to freezing. It is important that all fragment molecules be included in the cryo drop at the same concentration as the soak drop to prevent concentration-dependent diffusion and loss of fragments from the crystal. Typical cryogenic conditions are 25 % (v/v) ethylene glycol, glycerol, or low molecular weight polyethylene glycol mixed with the soaking solution. This cryo drop is used to resolubilize fragment pools previously added as dry film to separate wells or mixed with fragments in solution, before dipping and freezing crystals.

2.7. Data collection, data processing, and fragment-soaking analysis

In most cases, collection parameters for a fragment screening target are automatically selected using d*TREK software with our Rigaku X-ray system (Pflugrath, 1999). These parameters are based on the symmetry order determined from previous experiments and a few diffraction images of the first crystal tested. Diffraction data are reduced with XDS/XSCALE and customized in-house processing scripts (Kabsch, 1993). Initial structure solutions are achieved by molecular replacement methods, followed by rigid body refinement using a previous structure of the protein stripped of waters and other ligands, but retaining any catalytic metals or prosthetic groups. Crystals must diffract to 2.5 Å or better resolution for a robust analysis of fragment screening; if crystals dissolve, or diffract to weaker than 2.5 Å resolution after initial testing with fragment soaks, we reoptimize for new conditions before pursuing a full library screen. Unmodeled blobs with $(F_o - F_c) > 2.5\sigma$ and $(2F_o - F_c) > 1.0\sigma$ are first verified against the model protein structure using Coot, the Crystallographic Object-Oriented Toolkit (Emsley and Cowtan, 2004). Blobs are ignored which are likely to be unmodeled protein density, or identified as small molecules associated with crystallization conditions, cryogens, or the protein buffer. The remaining regions of unmodeled density are then examined for shape complimentarity with individual fragments in a given mixture. When a putative fragment hit is isolated, preliminary deconvolution is done to identify the ligand by modeling in the suspected fragment and refining the structure. The fragment identity is then confirmed by conducting additional soaks and data collection for protein crystals with individual fragments.

We utilize the Collaborative Computational Project Number 4 (CCP4) suite of programs, along with TLS motion determination, to refine crystallographic datasets into accurate structural models (CCP4, 1994; Painter and Merritt, 2006). Every protein residue, ligand, and water molecule in every SSGCID structure model that is deposited in the PDB is visually examined and compared to its local electron density. Every SSGCID structure solved by a crystallographer is then internally reviewed by another crystallographer for consistency between electron density and backbone, side chain, and solvent atoms in the model, prior to deposition. This procedure routinely catches small errors, such as side chain rotamer flips, poorly modeled waters, and novel ligand stereochemical or tautomeric centers. All structures released by the SSGCID must meet a series of criteria prior to deposition with the PDB. Datasets must be $\geq$80% complete in the highest resolution shell and $\geq$90% over the entire dataset. Resolution must be 2.5 Å or better, with a signal-to-noise ratio (I/Ω) $>$2.0 across the top 20 resolution shells. All models must have $>$95% of residues in favored Ramachandran regions as defined by MolProbity (Chen *et al.*, 2010; Davis *et al.*, 2007), with a structural rationale provided for any outliers.

3. Case Studies

3.1. Fragment screening for lead generation in structural genomics

Much interest has been garnered in the past decade on the methylerythritol isoprenoid (MEP) biosynthetic pathway as a point of intervention for antimicrobials (Akerley *et al.*, 2002; Brown and Parish, 2008; Campbell and Brown, 2002; Campos *et al.*, 2001; Freiberg *et al.*, 2001). This pathway is present and appears to be essential in *Plasmodium falciparum*, *Mycobacterium tuberculosis*, *Pseudomonas aeruginosa*, *Burkholderia pseudomallei*, *Escherichia coli*, and other infectious disease organisms (Brown and Parish, 2008; Freiberg *et al.*, 2001; Hunter, 2007; Jacobs *et al.*, 2003; Jomaa *et al.*, 1999). MEP enzymes are absent in mammalian species, all of which utilize the well-characterized mevalonate-dependent (MAD) pathway to synthesize isoprenoid precursors (Akerley *et al.*, 2002; Brown and Parish, 2008; Campbell and Brown, 2002; Campos *et al.*, 2001; Freiberg *et al.*, 2001). Early studies on fosmidomycin indicate clinical utility and minimal side effects in its use for treating *P. falciparum* infections by inhibiting IspC, one enzyme in the MEP pathway (Borrmann *et al.*, 2004a,b; Jomaa *et al.*, 1999; Wiesner *et al.*, 2002, 2003). This observation suggests that enzymes within the MEP pathway, such as IspC, can be targeted to treat infectious agents with minimal host toxicity. Recrudescence after 2 weeks of treatment revealed fosmidomycin alone allows rapid development of drug resistance, but if given in combination, fosmidomycin appears to retain its clinical activity (Missinou *et al.*, 2002; Wiesner *et al.*, 2003). Therefore, novel chemotherapies which target multiple enzymes within the MEP pathway should minimize the chance for resistance mutations to develop and survive in variant strains of pathogenic species for which it is essential.

Another enzyme in the MEP pathway is 2C-methyl-D-erythritol-2,4-cyclo-diphosphate synthase (IspF). Sequence analysis shows over 30% identity for IspF proteins across prokaryotic and eukaryotic species, and structural studies reveal high conservation for residues essential in substrate-binding, catalysis, and other aspects of IspF enzymatic function (Buetow *et al.*, 2007; Eisenreich *et al.*, 2004; Eoh *et al.*, 2007; Hunter, 2007; Lange *et al.*, 2000; Ni *et al.*, 2004a,b; Rohdich *et al.*, 2001; Steinbacher *et al.*, 2002; Ulaganathan *et al.*, 2007). Preliminary analysis of IspF from *B. pseudomallei* (BpIspF) showed the purified enzyme to be active *in vitro*, stable in low salt buffers, and capable of forming relatively large, solid crystals that are amenable to soaking in fragment-friendly drop conditions. Screening BpIspF with a subset of our library also revealed that approximately 15% of fragments to bind in solution by ligand-observe NMR spectroscopy. These results and success with preliminary crystal trials prompted a full screen of BpIspF against our complete FOL library (Begley *et al.*, manuscript in preparation).

Screening BpIspF with our FOL library generated an ensemble of fragment-bound structures (Fig. 21.1). One subset of BpIspF structures generated from fragment screening contain cytosine-based fragments, including cytosine, cytidine, CMP, CDP, and CTP. These nucleotide-derived fragments all bind to the same active site subpocket and mimic the cytidyl portion of the native substrate, 4-diphospho-cytidyl-2-C-methyl-D-erythritol 2-phosphate (Fig. 21.1). Results for this subset of fragments with BpIspF were somewhat anticipated, and the binding

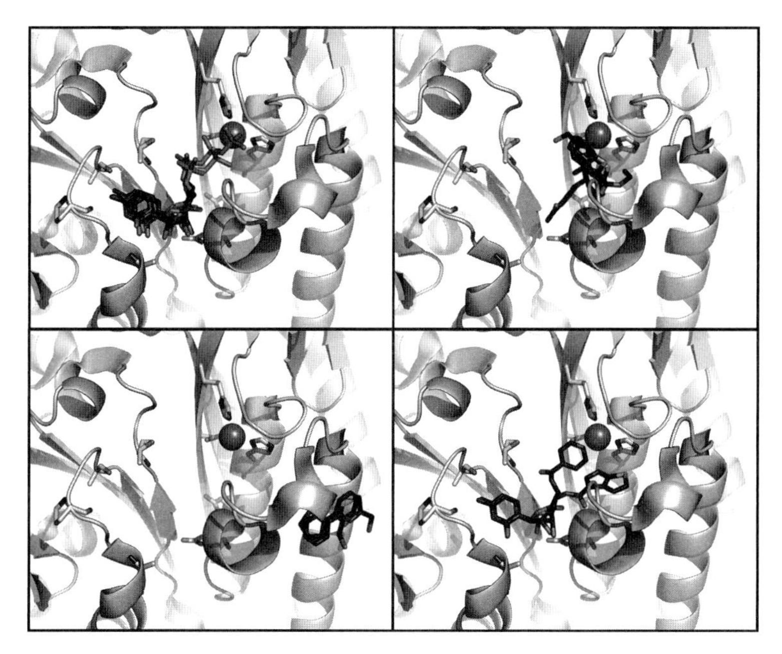

Figure 21.1 Small molecules from ligand-bound structures superimposed onto the *apo* structure of IspF from *B. pseudomallei*, with key side chains (lines) and the catalytic zinc ion (sphere) depicted. Fragment screening by NMR spectroscopy and X-ray crystallography led to three distinct subsets of ligands which bind in the cytosine subpocket (above left), coordinate to catalyic zinc (above right), and bind a hydrophobic cleft external to the active site (bottom left). Two lead compounds were designed by linking cytidine to zinc site-binding fragments (bottom right). While their cytosine moieties bind in the same manner, only one lead compound recapitulates the binding mode of its fragment precursor by coordinating to zinc through a heteroaromatic nitrogen. Figure generated using PyMol (DeLano, 2008), PDB IDs 3F0E, 3IKE, 3IEQ, 3F0F, 3IEW, 3IKF, 3JVH, 3K14, 3KE1, 3P0Z, 3P10 and 3Q8H).

modes of these cytosine-based molecules recapitulate those observed in IspF from other organisms (Kemp *et al.*, 2005). Fragment screening generated another subset of structures which contain heteroaromatic fragments bound to a zinc ion in the BpIspF active site (Fig. 21.1). Each of these complex structures reveals an aromatic nitrogen coordinated to zinc, displacing a crystallographically conserved water molecule that normally occupies the tetrahedral coordination site. These structures are unlike most ligand-bound complexes of IspF, as they do not require any substrate-derived moieties for binding. Finally, a third subset of structures resulted from a focused set of crystal soaks with NMR-derived fragment hits in the presence of cytidine (Fig. 21.1). These fragments bind in a small hydrophobic cleft outside the active site formed by a protein loop just above the catalytic zinc ion. Soaking trials for these fragments alone and in combination with cytosine did not result in ligand-bound structures, suggesting that the ribose moiety of cytidine is required for high occupancy binding (Begley *et al.*, manuscript in preparation). Thus for one target, fragment screening by two techniques with a single library has led to the deposition of at least a dozen complexes, which currently comprise nearly a third of all IspF structures from any species currently available in the PDB.

The ensemble of fragment-bound BpIspF structures generated by SSGCID provides several avenues for structure-based, fragment-assisted inhibitor design. Novel fragments bound in the zinc site serve as starting points for elaboration into higher affinity metal-chelating compounds. Fragment pairs bound to adjacent subpockets of the active site provide accurate intermolecular distances and orientations, information which can be used in merging or linking strategies to enlarge fragments into initial leads. Preliminary attempts to design and synthesize such leads led to a small series of cytidine derivatives (Begley *et al.*, manuscript in preparation). These pseudo-bidentate ligands were designed from adjacent binding fragment pairs of cytidine and several heteroaromatic zinc-binding fragment motifs. Our first cytidine fusion molecule (PDB ID 3KE1) shows the pyridine nitrogen coordinated to zinc in the same manner as was observed in its fragment precursor (Fig. 21.1). Our second cytidine derivative (PDB ID 3Q8H) recapitulates a different zinc-coordinated fragment, but its mode of binding does not reflect that of the fragment which inspired it. Instead of coordinating directly to zinc, this bicyclo-aryl octane moiety binds into a hydrophobic groove above the metal, on the opposite side of the external fragment-binding site loop (Fig. 21.1). This hydrophobic pocket helps position the erythritol portion of the substrate for catalysis, and although computationally predicted to bind fragments, none were observed through experimental fragment screening. Utilizing fragment hits to develop and test larger molecules in this manner expands our available knowledge of the ligand-binding active site landscape and shows that larger ligands do not always recapitulate the binding modes of their fragment precursors. Hence, the acquisition and interpretation of these and future datasets on BpIspF complexes provides the necessary details required for each iterative cycle of our design efforts, in pursuit of potent, novel inhibitors for this infectious disease target.

Investigations into the mechanism of IspF suggest that the loop above the catalytic zinc ion properly positions the 2-phosphate of the substrate for attack on its own β-phosphate, prior to cyclization and release of product (Hunter, 2007; Steinbacher *et al.*, 2002). This loop is well-ordered in the *apo* structure of IspF (PDB ID 3F0E) but appears too disordered to accurately model when complexed with a bidentate ligand (PDB ID 3KE1), suggesting a degree of flexibility which may be required for catalytic activity. External site-binding fragments appear to stabilize this loop, binding to it outside the active site, and result in complexes which recapitulate the *apo* structure (Begley *et al.*, manuscript in preparation). Currently, any real or potential biological effects of binding small molecules to this external site are unknown. Investigations with such external site fragment hits should always be undertaken cautiously, especially when there are no proven competitors for the site or any known or hypothesized allosteric mechanisms affecting enzymatic function. Nevertheless, this example of biophysical screening and structure ensemble generation reveal the potential to identify novel binding sites through fragment screening. Such complexes may also be of particular value in a structural genomics environment where annotated function is unconfirmed and supply biological research with additional chemical probes for functional characterization.

3.2. Ligand screening for functional annotation in structural genomics

Several genes from the SSGCID pipeline under investigation are annotated as arginine/ornithine transport system ATPases from *Mycobacterium* genomes. Our initial efforts were focused on the ATPases from *M. tuberculosis*, the organism which causes tuberculosis and which is estimated to infect one-third of the world's population (Jasmer *et al.*, 2002). Crystallization trials of this target in the *apo* form yielded needle clusters that failed to diffract better than 7 Å resolution, despite several attempts at gradient screening to optimize conditions. Due to the difficulty in obtaining well-diffracting *apo* crystals, this particular target was not a candidate for fragment screening. Our continued efforts to obtain any structures of this target then took a two-pronged approach: knowledge-based ligand additives to enhance crystallization conditions, and reengineering of the construct using genetic data from closely related nonpathogenic *M. smegmatis* and *M. thermoresistible* species (Edwards *et al.*, manuscript in preparation).

Although the *Mycobacteria* genes which code for this target family are annotated as ATPases, a BLAST search (Altschul *et al.*, 1990) against existing PDB entries yielded four hits, all of which belong to the RAS-like GTPase superfamily of proteins. Of these four entries, two structures were in the *apo* form, while the other two contained molecules of GDP bound to the protein. To maximize our chances for success, we attempted cocrystallization trials with the *M. tuberculosis* transporter protein in the presence of ATP, ADP, GTP, and GDP. Crystals grown in the presence of ATP or ADP yielded the same weakly diffracting needle-shaped clusters that were observed for the *apo* protein samples. In contrast, crystals grown in the presence of either GTP or GDP yielded block-shaped crystals. From a crystal grown in the presence of GDP, we determined a 2.45 Å resolution structure (PDB ID 3MD0). In this structure, the guanine moiety of GDP forms an N1/N3 interaction with the side chain carboxylate of Asp207, and stacks between the side chains of Lys205 and Val145. It is highly unlikely that an adenosine could form similar N6/N1 interactions with Asp207

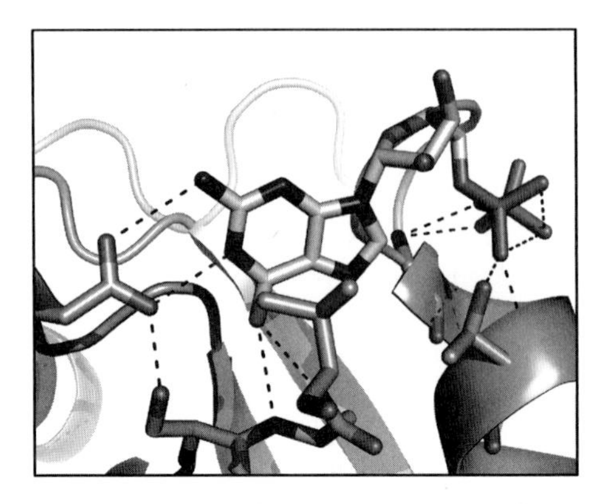

Figure 21.2 GDP-bound structure of a protein monomer from *M. tuberculosis* annotated as an arginine/ornithine transporter with ATPase activity. Diffraction-quality crystals were not obtainable in the presence of ADP or ATP, while hydrolysis of GTP to GDP was observed crystallographically. The N1/N3 interaction between GDP and the Asp207 side chain confirm a highly specific guanine-binding pocket. Figure generated using PyMol (DeLano, 2008) from PDB ID 3MD0 (Edwards *et al.*, manuscript in preparation).

without significantly shifting the nucleobase out of the binding pocket (Fig. 21.2). Using optimization grid screening to further enhance crystallization conditions, we then obtained a 1.9 Å resolution structure from a crystal grown in the presence of GTP (PDB ID 3P32). Upon refining the model against the data, it became clear that a single molecule of GDP, not GTP, was bound in the G-domain active site. Apparently, the enzymatic activity of this target protein is sufficient to hydrolyze GTP into GDP during crystallization, with enough efficiency to effectively remove the phosphate by-product from the binding pocket. This finding strongly suggests that the original ATPase annotation for this transporter is incorrect and demonstrates the power of structural genomics to revise such information when searching for and prioritizing ligand-bound structures (Edwards *et al.*, manuscript in preparation).

In further attempts to characterize this target family, we conducted cocrystallization trials of the transporter protein with GDP, arginine, and ornithine. Despite obtaining several 1.8 Å resolution or better datasets, we did not observe electron density for either arginine (50 m*M* in the crystallization drop) or ornithine (5 m*M* in the crystallization drop) in the *M. tuberculosis* target. We then determined crystal structures of the homologous transporter systems from *M. smegmatis* (2.3 Å resolution, PDB ID 3NXS) and *M. thermoresistible* (data not shown). In each case, we observed GDP in the highly conserved active site of each protein, but no arginine or ornithine in the unit cell. The sum total of this effort has led us to conclude that these proteins from closely related *Mycobacterium* species are not likely arginine/ornithine transport system ATPases. Not only do these proteins fail to bind to ATP, ADP, arginine, or ornithine in crystallization trials, they require the presence of GDP or GTP to obtain crystals capable of high-resolution X-ray diffraction. Furthermore, the target from *M. tuberculosis* is capable of hydrolyzing GTP to GDP in the crystallization drop, demonstrating enzymatic activity. More detailed sequence analyses show these proteins to align well with metallochaperones from the G3E family of P-loop GTPases, such as MeaB from *Methylobacterium extorquens* and human MMAA, a protein associated with methylmalonic aciduria (Dobson *et al.*, 2002; Korotkova and Lidstrom, 2004; Korotkova *et al.*, 2002). Therefore, these proteins, which are currently annotated as arginine/ornithine transport system ATPases, are most likely metallochaperone GTPases similar to MeaB.

4. Conclusions

To have the biggest impact on biological research, we believe that structural genomics must continue to increase and improve the diversity of available ligand-bound structures. We believe that structural genomics of infectious diseases garners more significance and better engages the scientific community by producing *ensembles* of complex structures, rather than *apo* structures alone. Such structures have already generated interest within the scientific community, whose requests for ligand-bound structures and fragment screening at SSGCID continue to grow in number. Utilization of novel fragments complexed with infectious disease proteins may also provide a much-desired avenue for the development of novel antimicrobial drugs in the wake of recent failures to generate such leads by other means. Historically, the great challenge in antibacterials is the generation of new chemical scaffolds. The path to registration has less risk when compared with other disease areas, once an effective scaffold is recognized which targets a particular bacterium. A recent survey noted over 125 antibacterial HTS campaigns at 34 different pharmaceutical companies in the past 15 years yielded no promising drug candidates (Chan *et al.*, 2004; Payne *et al.*, 2004, 2007). HTS studies typically employ larger, more developed small molecules in the hopes of immediate identification of a highly potent inhibitor, and insufficient chemical diversity across such libraries is a likely reason for this failure. The HTS assay format requires low micromolar activity for detection, which means that inherently tight-binding leads must already be present within the library before screening has begun (Keseru and Makara, 2006; Murray and Blundell, 2010; Payne *et al.*, 2007; Rees *et al.*, 2004). Conversely, fragment molecules are smaller and less complex than those used in HTS technologies, and therefore are inherently less biased toward any specific protein family. Thus, academic and industrial researchers alike can benefit from utilization of small molecular fragments in structural investigations of rare and structurally diverse drug targets, such as those present in pathogenic organisms.

Not every structural genomics protein is suitable for ligand-binding studies, as the biological functions of many proteins do not involve small molecules. However, any structural genomics pipeline equipped to rapidly generate protein structures by X-ray crystallography has the ability to generate ensembles of structures for targets which do bind small molecules. One reason for this is that the majority of backbone and side chain residues across such an ensemble (for most targets) will be identical or highly similar; in most cases, only a small portion of target atoms are altered to accommodate the binding entity. But it is precisely this structural data at the interface of protein–ligand binding interactions which can be of greatest value to researchers investigating their biological function. Adding the known (or putative) native substrate to a target for structural purposes can provide evidence of functional activity and help correctly annotate function when homology or literature assignment is incomplete. For chemical biologists, fragments which have been biophysically characterized can provide soluble starting points for experimental investigations. Perhaps the greatest impact fragment-bound structures can make is on the development of novel chemotherapeutics, whether to treat infectious diseases or to guide other structure-based drug design endeavors. We believe fragment screening furnishes a valuable resource to the infectious disease community by providing blueprints for structure-based drug design which would not be discovered by any other methodology.

ACKNOWLEDGMENTS

We would like to thank Alex S. Kiselyov and John M. McCall for helpful comments in reading this book chapter. We further acknowledge and greatly appreciate the efforts of the entire team at SSGCID, without whom our accomplishments would not be possible. Support for this research was funded by NIAID under Federal Contract No. HHSN272200700057C.

REFERENCES

Abad-Zapatero, C., and Metz, J. T. (2005). Ligand efficiency indices as guideposts for drug discovery. *Drug Discov. Today* **10,** 464–469.

Akerley, B. J., Rubin, E. J., Novick, V. L., Amaya, K., Judson, N., and Mekalanos, J. J. (2002). A genome-scale analysis for identification of genes required for growth or survival of Haemophilus influenzae. *Proc. Natl. Acad. Sci. USA* **99,** 966–971.

Altschul, S. F., Gish, W., Miller, W., Myers, E. W., and Lipman, D. J. (1990). Basic local alignment search tool. *J. Mol. Biol.* **215,** 403–410.

Andrews, P. R., Craik, D. J., and Martin, J. L. (1984). Functional group contributions to drug-receptor interactions. *J. Med. Chem.* **27,** 1648–1657.

Assadi-Porter, F. M., Tonelli, M., Maillet, E., Hallenga, K., Benard, O., Max, M., and Markley, J. L. (2008). Direct NMR detection of the binding of functional ligands to the human sweet receptor, a heterodimeric family 3 GPCR. *J. Am. Chem. Soc.* **130,** 7212–7213.
Begley, D. W., Hartley, R. C., Davies, D. R., Edwards, T. E., Leonard, J. T., Abendroth, J., Burris, C. A., Bhandari, J., Myler, P. J., and Stewart, L. J. Leveraging Structure Determination with Fragment Screening for Infectious Disease Drug Targets. *J. Struct. Funct. Genomics.* (manuscript in preparation).
Bembenek, S. D., Tounge, B. A., and Reynolds, C. H. (2009). Ligand efficiency and fragment-based drug discovery. *Drug Discov. Today* **14,** 278–283.
Bemis, G. W., and Murcko, M. A. (1996). The properties of known drugs. 1. Molecular frameworks. *J. Med. Chem.* **39,** 2887–2893.
Binkowski, T. A., and Joachimiak, A. (2008). Protein functional surfaces: Global shape matching and local spatial alignments of ligand binding sites. *BMC Struct. Biol.* **8,** 45.
Bornhop, D. J., Latham, J. C., Kussrow, A., Markov, D. A., Jones, R. D., and Sorensen, H. S. (2007). Free-solution, label-free molecular interactions studied by back-scattering interferometry. *Science* **317,** 1732–1736.
Borrmann, S., Adegnika, A. A., Matsiegui, P. B., Issifou, S., Schindler, A., Mawili-Mboumba, D. P., Baranek, T., Wiesner, J., Jomaa, H., and Kremsner, P. G. (2004a). Fosmidomycin-clindamycin for Plasmodium falciparum Infections in African children. *J. Infect. Dis.* **189,** 901–908.
Borrmann, S., Issifou, S., Esser, G., Adegnika, A. A., Ramharter, M., Matsiegui, P. B., Oyakhirome, S., Mawili-Mboumba, D. P., Missinou, M. A., Kun, J. F., Jomaa, H., and Kremsner, P. G. (2004b). Fosmidomycin-clindamycin for the treatment of Plasmodium falciparum malaria. *J. Infect. Dis.* **190,** 1534–1540.
Boucher, H. W., Talbot, G. H., Bradley, J. S., Edwards, J. E., Gilbert, D., Rice, L. B., Scheld, M., Spellberg, B., and Bartlett, J. (2009). Bad bugs, no drugs: No ESKAPE! An update from the Infectious Diseases Society of America. *Clin. Infect. Dis.* **48,** 1–12.
Brough, P. A., Barril, X., Borgognoni, J., Chene, P., Davies, N. G., Davis, B., Drysdale, M. J., Dymock, B., Eccles, S. A., Garcia-Echeverria, C., Fromont, C., Hayes, A., *et al.* (2009). Combining hit identification strategies: Fragment-based and in silico approaches to orally active 2-aminothieno[2, 3-d]pyrimidine inhibitors of the Hsp90 molecular chaperone. *J. Med. Chem.* **52,** 4794–4809.
Brown, A. C., and Parish, T. (2008). Dxr is essential in Mycobacterium tuberculosis and fosmidomycin resistance is due to a lack of uptake. *BMC Microbiol.* **8,** 78.
Buetow, L., Brown, A. C., Parish, T., and Hunter, W. N. (2007). The structure of Mycobacteria 2C-methyl-D-erythritol-2, 4-cyclodiphosphate synthase, an essential enzyme, provides a platform for drug discovery. *BMC Struct. Biol.* **7,** 68.
Burley, S. K., Joachimiak, A., Montelione, G. T., and Wilson, I. A. (2008). Contributions to the NIH-NIGMS Protein Structure Initiative from the PSI Production Centers. *Structure* **16,** 5–11.
Campbell, T. L., and Brown, E. D. (2002). Characterization of the depletion of 2-C-methyl-D-erythritol-2, 4-cyclodiphosphate synthase in Escherichia coli and Bacillus subtilis. *J. Bacteriol.* **184,** 5609–5618.
Campos, N., Rodriguez-Concepcion, M., Sauret-Gueto, S., Gallego, F., Lois, L. M., and Boronat, A. (2001). Escherichia coli engineered to synthesize isopentenyl diphosphate and dimethylallyl diphosphate from mevalonate: A novel system for the genetic analysis of the 2-C-methyl-d-erythritol 4-phosphate pathway for isoprenoid biosynthesis. *Biochem. J.* **353,** 59–67.
Card, G. L., Blasdel, L., England, B. P., Zhang, C., Suzuki, Y., Gillette, S., Fong, D., Ibrahim, P. N., Artis, D. R., Bollag, G., Milburn, M. V., Kim, S. H., *et al.* (2005). A family of phosphodiesterase inhibitors discovered by cocrystallography and scaffold-based drug design. *Nat. Biotechnol.* **23,** 201–207.

CCP4 (1994). The CCP4 suite: Programs for protein crystallography. *Acta Crystallogr. D Biol. Crystallogr.* **50,** 760–763.

Chan, P. F., Holmes, D. J., and Payne, D. J. (2004). Finding the gems using genomic discovery: Antibacterial drug discovery strategies–The successes and the challenges. *Drug Discov. Today Ther. Strateg.* **1,** 519–527.

Chen, V. B., Arendall, W. B., 3rd, Headd, J. J., Keedy, D. A., Immormino, R. M., Kapral, G. J., Murray, L. W., Richardson, J. S., and Richardson, D. C. (2010). MolProbity: All-atom structure validation for macromolecular crystallography. *Acta Crystallogr. D Biol. Crystallogr.* **66,** 12–21.

Ciulli, A., Scott, D. E., Ando, M., Reyes, F., Saldanha, S. A., Tuck, K. L., Chirgadze, D. Y., Blundell, T. L., and Abell, C. (2008). Inhibition of Mycobacterium tuberculosis pantothenate synthetase by analogues of the reaction intermediate. *Chembiochem* **9,** 2606–2611.

Congreve, M., Carr, R., Murray, C., and Jhoti, H. (2003). A 'rule of three' for fragment-based lead discovery? *Drug Discov. Today* **8,** 876–877.

Davies, D. R., Mamat, B., Magnusson, O. T., Christensen, J., Haraldsson, M. H., Mishra, R., Pease, B., Hansen, E., Singh, J., Zembower, D., Kim, H., Kiselyov, A. S., *et al.* (2009). Discovery of leukotriene A4 hydrolase inhibitors using metabolomics biased fragment crystallography. *J. Med. Chem.* **52,** 4694–4715.

Davis, I. W., Leaver-Fay, A., Chen, V. B., Block, J. N., Kapral, G. J., Wang, X., Murray, L. W., Arendall, W. B., 3rd, Snoeyink, J., Richardson, J. S., and Richardson, D. C. (2007). MolProbity: All-atom contacts and structure validation for proteins and nucleic acids. *Nucleic Acids Res.* **35,** W375–W383.

DeLano, W. L. (2002). Unraveling hot spots in binding interfaces: Progress and challenges. *Curr. Opin. Struct. Biol.* **12,** 14–20.

DeLano, W. L. (2008). The PyMOL Molecular Graphics System. DeLano Scientific LLC, Palo Alto, CA, USA.

Deng, J., Davies, D. R., Wisedchaisri, G., Wu, M., Hol, W. G., and Mehlin, C. (2004). An improved protocol for rapid freezing of protein samples for long-term storage. *Acta Crystallogr. D Biol. Crystallogr.* **60,** 203–204.

Dessailly, B. H., Nair, R., Jaroszewski, L., Fajardo, J. E., Kouranov, A., Lee, D., Fiser, A., Godzik, A., Rost, B., and Orengo, C. (2009). PSI-2: Structural genomics to cover protein domain family space. *Structure* **17,** 869–881.

Dobson, C. M., Wai, T., Leclerc, D., Kadir, H., Narang, M., Lerner-Ellis, J. P., Hudson, T. J., Rosenblatt, D. S., and Gravel, R. A. (2002). Identification of the gene responsible for the cblB complementation group of vitamin B12-dependent methylmalonic aciduria. *Hum. Mol. Genet.* **11,** 3361–3369.

Duarte, C. D., Barreiro, E. J., and Fraga, C. A. (2007). Privileged structures: A useful concept for the rational design of new lead drug candidates. *Mini Rev. Med. Chem.* **7,** 1108–1119.

Edwards, T. E., Bullen, J., Abendroth, J., Sankaran, B., Staker, B. L., Van Voorhis, W. C., Myler, P. J., and Stewart, L. J. Structures of putative *Mycobacterium* arginine/ornithine transport system NTPases. *Acta. Crystallogr. F* (manuscript in preparation).

Eisenreich, W., Bacher, A., Arigoni, D., and Rohdich, F. (2004). Biosynthesis of isoprenoids via the non-mevalonate pathway. *Cell. Mol. Life Sci.* **61,** 1401–1426.

Emsley, P., and Cowtan, K. (2004). Coot: Model-building tools for molecular graphics. *Acta Crystallogr. D Biol. Crystallogr.* **60,** 2126–2132.

Eoh, H., Brown, A. C., Buetow, L., Hunter, W. N., Parish, T., Kaur, D., Brennan, P. J., and Crick, D. C. (2007). Characterization of the Mycobacterium tuberculosis 4-diphosphocytidyl-2-C-methyl-D-erythritol synthase: Potential for drug development. *J. Bacteriol.* **189,** 8922–8927.

Erlanson, D. A., McDowell, R. S., and O'Brien, T. (2004). Fragment-based drug discovery. *J. Med. Chem.* **47,** 3463–3482.

Ertl, P., Jelfs, S., Muhlbacher, J., Schuffenhauer, A., and Selzer, P. (2006). Quest for the rings. In silico exploration of ring universe to identify novel bioactive heteroaromatic scaffolds. *J. Med. Chem.* **49,** 4568–4573.

Fejzo, J., Lepre, C. A., Peng, J. W., Bemis, G. W., Ajay, G. W., Murcko, M. A., and Moore, J. M. (1999). The SHAPES strategy: An NMR-based approach for lead generation in drug discovery. *Chem. Biol.* **6,** 755–769.

Fox, B. G., Goulding, C., Malkowski, M. G., Stewart, L., and Deacon, A. (2008). Structural genomics: From genes to structures with valuable materials and many questions in between. *Nat Methods* **5,** 129–132.

Freiberg, C., Wieland, B., Spaltmann, F., Ehlert, K., Brotz, H., and Labischinski, H. (2001). Identification of novel essential Escherichia coli genes conserved among pathogenic bacteria. *J. Mol. Microbiol. Biotechnol.* **3,** 483–489.

Gerlt, J. A. (2007). A protein structure (or function ?) initiative. *Structure* **15,** 1353–1356.

Hajduk, P. J. (2006a). Fragment-based drug design: How big is too big? *J. Med. Chem.* **49,** 6972–6976.

Hajduk, P. J. (2006b). Puzzling through fragment-based drug design. *Nat. Chem. Biol.* **2,** 658–659.

Hajduk, P. J., Huth, J. R., and Tse, C. (2005). Predicting protein druggability. *Drug Discov. Today* **10,** 1675–1682.

Hann, M. M., Leach, A. R., and Harper, G. (2001). Molecular complexity and its impact on the probability of finding leads for drug discovery. *J. Chem. Inf. Comput. Sci.* **41,** 856–864.

Hartshorn, M. J., Murray, C. W., Cleasby, A., Frederickson, M., Tickle, I. J., and Jhoti, H. (2005). Fragment-based lead discovery using X-ray crystallography. *J. Med. Chem.* **48,** 403–413.

Hermann, J. C., Marti-Arbona, R., Fedorov, A. A., Fedorov, E., Almo, S. C., Shoichet, B. K., and Raushel, F. M. (2007). Structure-based activity prediction for an enzyme of unknown function. *Nature* **448,** 775–779.

Hesterkamp, T., Barker, J., Davenport, A., and Whittaker, M. (2007). Fragment based drug discovery using fluorescence correlation: Spectroscopy techniques: Challenges and solutions. *Curr. Top. Med. Chem.* **7,** 1582–1591.

Hopkins, A. L., Groom, C. R., and Alex, A. (2004). Ligand efficiency: A useful metric for lead selection. *Drug Discov. Today* **9,** 430–431.

Hubbard, R. E., Davis, B., Chen, I., and Drysdale, M. J. (2007). The SeeDs approach: Integrating fragments into drug discovery. *Curr. Top. Med. Chem.* **7,** 1568–1581.

Hunter, W. N. (2007). The non-mevalonate pathway of isoprenoid precursor biosynthesis. *J. Biol. Chem.* **282,** 21573–21577.

Jacobs, M. A., Alwood, A., Thaipisuttikul, I., Spencer, D., Haugen, E., Ernst, S., Will, O., Kaul, R., Raymond, C., Levy, R., Chun-Rong, L., Guenthner, D., *et al.* (2003). Comprehensive transposon mutant library of Pseudomonas aeruginosa. *Proc. Natl. Acad. Sci. USA* **100,** 14339–14344.

Jasmer, R. M., Nahid, P., and Hopewell, P. C. (2002). Clinical practice. Latent tuberculosis infection. *N. Engl. J. Med.* **347,** 1860–1866.

Jomaa, H., Wiesner, J., Sanderbrand, S., Altincicek, B., Weidemeyer, C., Hintz, M., Turbachova, I., Eberl, M., Zeidler, J., Lichtenthaler, H. K., Soldati, D., and Beck, E. (1999). Inhibitors of the nonmevalonate pathway of isoprenoid biosynthesis as antimalarial drugs. *Science* **285,** 1573–1576.

Kabsch, W. (1993). Automatic processing of rotation diffraction data from crystals of initially unknown symmetry and cell constants. *J. Appl. Crystallogr.* **26,** 795–800.

Kemp, L. E., Alphey, M. S., Bond, C. S., Ferguson, M. A., Hecht, S., Bacher, A., Eisenreich, W., Rohdich, F., and Hunter, W. N. (2005). The identification of isoprenoids that bind in the intersubunit cavity of Escherichia coli 2C-methyl-D-erythritol-2,

4-cyclodiphosphate synthase by complementary biophysical methods. *Acta Crystallogr. D Biol. Crystallogr.* **61,** 45–52.
Keseru, G. M., and Makara, G. M. (2006). Hit discovery and hit-to-lead approaches. *Drug Discov. Today* **11,** 741–748.
Korotkova, N., and Lidstrom, M. E. (2004). MeaB is a component of the methylmalonyl-CoA mutase complex required for protection of the enzyme from inactivation. *J. Biol. Chem.* **279,** 13652–13658.
Korotkova, N., Chistoserdova, L., Kuksa, V., and Lidstrom, M. E. (2002). Glyoxylate regeneration pathway in the methylotroph Methylobacterium extorquens AM1. *J. Bacteriol.* **184,** 1750–1758.
Kuntz, I. D., Chen, K., Sharp, K. A., and Kollman, P. A. (1999). The maximal affinity of ligands. *Proc. Natl. Acad. Sci. USA* **96,** 9997–10002.
Ladbury, J. E., Klebe, G., and Freire, E. (2010). Adding calorimetric data to decision making in lead discovery: A hot tip. *Nat. Rev. Drug Discov.* **9,** 23–27.
Lange, B. M., Rujan, T., Martin, W., and Croteau, R. (2000). Isoprenoid biosynthesis: The evolution of two ancient and distinct pathways across genomes. *Proc. Natl. Acad. Sci. USA* **97,** 13172–13177.
Lepre, C. A., Moore, J. M., and Peng, J. W. (2004). Theory and applications of NMR-based screening in pharmaceutical research. *Chem. Rev.* **104,** 3641–3676.
Lipkus, A. H., Yuan, Q., Lucas, K. A., Funk, S. A., Bartelt, W. F., 3rd, Schenck, R. J., and Trippe, A. J. (2008). Structural diversity of organic chemistry. A scaffold analysis of the CAS Registry. *J. Org. Chem.* **73,** 4443–4451.
Makara, G. M. (2007). On sampling of fragment space. *J. Med. Chem.* **50,** 3214–3221.
Mayer, M., and James, T. L. (2004). NMR-based characterization of phenothiazines as a RNA binding scaffold. *J. Am. Chem. Soc.* **126,** 4453–4460.
Missinou, M. A., Borrmann, S., Schindler, A., Issifou, S., Adegnika, A. A., Matsiegui, P. B., Binder, R., Lell, B., Wiesner, J., Baranek, T., Jomaa, H., and Kremsner, P. G. (2002). Fosmidomycin for malaria. *Lancet* **360,** 1941–1942.
Mochalkin, I., Miller, J. R., Narasimhan, L., Thanabal, V., Erdman, P., Cox, P. B., Prasad, J. V., Lightle, S., Huband, M. D., and Stover, C. K. (2009). Discovery of antibacterial biotin carboxylase inhibitors by virtual screening and fragment-based approaches. *ACS Chem. Biol.* **4,** 473–483.
Moore, P. B. (2007). Let's call the whole thing off: Some thoughts on the protein structure initiative. *Structure* **15,** 1350–1352.
Mpamhanga, C. P., Spinks, D., Tulloch, L. B., Shanks, E. J., Robinson, D. A., Collie, I. T., Fairlamb, A. H., Wyatt, P. G., Frearson, J. A., Hunter, W. N., Gilbert, I. H., and Brenk, R. (2009). One scaffold, three binding modes: Novel and selective pteridine reductase 1 inhibitors derived from fragment hits discovered by virtual screening. *J. Med. Chem.* **52,** 4454–4465.
Muller, G. (2003). Medicinal chemistry of target family-directed masterkeys. *Drug Discov. Today* **8,** 681–691.
Murray, C. W., and Blundell, T. L. (2010). Structural biology in fragment-based drug design. *Curr. Opin. Struct. Biol.* **20,** 497–507.
Myler, P. J., Stacy, R., Stewart, L., Staker, B. L., Van Voorhis, W. C., Varani, G., and Buchko, G. W. (2009). The Seattle Structural Genomics Center for Infectious Disease (SSGCID). *Infect. Disord. Drug Targets* **9,** 493–506.
Nair, R., Liu, J., Soong, T. T., Acton, T. B., Everett, J. K., Kouranov, A., Fiser, A., Godzik, A., Jaroszewski, L., Orengo, C., Montelione, G. T., and Rost, B. (2009). Structural genomics is the largest contributor of novel structural leverage. *J. Struct. Funct. Genomics* **10,** 181–191.
Navratilova, I., and Hopkins, A. L. (2010). Fragment screening by surface plasmon resonance. *ACS Med. Chem. Lett.* **1,** 44–48.

Neumann, T., Junker, H. D., Schmidt, K., and Sekul, R. (2007). SPR-based fragment screening: Advantages and applications. *Curr. Top. Med. Chem.* **7,** 1630–1642.

Ni, S., McAteer, K., Bussiere, D. E., and Kennedy, M. A. (2004a). Crystallization and preliminary crystallographic analysis of an Enterococcus faecalis repressor protein, CylR2, involved in regulating cytolysin production through quorum-sensing. *Acta Crystallogr. D Biol. Crystallogr.* **60,** 1145–1148.

Ni, S., Robinson, H., Marsing, G. C., Bussiere, D. E., and Kennedy, M. A. (2004b). Structure of 2C-methyl-D-erythritol-2, 4-cyclodiphosphate synthase from Shewanella oneidensis at 1.6 A: Identification of farnesyl pyrophosphate trapped in a hydrophobic cavity. *Acta Crystallogr. D Biol. Crystallogr.* **60,** 1949–1957.

Nienaber, V. L., Richardson, P. L., Klighofer, V., Bouska, J. J., Giranda, V. L., and Greer, J. (2000). Discovering novel ligands for macromolecules using X-ray crystallographic screening. *Nat. Biotechnol.* **18,** 1105–1108.

Painter, J., and Merritt, E. A. (2006). Optimal description of a protein structure in terms of multiple groups undergoing TLS motion. *Acta Crystallogr. D Biol. Crystallogr.* **62,** 439–450.

Payne, D. J., Gwynn, M. N., Holmes, D. J., and Rosenberg, M. (2004). Genomic approaches to antibacterial discovery. *Methods Mol. Biol.* **266,** 231–259.

Payne, D. J., Gwynn, M. N., Holmes, D. J., and Pompliano, D. L. (2007). Drugs for bad bugs: Confronting the challenges of antibacterial discovery. *Nat. Rev. Drug Discov.* **6,** 29–40.

Pellecchia, M. (2005). Solution nuclear magnetic resonance spectroscopy techniques for probing intermolecular interactions. *Chem. Biol.* **12,** 961–971.

Petsko, G. A. (2007). An idea whose time has gone. *Genome Biol.* **8,** 107.

Pflugrath, J. W. (1999). The finer things in X-ray diffraction data collection. *Acta Crystallogr. D Biol. Crystallogr.* **55,** 1718–1725.

Rakus, J. F., Kalyanaraman, C., Fedorov, A. A., Fedorov, E. V., Mills-Groninger, F. P., Toro, R., Bonanno, J., Bain, K., Sauder, J. M., Burley, S. K., Almo, S. C., Jacobson, M. P., *et al.* (2009). Computation-facilitated assignment of the function in the enolase superfamily: A regiochemically distinct galactarate dehydratase from Oceanobacillus iheyensis. *Biochemistry* **48,** 11546–11558.

Rees, D. C., Congreve, M., Murray, C. W., and Carr, R. (2004). Fragment-based lead discovery. *Nat. Rev. Drug Discov.* **3,** 660–672.

Reitz, A. B., Smith, G. R., Tounge, B. A., and Reynolds, C. H. (2009). Hit triage using efficiency indices after screening of compound libraries in drug discovery. *Curr. Top. Med. Chem.* **9,** 1718–1724.

Rohdich, F., Eisenreich, W., Wungsintaweekul, J., Hecht, S., Schuhr, C. A., and Bacher, A. (2001). Biosynthesis of terpenoids. 2C-Methyl-D-erythritol 2, 4-cyclodiphosphate synthase (IspF) from Plasmodium falciparum. *Eur. J. Biochem.* **268,** 3190–3197.

Sanders, W. J., Nienaber, V. L., Lerner, C. G., McCall, J. O., Merrick, S. M., Swanson, S. J., Harlan, J. E., Stoll, V. S., Stamper, G. F., Betz, S. F., Condroski, K. R., Meadows, R. P., *et al.* (2004). Discovery of potent inhibitors of dihydroneopterin aldolase using CrystaLEAD high-throughput X-ray crystallographic screening and structure-directed lead optimization. *J. Med. Chem.* **47,** 1709–1718.

Service, R. F. (2008). Structural biology. Protein structure initiative: Phase 3 or phase out. *Science* **319,** 1610–1613.

Shuker, S. B., Hajduk, P. J., Meadows, R. P., and Fesik, S. W. (1996). Discovering high-affinity ligands for proteins: SAR by NMR. *Science* **274,** 1531–1534.

Siegel, M. G., and Vieth, M. (2007). Drugs in other drugs: A new look at drugs as fragments. *Drug Discov. Today* **12,** 71–79.

Smith, E., and Morowitz, H. J. (2004). Universality in intermediary metabolism. *Proc. Natl. Acad. Sci. USA* **101,** 13168–13173.

Song, L., Kalyanaraman, C., Fedorov, A. A., Fedorov, E. V., Glasner, M. E., Brown, S., Imker, H. J., Babbitt, P. C., Almo, S. C., Jacobson, M. P., and Gerlt, J. A. (2007). Prediction and assignment of function for a divergent N-succinyl amino acid racemase. *Nat. Chem. Biol.* **3,** 486–491.

Steinbacher, S., Kaiser, J., Wungsintaweekul, J., Hecht, S., Eisenreich, W., Gerhardt, S., Bacher, A., and Rohdich, F. (2002). Structure of 2C-methyl-d-erythritol-2, 4-cyclodiphosphate synthase involved in mevalonate-independent biosynthesis of isoprenoids. *J. Mol. Biol.* **316,** 79–88.

Swayze, E. E., Jefferson, E. A., Sannes-Lowery, K. A., Blyn, L. B., Risen, L. M., Arakawa, S., Osgood, S. A., Hofstadler, S. A., and Griffey, R. H. (2002). SAR by MS: A ligand based technique for drug lead discovery against structured RNA targets. *J. Med. Chem.* **45,** 3816–3819.

Ulaganathan, V., Agacan, M. F., Buetow, L., Tulloch, L. B., and Hunter, W. N. (2007). Structure of Staphylococcus aureus1, 4-dihydroxy-2-naphthoyl-CoA synthase (MenB) in complex with acetoacetyl-CoA. *Acta Crystallogr. F Struct. Biol. Cryst. Commun.* **63,** 908–913.

Van Voorhis, W. C., Hol, W. G., Myler, P. J., and Stewart, L. J. (2009). The role of medical structural genomics in discovering new drugs for infectious diseases. *PLoS Comput. Biol.* **5,** e1000530.

Waybright, T. J., Britt, J. R., and McCloud, T. G. (2009). Overcoming problems of compound storage in DMSO: Solvent and process alternatives. *J. Biomol. Screen.* **14,** 708–715.

Wells, J. A., and McClendon, C. L. (2007). Reaching for high-hanging fruit in drug discovery at protein-protein interfaces. *Nature* **450,** 1001–1009.

Wiesner, J., Henschker, D., Hutchinson, D. B., Beck, E., and Jomaa, H. (2002). In vitro and in vivo synergy of fosmidomycin, a novel antimalarial drug, with clindamycin. *Antimicrob. Agents Chemother.* **46,** 2889–2894.

Wiesner, J., Borrmann, S., and Jomaa, H. (2003). Fosmidomycin for the treatment of malaria. *Parasitol. Res.* **90**(Suppl 2), S71–S76.

Yuvaniyama, J., Chitnumsub, P., Kamchonwongpaisan, S., Vanichtanankul, J., Sirawaraporn, W., Taylor, P., Walkinshaw, M. D., and Yuthavong, Y. (2003). Insights into antifolate resistance from malarial DHFR-TS structures. *Nat. Struct. Biol.* **10,** 357–365.

Zartler, E., and Shapiro, M. (2008). Fragment-based drug discovery: A practical approach. John Wiley & Sons, Hoboken, N.J.

Author Index

B

C

D

E

F

G

I

J

L

M

N

Q

R

X

Y

Z

Subject Index

G

H

Q

R

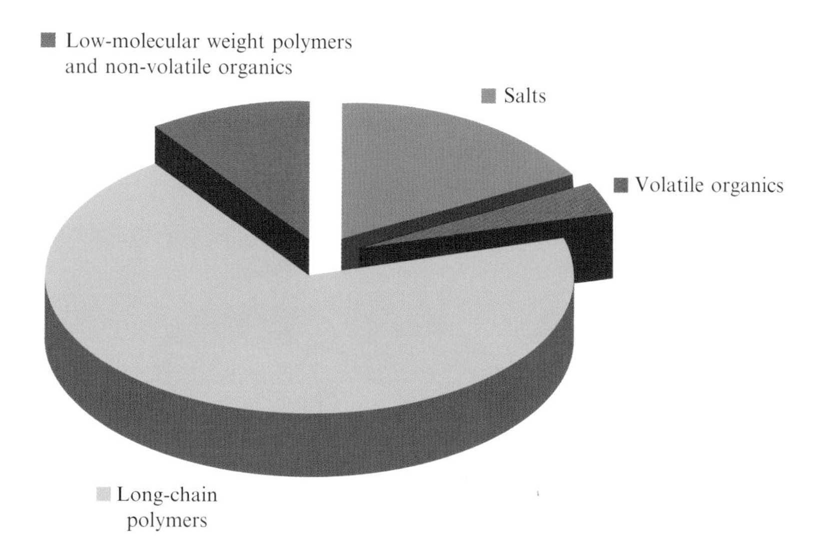

Jark Böttcher *et al.*, Figure 3.3 Usage of precipitants in successful protein–fragment structure determination.

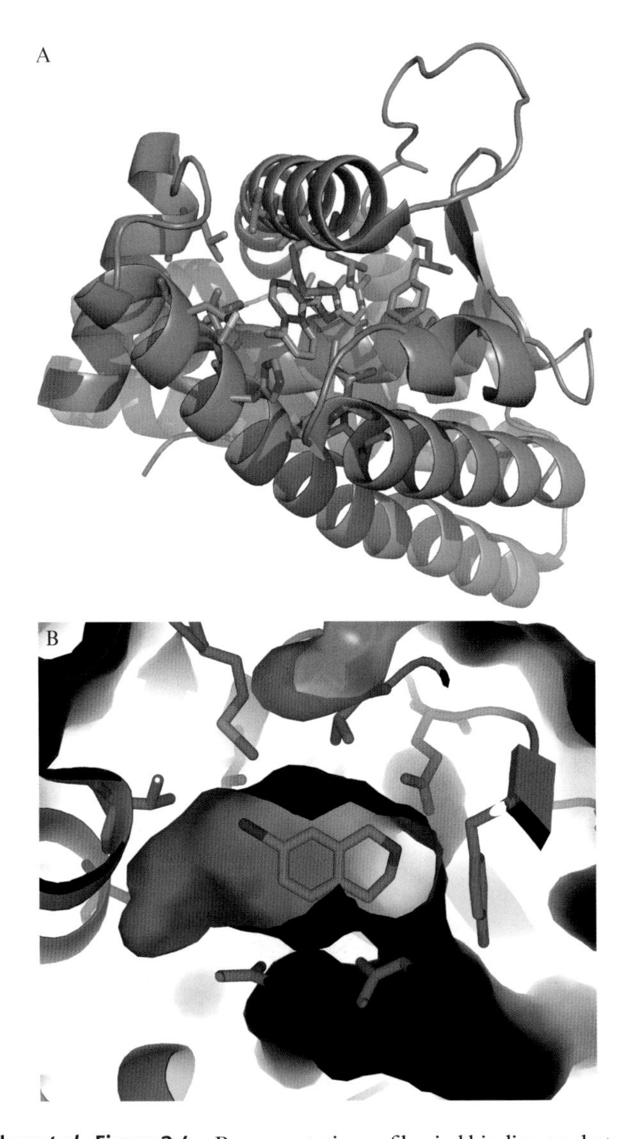

Jark Böttcher *et al.*, Figure 3.4 Representations of buried binding pockets not suitable for complex formation by soaking. (A) Estrogen receptor beta; (B) phenylethanolamin *N*-methyl transferase. The protein binding pocket is shown in blue, the buried ligand as orange sticks.

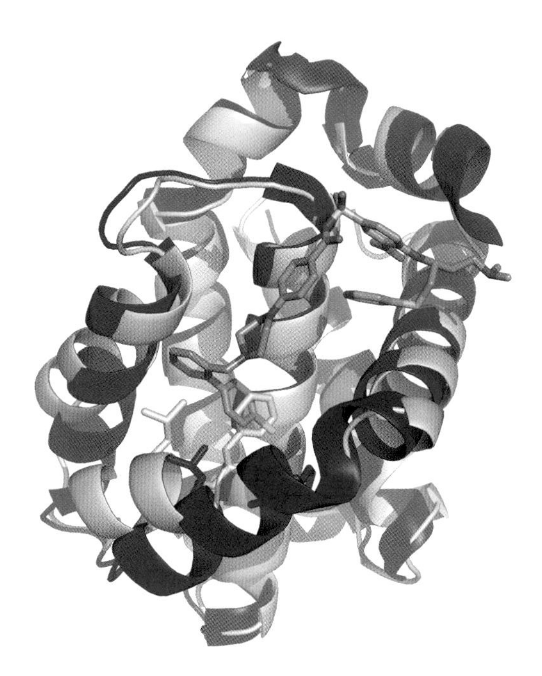

Jark Böttcher *et al.*, Figure 3.5 Superposition of Bcl-XL crystal structures in an unbound state (PDB number 1MAZ, gray) and a ligand-bound state (PDB number 2YXJ, blue, ligand shown as sticks in light blue). The conformation observed for the unbound protein in space group $P4_12_12$ strongly interferes with ligand binding, while the crystal form observed in space group $P2_12_12_1$ exhibits a pronounced binding capability.

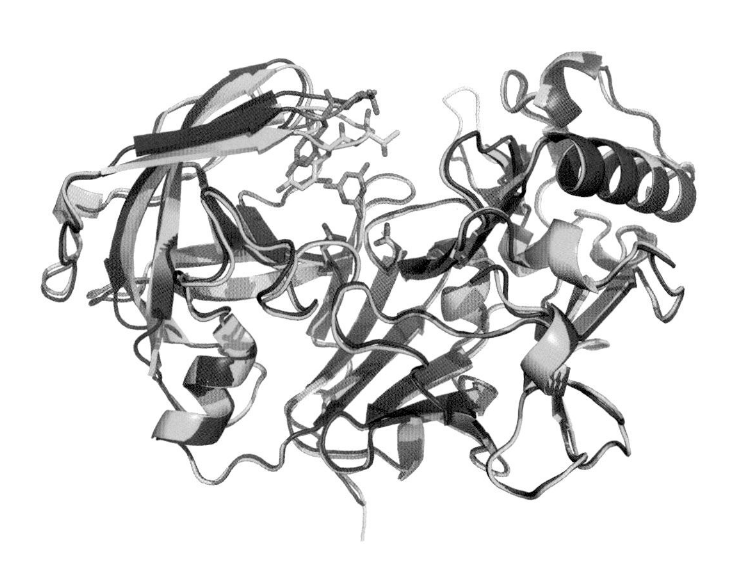

Jark Böttcher *et al.*, Figure 3.6 Superposition of monomers A and C of the BACE crystal structure (PDB number 3HVG; Godemann *et al.*, 2009). While monomer C (dark blue) represents the unbound state with the flap region in open conformation, monomer A (green) exhibits a closed conformation interacting with the bound fragment (orange).

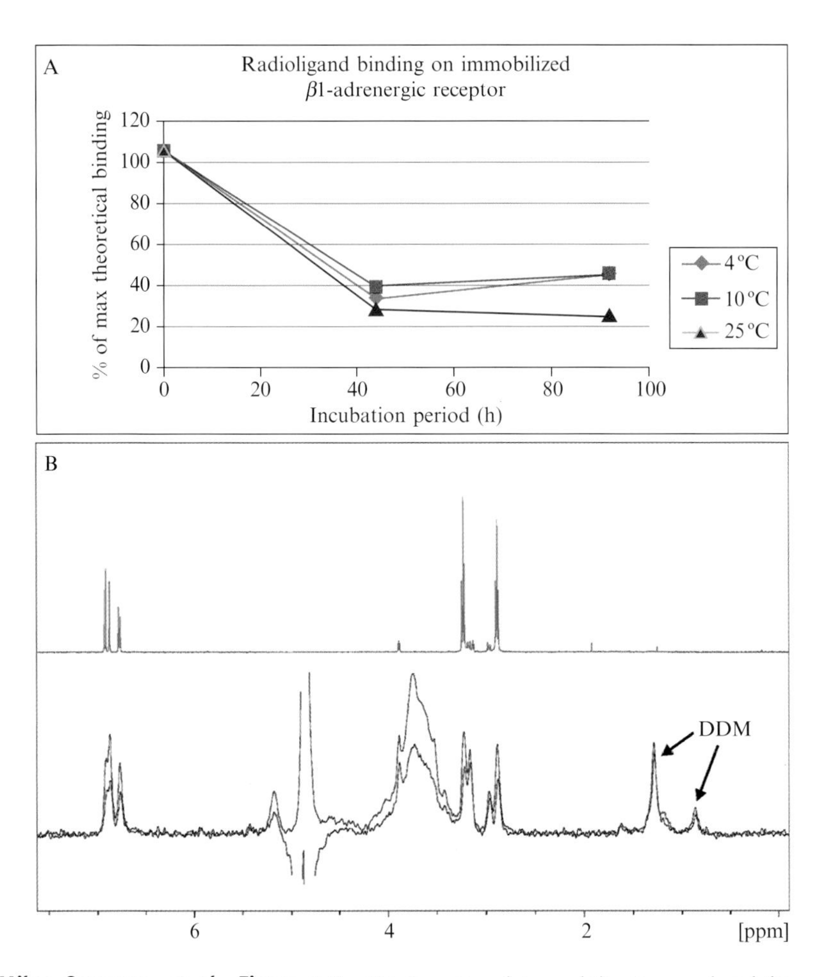

Miles Congreve *et al.*, Figure 5.7 (A) Functional immobilization and stability of micelle-solubilized β_1AR. β_1AR was covalently bound to Sepharose resin (see text for details). Functionality was assessed by binding of ^{3}H-dihydroalprenolol upon storage of the protein at the indicated temperature and for the indicated time periods. (B) Use of TINS to detect weakly binding ligands. A solution of 500 μ*M* dopamine was injected into the cell containing immobilized β_1AR (blue) and OmpA (red) and a ^{1}H spectrum of each is presented. The green spectrum is that of dopamine in solution. The NMR signals from DDM are indicated.

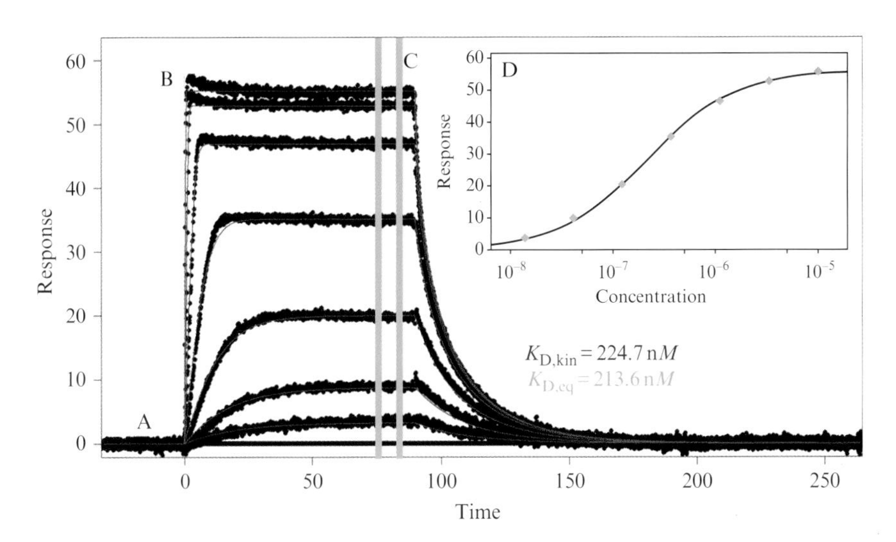

Anthony M. Giannetti, Figure 8.1 Experimental SPR sensorgram from a Biacore T100 with data, black points, overlaid with a 1:1 Langmuir binding model kinetic fit (red). The top concentration is 10 μ*M* and injections are related by threefold dilutions. (A) Preinjection baseline and (B) association phase. Compound is injected at $t = 0$ s and continues for 90 s. (C) Injection ends and dissociation is observed back to baseline levels. (D) Plot of the equilibrium response calculated as the average of the response values between the green bars versus the log of the sample concentration in molar. The $K_{D,kin}$ is taken as the ratio of the off-rate divided by the on-rate and $K_{D,eq}$ is derived from the fit in (D). The lowest concentration injection represents 6.2% surface occupancy and is at exactly three times the standard deviation of the noise.

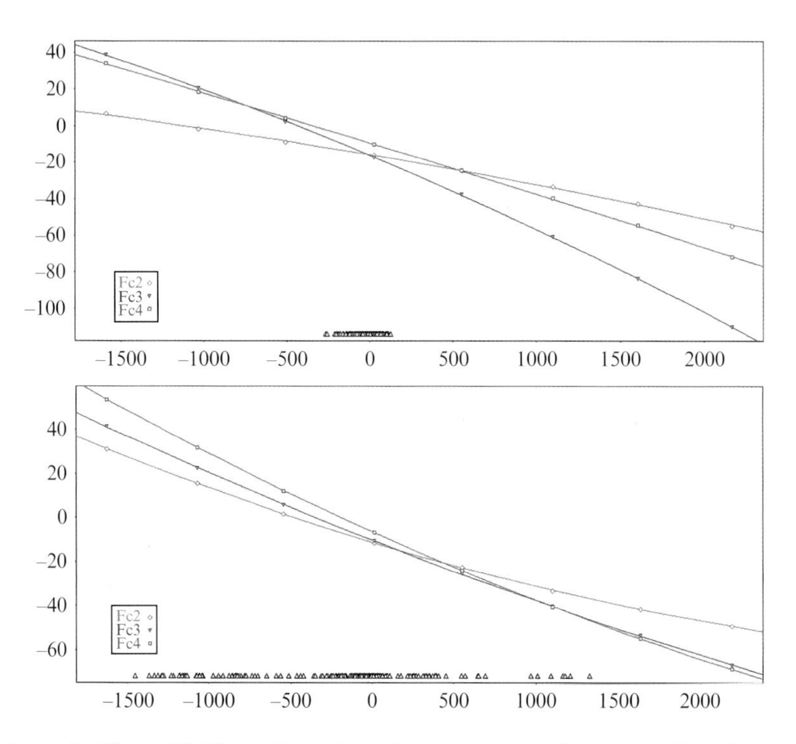

Anthony M. Giannetti, Figure 8.4 Example solvent correction curves from fragment screening. Each panel represents a screen of 320 fragments on three proteins. The *X*-axis is the bulk response value observed on the target flow cell and the *Y*-axis is the difference in response between the target and reference flow cells. Eight solvent correction solutions were used to construct the standard curve. Variation in sample DMSO levels is shown by plotting a triangle on the *X*-axis representing the bulk shift in response on the target surface due to solvent mismatch. Due to differences in capture levels and target weights, the shape of the correction curves varies between days and flow cells and is rarely linear. In (A), the dispersion of DMSO concentrations in the fragment plate is low but is very high in (B). All points are within the correction curve and can be retained for analysis.

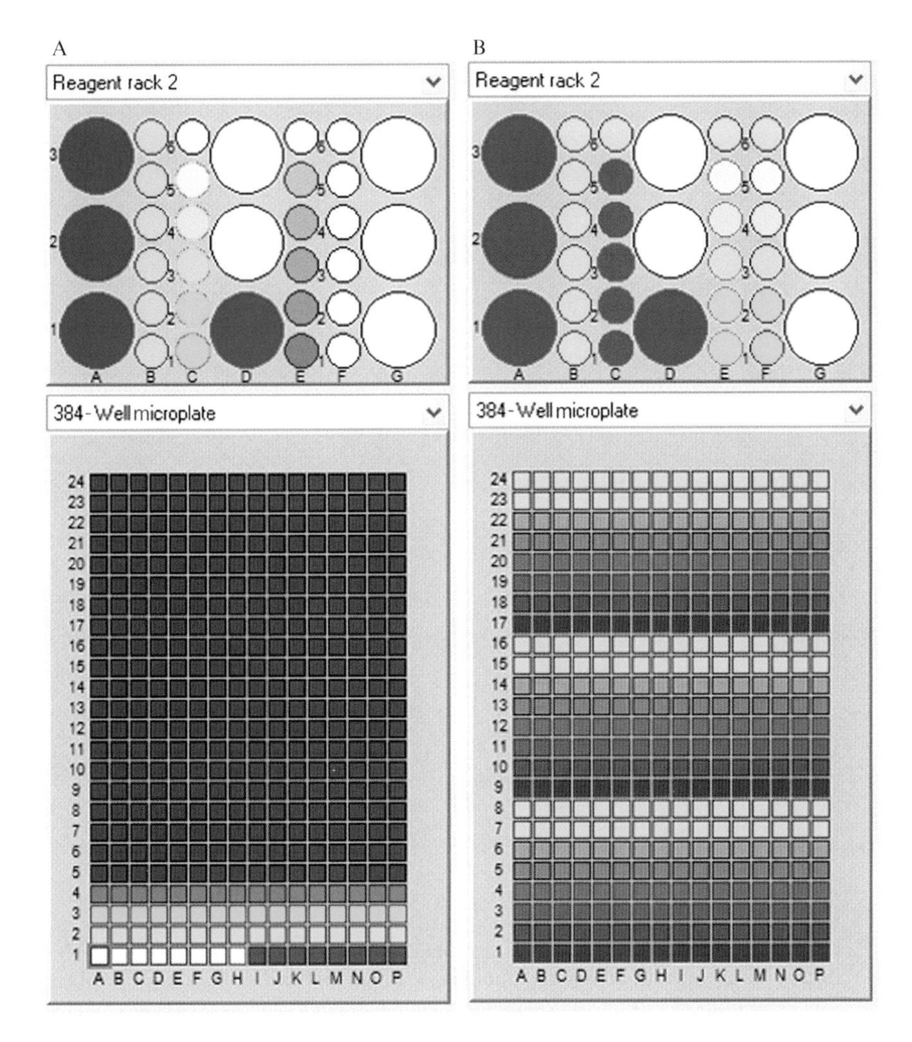

Anthony M. Giannetti, Figure 8.7 Example platemaps for fragment screening and characterization. Maps are specific to the Biacore T100 SPR platform but can be adapted to other systems. (A) Screening platemap with positions colored by sample type. Yellow: Startup cycles, magenta: solvent correction, cyan: running buffer used for double referencing, green: first control compounds, orange: second control compound, red: fragment, and blue: wash solution. Often the wash solution is simply the running buffer but can be different for various projects. In the reagent rack, dilution series of the controls are indicated by a color gradient in the green and orange samples. The blue samples are the blanks for double referencing the controls dilutions. (B) Colored as in (A) but with the layout for a dose–response plate. Forty-eight fragments (red) are shown in dilution with two buffer blanks for double referencing. There is only sufficient space for two dilution series of one control compound to be tested at the beginning and end of the run. If an additional control is required, then two fragment positions can be used for the second control reducing throughput to 46 fragments/machine/day.

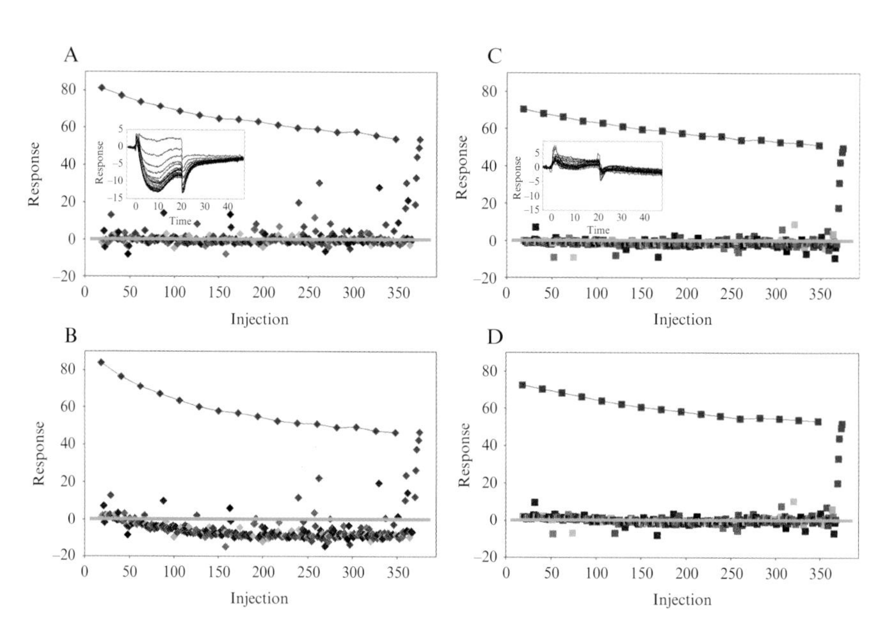

Anthony M. Giannetti, Figure 8.8 Effect of double referencing. Equilibrium response values for two different fragment plates on the same target and machine (but fresh surface) are plotted with (A) and (C) and without (B) and (D) double referencing. The 16 buffer blank injections from each screen are shown in the insets. The zero RU line is highlighted in green. Connected red dots are control replicates, magenta samples at the end are the control dilution series at the end of the run, and the rest of the points are individual fragment samples. The lack of double referencing in (B) significantly distorts the baseline and allows some fragment hits to have negative RU values, as compared with the double referenced data in (A). The improvements in the data in (C) are hard to visualize but statistical analysis reveals significant improvement in standard deviation and skew (see text).

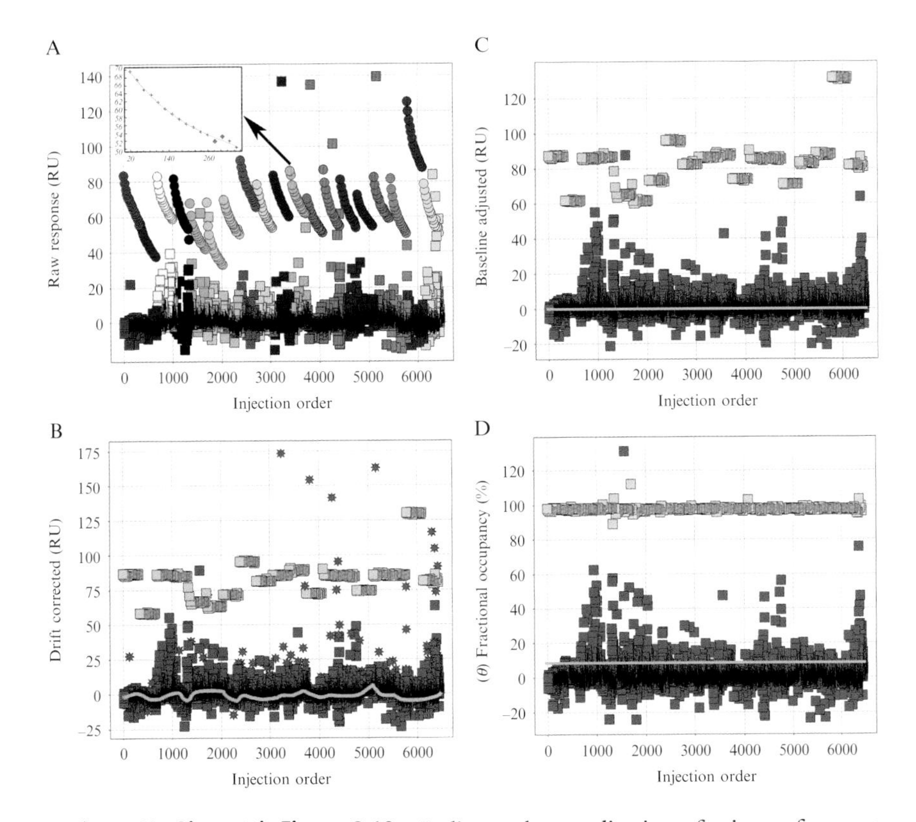

Anthony M. Giannetti, Figure 8.10 Scaling and normalization of primary fragment screening data. The plots represent a screen of ~6000 fragments (19 plates) on a single target. (A) Raw equilibrium response values (R_{eq}) from the instrument for all fragment and control replicate injections are shown versus the injection order. Points are colored by individual screening plate. Dose–response injections at the end of each plate are omitted for clarity. The inset shows a plot of the R_{eq} versus injection cycle number for the 16 replicate controls for the plate colored in gray. The green line represents the fit of a third-order polynomial to model the control decay. Two points highlighted in diamond boxes are statistical outliers and are excluded from the fit. (B) Screening data after correction for drift in the controls. Fragment samples shown in magenta, controls in cyan, and promiscuous binders identified from visual inspection of the sensorgrams are shown as stars and are excluded going forward. The green line trends the average of the baseline showing variance from day to day due to small buffer mismatches. (C) Data adjusted on a per-plate basis so the baseline scatters evenly around zero RU (green line). (D) Screen adjusted from RU to fractional occupancy, expressed as percentage, per the calculations described in Section 13. The average and standard deviation of the controls and baseline are 97.6 ± 1.33% ($N = 304$) and 0.17 ± 2.80% ($N = 5439$) yielding a Z′ factor (Zhang *et al.*, 1999) of 0.95. The three-sigma level used for initial hit selection is shown with a green line drawn at 8.4% and selects 344 compounds (5.7%). This initial hit rate is not officially quoted as only confirmed compounds from the dose–response analysis will qualify as true hits.

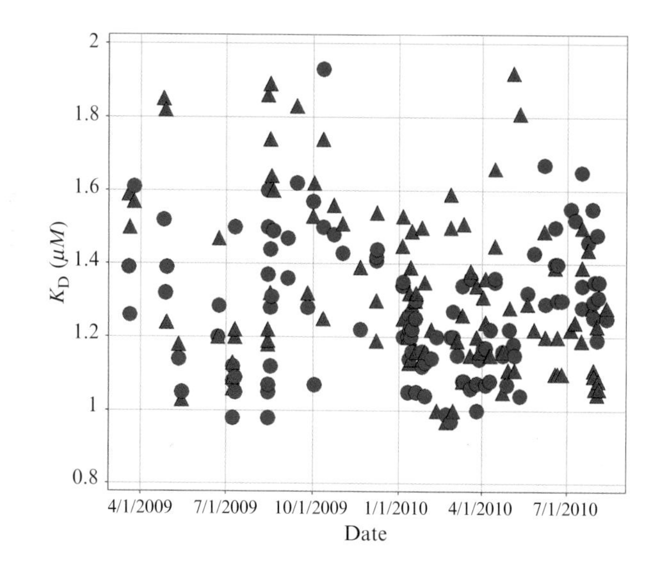

Anthony M. Giannetti, Figure 8.11 Reproducibility of "early" and "late" controls over ~18 months. The K_D values for 112 pairs of "early" and "late" controls used in a routine SPR-based SAR assay are plotted on a linear scale on the Y-axis versus the date of the collection. Note the Y-axis ranges from 0.8 to 2 μM. The early controls are shown as blue circles and the late controls, at the end of the plate, are depicted as red triangles. The average and standard deviation for the late and early controls are $1.27 \pm 0.18\ \mu M$ and $1.32 \pm 0.22\ \mu M$ ($N = 112$ for each set).

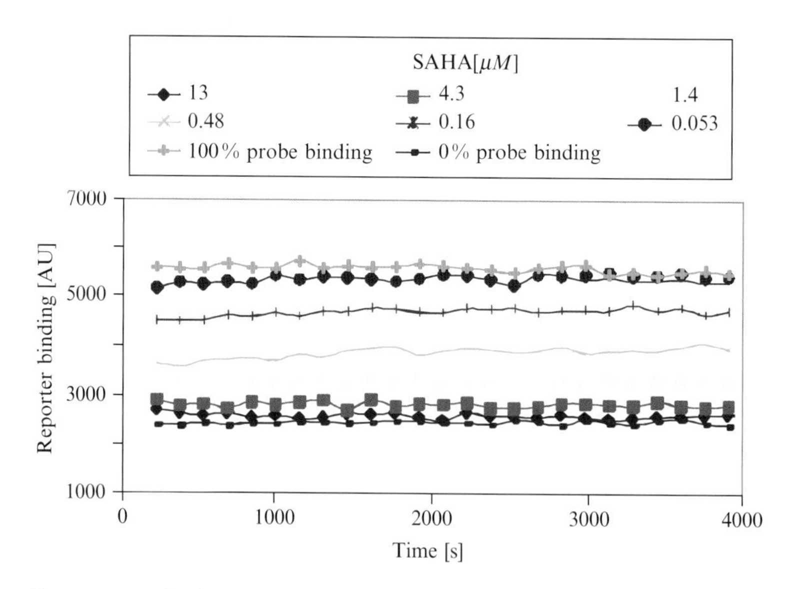

Lars Neumann *et al.*, Figure 12.4 Controlling reporter–target dissociation rate exemplified by the HDAC1–SAHA interaction. HDAC1 and reporter were incubated to form the reporter–target complex. Thereafter increasing concentrations of the fast binding reference inhibitor SAHA were added and reporter displacement was measured. Reporter displacement was completely finished within the 40 s that were required to transport the assay plate after compound addition to the plate reader, as indicated by the parallel run of the lines. Thus, reporter dissociation is very rapid and not the rate-limiting step in the measurements of compound with slow binding kinetics.

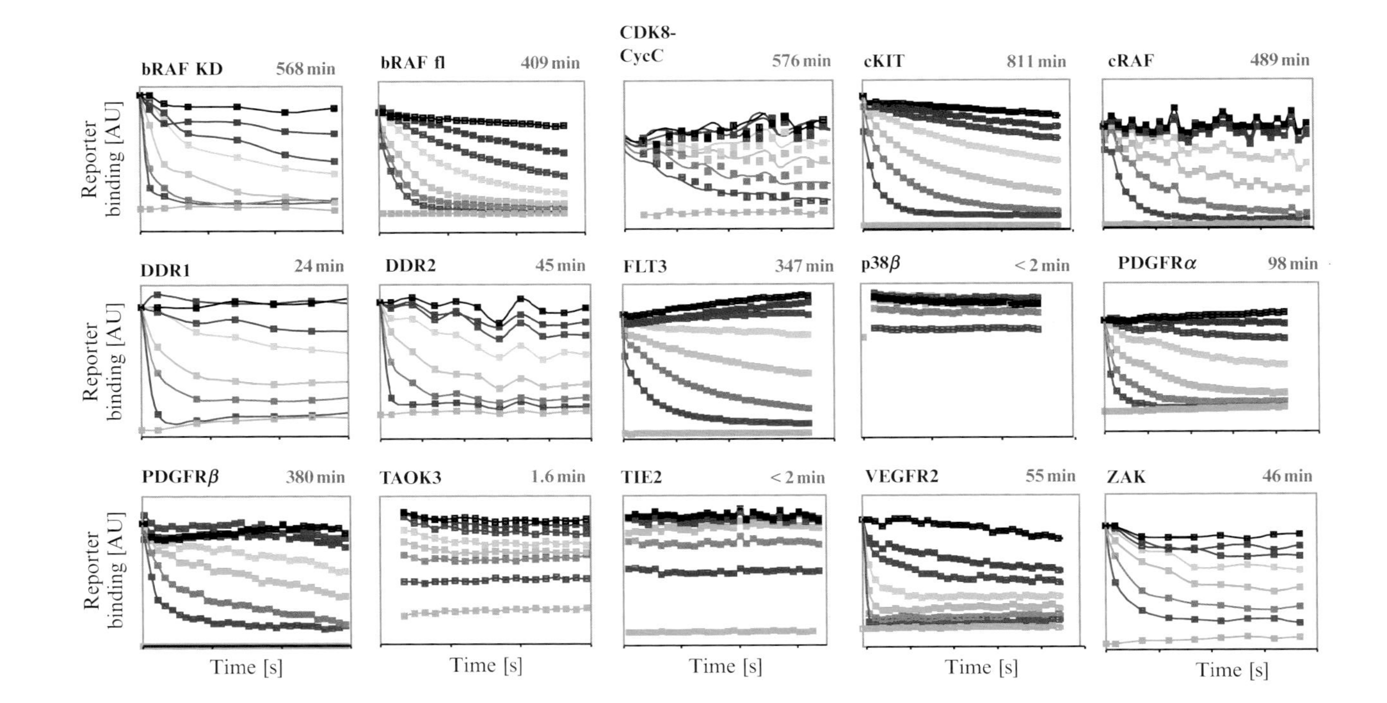

Lars Neumann *et al.*, Figure 12.6 Kinetic selectivity profiling of Sorafenib. Residence times of Sorafenib on 15 targets are measured using the reporter displacement assay. For each target reporter displacement is given over time in the presence of increasing Sorafenib concentrations. The color coding of the traces reflects the Sorafenib concentration (Sorafenib concentration: red > orange > green > light blue > dark blue > purple). Black traces represent reporter binding signal over time in the absence of Sorafenib and gray traces represent the signal in absence of target, indicating the signals for 100% and 0% reporter binding, respectively. Target names are given in black and residence times in blue. For illustration purposes, time scales differ between the different targets and are not shown.

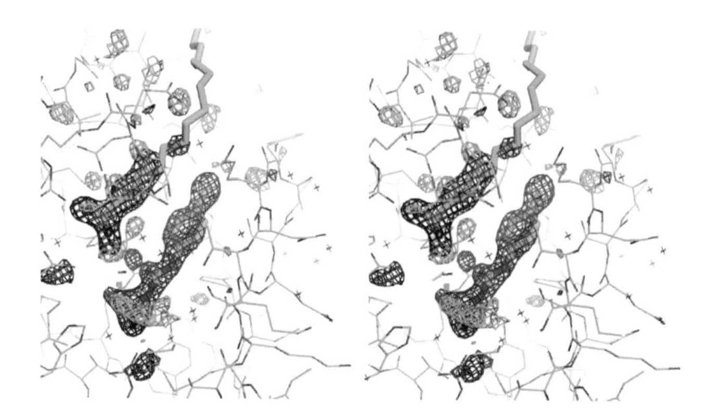

John C. Spurlino, Figure 13.8 A Stereo picture of electron density maps resulting from the refinement of fragment 3c with the $2F_o–F_c$ map (blue) contoured at 1.5s. The $2F_o–F_c$ maps are colored green (3.0s) and red (−3.0s). This view clearly shows the displacement of the palmitic acid by the fragment.

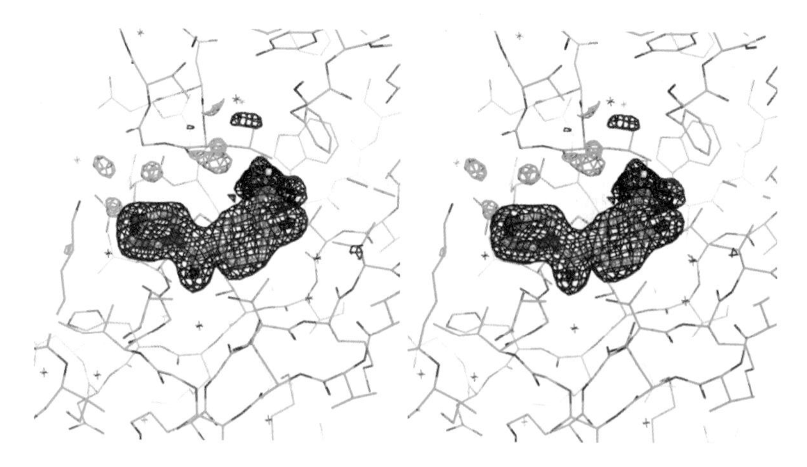

John C. Spurlino, Figure 13.14 Stereoviews of the refined electron density maps for fragment *4e* are shown. The $2F_o - F_c$ map (A) is contoured at 1.5 σ. The stereoviews for the $F_o - F_c$ maps for 3.0 σ (B) and −3.0 σ (C) are shown. The negative density around the trifluoro atoms suggests that they only partially occupy the site or that the *t*-butyl substituent would be better.

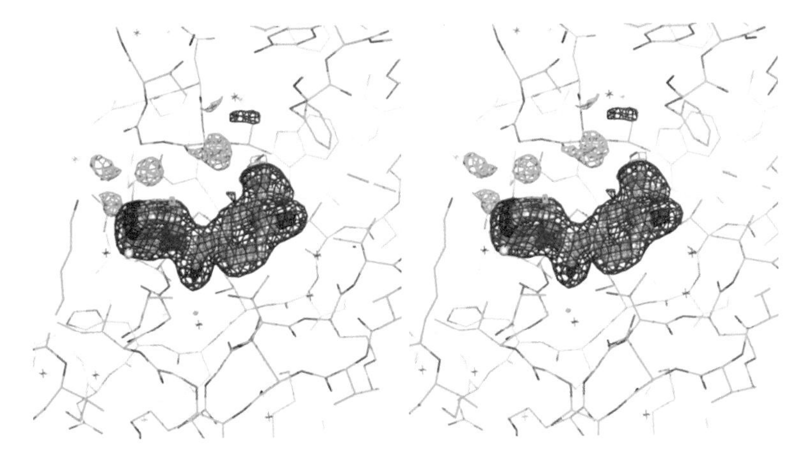

John C. Spurlino, Figure 13.15 Stereoviews of the refined electron density maps for co-refinement of all four fragments are shown. The $2F_o - F_c$ map (A) is contoured at 1.5 σ. The stereoviews for the $F_o - F_c$ maps for 3.0 σ (B) and -3.0 σ (C) are shown.

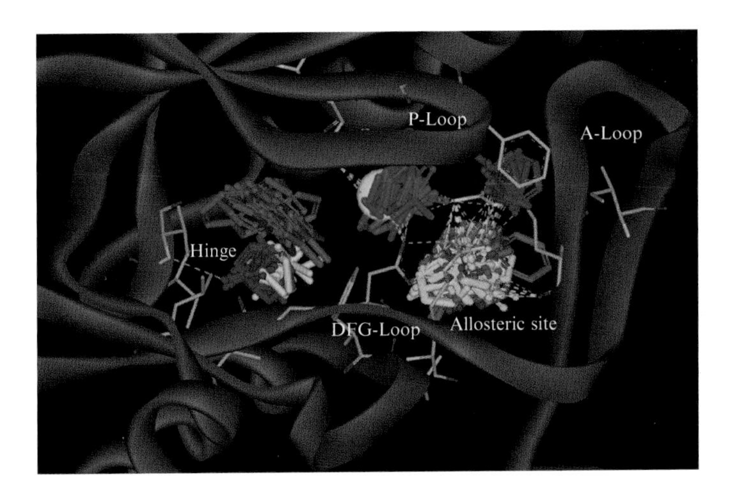

Zenon D. Konteatis *et al.*, Figure 14.7 A close-up view of the ATP and allosteric sites for the protein kinase in Fig. 14.6 showing an example of a druggable site. Note the different types of fragments within covalent bonding distance in the binding site and the presence of one high-affinity water molecule (cyan sphere).

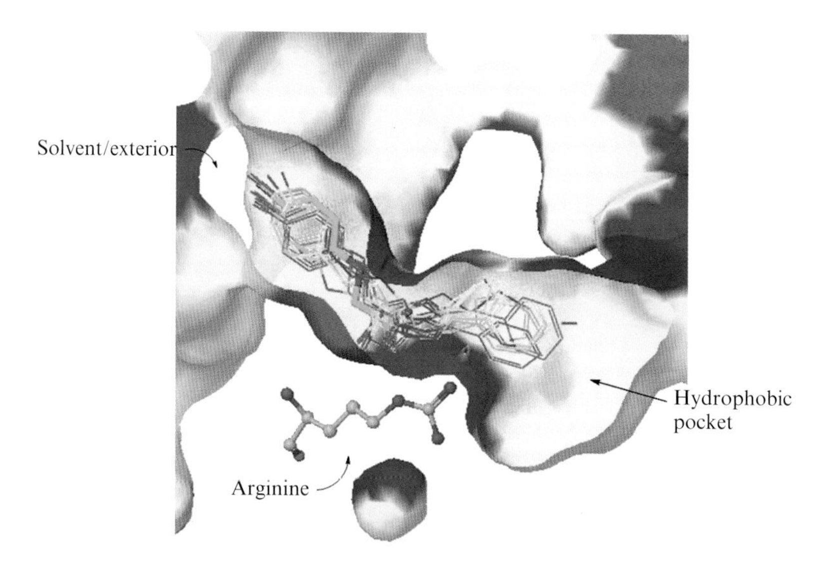

James Lanter *et al.*, Figure 16.9 A model of overlap between the primary library hit (green) and selected secondary library fragments in the Tert site.

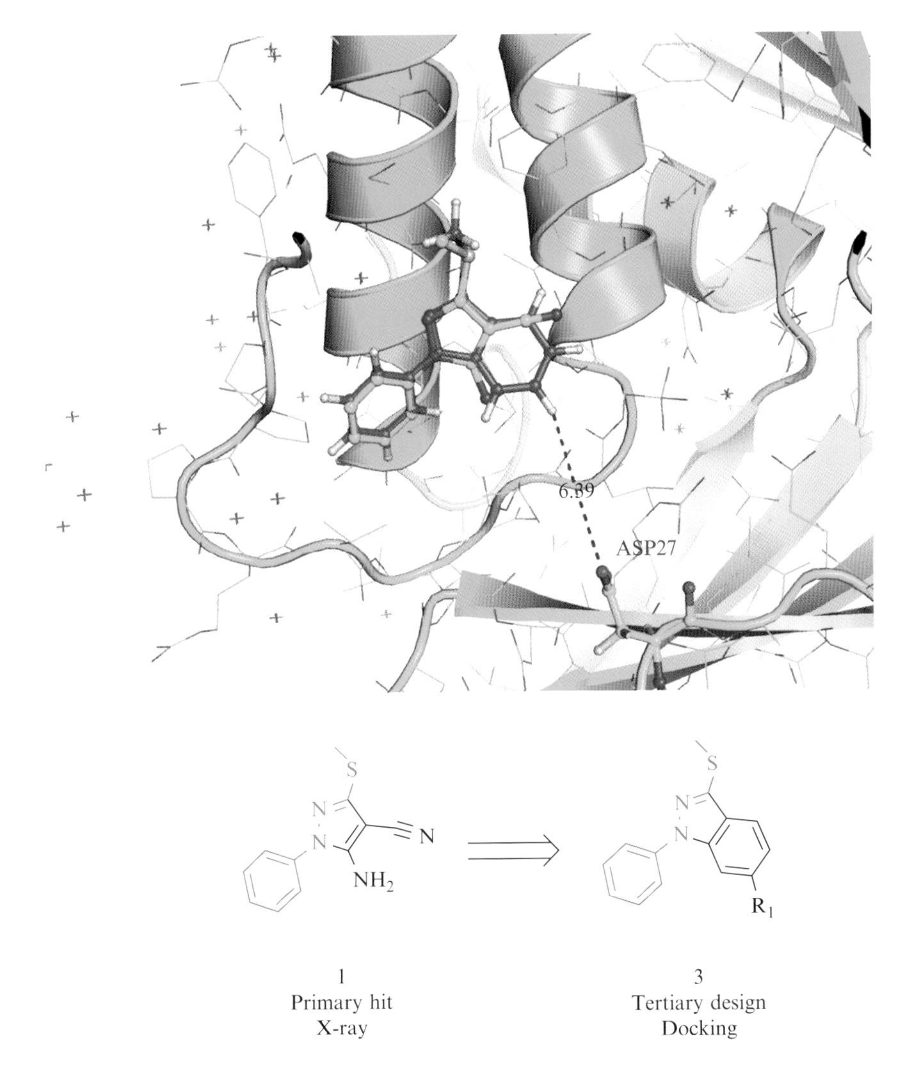

James Lanter *et al.*, Figure 16.10 A primary library hit pyrazole **1** (green) crystal structure with a modeled tertiary library structure indazole **3** (magenta).

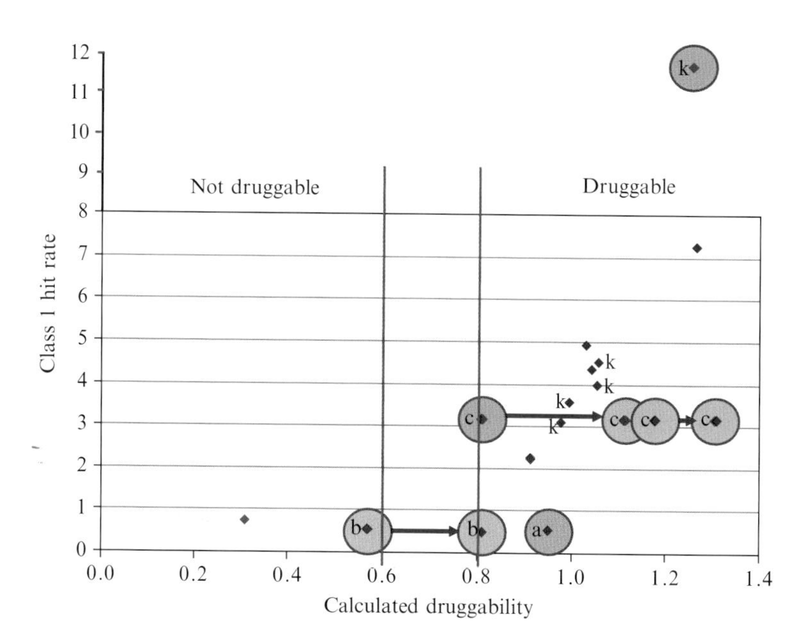

Roderick E. Hubbard and James B. Murray, Figure 20.6 The relationship between validated hit rate from a fragment screen and target druggability. k, kinases; a, Hsp70; b, Pin-1; c, a PPI target. The gray circles highlight particular data discussed in the text and the red arrows show that the notional druggability of the target active site alters as different ligands select different conformations of the protein.